What the Experts Are Saying About the Real Goods *Solar Living Sourcebook:*

At the corner of Wasteful Living and Solar Way sits a store called Real Goods. Take a tour of the store and walk down this road apiece with the *Solar Living Sourcebook.* This is a fabulous compendium of hows, whys, and widgets for making a sensible home under the Sun (and wind and micro-hydro and healthy living and sustainable transportation, too).

—**Amory B. Lovins,** CEO, Rocky Mountain Institute

Today, the most vital way to take control of our lives and feel "powerful" is to take responsibility for our energy footprint. The *Solar Living Sourcebook* continues to be the bible for everyone wanting to get off carbon and on the solar grid. It is scrupulously honest, technically flawless, and steeped in experience and practical wisdom. There is nothing like it in the world.

—**Paul Hawken,** author of *Natural Capitalism,*
The Ecology of Commerce, and *Blessed Unrest*

By 2025, most of what is being written today about energy will seem ridiculous. The *Solar Living Sourcebook* is one of the rare exceptions. John Schaeffer has had as much experience with the practicalities of sustainable living as anyone on the planet. With zest and intelligence, this book describes his vision, his tools, and his pragmatic problem-solving for those of us who expect to follow a parallel path.

—**Denis Hayes,** Chair of the Earth Day Network

The *Solar Living Sourcebook* is a critical resource, to supply both information and inspiration, in this time when the U.S. dependence on foreign oil and the U.S. foreign policy that supports that thirst are wreaking havoc across the globe. From the Timor Gap to the Niger Delta, oil is the source of so much pain in the world. Thanks to John Schaeffer and the Solar Living Institute for making energy alternatives real, viable, and accessible.

—**Amy Goodman,** Democracy Now! and author of *The Exception to the Rulers:*
Exposing Oily Politicians, War Profiteers, and the Media That Love Them and
Static: Government Liars, Media Cheerleaders, and the People Who Fight Back

Epic in scope, comprehensive in content, hopeful in spirit, the Real Goods *Solar Living Sourcebook* might be the most indispensable book in your library.

—**Thom Hartmann,** author, *The Last Hours of Ancient Sunlight*

More of What the Experts Are Saying About the Real Goods *Solar Living Sourcebook:*

The priceless *Solar Living Sourcebook* is a gold mine of information on products, renewable energy, and sustainable living. I spend hours browsing through this book, reading about ideas that will help us forge an enduring human presence.

—**Dan Chiras,** Ph.D., author of *The New Ecological Home, The Solar House, The Natural House, The Natural Plaster Book,* and *Eco Kids: Raising Children Who Care for the Earth*

As John Schaeffer points out in the Introduction to this extraordinarily useful book, solar living is not just an eco-groovy lifestyle choice for the environmentally hip; it is the only survivable option for humanity as fossil fuels dribble away during the remainder of this century. I bring every class of students from New College to the Solar Living Center in Hopland—a sustainability theme park—where they can see, hear, touch, and taste a way of life that uses dramatically less energy, leaves a reduced human footprint on the landscape, and yet opens space for enhanced human creativity and enjoyment. This book is the Solar Living Center on paper—an essential reference for anyone interested in practical sustainability. Study it carefully. And apply what you learn to your home, office, and daily routine.

—**Richard Heinberg,** author of *The Party's Over: Oil, War and the Fate of Industrial Societies; Powerdown: Options and Actions For a Post-Carbon World;* and *The Oil Depletion Protocol*

More than ever—the *Solar Living Sourcebook* is the world's best guide in the great transition to The Solar Age.

—**Hazel Henderson,** author of *The Politics of the Solar Age* and *Building a Win-Win World.*

As the world crests the peak of petroleum supply, the *Solar Living Sourcebook* offers both a wealth of tools and skills for our communities to be more self-reliant and ways we can use much less of the world's resources.

—**Julian Darley,** author of *High Noon for Natural Gas* and Founder of the Post Carbon Institute

The *Solar Living Sourcebook* is the one-step guide to how to live wisely, sustainably, and with optimism. A beacon to light our way in the 21st century—solar powered, of course.

—**David Pearson,** author of *The New Natural House Book, The Natural House Catalog, Earth to Spirit: In Search of Natural Architecture,* and *The House that Jack Built* series

THIRTIETH-ANNIVERSARY EDITION

Solar Living Source Book

 GAIAM    REAL GOODS

THE REAL GOODS SOLAR LIVING BOOK SERIES

Real Goods Trading Company in Hopland, California—now Gaiam Real Goods—was founded in 1978 to make available new tools to help people live self-sufficiently and sustainably. Through seasonal catalogs, a biannually updated *Solar Living Sourcebook,* a website (www.realgoods.com), and a flagship retail store located at the 12-acre permaculture oasis called the Solar Living Center, in Hopland, California, Gaiam Real Goods provides a broad range of tools for independent living. Real Goods is the original solar pioneer, having sold the very first photovoltaic (solar electric) module in the world at retail in 1979. Thirty years later, Real Goods can boast of having solarized over 60,000 homes and businesses in the United States.

"Knowledge is our most important product" is the Real Goods motto. To further its mission, Gaiam Real Goods is a strong promoter of books for renewable energy and sustainable living, and has pioneered a co-publishing venture with Chelsea Green Publishing Company to bring important writings by relatively unknown authors to a larger market. Many of these books (listed above) are quoted throughout this *Sourcebook.* The titles in this series are written by pioneering individuals who have firsthand experience in using innovative technologies to live lightly on the planet. Our copublished books are both practical and inspirational and they enlarge our view of what is possible. In 2004 we began partnering with New Society Publishers, who is distributing the *Sourcebook* for us and representing it at special events throughout North America to popularize the ideas, products, and practices of sustainable living to the nooks and crannies of America. The time has come.

John Schaeffer
President & Founder, Real Goods

GAIAM REAL GOODS

30th
ANNIVERSARY

SOLAR LIVING SOURCE BOOK

Your Complete Guide to
Renewable Energy Technologies
and Sustainable Living

Written and edited by **John Schaeffer**

Assistant Editor **Alan Berolzheimer**

Product Editor **Bill Giebler**

A REAL GOODS SOLAR LIVING BOOK

Distributed by

NEW SOCIETY PUBLISHERS

ISBN 978-0-916571-06-1
The Real Goods *Solar Living Sourcebook* is the 13th edition of the book originally published as the *Alternative Energy Sourcebook,* with over 600,000 in print, distributed in 44 English-speaking countries.

Distributed by:
NEW SOCIETY PUBLISHERS LIMITED
PO Box 189, Gabriola Island, BC V0R 1X0 Canada
Telephone: 800-567-6772
www.newsociety.com

Gaiam Real Goods
Real Goods Solar Living Center
13771 Highway 101
Hopland, CA 95449
Business office: 707-744-2100
To order: 800-919-2400 or fax 800-482-7602
International orders: 707-744-2100
International fax: 707-744-2104
For technical information: 800-919-2400
Renewable Energy email: techs@realgoods.com
Website: www.realgoods.com

NEW LEAF PAPER®
ENVIRONMENTAL BENEFITS STATEMENT
of using post-consumer waste fiber vs. virgin fiber

Real Goods saved the following resources by using New Leaf Frontier 100 (FSC), made with 100% post-consumer waste, and New Leaf Reincarnation Matte, made with 100% recycled fiber and 50% post-consumer waste, both manufactured with electricity offset with Green-e® certified renewable energy certificates and processed chlorine free.

1,317	Trees
561,932	Gallons of Water
950	Million Btu of Energy
63,112	Pounds of Solid Waste
124,099	Pounds of Greenhouse Gases

Calculations based on research by Environmental Defense and other members of the Paper Task Force.

©2007 New Leaf Paper   www.newleafpaper.com

 ANCIENT FOREST FRIENDLY™    NEW LEAF PAPER manufactured with wind power   Green-e

FOR THE EARTH

May we preserve and nurture it in our every action.

Contents

Acknowledgments

WHILE IT'S IMPOSSIBLE to adequately thank everyone who has contributed to our 30th-anniversary edition of this *Solar Living Sourcebook* and to Real Goods' extraordinary 30-year odyssey in general, several people stand out above the crowd. In the 26 years and 13 editions of this *Sourcebook*, more than 600,000 copies have been sold in 44 English-speaking countries. For this edition, thanks go first to Alan Berolzheimer, who toiled tirelessly to manage the new chapters (and many of the old) and whose masterful pen and mind often performed miracles of weaving straw draft copy into gold nuggets of final prose. Bill Giebler helped tremendously with products both old and new, with capable technical help from Norman Franks and merchandising assistance from Jenny Muchow and Mike Farhar. Jill Shaffer, once again, did a masterful job of designing the book and rose gracefully to our intense time pressures as we brought the book to the printer at its rapid conclusion. Ellen Bingham was a workhorse copyeditor and proofreader, and Karen Griffiths once again created an excellent index.

For the chapters new to this 30th-anniversary edition very hearty thanks go to Jason Bradford for his contributions to Relocalization, to Dan Chiras for his contributions to Land and Shelter, to Bob Ramlow for his contributions to the Solar Hot Water section, to Jim Fullmer and Hugh Courtney for their contributions to the Biodynamics section, to Benjamin Fahrer for his contributions to the Permaculture section, to Steve Heckeroth and Dave Blume for their help with "Sustainable Transportation," and to Cynthia Beal for her tremendous work on "Natural Burial."

Thanks also to Chris and Judith Plant and their capable staff at New Society Publishers for doing such an amazing job of distribution with our 12th edition and for encouraging us to continue. And for *Sourcebook*s past, there are many to thank who have nurtured the book over the years and whose work still remains current to this day. I especially thank Doug Pratt, Stephen Morris, and Jeff Oldham for their inspiration and dedication to Real Goods and the entire *Sourcebook* project. Thanks also go of course to Gaiam's management, particularly Lynn Powers and Vilia Valentine, whose financial support made this book possible.

And on a personal note, I thank my wife, Nancy, for her frequent insight and inspiration for this and most all of my writing, and my children, Sara, Ashley, and Cy, for their help over the years. Finally, for their unwavering financial support, I thank the original 10,000 Real Goods shareowners for directly bringing the Solar Living Center into fruition, and the directors of the Solar Living Institute, Bob Gragson, Doron Amiran, Coral Mills, and Lindsay Dailey, for nurturing that original vision to the present.

The Nonprofit Real Goods Solar Living Institute

The Solar Living Institute is a 501-C-3 nonprofit organization that promotes inspirational environmental education through hundreds of hands-on workshops; SolFest—an annual renewable energy educational event now in its 13th year; displays and interactive demonstrations at the solar- and wind-powered 12-acre permaculture oasis called the Solar Living Center in Hopland, California; and numerous children's educational programs. More information about the Institute can be found on pages 569-573 in this book or at its website: www.solarliving.org or by calling 707-744-2017. The Institute has an innovative membership program and encourages tax-deductible donations to further its programs.

Introduction

AS I LOOK BACK 30 YEARS to our humble beginnings at Real Goods in 1978, I am struck by how isolated we in the environmental movement really were. Our clientele at Real Goods was a cadre of young and idealistic hippies living in the woods of Mendocino County, California, having just fled the major urban centers of America for a simpler and more meaningful existence in the hills. Most of our customers got their light from kerosene, their heat from wood, their food from the garden, and their entertainment from books. Back then, there were no computers, and the Internet, Google, YouTube, and cell phones did not exist. Jimmy Carter was president, Jerry Brown was governor, and everyone was optimistic that, with the strength of our back-to-the-land movement, we could eventually overcome the misguided ways of our over-logging, over-consuming, shortsighted polluting peers. Global warming was not yet even a whisper, and there seemed to be plenty of time left to deal with oil depletion. I wrote an editorial in 1979 that declared: "According to the U.S. Congress's own Office of Technology Assessment, all known oil reserves will be exhausted by 2037." That was almost 60 years away, and if we could put a human on the Moon in only 10 years, fixing our fossil fuel habit sounded like a piece of cake—if we could only find the will to do it.

Today, 30 of those 60 years are gone, and we have made few, if any, major steps toward becoming fossil fuel-free. 2007 marked the first year when nearly all scientists, and even oil company executives, acknowledged that we have reached peak oil—or that we will reach it—within just a few years. Coincidentally, consensus about the reality of global climate change also finally coalesced in 2007, and it will probably be looked back on as the year when that consciousness reached the tipping point. From countless magazine covers to *An Inconvenient Truth* and hundreds of newspaper articles and media news clips, most of us now accept the reality that global warming is here to stay. California Governor Arnold Schwarzenegger, representing the sixth-biggest economy on the planet, has brought climate change to the forefront of politics, and he is partnering with many other governors and heads of state to take action despite the official U.S. position of denial. Across the world, we're beginning to realize that if we're to have any chance of mitigating global warming, we must lower our fossil fuel consumption—and the resulting carbon dioxide emissions— between 80% and 90% from today's levels as soon as 2030, or at the latest by 2050. There is simply no time left to begin acting decisively if humanity—and all other life on Planet Earth—is to have a chance to avoid implosion by the middle of the 21st century. Since 1980, we've had the warmest years on record (with 2006 and 2005 now ranking #1 and #2) since scientists began recording average annual temperatures in 1866. Parts-per-million carbon dioxide levels have increased from 280 ppm (for centuries) to 360 ppm today and are forecast to hit 560 ppm by 2050.

> Most of our customers got their light from kerosene, their heat from wood, their food from the garden, and their entertainment from books. Back then, there were no computers, and the Internet, Google, YouTube, and cell phones did not exist.

> There is simply no time left to begin acting decisively if humanity— and all other life on Planet Earth—is to have a chance to avoid implosion by the middle of the 21st century.

Modern society's addiction to fossil fuels has brought us to peak oil. Marion King Hubbert (the concept of peak oil), geologist Colin Campbell, and Richard Heinberg (author of *The Party's Over: Oil, War, and the Fate of Industrial Societies; Power Down;* and *The Oil Depletion Protocol*) have done exhaustive work to demonstrate that the "oil peak"—the point in time when 50% of all known oil reserves have been consumed—is upon us. Even the CEOs of Exxon-Mobil and Royal Dutch Shell, the world's largest oil companies, concur. The world currently uses about 80 million barrels of oil per day and is expected to need 120 million barrels per day in 2020. However, while 13 megaprojects (more than 500 million barrels of oil) were discovered in 2000, only two were found in 2002, and none since. Geologists believe we'll never find another one. The "age of oil," in the geologic history of the world, will prove to be only a small blip—from the time that Edwin L. Drake drilled the very first oil well in northwestern Pennsylvania in 1859 until that last drop is consumed sometime in the mid-21st century.

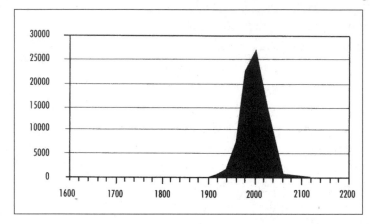

World oil production from 1600 to 2200, history and projection, in million of barrels per year. (Source: C. J. Campbell)

We will all be profoundly affected by peak oil, particularly in the industrialized countries where we've become accustomed—and addicted—to exponential energy growth. The first and most obvious consequence, which we've already begun to experience, is skyrocketing oil and gasoline prices. It is unlikely that these prices will ever again decline significantly from today's rates of $65/barrel and at least $3/gallon. Indeed, soon those prices may sound very cheap. Because fossil fuels permeate all sectors of our world economy, decreasing supplies and rising prices will lead to economic contractions. Perhaps most significantly, as oil and natural gas become more scarce and costly, so will chemical fertilizers, which will have a huge impact on our ability to feed ourselves.

In spite of the evidence, our government continues to subsidize fossil fuel technologies, and the energy playing field is not level. In 2003, the White House and Congress enacted incentives to give the owner of a 10-mpg Hummer a whopping tax deduction of $34,000 but the owner of a 50-mpg hybrid vehicle just $4,000. For every 100,000 SUVs sold this year, American taxpayers will pay a subsidy of $1 billion, which ironically is the same amount that the federal government spent in the 1990s to encourage American car companies to build a hybrid car, and about the same amount we're spending each week to keep our troops in Iraq. After the first Gulf War in 1991, both the Japanese and American governments launched major initiatives to reduce their dependence upon Middle East oil as an urgent matter of national security. The Japanese government aggressively helped develop the Toyota Prius and the Honda Insight, assisting the automakers to sell vehicles below cost to get them out to the public. Today, hybrid vehicles are a major success story, and Toyota expects to sell nearly half a million Priuses by 2008. Meanwhile, the American automobile industry is far behind the curve—and losing tons of money. President Bush has embraced ethanol, but he has also refused to push for higher fuel-efficiency standards that are more than feasible with current technologies. The U.S. fleet average (miles per gallon of cars and trucks) remains worse than it was 25 years ago, due to ideological stubbornness, multimillion-dollar lobbying campaigns, and duplicity by the government and the car companies.

The world currently uses about 80 million barrels of oil per day and is expected to need 120 million barrels per day in 2020.

We must face the prospect of changing our basic ways of living. This change will either be made on our own initiative in a planned way, or be forced on us with chaos and suffering by the inexorable laws of nature.

—President Jimmy Carter, 1976

Back in the good old days (30 years ago), we approached environmentalism by educating as many people as possible as fast as possible about new renewable energy technologies and the wisdom of simple processes like recycling, composting, natural building, and resource conservation. Through the Real Goods catalog, website, and 12-acre Solar Living Center, I think we've done a pretty good job of achieving that goal. When we opened our Solar Living Center in 1996, we aimed for 50,000 visitors a year within 10 years. We've greatly exceeded those projections and are now proud to host nearly 250,000 visitors annually and to be the largest tourist attraction north of San Francisco. But, as we celebrate our 30th anniversary, it's clear to me that there is so much more to do if we're going to make an impact on minimizing global climate change. We have to change human consciousness, thereby changing deeply ingrained habits. The two enormous challenges of our time—oil depletion and climate change—are inextricably connected and *must* be dealt with head on. This means addressing the systemic causes of these crises, not just the superficial symptoms.

Without a doubt, we're the most fortunate generation in the 2-million-year history of the genus *Homo*. And we may have it easier than any of the generations to come, given the sacrifices necessary on the horizon regarding energy and fossil fuel use. The abundance of new energy resources that became available during the last two centuries—primarily oil—sparked a population explosion (from 1.7 billion to 6.4 billion people during the 20th century), an awesome industrial revolution, and a level of mass prosperity never before imaginable (not to mention wars of a correspondingly horrific scale and scope). Agriculture is perhaps the area of human endeavor that has been most profoundly transformed by fossil fuels, through the invention of synthetic fertilizers and motorized transportation. In the U.S., the number of farmers has fallen from 50% of the population in the 1890s to 1% today. The average piece of food on an American plate travels 1,500 miles to get to your table, and changes hands six times along the way. The Swedish Food Institute discovered that growing and distributing a pound of frozen peas requires 10 times more energy than the peas themselves contain. A head of iceberg lettuce grown in California and shipped to London consumes 127 times more calories than it provides to the person who eats it. Because the current industrial agricultural system is totally dependent on fossil fuels (from fertilizers to fueling tractors to gassing up trucks that drive foods across country), and because it takes 400 gallons of oil equivalents annually to feed each American, it is unlikely that we will be able to feed the 8+ billion of us forecast to be alive in 2050 using our current industrial agricultural system—it is unsustainable.

And what has happened to our energy habits in the last half-century? At the turn of the 21st century, the average American family owned twice as many cars, drove two and a half times farther, used 21 times more plastic, and traveled 25 times farther by air than did the average family in 1951. The size of new houses has doubled since 1970 while the average number of people living in each home has shrunk dramatically. Even with all the additional space, we've not only managed to fill it up but have created a new and booming storage locker industry to hold all the stuff we can't fit in our homes. And finally, the enormous

The two enormous challenges of our time—oil depletion and climate change—are inextricably connected and *must* be dealt with head on. This means addressing the systemic causes of these crises, not just the superficial symptoms.

In the U.S., the number of farmers has fallen from 50% of the population in the 1890s to 1% today. The average piece of food on an American plate travels 1,500 miles to get to your table, and changes hands six times along the way.

The enormous irony in our acquisition of all this material abundance—and the billions of barrels of oil and millions of acres of trees it took to create it—is that, according to numerous surveys, many Americans are not happy people.

The average size of a house built in 2005 was nearly 2,500 square feet. If all the dimensional lumber used to build stick-frame houses in America were laid end to end, it would extend to the Moon and back 23 times—4.2 million miles!

irony in our acquisition of all this material abundance—and the billions of barrels of oil and millions of acres of trees it took to create it—is that, according to numerous surveys, many Americans are not happy people.

Something is very wrong if all this abundance can't make us happy. And with the "age of oil" rapidly coming to an end, it's now clear that our species cannot survive the continued loss of biodiversity, decimation of forests, shrinking ocean fisheries, falling water tables, foul water and air, and relentless increase in carbon dioxide emissions that is changing Earth's climate forever.

Maybe it's time to finally change course? That's always been the mission of Real Goods and the *Sourcebook*. It is time to make systemic changes to the ways we live. Every chapter in this *Sourcebook* will provide you with specific information and tools to help you plot your course, no matter where you are on the continuum of change.

We begin our 30th-anniversary edition with something new, something we believe to be the primary solution to our present dilemma—a very promising strategy called Relocalization. Evolving from the back-to-the-land movement that originally created Real Goods in 1978, Relocalization encompasses the best parts of the movements for environmental protection, sustainable living, regenerative design and natural building, new urbanism, Slow Food, voluntary simplicity, and the whole concept of "think globally and act locally." It's a strategic response to the problems of peak oil, climate change, and the overshoot of Earth's physical limits caused by exponential population and economic growth during the industrial era. It honors, encourages, and nurtures local businesses, farmer's markets, energy production, and community involvement, while rejecting the malign aspects of globalization and our current fossil fuel-based economy. We've devoted an entire chapter to Relocalization as a prelude to the nuts-and-bolts education you'll find in the balance of the *Sourcebook*.

We've completely rewritten and enlarged our "Land and Shelter" chapter because our land and homes are not only the cornerstone of our existence but also the area of our largest carbon-footprint impact. The choices we make about where and how we shelter ourselves are absolutely fundamental to any concept of sustainability. These choices pose a major environmental threat as well as a conservation opportunity. Americans build more than 1.8 million new houses every year. The average size of a house built in 2005 was nearly 2,500 square feet. If all the dimensional lumber used to build stick-frame houses in America were laid end to end, it would extend to the Moon and back 23 times—4.2 million miles! About 20% of the nation's fossil fuel energy consumption goes to heat, cool, light, and ventilate our homes. In the "Land and Shelter" chapter, you'll explore in depth a variety of innovative, fascinating, and environmentally low-impact ways to site, plan, build, and operate your new dream home—or retrofit your existing house—to make it as close to carbon neutral as possible.

Consistent with the last 12 editions, the heart of our *Sourcebook* focuses on the nuts and bolts of renewable energy—photovoltaics (a.k.a. PV or solar panels) and all the accouterments of living *off* the grid or *on* the grid with solar power. Renewable energy is the foundation of Real Goods. We made the very first retail sale in the world of a photovoltaic module in 1979. In the nearly 30 years since, we have solarized more than 60,000 homes and businesses. Worldwide, well over 1 million homes now get their electricity from solar cells. For the 1.7 billion people in the world still not connected to an electric grid, PV is by far the cheapest source of electricity: In the developing world, the monthly payment on a

3- to 5-year loan for a small PV system is less than what is typically spent on candles and kerosene for lamps.

Photovoltaic technology is totally reliable, and our solar industry is now fully mature. In what other industry can you find a 25-year warranty like the one that comes with almost every solar panel sold? Computers? Cars? I don't think so. And when building a new home, you can easily wrap the cost of the solar system into your 30-year mortgage, making PV cost effective from day one. For a retrofit on an existing home, payback comes in 6-12 years (depending on incentives and electric rates), which translates into a return on investment of between 8% and 16%—far better than what you can earn in the stock, bond, or long-term CD markets. In the commercial sector, the numbers are even better. With federal tax incentives and accelerated depreciation, payback typically comes in 2-5 years, delivering a return on investment of between 20% and 40%. Solar power substantially increases property value, too. According to the *Appraisal Journal* (October 1999) of the National Appraiser's Association, home value increases $20 for every $1 achieved in annual energy savings. This means that a 3kW system (the size of the average residential system Real Goods sells) can increase your property value by $20,000, while costing you only $18,000 (with California rebates). You're money ahead from day one. More important, a PV system allows you to lock in utility rates of less than $0.14 per kWh for 30 years—and many expect that utility electric rates will skyrocket over next three decades.

Solar is growing by leaps and bounds. In 2002, 525 megawatts (MW) of PV power was produced worldwide, up 82% from 2000. In 2004, production doubled to 1,040MW. And by 2006, it doubled again to over 2,000MW. The photovoltaic industry has experienced a compound annual growth rate approaching 50% annually in the last few years. Unfortunately, the United States' share of solar power is dwindling. In 1997, the U.S. accounted for 42% of all the PV power in the world, Europe for 18%, and Japan for 25%. By 2006, the U.S. share declined to only 7%, while Europe increased to 29% and Japan doubled its share to 51% of PV production worldwide. The good news is that the price of PV continues to tumble. Solar is following the same pattern as the computer, electronics, and cell phone industries: For every doubling of supply, the price declines about 20%. Forecasts call for worldwide production increases to over 10,000MW by 2010, up five times from 2006!

America is way behind in the PV race because Japan and Germany have spent considerably more money than the U.S. on PV research and development, on generous tariffs for solar power fed into the grid, and on rebates and incentives for the installation of PV systems. As a result, the German PV market (even with sunshine equivalent to only the northeastern U.S.) grew from 100MW in 2001 to more than 500MW in 2006. The Japanese PV market has been even more impressive, growing from just 2MW of residential PV in 1994 to over 900MW by 2006. And the installed cost of PV in that country was reduced by 62% in seven years—during a period of very poor performance by the Japanese economy. Meanwhile, net metering (where the utility has to buy power from your renewable grid-intertie system for at least the same price as the power you buy from them) pokes along in the U.S., a hodgepodge of nice, but inadequate, incentives now available in 41 states and the District of Columbia; and the nuclear industry still claims the lion's share of our government's largesse.

Programs to encourage the expansion of renewable energy in the U.S. would provide significant benefits: stimulating the economy by adding hundreds of thousands of jobs in

Photovoltaic technology is totally reliable, and our solar industry is now fully mature. In what other industry can you find a 25-year warranty like the one that comes with almost every solar panel sold? Computers? Cars? I don't think so.

America is way behind in the PV race because Japan and Germany have spent considerably more money than the U.S. on PV research and development.

manufacturing, sales, and installation of renewable energy systems; relieving pressure on our overtaxed electricity transmission grid; and reducing carbon emissions and pollution. Here's one concrete example: Our little company, Gaiam Real Goods, increased residential solar sales 15 times in one year (between 2002 and 2003), almost solely as a result of California Energy Commission incentives, and we almost doubled our workforce. Just imagine what a well-thought-out national program could do.

Solar hot water is another fantastic opportunity just begging for a chance, and it receives greatly expanded attention in this 30th-anniversary *Sourcebook* because the technology is a no-brainer that nearly everyone can use to reduce fossil fuel consumption. Domestic hot water accounts for between 20% and 40% of a typical home's annual energy budget, and a solar water heating system can offset up to 75% of that energy load. A standard, two-panel system will thus offset three-quarters of a ton of CO_2 per year, or the equivalent of driving 1,300-1,750 miles at 22 miles per gallon. Solar water heaters typically have a life span of more than 40 years and are one of the best investments available in America today, with paybacks often coming in less than five years, which equates to more than a 20% return on investment.

Given the realities of climate change and peak oil thrust upon us even more convincingly since we last published the *Sourcebook*, we've also expanded and rewritten our "Sustainable Transportation" chapter. For most of us Americans, our cars are responsible for between 25% and 50% of our total CO_2 emissions. What better place to begin reducing our carbon footprint than in our garages? By doubling the average fuel economy of American cars and trucks to 40 miles per gallon—which can easily be accomplished now with existing hybrid technology—we can save 5 million barrels of oil every day and eliminate our current Middle East oil imports in less than a decade. With that one simple but decisive step, we would offset nearly 12 billion tons of CO_2! The transportation chapter will give you the latest updates on biodiesel, ethanol, electric vehicles, and hydrogen fuel cells.

We bring the 30th anniversary edition of the *Solar Living Sourcebook* to a close with a brand new chapter (if not a new idea) whose time has come. Aptly subtitled *The Ultimate Back-to-the-Land Movement*, our "Natural Burial" chapter serves as a fitting conclusion to an environmentally responsible life and a righteous entrance into an afterlife, if you believe in that sort of thing. Think about what's buried along with our loved ones in American cemeteries every year: more than 800,000 pounds of embalming fluid, over 180 million pounds of steel, more than 5 million pounds of copper and bronze, and over 30 million board-feet of hardwoods. Contrast this to the U.K., where there's a burgeoning new movement in which people are burying loved ones in biodegradable containers—without embalming fluid or synthetics—and returning bodies to the Earth to compost into soil nutrients with a forest of trees marking the spot. We must question the "waste management" behavior of our society and the wisdom of leaving toxic burial chemicals and other synthetic substances in the ground (and atmosphere as a result of cremation) for future generations to clean up. As 80 million American baby boomers arrive at their end over the next 20-30 years (myself included!), the natural burial movement will undoubtedly gather lots of steam, and because many of us will be "dying to do the right thing," it makes sense to conclude our *Sourcebook* with this topic.

Actually, we figure that knowledge is even more powerful than death, so an updated, newly comprehensive "Sustainable Living Library" is the final chapter in the book. Enormously expanded from the previous *Sourcebook*, we're proud to offer you even more resources to help you fulfill your dreams.

Our little company, Gaiam Real Goods, increased residential solar sales 15 times in one year (between 2002 and 2003), almost solely as a result of California Energy Commission incentives, and we almost doubled our workforce.
Just imagine what a well-thought-out national program could do.

When we opened our first Real Goods store in Willits, California, in June 1978, our mission was to demonstrate and provide renewable energy alternatives—and it still is. After 30 years, we are better positioned than ever to help you realize your dreams of transforming your lifestyle toward sustainability, whether your goal is to buy land and build a totally self-sufficient solar home, or simply to modestly reduce your carbon footprint by buying some energy-efficient light bulbs and figuring out the best alternative fuel for your next car. Over the years, we've gathered an unbeatable team of renewable energy experts with more than 200 years of combined experience in solar—most of whom live with the products we sell. Our Real Goods residential solar division specializes in residential renewable energy design and installation. Our Real Goods catalog division produces two 100+-page color catalogs every year that feature the latest products for energy conservation, healthy living, renewable energy, and environmental education, as well as the most thorough sustainable living library on the planet.

We are headquartered at the Real Goods Solar Living Center in Hopland, California, our 12-acre permaculture oasis where all our products, ideas, and concepts come alive—not only in the interactive displays on site but in the 200,000 people who annually visit the site. The Solar Living Center is operated by the nonprofit Solar Living Institute, which nurtures the site and offers more than 1,500 class days per year on renewable energy, green building, permaculture, and other sustainable living workshops. The SLI's mission is providing inspirational environmental education. Since the flood that devastated the Solar Living Center in the winter of 2006, the site has been re-created with a brand new Intern Village complete with green buildings, greywater systems, and a gorgeous shower house. The site has lots of improved educational signage, and many new interactive displays are in the works for the 2008 season. If you haven't visited northern California's #1 tourist attraction, we invite you to stop by and see the future of sustainability. The Solar Living Center is 100% solar powered, and educational opportunities abound.

I've been a passionate adventurer in the solar industry and the sustainability movement my whole life. I try hard to walk my talk. My wife and I live in a home built of recycled and green materials, powered by solar (passive and active) and hydroelectric energy, with gorgeous gardens that provide most of our food, a 15-acre biodynamic olive orchard, an 8-acre biodynamic vineyard, and a new beehive positioned next to our lavender labyrinth. Our tractor and VW Jettas run on biodiesel. I'm tremendously gratified to see the fruits of all our labors. As the solar industry continues to grow and mature, and as our cultural consciousness continues to evolve, I remain hopeful that, once and for all, we will get things right in our homes, in our communities, in our country, and on our planet. Instead of forever being blamed for the excesses that put our planet on the dangerous path to destruction, we baby boomers can instead be viewed by our descendants as the generation that rose above our good fortune and decadence, finally saw the light, and embraced a vision that turned us all around while there was still a chance. We are indeed living on borrowed time. Let's turn it all around now, while we have this very last chance.

For the Earth,

John Schaeffer
Real Goods Founder and President

Solar Pioneers Reflect on Real Goods' First 30 Years

DENIS HAYES

John Schaeffer and Real Goods: Visionary, or Clear-Eyed Realist?

Starting up the Solar Living Center in 1978 was akin to starting a dot.com in 2000. All the omens looked good, but treacherous rapids lay just around the bend. The defeat of Jimmy Carter by Ronald Reagan in the 1980 election; the dramatic decline of world oil prices after the end of the Iran-Iraq war; and the expiration of a raft of federal incentives for renewable energy— all combined to swiftly turn much of the solar industry into hamburger.

Real Goods soldiered on. It is the kind of place that business writers envision when they praise the "can-do spirit of the American entrepreneur." John Schaeffer is someone I'd like at my side if I ever had to storm the beach at Iwo Jima. He just frigging doesn't know when to quit.

Working out of a remote location in northern California, beset by floods and fire, facing a national government that fluctuated between active hostility and benign neglect, Real Goods did more than just survive. It grew into a vibrant business serving a national base.

About the same time that Real Goods was founded—when some legitimate disagreements still existed among serious scholars about many of the details of global warming—the National Academy of Sciences issued a report surveying the field. It concluded: "Unfortunately, it will take a millennium for the effects of a century of use of fossil fuels to dissipate. If the decision [to reduce CO_2 emissions] is postponed until the impacts of man-made climate changes have been felt, then, for all practical purposes, the die will already have been cast."

America ignored its most distinguished scientists for 30 years. Now the "impacts of man-made climate change have been felt" by Americans from New Orleans to Alaska. Instead of pursuing the smart, cheap, relatively painless course that was available 30 years ago, we are now faced with choices that will be costly and painful. But at least we are finally starting to move.

California has adopted the most ambitious climate strategy of any major government in the world. Lying at the core of that strategy are the very products that Real Goods has been selling for the last three decades. (This must be what it would have been like to own the most experienced distillery when they finally lifted Prohibition.)

Some say that John Schaeffer was a visionary who was 30 years ahead of his time. In truth, John was, and is, a clear-eyed realist. The rest of the world is 30 years late.

Denis Hayes is president of the Bullitt Foundation in Seattle and the immediate past chair of the Energy Foundation. During the Carter Administration, he headed the federal Solar Energy Research Institute; from 1983 to 1988, he was a professor of energy engineering at Stanford University. Hayes chairs the board of trustees of the American Solar Energy Society.

STEPHEN MORRIS

What's in a Word?

The world has seen a succession of trendy environmental buzzwords. The first, and in many ways the most enduring, is "green," as in *The Greening of America*, the title of a 1970 bestseller by Charles A. Reich (Random House, 1970) that predicted, "There is a revolution coming. It will not be like revolutions of the past. It will originate with the individual and with culture, and it will change the political structure only as its final act."

John Schaeffer enlisted in the revolutionary "army" and headed north to the green and gold hills of Mendocino. For John, and many others, the necessary first step was backward to learn the lessons of the past that had been forgotten in our rush to modernity. By 1978, the buzzword had become "natural." But natural was eventually co-opted by Madison Avenue, which hung the word on everything from pesticides to cigarettes.

John Schaeffer came up with the phrase "real goods" to describe the essential items for a life on a rural commune. The real goods were the tools, the equipment, the articles of clothing, the information that made possible, and improved, the communal lifestyle. The folks at Real Goods have spent the last three decades redefining and further articulating the meaning of the phrase.

Real Goods soldiered on. It is the kind of place that business writers envision when they praise the "can-do spirit of the American entrepreneur."

California has adopted the most ambitious climate strategy of any major government in the world. Lying at the core of that strategy are the very products that Real Goods has been selling for the last three decades.

Meanwhile, other buzzwords eventually bobbed to the surface. "Ecological," "sustainable," "environmental," "reduce/reuse/recycle" . . . each has had its champions, each has had its moments. More recently, "organic" has been hot, but its power has also been diluted by corporate takeover, its place on the top 10 threatened by the more recent notion that "local" is the new organic.

But as the words come and go, what about that revolution? It's taking longer than we thought, but we're better equipped than ever to be successful, thanks to the real goods.

What are our victories? The Sun is the same, but our ability to harness its power, virtually nonexistent in 1970, is now universally available. The products we use to build our homes can now be compatible with nature, not designed to overcome it. Within our homes, we have the option of nontoxic products. We have access to fresh, organic, and increasingly local foods. We can choose from a selection of whole grain breads. Heck, even the beer is better!

Tools, technology, information, products, and wisdom . . . these are the real goods. The revolution still lies before us, but thanks to the continued evolution of Real Goods (the company) and real goods (the concept), we are in our best shape ever.

We aren't close to a declaration of victory, but our modest successes give us a taste of the ultimate outcome. Charles Reich described it back in 1970: "It promises a higher reason, a more human community, and a new and liberated individual. Its ultimate creation will be a new and enduring wholeness and beauty—a renewed relationship of man to himself, to other men, to society, to nature, and to the land."

Stephen Morris *is the editor/publisher of* Green Living: A Practical Journal for Friends of the Environment *and is the founder of The Public Press, an experimental book publishing company. His most recent book,* The New Village Green *(New Society Publishers, 2007), is an anthology that explores the new communities that have been created in the wake of new technologies.*

JAY BALDWIN

Real Goods Lights the Way

When the experimental 35-watt PV panel we'd had on our Airstream trailer since 1972 needed replacing, a budding Real Goods was ready to sell us a "real" one. More important, they could provide the experienced know-how needed to install it properly. At that time, the PV panels were hiding behind the wood stoves and grow lights in the store. Not now—they're out front, literally: There's a commercial PV power plant right outside the Solar Living Center in Hopland.

Unlike most catalogs, the *Solar Living Sourcebook* continues to serve as a unique textbook that provides customers with the understanding and knowledge necessary to select the energy-making and energy-using products they need for a conscientious, low-environmental-impact lifestyle. Classes, available experts, the inspiring Solar Living Center, and the exuberant annual Solfest celebration augment the *Sourcebook* and help to swell the ranks of solar enthusiasts. What a great way to run a business, and what a great way to spread the word! But . . .

But the outrageous building boom in the Bay Area and elsewhere has produced few new homes that reflect an appreciation of solar advantages by architects, builders, or buyers. California law forces builders to meet certain minimum insulation standards, and that's about it. PV panels, solar water heaters, and solar swimming pool heaters are rarely seen. Few new homes are passive solar or even oriented and fenestrated to use sunlight for day lighting. Most roofs are black, significantly increasing air-conditioner loads.

About 45 years ago, solar pioneer Steve Baer of Zomeworks fame warned that proponents of solar power can't claim important success until solar hardware is available at all major building supply stores, and erecting a nonsolar building is unthinkable. There's a long way to go yet, but there's hope for the future as more and more folks are going solar. At last.

Jay Baldwin *("JB") is an inventor, ex-*Whole Earth Catalog *senior editor, and author of* Bucky-Works *and three other books, as well as "JB" Professor of Sustainable Design, lecturer, land steward, and whitewater guide. He's now testing his radical RV that gets double the mpg of conventional designs, and writing a book on managing the dark side of any technology.*

DAVE BLUME

A Full Turn up the Spiral

When I think back to the late 1970s and solar energy, I remember revolution in the air. The catalysts that set it off were two separate streams:

Tools, technology, information, products, and wisdom . . . these are the real goods. The revolution still lies before us, but thanks to the continued evolution of Real Goods (the company) and real goods (the concept), we are in our best shape ever.

Classes, available experts, the inspiring Solar Living Center, and the exuberant annual Solfest celebration augment the *Sourcebook* and help to swell the ranks of solar enthusiasts. What a great way to run a business, and what a great way to spread the word!

the antinuclear power movement and the anger over MegaOilron's exploitation of the Arab oil embargos. Suddenly, everyone could see the disaster our national energy policy was creating for the planet and the economy. Jimmy Carter hit MegaOilron with a windfall profits tax and gave the money to alternative energy. Jerry Brown was governor of California, and we had huge solar tax credits and a revolving loan fund for farmers to build alcohol fuel distilleries.

The crises resulted in an amazing outpouring of creativity, as true American ingenuity was brought to bear. Everyone from *Mother Earth News* to Midwest farmers and every sort of technical tinkerer threw themselves at the renewable energy issue. Along with the *Whole Earth Catalog*, another beacon of sanity shone brightly, the dry goods store of the future, the Real Goods Company. In the early days, it seemed that Real Goods made tangible the ecotopian vision so many of us were trying to bring into existence.

Then, of course, came the very bad dark days of Reagan/Bush and the oily cloud eclipsing the Sun of the solar revolution. Real Goods hardly missed a beat with their inspired campaign, in the tradition of permaculture, to show folks how much power could be saved by changing lighting choices. They showed that overall system design could optimize the impact of small details. And Real Goods showed us a thousand small things we could change in our everyday lives. A switch from revolution to evolution, and the message kept evolving. Let's not forget, too, that Real Goods originally built itself not from Vulture Capital but from thousands of people personally investing small amounts to grow that business. Certainly no banks believed in what John Schaeffer was doing.

Today we are coming around, not full circle, but a full turn up the spiral, to a new solar revolution. But now we come at it with the fully professional infrastructure to make revolution happen in ways we never imagined in 1978. For example, look at the new alcohol fuel movement. Within six months of George Bush admitting that our nation "is addicted to oil" and officially giving the nod to biofuels, more than $10 billion dollars in private capital has flooded into the field. We couldn't have pulled that off in 1978. But with the foundation work diligently laid down by those pioneers and continued by die-hards like John during the thin years, we're more than ready to make the solar vision a Sun-drenched reality.

In the early days, it seemed that Real Goods made tangible the ecotopian vision so many of us were trying to bring into existence.

With the foundation work diligently laid down by those pioneers and continued by die-hards like John during the thin years, we're more than ready to make the solar vision a Sun-drenched reality.

Dave Blume *is a master permaculture designer and teacher. He has also been at the forefront of biofuels development since 1979, including serving as writer and host of the PBS television series,* Alcohol as Fuel, *and author of* Alcohol Can Be a Gas. *The U.S. Patent Office recently issued his "patent to destroy Monsanto," an organic herbicide/fertilizer derived from alcohol fuel by-products.*

JOEL DAVIDSON

Many Happy Returns Beyond the Tipping Point

I have always felt a kinship with Real Goods. We both started selling off-grid photovoltaic (PV) equipment by mail order in 1978 and were among the first to sell on-grid PV systems on the Internet. Also like me, some people at Real Goods have been living in PV-powered homes for many years.

After the 1978 oil crisis, some Americans turned their backs on the Sun and tried to exert further U.S. control over Middle East oil fields. During the 1980s and early 1990s, the off-grid PV market grew, but U.S. solar research and development funds were cut. American PV companies suffered, while Japan and Germany became solar powerhouses and PV grew into a multi-billion-dollar industry.

During the end of the last century, Real Goods and other solar advocates relentlessly promoted real alternatives to resource depletion and pollution. Finally, their ideas took root, and in 1998, the U.S. began the transition from fossil fuels to renewables. Californians passed the first PV net-metering and solar incentive laws thanks to Real Goods' and other people's grassroots work. During the next decade, solar arrays that had only dotted the rural landscape began to spread to cities and towns across America.

I think that the U.S. solar tipping point happened during the 2001 California energy crisis, which triggered laws that encouraged businesses to go solar in a big way. In rapid succession came solar programs in other states; the 30% federal solar tax credit; the $3.2 billion, 10-year California solar program; and more. Meanwhile, most European countries implemented or expanded solar and wind power use, and China entered the solar market.

All the while, Real Goods and other solar advocates led by example and worked hard to replace waste, pollution, and war with renewable energy and peace. Congratulations, Real

Goods, on your 30th anniversary, and many happy returns.

Joel Davidson *has worked at all levels in the photovoltaics and building industry. He wrote* The New Solar Electric Home: The Photovoltaics How-to Handbook *and lives in California in a solar-powered house.*

DAVID KATZ

Between Capitalism and Social Consciousness

John Schaeffer and Real Goods have been synonymous with sustainable living for nearly three decades. Over the years, John and I have been competitors, customers, and vendors of each other—and always good friends. John has done as much as anyone I can think of to put renewable energy front and center before everyday consumers and the public, an invaluable service to the cause of promoting renewable energy and green living.

John is an expert at walking the line between market capitalism and social consciousness. And whether selling recycled toilet paper, poopettes, or solar modules, he has always done so with sensitivity, style, and grace.

In the early 1980s, Real Goods and Alternative Energy Engineering were competitors in the retail mail-order renewable energy business. Since we changed our name to AEE Solar and decided to sell exclusively wholesale, Gaiam Real Goods has become one of our best customers.

One especially fond memory of time spent with John was in 1996 when he and I were arrested in the first wave of trespassers on Pacific Lumber property at a protest against the clear-cutting of Headwaters Forest, one of the few remaining large tracts of old-growth redwood trees. The two of us ended up in handcuffs on one of the first police buses to leave the site. But we never made it to jail. The bus dropped everyone off on a freeway exit ramp and went back to pick up more trespassers. Hundreds of people were arrested that day, and we were proud to be among them. And most important, much of Headwaters Forest was saved for future generations in a landmark agreement brokered by Senator Dianne Feinstein in 1999.

The alternative energy industry has changed so much in the 30 years John and I have been in the business that one can hardly recognize it anymore. Indeed, PV and other renewable energy sources cannot even be considered "alternative" any longer, as renewable energy is well on its way to becoming an accepted part of the mainstream energy supply—thanks in no small measure to the hard work, dedication, and vision of John Schaeffer and his team at Real Goods.

David Katz *founded Alternative Energy Engineering (AEE) in 1979, originally as an equipment supplier to homesteaders in northern California. AEE Solar, now a subsidiary of Mainstream Energy, is one of the largest wholesalers of renewable energy systems in the U.S. David has lived in solar homes for 25 years and authored numerous articles on renewable energy.*

PAUL MAYCOCK

Real Goods Spearheads the PV Revolution

I met John Schaeffer in the late 1970s while I was in Washington, D.C., managing the U.S. photovoltaics program. He gave me my first Real Goods catalog, a portent of things to come.

Solar photovoltaic production was a grand total of 320 kilowatts in 1976. All the products in the '70s were small power (5-200 watts) for off-grid applications; the modules cost about $15-$20 per peak watt. By 1985, the market had grown to 25MW per year with prices of $6-$7 per watt, still all for off-grid applications. The Real Goods catalog offered virtually all the consumer and household products in existence. In 2006, the world PV market reached nearly 2.3GW. The portion of the market serving homes and commercial buildings connected to the utility grid has grown from nearly zero in 1995 to 1.6GW in 2006. This historic growth is largely due to generous PV subsidy programs in Germany, Japan, and the United States (mainly California). The off-grid sector has continued to grow at its historic rate of 13%-15% per year to the 400MW level in 2006.

Let's consider some of the key products that spurred the early development of PV.

Consumer PV Applications. This key sector serves millions of people throughout the world with products that demand fewer than 50 watts: battery chargers for dozens of applications, including car batteries; solar-powered lights, calculators, watches, and other electronics; PV-powered lanterns and garden lights; marine modules, solar ventilators, fans on hats, DC

John is an expert at walking the line between market capitalism and social consciousness.

One especially fond memory of time spent with John was in 1996 when he and I were arrested in the first wave of trespassers on Pacific Lumber property at a protest against the clear-cutting of Headwaters Forest.

blenders, insect and mole repellents, electric fence chargers, ozone generators for pool purification, and, most recently, power for cell phones and other portable electronics. Real Goods led the way in the creation of this market.

PV for Habitat (50W-1kW). This most important sector provides electricity for over 500,000 remote, off-grid residences—vacation cottages, boats, and homes, especially in the developing world. Having volunteered with the Solar Electric Light Fund (SELF), Solar Light for Africa (SLA), and SELCO (the Solar Electric Light Company) in India, Sri Lanka, China, and Viet Nam, I can vouch that PV is life changing for the 100 million homes in the developing world without utility electricity. Real Goods and hundreds of "mom and pop" dealers spread the word on this vital service to the "outback" of the United States.

PV for Commercial Applications (50W-20kW). PV has served hundreds of off-grid commercial applications that prove its reliability and economic feasibility as a primary source of electricity. Most important in this regard is remote power for communications and signals, including buoys for navigation aids, repeaters and amplifiers for telecommunications, emergency call boxes on highways, and parking lot lighting. PV is also used for portable highway signs, railway crossings, bus stop shelters, remote sensors, and commercial lighting. Over 30% of the total off-grid PV market is installed in the commercial sector.

The ultimate success of PV is very likely attributable to the remarkable progress made in the off-grid sector. John Schaeffer, with his 30-year commitment to this PV market through the Real Goods catalog, the *Solar Living Sourcebook*, excellent training programs, and the Solar Living Institute in Hopland, California, has led the way.

The chart below shows the forecast for the world PV market to 2010. The off-grid market totals 400MW in 2006, accounting for 20% of applications worldwide. It is interesting to note that his key sector was 80MW, or 90% of the world market in 1996, and was nearly 100% from 1982 to 1996. John and his Real Goods crew have made a major contribution to the development of PV applications.

Paul Maycock, *president of Photovoltaic Energy Systems (pvenergy.com), is a physicist who has been involved in the PV field since the late 1950s. Along with three patents, he has authored over 100 publications on thermoelectricity, thermal transport, strategic planning, and all aspects of photovoltaic energy conversion.*

PV World Market Forecast (Installed Systems, MW)

Market Sector	1996	1998	2000	2002	2004	2005	2006**	2010
Consumer products	22	30	40	60	75	80	90	160
World off-grid rural	23	34	53	85	110	125	140	450
Communicate/signal	23	31	40	60	80	90	100	350
Off-grid/commercial	12	20	30	45	55	60	70	140
Grid-con. res./comcl.	7	35	120	270	700	1,375	1,600	4,500
Large>500k=W	2	2	5	5	20	30	100	400
Forecast (MW/yr.)						1,400*	2,100**	6,000**
Actual (MW/yr.)	89	152	288	525	1,040	1,550	2,200	
Forecast avg. price ($/W)	—	—				3.00*		2.50**
Actual avg. price ($/W)	4	4	3.5	3.25	3.25	3.50	3.75	—

*Forecast in 1995   **Forecast in 2005

30th ANNIVERSARY

Relocalization

A Strategic Response to Peak Oil and Climate Change

WE BEGIN THIS 30TH-ANNIVERSARY EDITION of the *Real Goods Solar Living Sourcebook* with an examination of what are likely the two most critical issues of our time and the rest of the 21st century: peak oil and global climate change. The end of the age of cheap and abundant fossil fuels, and the disruption, unprecedented in human history, of long-stable climatological and biological systems that industrialization has wrought, are destined to alter life on Earth as we know it. This statement is not made from an ideological stance. It is grounded in the overwhelming weight of scientific fact and the laws of physics and ecology as we understand them. Even the last remaining scientist skeptics—even the last holdouts among the oil companies themselves!—are beginning to grudgingly acknowledge the reality of these earthshaking interlocking trends. To guide us in the exploration of these issues and some promising strategies for grappling with them, we called upon one of our local heroes, who has done much groundbreaking work on relocalization and has developed into a key community leader, Dr. Jason Bradford.

The greatest hope, we think, rests in the ability to honestly accept the reality of a situation and then make the best of it.

People may indeed be scared or shocked by predictions of ensuing environmental and social chaos, but while their awareness and concern are definitely growing, many are still indifferent. How people respond emotionally to facts and deductions is important, but ultimately, if they are unable and unwilling to accept what is true because it makes them feel badly, then positive change is not possible. The greatest hope, we think, rests in the ability to honestly accept the reality of a situation and then make the best of it.

We also hear a lot these days about globalization, the economic dimension of the same interdependence between all corners of the Earth, which propels the peak oil and climate change crises. A new global movement is forming to challenge the existing economic system in light of energy constraints, threats from pollution, erosion of ecosystem services, the social costs of mass consumerism, and a built environment designed around automobiles. This movement draws from a rich history of activism related to environmental and social concerns and is termed "Relocalization" by the Post Carbon Institute, a think tank, media outlet, and networking and support organization for local groups around the world.[1] Relocalization is a promising strategy—perhaps the most potent strat-

egy—for effectively grappling with the enormous challenges posed by the probable impacts of both peak oil and climate change.

Relocalization may be a new term, but conceptually it has long roots. Some related recent precursors include E. F. Schumacher[2], Ted Trainer[3], Garrett Hardin[4], and Wendell Berry[5] as well as what are called the "antiglobalization" movement, the "slow food" movement, the "voluntary simplicity" movement, the "back to the land" movement, "new urbanism," and the "environmental movement." In general, common themes include decentralization of political and economic structures, less material consumption

One aspect of relocalization is buying locally grown produce, such as the fruits, vegetables, and other products available at Rosaly's Garden, an organic farm in Peterborough, New Hampshire.

Photo: © Ann Card

and pollution, a focus on the quality of relationships, culture and the environment as sources of fulfillment, and downscaling of infrastructural development.

Before describing some of the details of Relocalization, let's examine its basic premises. We believe that these premises are sound, being grounded in good science and common sense. By contrast, the assumptions underlying most of the economic and social models that have led to our current environmental and resource predicaments are essentially unsound rationalizations to justify short-term, often individual interests. Our society's obsession with growth and gain have blinded us to the real common good, the needs of future generations, and the welfare of nonhuman life on the planet. Changing these paradigms and reorienting the trajectory of our society is a major undertaking, but all of us—as individuals and working together collectively—*do* have the power to make a positive contribution.

Ecological Economics

During the era of cheap energy, the study of economics became divorced from an understanding of how human systems are connected to planetary systems. Not surprisingly, the nearly free energy available from fossil fuels, and the rapid technological advances they fostered, made people in modern industrialized societies believe they were no longer constrained by tangibles like food, energy, water, and the weather. But the hubris of our recent past is being revealed, and many are searching for a more honest and realistic reckoning of humanity's place on Earth.

A helpful place to look is the discipline called Ecological Economics.[6] A conceptual model based on Ecological Economics is useful both to understand the current economic system and its vulnerabilities and to guide the development of a sustainable alternative.

Mainstream economic thinking usually distorts or fails to fully understand the fundamental interconnectedness of "the economy" and "the environment." Only in recent decades have economists begun to consider the environmental or ecological dimensions of human productive activity. But even when economists do take account of these relationships, their formulations are typically partial or misguided from a vantage point that takes the global environment seriously. For example, wealthy and environmentally responsible nations are sometimes touted as examples of how economic growth and stewardship of the planet go hand in hand. However, while local measures of air quality, forest cover, and water cleanliness may be high, all wealthy nations are actually importers of much of their environmental carrying capacity, whether it is raw materials or finished industrial products, and these imports are possible because of fossil fuels used to mine, harvest, manufacture, and transport goods. Wealthy nations may protect their own environment—to some extent, anyway—but they do so by outsourcing the damage caused by overconsumption to other places.[7]

In the Ecological Economics model, the Human Economy is a subset of the Earth System, and therefore the *scale* of the Human Economy is ultimately limited. The Human Economy depends upon the *throughput* of materials from and back into the Earth System. Just pick up any trinket in your possession and ask, What is it made of? Where did these materials come from? How much energy was used? and What happens to the waste products?[8] Limits to the size of the Human Economy are determined by three related factors: 1) the capacity for the Earth System to supply inputs to the Human Economy (Sources), 2) the capacity of the Earth System to tolerate and process wastes from the Human Economy (Sinks), and 3) the negative

Fig. 1. The Ecological Economics model of the relationship between the Human Economy and the Earth System highlights the importance of sources, sinks, feedbacks, and scale.

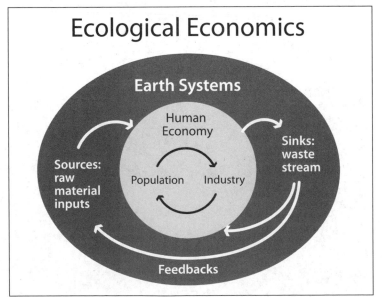

impacts on the Human Economy and the resources it relies on (Feedbacks) caused by too much pollution.

For example, mining coal makes available a "source" of energy for industry that produces pollution, including sulfur dioxide that causes acid rain. Too much acid rain degrades built infrastructure and overwhelms the capacity of natural "sinks," such as forests, killing them or slowing their growth. The loss of ecosystems also creates new costs to society for ecological services that were previously accomplished "free of charge" through ecological processes. Clean air and water, stable climates, and species interactions that moderate outbreaks of disease are all compromised when damage is done to ecosystems. The human economy then invests in expensive technologies to try and compensate for this damage, such as pollution control devices, flood control walls and canals, pesticides, medicines, and more.

The current Human Economy is clearly unsustainable because it relies heavily on nonrenewable raw-material sources. These are by definition finite, and using them produces tremendous pollution that leads to many negative "feedbacks" that impair ecosystems and disrupt climate. A sustainable economy would need to run on the income from solar energy and not degrade ecosystems through the buildup of wastes or the mining of nutrients.

Relocalization is based on an ethic of protecting the Earth System and its "natural capital," knowing that despite human cleverness, our well-being is fundamentally derived from the ecological and geological richness of Earth.

> A sustainable economy would need to run on the income from solar energy and not degrade ecosystems through the buildup of wastes or the mining of nutrients.

Overshoot

If the scale of the Human Economy (inner circle within the Ecological Economics model) is too large relative to the Earth System, the Human Economy is in a state of *overshoot*. This means that the environmental load of humanity on the planet is greater than the long-term ability of the planet to support it. Overshoot means we are *above carrying capacity*. This environmental load will eventually be reduced through declines in some combination of population, resource consumption, and pollution. Either we manage to reduce our environmental load, or resource constraints and pollution will limit it for us—with unpleasant and potentially catastrophic consequences.[9]

The concept of overshoot can be confusing. You may ask, How can a population go beyond the carrying capacity of the environment to support it? Won't a population simply increase until it reaches carrying capacity, and then stabilize? Isn't the human population projected to stabilize in this century? Sophisticated modeling of population, resource, and consumption dynamics provides answers to these questions that persuasively suggest the reality of overshoot.

Population overshoot may happen for several different reasons: 1) resource windfall and drawdown, 2) release from negative species interactions, 3) demographic momentum, and 4) fluctuating carrying capacity. These mechanisms of overshoot are not exclusive, and in fact they can feed positively on one another. Here is an example of how these mechanisms have interacted in modern human history.

In the middle of the 19th century, people discovered a dense and versatile energy source in fossil fuels, especially petroleum. The use of fossil energy freed up other resources, such as land and labor. Without the need to feed draft animals to power equipment, more land was available to grow food for humans (i.e., resource windfall and drawdown). With fossil fuel-powered equipment, fewer humans were needed for manual labor, enabling extended educational opportunities and a shift of resources into fields such as public health and medicine. Increased societal attention to health and medicine, and corresponding technologies like vaccines,

Fig. 2. Human demographic models of population show a plateau this century (solid line is approximate historic and demographic projected), whereas systems models show a decline (gray line). The difference exists because human demographic models do not include negative feedbacks from either resource scarcity or pollution, whereas systems models do.

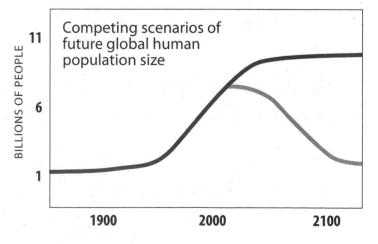

For the past 34 years, Stephen and Gloria Decater have been powering their 40-acre biodynamic farm in Covelo, California, with some form of solar energy—including draft horses and photovoltaic panels.

antibiotics, and sanitation, resulted in increased human life expectancy (i.e., reduced negative species interactions). A rapid increase in the human population increased the number of fertile women of childbearing age, leading to an even larger population (i.e., demographic momentum). As this population became very large, it began to impact the natural world around it substantially. Toxic emissions built up that harmed the basic life-support systems humans depend on, eventually making it more and more difficult to provide essentials, such as food (i.e., fluctuating carrying capacity).

Experts in the field of human demography project that the human population will stabilize around the middle of the 21st century.[10] Most people accept this analysis without knowing the underlying assumptions. Unfortunately, most studies of human population are akin to most studies of the human economy. The broader environment is not factored in to models of growth. If you have ever asked yourself, How are we going to feed 9 billion people when the soils are eroding and the aquifers are being depleted and the climate is changing and the deserts are expanding and oil and natural gas supplies are dwindling? then you have stumbled upon

this disconnect between most human population models and the physical world. Biologists studying any population would include those environmental factors in their models, whereas human demographers do not.

However, models do exist that contextualize the human population and our well-being within a dynamic study of resource availability, pollution levels, and even climate change and the fate of ecosystems. The classic example is the World3 model developed by the authors of *Limits to Growth*, where the baseline scenario shows human population declining after 2020.[11] Another model is GUMBO, from the University of Vermont's Gund Institute of Ecological Economics.[12] These models are not perfect, but they at least begin with the right premises and tell us what aspects of human civilization are likely pushing the boundaries of, or already exceeding, the physical and ecological capacities of Earth.

Relocalization starts from the premise that the world is a finite place and that humanity is in a state of overshoot. Perpetual growth of the economy and the population is neither possible nor desirable. It is wise to start planning now for a world with less available energy, not more.

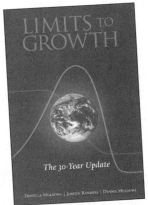

The 30-year updated edition of *Limits to Growth*.

Peak Oil and Implications for a Transportation-Dependent Economy

To a great extent, the Relocalization movement was sparked by concerns about "peak oil."[13] Given that the amount of oil on the planet is finite, at some point in time less is going to be available to us than in the past. Peak oil doesn't mean that oil "runs out," but it does mean that

the cheap and easy-to-get oil is gone and that what remains requires more energy, is more costly to produce, and is extracted at a progressively slower rate. The rate of decline of oil after peak is difficult to predict, but scenarios range from 1% to 8% per year. The peak may be somewhat "flat," with a slow initial decline that accelerates over time toward the higher end of the depletion-rate range. How human societies respond to the postpeak environment will likely be as important a factor as geology in determining what is available to societies. Do we cooperate, or do we fight like cats in a sack?

In the Ecological Economics model, peak oil is a "source" issue. Several source problems face the human economy, including peak natural gas and peak water.[14] Greater expansion of the human economy requires greater inputs, and aside from the ecosystem services provided by nature, oil is probably the single most important economic resource on the planet. Oil is critical for at least two reasons: energy density and versatility.

The energy output of a single person doing manual labor, averaged over a period of days, is equivalent to about 200-300 British Thermal Units (BTUs) per hour. A single gallon of gasoline contains about 150,000 BTUs of potential energy, roughly equivalent to 500-750 hours of hard human labor.[15] The energy density of oil has not simply permitted a life of leisure and travel for those with access to it—it has in fact greatly expanded the short-term carrying capacity of the human population. By harnessing the energy of oil (and other fossil fuels), our species has been able to outcompete others for space and resources. The expansion of industrial agriculture and "green revolution" technologies are based on oil and natural gas feedstocks and energy. Construction of large dams, water diversion systems, and pumps for groundwater and water delivery to fields and cities depends upon plentiful fuel. Land, water, and other resources that in the past had been available to a diversity of species are being funneled toward the appetite of only one species—hence the biodiversity crisis.

Oil is versatile because it is a liquid, making it easier to extract and transport than coal and natural gas. Oil is more readily available as a fuel for a global market because it can be put into pipelines and tankers without requiring special treatment. Natural gas, by contrast, needs to be cooled and pressurized for tanker travel, and coal needs to be pulverized into slurry to be piped or put onto freight cars or barges for long-distance transport.

Because oil can be delivered anywhere so efficiently, modern transportation systems have become reliant on it. A few buses and cars use natural gas. Some trains run on electricity. But the vast majority of transportation applications

on the planet—more than 90%—use oil in the form of gasoline, diesel, or kerosene (jet fuel). Consequently, modern economies are extremely vulnerable to shortages in transportation fuels. The relative stability of the oil market over the past several decades has led to the development of "just in time" delivery of products and to commercial linkages across the globe. Local and regional warehouses are uncommon now, with stores and businesses relying on frequent shipments to maintain a low overhead. Before the era of cheap transportation, each town and city had a full complement of craftspeople who relied on each other. Nowadays, businesses are connected through vast transportation networks, with a manufacturing company in California, for example, relying on components shipped in from Asia and Europe.

The food economy is perhaps the finest example of the insecurity that is now bred into normal societal infrastructures. Markets selling food are typically restocked daily with only a few days' supply available in the store. This fact leads many people concerned about peak oil to reason: no fuel, no trucks; no trucks, no food. The shifts in agricultural practices that have occurred in the past 30 or 40 years make it difficult to quickly switch to a less transportation-intensive food system. Many agricultural regions are overly specialized to serve global markets. For example, a place where 50 years ago granaries, dairies, vegetable farms, and ranches coexisted is now dominated by premium wine grapes.

These developments have been possible only because cheap oil has allowed us to overcome the limitations of local ecologies. And because oil possesses a unique combination of attributes, finding a suitable and equally effective substitute is no easy task, and perhaps [is] impossible.[16]

All proposed "substitutes" for cheap oil appear to fail the test of Energy Returned on Energy Invested (EROEI).[17] For an energy source to be useful to society, it must deliver more energy than it takes to find, harvest, and distribute it. Our economies have become addicted to energy sources with EROEIs of from 100:1 to 20:1, whereas biofuels, tar sands, and many renewable energy technologies range from about 10:1 to 1:1 or less. If a fuel has an EROEI of 1:1, it is essentially useless, because as much energy goes into producing the fuel as the fuel delivers. A complex and sustainable society will probably require substantial EROEI ratios, such as 5:1 or greater. Energy policies need to be devised based on sound EROEI analyses, which are currently difficult to find.

No fuel, no trucks; no trucks, no food.

Photo: © Jupiterimages, www.comstock.com

Highly industrialized agriculture requires about 10 times more energy to grow, harvest, process, and distribute the food than is contained in the food itself.

In the U.S., a high-EROEI energy source permits about 1% of the population to feed the other 99%. In places without access to fossil fuels, such as Afghanistan, more than 90% of the working population is engaged in growing food. Agriculture is, in essence, a means of capturing solar energy through investment in planting, maintenance, and harvesting. While the Afghani agricultural system looks inefficient from a labor point of view, it is actually far more efficient from an EROEI perspective than U.S. agriculture is. The extensive use of fossil fuels in industrialized food systems makes them energy sinks. Highly industrialized agriculture requires about 10 times more energy to grow, harvest, process, and distribute the food than is contained in the food itself—an EROEI of 1:10. Such a system is clearly unsustainable and begs for relocalization. With investment today in the right kind of equipment and training, industrialized nations could transition from fossil fuel-dependent agriculture and grow food with far less than 90% of their working population engaged in farming. According to some estimates, perhaps a third of the population would suffice.[18]

Climate Change and the Need to Eliminate Fossil Fuels

While peak oil is a source problem, climate change is a sink problem.

During the most recent ages of geologic history, Earth has cycled between ice ages and intervening warm periods. These cycles are primarily driven by orbital variations, both with respect to the angle of Earth's tilt toward the Sun and the shape of Earth's orbit around the Sun.[19] Carbon dioxide (CO_2) fluctuated as a result of how ecosystems responded to changes in Earth's temperature, and changes in temperature then amplified those ecosystem changes. In systems theory, which is the guiding paradigm for Ecological Economics and computer modeling, this process is known as a positive feedback loop.

Currently, CO_2 and other greenhouse gas concentrations are rising not because of orbital changes but from the use of fossil fuels. The pre-industrial level of CO_2 in Earth's atmosphere was 280 parts per million (ppm); now it's about 380 ppm. Consider that 100 ppm of CO_2 is what separated the Ice Age from the warm, stable climate of the past several thousand years, and that the corresponding temperature transition took about a thousand years. By comparison, half of all the energy used since the beginning of the industrial revolution has been consumed during the past 20 years, and global average temperatures are rising about 100 times faster than during transitions out of ice ages. Increased greenhouse gas concentrations are only partly responsible for the changes in temperature between an ice age and today. Much of the rise in temperature as an ice age ends is due to the loss of ice sheets that help cool the planet by reflecting sunlight. However, the current rate of

Photo: © Jupiterimages, www.comstock.com

change in the chemistry of Earth's atmosphere and oceans is comparable to only a few previous mass-extinction episodes over the past several hundred million years that appear to be related to radical, rapid climate change.[20]

The rate of change is perhaps more important to the climate system and life on Earth than is the amount of change. A slow rate of change is akin to gently applying the brakes to stop at a light, while a fast rate of change is akin to hitting a brick wall. Both take the vehicle and a passenger from 60 to 0 miles per hour, only one is faster.

Nobody really knows what this means for the climate system, the pH of the oceans, the physiology of plant growth, and other planetary processes. Policymakers ask scientists how much pollution can be tolerated before "dangerous interference" occurs. Unfortunately, answering how much is too much is not possible, and in all probability we have already passed some very dangerous thresholds that will become apparent only as the future unfolds.

There are many reasons for why a precise answer to "how much is too much" is not possible. Consider that for any factor built into a model, scientists 1) work with what they know, 2) try to incorporate plausible ranges for what they know they don't know, and 3) obviously exclude what they don't know they don't know. Some would argue that because we can't be sure climate models are correct, we should do nothing. Would "do nothing" skeptics be as cavalier about uncertain dangers if the food their children ate had *possibly* been contaminated by a deadly poison? What you don't know can kill you. Given the stakes, many advocates of energy policies leading to a curtailment of greenhouse gas emissions take a precautionary stance.[21] After all, if the U.S. is so concerned about security that it is willing to spend about half a trillion dollars a year on the military, what is it worth to help secure our climate?

Computer power limits the ability of models to capture many of the details of change. For example, models can't scale down to the future climate of a single town, which makes it difficult, perhaps, for local officials to understand the implications of global models. Nor can models usually identify critical thresholds in a complex system with much accuracy. Systems can remain remarkably stable over long periods under stress until something snaps, like a balloon expanding until it pops. The Earth System has been remarkably tolerant of the stresses it is under, but when something finally gives, it

will probably be "loud." It is very possible that current models actually underestimate the true threats of climate change.

Abundant evidence exists, both quantitative and qualitative, suggesting that climate change in the direction of global warming is an indisputable fact. The information, images, and data compiled by Al Gore in the movie and book *An Inconvenient Truth* are compelling to any open-minded observer. As we write at the end of 2006, the evidence continues to accumulate. The BBC recently reported that the Global Carbon Project has found that CO_2 concentrations increased by 1% a year from 1990 to 2000 but at 2.5% a year from 2000 to 2005. This finding corroborates conclusions recently issued by the World Meteorological Organization. Such rates of emission acceleration would push global temperature increases over the next century toward the high end of the range predicted by the Intergovernmental Panel on Climate Change, about 10 degrees Fahrenheit.[22] Scientists at the National Climatic Data Center have just concluded that *2006 was the warmest year on record in the U.S.*; each of the past 9 years has been among the 25 warmest years on record; and 8 of the past 10 years are the warmest on record worldwide.[23] Do you see a pattern?

Those of us living in the United States *do* have a special responsibility: While Americans make up about 5% of the world's population, we produce about 25% of all greenhouse gas emissions. A constellation of interrelated climate-change pressure points examined by Professor John Schellnhuber, former chief environmental advisor to the German government and currently research director at the Tyndall Centre for Climate Change Research in the U.K., reveal both the complexity of the problem and the shockingly realistic possibility of dramatic environmental damage.[24] Schellnhuber has identified 12 global ecological "tipping points," weak links that could, under the impact of global warming, trigger the catastrophic collapse of some of Earth's critical ecosystems. Because each of these phenomena operates on a planetary scale, disruption of one influences the behavior of others in chains of positive feedback loops. The dozen fragile systems and/or places in Schellnhuber's model are:

- Amazon Rain Forest
- North Atlantic Current
- Greenland Ice Sheet
- Ozone Hole
- Antarctic Circumpolar Current
- Sahara Desert

While Americans make up about 5% of the world's population, we produce about 25% of all greenhouse gas emissions.

- Tibetan Plateau
- Asian Monsoon
- Methane Clathrates within Siberian permafrost and ocean sediments
- Salinity Valves in the oceans, especially the Mediterranean Sea
- El Niño
- West Antarctic Ice Sheet

The ways in which these tipping points interact with one another are too numerous and complicated to detail here, but a couple of examples will give a good idea of what's at stake. What do the Amazon Rain Forest, the North Atlantic Current, and the Greenland Ice Sheet have in common? Well, we know that deforestation is occurring in the Amazon, and we know that the Greenland Ice Sheet is melting. In fact, studies show that its rate of melting unexpectedly doubled between 1996 and 2005. As it melts, it is discharging massive volumes of freshwater into the sea where the vast oceanic river known as the North Atlantic Current delivers warmth to the European continent via a mechanism called thermohaline circulation (THC). That new freshwater dilutes the salt water of the ocean, thereby disrupting the THC and potentially shutting it down—which probably would make many parts of coastal Europe considerably colder than they are now. Conversely, the warming ocean temperatures would drive accelerated melting of the ice sheet and could also result in drier air over the Amazon. Some climate models predict that even if clearcutting of the rain forest ceased today, the global warming already set in motion will change the Amazon basin from forest to savanna within the 21st century, potentially releasing as much CO_2 as all the fossil fuels burned during the 20th century. The circulation of warmer and drier air connected to the melting of the ice sheet and temperature increases in the ocean would only accelerate the transformation of the Amazonian Rain Forest.

The global warming already set in motion will change the Amazon basin from forest to savanna within the 21st century.

Another scenario, which probably has never occurred to most people, involves the Sahara Desert. Global warming is expected to increase rainfall along the southern edge of the Sahara, which means more plants will grow and the desert will shrink. As it stands, windblown dust from the Sahara seeds the ocean with nutrients that support the phytoplankton population, which is the foundation of the entire oceanic food chain. More rain, less dust, less food for the phytoplankton, decreasing phytoplankton populations, less food for fish. . . . You get the picture.

Of course, these and similar scenarios involving fragile environmental tipping points are somewhat hypothetical, especially when spun out to what some might see as worst-case scenarios. But the point is that the evidence that global warming and large-scale climate change are occurring is real, if not overwhelming, to those who will see it. And a systems view of how ecosystems and biological, hydrological, and climatological processes interact on a planetary scale is sensible. Therefore it is not a stretch—indeed, it only makes sense to take Schellnhuber's warnings seriously—to imagine that the changes being wrought by greenhouse gas emissions could, relatively suddenly, inflict even more-massive damage to Earth than we are already seeing.

And here's one more sobering consideration. Although climate models have limits, they also do an incredible job of accurately modeling the *past* climate. For example, when comparing images from weather satellites to the most advanced climate models, you can see how well models match the actual formation and movement of storm clouds around the globe.

One of the tests climate modelers perform to decide whether human-induced changes in the atmosphere are causing climate change is to run climate models for the 20th century *as if* we hadn't burned so much fossil fuel. The rise in global temperatures and the shifts in rainfall patterns seen during the 20th century can be accurately modeled only when fossil fuel-induced greenhouse gas emissions are included. Natural variations in solar radiation and the shape of Earth's orbit around the Sun do not account for recent climate change. Climate change is our problem.

While we can't know future threats precisely, scientists agree that creating a carbon cycle-neutral economy should be the dominant task occupying our minds. This is exactly what Relocalization aims to do.

Relocalization:
A Strategic Response to Overshoot

Economic and population growth were made possible by the synergies permitted by cheap energy. The limits of productivity in one locality could be overcome by importing something that was produced in excess elsewhere. A global economy emerged, propelled by an imperative that each place seek its comparative advantage and specialize in the marketplace. The recent trend toward "free trade" agreements is further indication that most economists, policymakers, and political leaders see only the benefits of globalized commerce and ignore or minimize the long-term liabilities.

One particular flawed assumption behind globalization is especially glaring, i.e., that transportation costs will always be low, both in terms of fuel availability and the environmental problems associated with their use.[25] If that assumption is false—and certainly peak oil and climate change make it appear false—then localities should not be specializing to trade globally. Take the example of California wine country again. That place grows far more grapes than the local population can eat, but it lacks just about every other kind of food production in sufficient quantity. As long as the region can sell its wine to a global market and buy the other stuff people need, this situation seems reasonable. But a "peak oil" perspective reveals the region's vulnerability, and a "climate change" perspective calls this entire socioeconomic system irresponsible.

Relocalization advocates rebuilding more-balanced local economies that emphasize securing basic needs. ***Local food, energy, and water systems are perhaps the most critical to build.***[26] The movement toward relocalizing food networks is perhaps the most advanced today. A new book by Sandor Ellix Katz, *The Revolution Will Not Be Microwaved: Inside America's Underground Food Movements* (Chelsea Green, 2006), documents the growth of initiatives that are working to restore traditional food production and distribution methods and revive local economies. Katz's analysis speaks directly to the issues we've been discussing and the promise of relocalization:

> [F]ood-related political activism . . . seeks to revive local food production and exchange and to redevelop community food sovereignty. There is no sacrifice required for this agenda because, generally speaking, the food closest at hand is the freshest, most delicious, and most nutritious. This revolution will not be genetically engineered, pumped up with hormones, covered in pesticides, individually wrapped, or microwaved. This is a revolution of the everyday, and it's already happening. It's a practice more of us can build into our mundane daily realities and into a grassroots groundswell. This revolution is wholesome, nurturing, and sensual. This revolution reinvigorates local economies. This revolution rescues traditional foods that are in danger of extinction and revives skills that will enable people to survive the inevitable collapse of the unsustainable, globalized, industrial food system.[27]

In the absence of reliable trade partners, whether from peak oil, natural disaster, or political instability, a local or regional economy that at least takes care to produce the essentials will have a true comparative advantage. Relocalization will promote local and regional stability. Because it reduces the distances that goods travel between production and consumption, it will also significantly reduce pollution and greenhouse gas emissions. Ideally, relocalization is grounded in the principles of Ecological Economics and is developed around renewable energy inputs and cycling of nutrients. The more we can do, the greater the positive impact it will have on reducing the potential consequences of these impending crises.

> *Relocalization advocates rebuilding more-balanced local economies that emphasize securing basic needs.*

Approaching Social Change

A local economy that takes care of its basic needs is also a very interesting place to be.

The problems we face tempt many to drop out of society as much as possible and live a simple life, semi-isolated from the horrors "out there." However appealing this may be, global traumas will most likely catch up with everyone. Reversing course and implementing a complete overhaul of our collective lives requires massive cooperative action.

History shows some instances when societies have responded wisely and plenty of other instances when they didn't change in time.[28] There's no guarantee that relocalization will be successful, but we can hope that our work will improve the odds. And the journey has many inherent benefits in any event.

When approaching others, it's useful to keep an old sales adage in mind: Sell the benefits, not the features. Truthfully, nobody really knows "how" to relocalize economies in any intimate detail. Many examples of the components of a local, sustainable economy can be found, but nowhere can we point to an example of a place where it has all been put together harmoniously. This is something we will have to learn how to do together. So don't get too hung up on the "features" of a local economy beyond some broad principles and working examples. In the meantime, sell the benefits to enroll people in the vision.[29]

The benefits of a local, sustainable economy would be extensive, beyond the environmental pluses already noted. Such an economy is more responsible, secure, and potentially more "free" in some ways. With greater self-reliance comes greater political autonomy and less vulnerability to instabilities elsewhere. A local economy that takes care of its basic needs is also a very interesting place to be. The diversity of goods, services, and skills required is much greater than what many in the U.S. are accustomed to in their communities. Such a place to live would be attractive to all generations, as each individual would be more likely to find a role suited to his or her talents and interests. The stability of a locally focused economy enhances community bonds. Instead of a hyper-mobile society in which anonymity is prevalent, people would have the time to get to know each other and work together based on mutual understanding.[30] This social lubricant lowers what economists call transaction costs, which can be very important for getting work done together efficiently and responding cohesively during crises. It is hard for people nowadays to "love thy neighbor," when in many cases they wouldn't even recognize their neighbor.[31]

Beyond the personal level, the challenge is to engage our neighbors and communities in the project of shifting public investment strategies and creating laws that can lead to significant behavioral changes at the societal level. Changing a behavior required to get by on a daily basis, such as driving a car, requires that the built environment make alternatives relatively convenient. That means that tax dollars going toward highway projects and airport expansion need to go instead toward a locally scaled, non-fossil fuel-dependent transportation system.

Activists will often hear this response: "Sounds great, but we have no money." In the United States, however, vast material resources are devoted to expanding freeways, building cities in deserts, generating electrical power using coal and natural gas, and producing military hardware. In truth, shifting public investment resources is mainly a matter of priorities. But that requires involvement in one of the most difficult social environments of all: politics. Bumping up against institutional norms may sound daunting, but the good news is that reality is conflicting with dominant belief systems. This conflict sets up an opportunity to help the disillusioned or confused by offering a coherent explanation of what is happening, and pathways to realign their thinking with the new realities.

Members of the Peterborough, New Hampshire, community garden gather to share the bounty.

Photo: © Ann Card

Being politically active doesn't require running for office or joining a political party, though those are fine options. Because relatively few Americans are actually civically engaged, a small group of well-organized and thoughtful people can wield great power. After all, this is what professional lobbyists do.

The message that we need to develop local economies is an easy sell because most people readily understand the advantages of greater self-reliance and strengthened communities. Job loss trends related to the dissipation of local manufacturing and agriculture are sore spots that lead many to question the wisdom of globalization. In this context, people are receptive to cogent arguments that reveal the insecurity and environmental problems of our dependence on fossil fuels and other imports, and they are more willing to see the vision of an interesting and beautiful alternative.

Shifting established patterns of behavior and generating the same sense of urgency many activists feel is more difficult. Two parallel strategies can help us to move forward.

The first is a strategy of earnest but careful dialog. Developing a relationship with leaders in your community based on mutual respect and trust is crucial. Establishing such personal connections will improve the likelihood that decisions will be made from the perspectives and priorities of relocalization. However, be aware that because the message of overshoot challenges cherished assumptions and intrudes on many people's comfort zones, it can be a somewhat awkward dance getting to know someone and communicating your purpose.

Understanding the art of conversation is a useful skill when navigating interpersonal relationships. Being able to listen carefully to understand others' perspectives, and finding avenues of shared concern, is a great way to start. Build from an initial foundation of respect. Pointing fingers and demanding that people "get on board" is less effective than asking friends to assist in dealing with a mutual problem. Strive for understanding and agreement, but learn how to live with differences and to tolerate tension and conflict. We don't always get along with those we care most about, but we still try to stick together.

The second strategy is modeling through tangible examples the kinds of changes being advocated. Those of us "ahead of the curve," so to speak, will need to pull up our sleeves and create some of the alternatives we're talking about. Examples abound, including local monetary systems, renewable energy devices, community gardens and farms, farmer's markets, bike clubs, and small businesses or nonprofit organizations helping people do all of the above and more. Ideally, as these activities demonstrate success, gain credibility, and become more cost-effective, more and more people will rally to support them.[32]

If you want to start a new group in your area to work on peak oil, climate change, renewable energy, or relocalization issues, consider joining the Relocalization Network.[33] The network is valuable as a distributor of information resources, a way to learn from others on the same path, and, as it grows, a potentially influential voice.

Most important, don't be paralyzed by indecision and fear. Doing nothing is a capitulation to disaster, while doing something is empowering and potentially transformational.

> Those of us "ahead of the curve," so to speak, will need to pull up our sleeves and create some of the alternatives we're talking about.

End Notes

1. www.postcarbon.org/
2. www.schumachersociety.org/
3. http://socialwork.arts.unsw.edu.au/tsw/
4. www.garretthardinsociety.org/
5. www.brtom.org/wb/berry.html
6. A college-level textbook by Herman E. Daly and Joshua Farley titled *Ecological Economics: Principles and Applications* (Island Press, 2004) exists. Also look for popular books by Herman Daly, Brian Czech, and Richard Douthwaite.
7. See measures like the Ecological Footprint (www.footprintnetwork.org) and the Genuine Progress Indicator (www.redefiningprogress.org/projects/gpi/).
8. A great book that leads the reader through this process for several consumer items is John C. Ryan and Alan Thein Durning, *Stuff: The Secret Lives of Everyday Things* (New Report No. 4, January 1997, Northwest Environment Watch, Seattle).
9. The book by William R. Catton Jr. gives a thorough overview of ecological and social mechanisms and consequences of overshoot.
10. A great place to review standard population projections and the underlying assumptions is through the United Nations Population Division websites: www.un.org/esa/population/unpop.htm and http://esa.un.org/unpp/.
11. Donella Meadows, Jorgen Randers, and Dennis Meadows, *Limits to Growth: The 30-Year Update* (Chelsea Green Publishing, 2004).

12. www.uvm.edu/giee/research/publications/Boumans_et_al.pdf

13. Literally dozens of books, websites, and articles about peak oil exist. Richard Heinberg, *The Party's Over: Oil, War, and the Fate of Industrial Societies*, 2nd Ed. (New Society Publishers, 2005) is hignhly recommended. On the Internet, try www.energybulletin.net/ and www.theoildrum.com/.

14. Much less has been written specifically about natural gas, but see Julian Darley, *High Noon for Natural Gas: The New Energy Crisis* (Chelsea Green, 2004).

15. For a slim but comprehensive book on energy and conversion factors, see John G. Howe, *The End of Fossil Energy and the Last Chance for Sustainability*, 2nd Ed. (McIntire Publishing Services, 2005).

16. A fundamental concept of ecology is Liebig's Law of the Minimum, which states that the growth of a population will be limited by whatever single factor of production is in short supply, not the total amount of resources. The expression "for the want of a nail" captures Liebig's Law.

17. An important book covering EROEI and agriculture is John Gever, Robert Kaufmann, David Skole, and Charles Vorosmarty, *Beyond Oil: The Threat to Food and Fuel in the Coming Decades* (Ballinger Publishing Company, 1986).

18. A comparison of the energy balance of different food systems is provided by David Pimental and Marcia Pimental, Eds., *Food Energy and Society* (University Press of Colorado, 1996). For a view of how to transform U.S. agriculture, see Richard Heinberg, "Fifty Million Farmers," www.energybulletin.net/22584.html.

19. http://en.wikipedia.org/wiki/Milankovitch_cycles

20. Dozens of references are possible for climate change. A good recent book, written by a scientist, is Tim Flannery, *The Weather Makers: How Man Is Changing the Climate and What It Means for Life on Earth* (Atlantic Monthly Press, 2005). On the Internet, see this site run by climatologists: www.realclimate.org/.

21. http://en.wikipedia.org/wiki/Precautionary_principle

22. Al Gore, *An Inconvenient Truth: The Planetary Emergency of Global Warming and What We Can Do About It* (Rodale, 2006), www.climatecrisis.net/. Richard Black, "Carbon Emissions Show Sharp Rise," http://news.bbc.co.uk/2/hi/science/nature/6189600.stm.

23. Robert Lee Hotz, "Record Warmth (Again) in 2006," *Los Angeles Times*, January 10, 2007.

24. See Julia Whitty, "The Thirteenth Tipping Point," *Mother Jones* (November/December 2006), www.motherjones.com/news/feature/2006/11/13th_tipping_point.html. For more on Schellnhuber, see Ian Sample, "Pressure Points," *The Guardian*, October 14, 2004, www.guardian.co.uk/guardianweekly/story/0,,1333874,00.html.

25. A few books do a fine job discussing both "source" and "sink" problems with fossil fuels, including Thom Hartmann, *The Last Hours of Ancient Sunlight: Waking Up to Personal and Global Transformation* (rev. ed., Three Rivers Press, 2004); Jeremy Leggett, *The Empty Tank: Oil, Gas, Hot Air, and the Coming Global Financial Catastrophe* (Random House, 2005); and James Howard Kunstler, *The Long Emergency: Surviving the Converging Catastrophes of the Twenty-First Century* (Grove Press, 2006).

26. Books addressing the benefits of a local economy focused on basic needs include Richard Douthwaite, *Short Circuit: Strengthening Local Economies for Security in an Unstable World* (Green Books, 1998), and Michael Shuman, *Going Local: Creating Self-Reliant Communities in a Global Age* (Routledge, 2000).

27. Sandor Elliz Katz, *The Revolution Will Not Be Microwaved: Inside America's Underground Food Movements* (Chelsea Green, 2006), Introduction.

28. Jared Diamond, *Collapse: How Societies Choose to Fail or Succeed* (Viking Penguin, 2005).

29. An example of how to sell relocalization to a community is the document called "Joint Statement toward a Sustainable, Healthy Willits" by Willits Economic Localization, http://willits.postcarbon.org/files/well/JointStatement.pdf.

30. Intriguing insights into the social benefits of a more cooperative local economy can be found in Eric Brende, *Better Off: Flipping the Switch on Technology* (Harper Collins, 2004), where the author joins a traditional community in America's heartland that uses no motors, electricity, or fossil fuels.

31. See Robert D. Putnam, *Bowling Alone: The Collapse and Revival of American Community* (Simon and Schuster, 2000) for a detailed review of the loss of "social capital" in the U.S. and what this costs us.

32. For a more in-depth discussion of what it is like to develop a community group with the goal of relocalization, see http://willits.postcarbon.org/files/well/OutpostGuideWillits.pdf.

33. www.relocalize.net/

Pioneering Solar down on the Farm

Left: The Decaters plow their fields with "original" solar farming technology—draft horses that eat grass fed by the Sun. **Above:** Tended by apprentices, compost piles at Live Power Community Farm recycle onsite waste. **Below:** John Schaeffer visits with Steve and Gloria Decater down on the farm.

Live Power Community Farm has been generating its electricity via some form of solar energy for decades. Now the farm is set to add another leg to its legendary system.

Talk about your renewable energy pioneers. For the past 34 years, Stephen and Gloria Decater have been powering Live Power Community Farm, their 40-acre Demeter-certified biodynamic operation in Covelo, Calif., with some form of solar energy.

Whether it be from the draft horses that plow their fields ("Horses are eating grass and forage, and their energy source is the Sun, so, in effect, horses are solar powered," Stephen says) or from photovoltaics (PV), the Decaters have been committed to solar energy for nearly as long as the term *solar energy* has been around.

"Our goal has always been to produce food from solar energy rather than fossil fuel. We designed the farm on that principle," Stephen says.

Right now the farm is on its third leg of solar power development. In addition to their draft horses, the Decaters currently have a 13kW PV system that provides much of the farm's power. For help designing and installing the third phase, they contacted Real Goods. "The system is

getting pretty involved," Stephen says. "The people at Real Goods were extremely knowledgeable in helping us understand the next leg of it." When phase three is complete, an additional 30 panels will provide another 5kW of electricity for the farm. And the Decaters are also developing a separate thermal system for all the farm's hot-water needs, including radiant heat, and they hope to have it online this year.

POWERING COMMUNITY SUPPORTED AGRICULTURE

If there's a quintessential statement to be made about reverence for the Earth, it's being made every day at Live Power Community Farm. The farm was one of the first Community Sup-

ported Agriculture (CSA) projects in California. CSAs now number between 1,200 and 1,500 in North America and are based on an economic paradigm vastly different from traditional market-based systems.

Essentially, CSA is a social and economic contract between growers and consumers. "For us, CSA means 100% community-based. Members support our annual operating budget, and all the food we grow is distributed to members," Gloria says. "It's an associative economy rather than a market economy. The association is between growers, consumers, and the Earth."

TECH SPECS	
Solar system size	13kW currently; 18kW after new panels installed
Est. average annual savings	$1,600+
Solar panels	32 Siemens SR 100; 72 Shell SP 140; 30 Sharp 165W modules
Controller	2 Xantrex C40; 3 Outback MX60
Inverters	SMA SB6000U 6kW inverter for Sharp panels. Siemens and Shell panels feed into 2 Xantrex SW5548 inverters
Batteries	9,220 amp-hour, 6V Trojan T105 batteries

RELOCALIZATION BOOKS

Peak Oil and Climate Change

An Inconvenient Truth

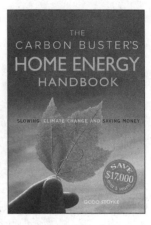

A relatively easy and entertaining read, *An Inconvenient Truth* is a follow-up to Al Gore's best-selling *Earth in Balance: Ecology and the Human Spirit*, and is inspired by a series of multimedia presentations on global warming that Gore created and delivers to groups around the world. Highly recommended for your personal collection and as a gift that gives to the environment by educating and entertaining your friends and family. Includes 330 photographs, illustrations, and charts. 328 pages, softcover. DVD run time 100 minutes.

21-0389 An Inconvenient Truth (book) $21.95

90-0169 An Inconvenient Truth (DVD) $29.95

Heat: How to Stop the Planet from Burning

We don't think it's exaggerating to call this the most important book of our time. With fierce intelligence and optimistic wit, activist/journalist George Monbiot looks at climate crisis science and comes to one conclusion: To avoid catastrophe, we must cut global CO_2 emissions 90% by 2030. Sound impossible? Guess again. Monbiot shows us how we can do it—without sending humanity back to the Stone Age. Here's what will work, what won't, and how it all fits together to create a solution whose time has come. Read it and act. 304 pages, softcover.

21-0420 Heat: How to Stop the Planet From Burning $22

The Party's Over

Oil, War, and the Fate of Industrial Societies

Without oil, what would you do? How would you travel? How would you eat? The world is about to change dramatically and forever as the global production of oil reaches its peak. Thereafter, even with a switch to alternative energy sources, industrial societies will have less energy available to do all the things essential to their survival. Author Richard Heinberg deals head-on with this imminent decline of cheap oil in *The Party's Over*, a riveting wake-up call that does for oil depletion what Rachel Carson's *Silent Spring* did for the issue of chemical pollution—raise to consciousness a previously ignored global problem of immense proportions. 288 pages, softcover.

92-0267 The Party's Over $17.95

The Carbon Buster's Home Energy Handbook

There's no need to wait for governmental leadership or the Kyoto Protocol. Start your own grassroots carbon reduction program and potentially save $15,000 in energy costs over five years. This empowering book helps you assess a detailed carbon accounting of your own emissions and prioritizes solutions based on your highest reductions and returns—often with surprising results! 170 pages, paperback.

21-0418 The Carbon Buster's Home Energy Handbook $12.95

Peak Oil Survival

Preparation for Life After Gridcrash

Oil and energy are not limitless resources, and someday the supply will be depleted. *Peak Oil Survival* shows readers how to plan for the future: how to survive and thrive when the food, transport, and energy industries sputter out. Author Aric McBay gives an essential crash course complete with clear, simple instructions and easy-to-read diagrams. *Peak Oil Survival* will explain how people can protect their families and strengthen their communities in the event of a crisis—and live comfortably off the grid. 128 pages, softcover.

21-0580 Peak Oil Survival $12.95

The Empty Tank

Oil, Gas, Hot Air, and the Coming Global Financial Catastrophe

Geologist and ex-Greenpeace official Jeremy Leggett, author of *The Carbon War*, argues incisively that oil production has peaked, with dwindling supplies and

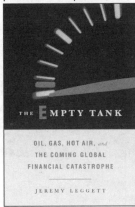

soaring prices in the offing. News? No. The more pertinent story told here is his suggestion that, worse than the possibility that the world cannot cope, the threat is that it will cope all too well by burning more coal and coal-derived synthetic fuels, thus exacerbating atmospheric carbon dioxide buildup and global warming. This will lead the planet down "the road to horror," illustrated in a sketchy montage of the usual environmental doomsday scenarios: rising sea levels, extreme weather, famine, and war. 265 pages, hardcover.

92-0269 The Empty Tank   **$24.95**

The Weather Makers

How Man Is Changing the Climate and What It Means for Life on Earth

By Tim Flannery. Sometime this century, the day will arrive when the human influence on the climate will overwhelm all other natural factors. Over the past decade, the world has seen the most powerful El Niño ever recorded, the most

devastating hurricane in 200 years, the hottest European summer on record, and one of the worst storm seasons ever experienced in Florida. With one out of every five living things on this planet committed to extinction by the levels of greenhouse gases that will accumulate in the next few

decades, we are reaching a global climatic tipping point. *The Weather Makers* is both an urgent warning and a call to arms, outlining the history of climate change, how it will unfold over the next century, and what we can do to prevent a cataclysmic future. 384pages, hardcover.

21-0500 Life on Earth   **$24**

The Last Hours of Ancient Sunlight

Revised and Updated: The Fate of the World and What We Can Do Before It's Too Late

The Last Hours of Ancient Sunlight details what is happening to our planet, the reasons for our culture's blind behavior, and how we can fix the problem. Thom Hartmann's comprehensive book, originally published in 1998, has become one of the fundamental handbooks of the environmental activist movement. Now, with fresh, updated material and a focus on political activism and its effect on corporate behavior, *The Last Hours of Ancient Sunlight* helps us understand—and heal—our relationship to the world, to each other, and to our natural resources. 400 pages, paperback.

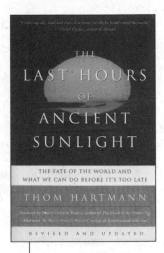

21-0501 The Last Hours of Ancient Sunlight   **$14.95**

Relocalization, Growth, Technology, and Social Change

Deep Economy

The Wealth of Communities and the Durable Future

How can we live successfully in a fossil-fuel scarce and climate changed world? Author, thinker and environmentalist Bill McKibben's provocative and highly engaging new book provides a promising answer. We can "relocalize" and, in the process become happier. In *Deep Economy*, McKibben puts forth a compelling new way to think about what we buy, what we eat, the energy we use, and the money that pays for it all. His book may be one of the first to fully explore the topic of climate change and peak oil from an economic perspective. His argument: More is no longer better. Economic growth can no longer be the definition of economic success. 261 pages, hardcover.

21-0413 Deep Economy   **$25**

The Small-Mart Revolution

How Local Businesses Are Beating the Global Competition

Author Michael Shuman (*Going Local*) once again explores the issue of relocalization and the benefits—

from the nostalgic return of small-town America's "main street," through the economic stability of our nation, to the very environmental stability of our planet—of this growing movement. In *Small-Mart*, his analysis turns to the realm of mom-and-pop in a world of big-box giants as he presents a thorough analysis and how-to guide for consumers, investors, entrepreneurs, policymakers, and more. Refreshing in tone, this is an inspired guide to the revolution that is the resurgence of locally owned business and is not another big-box-bashing tome. A must-read for anyone interested in prioritizing real actions toward reclaiming a unique and vibrant local landscape. 200 pages, hardcover.

21-0400 The Small-Mart Revolution $24

It's a Sprawl World After All

The Human Cost of Unplanned Growth—and Visions of a Better Future

It's a Sprawl World After All is the first book to link America's increase in violence and the corresponding breakdown in society with the post-World War II development of suburban sprawl. Without small towns to bring people together, the unplanned growth of sprawl has left Americans isolated, alienated, and afraid of the strangers that surround them. Suburbia has substituted cars for conversation, malls for main streets, and the artificial community of television for authentic social interaction. 245 pages, paperback.

21-0502 It's a Sprawl World $17.95

Limits to Growth: The 30-Year Update

Updated for the second time since 1992, this book by a trio of professors and systems analysts (Donella Meadows, Jorgen Randers, and Dennis Meadows) offers a dark view of the natural resources available for the world's population. Using extensive computer models based on population, food production, pollution, and other data, the authors demonstrate why the world is in a potentially dangerous "overshoot" situation. The consequences may be catastrophic: "We . . . believe that if a profound correction is not made soon, a crash of some sort is certain. And it will occur within the lifetimes of many who are alive today." The book discusses population and industrial growth, the limits on available resources, pollution, technology, and particularly, ways to avoid overshoot. The authors do an excellent job of summarizing their extensive research with clear writing and helpful charts illustrating trends in food consumption, population increases, grain production, etc., in a serious tome likely to appeal to environmentalists, government employees, and public policy experts. 368 pages, paperback.

21-0503 Limits to Growth $22.50

Collapse: How Societies Choose to Fail or Succeed

In his Pulitzer Prize-winning bestseller *Guns, Germs, and Steel*, geographer Jared Diamond laid out a grand view of the organic roots of human civilizations. That vision takes on apocalyptic overtones in this fascinating comparative study of societies that have undermined their own ecological foundations. Diamond examines storied examples of human economic and social collapse, and even extinction. He explores patterns of population growth, overfarming, overgrazing, and overhunting—often abetted by drought, cold, rigid social mores, and warfare—that lead inexorably to vicious cycles of deforestation, erosion, and starvation. Extending his treatment to contemporary environmental trouble spots from Montana to China and Australia, he finds today's global, technologically advanced civilization far from solving the problems that plagued primitive, isolated communities in the remote past. Diamond is a brilliant expositor of everything from anthropology to zoology, providing a lucid background of scientific lore to support a stimulating, incisive historical account of many declines and falls. Readers will find his book an enthralling—and disturbing—reminder of the indissoluble links that bind humans to nature. 592 pages, paperback.

21-0504 Collapse $17

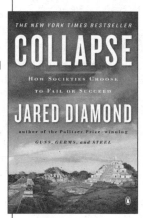

Better Off:
Flipping the Switch on Technology

A decade ago, author Eric Brende was pursuing a graduate degree at MIT studying technology's influence on society, and he reached conclusions that disturbed both him and his faculty mentors. Brende and his new wife moved to a religious "Mennonite-type" community (that in many respects makes the Amish seem worldly), where he hoped to pare his environment down to "a baseline of minimal machinery" that could sustain human comfort while allowing him to stay off the power grid. The pervasive back-to-basics sentiment will surprise few, but Brende's nostalgia for a simpler way of life is far from rabid. His gentle case for simple living will easily resonate with the converted and may inspire skeptics to grapple more intimately with the issue. 272 pages, paperback.

21-0505 Better Off   **$13.95**

Future Hype

Future Hype debunks the popular myth of technology growth and teaches that technology does not have to control you, nor do you need to feel overwhelmed by the ubiquitous white wires streaming from the ears of nearly everybody. The key lies in understanding how technology works and allowing yourself to better evaluate, anticipate, and control it. New technology comes out every day, and yet, as Bob Seidensticker shows us, the popular view of technology is wrong and the future won't be so shocking. 240 pages, softcover.

21-0388 Future Hype   **$15.95**

Blessed Unrest

How the Largest Movement in the World Came into Being and Why No One Saw It Coming

What will save the world is already here in the form of countless efforts large and small each taking care of a tiny piece of the social and environmental puzzle in their part of the world. That's what visionary Paul Hawken sees in this inspiring exploration of the unstoppable movement whose ancient history, deep diversity, and ingeniously innovative ideas are leading us to a new era. It's a grass-roots effort with no name or leader, and as it re-imagines our relationships to each other and the Earth, Hawken finds the lasting hope we seek. 352 pages, hardcover.

21-0412 Blessed Unrest   **$24.95**

Food and Food Security

Plenty: One Man, One Woman, and a Raucous Year of Eating Locally

It began with a spontaneous dinner that circumstance demanded be created completely from foraged edibles. It ended with a question: Is it possible to defy the industrial food chain and live only on foods produced within 100 miles of home? With that, a year-long experiment to try began, one that asked vital questions about globalization, agribusiness, the oil economy, ecological collapse, and community. This is the inspiring story

of that thought-provoking journey. From initial struggle to deep contentment, follow as the authors learn about themselves, our food, and our world, and start a movement that's changing how we eat and live. 272 pages, hardcover.

21-0414 Plenty   **$24**

Animal, Vegetable, Miracle
A Year of Food Life

Overflowing with the same illuminating insight and graceful beauty that mark her award-winning fiction, acclaimed author Barbara Kingsolver's first foray into nonfiction details her family's year-long quest to eat only locally-produced foods. It's an endlessly fascinating, frequently funny rumination on everything from the industrial food chain to turkey mating. Its fact- and recipe-filled

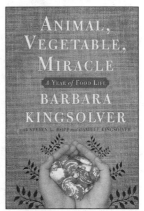

pages are part memoir, part exposé, and part passionate call to restore the American kitchen and the farms that feed it to sustainable health. 384 pages, hardcover.

21-0415 Animal, Vegetable, Miracle   **$26.99**

The Revolution Will Not Be Microwaved

Inside America's Underground Food Movements

The struggle between constant convenience and natural, vibrant food is one that plagues many of us. In his

latest book, Sandor Ellix Katz explores this topic far beyond the organic label and the lack of GM (genetically modified) food labeling, and challenges the way we think about food. At the same time, he offers a rich history and analysis of profit-driven policy and a delightful recipe book of individual actions, from seed saving and sprouting to gardening and foraging. An inspired read that profiles the heroes of the growing renaissance of healthy, living food. 378 pages, softcover.

21-0393 The Revolution Not Microwaved $20

The Way We Eat:
Why Our Food Choices Matter

Eating is a lot more complicated than it used to be. Is it better to buy fair trade? Go organic? Eat local? Avoid animal products? Meticulously researched, this landmark book resolves today's dietary dilemmas. Dissecting the eating habits of three typical American

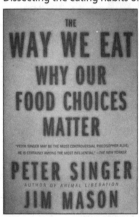

families, renowned ethicist Peter Singer and ag expert Jim Mason follow our food from farm to table and explore all the moral considerations involved, from animal welfare to environmental stewardship. Eating ethically needn't involve fanaticism. Here's how it's done and why we all should do it. 328 pages, hardcover.

21-0406 The Way We Eat $29.95

The Omnivore's Dilemma

Follow award-winning writer Michael Pollan on this fascinating journey through the American food chain. From the pitfalls of industrial agriculture and the promise of organic farming to strange but true supermarket tales and his experiences foraging in the wild, Pollan looks at what we eat, where it comes from, and the good, the bad, and the unappetizing hiding in between. It's a thoughtful and lucid, yet hugely entertaining volume that leaves no plate unturned. 288 pages, hardcover.

21-0401 The Omnivore's Dilemma $26.95

Recipes from America's Small Farms

Nothing beats a farmhouse meal, and with this incredible cookbook in your kitchen, you'll enjoy them every day. It's overflowing with the favorite recipes of farmers, CSA members, and renowned chefs. Chapters are organized by food type and discuss history, characteristics, and nutrition while providing cooking techniques, useful sidebars, farmer profiles, and (of course!) dozens of delicious original dishes you have to eat to believe, including vegan, meat, and dairy recipes. Edited by Joanne Hayes and Lori Stein. 304 pages, softcover.

21-0399 Recipes from America's Farms $17.95

Food Not Lawns

Forget sterile lawns! Grow a vibrant garden filled with food and beauty instead. That's the advice of one of our favorite new books. It's a virtual encyclopedia of wisdom and how-to advice about ecological design, organic gardening, food and eating, health, self-reliance, community building, habitat, and biodiversity. Learn how city dwellers and rural denizens alike can use a nine-step permaculture strategy to create a piece of paradise that feeds both body and soul. By H. C. Flores. 344 pages, softcover.

21-0392 Food Not Lawns $25

Land and Shelter

Designing and Building Your Green Dream Home

IT ALL BEGINS AND ENDS WITH LAND. The choices we make about how we treat the land upon which we live, and how we create our sheltering places, are absolutely fundamental to any concept of sustainability. For the "Land and Shelter" chapter of this 30th-anniversary *Solar Living Sourcebook*, we have turned to Dan Chiras, sustainable-design consultant and author of numerous books, including *The Natural House*, *The New Ecological Home*, and *The Solar House*.

Shelter is a fundamental requirement of all living beings. There's hardly an animal on the planet that doesn't seek shelter at some point in its daily life, and humans are no exception. In fact, we humans are rather poorly equipped to survive without shelter, even in the most hospitable regions.

Human shelter, of course, is rooted in the land. We build our homes upon the thin layer of topsoil that nourishes the plants that make all life on Earth possible. And all the materials we use to build our homes come from the land. Wood, metals, glass, and even synthetic materials such as plastics are harvested from the world's forests and fields or deep within Earth's crust. (Plastics come from oil produced from ancient marine algae that were buried in the sediments of the world's oceans many millions of years ago.)

Few human activities so profoundly influence Earth and its ecosystems as creating shelter.

Because our homes take up space that other organisms once occupied, because the resources needed to build and maintain our shelters come from the Earth in vast quantities, and because there are so many of our kind on the planet, few human activities so profoundly influence Earth and its ecosystems as creating shelter. Through intelligent placement and smart design, however, we can create shelter—even entire communities—that honor and respect the Earth and its ecosystems. It's up to us. The choices we make about shelter can help us create a sustainable future.

For years, however, the culture of consumption, predicated on convenience, comfort, and

Passive solar home of residential renewable energy and green building author and consultant Dan Chiras in Evergreen, Colorado.

immediate personal gratification, has wreaked havoc on Earth's biological life-support system. Although most *Sourcebook* readers understand the fundamental incompatibility of modern human culture and its many unsustainable practices, many only barely grasp the full potential of a sustainable response in the building sector—how profoundly we can change our future by building sustainably. Fortunately for all of us, many people are beginning to explore new ways of building shelter, using techniques and materials that could, if widely used in the future, help steer society onto a sustainable path.

Changes in house-building practices begin with shifts in the ways we think. One of those shifts, now embraced by many forward-thinking people the world over, harkens back to the Iroquois Indians of the northeastern United States. When tribal elders pondered important issues, a key element in their decision making considered the potential effects, both positive and negative, of contemplated actions on their seventh generation of descendants (their grand-children's great-grandchildren).

But contemplating the impacts of our actions on future stakeholders, while vital to our long-term success, is not enough to reshape our society or the way we build houses. We need an ethics of sustainability to guide us. Sustainable ethics consists of three basic tenets. (These are discussed in greater detail in *Lessons from Nature: Learning to Live Sustainably on the Earth*, by Dan Chiras.) The first is that humans are a part of nature. We are not superior to nature, nor are we the crowning achievement of nature. A tree manages its affairs much better. We are members of the intricately linked web of life, and we are highly dependent on nature.

Although humans have accomplished amazing feats and created astonishing civilizations, we mustn't lose track of the fact that, like all other species, we depend on nature for the resources that make our lives and our economies possible. In short, nature is the source of all the goods and services we consume, and it is also the sink for all our wastes. Intricately connected as we are, nature has also become dependent on humans. What we do or don't do will profoundly affect the future of Earth's biological life-support system and the fate of humanity.

The second tenet of sustainable ethics is

that the Earth has limits. There are limits to freshwater supplies and limits to the amount of pollution the atmosphere can accept without serious consequence. There are limits to the amount of food we can produce and limits to the amount of oil and natural gas in Earth's crust. To meet our needs and prosper, we must learn to live within those limits. We must design houses, even entire communities, that operate within nature's limits.

The third tenet of sustainable ethics holds that the key to human success comes not from the domination and control of nature but from cooperation with it. Brute force, the old *modus operandi* of industrial society, is a prescription for disaster. Cooperation is the key to success. Cooperation means learning to rely on renewable resources to meet our needs, and conducting our economic affairs sustainably in accordance with patterns laid down by nature.

Our exploration of sustainable building begins with the land: what to look for when thinking about building a sustainable house. We'll then examine ways to heat and cool our houses naturally. This strategy, known as passive solar heating and cooling, is a great example of how we can ensure a comfortable existence for ourselves without bankrupting the Earth and catapulting ourselves further down the path to extinction. It epitomizes how humans can cooperate with nature and honor its limits. We then turn to building materials and examine ways to dramatically reduce the environmental and health impacts of shelter.

Before we embark on this journey, however, let's pause to consider the impacts of conventional housebuilding in order to set the stage for a new way of thinking about shelter.

> Although humans have accomplished amazing feats and created astonishing civilizations, we mustn't lose track of the fact that, like all other species, we depend on nature for the resources that make our lives and our economies possible.

The Impacts of Modern Building

When most of us think about environmental damage that threatens our long-term future, such as global climate change and ozone depletion, the last thing we think about is our homes. Our homes are supposed to be places of refuge, sanctuaries in a maddeningly hectic world. We think that pollution and other significant environmental threats come not from our homes but from factories, cars, trucks, smelters, corporate farming, and other activities.

Unfortunately, shelter is anything but environmentally benign. In fact, shelter poses a major environmental threat. The environmental damage from homes arises from two aspects

of housing: construction and day-to-day living. Both require a huge investment in energy and natural resources, and both are responsible for an enormous output of waste. Both aspects of shelter contribute significantly to local, regional, and global decline in environmental quality. Let's take a closer look at this bold allegation.

Each year, Americans build more than 1.8 million new houses. Each house requires a massive amount of energy and materials. Consider, for example, the amount of wood that goes into a typical new house.

Wood is used to build the majority of new houses in America. It is a primary component

Wood is a renewable yet endangered resource. Huge amounts of wood are used to frame walls, floors, and roofs in new construction and renovations.

of walls, floors, ceilings, and roofs. It is used to build decks, cabinets, shelves, fences, and furniture. Today, about 60% of all timber cut in the U.S. is used to build houses.

With the average house size increasing each year, demand for wood continues to skyrocket. The new houses built in 2005 averaged 2,434 square feet, up from 2,200 square feet just five years earlier. The lumber used to frame a house of this size, if laid end to end, would stretch nearly 2.8 miles. If all the dimensional lumber used to build stick-frame houses (85% of all houses) in America were laid end to end, it would extend to the Moon and back 23 times—or about 4.2 million miles.

Although wood is a renewable resource and forests can be managed sustainably, humankind's insatiable demand for this material is leading to ever-greater loss of native forests. Deforestation can lead to serious soil erosion that damages nearby rivers, streams, and lakes. Deforestation significantly decreases the planet's ability to absorb carbon dioxide (CO_2). According to one estimate, about one-fourth of the annual global increase in the greenhouse gas CO_2 results from deforestation.

Deforestation is also responsible for the loss of valuable wildlife habitat that, in turn, leads to species extinction. Nowhere are these problems more acute than in the tropical rain forests of the world. In this narrow band of land that straddles Earth's equator, forests continue to be leveled in mind-boggling fashion to provide wood and wood products for a fast-growing human population and the rapacious appetite of a continually expanding global economy. These and an assortment of additional impacts are expected to only worsen in the near and long-term future as this growth continues. In addition to wood, housebuilding consumes a huge amount of minerals and metals, among them concrete, granite, a variety of other kinds of stone, aluminum, copper, and steel. Supplying the demand for these materials requires extensive mining. The mining of minerals and metals for each new home built in the United States is responsible for a hole approximately the size of that home.

Construction also requires a huge input of energy—electricity, oil, natural gas, and gasoline—to manufacture materials, assemble products, transport products from the factory to the building site, and, finally, to assemble them to create a sturdy, weatherproof structure.

Building a house creates enormous quantities of solid waste. Scrap wood, glass, dry wall, stone, concrete block, packaging materials, and other waste is generated at building sites in astronomical amounts. In fact, a typical new stick-frame house produces three to seven tons of solid waste. Nationwide, waste from new construction and remodeling constitutes 25%-30% of the municipal solid waste stream. In regions where homebuilding is occurring at a rapid rate, the percentage can be much higher. Hauling this waste to the landfill consumes valuable resources, especially energy, and adds to the cost of building. By comparison, consider the construction of our Real Goods Solar Living Center in 1995, which sent less than 3% of building materials to the landfill! (Learn more about the Solar Living Center on pages 564-568.)

But while construction accounts for an enormous amount of resource consumption, it is responsible for only part of the environmental impact of modern housing. Many impacts continue long after a home is finished: the occupation or habitation impacts. These stem from the resource demands of the people living in a home over long years of service. To keep our homes running, we use enormous quantities of electricity, oil, natural gas, and water. Food and countless household items also stream in. Let's take energy as an example.

According to the U.S. Department of Energy, America's homes consume about 20% of the nation's fossil fuel energy. This energy is used to heat, cool, light, and ventilate our homes and to power a growing assortment of appliances and electronic devices meant to make our lives more comfortable, enjoyable, and convenient. Residential energy demand, in turn, is responsible for approximately one-fifth of America's enormous annual release of CO_2, a greenhouse gas that, when combined with other pollutants and

The mining of minerals and metals for each new home built in the United States is responsible for a hole approximately the size of that home

trends such as deforestation, is unleashing the greatest global catastrophe in human history, one that will profoundly affect each and every citizen on the planet, one that clearly threatens our long-term economic prosperity. (See Chapter 1, "Relocalization: A Strategic Response to Peak Oil and Climate Change.")

Our homes are also a source of tremendous amounts of waste, including food scraps, empty cans and bottles, junk mail, packaging materials, and yard waste. To get an idea of the volume generated each year, take a look at the garbage cans lined up on the street of a typical suburb on garbage day and multiply that by the 112 million homes throughout the country and multiply that by 52 weeks in a year.

Less obvious are the billions of gallons of sewage, containing human excrement, potentially hazardous cleaning chemicals, and excreted medications, among them cholesterol-lowering drugs, hormones from birth control pills, and heart medication, which have a deleterious effect on fish and other aquatic organisms. Each day, these billions of gallons of sewage flow to wastewater treatment plants that attempt to remove as much potentially harmful materials as they can before the water is released back into waterways where it is used for crop irrigation, industrial production, and drinking water and where it also serves as habitat for many aquatic species. In rural areas, waste from cities ends up

> Not only are our homes poisoning the many species that share this planet with us, they're also poisoning the people they're meant to serve.

in septic tanks. From the leach fields that drain the septic tanks, the wastes seep into the ground and, in some cases, may slowly percolate into groundwater.

Modern homes often poison their occupants through toxic chemicals found in numerous building materials. Certain manufactured wood products such as plywood, particleboard, and oriented strand board (OSB, also known as wafer board or chip board) contain formaldehyde resins that are used to bind the wood fibers together. (Formaldehyde is the chemical biologists use to preserve biological specimens.) Studies show that these materials release formaldehyde and other toxic chemicals into room air long after a building is completed.

Potentially toxic chemicals are also found in furniture made from particleboard, furnishings such as carpeting and curtains, as well as paints, stains, and finishes. Appliances like water heaters and furnaces can also release toxins (in this case, carbon monoxide). So not only are our homes poisoning the many species that share this planet with us, they're also poisoning the people they're meant to serve.

Another way to think about the environmental impact of our homes is to realize that they are like patients in an intensive care unit. They're supplied by electrical, gas, and water lines. Their wastes are removed by sewage lines and garbage companies. Just like an intensive care patient,

Can You Afford a Green-Built House?

The future of green building depends, in large part, on green houses being affordable to build, furnish, and operate. Contrary to the myth that building an environmental house is a costly venture reserved only for the well-to-do, this option can be quite cost-competitive. Even with some unfortunate cost overruns, my house cost $5 to $15 per square foot less than most other new spec houses in my area. And its operating costs are much lower. But that is not to say that every aspect of building and furnishing the house was cheaper. A few environmentally friendly products, such as my superefficient Sun Frost refrigerator and compact

fluorescent light bulbs, cost more, sometimes a lot more, than their less-efficient counterparts. However, many other green building products, such as recycled tile, carpeting, carpet pad, and insulation, cost the same or less than their counterparts.

Green builders I've talked with say that green building costs from 0% to 3% more. But remember that investments in some green products and materials may end up providing additional comfort and saving money right away. Adding extra insulation, for example, keeps a house warmer in the winter and cooler in the summer. This not only makes living space more comfortable, it dramatically lowers fuel bills, potentially saving tens of thousands of dollars over the lifetime of a house.

Furthermore, because a well-insulated house requires less heat, you'll be able to install a smaller and less expensive furnace, saving money right from the start. Smaller air-conditioning units are also needed, and in some cases measures to make a house more energy efficient make air conditioning unnecessary. Installing water-efficient toilets and showerheads reduces the demand for water, decreases electricity used for pumping water from wells, and could reduce the size of the leach field required for houses on septic systems. Because these options are no more expensive than standard fixtures, you end up saving money in the short and the long run.

Used with permission from *The Natural House*, by Dan Chiras.

our homes are highly vulnerable to disruptions in the flow of inputs and outputs: Cut off the supply of resources such as electricity, even for a short period, and they cease to function.

In sum, then, when most of us think about our homes, the last thing that comes to mind is resource depletion or pollution. We don't think of our homes as a major source of greenhouse gases that contribute to global warming. Nor do we think about the loss of wildlife habitat, species extinction, dangerous exposure to toxic chemicals, and our extreme dependence on outside resources. We think of our homes as sanctuaries, not sources of personal and environmental damage.

The bad news is that our homes are anything but environmentally friendly. The good news is that they don't have to be.

Houses can be built and operated at a fraction of the impact of a conventional modern home. They can be designed and built in ways that require a fraction of Earth's natural resources. They can be healthful places to live. If we're smart, our homes can work for us. They can heat and cool themselves, and they can even be a source of energy—not just for us, but for our neighbors as well, generating tax-free income by reducing monthly fuel bills. Shelter can be part of the solution to creating a sustainable future, if only we rethink our strategies.

Fortunately, many architects, builders, and consultants are rethinking their approach to

Photo: Bob Ramlow

Good solar exposure can take advantage of both solar hot-water and photovoltaic energy systems.

building. They are embracing the idea of "green building," which has assumed the status of a social movement. Green building means producing houses that satisfy a triple bottom line. They are good for people, good for the economy, and good for the environment. Equally important, green building is affordable. It is one strategy that will help us build a sustainable future.

Creating a green house begins with the selection of the land upon which you want to build. This decision has many ramifications. If you choose land wisely, you have a much better chance of creating shelter that contributes to a sustainable human presence on our Earth.

Houses can be designed and built in ways that require a fraction of Earth's natural resources. They can be healthful places to live.

Choosing Land

Choosing land is one of the most important challenges facing those who want to build a house. For most people, land is selected by proximity factors, for example, how close it is to work, grocery stores, schools, recreational opportunities, doctors, and services such as fire protection. In rural areas, proximity to the electrical grid is an important factor for some (for others, it's important to be far away from electricity!). In urban areas, proximity to mass transit may be important.

Although proximity factors may seem to have no bearing on sustainability, they are relevant. They involve more than mere lifestyle and convenience choices. Choosing a site near a major transit line or a place of work, for instance, helps reduce travel time. Fewer miles on the road each day means fewer gallons of gas each week, less money spent on gas, and less air pollution. Proximity might also permit you to commute by bicycle, a lifestyle change that has

many health benefits. Reducing the time spent commuting also means more time spent with loved ones.

We tend to associate buying land with rural living, but if you're interested in an urban location, you can search for a vacant lot to build on. You'll be amazed at how many lots are available, and one just might suit your needs and goals.

If you are dreaming of escaping to the country, choose your land wisely. Watch out for easements, zoning restrictions, and covenants in rural subdivisions. Check into future land development plans in the area. You don't want to build a house only to find that a landfill is slated to go in next door next year. All in all, you want to be sure that you can use and enjoy the land in the manner you anticipated when you signed on the dotted line.

Wherever you're looking for land, don't forget to shop for the many "free services" a piece of property can offer, such as a good source of water, hydroelectric potential, natural drainage, or access to the Sun. And wherever you choose to settle, consider buying an old house and fixing it up. Remodeling uses existing resources and requires less consumption of virgin resources and energy than new construction. Although fixer-uppers can be a challenge, you can make significant improvements that contribute to a sustainable future.

Here are some things to look for—and avoid—when shopping for land for building a sustainable house:

1. Choose a site with good solar access. Sunlight can provide or contribute to your space heating and your domestic hot water. It can also provide free lighting during the day and be used to generate electricity. When selecting a site, be

A Humboldt County homestead with great southern exposure.

sure that it provides good unobstructed access to the Sun from 9 a.m. to 3 p.m. each day.

2. Choose a south-sloping site for earth sheltering. "Backing" a house into a south-facing slope (in the Northern Hemisphere) so that it is sheltered by the earth dramatically reduces energy consumption. Houses built this way stay cool in the summer and warm in the winter with little outside energy input. When combined with passive solar design, earth sheltering can virtually eliminate the need for heating and cooling equipment.

3. Choose a site that offers unobstructed access to winds. Wind can provide a significant amount of electrical energy in many areas. Wind can also help cool a house in the summer. If you are building in an area with a good wind resource and want to tap into the potential of wind energy, be sure to select a site that is free from potential obstructions if you are thinking about installing a wind turbine.

4. Seek favorable microclimates. Climate can vary dramatically from one location to another within a state, a region, and even a neigh-

borhood. Some microclimates may be suitable for self-sufficiency; others may make it difficult to live on renewable energy.

5. Select a dry, well-drained site. Moisture is becoming a major issue of concern among homebuilders throughout the world. It can seep into basements and walls through the foundation. To prevent moisture problems, select a well-drained site. This will protect your house from water damage, reduce the cost of grading, and decrease construction costs.

6. Choose a site with stable subsoils. Some subsoils are extremely unstable. They expand and contract in response to changing moisture levels, which can cause a considerable amount of damage to foundations and overlying walls. Be sure to select a site in an area with stable subsoils or design your house accordingly.

7. Avoid marshy areas. Wetlands are a precious natural resource that support many species and provide many free services (for example, they recharge groundwater and reduce flooding after rainstorms). Building near wetlands can damage them and also expose inhabitants to pesky mosquitoes that carry the dangerous West Nile virus. Stay clear of wetlands.

8. Select land that has a suitable site for growing food. Gardens and orchards can provide a large portion or even all of a family's food in favorable climates with good topsoil. Growing food provides exercise, time outdoors, and substantial economic savings. It also offers many environmental benefits, especially if you grow with organic methods. Choose land that has good topsoil somewhere, or be prepared to spend years building up the organic matter in the soil. Always do a simple soil test before buying your land.

9. Select land that offers building resources. Land can provide many building materials, among them clay-rich subsoils for making natural plaster, wood for framing walls, and stones for building pathways or garden walls. Using materials harvested from a site dramatically decreases the energy required to build a house. Determine the type of house you want to build and then select a site that offers abundant free materials, or better yet, design the house around the site you've selected and the building materials it offers.

10. Choose a site with a good water supply. It's important to select land that has a reliable supply of clean water, either in the form of groundwater or a stream or spring you can tap.

11. Don't destroy beauty in your search for it. Be sure that the property you buy will allow you to build a house that looks out over the beauty of the place, instead of being dropped smack dab in the middle of it. As Christopher Alexander, coauthor of *A Pattern Language*, advises, "Leave those areas that are the most precious, beautiful, comfortable, and healthy as they are, and build new structures in those parts of the site which are least pleasant now." Create beauty; don't be an agent of its destruction.

Once you have purchased land, you'll have to determine exactly where your house should be situated. It is highly advisable that you become intimately familiar with your building site. Visit it often during different seasons and different times of the day. Plan on camping on the land on several different spots during different seasons to see how it feels at night and early in the morning. Study wind patterns. Determine the path of the Sun. Observe the land during rainstorms and snowstorms. Study natural drainage patterns. Notice where snow accumulates in the winter. If you're building on property with

a stream, look for evidence of flooding. Be on the lookout for wildlife corridors or paths. Locate the best views. Check for hot spots and cold spots that exist because of topography or other reasons. Determine how you can place a driveway so that it does the least amount of damage.

After you are intimately familiar with the land, you can select a site for your house that affords the best views; does the least damage; harvests the energy of the Sun, wind, and water (hydroelectric); avoids natural drainage paths; and provides optimal year-round comfort. Selecting the wrong site is a decision that will haunt you for a lifetime.

After you've settled on a house site, you can begin to design your home. Don't make the all-too-common mistake of imposing your dream home on a site. Design with the site in mind, and remember that your design should enhance the natural beauty of a site, not become an eyesore. To learn more about site selection, you may want to consult *The New Ecological Home*, by Dan Chiras, see page 518).

> Don't make the all-too-common mistake of imposing your dream home on a site. Design with the site in mind, and remember that your design should enhance the natural beauty of a site, not become an eyesore.

Creating Sustainable Shelter

Creating a sustainable society will require many changes in our current practices, in practically everything we do. As noted above, a sustainable human society will have to comply with ecological principles, which means that we must pattern human society according to rules set down by nature. For example, we must learn to use resources such as energy, wood, masonry materials, and water much more efficiently than we do at present. We must recycle and compost as much of the waste from modern society as we can and use recycled materials made from those waste products whenever possible to meet our needs, just as nature does. We must turn to renewable resources such as wind and solar and biomass energy to fuel our society. We must also restore land disturbed by human activities—for example, land that provides wood to build houses—and bring it back to its full po-

tential. Finally, we have to learn to control human population growth to prevent humankind from overrunning the planet. All these actions help reduce pollution and slash resource demand, which help us live within Earth's limits, a step that is vital to our efforts to create a sustainable future.

The rules of nature can serve as a guide for restructuring every human system from agriculture to industry. Because homebuilding has such an enormous impact on Earth and its ecosystems, efforts to create sustainable shelter can go a long way toward creating a more sustainable human presence. How do the ecological principles of sustainability apply to homebuilding?

Creating sustainable shelter requires efforts to conserve natural resources—to use what we need and use them efficiently. We can reduce our demand for natural resources, for example,

by building smaller and more energy-efficient houses. In arid climates or any area suffering from water shortages, homeowners can help conserve this rapidly dwindling resource by planting xeric vegetation—that is, grasses and colorful flowers, shrubs, and trees that require little water to thrive.

We can conserve energy and resources by using recycled materials to build houses and recycling all waste from the building site and the house once we move in. We can promote a healthy living environment by building with materials that are free of toxic chemicals that pollute indoor air and damage the health of a home's inhabitants.

We can design and build houses that make use of Earth's generous supply of clean, affordable, and reliable renewable energy. Many homeowners have taken this idea to heart and generate all their energy from renewable sources using passive solar design, solar electric panels, wind generators, hydroelectric systems, and solar hot water collectors. In effect, they've become their own sources of heat and power. Many families are no longer connected to the power grid at all—in fact, there are estimated to be over 25,000 folks living off the grid in northern California alone (nothing like the nearly 2 billion people in the world who have no electricity whatsoever). There is a huge growing trend of homeowners with solar systems sending surplus electricity back onto the grid, supplying part of their neighbors' electrical energy.

When construction is completed, we can restore the site, planting native vegetation that helps support some of the wildlife we have displaced by building our house in their habitat.

Creating sustainable shelter requires a new way of thinking about building that permeates every aspect of construction, starting at the beginning with careful planning and design. Success in achieving our goals will be enhanced by collaboration—working with a team of professionals, among them the architect, builder, engineers, and subcontractors. Before we explore planning and collaboration, let's take a look at some specific aspects of sustainable building that deserve special attention: green building materials, healthy homebuilding, and off-grid living.

Green Building Materials

One element of green building that receives a great deal of attention, and rightly so, is the use of environmentally friendly and people-friendly building materials, commonly referred to as *green building materials*. What exactly is a green building material?

Green building materials defy easy definition. In general, building materials are called *green* because they are good for the environment. For example, they may be produced from recycled materials, which puts waste to good use and reduces the energy required to make building materials. Three good examples include counter tiles made from recycled automobile windshields, carpeting made from recycled soda bottles, and decking made from recycled plastic bags or milk jugs.

Some building materials are green because they can be recycled once their useful lifespan is over, for example, aluminum roofing shingles. If that product itself is made from recycled materials, the benefit of recyclability is multiplied.

Some building materials are considered green because they're durable. A durable form of sid-

Photo: Trex Company

ing, for instance, outlasts less-durable products, resulting a significant savings in energy and materials over the lifetime of a house. If a durable product is made from environmentally friendly materials, for instance, recycled waste, it offers even greater benefits. Siding made from cement and recycled wood fiber (wood waste) is a good example of a green building material that offers at least two significant advantages over less environmentally friendly materials. It will very likely outlast many homeowners, and because it is durable, it saves on costly maintenance.

Other building materials are considered green because they are made from renewable resources that are sustainably harvested. Flooring made from sustainably grown and harvested lumber or bamboo is a good example.

Efficient manufacturing is another mark of a green building material. In this case, production requires much less energy and fewer raw materials than those of a conventional counterpart. Examples include engineered lumber such as glue lams and wafer board or wooden I joists. Engineered lumber is more efficient than conventional wood products because floor and ceiling joists made from engineered lumber use much less wood than a solid 2x10 framing timber. Moreover, you don't have to cut down an old-growth tree to create a 2x10 engineered wood joist. Engineered lumber is typically made from small, fast-growing trees.

Another group of green building products of considerable importance is the low- or no-VOC

Courtesy Classic Products, Inc.

Courtesy Re-New Wood, Inc.

(volatile organic compound) paints, stains, finishes, and adhesives. These products are made without toxic resins that can cause cancer or other health problems. They're safer for the factory workers who produce them, the subcontractors who install them when building a house, and the family who lives in the house. In other words, they're healthier for all who come in contact with them.

Ideally, green building products should be manufactured by companies that embrace environmental protection and resource efficiency, and demonstrate a strong commitment to protecting workers from exposure to toxic substances. Pollution prevention programs are an important sign of a company's environmental responsibility and its desire to safeguard the health of its workers. Energy-efficient operations, factory recycling programs, and the use of recycled waste or sustainably harvested wood are additional signs of a company's commitment to the environment.

In recent years, many companies have begun to produce green building materials. In fact, these days there's not a material or appliance or a piece of furniture or a furnishing that doesn't have a green alternative—sometimes several of them. The number of sources has also skyrocketed. Online mail-order companies that can drop ship hundreds of products to a building site are now operating nationwide. Some products, such as recycled plastic deck material and certified lumber, can even be purchased at local building supply stores. As green building continues to grow, you can expect wider availability. The industry is skyrocketing.

Characteristics of Green Building Materials

- Produced by socially and environmentally responsible companies
- Produced sustainably—harvested, extracted, processed, and transported efficiently and cleanly
- Low embodied energy
- Locally produced
- Made from recycled waste
- Made from natural or renewable materials
- Durable
- Recyclable
- Nontoxic
- Efficient in their use of resources
- Reliant on renewable resources
- Nonpolluting

There's not a material or appliance or a piece of furniture or a furnishing that doesn't have a green alternative—sometimes several of them.

All Photos: Dan Chiras

Green building materials help builders create more energy-efficient and environmentally friendly homes. Hardie plank (far left and left), made from recycled wood fiber and cement, creates a durable siding that will outlast cheaper alternatives. Structural insulated panels (right), used to build energy-efficient walls, roofs, and floors, reduce labor and lumber use. Wooden I-beams (far right), for framing floors and roofs, use substantially less wood than solid dimensional lumber and help protect old-growth trees.

Five years ago, no common definition existed for a high-performance green building, yet today more than 2,000 building projects are LEED (Leadership in Energy and Environmental Design)-registered, representing over 86 million square feet. In 2007, a majority of the builder population reported involvement with green building. The green home market now exceeds $7 billion, which represents a 40% growth rate. This market is projected to increase to $20-$38 billion by 2010.

Surely you're asking yourself, "But what about the cost of green building materials? Aren't they terribly expensive?"

The good news about green building materials is that many of them are cost competitive with conventional materials. Other green products may cost more—sometimes a lot more—but they provide peace of mind and satisfaction that you're doing the right thing as a homeowner, including helping this critically important industry gain a foothold in the marketplace. In time, as more and more builders go green and the demand for green building materials rises, prices of materials that cost more than conventional materials could start to tumble. And don't forget, a product that is healthier may be expensive, but so is cancer.

Choosing a green building material is not always easy. As just noted, some materials may cost more. Other products may have inherent

shortcomings; for example, they may be produced from a renewable resource that is sustainably harvested halfway around the world. To ship them to your building site requires an enormous amount of energy. Should you use a product like this?

That's your decision. You may have to compromise on some energy or cost factor, but you may still decide that the choice of a certain green building product makes sense anyway, even if isn't quite perfect. By using these environmentally friendly products and helping to stimulate markets needed to drive this important new endeavor, you can become an agent of change.

To achieve the greatest benefit, you probably want to choose green building materials that offer the greatest gains. You may also want to concentrate on big-ticket items—that is, products that are used in greatest quantity, such as framing lumber, insulation, roofing materials, floor coverings, wall finishes, and foundation materials. To learn more about green building materials, a helpful resource is Dan Chiras's book, *The New Ecological Home*. In addition to describing your options, it contains a list of many online suppliers of green building materials. Also check out *GreenSpec: Product Directory with Guideline Specifications*, published by Building Green. It lists hundreds of green building products, including contact information for the companies that manufacture them.

The Healthy Home

Green building materials reduce the impact homebuilding has on the environment. They help create a healthier environment, too. As noted briefly above, some products even improve indoor air quality, making our homes more healthful places in which to live.

Healthy home construction is essential today because many new homes are built to more exacting energy standards—that is, they are

better insulated and much more tightly sealed. Although tight construction is a good idea from an energy standpoint, it creates one significant problem: Many modern (non-green) building materials are manufactured with potentially toxic chemicals. Oriented strand board (OSB), for example, which is used to sheath walls, floors, and roofs, is an engineered wood product that is held together by a resin that contains formaldehyde, which has been shown to be hazardous to human health.

Combine air tightness with toxic chemicals and you could be creating a health nightmare. Occupants may have only minor complaints, such as headaches or allergies, or they may suffer more serious illnesses such as chronic bronchitis and asthma. Unhealthy indoor air can even lead to debilitating illnesses such as multiple chemical sensitivity and cancer.

Using healthy green building materials is a good idea but won't eliminate all health problems created by indoor air pollution. That's because indoor air pollution comes from many sources besides building materials. Prominent among those additional sources of pollution in homes are the many combustion technologies we rely on, such as gas stoves, ovens, water heaters, and furnaces. Wood stoves and fireplaces also release potentially harmful pollutants, such as carbon monoxide, into the air inside a home. But that's not all.

Conventional cleaning products and disinfectants, household pesticides, personal care products, and pets also generate indoor air pollution. That shouldn't be too surprising, since you're probably aware that these items contain volatile chemicals. But indoor air pollutants also come from sources you'd never suspect, such as carpeting, window coverings, furniture and cabinetry made from particleboard, and the stains and finishes that coat their surfaces. Vinyl floor coverings and vinyl wallpaper may also release potentially toxic chemicals. These and

other household furnishings may release dangerous substances for six months to a year after application or installation, sometimes longer. The more airtight a home is, the more problems these sources create.

To create a healthy home, it's best to eliminate as many of these sources as possible. Choose healthy alternatives—for example, nontoxic cleaning agents or water heaters with closed-combustion chambers. Passive and active solar heating reduce the need for furnaces or boilers that burn fossil fuel.

Eliminating sources of pollution is the first line of defense against indoor air pollution. The second line of attack is ventilation. An airtight energy-efficient home requires adequate ventilation. In the summer, open windows can purge air of toxic pollutants. When the house is closed up, however, you may want to provide mechanical ventilation.

These days, many contractors attentive to energy efficiency install heat-recovery ventilators in the houses they build. These devices continuously remove stale indoor air when a house is closed up and replace it with fresh outdoor air. The opposing air streams pass through a heat exchanger, a device that transfers heat from the warm outgoing air to the incoming cool air, thereby saving energy. There's no sense in losing all that hard-earned heat.

For more information about indoor air quality and cleansing or purification techniques, see Chapter 9, "Water and Air Purification."

> Eliminating sources of pollution is the first line of defense against indoor air pollution. The second line of attack is ventilation.

Living off the Grid

For many of us, creating sustainable shelter means living off the grid, disconnecting our homes from the conventional energy systems that are, for the most part, powered by fossil fuels (especially coal) and nuclear energy. Living off the grid means producing your own energy from clean, reliable, and renewable resources.

True off-grid living means supplying all your energy yourself. For example, you can generate all your electricity from solar energy and wind or hydroelectric. It also means heating and cooling your house naturally, relying on passive solar design, and heating water for showers and baths and washing dishes without fossil fuels. Although it is difficult to achieve, some off-grid families find ways to cook their food without fossil fuels, typically with a wood stove supplemented by a solar oven. (Most off-grid houses,

however, utilize some natural gas or propane for cooking.)

For many people, the goal of a 100% off-grid lifestyle is unattainable, but for the thousands generating their own electricity, they wouldn't have it any other way. While homeowners may be able to produce almost all their electricity from renewable sources, they may still need a back-up generator to supply their needs during prolonged cloudy periods. In northern California, for example, many off-grid houses run on a combination of solar and hydroelectric power. When the Sun doesn't shine in the winter, the rain provides ample hydropower. However, for a brief period in the fall before the rains have begun, and while the Sun is waning and becoming overcast, the generator may become necessary for short periods. Back-up generators are

> For many of us, creating sustainable shelter means living off the grid, disconnecting our homes from the conventional energy systems that are, for the most part, powered by fossil fuels (especially coal) and nuclear energy.

Photo: Bob Ramlow

Off the grid and loving it!

on investment can easily be 8%-16% or more depending on the incentive, which equates to payback times of only 6-12 years.

Generating your own power is rewarding and fun, but it can be challenging at times, especially early on. On days when clouds hide the Sun or winds do not blow, you may have to curtail activities that require a lot of electricity. And while renewable energy equipment is reliable, you may have to contend with an occasional mechanical problem, like a broken pump on a solar hot-water system. Renewable energy systems will require some maintenance, for example, batteries in an off-grid solar electric system need attention from time to time (monthly to quarterly in most instances) for optimal performance and to ensure the greatest return on your investment. But these maintenance issues are minor and differ little from maintaining conventional household systems. (The subsequent chapters in this *Sourcebook* provide tons of information about renewable energy systems and off-grid living.)

Many of us at Real Goods have lived off the grid for many years, and we all find it greatly rewarding. When lights are on or an appliance is operating, it makes us smile to think that the energy running them comes from the Sun 93 million miles away. Relying on wind and solar energy keeps us in touch with the weather and makes us acutely aware of our household's daily energy use. It also forces us to never take energy—any energy—for granted. Being your own power company makes you forever cognizant of waste. You even become aware of electricity that's being used by appliances and electronic devices that aren't operating but are in the standby mode. These demands, known as "phantom loads," can drain a system pretty quickly on cloudy days. Probably the most rewarding aspect of off-the-grid living is being a lone bright home in a sea of darkness when the electrical company has a power outage.

Monitoring energy use is critical to successful off-grid living. After many years of off-the-grid living, most of us at Real Goods still find ourselves checking the meters in our systems daily to see how much electricity our photovoltaic panels, wind generators, and hydroelectric plants are generating at any given time and how much electricity is being used in our homes. We watch our batteries, too, monitoring the voltage to be sure they don't get drained. You need to be sure that batteries get fully recharged after long, cloudy spells to ensure optimal performance over the long run.

typically powered by natural gas, diesel (which can run on biodiesel with no conversion), or gasoline. They may also use propane for cooking and to provide backup to their solar hot water or heating system. And most Americans, off grid or not, still rely on gasoline or diesel fuel (biodiesel works) to power their vehicles. (See Chapter 12 for sustainable transportation alternatives.)

Even though you may not be able to achieve 100% energy independence, any and all efforts to reduce our nation's reliance on fossil fuels and nuclear energy are beneficial. Renewable energy resources help to free us from burning coal, oil, and natural gas, thereby reducing pollution and environmental destruction and combating global warming. Every step we take toward using renewable energy, as individuals and as a society, builds a more secure energy future and a more prosperous and safer nation.

In addition to the personal satisfactions, going off grid has real economic benefits. It frees you from costly utility bills and provides a hedge against inflation. Renewable energy system costs may go up, but the fuel cost is the same today as it has been for all of human history: free. Unless someone finds a way to meter the Sun or charge for the wind that blows by our houses, those who rely on renewable energy won't ever have to worry about rising fuel costs—or even monthly bills for that matter! Studies also show that residential renewable energy is a good economic investment, one that equals or exceeds rates of return available through conventional means. In California, Colorado, New Jersey, New York, and any other states with solar incentives, the return

Residential renewable energy is a good economic investment, one that equals or exceeds rates of return available through conventional means.

If all this seems like too much, don't despair. You don't have to go off grid to live sustainably. In fact, many homeowners use the electrical grid as their storage battery, which is another way to achieve a measure of energy independence and also save money. When their renewable energy systems produce a surplus of electricity, it flows into the electrical grid, and their electric meters literally spin backward, subtracting consumption from their monthly bill. Their surplus is consumed by their neighbors. At night or on days when the output of their system can't meet demand, electricity is drawn from the grid and their electric meter runs forward.

If a grid-connected house sends as much electricity to the grid as it draws, there's no monthly bill, other than a meter-reading fee. If it produces a surplus of electricity beyond its own demand, the utility may send the homeowner a check, although most simply credit the surplus to the next month's bill.

Using the grid as a storage battery offers a unique advantage that few people realize: It yields a 100% return on the electricity a home generates. Every kilowatt-hour of electricity you put into the grid can be drawn off at no charge. Batteries, on the other hand, give back only about 80% of the electricity they receive due to inefficiencies in the conversion of electrical energy to chemical energy (for storage) and the reverse reaction (when electricity is drawn from a battery).

Feeling connected to your energy sources changes you forever. The thought of ever reverting to grid power is unfathomable. To learn more about off-grid living, you may want to consult *The Solar House* and *The Homeowner's Guide to Renewable Energy*, both by Dan Chiras (see pages 83 and 84).

Careful planning, especially early on, and collaboration among all the participants—you, your architect, your builder, and the subcontractors—will increase your chances of reaching your goals and building a house that provides healthful shelter with the least impact on Earth and its ecosystems.

Planning, Design, and Collaboration

Planning and collaboration are two keys to successfully building a sustainable house. Careful planning, especially early on, and collaboration among all the participants—you, your architect, your builder, and the subcontractors—will increase your chances of reaching your goals and building a house that provides healthful shelter with the least impact on Earth and its ecosystems.

Careful planning and collaboration also helps to control costs and can even save you money. Consistent communication among the various partners from the outset helps a homeowner develop realistic expectations of the cost of building and the time required to complete a project. It also tends to make the process more efficient, in terms of both time and money. Collaboration makes the whole process more enjoyable for everyone involved.

If you have selected a suitable piece of property and become familiar with it, your first task is to create a site map, a drawing that indicates where the house will be located, where gardens can be placed, and where the driveway and outbuildings will be situated. It should also indicate where wells or water lines and septic tanks or sewer lines can be safely sited.

Next comes the house design. While many people design a home based on the layout of rooms, a sustainable home should be designed around room layout *and* passive solar heating and cooling. Indeed, passive solar heating and cooling should be the central unifying concepts that dictate the floor plan and elevations. Hire an architect or an experienced solar builder with a knack for drawing to create plans. Use scale drawings to determine if the rooms are sized appropriately so that your furniture will fit. Compare the size of the rooms in your new home with the rooms in your existing residence. If you can, create three-dimensional renderings of the structure on a computer and do many virtual walk-throughs to be sure traffic flows nicely and the Sun will heat it comfortably.

The beginning of a site plan.

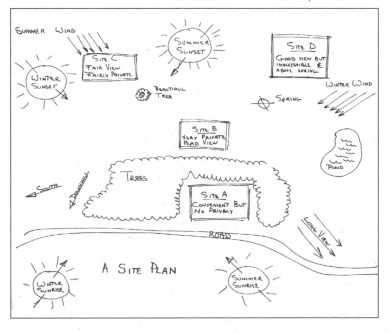

Energy Costs and Savings Resulting from Passive Solar Design

Location (date construction was completed)	Additional cost	Reduction in heating and cooling cost	Annual savings in fuel bill	30-year savings based on fuel costs at time of completion	30-year savings based on projected 5% and 10% increases in fuel costs per year
Jonesport, Maine (1988)	$1,000	70%	$300	$9,000	$19,900-$48,700
Falmouth, Massachusetts (1995)	$3,500	82%	$1,260 with PV system for electricity	$37,800	$76,894-$207,750
Burlington, North Carolina (1990)	$5,000	64%	$840	$25,200	$55,339-$138,160
Naperville, Illinois (1984)	$3,000	72%	$550	$16,500	$36,540-$90,430
Stevens Point, Wisconsin (1995)	$6,000	70%	$600	$18,000	$42,179-$98,690
Hanover, New Hampshire (1994)	0	95%	$2,255	$67,650	$141,390-$403,976
Andover, Connecticut (1981)	0	58%	$958	$28,740	$63,650-$157,675
Santa Fe, New Mexico (1985)	0	81%	$220	$6,600	$14,614-$36,196

From Daniel D. Chiras, *The Solar House: Passive Heating and Cooling*, Chelsea Green, White River Junction, Vermont, 2002.

> It's better to spot problems and make the necessary changes at the drawing stage than to make changes once the walls are up. And it is a lot cheaper to fix problems when all that's required is an eraser and a sharp pencil.

Pore over your plans many times and try to imagine living in the space. Ask others to give you their thoughts on the functionality of the design and the appearance of your house. Are the windows in the right place? Do the windows look good? Are they too small or too large? Will they afford a high level of privacy? Will rooms be useful? Can you haul wood to the wood stove without traipsing through the entire house? Will it be convenient to bring the groceries in? Will it be convenient to recycle and compost kitchen wastes? Do you have easy access to outdoor living spaces? How will air flow? Is there sufficient storage space? Will the design create private areas? How will sound carry? Will your privacy needs be met?

Revise the plans as required to meet your needs. Remember: It's better to spot problems and make the necessary changes at the drawing stage than to make changes once the walls are up. And it is a lot cheaper to fix problems when all that's required is an eraser and a sharp pencil.

Sam Clark, author of *The Independent Builder*, notes, "The builder and designer should both be hired early." Moreover, he adds, "They should be asked to collaborate from start to finish." During the design phase, he asserts that the builder should provide support and consultation to the designer or the architect, giving input on costs and suggesting changes that make the building affordable. "Later," Clark notes, "the roles reverse: During construction, the designer supports and consults with the builder on detailing and other problems that arise."

As noted earlier, the team should include a number of additional individuals as well, such as an energy and green building consultant, HVAC subcontractor, plumber, electrician, and perhaps solar installers. They should be involved early in the process, providing input to the designer and the contractor. As owner, you are a key member of the team. You set the goals. You set the tone. You make suggestions and demands. And don't forget, you pay the bills and everyone's salaries. But don't get too carried away with your newfound position of importance and forget to listen to the experts. They know their business and can help you avoid costly stupid mistakes.

Passive Solar Design: A Primer

Creating sustainable shelter is about homes that provide comfort during all seasons, have a small environmental footprint, and are easy on the pocketbook. Many such homes rely heavily on a design strategy known as *passive solar heating*. Simply defined, passive solar is a means of providing space heating without the use, or at most with minimal use, of costly outside energy sources or mechanical systems—hence the term *passive*.

Passive solar heating is not a new idea. It's been around for centuries. It is tried and tested and was practiced by the ancient Greeks. (The Greeks considered anyone who didn't heat their homes passively to be barbarians!) The ancient Anasazi of the Desert Southwest also built some of their communities using passive solar design principles that allowed them to take full advantage of the low-angled winter Sun for warmth. Today it is practically the first principle of sustainable shelter.

As shown in column 5 in the accompanying table (from 2002), the savings over a 30-year period from passive solar design can be substantial. Column 6 gives estimates of future savings based on 5% and 10% increases in the cost of natural gas over a 30-year period. As you can see, the savings are impressive. Gas prices actually skyrocketed by more than 300% in the subsequent four years, so future savings based on the likelihood of rising fuel prices could be even greater.

Passive solar design is intelligent, climate-sensitive design. If properly executed, it ensures that homes stay warm in the winter and cool in the summer with little, if any, outside energy and with little, if any, mechanical support from air conditioners or evaporative coolers. Year-round comfort happens because many of the measures that permit a house to be passively heated, such as insulation and proper orientation, also passively cool it. But that's not all.

Passive solar houses also provide a significant amount of light during the daytime, saving even more money on energy bills. This strategy, known as daylighting, creates a more healthful environment and for this reason is often incorporated into offices, warehouses, schools, and factories. Daylighting has been shown to increase test scores in schoolchildren and productivity in the work environment.

A Heating System with Only One Moving Part

We like to think of passive solar as a heating system with only one moving part, the Sun. In this design, the house constitutes a huge solar collector. It collects, stores, and releases heat internally to maintain comfortable indoor temperatures in the winter. Passive solar houses rely

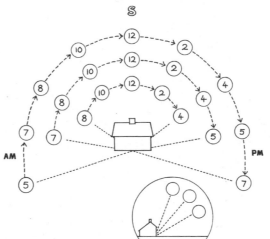

This drawing shows the relation of the Sun to true north at different times of the year and different times of day. The insert shows the angle of the Sun from the ground at the equinoxes and solstices.

on south-facing glass that is designed to allow the low-angled winter Sun to enter the building, where it is converted into heat that warms the interior through the day and into the night. What's more, passive solar houses rely on standard building components such as windows, walls, floors, insulation, and overhangs. You won't need to install photovoltaic panels or expensive mechanical systems to collect, store, and release solar heat. Properly designed, the house itself performs these functions marvelously.

Before we look more closely at passive solar design options and principles, let's examine the ordinary building components of a passive solar house that are responsible for its extraordinary function.

South-Facing Windows

South-facing windows are vital to all passive solar buildings. They allow the low-angled winter Sun to enter the interior of a structure. Inside, sunlight consisting of visible light, near-infrared radiation, and ultraviolet light is converted into heat, warming the interior. Heat is stored in the next component of a passive solar house, the thermal mass.

Thermal Mass

Thermal mass consists of heat-absorbing material in walls, floors, and ceilings. Tile on a concrete slab, for example, serves as thermal mass, as do walls and floors made of earthen material. Even framing and dry wall in a conventional house absorb heat from solar energy.

Thermal mass performs three important functions: 1) It converts sunlight into heat; 2) It radiates heat into the room of a house to provide warmth; and 3) It stores excess heat for later use at night or during long, cloudy periods.

Overhangs

Overhangs, or eaves, protect the walls of a house from rain and thus ensure longer life. In a solar house, however, they also regulate solar gain. To understand this phenomenon, you must first understand the path the Sun takes during the year.

As most readers know, the angle of the Sun from the horizon changes throughout the year. In the summer, the Sun angle (also known as altitude angle) is greatest. That is to say, the Sun cuts a high arc through the sky. In the winter, the altitude angle is lowest. The Sun cuts a low arc through the sky.

The angle changes every day. On June 21, the longest day of the year, the Sun angle is greatest. From that time until December 21, the shortest day of the year, the Sun angle decreases. After December 21, the Sun angle increases again in a cycle that repeats itself year after year after year. What does all this have to do with eaves or overhangs on a passive solar house?

The eaves of a house are the on-off switch of a passive solar system. They determine when sunlight begins to enter the house in the fall and when it exits in the spring. In the summer, the eaves shade walls and windows from the high-angled Sun, facilitating passive cooling. As the Sun descends in the sky after the summer solstice, sunlight is gradually able to enter the house below the overhang. As the Sun falls lower and lower in the sky, more and more sun-

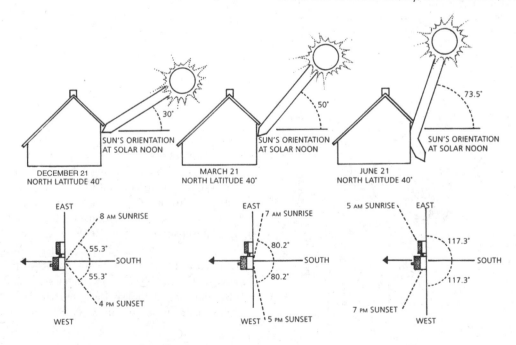

SUN'S ORIENTATION AT SOLAR NOON   30°

SUN'S ORIENTATION AT SOLAR NOON   50°

73.5°

SUN'S ORIENTATION AT SOLAR NOON

DECEMBER 21
NORTH LATITUDE 40°

MARCH 21
NORTH LATITUDE 40°

JUNE 21
NORTH LATITUDE 40°

EAST

8 AM SUNRISE

55.3°

SOUTH

55.3°

4 PM SUNSET

WEST

EAST

7 AM SUNRISE

80.2°

SOUTH

80.2°

WEST   5 PM SUNSET

5 AM SUNRISE   EAST

117.3°

SOUTH

117.3°

7 PM SUNSET

WEST

light enters, creating more heat just when it is needed most, during the winter heating season. As the heating season winds down, however, the Sun is on the rise again, providing less and less heat just as the days grow longer and warmer. By June 21, the Sun is at its peak again, beating down on the roof and overhangs, unable to penetrate the windows.

Overhangs on solar houses must be custom-tailored to one's heating needs. If they are too small, a house could overheat in the spring and fall. If they're too long, the house may not heat up soon enough and the heating system may shut off too soon as well.

Insulation and Window Coverings

Solar energy is a diffuse form of energy. It's not concentrated like gasoline or coal. Even so, small amounts of heat can be used to warm a home, even in the chilliest climate zones. To do so, you need thermal mass to store heat, and insulation to conserve the heat you collect.

In a well-designed solar house, insulation will form a complete and uninterrupted layer that courses through the walls and ceilings, under floors, around foundations, and over windows and skylights. During the summer, the insulation does double duty: It helps to maintain cool temperatures inside the house, blocking heat gain through walls, ceilings, and windows.

Passive Solar Design Options

Keeping the preceding background information in mind, let's look more closely at various passive solar design options. Passive solar design falls into two broad categories: sun-tempered and true passive solar.

Sun-tempered design is the simplest of all passive solar heating systems. It is achieved by placing the house on a lot so that the long axis (if there is one) is oriented along a line running east and west. The number of windows on the south side of the house (in the Northern Hemisphere) is increased to capture solar energy. Slightly higher than normal levels of insulation in walls, ceilings, and floors are usually installed. Unlike more advanced designs, no additional thermal mass is necessary. A sun-tempered house functions well without extra thermal mass, relying only on the "free," or "incidental," mass in framing, floors, walls, surfaces, and furnishings.

Sun-tempered designs are simple to conceptualize and build. The small effort required to convert an ordinary house into a sun-tempered house pays fairly huge benefits. It can reduce heating and cooling requirements by 20%-30%. The amount of savings depends on many factors, such as local conditions (sunlight availability and ambient temperature), house design, and construction choices (including insulation).

While the energy savings in a sun-tempered house are modest compared with a true passive solar house, the cost is equally modest. As a rule, building a sun-tempered house doesn't cost a nickel more than any other house—but a homeowner will be rewarded with a substantial decrease in annual heating bills for the life of the house.

Those who seek greater energy independence build true passive solar houses. This type of house can reduce heating and cooling requirements by 30%-80% or more. To do so, designers place more windows on the south side of the house and install additional thermal mass to accommodate the additional solar gain. They often add additional insulation and take other steps as well (outlined below).

True passive solar houses rely on three design features: 1) direct gain, 2) attached sunspaces, and 3) thermal storage walls.

Direct Gain Systems. The most popular passive solar design is called direct gain. It is probably what you think of as the classic solar house with lots of south-facing windows that allow the space to be heated directly by sunlight. Excess heat is stored in thermal mass within the living space and released into the interior at night or on cold days—anytime the indoor air temperature falls below the surface temperature of the thermal mass.

Sun-tempered designs are simple to conceptualize and build. The small effort required to convert an ordinary house into a sun-tempered house pays fairly huge benefits. It can reduce heating and cooling requirements by 20%-30%.

An illustration of direct-gain passive solar design.

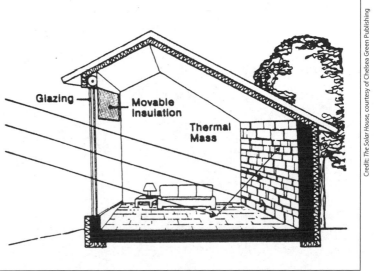

Credit: *The Solar House*, courtesy of Chelsea Green Publishing

Glazing
Movable Insulation
Thermal Mass

Attached sunspaces

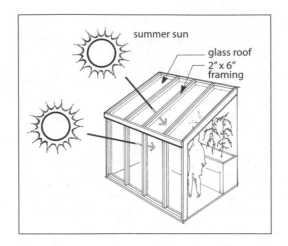

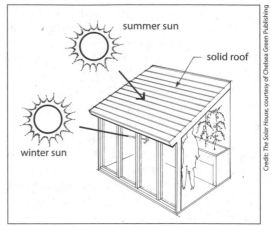

A direct gain house must be carefully thought out, however. If not designed correctly, sunlight can overheat and damage carpeting and furnishings and generally make the living space very uncomfortable.

Attached Sunspaces. The second most popular passive solar design strategy is the attached sunspace or solar greenhouse. This design is most commonly used to retrofit homes, though it can be used in new construction. Built on the south side of a house, an attached sunspace collects heat from sunlight that streams in through the glazing, which is transferred to the house through a door or window in the wall between the sunspace and the house. Fans are sometimes used to facilitate the flow of warm air into the living space. This concept is sometimes referred to as *isolated gain passive solar* because the attached sunspace is a solar collector that is isolated from the main living space.

Although sunspaces are relatively easy to attach to an existing home or incorporate into new construction, they must be designed care-

fully. All-glass attached sunspaces, for example, tend to overheat in the spring, summer, and early fall, sometimes even in the winter. They also tend to get very cold at night in the winter. If you are expecting year-round use, forget it. They'll be terribly disappointing. For most applications, an attached sunspace with a solid roof works best. The south-facing windows suffice to heat the structure during the day.

Thermal Storage Walls. The third type of passive solar design is called a thermal storage wall (TSW), also known as a *Trombe wall* (pronounced TROM) in honor of the French engineer who came up with the idea. It is a mass wall located on the south side of a house (just the opposite in the Southern Hemisphere). As their name implies, TSWs consist of some form of thermal mass that effectively absorbs heat, such as rammed earth, adobe, poured concrete, or cement block. The surface exposed to the Sun is typically painted black, and glass is placed 1-6 inches from the wall, creating a small air space.

Thermal storage (Trombe) walls use mass and convection to heat and cool a space.

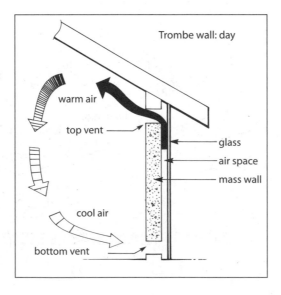

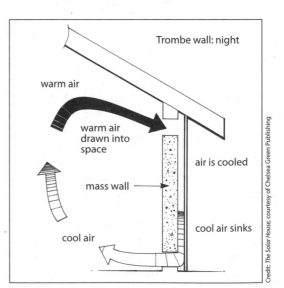

Passive Solar Checklist

- ✔ Small is beautiful
- ✔ East-west axis
- ✔ South-facing glazing
- ✔ Overhangs
- ✔ North-side earth berming
- ✔ Thermal mass inside building envelope
- ✔ High insulation levels
- ✔ Radiant barriers in roof
- ✔ Open airways to promote internal circulation
- ✔ Tight construction to reduce air infiltration
- ✔ Air-to-air heat exchanger
- ✔ Best high-tech windows available
- ✔ Reduced glazing on north and west sides
- ✔ Daylighting
- ✔ Investment in any energy-saving features possible
- ✔ Attention to the little details
- ✔ K.I.S.S. (keep it simple, stupid)

In a TSW system, sunlight penetrates the glass, strikes the mass wall, and heats up its surface. The heat then slowly migrates into and through the thermal mass, moving from the warm surface to the cooler interior of the home. By the time the Sun sets, the heat absorbed by the wall begins to radiate into the adjoining room, providing gentle radiant heat.

For daytime heating, some builders put vents in the wall. As illustrated in the accompanying drawing, cool air enters the lower vents and is warmed by sunlight. As it heats up, the air expands, becomes lighter, and rises in the cavity between the glass and the surface of the mass wall, exiting through the upper vents. More cool air flows in through the bottom vents, creating a thermal convection loop.

Thermal storage walls heat homes indirectly and are therefore referred to as a form of *indirect gain*. So that you're not living behind solid concrete walls, many designers install windows in their thermal storage walls to provide light during the day and a view of the outside world.

Designers often combine passive solar features. For example, they might choose to heat some rooms via direct gain and others, such as bedrooms or a home office, using thermal storage walls. Why? A thermal storage wall reduces the amount of direct sunlight in a room, a big plus

for an office where computers are used, because glare on the computer screen and eyestrain will be lessened. A bedroom with a thermal storage wall may also be darker at night, ensuring a better night's rest.

Choose your system or combination of systems carefully. Be sure to read more about each design, each of which offers a unique set of advantages and disadvantages.

Passive Solar Design Principles

Passive solar design, whether for a new house or a retrofit, requires a working knowledge of a dozen key principles. We highlight them here with a bit of friendly advice, which is to explore each one in much more detail before you design your home. *The Solar House*, by Dan Chiras, is a useful guide (see page 83).

PRINCIPLE 1:
Select a Site with Good Solar Exposure

It may seem silly to point out the obvious—that a solar home needs sunlight, and a lot of it, to work well—but you'd be amazed how many passive solar houses suffer from a chronic lack of sunlight. In many instances, they become shaded over time by trees or neighbors whose houses block the Sun.

The first principle of solar house design is to select a site with unobstructed solar access, now and in the long-term future. Ideally, the south side of the house should be exposed to the Sun from 9 a.m. to 3 p.m. during the heating season. In most locations, the heating season occurs in the late fall, winter, and early spring. In northern latitudes, it is even longer, around eight or nine months of the year, beginning in the early fall and ending in the late spring. In warmer areas, like the southeastern United States, the heating season may be a brief two-month period in the "dead of winter."

> The first principle of solar house design is to select a site with unobstructed solar access, now and in the long-term future. Ideally, the south side of the house should be exposed to the Sun from 9 a.m. to 3 p.m. during the heating season.

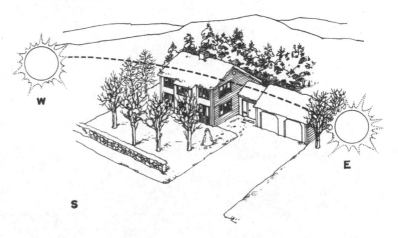

Choose a site carefully. Avoid wooded lots, or be prepared to remove trees on the south side of the house to open it up to the Sun. And avoid lots where the house will be shaded by large obstructions, such as hills or nearby buildings that will block the low-angled winter Sun.

When in doubt, visit the site in the winter, especially around December 21 (the shortest day of the year when the Sun is the lowest in the sky), to be certain that the house will receive a good dose of sunshine.

After years of experience in an area, many solar designers can tell whether a house site is suitable just by tracing the approximate path of the Sun through the sky. Others use a handy little device known as a Solar Pathfinder. It permits a year-round assessment of the solar potential of a site in about 10-15 minutes.

If you can afford only a small lot, select one that is deep from north to south to ensure good solar access. That way, it is less likely that a yet-to-be-built neighboring house will block the Sun at some future point. (Be sure to orient the long axis of the house from east to west.) In rural settings, locating your septic drainage field within the solar access zone is a good strategy to maintain good solar access, because that area will need to be kept clear of trees that could obstruct the southern Sun.

PRINCIPLE 2: Orient the Long Axis of the House from East to West

To ensure maximum solar gain, be sure to orient your house so that its long axis (if any) lies on an east-west plane. The east-west axis should be oriented precisely so that the house points within 10 degrees (east or west) of true south. These steps ensure that the maximum surface area is directed south for optimal solar gain. (Either the front, the back, or the side of the house can serve as the "solar face.")

When staking out a house site, bear in mind that true north and south are not the same as magnetic north and south. True north and south correspond to the lines of longitude, running from the North Pole to the South Pole. Magnetic north and south are determined by magnetic fields created by iron-containing minerals in Earth's core and they rarely run true north and south. In most places, however, true north and south vary slightly from magnetic north and south, from a few to as much as 10-15 degrees.

To determine whether and how much magnetic north and south deviate from true north and south in your location, contact a local surveyor and ask for the magnetic declination or simply consult our chart on page 583 of U.S. magnetic declinations. The magnetic declination is given in degrees and direction. If, for instance, the answer is 10 degrees east, true south is 10 degrees east of magnetic south on your property.

For optimal performance, it's best to orient a house exactly toward true south (in the Northern Hemisphere). Although you can orient a house so that it faces slightly east or west of true south—in fact, up to 20 degrees east or

For ideal solar gain, orient a house so its long axis is aligned with true east and west. This places the maximum amount of south-facing wall toward true south.

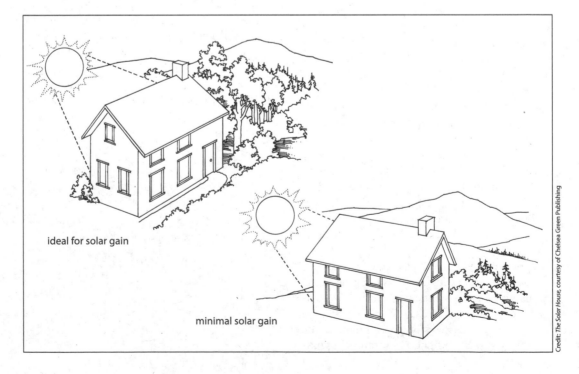

ideal for solar gain

minimal solar gain

Credit: *The Solar House*, courtesy of Chelsea Green Publishing

west of true south—without much reduction in solar heat gain, it's not advisable to deviate any more than you absolutely must. This is because deviating from a true north-south orientation only slightly decreases wintertime solar gain but increases solar gain in the summer, often a lot. Excess solar gain can result in substantial overheating and discomfort and much higher cooling costs.

PRINCIPLE 3: Concentrate Windows on the South Side of the House

As should be perfectly clear by now, a passive solar house relies on south-facing windows to capture sunlight energy during the heating season. Concentrating windows on the south side of the house therefore helps ensure sufficient solar gain.

As a general rule, true passive solar houses require south-facing glass whose cumulative square footage ranges between 7% and 12% of the total square footage of the house. If you're designing a 1,000-square-foot house, for example, the square footage of the south-facing window surface should be between 70 and 120 square feet. The more heat you want, the greater the square footage of south-facing glass you need.

PRINCIPLE 4: Minimize Windows on the North, East, and West Sides

While increasing the amount of glass on the south side increases solar gain in the winter, designers typically decrease window square footage elsewhere. As a rule, north- and east-facing glass should not exceed 4% of the total square footage. West-facing glass should not exceed 2%. Limits on glazing are imposed for several reasons. North-facing glass is restricted to 4% primarily to reduce wintertime heat loss. Limits on east- and west-facing glass are imposed to reduce heat loss as well, but also to reduce summertime heat gain to prevent overheating.

PRINCIPLE 5: Install Adequate Overhangs

As noted earlier, overhangs on the south side are the on-off switches of a passive solar house. The length of the overhang determines when solar heating begins and when it ends. They are designed to suit each climate and location.

Overhangs are most important on the south side of a passive solar house. Contrary to popular belief, eaves provide very little shade on the east and west sides of a house because the angle of the Sun in the morning and late afternoon is fairly low. That doesn't mean that you should skimp on this detail, however. Overhangs may not help to protect the walls and windows from Sun, but they do protect walls from rain.

In most northern locations, a 2-foot overhang on south-facing walls works well. That will shade an 8- to 9-foot wall well in most locations during most of the summer and ensures solar heating as the heating season begins. In south-

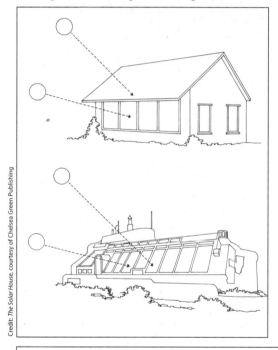

Credit: *The Solar House*, courtesy of Chelsea Green Publishing

Overhangs regulate solar gain—that is, when sunlight can begin to enter a solar home and when solar gain ends each year.

Insolation year-round for latitude 40 degrees north.

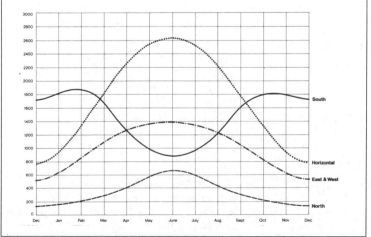

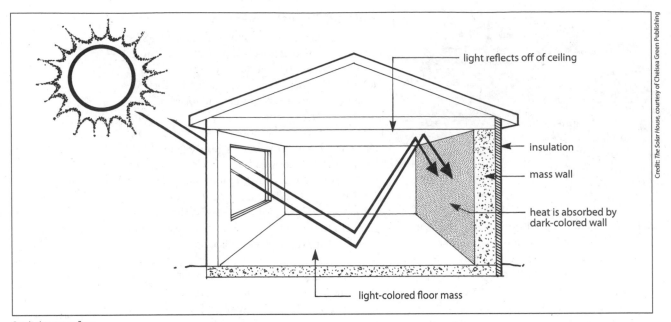

Credit: *The Solar House*, courtesy of Chelsea Green Publishing

light reflects off of ceiling

insulation

mass wall

heat is absorbed by dark-colored wall

light-colored floor mass

Sunlight can reflect off of light-colored mass onto darker, heat-absorbing mass situated deeper in the house.

ern locations, less heat is required and a longer overhang is advisable. Generally, the warmer and sunnier the climate, the greater the overhang. Log on to the website www.susdesign .com/overhang/index.html for help designing overhangs for a house in your region.

While we are on the subject, be sure to choose a house design with few projections such as decks and porches on the south side. These features typically shade windows beneath them, significantly reducing—often practically eliminating—their solar gain. Porches on the east and west sides of a house, however, are often beneficial, as they shade windows and exterior walls from the hot summer Sun.

PRINCIPLE 6:
Incorporate Sufficient Thermal Mass

Thermal mass is essential to all passive solar houses, regardless of type or design. For sun-tempered houses, meaning houses with less than 7% solar glazing, incidental mass found in tile floors, wood framing, wallboard (dry wall), and furniture is generally sufficient to accommodate the heat produced by the Sun. When south-facing glass falls within the 7%-12% range, additional mass is required to prevent overheating. Concrete slabs, adobe floors, masonry walls, and planters all work well.

Thermal mass is costly and tricky. It is tricky because placement of thermal mass is directed by two contradictory principles. For optimal performance, mass should be located in direct contact with incoming solar radiation. For greatest comfort, however, thermal mass should be distributed throughout the house for even heat. That way, the occupants of a home are sur-

rounded by heat at night and on cold winter days as the thermal mass relinquishes its stored solar energy. To achieve these contradictory goals, designers can use a lighter-colored floor mass near the windows to bounce light into the back of the home where it strikes darker-colored wall mass. Another innovative way to deposit sunlight energy into thermal mass is the Solar Slab, described in the accompanying sidebar.

PRINCIPLE 7:
Insulate, Insulate, and Insulate!

Many passive solar houses built in the 1970s and early 1980s gathered a significant amount of heat during the day but lost much of that heat through leaky, poorly insulated walls. Modern passive solar home designers don't make that mistake; in fact, smart designers tend to super-insulate, knowing that the better a home retains heat in the winter, the better it will perform and the more comfortable it will be.

A passive solar house requires superior levels of insulation in ceilings, walls, floors, and foundations to conserve hard-earned solar energy and maintain comfortable interior temperatures despite wide swings in outdoor temperature. Ron Judkoff, of the National Renewable Energy Lab, recommends insulating at least to the level prescribed by the International Energy Conservation Code for residential structures—but preferably beyond them. EPA and DOE's Energy Star program prescribe superb levels of insulation, too. Even so, many energy-smart builders prefer to insulate at levels that exceed these recommendations, often 30% higher than recommended.

A passive solar house requires superior levels of insulation in ceilings, walls, floors, and foundations to conserve hard-earned solar energy and maintain comfortable interior temperatures despite wide swings in outdoor temperature.

The Solar Slab

Thermal mass is vital to the function of a passive solar home, and its absence is often the Achilles' heel of passive solar houses. As noted in the accompanying text, thermal mass is tricky. It needs to be in direct contact with the Sun as much as possible and needs to be widely dispersed throughout a building for maximum comfort. Doing so is not easy.

One solution to the thermal mass dilemma that faces most designers is the Solar Slab, a technique invented by James Kachadorian and detailed in his book, *The Passive Solar House.*

As shown in the accompanying drawing, a Solar Slab is a concrete slab poured over cement blocks. The blocks are laid in parallel rows that are oriented north-south. This ensures that the cavities in each block line up to create a continuous path for air to move from the north side of the house to the south side. How is air propelled through the subslab channels?

According to Kachadorian, air movement through the subslab air channels is powered as a result of natural airflow—more specifically, a natural convection current. As you can see, sunlight streaming through south-facing windows warms air in the vicinity. Because heat causes air to expand, warm air along the south side of the house rises. Cool air flows in to replace it. That cool air comes out of the ducts under the slab.

Warm air from the south side then moves slowly along the ceiling to the back of the house. As it flows northward through the interior of the house, it is cooled. Along the north wall, heavier cooler air sinks. It then enters the vents on the north side of the house that connect to the subslab channels. The movement of air into the system at this end and the movement of air out of the channels along the south side are the drivers for the natural convection loop, also known as a thermosiphon. As the air flows through the channels in the cement blocks, it gives off additional heat. Heat absorbed by the blocks then migrates upward to warm the slab. The slab stores heat and radiates it into the room at night or on cold days.

As illustrated, the natural convection cycle is energized by incoming solar radiation. However elegant the system appears, though, it often needs a little boost in the form of fans to increase airflow through the slab.

Concerns have been raised about the potential for moisture to build up in the block cavities and promote the growth of mold and mildew. Accordingly, some builders use plastic pipes instead of cement blocks. They're very likely most important in areas that experience higher levels of humidity, according to Kachadorian. Although such modifications may be required, the Solar Slab remains a simple and effective means of storing solar heat in a location that permits the widest dispersal.

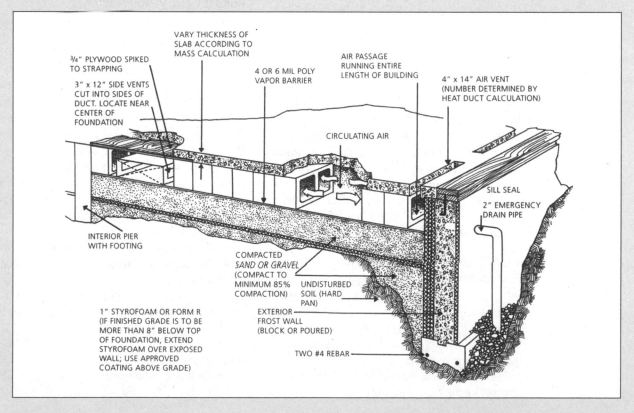

VARY THICKNESS OF SLAB ACCORDING TO MASS CALCULATION

3/4" PLYWOOD SPIKED TO STRAPPING

3" x 12" SIDE VENTS CUT INTO SIDES OF DUCT. LOCATE NEAR CENTER OF FOUNDATION

4 OR 6 MIL POLY VAPOR BARRIER

AIR PASSAGE RUNNING ENTIRE LENGTH OF BUILDING

4" x 14" AIR VENT (NUMBER DETERMINED BY HEAT DUCT CALCULATION)

CIRCULATING AIR

SILL SEAL

2" EMERGENCY DRAIN PIPE

INTERIOR PIER WITH FOOTING

COMPACTED *SAND OR GRAVEL* (COMPACT TO MINIMUM 85% COMPACTION)

UNDISTURBED SOIL (HARD PAN)

1" STYROFOAM OR FORM R (IF FINISHED GRADE IS TO BE MORE THAN 8" BELOW TOP OF FOUNDATION, EXTEND STYROFOAM OVER EXPOSED WALL; USE APPROVED COATING ABOVE GRADE)

EXTERIOR FROST WALL (BLOCK OR POURED)

TWO #4 REBAR

General insulation guidelines for superefficient passive solar houses are R-30 walls and R-60 roofs in temperate climates, and R-40 walls and R-80 roofs in extremely hot or cold climates, according to Ken Olson and Joe Schwartz, authors of "A Home Sweet Solar Home: A Passive Solar Design Primer," published in *Home Power* magazine.

When designing passive solar houses in sun-challenged climates, remember that adding insulation can offset the lack of sunshine. Even in good solar regions, in cases where builders can't add as much south-facing glass, say for aesthetic reasons, as is required for optimal solar gain, extra insulation may offset the reduced solar gain.

Bear in mind that insulation has to be installed correctly, too. Pay careful attention to detail, or hire pros who know how to do the job right. Batt insulation should not be compressed and should be flush against studs and ceiling joists so air cannot flow around it. Some builders are fond of the liquid foam insulation products. They are sprayed in wall and ceiling cavities as a liquid but quickly expand and convert to a solid. Liquid foam products create an airtight seal and also repel water, which has a devastating effect on many other types of insulation such as cellulose and fiberglass. Even a tiny amount of water in standard insulation products reduces their resistance to heat flow and thus lowers their R-value (ability to insulate). Liquid foam insulation nearly doubles the R-value per inch but tends to cost a lot more.

Although wall, roof, floor, and foundation insulation are important, don't forget to install well-insulated windows. Insulated windows are especially important in north-, east-, and west-facing walls. To make a house even more energy efficient, you can cover windows at night with insulated shades or rigid thermal shutters.

PRINCIPLE 8:
Protect Insulation from Moisture

As just noted, most commonly used forms of insulation lose most of their thermal resistance (their ability to block heat flow) when they contain even a small amount of moisture. Even a tiny increase in moisture level can reduce the R-value of insulation by half. So it is important to protect insulation from moisture.

In conventional stick-frame construction, builders frequently install a plastic vapor barrier on the warm side of the wall. (In cold climates, the vapor barrier is installed on the inside of the wall; in warm, humid climates, it is installed on the outside.) Placing a plastic house wrap on the exterior sheathing may provide additional protection. (Natural building materials typically don't require moisture barriers—see below.) Moisture barriers are meant to retard the movement of moisture into and through walls.

While vapor barriers are a good idea, they are not as important as many would believe. That's because most moisture enters walls elsewhere. Very little actually penetrates a wall surface under most circumstances. The number-one source of moisture in walls is through improper flashing around doors, windows, and roofs. The second major source of moisture is through penetrations in walls, for example, electrical outlets, light switches, recessed lighting, and other electrical and plumbing penetrations. When building a new house or energy-retrofitting an existing one, pay close attention to these areas to help protect the insulation in your walls and ceilings from moisture. Careful flashing and caulking of penetrations can pay a huge dividend.

PRINCIPLE 9:
Build It Tight, Ventilate It Right

Energy-efficient passive solar houses are built airtight to prevent moisture from entering the walls, as just noted, but also to prevent infiltration and exfiltration. Infiltration, the movement of air into a building, and exfiltration, the movement of air in the opposite direction, can rob a home of heat in the winter. On windy winter days, for instance, cold air may leak into a house through openings in the building envelope, creating uncomfortable drafts. On days when the wind isn't blowing, warm interior air may escape through leaks in the building envelope, reducing your heating efficiency and comfort levels.

In the summer, leaks in the building envelope also decrease comfort and raise fuel bills. Cool air created through passive or active systems may escape through those openings. If the wind is blowing, warm outside air may pour in.

Competent solar contractors seal all cracks in a building envelope around windows and doors and wherever else they occur, in order to reduce infiltration and exfiltration. Designers may also include entryways and mudrooms that can be

tor. As noted earlier, heat-recovery ventilators remove stale air and replace it with fresh outside air, and thus help to maintain indoor air quality while reducing heat loss in the winter and heat gain in the summer.

Good insulation, protecting the insulation from moisture, and effective sealing—principles 7, 8, and 9—all make a residence more energy efficient. Energy efficiency is integral to passive solar design, almost as important as the sunlight that provides the energy to heat a house. Without efficiency measures, passive solar design won't work. Efficiency pays huge dividends in the summer months as well by helping to keep a home comfortable when outside temperatures are in the 90s.

PRINCIPLE 10:
Design Your House So Most of the Rooms Are Heated Directly by Incoming Sunlight

Most passive solar houses are rectangular designs. This shape is thermally superior to others for several reasons. First, rectangular designs result in the largest possible amount of exterior wall space and window area on the south side of the building—the surface that's exposed to the low-angled winter Sun. Second, rectangular designs permit sunlight to penetrate more deeply into the interior of a building than square floor plans. In fact, if the house is narrow enough, a rectangular design can ensure that each room is heated independently by the Sun. This, in turn, eliminates the need for fans and ducts to move warm air from one part of the house to another. Third, rectangular floor plans minimize the exposure of east and west walls to summer Sun, thereby reducing heat gain, a feature that is especially important in hot climates.

Some solar houses are designed as long, narrow rectangles often only one room deep. Al-

closed off from the main living areas, especially in cold climates. In the winter, sealed entryways (sometimes referred to as airlocks) prevent cold air from rushing in when the outside door is opened. Another effective way of reducing air infiltration is earth sheltering.

Earth sheltering helps to create a much more airtight house because it dramatically reduces the exterior wall space exposed to the elements. Earth sheltering also reduces exterior wall maintenance (painting and staining siding). In addition, this technique helps to reduce heat loss through exterior walls and roofs because the earth surrounding the house stays at a constant temperature, usually around 50°F, below the frost line. This constant and relatively warm thermal blanket around the earth-sheltered portion of a house helps to keep it warm in the winter and cool in the summer.

But how airtight should a house be? An airtight house should permit 0.35-0.5 total air changes per hour. That's sufficient to permit fresh air inside and purge stale air. If your house is tighter, you may need to install a whole-house ventilation system with a heat-recovery ventila-

This floor plan of Dan Chiras's passive solar rammed earth tire/strawbale house shows various design features that reduce sun drenching and create usable living space. The front air lock (an attached sunspace) shields the living room; a wide hallway down the south side protects the kitchen and living room; partition walls shield bedrooms; and the bathroom shields the office.

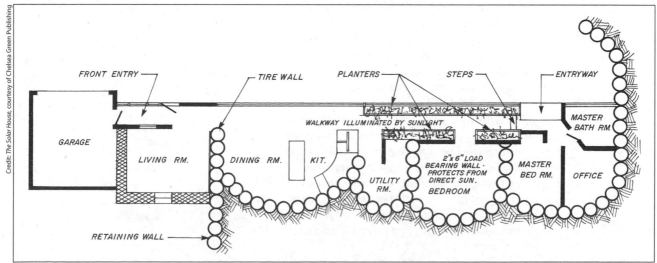

Credit: *The Solar House*, courtesy of Chelsea Green Publishing

though this design ensures maximum solar gain in each room, houses can, as solar expert Johnny Weiss at Solar Energy International notes, end up looking like "solar trailers."

If a rectangular design is not possible or desirable, it is best to place rooms that require less heat, such as bedrooms—or rooms that have their own source of heat, such as kitchens and utility rooms—on the north side of the house. Place frequently used rooms, such as home offices and living rooms, on the south side where they can be warmed by sunlight. Dining rooms are often best placed on the south side, too. Kitchens shouldn't be located on the south side, as wall and base cabinets often limit solar glazing, and intense sunlight can make meal preparation rather uncomfortable.

PRINCIPLE 11: Create Sun-Free Zones

Unless you live in an area that is starved for sunlight—that is, regions solar designers refer to as the nation's gloom belt, it is wise to create sun-free zones where family members can watch TV, work on computers, or just lie on a couch without being baked by the sunlight. Too much sunlight in the wrong place can make sun-drenched rooms extremely hot, uncomfortable, and relatively useless during daylight hours

Unfortunately, many passive solar designs are little more than rectangular boxes with south-facing windows that permit maximum sunlight penetration with little thought given to habitability. During the day, many of these homes are bathed in blinding sunlight so that baseball caps and sunglasses become required gear for occupants. Glare on computer screens and televisions can create eyestrain and headaches and be a major nuisance. If you're not careful, you may create a solar oven that bakes you and your pets and bleaches furniture and carpeting on sunny winter days.

Creating a design that permits sunlight to enter a home yet doesn't render living space useless on sunny days can be an enormous challenge. Consider placing hallways and stairwells that lead to bedrooms and offices in locations that receive maximum solar gain. Although they'll be bathed in sunlight during the day, very few people linger in them. Dining rooms can also be used to collect sunlight, as they are rarely used during daylight hours. Thermal storage walls can also be installed to eliminate sun drenching without sacrificing solar gain. Home offices and bedrooms are good candidates for the placement of thermal storage walls.

PRINCIPLE 12: Provide Back-Up Heat

Although passive solar houses often receive the majority of their heat from the Sun, most of them still require back-up heat—a furnace, boiler, or wall heater that meets a family's needs during long, cold, cloudy periods when all the solar energy stored in the thermal mass has been dissipated. Back-up heating systems are required by code and do come in handy from time to time. To meet code, they must be automatic—that is, able to function on their own, turning on and off as needed when a family is not home.

Remember, however, that because most of your home's heat will come from the Sun, back-up heating systems must generally be downsized, often substantially. Downsizing—or more accurately, rightsizing—means installing a smaller-than-usual heater, wood stove, or boiler. This reduces the initial costs, thus offsetting higher costs elsewhere—for example, higher insulation costs. It also saves fuel and money over the long haul because an oversize heating system tends to operate inefficiently. (The same pertains to mechanical back-up cooling systems.)

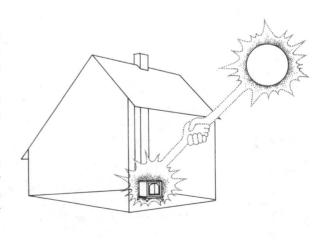

Radiant floor heating has become popular over the last several decades and is an excellent backup for passive solar houses. Typically powered by propane or natural gas, these systems have tubes that circulate hot water through concrete floor slabs or under conventional wooden floors. They keep a home at a fairly constant temperature throughout the winter. However, with sufficient passive solar gain, radiant floor heat will be required infrequently. If you are building an energy-efficient passive solar house in a sunny climate, radiant floor heat is a waste of money. When selecting a back-up heating (or cooling) system, it's best to purchase one that is efficient and nonpolluting. Not only does this help to make a passive solar house more environmentally friendly, it cuts fuel bills and may qualify for tax incentives from various governmental agencies or rebates from local utilities. Among the most environmentally friendly back-up heating systems are solar hot-water systems, masonry heaters, clean-burning wood stoves, and heat pumps. For further information, check out *The Homeowner's Guide to Renewable Energy*, by Dan Chiras (see page 84).

While the passive design principles outlined above can be useful in designing your passive solar house, take the time to learn more. If you want further information, consult *The Solar House: Passive Heating and Cooling*, by Dan Chiras (page 83) or *The Passive Solar House*, by James Kachadorian (page 83).

Useful Design Tools

For many years, solar designers have based their work on general principles, hunches, theory, and a little wishful thinking. As a result, many early designs performed poorly. They either overheated and were extremely uncomfortable, or they didn't provide as much heat as hoped.

To create designs that function closer to expectations, many professionals use powerful new energy design software. One of our favorites is BWG2004 (formerly known as Builder Guide for Windows). Written by Fred Roberts, a Colorado architect, this program is available from Solaequis (www.solaequis.com). Another is Energy-10, a program available from the Sustainable Buildings Industry Council in Washington, D.C. Both computer programs allow designers to predict the energy performance of a building early in the process, long before costly blueprints have been generated. Both programs predict the amount of solar gain and passive cooling a house design will provide in a particular location.

This information facilitates the fine-tuning of a house design. If the projected performance falls short of expectations, for example, the designer may add windows on the south side, beef up insulation, and/or tighten up the building envelope. After these adjustments have been made, BGW2004 and Energy-10 provide new energy estimates.

Of the two programs, BWG2004 is the most user-friendly. Energy-10, while sophisticated and amazingly useful, requires a considerable amount of training and a lot of time to input data.

Passive Solar Design Challenges

Passive solar is simple in theory. Find a sunny lot, orient the house to true south, concentrate windows on the south side, provide overhang, insulate well, and add a sufficient amount of thermal mass. After the house is done, sit back and let the Sun shine in and enjoy the free solar heat while your neighbors pay extravagant fuel bills to maintain some degree of comfort.

But even though the rules of passive design are elementary, designing a passive solar house that works well year-round can be a challenge. Many houses can't be oriented to true south for one reason or another. For example, the view may be in another direction. Floor plans may create obstacles to optimal solar gain. Cabinets in kitchens on the south side, for instance, may reduce solar gain. Even entryways, if designed improperly, can block solar gain. Decks and porches on the south side have the same effect and sometimes negate solar gain entirely. Trees may shade the site. Budgetary constraints may make certain desirable design features impractical.

In our experience, very few passive designs offer optimal solar gain. Most designs involve compromises, which often lead to substantial decreases in energy performance and comfort. Too many west-facing windows included in a design to avail oneself of a killer view, for instance, may lead to summertime overheating. Too many north-facing windows to take in a breathtaking view of a mountain, ocean, or butte will result in significant heat loss in the winter and the resulting discomfort. Even many contemporary building fashions, such as tall ceilings and skylights, can compromise the year-round function, utility, and comfort of a passive solar house.

If you design correctly, however, passive solar will result in a lifetime of low-cost comfort.

To dramatically reduce the environmental impact of housing, many contractors are turning to locally available natural materials such as straw bale, cob, straw-clay, rammed earth, and adobe.

Guaranteed. Contrary to popular belief, passive solar can work well in most climates. Don't exclude passive solar from your plans simply because your house site is not as sunny as that of a friend who lives in Colorado, New Mexico, or California. Plan to obtain as much free solar energy as possible.

Even if you live in a sun-challenged area, such as the northeastern United States, solar energy can provide a significant amount of heat to an energy-efficient household. You may acquire most of that heat early and late in the heating season, but the gains can be impressive.

You will be amazed at how little passive solar design actually costs. In fact, if you plan well, a passive solar house may not cost a penny more than a conventional house. Proper orientation and concentrating the windows on the south side, for example, won't add to the cost of your new home. Adding insulation beyond levels re-quired by code will increase the price a bit, but higher costs on insulation are often offset by installing a smaller heating and cooling system. You will certainly save huge amounts of money over the lifetime of your home in lower energy bills that will pay back any additional initial investment many times over.

In recent years, passive solar house design has become even more cost competitive with conventional construction, because many building departments have increased energy-efficiency requirements. As a result, many measures such as energy-efficient windows that were once unique to passive solar houses are now required by code. Bottom line: It isn't going to cost you an arm and a leg to go solar. And over the lifetime of your home, passive solar design could save you thousands, even tens of thousands, of dollars in fuel bills.

Architecture and Building Materials

For years, most of the interest in green building focused on energy-efficient and solar construction. Today, however, green builders have widened their scope to include many other aspects of homebuilding. One area where significant change has occurred is in the materials used to build houses. To dramatically reduce the environmental impact of housing, many contractors are turning to locally available natural materials such as straw bale, cob, straw-clay, rammed earth, and adobe.

Using locally available natural materials can save money and a considerable amount of energy. In this section, we take a look at a number of natural building materials, including wood. We'll explore the pros and cons of each one to help you choose wisely among the many options.

When choosing a building material, two considerations rise to the forefront: 1) how well it functions in a particular climate, and 2) its availability. As noted earlier, many natural building materials may even be found on the site itself, which makes them a potentially great choice.

Building with Wood

Wood is a versatile material that is used to build houses throughout the world. In fact, most new houses in more developed countries are made from wood harvested from the world's forests. It is often shipped hundreds, even thousands, of miles to house sites. Even many natural homes whose walls are made out of earth, straw, or combinations of the two, require a significant amount of wood to build roofs, floors, interior walls, cabinets, and decks.

STICK-FRAME HOUSES

Although wood-frame houses are often criticized in natural building circles, they can provide many years of service if they are well designed and well built. Roofs and foundations require special consideration. The problem is that some builders cut corners to save money up front. Cost-cutting measures may lead to potentially devastating problems, including mold buildup and rot, that dramatically reduce the useful life of a stick-frame house.

Wood does offer many advantages. It is widely available and well understood, and stick-frame homes can be remodeled with relative ease, although it can be a pain in the neck, even under the most ideal circumstances. Moreover, wood is a renewable resource and can be grown and harvested sustainably.

Unfortunately, sustainable wood production is only just beginning to emerge in many parts of the world. Although more and more wood is grown and harvested intelligently, abuses continue that cause significant damage to forests and the surrounding environment.

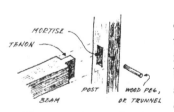

Mortise and tenon joint. From *A Shelter Sketchbook*. Used with permission.

If you are thinking about building a house using wood, you can take many steps to reduce the environmental impact of this ubiquitous and potentially environmentally friendly material. One way is to harvest wood locally on your land or on a woodlot nearby (with permission, of course!). Locally harvested wood can be milled into usable lumber on site using a portable sawmill like an Alaskan Mill or "mobile dimension saw" or hauled to a local sawmill for processing.

Another way to reduce the impact of wood is to purchase reclaimed lumber—that is, lumber salvaged from demolition sites or old redwoods submerged for years in a river. Salvaged lumber is widely available in many parts of the country. You may also want to use engineered lumber—that is, sheathing and dimensional lumber made from wood chips that are glued together. Engineered lumber provides sufficient strength and uses more fiber from each tree than dimensional lumber uses. In other words, there's less waste. In addition, in many instances, engineered lumber is manufactured from smaller-diameter trees.

Yet another way to reduce the impact of your new home—and one that's rarely considered—is to build a smaller house. Smaller houses use fewer materials, including wood. A small house needs not just fewer timbers but smaller ones, since the spans are shorter. This not only reduces building costs, it reduces the strain on the world's already taxed forests. Check out *The Not So Big House*, by Sarah Susanka (page 85).

POST AND BEAM

Stick-frame houses, erected by the millions each year, are the most popular type of construction in more-developed countries. Another technique that uses wood for framing and is popular among the natural building community, especially straw bale builders, is post-and-beam construction. Post and beam, which has been around for centuries, produces amazingly beautiful houses. In the hands of a skilled craftsman, timber framing results in works of art. Bear in mind, however, that it requires heavy lifting, as timbers weigh a great deal. Teams must be assembled on days when the frame is to be assembled. Timber framing also requires special tools; some builders perform all the work using hand tools. As Sam Clark notes in *Independent Builder*, the frame can be fabricated indoors during the off-season and then hauled to a site and assembled quickly and efficiently.

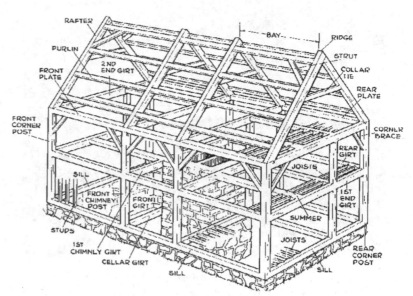

Framing nomenclature.

Unfortunately, say critics, timber-frame construction is fairly wood-intensive. It also often requires fairly large timbers that come from much older and much higher-quality trees that are increasingly more difficult to find and often quite costly. They may need to be shipped hundreds or thousands of miles. But there are ways to minimize the impact. Some timber framers cut their own wood or purchase timbers that have been hauled out of the woods behind a team of horses, rather than with heavy machinery that can cause major environmental damage. Fortunately, there are good local sources of timbers for post-and-beam construction in many parts of the world. Many building sites themselves have sufficient timber to build a timber-frame house. If the trees are cut from a sustainably managed woodlot, that's all the better!

While traditional post-and-beam construction requires considerable skill and experience, there are techniques that open up this method to a larger, less skilled audience. Rob Roy's book, *Timber Framing for the Rest of Us*, describes a simpler, more user-friendly approach.

LOG HOUSES

Although post-and-beam and timber-frame construction are especially popular among natural builders these days, log houses are far more prevalent. In the United States, over 70,000 log houses are built each year.

Log houses have been around for centuries. Early American settlers built their cabins in the woods from logs harvested from the lush forests of the eastern United States. Our early ancestors didn't invent log construction, however; they brought the knowledge and skills from their

Log homes can be energy efficient and built from sustainably harvested forests.

homelands in Europe, especially Finland, Sweden, and Germany. Today's log homes greatly surpass the survival shelter of the early pioneers. Many are elegant structures of unrivaled beauty.

Log house construction begins with a solid foundation to elevate the logs off the ground. Logs are debarked and then laid down one at a time. In olden days, logs were joined only by interlocking corner notches. The spaces between adjacent logs were filled with a material known as *chinking* in an attempt to create airtight and energy-efficient structures. Mud was used in early days; a cement mix is now more commonly employed.

To create an even more airtight seal, the Swedes developed chinkless log construction. In this technique, logs are notched lengthwise. The long notch, or groove, is placed down onto the log below it so they fit tightly together, thus eliminating the gap between logs and the need to apply chinking. Although this construction method is time-consuming, it results in a much more energy-efficient and comfortable building. It also reduces maintenance, as chinking is prone to cracking and needs periodic repair.

Log houses offer a rustic beauty that is unparalleled in modern building. If your heart is set on building a log house, you'll likely be able to find a company or two in your area that can help make your dream become reality. If you want to build your own house, you can purchase a log home kit, or if you're really ambitious, you can build it from scratch, cutting and notching logs yourself!

It's important to note, however, that while log houses are a form of natural building, they are probably the least sustainable of the natural building options. Why is that—aren't logs a natural byproduct of nature?

Logs are one of nature's many renewable resources. Moreover, trees harvested for log house construction require far less processing (milling) than trees that are converted to standard dimensional lumber (2x4s and 2x6s) and engineered lumber that is currently used in conventional stick-frame construction. The problem with log house construction, from an environmental standpoint, is that most of the 70,000 log homes built each year in the U.S. are made with trees harvested from distant and unsustainably managed forests. The logs must be transported hundreds of miles to the job site, consuming huge amounts of energy. If you want to build a log house, be sure to obtain locally harvested trees from sustainably managed forests and woodlots. Potentially toxic wood preservatives are also used to protect the logs and must be applied periodically to ensure long-term protection.

If you can build a log house sustainably, you will be rewarded many times over. For one thing, its thermal performance can be excellent. In fact, the U.S. National Bureau of Standards performed tests on log walls and found that a 6-inch-thick log wall equaled or exceeded the energy performance of any other type of (conventional) exterior wall during all seasons tested except the dead of winter. In the winter, conventionally insulated wood-frame walls won by a small margin.

CORDWOOD HOUSES

Cordwood construction represents yet another way that wood can be used to build houses (and other buildings). Cordwood houses, as their name implies, are made from a most unlikely material, firewood.

Here is how a cordwood house is built: To begin, a cement or lime-sand mortar is placed on a concrete foundation in two parallel bands—one

Details of cordwood construction.

Courtesy of Rob Roy

Photo : Bill Tishler

Cordwood walls such as this one under construction were used to build the Mecikalski general store in Jennings, Wisconsin.

on the inside and the other on the outside—with about 6 inches separating the inner and outer joints. Cordwood is then placed directly on the mortar joints, perpendicular to the joints and spanning them. The cordwood, which can be either round or split, is then worked into the mortar to create a strong bond. After the logs are embedded in the mortar, insulation is poured into the space between mortar joints; sawdust treated with lime is often used. The lime protects the insulation from moisture and insects. After the first course is complete, subsequent courses are laid down in similar fashion until the wall is complete.

Cordwood construction has been around for a long time. In fact, the first cordwood homes in North America were built in the mid-1800s. Like log house construction, the technique probably originated in Europe. In recent times, cordwood has gained popularity among owner-builders largely as a result of the work of Rob and Jaki Roy from upper New York State and others. Rob's books on the subject provide great guidance. (See page 89).

Cordwood offers many advantages over conventional construction. It uses a material that is locally available and abundant in many parts of the world. The technique is also relatively easy to learn and hence great for owner-builders. Moreover, cordwood walls are very strong and

durable. Buildings erected more than a hundred years ago are still standing in good condition.

Cordwood construction produces homes of rustic beauty, resembling stone masonry. The walls also offer thermal mass and insulation and therefore are ideal for passive solar designs. Like several other natural building techniques, cordwood construction is pretty labor-intensive. It also uses lime or cement to make mortar, both of which are energy intensive to produce.

Building with Earth and Straw

While wood has been a popular building material throughout much of human history, and will remain the mainstay for many years to come, it is not the only building material and it is not among the original materials used to erect shelter. Earthen materials have been used far longer, and continue to be used, to build shelter throughout the world. Today, at least half the world's population lives in homes built from earth. Many of them are beautiful and durable structures, and many have been continuously occupied for hundreds of years. Grasses and other fibrous materials have also played a huge role in building and are being used more and more by a growing legion of builders who want to create low-impact, energy-efficient buildings. In this section, we'll examine many of these options, starting with straw bale construction.

STRAW BALE

Straw bale building is the most popular of the natural homebuilding techniques today, second only to log house construction. The popularity of straw bale construction is largely due to the fact that straw bale homes are energy efficient and suitable for a wide range of climates, even humid ones.

Straw is the stem of harvested grains such as wheat, barley, oats, and rice. Harvesting machines remove the seeds and spit out straw, which is baled and used as animal bedding or left on the fields and either plowed under or burned to return the nutrients to the soil.

Although houses can be built using bales of hay, they are not as desirable. Hay consists of meadow grasses that are harvested green to retain as many nutrients as possible. It is used primarily as feed for livestock. Although hay bales can be used to build houses, builders generally avoid them because they could attract rats and mice. In contrast, straw has no nutritional value and is quite unattractive to rodents, except as a place to nest. The high nitrogen-to-carbon ratio

While wood has been a popular building material throughout much of human history, and will remain the mainstay for many years to come, it is not the only building material and it is not among the original materials used to erect shelter. Earthen materials have been used far longer, and continue to be used, to build shelter throughout the world.

Photo: Kelly Lerner, OneWorldDesign.com

The Drew residence is a beautiful straw bale home located in Lake County, California, designed by Pete Gang of Common Sense Design and Kelly Lerner of One World Design.

A straw bale wall under construction; the box frames out a window opening.

Photo: Dan Chiras

of hay also makes it more susceptible to composting.

Straw bales are used to build exterior walls. They are often stacked between a post-and-beam framing to provide insulation in a technique known as the in-fill method. But sometimes no framing is used. Straw bales form the walls and support the roof and upper stories, if any. Such structures are known as load-bearing straw bale houses.

Straw is popular for many reasons. It can reduce wood use and produces super-energy-efficient walls. It is inexpensive and fairly easy to work with. Straw is also a renewable resource grown throughout the world and is available in abundance anywhere grain crops are produced. With careful attention to detail, straw bales can create a house that is extremely attractive and durable. Moreover, as recent tests show, straw bale is quite fire resistant. Plastered straw bale walls hold up extremely well against fires—much better than wood-framed walls. Even

unplastered walls burn very slowly because the density of the bales means that little oxygen is available to support combustion. If walls are built correctly, they are also resistant to rodents and insects.

Although the idea of building the walls of a house out of straw bales may seem absurd to the uninitiated, these homes perform very well indeed. Like any building technique, however, straw bale must be done correctly. To endure weather over the long haul, straw bale houses must be equipped with good roofs with adequate overhang and solid foundations that prevent moisture from seeping into the walls. Builders must pay careful attention to details such as flashing around windows and doors and on roofs. They must carefully seal all penetrations in the building envelope to prevent moisture from entering the walls. They must also apply wall finishes (like earthen or lime plasters) that permit the escape of moisture that can creep into bale walls. If careful attention is paid to preventing moisture incursion, the walls will be free of mold and decay.

Straw bale building results in an energy-efficient structure. Although three-string bales laid flat in walls were once thought to provide an R-value (resistance to heat) of slightly more than 50, newer studies suggest a much lower R-value, about R-32. Even though the R-value is not what it was once thought to be, it still is considerably higher (about 30%) than a 2x6 framed wall insulated with wet-blown cellulose or fiberglass batts.

Straw bale building got its start in the late 1800s in the windswept plains of Nebraska in an area known as the Sand Hills. Because the soil was sandy, sod homes were out of the question. Because trees were rare, stick-frame homes were beyond the reach of the early settlers.

Shortly after the invention of mechanical baling equipment in the late 1800s, however, early settlers began to experiment with hay bales, using them to create walls of temporary housing. Much to their surprise, their temporary homes turned out to be exceedingly comfortable, both in the winter and the summer. Many settlers plastered the walls and forsook their dreams of building more expensive stick-frame houses like their eastern relatives had. As testimony to the durability of the early straw bale structures, a number are still standing today in good shape.

Straw bale building died out but experienced a resurgence in the 1970s. Although initially touted as a low-cost building method, experience shows that in most cases, it isn't necessarily

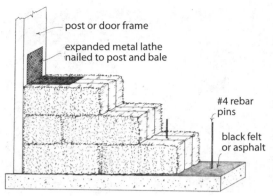

post or door frame

expanded metal lathe nailed to post and bale

#4 rebar pins

black felt or asphalt

Credit: *The Straw Bale House*, courtesy of Chelsea Green Publishing

Left: Typical straw bale construction, like bricks only bigger. **Center:** A straw bale foundation and door frame ready to begin stacking. **Right:** Stacking walls is the easy part! Stucco application takes a bit of elbow grease.

cheaper than standard construction, especially if you hire outside help. If you hire a contractor to build a straw bale house, expect to pay as much as for a conventional spec or custom-built house, perhaps more. Although straw is less expensive than wood, building the walls of a house represents only 15%-20% of the total cost of construction. Slight savings on wall material are often offset by higher costs of plastering. But with straw bale as with any other method, the cost of building will depend on the amount of work performed by the homeowners. The more work you do with your friends, the less you'll spend. Straw bale building is touted as environmentally friendly because the building material comes from a highly renewable resource grown locally in many instances. Straw bale construction generally reduces wood use, if done right. And as noted above, it results in walls with a high level of insulation that provide long-term economical comfort with little, if any, fuel consumption, especially when coupled with passive solar design.

Using straw bales to build houses also helps clean up the air. For years, farmers in California, Oregon, and other states burned straw in the fields after the grain harvest to return the nutrients to the soil. Because this practice contributes significantly to air pollution, it has been banned in many places. If the straw is baled and used to build houses, however, this waste product can be turned into a building material.

In spite of these environmentally friendly qualities, however, we must not forget that huge quantities of water, fertilizer, and, in many cases, pesticides and herbicides are used to grow cereal crops. If at all possible, use bales from organic farms.

Straw bale construction is promoted by advocates as easy to learn and fast. Walls for an entire house can be stacked in just a day or two. However, while building walls from straw is pretty easy to learn, and walls can be erected in a flash, house construction still requires a considerable amount of knowledge and skill. Completing the walls may take a long time. You'll need to install windows and doors, prepare walls for plaster, and apply several coats of plaster. Roof framing, electrical wiring, plumbing, and back-up heating system installation require expertise. Don't be lulled into thinking that building your straw bale house is going to occur on a single weekend with a few of your friends. Unless the house is small and the design is simple, expect to spend a lot of time learning about building, conferring with or hiring experts to help, and building. It all takes time.

When all is said and done, straw bale building, while not for the faint of heart, does produce dramatic results. The walls are impressively thick, and many people are drawn to the rounded walls that bales make relatively simple. Window wells can be beveled. Doorways can be arched. Niches can be carved into the plaster.

Straw bale interior finished in traditional Southwestern style. .

Credit: *The Straw Bale House*, courtesy of Chelsea Green Publishing

If you want to drastically reduce your heating and cooling bills, reduce stress on the world's forests caused by overharvesting, and help build a sustainable future, straw bale building may be just what the doctor ordered. Don't forget to design for passive solar, insulate ceilings well, install high-quality windows, and make sure the house is airtight. Real Goods offers numerous books on the subject as well as the hardware and specialized tools that make straw bale building easier. See pages 87 and 88 for several books on straw bale. *The Natural House*, by Dan Chiras, provides a great deal of information on all natural building materials for those seeking an accurate and honest overview of options.

Adobe

Adobe is one of the oldest natural building materials in use today. It has been used to build houses and many other structures such as churches for centuries throughout the world. So durable is it that many thousands of adobe structures remain standing today in China, the Middle East, northern Africa, South America, Central America, and the United States. Many of these buildings are still occupied.

Adobe buildings are made from bricks fashioned from local subsoils containing a mixture of clay and sand, with chopped straw or horse manure added to increase their strength. This material is wetted, mixed, and then poured into wooden forms the shape and size of the desired bricks. In arid climates, the forms are often left to bake in the sun. After the bricks have dried a bit, the forms are removed and the blocks are left in the sun to complete the process. The forms are reused to make more blocks.

To build walls, adobe bricks are laid in a running bond (overlapping pattern), just like modern bricks or stones in a stone wall, directly on

Adobe lends itself to curved walls and other organic architectural features.

Courtesy of Paul G. McHenry, Adobe and Rammed Earth Buildings, University of Arizona Press

a foundation. A mortar made of the same dirt is used to hold the bricks in place.

When completed, adobe walls are often coated with an earthen plaster that protects them from the weather, especially wind and rain. Although cement stuccos have been used in recent years, the results have proved disastrous and the requirements have been quietly dropped from building codes. Experience shows that earthen plasters are much better suited for use with adobe because, unlike cement stuccos, they expand and contract at the same rate as the adobe bricks. This reduces cracking and the need for repair. Earthen plasters also allow moisture that enters the wall to escape. Cement stuccos trap moisture that may erode walls internally.

Like other natural building materials, adobe blocks are relatively easy to make and inexpensive, if made on site. They can also be purchased from local suppliers in some parts of the world. If not, you can purchase or rent a machine that makes adobelike blocks, known as compressed earth blocks. This machine speeds up the process and makes it possible to build with adobe in cloudy and rainy areas where sun baking is out of the question.

Adobe blocks are made in wooden forms and dried in the sun. If the weather cooperates, they're usually ready to use in 11-14 days.

Courtesy of Paul G. McHenry, Adobe: Build It Yourself, University of Arizona Press

Wall building is relatively easy to master as well, and walls go up pretty quickly. Beginners find it quite forgiving. If you make a mistake, it's easy to tear a section of a wall down and start all over.

Historically, most adobe building has occurred in warmer regions of the world—for example, the Middle East, Central and South America, and the southwestern United States. Adobe performs well in such regions and functions especially well in deserts with hot, dry summer days and mild winters. In regions that have hot, dry summers and cold winters, adobe can be used without insulation, since its mass provides a "thermal flywheel" that moderates temperature swings and ensures comfort. In colder regions, adobe homes can perform well but only if the walls are well insulated.

Like other natural building materials, adobe construction is ideally suited to passive heating and cooling. However, unlike straw bales, adobe mainly provides thermal mass to absorb solar heat. Insulation must be incorporated inside the walls or on the exterior of the walls to achieve maximum comfort in many climates.

Despite its many positive attributes, adobe has a few disadvantages. For example, adobe construction is slow and labor intensive and can be costly. Creating adobe bricks and laying up walls requires a considerable amount of time. If you hire someone to build for you, don't expect to get by cheaply if labor costs are high in your area. In the U.S., where labor costs are notoriously high, custom-built adobe houses are expensive and tend to be built for a fairly wealthy clientele. That said, if you make your own bricks and lay up the walls yourself, an adobe home can be a fairly economical option.

Rammed Earth

Another ancient natural building technique that is gaining popularity in some areas of the world is rammed earth construction. Rammed earth building has been around for thousands of years—in fact, parts of the Great Wall of China were made from rammed earth.

In rammed earth building, a mixture of clay and sand is compacted (rammed) in forms, like those used to create concrete walls. Some rammed earth builders add cement to their mix while others use sand stabilized with a small amount of cement, rather than clay.

Workers begin by erecting wooden or steel forms on the foundation. Moistened subsoil containing approximately 70% sand and 30%

Adobe lends itself to fantastic and wonderful building shapes. From *Build with Adobe.* Used with permission.

clay is shoveled into the forms, about 6-8 inches at a time. The mix is then compacted, typically with a pneumatic tamping device powered by a compressor. Once compacted, additional soil is added, tamped down, and the process continues until the forms are filled. When a section is completed, the forms can be removed. The result is a solid block of earthen material, typically 12-18 inches thick and 6-8 feet long. A new form is then placed next to it, filled with dirt. After the walls are formed, the roof, windows, and door frames are installed.

David Easton, a California builder and author of *The Rammed Earth House*, thinks of rammed earth construction as a means of creating "instant rock." "To me," he says, "rammed earth construction is magic . . . watching soil become stone beneath your feet, and knowing that, when the forms are removed, a well-built wall will be here that will survive the test of centuries."

Exterior surfaces of rammed earth walls may be plastered to protect them from the weather or left as is in areas where rain is scarce. Interior surfaces are typically left unplastered so the occupants of the home can enjoy the raw natural beauty of this material.

A rammed earth, passive solar home in Buena Vista, Colorado.

Photo: Dan Chiras

Credit: The Rammed Earth House, courtesy of Chelsea Green Publishing

Rammed earth requires serious form work during construction.

Rammed earth building is growing in popularity in the United States thanks to the pioneering work of David Easton. In recent years, other rammed earth builders have emerged in California, Arizona, and New Mexico.

To reduce costs, Easton is using a modified building technique, known as PISE, which stands for Pneumatically Impacted Stabilized Earth. This technique requires much less formwork and no tamping. Rather, the earthen mix material is sprayed from the outside, using a gunite spray rig against a plywood form, to a thickness of 18-24 inches. In other words, the wall is built out laterally instead of vertically, and much more quickly. In seismically active regions, the walls are reinforced with rebar for protection. The Real Goods store at the Solar Living Center is built of 600 rice straw bales all covered with PISE, applied by a machine similar to the one that affixes gunite to a swimming pool.

Like other building materials derived from the earth, rammed earth produces solid, fireproof walls. These massive walls can withstand extremes of weather, including hurricanes and tornados. They stand up to wind much like concrete or concrete block walls and far better than the walls of stick-frame homes.

Rammed earth, like adobe, is ideal for passive solar heating and cooling in desert climates and works well in colder climates as well, so long as exterior insulation is applied to prevent heat loss. These homes stay unbelievably cool in the summer and warm in the winter, and they look great, too!

Although rammed earth is a great building material that has been around for centuries, it has some disadvantages. One of the most significant is that construction requires numerous forms. It also requires heavy equipment to transfer the raw mix into the forms. For the most part, rammed earth construction is best reserved for contractors who build a number of houses each year and thus can afford the forms and heavy machinery.

Rammed Earth Tire Houses ("Earthships")

Americans throw away well over 250 million automobile tires each year. Although more and more tires are recycled into useful products, such as doormats, highly durable roof shingles, the spongy surfacing of high school tracks, spongy mats placed around playground equipment, mulch for trees and shrubs, and (more recently) sidewalks, millions of tires are still thrown away. They are, for the most part, tossed into special tire dumps that are notorious for catching fire spontaneously and resisting attempts to extinguish them. Today, however, a small percentage of used automobile tires are being salvaged to build houses.

Building houses with tires, a technique known as *rammed earth tire construction*, is not entirely natural, but it has many redeeming qualities worth serious consideration. Before we examine the pros and cons of this technique, let's see how a house is built from tires.

Rammed earth tire houses are typically built into south-facing slopes for earth sheltering and passive solar design, so excavation must precede building. Once the site is ready, the tires are laid on the compacted subsoil. Tires are typically laid in large U-shapes, one for each room of the house. Next, a small piece of cardboard is placed over the center hole at the bottom to prevent dirt that is shoveled in from escaping. One worker begins shoveling dirt from the site into the tires while another compacts the dirt using a sledgehammer or a pneumatic tamping device like those used in rammed earth construction. After extensive pounding or tamping, the tire is fully compacted and easily weighs 300 pounds.

A passive solar rammed earth tire home, also known as an Earthship, in Taos, New Mexico. Although angled glass increases solar gain in the winter, it tends to lead to overheating in the summer months.

Photo: Dan Chiras

Photo: Dan Chiras

This Earthship is made from used automobile tires, collects rainwater from its roof, and heats and cools itself naturally.

After the first row is compacted, a second row is placed on it in a running bond (overlapping) pattern, and the process continues until the wall, typically consisting of six to eight rows of tires, is completed. After the wall is finished, a roof is attached, usually via a top plate. The tire walls are then mud-plastered or cement-stuccoed.

Rammed earth tire construction is the creation of forward-thinking architect and builder Michael Reynolds, who resides in the small artist community of Taos, New Mexico. Reynolds has been building rammed earth tire houses, which he calls *Earthships*, since the mid-1970s. His ever-evolving designs are meant to be largely independent living structures. They're earth sheltered and passively heated and cooled. They generate their own electricity using photovoltaic panels or wind turbines. They recycle greywater and blackwater to nourish gardens. These buildings even capture rainwater from the roof, which is purified and used for showering, cooking, and drinking. This building technique is not entirely natural, of course, because tires made from synthetic rubber are derived from petroleum and/or coal tar. (Interestingly, radial tires are made from a fair amount of natural rubber.)

Rammed earth tire houses have many redeeming qualities, however. For one, they put an abundant and troublesome waste material to good use. Like other natural building materials, they reduce wood use dramatically and they use local resources—both tires and dirt from the site. Using local resources helps to reduce the embodied energy of a building.

Rammed earth tire homes are also very comfortable thanks to the fact that they are typically earth sheltered and rely on passive heating and cooling. They stay warm in the winter and cool in the summer with very little, if any, outside energy. They also tend to have an open design, which makes for a bright and cheery interior.

Like all building techniques, rammed earth tire construction has its downsides. For one,

packing tires can be backbreaking work, if done with sledgehammers. (You can speed up the process by using a pneumatic tamper.) Earthships can also be a bit damp, like a basement, in some locations. To avoid this problem, special precautions must be taken to protect the buried (earth-sheltered) walls from ground moisture, usually waterproofing the walls and installing a French drain. Like other natural building techniques, Earthships may not be looked on favorably by local building officials. Some people, official or not, also have trouble with the appearance of Earthships, asserting that they appear a bit too funky for their tastes. With a little ingenuity, however, this unusual building technique can be used to produce reasonably conventional-looking houses.

Another disadvantage is that the open design creates a somewhat noisy interior. Sound travels well, and the open design is not conducive to privacy, although measures can be taken to ensure greater privacy.

Recently, Michael Reynolds and his colleagues at Earthship Biotecture (formerly Solar Survival Architecture) have created smaller types of Earthships (the Nest and the Woodless Hut) that lend themselves to modular or kit construction and are less expensive. Many natural builders have also been experimenting with rammed tires as part of the foundation for other natural wall systems—their width in particular makes them appropriate for a stem wall for straw bale or cob construction (see below), with appropriate moisture detailing.

Cob

If you like the idea of building a house using earth but aren't inclined to wield a sledgehammer or a pneumatic tamper or spend long hours making adobe blocks, you may want to consider cob. *Cob* is the English word for "a lump" or "a rounded mass." In this context, it refers to

Cob lends itself to organic forms. The house at the left is finished with lime plaster.

Photo left: Becky Bee; right: Robert Bolman

a lump of earthen material, and cob houses are built from lumps of mud.

Cob contains a mixture of sand, clay, and straw that is fairly similar to adobe, although many builders are now using long strands of straw and a carefully adjusted amount of sand with just enough clay to bond the mix together. This rendition, known as Oregon cob, shrinks very little as it dries.

Rather than making blocks from this mix, builders apply the mud directly to the foundation, most often by hand or, occasionally, by the shovelful. The walls of mud are then massaged into shape much like a clay bowl made in a ceramics class. As a result, cob construction lends itself to sensuous curved walls, arches, and niches. Cob builders Ianto Evans, Michael Smith, and Linda Smiley note in their book, *The Hand-Sculpted House: A Practical and Philosophical Guide to Building a Cob Cottage* (available on page 90), that cob homes are cozy and delightful places to live.

Cob building is great fun and permits extraordinary freedom of expression. Cob construction is also an ideal option for those who want to build their own house, as most of the work is done by hand or with simple hand tools.

If the thought of building a house from mud seems absurd, rest assured; cob is a highly durable building material. Cob walls are strong and thick, up to 36 inches in some homes, and are as solid as a rock. To protect against the weather, cob walls are either whitewashed, lime-plastered, or coated with an earthen plaster. The proof is in the pudding (or the mud pie, as the case may be): Many thousands of cob homes can be found in southern England and many have been continuously occupied for the past 500 years!

Cob is suitable for many climates, although insulation must be incorporated in cold climates to maintain comfort. Exceedingly thick walls may also suffice. In rainy climates, exterior walls must be protected from rain by large overhangs or highly durable plasters or both. Lime-sand plasters are often the product of choice in such instances. In general, though, cob erodes slowly and is much less vulnerable to moisture than straw-based wall systems.

Earthbags

In recent years, many new building techniques have emerged to try out new ideas, provide greater flexibility, and create greater diversity in outcome. One recent technique is earthbags.

Earthbag construction is a type of rammed earth building. It was pioneered by Iranian architect Nader Khalili, who now resides in California, and popularized and further perfected by two maverick builders in Moab, Utah, Doni Kiffmeyer and Kaki Hunter. Rather than using wooden or steel or tire forms, earthbaggers use polypropylene bags. These bags, used to hold rice and other grains, are widely available and fairly inexpensive. They're also pretty easy to work with.

Children like "working" with cob, too!

Credit: *The Natural House*, courtesy of Chelsea Green Publishing

An earthbag wall under construction, above left and below, and the finished product, above right. Built by Doni Kiffmeyer and Kaki Hunter of Moab, Utah.

To begin, bags are filled with a slightly moistened mixture of dirt from the site—typically subsoil containing a mixture of sand and clay. The bags are filled on the foundation one by one and laid flat. When a course is completed, the bags are tamped. Tamping flattens and compresses the earthen material within the bags, transforming them into solid blocks. Before the next course is laid, two strands of four-point barbed wire are laid down on top of the bags. The wire provides continuous tensile strength around the building, and the barbed points dig into the bags above and below and "Velcro" the courses together. The next course is laid on top in a running bond pattern. Once filled, the second course is hand-tamped. Slowly but surely, the walls are created.

When completed, the walls are plastered, earthen and lime-sand plasters being preferable. Both adhere admirably well to the earthbags without stucco netting or diamond lath.

Earthbag construction is a great choice for owner-builders. It's fairly easy to master and requires basic tools. Earthbag construction lends itself to round structures, small structures such as sheds and privacy walls, as well as features such as vaults and domes. Like cob, adobe, and rammed earth, it yields homes with excellent thermal mass but very little insulation. As a result, most earthbag building tends to take place in desert climates.

On the downside, earthbag construction is fairly slow and very labor intensive. There's a lot of pounding required! This new technology is also largely untested, so obtaining approval by your local building department may be difficult.

CAST EARTH

Cast earth is the newest of all earthen building technologies. It was invented in 1993 by Harris Lowenhaupt, a Phoenix-based metallurgist.

In cast earth building, houses are made from a slurry consisting of water, soil, and 10%-15% heated (calcined) gypsum. The slurry is poured into forms set on a concrete foundation (as in rammed earth construction). The slurry sets up fairly quickly after being poured, thanks to the gypsum. The forms are then removed and voilà! The exterior walls are typically coated with plaster or stucco for protection and aesthetics.

Cast earth houses look a lot like rammed earth houses and are ideal for passive solar heating and cooling, especially in hot, dry climates. In colder climates, the walls must be insulated externally to reduce heat loss and ensure maximum comfort.

Cast earth is one of the quickest ways of building a house. Unfortunately, this technology is not available to owner-builders. Unlike all other natural building methods, cast earth building is a proprietary process. If you want a cast earth home, you will need to contract with a certified builder or figure out the secret ingredients that make this technology work. The inventor, Harris Lowenhaupt, holds the patent and guards the secrets of this process.

STRAW-CLAY

Straw-clay is a relative newcomer to the natural building field in North America, although it has been used in Europe (especially Germany) for at least 500 years. Introduced by New Mexico-based architect and builder Robert LaPorte, straw-clay produces homes of extraordinary beauty.

Straw-clay is made from straw with a little clay slip, a mixture of clay and water that serves as a binder. Clay slip is poured onto the straw and mixed in until it slightly wets the surface of all the straw. The mixture is then packed into 2-foot-high wooden forms attached to the wall framing, either by lumber or posts. The slightly moist straw-clay mix is then tamped by hand.

Build It with Bales—Paper Bales, That Is!

All photos: Dan Chiras

When you think of a home built from bales, you probably think straw bales. Interestingly, a few maverick builders are creating homes from another kind of bale: bales of paper or paperboard they collect from recycling centers.

Although there are only a few in North America, paper bale houses could increase in popularity. They perform well and put to use an abundant waste product.

In the tiny town of Fraser, Colorado, near the Winter Park ski resort, Rich Messer and his partner, Ann Douden, built the walls and foundation of their energy-efficient mountain home from huge bales containing laundry soapboxes. It's known as poly-coated Kraft carrier board, but any type of paper would work.

The paperboard was collected by Tri-R, a major recycling company in Denver, about 100 miles away. This material is quite clean, but because it is coated with a thin layer of plastic, it is relatively difficult to recycle. Much of it ends up in the landfills. Messer points out that for this reason, paperboard is abundantly available in cities and towns throughout North America. Moreover, it's inexpensive: Delighted to get rid of the stuff, Tri-R gave him the bales for free!

The huge bales of waste paperboard produce 36-inch-thick, heat-resistant walls that keep Rich and Ann warm and cozy in this cold, wintry climate. When covered with stucco, they appear a bit bumpy but otherwise pretty normal.

Inspired by the idea of using waste to build a home, Messer decided to make the foundation out of another abundant waste material: huge bales of waste plastic. This required 28 bales containing postconsumer PVC plastic—mostly old toys, laundry baskets, and shampoo bottles—but any type of plastic would work. Rich paid $20 per bale for them.

The plastic bales were laid into a 5½-foot-wide foundation trench with compacted road base providing the

This house is being built by Rich Messer with huge bales made of hard-to-recycle paper, near Winter Park, Colorado.

underlying support. He then applied 3½ inches of foam insulation inside the plastic bale foundation to reduce heat loss.

Rich and a helper secured the roof to the paperboard bales by installing a concrete bond beam poured on top of the bales. (It is attached to the bales by rebar.) After the roof was installed, Rich blew recycled cellulose insulation into the ceiling cavity, yielding an R-50 layer of heat-resistant material. They then sealed air cracks with leftover paper.

In addition to enjoying low fuel bills, the house is quiet and warm. It creates "a serenity we didn't think was part of what we were building until we lived in it,"

says Messer. The only obstacle to getting more of them built is financing, he says. "Although many people have expressed interest in the building method, the biggest hurdle is bankers, who just don't like to hear about building houses with trash."

"To do so requires a willingness to take a risk," adds Messer. As forests decline and people look for other options, wastepaper may become yet another source of building materials. Who knows, maybe someday you too will live in a house made of recycled trash.

Adapted from *The New Ecological Home*, by Dan Chiras.

All photos: Gordon Solberg

Bamboo reinforcing is installed to ensure lateral stability, usually attached to holes drilled in the framing members every few feet along the height of the wall.

When a form is filled, a new one is added. After two forms are filled, the lower one is removed so the walls can start to dry. When the walls have dried, they can be coated with plaster.

Straw-clay construction is easy to learn, and walls are some of the easiest, if not *the* easiest, of all natural building materials to plaster. For these reasons, straw-clay is an ideal building material for owner-builders and neophytes. When covered with plaster, the walls are fireproof, fairly soundproof, and look marvelous. Because it provides a lot of insulation, straw-clay is an excellent choice for those interested in building an energy-efficient, passive solar house. Likewise, when interior surfaces of exterior walls are coated with a 1- to 2-inch layer of earthen or lime-sand plaster, they provide a fairly good source of thermal mass. Along with straw bale, straw-clay homes are suitable for a wide range of climates.

Unfortunately, building with straw-clay is a fairly slow and labor-intensive process. The drying time is the slowest of all the natural wall systems—about a week per inch of wall thickness—so they should be built when a lot of good drying weather is expected. Generally, walls do not exceed a foot in thickness; otherwise they dry so slowly that mold becomes a concern.

Papercrete

Papercrete is another newcomer to the natural building field. Unlike straw-clay, it is brand new. It has no long European history to draw on.

Papercrete is a mixture of newspaper, sand, cement, and water. Newspaper is cut into small pieces that are soaked in water for a day or two until they begin to break down. At this point, sand and cement are added. The mixture is stirred, and the result is a thick, gray slurry that can be poured into block forms (similar to those used to make adobe blocks) or directly into wall forms on the foundation (like those used in rammed earth or straw-clay construction).

Once poured, the papercrete slurry slowly dries and sets up, forming a solid block that offers thermal mass as well as insulation. When dry, the blocks are placed on a wall on a foundation in a running bond pattern, mortared in place (using wet papercrete slurry), and then plastered.

Papercrete is relatively easy to make and, like rammed earth tire construction, helps to put a ubiquitous waste material—in this case, newspaper—to good use. Materials are readily available almost anywhere. The blocks are lightweight and easy to work with.

A near relative of papercrete, fiber adobe, uses clay instead of cement as the binder. It dries extremely slowly but has a lower embodied energy and is less expensive than papercrete.

Bear in mind, however, that these are new building techniques, so there's little experience to draw on. Building departments may be skeptical of these newcomers. Although it appears that papercrete and fiber adobe would perform well in many climates, the jury is still out.

Sean Sands's papercrete house in Columbus, New Mexico. Papercrete blocks, which put waste paper to good use, are lightweight and easy to work with.

Stone Houses

Another oldie but goodie in the natural building field is stone. One of the most ancient of building materials, stone has been used to build massive castles in Europe, numerous cathedrals, and soaring aqueducts in Italy but also less lofty structures such as barns, houses, garden walls, roads, and walkways. Except for the more recent use of cement mortar to hold the stones together, stone building has changed little over the centuries.

Like rammed earth construction, walls made of stone must be laid on very sturdy foundations typically made from stone or concrete. Rocks are typically laid in a running bond pattern with cement mortar. Mortar holds the carefully stacked stones in place and reduces air infiltration.

Homes built from natural stone are beautiful and extremely durable: A well-built home could last for many centuries. Stone is also a locally available material, abundant in many areas. Like earthen natural building techniques, stone building is ideal for passive solar heating and cooling. As a rule, thick, uninsulated exterior stone walls work best in hot, arid climates. In

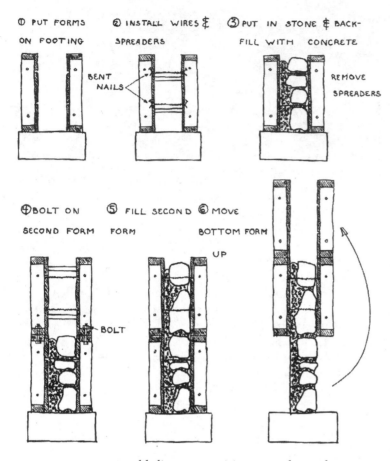

NEARING METHOD

① PUT FORMS ON FOOTING

② INSTALL WIRES & SPREADERS

BENT NAILS

③ PUT IN STONE & BACK-FILL WITH CONCRETE

REMOVE SPREADERS

④ BOLT ON SECOND FORM

⑤ FILL SECOND FORM

⑥ MOVE BOTTOM FORM UP

BOLT

Formed-wall stone construction. From *Independent Builder.* Used with permission.

for occupants, as moisture promotes the growth of mold and mildew. Both of these problems can be solved by insulating the interior walls or building double stone walls—two walls on the foundation separated by an airspace. Because air is a poor conductor of heat, it blocks heat flow out of the building in the winter and into the building in the summer. Insulation can also be inserted in the cavity between the two stone walls to further reduce heat loss.

Many good books on building with stone are available. One is *Stone House*, by Tomm Stanley. Another is Charles Long's *The Stone Builder's Primer: A Step-by-Step Guide for Owner-Builders* (Firefly Books, 1998).

SLIPFORM STONE BUILDING

Slipform construction is the lazy person's way of building stone walls, although it still requires a lot of work. A slipform is a wooden form that is placed on a foundation. Forms are squared and plumbed, and stones are selected. Rather than stacking stones on top of one another, as in conventional stone construction, the stones are laid along the outer face of the form, one course at a time. Concrete is then poured into the form after each course to fill the remaining space. Walls are laid up in 1½-foot sections. After a section is completed and the mortar has set up, the form is raised, or slipped up, and work continues (hence the name slipform). Those who want to learn more about this technique should refer to Karl and Sue Schwenke's book, *Build Your Own Stone House Using the Easy Slipform Method.*

Although the slipform technique greatly reduces the labor of building with stone, it requires construction of wooden forms—at the very minimum eight, but usually more. This, in turn, means more wood, more labor, and higher costs—all of which may be justifiable if the method helps you build better walls. Slipform stone walls look quite different from stones laid up conventionally. Moreover, the inside surface of the wall is concrete and must be plastered or covered with dry wall.

cold climates, provisions must be made to insulate the walls to ensure maximum comfort.

Although stone is an excellent material, it creates some challenges. One of the most significant is that stone is extremely heavy and requires a level of fitness beyond most, if not all, other building techniques. For this same reason, building with stone can cause injuries such as hernias and back strain. Stonework is slow and tedious, although there are some newer ways of building with stone, notably the slipform method, that reduce both labor and time.

Like concrete, stone exhibits great compressive strength but offers little tensile strength. In seismically active areas, steel reinforcement is required to compensate for this weakness. (Steel rods are laid in the mortar joints between adjacent courses.)

Another problem with stone homes is that they can be quite cold if not built right. Like earthen building materials, stone is a good conductor of heat. On cold winter days, heat passes right through stone walls, making it difficult and costly to heat the home. In addition, moisture in the indoor air often condenses on cold stone, creating a potential health problem

Hybrid Natural Houses

There was a time when a straw bale homebuilder built entirely with straw bales and a cob builder built exclusively out of cob. He or she may have viewed other natural building materials with undeserved suspicion. As more and more natural builders have learned about the benefits of the wide array of natural building materials, however, the field has opened up considerably.

Most, if not all, natural builders today utilize two or more natural building materials in any given project. They may, for instance, build a straw bale house with rammed earth interior walls for thermal mass. They might build a cob bench and shed. They might even build interior straw-clay walls in places where acoustic insulation is required. Houses may be placed on a stone or rammed earth tire foundation. Many builders use cob and straw-clay to fill in unusual places—nooks and crannies that are hard to fill with straw bales. They may use cob to sculpt window and door openings as well as shelves and even freeform fireplaces. And, finally, many natural builders use earthen plasters or an earthen plaster base coat with a lime-sand plaster finish coat on their creations. In many cases, they even use conventional building materials and techniques like stick framing for interior walls, roofs, and floors, or green building materials like Rastra blocks or insulated concrete forms (discussed below). All this is to say that contemporary natural builders frequently mix and match natural and conventional building techniques to achieve the best results. The result is structures that are a hybrid of several natural and conventional building materials and techniques.

"By viewing building as a process of combining different, but complementary, materials rather than adhering to a particular building system," note Bill and Athena Steen, two extraordinary veteran natural builders, "we have given ourselves the freedom to create structures that respond to a wide variety of contexts and circumstances. They can be elegant or simple, quick or detailed, inexpensive or costly, and probably most important, they can be built from predominantly local materials in whatever combination best matches the local climate."

Utilizing a number of materials does create challenges, so be sure to heed the advice of two veteran straw bale builders, Matts Myhrman and S. O. MacDonald, authors of another classic book, *Build It with Bales*: "By its nature, a hybrid structure often requires extra thought during the design process. Draw it, model it, get a 'second opinion,' and still expect to have to think on your feet once you get started."

Insulated Concrete Forms

As just noted, in the process of hybridization that currently characterizes the natural building movement, builders sometimes use green building products that, while not natural, offer substantial benefits. One of them is the insulated concrete form.

Insulated concrete forms, or ICFs, are hollow blocks of varying length made from rigid foam insulation (beadboard, also known as expanded polystyrene). One common brand, known as Rastra, is made of 85% recycled polystyrene (Styrofoam) and 15% cement, making it very environmentally friendly. Another, known as Durisol, is made of recycled wood and cement.

ICFs are used to build foundations and exterior walls and often interior walls as well. They are reinforced with rebar (internally) to increase tensile strength and then filled with concrete.

Unlike wood or steel or concrete forms used to build walls and foundations, ICFs are lightweight and easy to assemble. They pretty much snap in place like pieces of a three-dimensional puzzle. To prevent blowout when concrete is poured into them, manufacturers install plastic or steel cross bridges that attach one side of the form to the other. Products like Durisol and Rastra, however, are more like concrete blocks in their structure. After concrete is poured, the ICFs are left in place. The foam sandwich produces a foundation or external wall with an insulation value of around R-35, depending on the width and the manufacturer. Tech Block, a relative newcomer, boasts an R-value of over 45!

ICFs significantly reduce both the amount of concrete needed to build foundations and exterior walls and the amount of lumber used throughout a structure. Although expanded polystyrene is made from a chemical extracted from oil, bear in mind that the foam is 98% air bubbles. (Air bubbles block heat flow through walls.) Also, no ozone-depleting chemicals are used in the production of expanded polystyrene.

In the process of hybridization that currently characterizes the natural building movement, builders sometimes use green building products that, while not natural, offer substantial benefits.

Interlocking insulated concrete forms (here made by American Polysteel) are placed, reinforced with rebar, then filled with concrete. They're great for foundations and walls, providing superior insulation.

Photo: American Polysteel

Although ICFs are, for the most part, made from virgin materials derived from oil, there's a notable exception: the Rastra block. Rastra blocks, which Real Goods founder John Schaeffer used to build his house (see page 78), are made from a small amount of cement (about 15%) and 100% recycled expanded polystyrene (EPS) foam (about 85%). Like other ICFs, Rastra blocks are lightweight and easy to handle—they can even be cut with a saw. Another huge advantage to Rastra, then, is that it lends itself beautifully to being shaped and sculpted into just about any design. As noted above, Durisol blocks are made from recycled wood fiber.

According to *Environmental Building News*, Rastra "is insect-proof, creates a very strong wall, and is extremely fire-resistant. Recent structural testing found a Rastra wall to be seven times better under earthquake-type stresses than a wood-framed shear wall, according to marketing director Richard Wilcox" (July/August 1996).

Yurts

If you need temporary shelter, say to live in for five years while you build your house, or you want an inexpensive, portable home to live in for a long time, you may want to look into a yurt. Yurts can also make nice playhouses, studios, storage sheds, workshops, or meeting rooms.

Yurts are relatively simple, tent-like circular structures invented by the nomads of central Asia. Modern yurts come in three basic varieties, according to Becky Kemery, author of *Yurts: Living in the Round* (Gibbs Smith, 2006): the tapered-wall yurt, the modern fabric yurt, and the frame-panel yurt. Tapered-wall and frame-panel yurts are typically made from wood. As its name implies, the modern fabric yurt is made from fabric. Regardless of its composition, the modern yurt is a "spacious, economical structure—one of the most efficient surface-to-volume structures ever devised," according to Bill Coperthwaite, American yurt pioneer and advocate.

Photo: Colorado Yurt Co.

In addition to their efficient use of space, many yurts are portable. They can be disassembled and reassembled fairly quickly. They can also be made from locally available materials or from kits that can be assembled on site.

Yurts are placed on a wood platform. The top cover of Pacific Yurts, the original designer and manufacturer of the modern yurt, is made from a flame-retardant vinyl laminate that, according to the company literature, provides excellent durability, low maintenance, and protection from the elements. The side cover is made from an acrylic-coated polyester fabric that, likewise, provides exceptional strength, durability, and low maintenance. The walls are supported by an expandable Douglas fir lath. Rafters to support the roof are made of structural-grade Douglas fir. Windows are made from clear vinyl, although glass windows can be installed if wood frames are built.

Although these yurts are not natural, because they're made with synthetic materials, they manifest other aspects of green shelter and deserve consideration in this context.

To learn more about yurts, consult one of the excellent books on the subject, such as the newest entry into the field by Becky Kemery, referred to above.

Yurts provide temporary or permanent shelter at a fraction of the cost of conventional materials. With insulation and a heat source, they can provide year-round shelter in cold climates.

Photo: Rainier Yurts

Photo: Scott Vlaun

Build Your Passive Solar Natural House

Building a house is a major undertaking, unless all you are building is a small shack to live in. So many people elect to hire an architect to design their house and a contractor to build it. Contractors typically hire a number of subcontractors to carry out specialized functions—for example, pouring concrete foundations and floors, applying roofing materials, and installing electrical service and plumbing.

Another option is to act as your own general contractor. You then coordinate the process, hiring people who carry out specific tasks. Doing so can save you a fair amount of money, perhaps as much as 25%.

While serving as a general contractor may seem preposterous, many building departments permit this approach. You don't have to be a licensed builder, either. Any ordinary citizen can serve as a general contractor. And it's not as daunting as you may imagine. It does, however, require some knowledge and a willingness to learn a lot and quickly. Before you start, read everything you can on building and don't be afraid to ask for help. A competent green building consultant or a good builder who is willing to consult on your project will be worth his or her weight in gold!

Yet another approach is to build your house yourself. That is, you can build as much as you can. Many people handle their own basics like pouring a foundation, building walls, stacking bales, installing windows, and putting in electrical wiring, and hire others for more specialized jobs, like heating system installation or plastering. Or, if those projects seem too daunting, you can tackle smaller ones and let the professionals do the rest and perhaps show you ways you can help out.

Building your own house requires courage, patience, and a willingness to climb a steep learning curve, unless, of course, you come to the task with a great deal of knowledge and experience. If you aren't already versed in construction, expect to spend many hours thinking about the project, planning, buying materials, returning materials, asking questions, trying new things, consulting with experts, and building. Don't be discouraged if your experience is limited. You can succeed if you're committed and willing to learn, as Carolyn Roberts documents in her wonderful book, *A House of Straw: A Natural Building Odyssey*.

Building your own house will save you a lot of money and can be tremendously gratifying. There's nothing better than admiring your handiwork at the end of each day. And there's nothing more gratifying than telling friends and family, "I built my house!"

The natural building techniques discussed in this chapter are especially well suited to owner building, but don't be lulled into complacency by experts who proclaim that their building techniques are easy. Some are, some aren't. Even if a natural building technique is easy to master, a home is much more than its walls, and building an entire house is enormously challenging. So remember: Building walls from cob or straw bales or straw-clay may be relatively simple, but that's just part of a very complex project.

Words of Wisdom from Real Goods Staffers and Colleagues

Many of the staffers at Real Goods have built their own houses and acquired some good down-to-earth advice from the school of hard knocks in the process. The following tips are a distillation of lessons they and others have learned along the way. We hope that some of the mistakes we have made and some of the advice we offer can help you.

On Where to Build

■ "Stay away from north-facing slopes. They're very cold."
—Terry Hamor

■ "I would have chosen a building site nearer to the main road and with better gravity flow for water instead of building a water tower."
—Debbie Robertson

■ "Don't buy too far out in the boonies. Elbowroom is great, but there are practical limits. My property was five miles out a steep, rough dirt road. That's fine if you're independently wealthy and don't have kids or friends. The difficult commute was my primary reason for eventually selling the property."
—Doug Pratt

■ "When looking for land, consider in-town lots. You'll ride your bike more, shop more locally, and help strengthen the web of your community. After building my first straw bale home in a beautiful but remote location, I love the options I have living in town and not needing my car all the time. My life is far more sustainable now."

—Laura Bartels

On Being Realistic with Time and Money Plans

■ "Be patient. You're going to be living in your house hopefully for many decades. If you can do it right by spending a few extra months, stop rushing, and pay attention to detail."

—John Schaeffer

■ "Don't underestimate the value of the little things you can do first, before you design and before you build. Educate yourself, take a hands-on workshop, or hire a consultant to review even your most preliminary plans. While relatively inexpensive, these steps can save you thousands of dollars and much anguish."

—Laura Bartels

■ "If I had to do it over again, I'd realize that it takes three times as long and costs three times as much as expected. I'd have my finances together so it could be built within a year's time instead of 15 years!"

—Debbie Robertson

On Energy and Size

■ "To strengthen our communities for the coming years, we need to localize our energy sources. Will everyone need to have solar panels on their roof? We might instead come together as neighbors and communities to create small-scale renewable energy systems, reducing our dependency on remote sources of energy while building community and providing a potential source of income for the long haul."

—Laura Bartels

■ "Small is beautiful in more ways than you think. Five hundred fewer square feet at $100-$150 per square foot is $50,000-$75,000 that you can invest in higher-quality and more sustainable materials, higher-performance systems,

and better design. The smaller house also makes it easier to achieve zero-energy with renewable energy systems. Whatever you build, you're going to pay for initially and then pay for over and over as you pay taxes on it and clean, maintain, repair, and heat and cool it. Building a smaller house will have a smaller ecological footprint—an even bigger benefit."

—David Eisenberg

■ "Before you even think about installing a solar electric system or a wind machine or a solar hot-water system, be sure to make your home as energy-efficient as possible. Each dollar you spend on energy-efficiency measures will save $3-$5 in renewable energy systems cost. Remember: Efficiency first!"

—Dan Chiras

On When to Move In

■ "Get a cheap, portable living space initially. I had a refurbished school bus that allowed me to move onto the property with a minimum of development. This saved rent and allowed me to check out solar access and weather patterns before choosing a building site and designing a house. A house trailer can do the same thing. Don't move in until it's finished. It's real tempting to move in once the walls and roof are up. Resist if at all possible."

—Doug Pratt

■ "I lived in a school bus until it was finished. It was good to be able to move into the shade in the summer and into the sun in the winter. Buses are cheaper than trailers and come with a motor and charging system. One of my biggest mistakes was moving into the house before it was 100% completed. Most plugs and switches are still not done 13 years later! Finish it before you move in, or you never will."

—Jeff Oldham

■ "I wouldn't have moved into an unfinished structure."

—Debbie Robertson

On General Building Plans

■ "Use passive solar design. This was one of the things I did right. My house was cool in the summer and warm in the winter, and, with an intelligent design, yours can be too. Avoid the temptation to overglaze on the south side. You'll end up with a house that's too warm in the winter and cools off too quickly at night. Buy quality stuff. Cheap equipment will drive you crazy with frustration and eat up your valuable time."

—Doug Pratt

■ "Hire a carpenter and become his apprentice."

—Terry Hamor

■ "Take an honest assessment of your skills. Decide up front which tasks you can accomplish through your own efforts and which will require assistance. Assign a dollar amount and time required for each. Establish a working budget, which you should update throughout the project."

—Robert Klayman

■ "Money is no object when it comes to insulation! Don't use aluminum door or window frames. Aluminum is too good a conductor of heat. Use wood, vinyl, or fiberglass instead. Pay attention to little details. It's not a matter of which roofing or siding system you choose; it's the small details like flashing, corner joints, and drainage runoffs that really count."

—Jeff Oldham

■ "Design a compost bucket into your kitchen counters that's flush with the top. Design recycling chutes into the kitchen walls with 6-inch PVC so you can use it for bottles and cans—it makes recycling fun for kids and a breeze to keep clean and organized."

—John Schaeffer

■ "Design your wood storage with the ability to load from the outside and retrieve from the inside. Build a laundry chute and a dumbwaiter. Have a small utility bathroom accessible from a back door. Plan for, or install, a dishwasher."

—Debbie Robertson

■ "If you're hiring a contractor, get him or her involved from the beginning of the design phase. Get his/her buy-in for the project before you start. Always find a contractor you can trust and pay him/her time and materials instead of doing a contract. With a contract, one of you loses every time—with time and materials, it's a win-win."

—John Schaeffer

Spirit of the Sun

Green living pioneers John Schaeffer and Nancy Hensley have built a home that honors the land, wildlife, and pioneering spirit of their company, Real Goods.

Between the two of them, John Schaeffer and Nancy Hensley have 40 years of experience researching green building, off-the-grid living, and permaculture. John founded Real Goods, the nation's first solar retail business, in 1978 in Hopland, California. Nancy has lived off the grid since 1973 and joined Real Goods in 1989. The couple has realized their dream of employing their collective wisdom to create an energy-independent, nontoxic, environmentally gentle home that promotes sustainability—while also being tastefully beautiful and soul soothing.

They were right on track with that vision as construction on their 2,900-square-foot roundhouse—oriented to the cardinal directions and patterned after a red-tailed hawk ready to take flight—began in early 2001. The house is set on 320 acres that are richly landscaped with gardens, orchards, ponds, a lake, and a grotto with a waterfall, and that overlook the Hopland Valley. In 2002, John and Nancy moved into a barn on their property where they could watch the building progress. That summer, they

noticed several large black birds pecking mercilessly at the windows. How cute, they thought, until they realized that weeks later, the birds—ravens—continued to attack the house's 69 windows with a vengeance.

"We went into major raven research mode," Nancy says. "We talked to biologists, ornithologists, and shamans. People told us to dance around in circles with corn, build altars, give them offerings. We eventually learned that when the birds are nesting, they're very territorial. They saw their reflections and tried to scare off those 'other ravens.'"

"A shaman told us the ravens were upset because we were calling the house Sunhawk—they have a natural animosity toward hawks," John says. "So we made an altar, and for 30 days we brought them tobacco, fish, and meat." In marked contrast to locals who thought a shotgun was the answer to their raven problem, John and Nancy's attitude was that the birds had inhabited this piece of land first and that they, as the human "intruders," should strive to live in harmony with the ravens. That open-mindedness

translated into every step of building their ecologically friendly home.

BETTER LIVING THROUGH TECHNOLOGY

Built from Rastra blocks, which are made from 85% recycled polystyrene beads and 15% cement, Sunhawk is a masterful example of sustainability. Nancy spent months researching building materials and appliances. Her finds include recycled-tire roof shingles and repurposed granite countertops from a Berkeley café. Roof decking, fascia, barge rafters, and beams were made from reclaimed redwood, Douglas fir, and walnut from an area winery, vinegar plant, warehouse, and converted orchard.

John and Nancy's house is, of course, entirely off the grid. (How could you expect less from a renewable energy pioneer?) A 17-kilowatt solar system—recycled from a Gaiam Real Goods installation in Belize blown down in a hurricane—provides ample power in summer months, and a hydroelectric turbine produces 1.5 kilowatts per hour from a seasonal creek that runs through the property from December through May. "Our hydro system provides 36 kilowatt-hours per day—almost twice the national average for electricity use," John points out. "And it cost only $1,500—it's way more cost-effective than the electric company or any other source of electricity. It's a powerful feeling knowing we'll never have to pay an electric bill for the rest of our lives."

John is most proud of the home's innovative heating and cooling systems, which work so well largely because of the passive heating and cooling design. Cooling the house without air conditioning is no small feat in Hopland, where summer temperatures consistently rise to three digits. Sunhawk is cooled by earth-cooling tubes, two 150-foot-long pipes (12-inch culverts) that draw outside air 10 feet below the surface of the ground, where it is cooled naturally. The culverts empty into a 9-foot-deep rock storage chamber below the central portion of the house. Two solar-powered fans pull cool air into the central rock core, which remains at 67°F. From there, it travels by convection to the rest of the house. Even on the hottest day, the home's interior has never exceeded 76°F. The home is heated primarily by radiant floor tubing powered by rooftop solar hot-water panels and water heated by the excess voltage from the solar and hydroelectric systems.

Living off the grid entails absolutely no sacrifice, John quickly points out. In fact, there are unexpected bonuses. "We got a satellite system for Internet access and it's two to three times as fast as DSL or cable," he says. "The fun part is we get to use our house as a laboratory for the technology and products we order for Gaiam Real Goods."

NOW *THIS* IS PERMACULTURE

After John and Nancy bought their acreage in 1998, they spent numerous nights camping in various places around their property to find the ideal spot for their house—in the end, exactly the location where Nancy's intuition had initially told her it should be.

Building the Real Goods Solar Living Center in Hopland had taught the couple the importance of landscaping as an integral part of designing a homestead, so they made that a priority—even before finding a designer for the house. "We knew early on that this would be much more than just a house-building project," John says.

Their first major project after grading the road was to dig a 10-acre-foot lake flanked by a 30-foot-wide grotto overflowing with waterfalls from the property's three natural springs. Three and a half acres of lush permaculture landscaping include native grasses, a coastal redwood grove, Mediterranean foliage, lavenders, and a corridor of swamp cypress by the pond, which attracts a variety of wildlife including herons, egrets, ducks, coots, and giant bullfrogs. To complement the landscaping, the couple added fruit and nut trees and Italian olive trees from which they hope to make their own olive oil. The orchards and the

Stone "tree" in the living room from local rocks.

pond are key to the home's comfort; prevailing winds from the northwest flow across them and bring evaporative cooling inside.

The grounds continue to be a work in progress. Four Rastra window boxes on the home's south side provide enough growing area to keep the couple well fed. Vegetable beds are

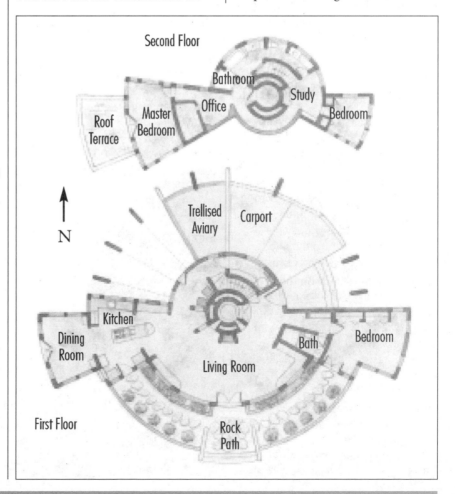

The equinoxes and soltices are marked by the stained glass hawk aligning with two ceramic hawks inside.

located 6 feet from the kitchen door so John and Nancy can step outside in any kind of weather to harvest arugula, cilantro, and exotic lettuces. Recently, as part of a Real Goods Solar Living Center permaculture workshop, students spent four days building an herb spiral on the home's north side and installing a composting system, worm bin, and drip irrigation systems to water the vegetation.

LETTING THE HAWK SOAR

With the major landscaping in place, John and Nancy faced the task—and privilege—of creating a house that could live up to their ideals. "A home is far more than a shelter," John says. "It's an expression of our values and commitment, and it enables us to put our convictions into action. We wanted ours not just to promote the principles of sustainability but to engender restoration and regeneration of the environment, while also nourishing the spirit."

They searched until they found architect Craig Henritzy of Berkeley, who understood that vision. However, they were taken aback when he showed them a set of plans for a Rastra house based loosely on the California Native American roundhouse (in the style of the indigenous Pomo Indians) and symbolic of the hawk—which he declared John and

Nancy's "house totem." "We thought it was visionary and unique but a real challenge to pull off in a practical sense," John admits. Being open-minded, John and Nancy visited another roundhouse Henritzy had designed in Napa, and they knew they'd found their home. "That house was unlike any other we'd ever seen," John says. "It felt Native American yet 22nd century—both ancient and futuristic."

"For Native Americans, the hawk symbolizes 'vision,' which has been important in John and Nancy's work," Henritzy explains. "Also, in the 'green architecture' field, we often use just shed-type designs, which I feel has limited the progress of alternative designs being accepted in the housing market. This design explores passive solar with a geometry that celebrates the Sun's cycles and playfully and beautifully assumes a hawk shape."

Because Rastra's polystyrene beads give it a fluid quality, it's easily cut and sculpted, making the hawk shape and orientation with the cardinal directions possible. Sunlight falling on a solar calendar running from north to south on the living room floor marks the passing seasons. On the winter solstice, sunbeams stream through a stained-glass hawk above the south-facing French doors, causing the bird to "fly" across the floor from west to east. At exactly solar noon, the Sun

Master bedroom

All cabinets are made from recycled walnut from a tree that came down at the Solar Living Center.

illuminates a slate hawk in the floor in front of the living room woodstove.

"I appreciate always knowing the position of the Sun," John says. And that, Nancy adds, is really just a fringe benefit of a good passive solar design. "The very basic, most important thing is good southern exposure—taking advantage of Sun and light," she points out. "What I love most is that the Sun comes in at the right time and doesn't come in at the wrong time."

Reprinted from *Natural Home* magazine, November/December 2004, by Robin Griggs Lawrence, photography by Barbara Bourne.

Sunhawk: The Basics

HOUSE CONSTRUCTION:
Rastra block (85% recycled Styrofoam, 15% cement)
Foundation and Rastra block grout: 6-sack concrete and fly ash mix reinforced with rebar

ENERGY SYSTEM:
—17kW photovoltaics (4kW AstroPower 110W modules and 13kW Siemens 75W modules)
—Harris hydroelectric turbine

A Conversation with the Homeowners

What do you love most about this house?

NANCY: I love the feeling of permanence and belonging. European houses are frequently homes to families for a thousand years. I see no reason ours won't be here to see our newly planted redwood trees turn 1,000 years old.

JOHN: There's nothing better than walking with bare feet on the warm floors on a cold winter day and curling up in the cushions in the south-facing window seats.

What's your favorite room?

NANCY: In the dining room, we can sit at the table and enjoy 360-degree views of the pond and the wildlife. We keep a bird guide and binoculars on the dining room table.

JOHN: I love having a great "dishwashing" window in the kitchen, where I can look out over the valley while I work. And, of course, there's the tower where we hold great solstice rituals.

How does the home perform? How comfortable is the house throughout the seasons?

JOHN: The house has exceeded our expectations. The telling moment was the Solar Living Institute's board meeting in July 2006. The meeting took place in the midst of the worst heat wave I've ever experienced in the 35 years I've lived in Mendocino County. The temperature topped out at 110°F on Friday, 112°F on Saturday, and 110°F on Sunday and never fell below 85°F, even at night.

In the board meeting on the 112°F Saturday, 12 people met in our dining room. We were all totally comfortable, and the temperature never got over 76°F inside! The passive cooling measures (high levels of insulation) and the earth-cooling tubes worked admirably. All we used was a small fan in the meeting room.

In the winter, the passive solar and radiant floor provide all the heat we need. We have found that our wood stove (Tulikivi) is oversized, as the R-65 insulation on the roof and R-35 Rastra walls ensure that a minimum amount of radiant floor heat keeps us toasty, even on 25°F winter days. The woodstove is more of a fireplace with a glass door and is wonderfully cozy but not really needed to heat the house.

What would you do differently during the construction phase?

NANCY: I would have created a hazardous materials collection station when we began building and introduced all the subcontractors to it. You don't want even nontoxic paints and sealers ending up in the landfill or your future garden.

JOHN: I would be more patient. When you're going to live in a home for decades, who cares if it takes an extra six months to complete? I'd spend more time and think out every detail. It's easier to change things before the house is done.

What advice can you offer new homebuilders?

NANCY: Include the outside areas in your planning. We just recently poured the sidewalks around the house, so we had a year of mud and dust. Don't wait until the house is done—do it at the same time.

JOHN: Erect your solar system before you start construction. That way, you build with solar energy and don't have to listen to and smell generators—and you save money in the long run.

Now that you have lived in the house for a few years, do you think you would change anything about the energy design?

JOHN: The earth-cooling tubes work great in the immediate vicinity of the central core of the house, and it's a great place to hang out when it's really hot outside, but we've found that the insulation and passive cooling account for 95% of the cooling and the cooling tubes probably only 5%.

The only thing I can think of would be to extend pipes carrying cool air (from the earth-cooling tube system) throughout the house to cool the upstairs better. On the hottest of days, the temperature downstairs climbs to 76°-78°F, but upstairs can sometimes reach 84°F. We have found that window blinds significantly help to cool down the upstairs on the west side. I think awnings or overhangs on the west side of the house would have helped reduce the late-afternoon heat gain in the summer.

LAND AND SHELTER PRODUCTS

Colorado Yurt Company: Making tipis and yurts into hip housing solutions

For former ski bums Dan and Emma Kigar, living in a yurt or tipi was a way to be part of the back-to-the-land movement of the 1970s. Thirty years later, the rest of us are catching up.

The husband-and-wife team—Emma, a seamstress by trade, and Dan, a cabinetmaker—built their first tipi at 11,000 feet in 1976. Today, their Colorado Yurt Company sells tipis and yurts for use around the world as corporate retreats, yoga studios, vacation homes, and beachside resorts.

"Tipis and yurts have become almost symbolic of a cool, hip, quality-of-life choice for people who are environmentally conscious and are seeking a temporary getaway from the rat race in the city," says Dan from his yurt studio in Montrose, Colorado, adding that many retirement-age couples are buying yurts for backcountry land they bought years ago.

Their rising popularity is part of a growing mission to remove the "beeps and buzzes" of everyday life and live simpler, he explains. Yurts and tipis let people easily do that—it can take just a few hours to raise a tipi, and only one to two days to erect a yurt. And while people use them primarily as temporary getaways, Dan says, "with proper care and maintenance, yurts can last a lifetime. Our list of options allows our customers to design a yurt for their specific conditions, from subarctic to tropical."

Colorado Yurt Co.

Colorado Yurt and Tipi Kits

Our Standard Yurt Kit is available in three sizes and includes frame with engineered compression ring; three heavy-duty zippered screen/storm windows; handcrafted wood window door; Pro-Structure industrial-duty fabric roof; Pro-Tech® water-, UV-, and mildew-resistant fabric walls; roof and wall insulation; and stovepipe outlet (specify diameter when ordering). Additional door and windows available as extra options. USA.

51-0107	**Standard Yurt Kit (16')**	**$6,999**
51-0107	**Standard Yurt Kit (24')**	**$10,399**
51-0107	**Standard Yurt Kit (30')**	**$12,999**
51-0108	**Big Wind Package (16')**	**$495**

(Includes wind load columns and brackets for high wind exposure)

51-0108	**Full Snow and Wind Load Package (24')**	**$1,199**
51-0108	**Full Snow and Wind Load Package (30')**	**$1,549**

(Designed for alpine conditions with high wind and snow loads)

Ideal for both the backcountry and the backyard, our Tipi Pack zage is available in six sizes and includes Tuff Star® water-, UV-, and mildew-resistant fabric cover, poles, door cover, rope, stakes, and lacing pins. Options include flame-retardant liner and flame-retardant floor covers. Easy to set up in a few hours. USA.

51-0112	**Tipi Package**	**$759-$1,249**
51-0113	**Tipi Package, with lodgepole pine poles**	**$1,099-$1,799**
51-0114	**Tipi Liner**	**$349-$479**

Earth Plaster and Pigment

Our completely natural, interior-wall plaster contains no VOCs, fillers, or additives. Create a beautiful, durable, and breathable adobe-like surface that naturally resists mold and helps regulate temperature and humidity. Just mix with water and apply to almost any material, including wood, dry wall, painted surfaces, and masonry. One 50-lb. bag covers 80 to 120 sq. ft. with two coats. Most surfaces require a precoat of sanded primer. USA.

51-0101 Earth Plaster (50-lb. bag of Snow Canyon)   **$74 ($10)**

51-9102 Colored Earth Plaster (50-lb. bag of Snow Canyon Base plus choose one pigment)   **$86 ($10)**

51-0105 Earth Plaster Sanded Primer (1 gal. per 180 sq. ft.)   **$46**

51-0103 Demonstration Video (14 min.)   **$18**

51-0104 Earth Plaster Samples (6 chips of actual plaster)   **$10**

Cork Flooring Planks

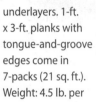

Real Goods' cork flooring planks are made from the bark of the cork oak tree, found in Europe's Mediterranean region and harvested under strict regulatory oversight. Tongue-and-groove construction makes the flooring easy to install without the use of toxic adhesives. Slight color variations are inherent within this natural product; slight fading can occur with extreme sun exposure. Cork veneer planks have high-density fiberboard cores and acoustical cork

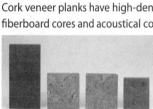

underlayers. 1-ft. x 3-ft. planks with tongue-and-groove edges come in 7-packs (21 sq. ft.). Weight: 4.5 lb. per plank. Cost is $6 to $6.50 per sq. ft. 10-yr. limited warranty. Portugal.

51-0106 Patchwork, natural finish (7-pk.)   **$126**

51-0106 Patchwork, stained finish (7-pk.)   **$136**

51-0106 Burl, natural finish (7-pk.)   **$126**

51-0106 Burl, stained finish (7-pk.)   **$136**

51-9106 Samples (4"x4" squares, 1 of each color)   **$10**

BOOKS

Sensible Design

The Solar House: Passive Heating and Cooling

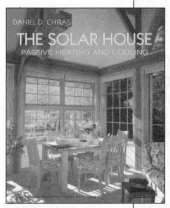

Passive solar design is basically simple—the devil is in the details. This book admits, even highlights, ideas from the '70s that didn't work, and mostly highlights the good design ideas that DO work. Dan Chiras explains in detail how homebuilders can design excellent solar buildings attuned to regional climates—buildings that are warm in the winter and cool in the summer, and require a minimum of energy to stay that way. This is simply the best passive solar design book that's ever been available. Includes experienced reviews of current software programs for solar design, and extensive resource lists and appendixes. 284 pages, softcover.

21-0335 The Solar House   **$30**

The Passive Solar House

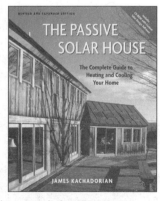

In the expanded second edition of the best-selling guide to building homes that heat and cool themselves, James Kachadorian delivers a book that does not fail to impress. Kachadorian utilizes techniques that translate the essentials of solar design into practical, on-the-ground wisdom for builders of all types. Includes CSOL passive solar design software to help you analyze the efficiency of your passive solar house design and the passive solar potential of your current home. Includes CD-ROM. 240 pages, hardcover.

80372 The Passive Solar House   **$40**

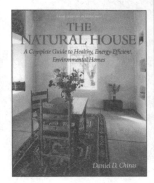

The Natural House

A Complete Guide to Healthy, Energy-Efficient, Environmental Homes

By Daniel D. Chiras. Aimed at both builders and dreamers, this extraordinary book explores environmentally sound, natural, and sustainable building methods, including rammed earth, straw bale, Earthship, adobe, cob, and stone. Besides sanctuary to shelter our families, today we want something more—we want to live free of toxic materials. Chiras is an unabashed advocate of these new technologies, yet he openly addresses the pitfalls of each while exposing the principles that make them really work for natural, humane living. This book is devoted to analysis of the sustainable systems that make these new homes habitable. This is the short course in passive solar technology, green building material, and site selection, addressing more-practical matters—such as scaling your dreams to fit your resources—as well. 469 pages, softcover.

82-474   The Natural House $35

The New Ecological Home

By Daniel D. Chiras. Shelter, while crucial to our survival, has come at a great cost to the land. But, as author Dan Chiras shows in his follow-up book to *The Natural House*, this doesn't have to be the case. With chapters ranging from green building materials and green power to landscaping, he covers every facet of today's homes. A good starting point for decreasing the environmental impact of new construction. 352 pages, softcover.

21-0346 The New Ecological Home $35

The Homeowner's Guide to Renewable Energy

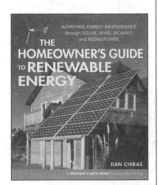

The Homeowner's Guide to Renewable Energy outlines the likely impacts of fossil fuel shortages and some basic facts about energy, and discusses energy conservation as a way to slash energy bills and prepare for renewable energy options. Focusing on strategies needed to replace fuels, the book examines each practical energy option available to homeowners: solar hot water; cooking and water purification; space heating (solar and wood); passive cooling; solar, wind, and microhydroelectricity; and hydrogen, fuel cells, methane digesters, and biodiesel. The book gives readers sufficient knowledge to hire and communicate effectively with contractors, and for those wanting to do installations themselves, it recommends more-detailed manuals. With a complete resource listing, this well-illustrated and accessible guide is a perfect companion for illuminating the coming dark age. 352 pages, softcover.

21-0506 The Homeowner's Guide $27.95

Good Green Homes

Good Green Homes is a guide for people who want to live in a comfortable, healthy, environmentally conscious home. With some simple steps outlined in this book, you can save money and do your part to help save the environment. Perfect for homeowners, remodelers, renters, architects, builders, and interior designers, this book lays out seven fundamental principles of green building, illustrated with more than 150 color and 20 black-and-white photographs of more than 25 homes. 160pages, hardcover.

21-0507 Good Green Homes $39.95

Independent Builder

Designing and Building a House Your Own Way
(REAL GOODS INDEPENDENT LIVING BOOKS)

Subtitled "Designing and Building a House & Own Way," this is *the* book for anyone thinking about building their own home. It is comprehensive and detailed and covers subjects not often covered in homebuilding books, like how to make a small house seem bigger, incorporating ergonomics and accessibility, doing your own drawings and scale models, making contracts that work, and working effectively with professional designers and builders. 528 pages, paperback.

21-0508 Independent Builder $40

Green Remodeling: Changing the World One Room at a Time

By David Johnston and Kim Master. Millions of North Americans are renovating their homes every year. How do you remodel in a healthy, environmentally friendly way? *Green Remodeling* is a comprehensive guide. Buildings are responsible for 40% of worldwide energy flow and material use; so how you remodel can make a

difference. Green remodeling is more energy-efficient, more resource-conserving, healthier for occupants, and more affordable to create, operate, and maintain. The book discusses simple, green renovation

solutions for homeowners, focusing on key aspects of the building, including foundations, framing, plumbing, windows, heating, and finishes. Room by room, it outlines the intricate connections that make the house work as a system. Then, in an easy-to-read format complete with checklists, personal stories, expert insights, and an extensive resource list, it covers easy ways to save energy, conserve natural resources, and protect the health of loved ones. Addressing all climates, this is a perfect resource for conventional homeowners, as well as architects and remodeling contractors. 368 pages, softcover.

21-0365 Green Remodeling **$29.95**

More Small Houses

In *More Small Houses*, you'll find that smaller is beautiful—and more intimate and affordable. Twenty homes, each less than 2,000 square feet, each a study in craft and efficiency, are used to explore space-saving design ideas. 159 pages, hardcover.

21-0217 More Small Houses **$24.95**

The Not So Big House
A Blueprint for the Way We Really Live

By Sarah Susanka, with Kira Obolensky. This is one of the best, most inspiring home design books that's ever been written. Sarah's sense of what delights us, what makes us feel secure, and what simply works well for all humans is flawless. Is it a surprise that people gather in the kitchen? It shouldn't be; human beings are drawn to intimacy, and this book proposes housing ideas that serve our spiritual needs while protecting our pocketbook. Commonsense building ideas are combined with gorgeous home design and dozens of small, simple touches that delight the senses. This is an inspiring book on building houses that make us feel safe, protected, and comfortably at home. 199 pages, softcover.

21-0132 The Not So Big House **$23**

XS: Small Structures, Green Architecture

A follow-up to the highly successful *XS: Big Ideas, Small Buildings*, this book features contemporary solutions to two of today's most challenging problems: how to conserve space and how to help save the environment. The design goals of the 40 houses included here are to build as small as possible, to harmonize with the site, to use natural heating and cooling techniques, and, above all, to combine aesthetic beauty with ecological sensitivity. The houses are striking in appearance, inexpensive to build, and totally functional and will serve as inspiration for architects and potential owners. 224 pages, hardcover.

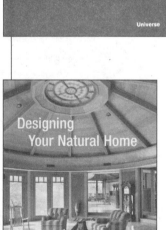

21-0509 XS: Small Structures **$29.95**

Designing Your Natural Home

From studs to refrigerators, this book offers a practical, how-to guide on creating an eco-home for the do-it-yourselfer. Handy for remodeling, expanding, starting from scratch, or even just coffee table fodder. Photo essays of 10 homes exemplify eco-living for every style and budget. Architect and author David Pearson has been at the forefront of integrating home design with the green living movement. 160 pages, softcover.

21-0381 Designing Your Natural Home $29.95

The Art of Natural Building

Design, Construction, Resources

For architects, designers, and the rest of us, this introduction to natural building covers how to build economically and environmentally sound houses that are as beautiful as they are comfortable. Organized as an anthology of articles from leaders in various building techniques—from straw bale and cob to recycled concrete and salvaged materials. Five sections take you from an overview of natural building through planning and design, specific techniques, example structures, and complementary systems for completing your natural home. Edited by Joseph F. Kennedy, Michael G. Smith, and Catherine Wanek. 288 pages, softcover.

82669   The Art of Natural Building $27

Building with Awareness

The Construction of a Hybrid Home
DVD & Guidebook

Constructing a straw bale solar home requires the merging of solar design with alternative building techniques. Books are useful for accessing information on these topics quickly and easily, while videos are a great way to show actual building methods and techniques firsthand. The *Building with Awareness* DVD/guidebook combination brings you the best of both worlds. The award-winning DVD is inspiring and informative, with over five hours of material on every aspect of building a green home. The handy reference guidebook complements the DVD, with color photographs, diagrams, suggestions, and step-by-step methods, all condensed into a nuts-and-bolts format. Used together, these two valuable resources provide a visually dynamic and easy-to-understand library of information. 152 pages, paperback.

21-0378 Building with Awareness $35

Building with Vision

Optimizing and Finding Alternatives to Wood (Wood Reduction Trilogy)

Part green building primer, part architectural photo essay, this is an essential resource for professionals and homeowners interested in the leading edge of environmental building. Author Dan Imhoff traveled extensively to document and photograph beautiful and novel alternatives to wood-intensive building. 136 pages, paperback.

21-0511 Building with Vision $22

Building Types and Materials

The Complete Yurt Handbook

The yurt is a low-impact structure that causes no permanent damage to the land on which it is pitched. It is easy to erect and can be taken down in an hour. This book gives fully illustrated and detailed instructions on how to make several of the most popular types of yurts, including the "weekend yurt." It also provides a thorough history of the yurt and the principles behind its construction, exploring modern life in a Mongolian ger (yurt) and the culture and etiquette of ger living. For centuries, people throughout Central Asia have made robust and versatile yurts their homes. It is the ultimate portable dwelling—perfect for offices, summerhouses, meditation spaces, and spare rooms, or just as beautifully satisfying spaces to be in! With a few common woodworking tools, even an absolute beginner could build the frame for this simple, elegant structure. This handbook shows you how. 121 pages, paperback.

The complete Yurt handbook

Paul King

21-0512 The Complete Yurt Handbook $21.95

Mongolian Cloud Houses: How to Make a Yurt and Live Comfortably

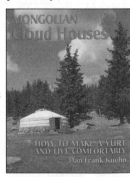

Written for those interested in alternative lifestyles, outdoor living, camping, and do-it-yourself projects, this lively book recounts the author's experiences building his first yurt. Dan Frank Kuehn carefully guides readers through every step of the creation of a 13-ft.-diameter, 10-ft.-tall model. He covers everything from the poles and lattice that form the basic structure to the pluses and minuses of various materials and the "luxuries" like solar windows and lofts. The book highlights new building techniques and contains detailed lists of commercial yurt manufacturers, tools, and materials. 152 pages, paperback.

21-0513 Mongolian Cloud Houses $16.95

The Straw Bale House

(REAL GOODS INDEPENDENT LIVING BOOKS)

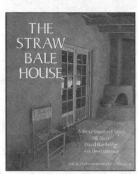

Authors Athena Swentzell Steen and Bill Steen founded the Canelo Project, which promotes innovative building; David Bainbridge is a California restoration ecologist; and David Eisenberg is an alternative-materials builder who pioneered straw bale wall testing. Between them, they have encyclopedic knowledge of their subject. This book is comprehensive, broadly covering why and how to build with straw and then focusing on the details, which are both intellectually and aesthetically delightful. Besides being cheap, clean, and lightweight, straw also provides advantages like energy efficiency and resistance to seismic stresses. 336 pages, paperback.

21-0514 The Straw Bale House $30

Design of Straw Bale Buildings: The State of the Art

Certainly the must-own book for anyone embarking on a straw bale project. Bruce King details straw bale construction—including invaluable information dealing with technical issues and codes—in a writing style that's accessible to everyone. Written by the author of *Buildings of Earth and Straw: Structural Design for Rammed Earth and Straw Bale Architecture*, as well as other useful references on sustainable building, *Design of Straw Bale Buildings* is long overdue and a welcome resource for anyone interested in constructing a living space that emulates the natural world. 296 pages, softcover.

21-0391 Design of Straw Bale Buildings $40

Serious Straw Bale

Until the publication of *Serious Straw Bale: A Home Construction Guide for All Climates*, straw bale was assumed to be appropriate only for dry, arid climates. This book does not advocate universal use of straw bale but provides practical solutions to the challenges brought on by harsh climates. This is the most comprehensive, up-to-date guide to building with bales ever published. Builders, designers, regulators, architects, and owner/builders will benefit from learning bale building techniques, even for climates and locales once thought impossible. 371 pages, softcover.

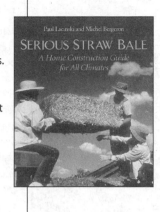

21-0180 Serious Straw Bale $29.95

The Beauty of Straw Bale Homes

By Athena and Bill Steen. This book is a visual treat, a gorgeous pictorial celebration of the tactile, timeless charm of straw bale dwellings. A diverse selection of building types is showcased—from personable and inviting smaller homes to elegant large homes and contemporary institutional buildings. The lavish photographs are accompanied by a brief description of each building's unique features. Interspersed throughout the book are insightful essays on key lessons the authors have learned in their many years as straw bale pioneers. 128 pages, softcover.

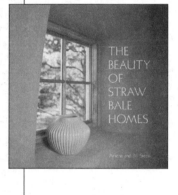

**21-0158 The Beauty of
           Straw Bale Homes $22.95**

Straw Bale Details

By Chris Magwood and Chris Walker. Nothing conveys a wealth of information like a good drawing by a professional architect, which is what this book is all about. It's the nitty-gritty of how to assemble a well-built, long-lasting straw bale structure. All practical foundation, wall, door and window, and roofing options are illustrated, along with notes and possible modifications. Has a clever ring binding that lies flat for easy worksite reference. 68 pages, softcover.

21-0341 Straw Bale Details $32.95

Building a
Straw Bale House:
The Red Feather
Construction
Handbook

Building a Straw Bale House

The Red Feather Construction Handbook

This book is a timely and important tool for the empowerment of communities facing housing deficits. For more than a decade, the Red Feather Development Group, a volunteer-based organization, has built and repaired straw bale houses for Native Americans. This inexpensive, environmentally sound, easily constructed, and downright beautiful form of building has, for good reason, caught the public's imagination. Here, Red Feather provides a step-by-step, easy-to-follow manual for would-be straw bale builders. Informative sections on safety, design, tools, and materials and case studies picked from over 35 Red Feather projects give a comprehensive overview to straw bale building. But this book is much more than a construction manual. It is also the inspiring story of Red Feather itself, a tale of community action and cooperation that suggests a can-do solution to the growing housing crisis on America's Native American reservations. 192 pages, paperback.

21-0515 Building a Straw Bale House $24.95

More Straw Bale Building

A Complete Guide to Designing and Building with Straw

Straw bale houses are easy to build, affordable, super energy efficient, environmentally friendly, and attractive and can be designed to match the builder's personal space needs, esthetics, and budget. Despite mushrooming interest in the technique, however, most straw bale books focus on "selling" the dream of straw bale building but don't adequately address the most critical issues faced by bale house builders. Moreover, since many developments in this field are recent, few books are completely up to date with the latest techniques. *More Straw Bale Building* is designed to fill this gap. A completely rewritten edition of the 20,000-copy best-selling original, it leads the potential builder through the entire process of building a bale structure, tackling all the practical issues: finding and choosing bales; developing sound building plans; roofing; electrical, plumbing, and heating systems; building code compliance; and special concerns for builders in northern climates. 276 pages, paperback.

21-0516 More Straw Bale Building $32.95

The New Strawbale Home

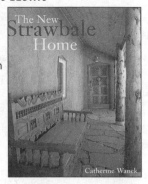

The New Strawbale Home compiles floor plans and images from 30 cutting-edge homes across North America, from California to Quebec, from New Mexico to New England, showcasing a spectrum of regional styles and personal aesthetic choices. This practical guide discusses varying climate considerations and essential design details for problem-free construction and low maintenance, and also points out the ecologically friendly, energy-saving aspects of straw bale construction. 176 pages, hardcover.

21-0517 The New Strawbale $39.95

Small Strawbale

Natural Homes, Projects & Designs

This practical guide is filled with rich photos of homes, greenhouses, studios, sheds, open-air structures, and more, each pulsating with unique yet subtle creativity. Both a pragmatic construction manual and a philosophical, artistic guidebook, *Small Strawbale* is an inspirational starting point for a straw bale dreamer and a great source of information for those who are ready to get baling. 240 pages, paperback.

21-0518 Small Strawbale $29.95

Buildings of Earth and Straw
Structural Design for Rammed Earth and Straw Bale

By Bruce King. Straw bale and rammed earth construction are enjoying a fantastic growth spurt in the United States and abroad. When interest turns to action, however, builders can encounter resistance from mainstream construction and lending communities unfamiliar with these materials. *Buildings of Earth and Straw* is written by structural engineer Bruce King and provides technical data from an engineer's perspective. Information includes special construction requirements of earth and straw, design capabilities and limitations of these materials, and most important, the documentation of testing that building officials often require. This book will be an invaluable design aid for structural engineers and a source of insight and understanding for builders of straw bale or rammed earth. Includes photos, illustrations, and appendixes. 190 pages, softcover.

80-041   Buildings of Earth and Straw $25

Building with Earth
A Guide to "Dirt Cheap" and "Earth Beautiful"

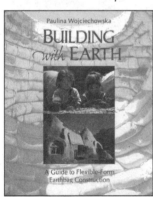

Building with Earth is the first comprehensive guide to the re-emergence of earthen architecture in North America. Even inexperienced builders can construct an essentially tree-free building, from foundation to curved roof, using recycled tubing or textile grain sacks. Featuring beautifully textured earth- and lime-based finish plasters for weather protection, earthbag buildings are being used for retreats, studios, and full-time homes in a wide variety of climates and conditions. This book tells (and shows in breathtaking detail) how to plan and build beautiful, energy-efficient earthen structures. Author Paulina Wojciechowska is founder of Earth, Hands, and Houses, a nonprofit that supports building projects that empower indigenous people to build shelters from locally available materials. 200 pages, softcover.

21-0323   Building with Earth $24.95

Earthbag Building
The Tools, Tricks, and Techniques

By Kaki Hunter and Donald Kiffmeyer. *Earthbag Building* is the first comprehensive guide for building with bags filled with earth—or earthbags. The authors developed this "flexible form rammed earth technique" over the last decade. A reliable method for constructing homes, outbuildings, garden walls, and much more, this enduring, tree-free architecture can also be used to create arched and domed structures of great beauty in any climate. This profusely illustrated guide discusses the many merits of earthbag construction and leads the reader through the key elements of an earthbag building, including special design considerations; foundations, walls, and floors; electricity, plumbing, and shelving; lintels, windows, and door installations; roofs, arches, and domes; and exterior and interior plasters. With dedicated sections on costs, making your own specialized tools, and building code considerations, as well as a complete resource guide, *Earthbag Building* is the long-awaited, definitive guide to this uniquely pleasing construction style. 288 pages, softcover.

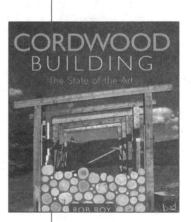

21-0367   Earthbag Building $29

Cordwood Building:
The State of the Art

By Rob Roy. Cordwood masonry is an ancient building technique whereby walls are constructed from "log ends" laid transversely in the wall. It is easy, economical, aesthetically striking, energy-efficient, and environmentally sound. *Cordwood Building* collects the wisdom of more than 25 of the world's best practitioners, detailing the long history of the method and demonstrating how to build a cordwood house using the latest and most up-to-date techniques, with a special focus on building code issues. 240 pages, paperback.

21-0519   Cordwood Building $26.95

Logs, Wind, and Sun

Handcraft Your Own Log Home... Then Power It with Nature

Many dream of getting back to nature and living self-sufficiently in a house built with their own hands. The Ewings show readers how in this account of how they built a log house and then powered it using sun and wind. This monumental undertaking is pragmatically described by the Ewings in an easygoing prose style. They take a clear, step-by-step approach to log building—a project one should not undertake without considerable building experience. Fully a third of the text is devoted to explaining how to run a home completely off the power grid. Readers are offered a wealth of information about solar modules and wind generators, charge controllers, batteries, and inverters. This information makes their work stand out from other log-building books. 304 pages, paperback.

21-0520 Logs, Wind, and Sun **$28.95**

Building with Structural Insulated Panels

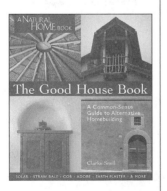

By Michael Morley. Structural insulated panels (SIPs) are replacing the postwar norm of stick-framed, fiberglass-insulated houses and light commercial buildings. SIPs produce a structurally superior, better insulated, faster to erect, and more environmentally friendly house than ever before possible. Experienced SIP builder Michael Morley covers choosing the right panels for the job, tools and equipment, wall and roof systems, and mechanical systems. This is an indispensable reference for anyone who wants to work with this innovative, field-tested construction system. 186 pages, hardcover.

82621   Building with SIPs **$35**

The Good House Book

A Common-Sense Guide to Alternative Home-building—Solar • Straw Bale • Cob • Adobe • Earth Plaster • & More

An intelligent look at how a home is supposed to function and a variety of different building approaches. What's important is finding the right solution to fit your individual needs, local climate, and natural resources. The broad range of topics covered include choosing a site; selecting materials; building with straw bale, cob, adobe, or rammed earth; and plugging into alternative home power systems. Interviews with six homeowners, and photos of the dream homes they built, provide invaluable insight. 240 pages, paperback.

21-0521 The Good House Book **$19.95**

The Hand-Sculpted House

Learn how to build your own home from a simple earthen mixture. Derived from the Old English word meaning "lump," cob has been a traditional building method for millennia. Authors Ianto Evans, Michael G. Smith, and Linda Smiley teach you how to sculpt a cozy, energy-efficient home by hand without forms, cement, or machinery. Includes illustrations and photography. 384 pages, softcover.

21-0331 The Hand-Sculpted House **$35**

Building with Cob: A Step-by-Step Guide

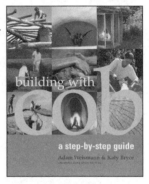

Learn how to create your dream home with clay sub-soil, aggregate, straw, and water in *Building with Cob*, one of the most practical and well-illustrated books on earth building ever published. From designing, planning, and siting your home to adding the roof, insulation, and floors, authors Adam Weismann and Katy Bryce impart wisdom gained from having built and restored many cob buildings and run their own building company for years. More than 400 photographs and illustrations make it easy for even a novice to understand and construct a cob home. 256 pages, paperback.

21-0387 Building with Cob **$45**

The Cob Builders Handbook

You Can Hand-Sculpt Your Own Home

This how-to book includes information on all aspects of cob building. Topics range from choosing a building site to drainage, floors, plaster, and finishing touches. An informal style and numerous hand-drawn illustrations make this book easy to read. 173 pages, paperback.

21-0522 The Cob Builders Handbook **$23.95**

Concrete at Home

In his follow-up book to *Concrete Countertops*, Fu-Tung Cheng expands from countertops to floors, walls, fireplaces, and more, showing us that truly anything is possible with concrete. His beautifully designed book includes step-by-step guidance in forming, pouring, and polishing concrete throughout your home. You'll find yourself amazed to discover how well concrete can emulate natural materials and create a feeling of warmth and beauty. Packed with hundreds of color photos and drawings, this book is part art, part instruction manual. 224 pages, softcover.

21-0376 Concrete at Home   **$32**

Concrete Countertops

Design, Form, and Finishes for the New Kitchen and Bath

By Fu-Tung Cheng and Eric Olsen. Here at last is a start-to-finish book on creating concrete countertops for your kitchen and bathroom. This sustainable material works equally well in traditional and modern homes. Cheng and Olsen take you step by step through the entire process. You'll be inspired by the 350 color photos that showcase this versatile medium and the various creative options for customizing with color, polish, stain, stamp, and embedded objects. Throughout this well-organized book, you'll find valuable troubleshooting advice and useful tips on maintaining your countertop. 208 pages, softcover.

21-0361 Concrete Countertops   **$30**

Green Building Products

The GreenSpec Guide to Residential Building Materials

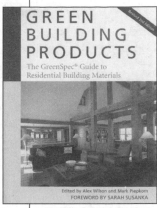

Green Building Products includes more than 1,600 products from the GreenSpec database, which the company has maintained since 1998. Includes everything from precast concrete foundation systems to recycled-plastic roofing shingles and top-efficiency heating equipment. Products are organized according to building component and indexed under manufacturer or product name. Photos are included for about 300 products. A must for anyone—builder, designer, architect, or homeowner—who wants to know the unbiased and unembellished truth about what's really green. Products are selected on criteria that *Environmental Building News* editors have developed over more than 10 years, including recycled content, FSC-certified wood, avoidance of toxic constituents, reduction of construction impacts, energy or water savings, and contribution to a safe, healthy indoor environment. Not a paid directory, they "base selections on careful in-house review by [their] editorial staff." Details on the product selection process and the full list of criteria are included. 338 pages, paperback.

21-0523 Green Building Products   **$34.95**

Pocket Ref

Almost every conversion factor and bit of technical data you could ever imagine wanting to know packed into a pocket-size book you can easily take with you anywhere. The BTUs in a gallon of propane or a kilowatt-hour, wire and lumber sizing, area codes and zip codes, mixing ratios, wind and water formulas, exchange rates, maps, charts—it goes on and on. 3rd edition. By Thomas J. Glover. 544 pages, softcover.

21-0312 Pocket Ref   **$12.95**

Other

Rustic Retreats

This inspiring book presents 20 step-by-step plans for buildings, including a log cabin, floating water gazebo, yurt, grape arbor, sauna hut, wigwam, river raft, and foldup tree house. Many of these low-cost outdoor buildings can be built in a few hours or in less than a day and do not require a high level of building skills. Individual creativity can make any of the projects unique personal statements. Includes detailed illustrations. By David and Jeanie Stiles. 160 pages, softcover.

21-0362 Rustic Retreats   **$20**

Rocket Mass Heaters

Newly updated (a companion volume to the very popular *Rocket Stoves*), *Rocket Mass Heaters*, by Ianto Evans and Leslie Jackson, teaches readers how to construct a rocket stove in a single weekend for less than $100. As heating costs spiral upward, learn how to use locally procured materials to build and warm your home with this extra-efficient heating system. Expanded case studies and color photographs, extended troubleshooting, and a Q&A section help explain the smallest details. 100 pages, paperback.

21-0373 Rocket Mass Heaters   **$18**

The Book of Masonry Stoves: Rediscovering an Old Way of Warming

The Book of Masonry Stoves represents the first comprehensive survey ever published of all the major types of masonry heating systems, ancient and modern. Detailed plans and building information are included in the book. As a complete introduction to masonry stoves, it will help many people, as the title states, rediscover an old way of warming. Within the past decades, millions  of Americans have discovered the economic benefits and personal pleasures of heating with wood, yet many have discovered that there are serious problems associated with wood heat and iron stoves: chimney fires, air pollution, and structural fires. Masonry stoves offer good solutions to many of the problems associated with wood burning. They provide clean combustion at a high temperature, good efficiency, a high degree of safety, and little or no pollution, while requiring little care. They come in a wide variety of shapes and sizes, from simple to elegant and from austere to gothic, and are easily adapted to a variety of structures including solar designs. 192 pages, paperback.

21-0524 The Book of Masonry Stoves   **$30**

CHAPTER 3

Sunshine to Electricity

Renewable Energy 101—Solar, Wind, and Hydroelectric

BETTER THAN 99% OF THE WORLD'S ENERGY comes from the Sun. Some is harvested directly by plants, trees, and solar panels; much is used indirectly in the form of wood, coal, or oil; and a tiny bit is supplied by the nonsolar sources, geothermal and nuclear power. Solar panels receive this energy directly. Both wind and hydro power sources use solar energy indirectly. The coal and petroleum resources that we're so busy burning up now represent stored solar energy from the distant past, yet every single day, enough free sunlight energy falls on Earth to supply our energy needs for four to five years at our present rate of consumption. Best of all, with this energy source, there are no hidden costs and no borrowing or dumping on our children's future. The amount of solar energy we take today in no way diminishes or reduces the amount we can take tomorrow or at any time in the future.

Solar energy can be directly harnessed in a variety of ways. One of the oldest uses of solar is heating domestic water for showering, dishwashing, or space heating. At the turn of the century, solar hot-water panels were an integral part of 80% of homes in southern California and Florida until gas companies, sensing a serious low-cost threat to their businesses, started offering free water heaters and installation. (See pages 388-401 for the full story of solar hot-water heating.) What we're going to cover in this chapter is one of the more recent, cleanest, and most direct ways to harvest the Sun's energy: photovoltaics, or PV.

> Every single day, enough free sunlight energy falls on Earth to supply our energy needs for four to five years at our present rate of consumption.

What Are Photovoltaic Cells?

Photovoltaic cells were developed at Bell Laboratories in the early 1950s as a spinoff of transistor technology. Very thin layers of pure silicon are impregnated with tiny amounts of other elements. When exposed to sunlight, small amounts of electricity are produced. They were mainly a laboratory curiosity until the advent of spaceflight in the 1950s, when they were found to be an efficient and long-lived, although staggeringly expensive, power source for satellites. Also, the utility companies couldn't figure out how to get their wires out into space, so PV was really the only option! Since the early '60s, PV cells have slowly but steadily come down from prices of over $40,000 per watt to current retail prices of around $5 per watt, or in some cases as low as $3 per watt for distributors or in very large quantities. Using the technology available today, we could equal the entire electric production of the United States with photovoltaic power plants using about 10,000 square miles, or less than 12% of the state of Nevada. (See the map on page 103 for details.) As the true environmental and societal costs of coal and petroleum become more apparent, PVs promise to be a major power resource in the future. And with worldwide petroleum sources close to peak, we are rapidly running out of "cheap" oil, making renewable energy all the more critical. Who says that space programs have no benefits for society at large?

In 1954, Bell Telephone Systems announced the invention of the Bell Solar Battery, a "forward step in putting the energy of the Sun to practical use."

Fifty Years of Photovoltaics

Solar Energy Still Strikes a Powerful Chord

Hearing the hit songs of Rosemary Clooney or Perry Como was an ordinary occurrence in the mid-1950s. But it turned extraordinary on Sunday, April 25, 1954.

That was the day that Bell Laboratories executives electrified members of the press with music broadcast from a transistor radio—powered by the first silicon solar cell. Bell called its invention "the first successful device to convert useful amounts of the Sun's energy directly and efficiently into electricity."

The discovery was, indeed, music to people's ears. The *New York Times* heralded it as "the beginning of a new era, leading eventually to the realization of one of mankind's most cherished dreams—the harnessing of the almost limitless energy of the Sun for the uses of civilization."

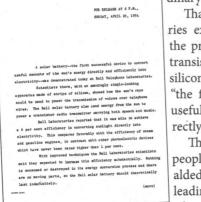

Single Cell Gives Birth to Millions

This 50-year-old prophecy has, in many ways, come to fruition, with a billion watts of solar-electric modules now powering satellites, telescopes, homes, water pumps, and even Antarctic research stations.

Today's solar cells are a vast improvement on their ancestors—achieving 16%-19% efficiency compared with the 6% efficiency of Bell Labs' foray into the solar industry. Silicon is still used—though in slightly different forms—to achieve higher efficiency rates as well as lower costs. Lab tests using single-crystal silicon modules, for example, have produced efficiency rates of up to 24%.

The two- to threefold increase in efficiency is only part of the equation in the success of photovoltaics—an industry that has grown more than 20% compounded per year since 1980 and 50% per year recently. A thousandfold price drop since 1954, along with tax incentives and rebates, have made the move to solar power economically feasible for many homeowners and businesses, with typical payback periods of less than 10 years. PV seems to follow the same economic rules as other new technologies, like

Chemist Calvin Fuller gets ready to diffuse boron into negative silicon. The addition of boron resulted in the first solar cell capable of producing useful amounts of electricity from the Sun.

computers and cell telephony—that is, for every doubling of production, the price drops by 15%-20%. At this rate, PV should be competitive with fossil fuel-generated power company electricity with no rebates or incentives whatsoever within the next few years.

Where Will Solar Energy Be After Its First 100 Years— in 2054?

Of course, the satisfaction of using a renewable energy source may be the biggest motivator of all—and that's what fuels our excitement most. As Margaret Mead said so eloquently, "Never doubt that a small group of thoughtful, committed citizens can change the world. Indeed, it is the only thing that ever has."

Real Goods has now solarized more than 60,000 homes and businesses, and we can't wait to see what will happen as that number gets bigger over the next 50 years. You can count on Real Goods to continue leading the way.

A Brief Technical Explanation

A single PV cell is a thin semiconductor sandwich, with a layer of highly purified silicon. The silicon has been slightly doped with boron on one side and phosphorus on the other side. Doping produces either a surplus or a deficit of electrons, depending on which side we're looking at. Electronics-savvy folks will recognize these as P- and N-layers, the same as transistors use. When our sandwich is bombarded by sunlight, photons knock off some of the excess electrons. This creates a voltage difference between the two sides of the wafer, as the excess electrons try to migrate to the deficit side. In silicon, this voltage difference is just under half a volt. Metallic contacts are made to both sides of the wafer. If an external circuit is attached to the contacts, the electrons find it easier to take the long way around through our metallic conductors than to struggle through the thin silicon layer. We have a complete circuit and a current flows. The PV cell acts like an electron pump. There is no storage capacity in a PV cell; it's simply an electron pump. Each cell makes just under half a volt regardless of size. The amount of current is determined by the number of electrons that the solar photons knock off. We can get more electrons by using bigger cells, or by using more efficient cells, or by exposing our cells to more intense sunlight. There are practi-

cal limits, however, to size, efficiency, and how much sunlight a cell can tolerate.

Since 0.5-volt solar panels won't often do us much good, we usually assemble a number of PV cells for higher voltage output. A PV "module" consists of many cells wired in series to produce a higher voltage. Modules consisting of about 36 cells in series have become the industry standard for large power production. This makes a module that delivers power at 17 to 18 volts, a handy level for 12-volt battery charging. In recent years, as PV modules and systems have grown larger, 24-volt modules consisting of 72 cells have also become standardized. The module is encapsulated with tempered glass (or some other transparent material) on the

A Technical Step Back in Time

Before the invention of the silicon solar cell, scientists were skeptical about the success of solar as a renewable energy source. Bell Labs' Daryl Chapin, assigned to research wind, thermoelectric, and solar energy, found that existing selenium solar cells could not generate enough power. They were able to muster only 5 watts per square meter—converting less than 0.5% of incoming sunlight into electricity.

The Solar Cell That Almost Never Was

Chapin's investigation may have ended there if not for colleagues Calvin Fuller and Gerald Pearson, who discovered a way to transform silicon into a superior conductor of electricity. Chapin was encouraged to find that the silicon solar cell was five times more efficient than selenium. He theorized that an ideal silicon solar cell could convert 23% of incoming solar energy into electricity. But he set his sights on an amount that would rank solar energy as a primary power source: 6% efficiency.

Nuclear Cell Trumps Solar Scientists

While Chapin, Pearson, and Fuller faced challenges in increasing the efficiency of the silicon cell, archrival RCA Laboratories announced its invention of the atomic battery—a nuclear-powered silicon cell. Running on photons emitted from Strontium-90, the battery was touted as having the potential to run hearing aids and wristwatches for a lifetime.

That turned up the heat for Bell's solar scientists, who were making strides of their own by identifying the ideal location for the P-N junction—the foundation of any semiconductor. The Bell trio found that shifting the junction to the surface of the cell enhanced conductivity, so they experimented with substances that could

The three inventors of the first silicon solar cell, Gerald Pearson, Daryl Chapin, and Calvin Fuller, examine their cells at their Bell lab.

permanently fix the P-N junction at the top of the cell. When Fuller added arsenic to the silicon and coated it with an ultrathin layer of boron, it allowed the team to make the electrical contacts they had hoped for. Cells were built using this mixture until one performed at the benchmark efficiency of 6%.

Solar Bursts the Atomic Bubble

After releasing the invention to the public on April 25, 1954, one journalist noted, "Linked together electrically, the Bell solar cells deliver power from the Sun at the rate of 50 watts per square yard, while the atomic cell recently announced by the RCA Corporation merely delivers a millionth of a watt" over the same area. The solar revolution was born.

Special thanks to John Perlin, author of *From Space to Earth: The Story of Solar Electricity*.

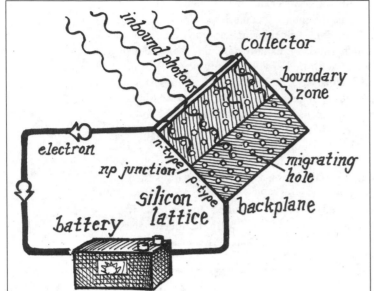

front surface and with a protective and water-proof material on the back surface. The edges are sealed for weatherproofing, and there is often an aluminum frame holding everything together in a mountable unit. A junction box, or wire leads, providing electrical connections is usually found on the module's back. Truly weatherproof encapsulation was a problem with the early modules assembled 20 years ago. We have not seen any encapsulation problems with glass-faced modules in many years.

Many applications need more than a single PV module, so we build an "array." A PV array consists of a number of individual PV modules that have been wired together in series and/or parallel to deliver the voltage and amperage a particular system requires. An array can be as small as a single pair of modules, or large enough to cover acres.

PV costs are down to a level that makes them the clear choice for remote *and* grid-intertie power applications. They are routinely used for roadside emergency phones and most temporary construction signs, where the cost and trouble of bringing in utility power outweighs the higher initial expense of PV, and where mobile generator sets present more fueling and maintenance trouble. It's hard to find new gate opener hardware that isn't solar powered. Solar with battery backup has proven to be a far more reliable power source, and it's usually easier to obtain at the gate site. More than 150,000 homes in the United States, largely in rural sites, depend on PVs as a primary power source, and this figure is growing rapidly as people begin to understand how clean, reliable, and mainte-nance-free this power source is, and how deeply

Sharp 123 PV Module

our current energy practices are borrowing from our children. Worldwide, there are currently over 1 million homes that derive their primary power from PV. Because they don't rely on miles of exposed wires, residential PV systems are more reliable than utilities, particularly when the weather gets nasty. PV modules have no moving parts; degrade very, very slowly; and boast a lifespan that isn't fully known yet but will be measured in multiple decades. Standard factory PV warranties are 25 years. Compare this with any other power generation technology or consumer goods. Could you find a car or truck or computer with a 25-year warranty? If you could, you'd probably buy it!

Construction Types

There are currently four commercial production technologies for PV cells.

SINGLE CRYSTALLINE

This is the oldest and most expensive production technique, but it's also the most efficient sunlight conversion technology commercially available. Complete modules have sunlight-to-wire output efficiency averages of about 10%-12%. Efficiencies up to 22% have been achieved in the lab, but these are single cells using highly exotic components that cannot economically be used in commercial production.

Boules (large cylindrical cylinders) of pure single-crystal silicon are grown in an oven, then sliced into wafers, doped, and assembled. This is the same process used in manufacturing transistors and integrated circuits, so it is well-developed, efficient, and clean. Degradation is very slow with this technology, typically 0.25%-0.5% per year. Silicon crystals are characteristically blue, and single crystalline cells look like deep blue glass. Examples are Sunpower, Solar World, and Sharp single-crystalline products.

POLYCRYSTALLINE OR MULTICRYSTALLINE

In this technique, pure, molten silicon is cast into cylinders, then sliced into wafers off the large block of multicrystalline silicon. Poly-crystal is slightly lower in conversion efficiency compared with single crystal, but the manufacturing process is less exacting, so costs are a bit lower. Module efficiency averages about 10%-11%, sunlight to wire. Degradation is very slow and gradual, similar to that of single crystal, discussed above. Crystals measure approximately 1 centimeter (two-fifths of an inch), and the multicrystal patterns can be clearly seen in the cell's deep blue surface. Doping and assembly are the same as for single-crystal modules. Examples are Sharp, Sanyo, and Kyocera polycrystalline products.

STRING RIBBON

This clever technique is a refinement of polycrystalline production. A pair of strings are drawn up through molten silicon, pulling up between them a thin film of silicon like a soap bubble. It cools and crystallizes, and you've got ready-to-dope wafers. The ribbon width and thickness can be controlled, so there's far less slicing, dicing, or waste, and production costs are lower. Sunlight-to-wire conversion efficiency is about 8%-9%. Degradation is the same as for ordinary slice-and-dice polycrystal. Examples are Evergreen modules.

AMORPHOUS OR THIN FILM

In this technique, silicon material is vaporized and deposited on glass or stainless steel. This production technology costs less than any other method, but the cells are less efficient, so more surface area is required. Early production methods in the 1980s produced a product that faded up to 50% in output over the first three to five years. Present day thin-film technology has dramatically reduced power fading, although it's still a long-term uncertainty. Uni-Solar has a "within 20% of rated power at 20 years" warranty, which relieves much nervousness, but we honestly don't know how these cells will fare with time. Sunlight-to-wire efficiency averages about 5%-7%. These cells are often almost black in color. Unlike other modules, if glass is used on amorphous modules, it is not tempered, so breakage is more of a problem. Tempered glass can't be used with this high-temperature deposition process. If the deposition layer is stainless steel and a flexible glazing is used, the resulting modules will be somewhat flexible. These are often used as marine or RV modules. In the mid '90s, it appeared that amorphous modules could deliver the magic $2 per watt that would make solar sprout on every rooftop. There was a rush to build assembly plants. Oddly, it turned out that few homeowners wanted to cover every square foot of their roof with an unproven and still fairly expensive solar product. We're seeing fewer examples of thin-film technology. Uni-Solar makes flexible, unbreakable modules.

Putting It All Together

The PV industry has standardized on 12-, 24-, or 48-volts for battery systems. These are moderately low voltages, which are relatively safe

Single crystalline module

Multicrystalline module

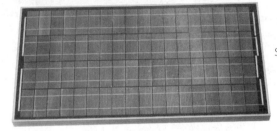

String ribbon module

Thin film

to work with. Under most circumstances, this isn't enough voltage to pass through your body. While not impossible, it's pretty difficult to hurt yourself on such low voltage. Still, whenever working with electricity, make sure you take the necessary safety precautions. Batteries however, where enormous quantities of accumulated energy are stored, can be very dangerous if mishandled or miswired. Please see Chapter 4, which discusses batteries and safety equipment, for more information.

Multiple modules can be wired in parallel or series to achieve any desired output. As systems get bigger, we usually run collection and storage at higher DC voltages because transmission is easier. Small systems processing up to about 2,000 watt-hours are fine at 12 volts. Systems processing 2,000-7,000 watt-hours will function better at 24 volts, and systems running more than 7,000 watt-hours should probably be

What's a Watt?

A Watt (W) is a standard metric measurement of electrical power. It is a rate of doing work.

A Watt-hour (Wh) is a unit of energy measuring the total amount of work done during a period of time. (This is the measurement that utility companies make to charge us for the electricity we consume.)

An Amp (A) is a unit measuring the amount of electrical current passing a point on a circuit. It is the rate of flow of electrons through a conductor such as copper wire: 1 Amp = 6.28 billion billion electrons moving past a point in one second. (Amps are analogous to the water-flow rate in a water pipe.)

A Volt (V) is a unit measuring the potential difference in electrical force, or pressure, between two points on a circuit. This force on the electrons in a wire causes the current to flow. (Volts are analogous to water pressure in a pipe.)

In summary, a Watt measures power, or the rate of doing work, and a Watt-hour measures energy, or the amount of work done. Watts can be calculated if you know the voltage and the amperage: **Watts = Volts x Amps**. More pressure or more flow means more power.

PV modules do not convert 100% of the energy that strikes them into electricity (we wish!). Current commercial technology averages about 10%-12% conversion efficiency for single- and multicrystalline modules, and 5%-7% for amorphous modules. Conversion rates slightly over 20% have been achieved in the laboratory by using experimental cells made with esoteric and rare elements. But these elements are far too expensive to ever see commercial production. Conversion efficiency for commercial single- and multicrystalline modules is not expected to improve; this is a mature technology. There's better hope for increased efficiency with amorphous technology, and much research is currently underway.

How Long Do PV Modules Last?

PV modules last a long, long time. How long we honestly don't yet know, as the oldest terrestrial modules are barely 45 years old and still going strong. In decades-long tests, the fully developed technology of single- and polycrystal modules has shown to degrade at fairly steady rates of 0.25%-0.5% per year. First-generation amorphous modules degraded faster, but there are so many new wrinkles and improvements in amorphous production that we can't draw any blanket generalizations for this module type. The best amorphous products now seem to closely match the degradation of single-crystal products, but there is little long-term data. Most full-size modules carry 25-year warranties, reflecting their manufacturers' faith in the

---

running at 48 volts. These are guidelines, not hard and fast rules! The modular design of PV panels allows systems to grow and change as system needs change. Modules from different manufacturers, different wattages, and various ages can be intermixed with no problems, so long as all modules have a rated voltage output within about 1.0 volt of each other. Buy what you can afford now, then add to it in a few years when you can afford to expand.

> Conversion efficiency for commercial single- and multicrystalline modules is not expected to improve; this is a mature technology. There's better hope for increased efficiency with amorphous technology, and much research is currently underway.

Efficiency

By scientific definition, the Sun delivers 1,000 watts (1 kilowatt) per square meter at noon on a clear day at sea level. This is defined as a "full Sun" and is the benchmark by which modules are rated and compared. That is certainly a nice round figure, but it is not what most of us actually see. Dust, water vapor, air pollution, seasonal variations, altitude, and temperature all affect how much solar energy your modules actually receive. For instance, the 1991 eruption of Mt. Pinatubo in the Philippines reduced available sunlight worldwide by 10%-20% for a couple of years. It is reasonable to assume that most sites will actually average about 85% of full Sun, unless they are over 7,000 feet in elevation, in which case they'll probably receive more than 100% of full Sun.

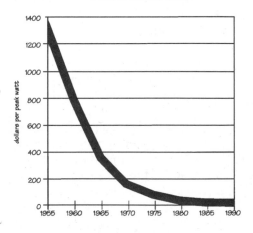

Photovoltaic Prices

dollars per peak watt (y-axis: 0, 200, 400, 600, 800, 1000, 1200, 1400)

(x-axis: 1955, 1960, 1965, 1970, 1975, 1980, 1985, 1990)

Photovoltaic prices have decreased dramatically since 1955. Prices continue to drop 10%-15% per year as demand and production increase. We have seen prices as low as $3 per watt for very large systems in late 2007.

durability of these products. PV technology is closely related to transistor technology. Based on our experience with transistors, which just fade away after 20 years of constant use, most manufacturers have been confidently predicting 20-year or longer life spans. However, keep in mind that PV modules are seeing only six to eight hours of active use per day, so we may find that life spans of 60-80 years are normal. Cells that were put into the truly nasty environment of space in the late 1960s are still functioning well. The bottom line? We're going to measure the life expectancy of PV modules in decades—how many, we don't yet know.

Payback Time for Photovoltaic Manufacturing Energy Investment

In the early years of the PV industry, there was a nasty rumor circulating that said PV modules would never produce as much power over their lifetimes as it took to manufacture them. During the early years of development, when transistors were a novelty, and handmade PV modules costing as much as $40,000 per watt were being used exclusively for spacecraft, this was true. The truth now is that PV modules pay back their manufacturing energy investment in about 1.5 years' time (only a fraction of the typical warranty period), depending on module type, installation climate, and other conditions. Now, in all honesty, this information comes to us courtesy of the module manufacturers. The National Renewable Energy Laboratory has done some impartial studies on payback time (see the results at www.nrel.gov/ncpv/pdfs/24596.pdf). It concludes that modules installed under average U.S. conditions reach energy payback in three to four years, depending on construction type. The aluminum frame all by itself can account for six months to one year of that time. Quicker energy paybacks, down to one to two years, are expected in the future, as more "solar grade" silicon feedstock becomes available, and simpler standardized mounting frames are developed.

Maintenance

It's almost laughable how easy the maintenance is for PV modules. Because they have no moving parts, they are virtually maintenance free. Basically, you keep them clean. If it rains irregularly or if the birds leave their calling cards, hose the modules down. Do not hose them off when they're hot, since uneven thermal shock could theoretically break the glass. Wash them in the morning or evening. For PV maintenance, that's it.

Control Systems

Controls for PV systems are usually simple. When the battery reaches a full-charge voltage, the charging current can be turned off or directed elsewhere. Open-circuited PV module voltage rises 5-10 volts and stabilizes harmlessly. It does no harm to the modules to sit at open-circuit voltages, but they aren't doing any work for you either. When the battery voltage drops to a certain set-point, the charging circuit is closed and the modules go back to charging. With the solid-state PWM (Pulse Width Modulated) controllers, this opening and closing of the circuit happens so rapidly that you'll simply see a stable voltage. The most recent addition to PV

It's almost laughable how easy the maintenance is for PV modules. Because they have no moving parts, they are virtually maintenance free. Basically, you keep them clean.

A Mercifully Brief Glossary of PV System Terminology

AC (alternating current)—This refers to the standard utility-supplied power, which alternates its direction of flow 60 times per second, and for normal household use has a voltage of approximately 120 or 240 (in the USA). AC is easy to transmit over long distances, but it is impossible to store. Most household appliances require this kind of electricity.

DC (direct current)—This is electricity that flows in one direction only. PV modules, small wind turbines, and small hydroelectric turbines produce DC power, and batteries of all kinds store it. Appliances that operate on DC very rarely will operate directly on AC, and vice versa. Conversion devices are necessary.

Inverter—An electronic device that converts (transforms) the low-voltage DC power we can store in batteries to conventional 120-volt AC power as needed by lights and appliances. This makes it possible to utilize the lower-cost (and often higher-quality) mass-produced appliances made for the conventional grid-supplied market. Inverters are available in a wide range of wattage capabilities. We commonly deal with inverters that have a capacity of anywhere between 150 and 6,000 watts.

PV Module—A "solar panel" that makes electricity when exposed to direct sunlight. PV is shorthand for photovoltaic. We call these panels PV modules to differentiate them from solar hot-water panels or collectors, which are a completely different technology and are often what folks think of when we say "solar panel." PV modules do not make hot water.

control technology is Maximum Power Point Tracking, or MPPT controls. These sophisticated solid-state controllers allow the modules to run at whatever voltage produces the maximum wattage. This is usually a higher voltage than batteries will tolerate. The extra voltage is down-converted to amperage the batteries can digest comfortably. MPPT controls extract an average of about 15% more energy from your PV array and do their best work in the wintertime when most residential systems need all the help they can get. Most controllers offer a few other bells and whistles, like nighttime disconnect and LED indicator lights. See the Controls and Monitors section in Chapter 4 for a complete discussion of controllers.

Powering Down

The downside to all this good news is that the initial cost of a PV system is still high. After decades of cheap, plentiful utility power, we've turned into a nation of power hogs. The typical American home consumes about 20-30 kilowatt-hours daily. Supplying this demand with PV-generated electricity can be costly; however, it makes perfect economic sense as a long-term investment. Fortunately, at the same time that PV-generated power started to become affordable and useful, conservation technologies for electricity started to become popular, and given the steadily rising cost of utility power, even necessary. The two emerging technologies dovetail together beautifully. Every kilowatt-hour you can trim off your projected power use in a standalone (off-grid) PV-based system will reduce your initial setup cost by as much as $3,500. Using a bit of intelligence and care in your lighting and appliance selection will allow

> The typical American home consumes about 20-30 kilowatt-hours daily. Supplying this demand with PV-generated electricity can be costly; however, it makes perfect economic sense as a long-term investment.

you all the conveniences of the typical 20-kWh-per-day California home, while consuming less than 10 kWh per day. At $3,500 per installed kilowatt-hour, that's $35,000 shaved off the initial system cost! With this kind of careful analysis applied to electrical use, most of the full-size home electrical systems we design come in between $15,000 and $30,000, depending on the number of people and intended lifestyle. Simple weekend cabins with a couple of lights and a boom box can be set up for $1,500 or less. With the renewable energy rebates and buydowns available in an increasing number of states, grid-tie PV can be very cost effective. Typical payback times in California run 6-12 years (an 8%-17% return on investment!). Commercial paybacks with tax incentives typically pay back in half that time.

Other chapters in this *Sourcebook* present an extensive discussion of electrical conservation, for both off and on grid (utility power), and offer many of the lights and appliances discussed. We strongly recommend reading these sections before beginning system sizing. We are not proposing any substantial lifestyle changes, just the application of appropriate technology and common sense. Stay away from 240-volt watt hogs, electric space heaters, cordless electric appliances, standard incandescent light bulbs, instant-on TVs, and monster side-by-side refrigerators, and our friendly technical staff can work out the rest with you.

PV Performance in the Real World

Okay, here's the dirt under the rug. Skeptics and pessimists knew it all along: PV modules could not possibly be all that perfect and simple. Even the most elegant technology is never quite perfect. There are a few things to watch out for, beginning with . . . Wattage ratings on PV modules are given under ideal laboratory conditions. Assuming you can avoid or eliminate shadows, the two most important factors that affect module performance out in the real world are percentage of full Sun and operating temperature.

SHADOWS

Short of outright physical destruction, hard shadows are the worst possible thing you can do to a PV module. Even a tiny amount of shading dramatically affects module output. Electron flow is like water flow. It flows from high voltage to low voltage. Normally the module is high

Photovoltaic Summary

ADVANTAGES	DISADVANTAGES
1. *No moving parts*	1. *High initial cost*
2. *Ultralow maintenance*	2. *Works only in direct sunlight*
3. *Extremely long life*	3. *Sensitive to shading*
4. *Noncorroding parts*	4. *Lowest output during shortest days*
5. *Easy installation*	5. *Low-voltage output difficult to transmit*
6. *Modular design*	
7. *Universal application*	
8. *Safe low-voltage output*	
9. *Simple controls*	
10. *Long-term economic payback*	

and the battery, or load, is lower. A shaded portion of the module drops to very low voltage. Electrons from other portions of the module and even from other modules in the array will find it easier to flow into the low-voltage shaded area than into the battery. These electrons just end up making heat and are lost to us. This is why bird droppings are a bad thing on your PV module. A fist-size shadow will effectively shut off a PV module. Don't intentionally install your modules where they will get shadows during the prime midday generating time from 10 a.m. to 3 p.m. Early or late in the day, when the Sun is at extreme angles, little power is being generated anyway, so don't sweat shadows then. Sailors may find shadows unavoidable at times, but just keep them clear as much as practical.

FULL SUN

As mentioned above, most of us seldom see 100% full-Sun conditions. If you are not getting full, bright, shadow-free sunlight, then your PV output will be reduced. If you are not getting bright enough sunlight to cast fairly sharp-edged shadows, then you do not have enough sunlight to harvest much useful electricity. Most of us actually receive 80%-85% of a "full Sun" (defined as 1,000 watts per square meter) on a clear sunny day. High altitudes and desert locations will do better on sunlight availability. On the high desert plateaus, 105%-110% of full Sun is normal. They don't call it the "sunbelt" for nothing.

TEMPERATURE

The power output from all PV module types fades somewhat at higher temperatures. This is not a serious consideration until ambient temperatures climb above 80°F, but that's not uncommon in full Sun. The backs of modules should be as well ventilated as practical. Always allow some airspace behind the modules if you want decent output in hot weather. On the positive side of this same issue, all modules increase output at lower temperatures, as in the wintertime, when most residential applications can use a boost. We have seen cases when modules were producing 30%-40% over specs on a clear, cold winter morning with a fresh reflective snow cover and hungry batteries.

As a general rule of thumb, we usually derate official manufacturer-specified "nameplate" PV module output by about 25%-30% (grid-tie systems) to 40%-50% (off-grid, battery-based systems) for the real world. For panel-direct systems (where the modules are connected directly to the pump without any batteries), derate by 20%, or even by 30% for really hot summer climates if you want to make sure the pump will run strongly in hot weather.

Module Mounting

Modules will catch the maximum sunlight, and therefore have the maximum output, when they are perpendicular (at right angles) to the Sun. This means that tracking the Sun across the sky from east to west will give you more power output. But tracking mounts are expensive and prone to mechanical and/or electrical problems, and PV prices have been coming down. Unless you've got a summertime high-power application, like water pumping, tracking mounts don't make a good investment anymore.

PV systems are most productive if the modules are approximately perpendicular to the Sun at solar noon, the most energy-rich time of day for a PV module. The best year-round angle for your modules is approximately equal to your latitude. Because the angle of the Sun changes seasonally, you may want to adjust the angle of your mounting rack seasonally. In the winter, modules should be at the angle of your latitude plus approximately 10 degrees; in the summer, your latitude minus a 10-degree angle is ideal. On a practical level, many residential systems will have power to burn in the summer, and seasonal adjustment may be unnecessary.

Generally speaking, most PV arrays end up on fixed mounts of some type. Tracking mounts are rarely used for residential systems anymore. Small water-pumping arrays are the most common use of tracking mounts now. This rule of thumb is far from ironclad; there are many good reasons to use either kind of mounting. For a more thorough examination, see the PV Mounting section, which includes a large selection of mounting technologies.

> The best year-round angle for your modules is approximately equal to your latitude.

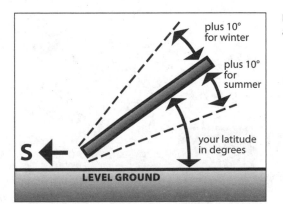

plus 10°
for winter

plus 10°
for summer

your latitude
in degrees

S ←

LEVEL GROUND

Proper PV mounting angle.

System Examples

Following are several examples of photovoltaic-based electrical systems, starting from simple and working up to complex. All the systems that use batteries can also accept power input from wind or hydro sources as a supplement or as the primary power source. PV-based systems constitute better than 95% of Gaiam Real Goods' renewable energy sales, so the focus here will be mostly on them.

A SIMPLE SOLAR PUMPING SYSTEM

In this simple system, all energy produced by the PV module goes directly to the water pump. No electrical energy is stored; it's used immedi-

PV-direct water pumping

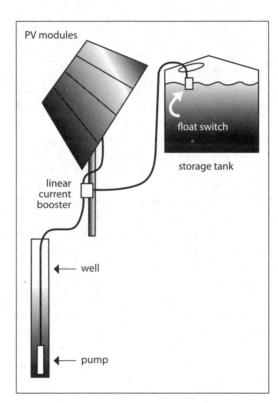

A utility intertie without batteries

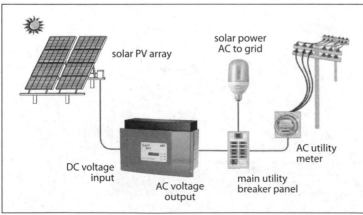

ately. Water delivered to the raised storage tank is your stored energy. The brighter the Sun, the faster the pump will run. This kind of system (without battery storage) is called PV-direct and is the most efficient way to utilize PV energy. Eliminating the electrochemical conversion of the battery saves about 20%-25% of the energy, a very significant chunk! However, PV-direct systems work only with DC motors that can use the variable power output of the PV module, and of course this simple system works only when the Sun shines.

There's one component of a PV-direct system you won't find in other systems. The PV-direct controller, or Linear Current Booster (LCB), is unique to systems without batteries. This solid-state device will down-convert excess voltage into amperage that will keep the pump running under low-light conditions when it would otherwise stall. An LCB can boost pump output by as much as 40%, depending on climate and load conditions. We usually recommend them for PV-direct pumping systems.

For more information about solar pumping, see Chapter 7, "Water Development."

A UTILITY INTERTIE SYSTEM WITHOUT BATTERIES

This is the simplest and most cost effective way to connect PV modules to regular utility power. All incoming PV-generated electrons are converted to household AC power by the intertie inverter and delivered to the main household circuit breaker panel, where they displace an equal number of utility-generated electrons. That's power you didn't have to buy from the utility company. If the incoming PV power exceeds what your house can use at the moment, the excess electrons will be forced out through your electric meter, turning it backward. If the PV power is insufficient, that shortfall is automatically and seamlessly made up by utility power. It's like water seeking its own level (except it's really fast water!). When your intertie system is pushing excess power out through the meter, the utility is paying you regular electric rates for your excess power. You sell power to the utility during the daytime; it sells power back to you at night. This treats the utility grid like a big 100%-efficient battery. However, if utility power fails, even if it's sunny, your PV system will be shut off for the safety of utility workers.

How Much PV Area to Equal U.S. Electric Production?

Here's the simple answer:

A solar-electric array, using today's off-the-shelf technology, sited in sunny, largely empty Nevada, that's big enough to deliver all the electricity the U.S. currently uses, would cover a square almost exactly 100 miles per side.

Here's the proof in more detail:

According to the Energy Information Administration of the U.S. Department of Energy, www.eia.doe.gov/emeu/aer/txt/ptb0802a.html, the U.S. produced 4,038 billion kilowatt-hours of electricity in 2005. Note that this is "production," not "use." Transmission inefficiencies and other losses are covered.

We'll want our PV modules in a sunny area to make the best of our investment. Nevada, thanks to climate and military/government activities, has a great deal of almost empty and very sunny land. So looking at the National Solar Radiation Data Base for Tonopah, Nevada http://rredc.nrel.gov/solar/pubs/redbook/, a flat-plate collector on a fixed mount facing south at a fixed tilt equal to the latitude, 38.07° in this case, saw a yearly average of 6.1 hours of "full-sun" per day in the years 1961 through 1990. A "full-sun" is defined as 1,000 watts per square meter.

For PV modules, we'll use the large Sharp 208-watt module, which the California Energy Commission, www.energy.ca.gov/ greengrid/certified_pv_modules .html, rates at 183.3 watts output, based on lab-tested performance. 183.3 watts times 6.1 hours equals 1,118 watt-hours or 1.118 kilowatt-hours per day per module at our Tonopah site. At 65 x 39 inches this module presents 17.6 square feet of surface area. We'll allow some space between rows of modules for maintenance access and for sloping wintertime Sun, so let's say that each module will need 23 square feet.

Conversion from PV module DC output to conventional AC power isn't perfectly efficient. Looking at the real-world performance figures from the California Energy Commission, www.energy.ca.gov/greengrid/certified_invert ers.html, we see that the SatCon Power Systems 75kW model AE-75-60-PV-D is rated at 96% efficiency. We'll probably be using larger inverters, but this is a typical efficiency for large intertie inverters. We'd better also deduct about 10% for whatever other losses might occur—dirty modules, etc. So our 1.118kWh per module per day becomes 0.966kWh by the time it hits the AC grid.

A square mile (5,280 x 5,280 feet) equals 27,878,400 square feet. Divided by 23 square feet per module, we can fit 1,212,104 modules per square mile. At 0.966 kilowatt-hours per module per day, our square mile will deliver 1,170,971kWh per day on average, or 427,404,328kWh per year. Back to our goal of 4,038,000,000,000kWh divided by 427,404,328kWh per year per square mile, it looks like we need about 9,448 square miles of surface to meet the electrical needs of the United States. That's a square area about 97 miles on a side. This is about 60% of the approximately 16,000 square miles currently occupied by the Nevada Test Site and the surrounding Nellis Air Force Range (www.nv.doe.gov/nts/ and www.nellis.af.mil/ environmental/default.htm).

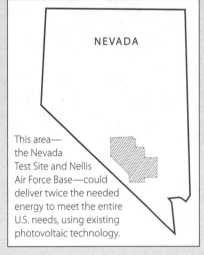

NEVADA

This area—the Nevada Test Site and Nellis Air Force Base—could deliver twice the needed energy to meet the entire U.S. needs, using existing photovoltaic technology.

What about the cost/benefit of such a project? Let's look at the cost of an array that would produce just one-quarter of the required electricity that is produced in the U.S. Currently small commercial or residential systems cost about $5-$6 per peak watt including typical government and utility incentives. If economies of scale, advances in efficiencies, and government subsidies are considered, the job might get done for $2/watt (just guessing here, but you get the idea). Therefore, our quarter-sized 596,000 megawatt array would cost a cool $1.19 trillion. This certainly is a lot of money; but then the potential benefits can be enormous. For one, think of the jobs created. The Apollo Alliance (www.apolloalliance.org) says that "Renewable power production is labor intensive. . . . Solar PV creates 7.24 jobs per MW." However, worldwide PV production in 2005 was only 1,652 megawatts (www.solarbuzz.com), so we've got a ways to go to meet this potential demand.

As a practical measure, PV power production happens during the daytime, and so long as we use lights at night, we will continue to use substantial power at night. Also, out in the desert, solar thermal collection may be a more efficient power generation technology. But however you run the energy collection system, large solar-electric farms on what is otherwise fairly useless desert land could add substantially to the electrical independence and security of any country. The existing infrastructure of coal, nuclear, and hydro power plants could continue to provide reliable power at night, but nonrenewable resource use and carbon dioxide production would be greatly reduced.

Small cabin systems
to run a few lights and
an appliance
or two can start at
under $1,000.

Batteries are the
most cost-effective
energy storage
technology available so
far, but batteries are a
mixed blessing.

There is a minimum of hardware for these intertie systems; a power source (the PV modules), an intertie inverter, a circuit breaker, and some wiring to connect everything. See the separate section specifically on Utility Intertie for more information.

A SMALL CABIN PV SYSTEM WITH BATTERIES

Most PV systems are designed to store some of the collected energy for later use. This allows you to run lights and entertainment equipment at night, or to temporarily run an appliance that takes more energy than the PV system is delivering. Batteries are the most cost-effective energy storage technology available so far, but batteries are a mixed blessing. The electrical/chemical conversion process isn't 100% efficient, so you have to put back about 20% more energy than you took out of the battery, and the storage capacity is finite. Batteries are like buckets, they can get only so full and can empty just so far. A charge controller becomes a necessary part of your system to prevent over- (and sometimes under-) charging. Batteries can also be dangerous. Although the lower battery voltage is generally safer to work around than conventional AC house current, it is capable of truly awe-inspiring power discharges if accidentally short-circuited. So fuses and safety equipment also become necessary whenever you add batteries to a system. Fusing ensures that no youngster, probing with a screwdriver into some unfortunate place, can burn the house down. And finally, a monitoring system that displays the battery's approximate state of charge is essential for reliable performance and long battery life. Monitoring could be done without—just as you

could drive a car without any gauges, warning lights, or speedometer—but this doesn't encourage system reliability or longevity.

The Real Goods Weekend Getaway Kit is an example of a small-cabin or weekend retreat system. It has all the basic components of a residential power system: A power source (the PV module), a storage system (the deep-cycle battery), a controller to prevent overcharging, safety equipment (fuses), and monitoring equipment. The Weekend Getaway Kit (see p. 115) is supplied as a simple DC-only system. It will run 12-volt DC equipment, such as RV lights and appliances. An optional inverter can be added at any time to provide conventional AC power, and that takes us to our next example.

A FULL-SIZE HOUSEHOLD SYSTEM

Let's look at an example of a full-size residential system to support a family of three or more. The power source, storage, control, monitoring, and safety components have all been increased in size from the small-cabin system, but most important, we've added an inverter for conventional AC power output. The majority of household electrical needs are run by the inverter, allowing conventional household wiring and a greater selection in lighting and appliance choices. We've found that when the number of lights gets above five, AC power is much easier to wire, plus fixtures and lamps cost significantly less due to mass production, so the inverter pays for itself in appliance savings.

Often, with larger systems like this, we combine and preassemble all the safety, control, and inverter functions using an engineered, UL-approved power center. This isn't a necessity, but we've found that most folks appreciate the

The typical Real Goods full-time system takes only 4'x8' of floor space inside your utility room.

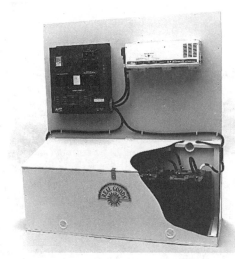

Handy Formulas for Estimating Household Renewable Energy Installations

[Photovoltaic (PV) array size (watts)] x [solar radiation (hours/day)] x [system efficiency] = [system output (watt-hours/day)]

Off-Grid Solar

[Average daily electric usage (watt-hours/day)] ÷ [solar radiation (hours/day)] ÷ [55% off-grid system efficiency] = [PV watts required]

Ballpark estimate:
[PV array size (W)] x 3 = [Output (Wh/day)]
[Output (Wh/day)] x ⅓ = [PV array size (W)]

On-Grid Solar

[Average daily electric usage (kilowatt-hours/day)] ÷ [solar radiation (hours/day)] ÷ [70% on-grid system efficiency] = [PV kilowatts required]

Ballpark estimate:
[PV array size (kW)] x 4 = [Output (kWh/day)]
[Output (kWh/day)] x ¼ = [PV array size (kW)]

1kW = 75 sq. ft. of PV panels

1MW (system rating) of PV energy powers 130 homes at the U.S. average of 31kWh/day (220 homes in California at 18kWh/day average)

1MW (system rating) of wind energy powers 250 average U.S. homes (450 homes in California)

Wholesale Cost of Producing Electricity

Coal = 4¢/kWh
Natural gas = 6¢/kWh
Wind = 7¢/kWh
PV = 14¢/kWh (Real Goods residential cost)
Hydro = 11¢/kWh
Geothermal = 11¢/kWh
Nuclear = 14¢/kWh
Centralized PV = 15¢/kWh
(Source: "Solar Revolution" by Travis Bradford)

In 2005, about 50% of the electricity produced in the U.S. was from coal, 20% from nuclear, 16% from natural gas, and 10% from renewables (mostly hydro). Only about one-third of this power actually reached the consumer—the rest was lost along the way (due to conversion, transmission, distribution, etc.). Source: eia.doe.gov.

Battery Bank Sizing

[kWh/day] x [3-5 days of storage] x 3 = [kWh size for battery bank]

Charge Controller Sizing

[PV Short-circuit current amps] x 1.56 = [Total amp size]

Fusing/Breakers Sizing

[Short-circuit current amps] x 1.56 = [Fuse/breaker amp size]

---

tidy appearance, fast installation, UL approval, and ease of future upgrades that preassembled power centers bring to the system.

Because family sizes, lifestyles, local climates, and available budgets vary widely, the size and components that make up a larger residential system are customized for each individual application. System sizing is based on the customer's estimate of needs and an interview with one of our friendly technical staff. See the "System Sizing" hints and worksheets in Chapter 4 or in the Appendix.

A UTILITY INTERTIE SYSTEM WITH BATTERIES

In this type of intertie system, the customer has both a renewable energy system and conventional utility-supplied grid power. Any renewable energy beyond what is needed to run the household and maintain full charge on an emergency back-up battery bank is fed back into the utility grid, earning money for the system owner. If household power requirements exceed the PV input, e.g., at night or on a cloudy day, the shortfall is automatically and seamlessly made up by the grid. If the grid power fails, power will be drawn instantly from the back-up batteries to support the household. Switching time in case of grid failure is so fast that only your home computer may notice. This is the primary difference between intertie systems with and without batteries. Batteries will allow continued operation if the utility fails. They'll provide back-up power for your essential loads and will allow you to store and use any incoming PV energy.

A number of federal and state programs exist to hasten this emerging technology, and an increasing number of them have real dollars to

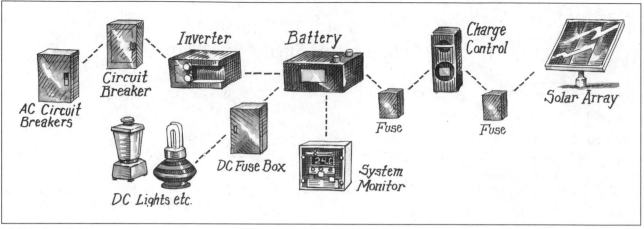

AC Circuit Breakers — Circuit Breaker — Inverter — Battery — Charge Control — Solar Array — Fuse — Fuse — System Monitor — DC Fuse Box — DC Lights etc.

A full-size household system has all these parts.

spend! These dollars usually appear as refunds, buydowns, or tax credits to the consumer—that's you. Programs and available funds vary with time and state. For the latest information, call your State Energy Office, listed in the Appendix, or check the Database of State Incentives for Renewable Energy on the Internet at www.dsireusa.org.

System Sizing

We've found from experience that there's no such thing as "one size fits all" when it comes to energy systems. Everyone's needs, expectations, budget, site, and climate are individual, and your power system, in order to function reliably, must be designed with these individual factors accounted for. Our friendly and helpful technical staff, with over 60,000 solar systems under its collective belt, has become rather good at this. We don't charge for this personal service, so long as you purchase your system components from us. We do need to know what makes your house, site, and lifestyle unique. So filling out the household electrical demands portion of our sizing worksheets is the first very necessary step, usually followed by a phone call (whenever possible) and a customized system quote. Worksheets, wattage charts, and other helpful information for system sizing are included at the end of Chapter 4, our "panel to plug" chapter—which also just happens to cover batteries, safety equipment, controls, monitors, and all the other bits and pieces you need to know a little about to assemble a safe, reliable renewable energy system.

What's It Going to Cost Me to Go Solar?

Three easy steps to get a ballpark calculation for utility-tied systems:*

1. Find your daily utility usage by dividing the kilowatt-hours (kWh) used on an average month's utility bill by 30.

2. Divide that number by 5 (the average number of peak Sun hours in the U.S.) and multiply by 1.43 to account for system losses. This is the size of the solar system, in kilowatts, that you will need for taking care of 100% of your electrical needs.

3. Multiply that number by $9,000 ($9/watt installed) for a good ballpark idea of the gross installed cost.

Can state rebate incentives take a chunk out of that price? Go to www.dsireusa.org to find out what grants or incentives are available in your state. For instance, in California, you can multiply your gross installed cost by 0.65 to account for rebates and tax credits. In New York or New Jersey, multiply by 0.5. For commercial systems, multiply by 0.3.

What ongoing savings can I expect?
Whatever you're now paying the utility for electricity will change to $0 (service charges will still apply).

Call our techs at 800-919-2400 for more information on how solar can work for your house.

For off-grid systems, roll up an estimated watt-hour calculation using our system sizing worksheet on pages 173-175 or page 579.

PHOTOVOLTAIC PRODUCTS

System Design Tools

Solar Pathfinder (Rent or Buy)

It takes only a fist-size chunk of shade to effectively turn off a solar module. Wondering if that tree or building will shade your array during certain times of the year? In less than two minutes on site, the Pathfinder will show exactly what times of day and months of the year

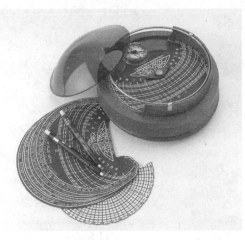

shade will fall on any site. This wonderfully simple tool requires only that you stand on the site, face south (there's a built in compass with magnetic declination adjustment), level the Pathfinder (built-in bubble level), and read the shade reflections on the dome. The chart under the dome shows hours and months that any bit of shade will be a problem. Every professional needs to own one of these. Homeowners probably need it only once, so we've got a nice $25/week rental program. Credit card required for deposit. The handheld model is just the Pathfinder head. The professional model adds an adjustable tripod mount and protective custom-metal carrying case. The Pathfinder Assistant software allows you to use your PC to do the somewhat tedious calculations automatically. Just take an overhead picture of the dome, download it to your PC, and the software does the rest. USA.

11605	Weekly Pathfinder rental	$25
11603	Handheld Pathfinder	$250
11604	Professional Pathfinder w/ tripod & metal case	$360
90-0191	Pathfinder Assistant software	$89

Solmetric SunEye™ Solar Access and Shade Analysis Tool

Similar in function to the Solar Pathfinder, the Solmetric SunEye adds a built-in fisheye lens, digital camera, and touch-screen display. This handheld tool gives the professional installer the ability to perform on-site data collection and analysis. The site readings are stored for later transfer to a PC. The SunEye automatically captures the horizon, detects unobstructed sky and shading obstacles, maps Sun paths onto a skyline based on location, calculates shading, accounts for latitude/longitude and magnetic declination, and stores data for more than 50 different site readings. The included desktop companion software allows you to review and export data and automatically generate complete reports for job quotes or rebate applications. 6"x8½". 2 lb. 30-day money-back guarantee. 1-yr. warranty. Made from recycled and refurbished materials. Case included. USA.

90-0500 Solmetric SunEye   **$1,255**

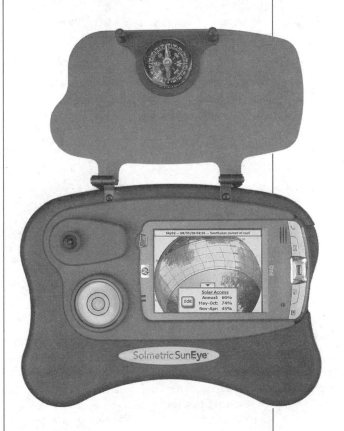

Photovoltaic Products

Evergreen Spruce Series PV Modules

Evergreen's newest product, the Spruce series, is optimized for grid-tie systems. These modules also work well with high-voltage-input MPPT charge controllers such as the OutBack Power MX60. Each Spruce module features 108 cells per panel, antireflection cover glass, clear anodized-aluminum frames with a 37.5-inch mounting width and 62-inch length, and watertight junction boxes that never need maintenance. Panels come in 170-, 180-, and 190-watt capacity. Maximum power up to 4% above rated; minimum power only 2% below rated. And Evergreen Spruce modules have multicontact cables already installed for easy installation and series wiring.

When it comes to environmental credentials, each Evergreen Spruce module features low-energy payback (its manufacturing impact is recouped in 18 months), low carbon output because it emits only 30g of carbon dioxide per equivalent kWh, and low lead construction because lead-free solder is used in the inter-cell connections. All Evergreen modules are UL-listed. 25-yr. limited power warranty; 2-yr. workmanship warranty. USA/Germany.

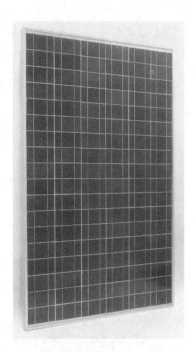
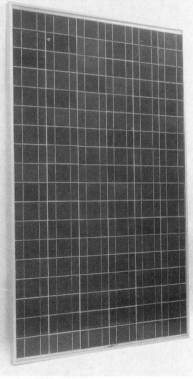
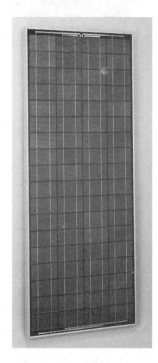

Evergreen	170	180	190
Rated power	170W	180W	190W
Voltage @ max. pwr.	25.3V	25.9V	26.7V
Current @ max. pwr.	6.72A	6.95A	7.12A
Open-circuit voltage	32.4V	32.6V	32.8V
Short-circuit current	7.55A	7.78A	8.05A
Series fuse	15A	15A	15A
LxWxD (inches)	61.8x 37.5x1.6	61.8x 37.5x1.6	61.8x 37.5x1.6
Weight	40.1 lb.	40.1 lb.	40.1 lb.
Price per watt	**$5.76**	**$5.72**	**$5.68**
Item #	**41-0255**	**41-0256**	**41-0257**
Price	**$979**	**$1,029**	**$1,079**

Evergreen Cedar-Series PV Modules

Evergreen Solar is a rapidly growing company with an innovative String Ribbon™ polycrystalline production technology. String Ribbon produces a silicon wafer by drawing up a film of molten silicon between two strings, like a soap bubble. Production is a simple, nearly continuous process. With none of the usual slicing or dicing required to get thin blanks of silicon to build PV cells on, there's very little waste, just fast, efficient cell production. That results in PV modules that use less energy and waste fewer raw materials in the process. Like all polycrystalline modules, this is a very stable power production material with a life expectancy measured in decades. As they've gained experience with String Ribbon production, Evergreen has been producing larger cells, resulting in higher wattage modules. Features include standard 36 cells in series configuration, tempered-glass covers, clear-anodized aluminum frames with standardized 24" mounting width, Tedlar backsheets and roomy watertight junction boxes. All Evergreen modules are UL-listed and feature a great 25-yr. warranty. USA.

Evergreen	EC-120
Rated power	120 watts
Voltage (@ max. pwr.)	17.6 volts
Current (@ max pwr)	6.782 amps
Open circuit voltage	21.5 volts
Short circuit current	7.68 amps
Series fuse	15 amps
LxWxD (inches)	62.4x25.7x1.4
Weight	28.0 lb./12.7kg
Price per watt	$5.79/W
Item #	**41-0205**
Price	**$695**

SHARP PV MODULES

Low prices and now made in the USA.

Since 2000, Sharp Solar has been the market leader in worldwide solar cell production. Sharp PV modules for all U.S. sales are produced at their new Tennessee plant, which went into production in 2003. When it comes to grid-tie PV systems, our most popular choices are the Sharp polycrystalline 170- and 208-watt modules and the monocrystalline 180-watt module because all of them feature watertight MC output cables for quick series connections and high power output. The 187-, 140-, and 70-watt modules are all polycrystalline with black anodized frames for lower visibility. The 70-watt modules are the PV industry's first triangular modules and come in left or right configurations. The 70- and 140-watt modules match and are designed to be used together. All Sharp modules feature extremely stout frames; tempered, low-reflection glass covers; built-in bypass diodes; and 25-year warranties. The 187-watt modules (with low visibility, black anodized frame construction) are available only in conjunction with specific mounting gear—call for details. If you've got a smaller, battery-based system, the 208-, 187-, 70-, and 140-watt modules aren't the first choice unless you use them with a combiner box and an MPPT charge controller. This is because they have lower voltages (no nominal rating) and come without junction boxes. 25-yr. mfr. warranties. USA.

Sharp	170	180	187	208
Rated power	170W	180W	187W	208W
Voltage @ max. pwr.	34.8V	35.9V	25.6V	28.5V
Current @ max. pwr.	4.9A	5.02A	7.3A	7.3A
Open-circuit voltage	43.2V	44.8V	32.5V	36.1V
Short-circuit current	5.47A	5.6A	8.13A	8.13A
Series fuse	10A	10A	15A	15A
LxWxD (inches)	62x32.5x1.8	62x32.5x1.8	58.3x39.1x2.3	64.6x39.1x1.8
Weight	37.5 lb./17kg	37.5 lb./17kg	39.6 lb./18kg	46.3 lb./21kg
Price per watt	$5.64	$6.05	$5.48	$5.52
Item #	41-0194	41-0221	41-0225	41-0195
Price	$959	$1,089	$1,025	$1,149

Model	140	70
Rated power	140W	70W
Voltage @ max. pwr.	19.95V	9.98V
Current @ max. pwr.	0.02A	7.02A
Open-circuit voltage	24.85V	12.43V
Short-circuit current	7.81A	7.81A
Series fuse	15A	15A
LxWxD (inches)	45.9x39x1.8	45.9x39x1.8
Weight	32 lb.	26.9 lb.
Price per watt	$4.89	$6.64
Item #	41-0116	41-0115 (L or R)
Price	$685	$465

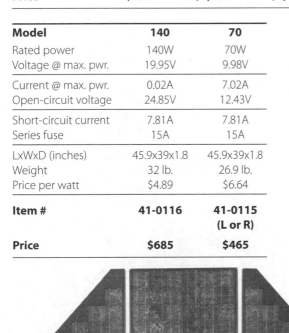

Sharp 80- and 123-Watt PV Modules

Using weathertight J-boxes, these high-efficiency polycrystalline modules are a great choice for battery-based systems. Featuring over 14% cell efficiency, which keeps the footprint smaller and the wattage higher, these 12-volt nominal modules also have an antireflective coating to increase light absorption, a textured cell surface to reduce reflection, terminal strips for easy electrical connections, and built-in bypass diodes. Construction is conventional with tempered glass, EVA encapsulation, and full aluminum frames. Uses the standard 36 cells in series, so these will intermix with any other 36-cell modules. UL-listed. 25-yr. mfr. warranty. Assembled in USA.

Sharp	80	123
Rated power	80W	123W
Voltage @ max. pwr.	17.1V	17.2V
Current @ max. pwr.	4.67A	7.16A
Open-circuit voltage	21.3V	21.3V
Short-circuit current	5.3A	8.1A
Series fuse	10A	15A
LxWxD (inches)	47.3x21.4x1.4	59x26.1x1.8
Weight	18.7 lb./8.5kg	30.9 lb./14kg
Price per watt	$6.19	$5.68
Item #	41-0177	41-0117
Price	$519	$729

SANYO PHOTOVOLTAIC MODULES

Sanyo 190- to 205-Watt PV Modules

Sanyo HIT (Heterojunction with Intrinsic Thin layer) solar cells are hybrids of single-crystalline silicon surrounded by ultrathin amorphous silicon layers. As a result, these Sanyo solar panels are a leader in cell and module efficiency. With up to 16.2 watts per square foot (17.4% module efficiency), you obtain maximum power within a fixed amount of space. You save costs for using fewer support materials and wiring, and you spend less time installing. The powerful modules are ideal for grid-connected solar systems, areas with performance-based incentives, and renewable energy credits. Another advantage to the HIT design is lower de-rating due to temperature. As temperatures rise, Sanyo HIT solar panels produce more electricity than conventional crystalline silicon solar panels at the same temperature. The modules are lightweight (31 lb.) and come with a 20-yr. mfr. warranty. USA/Mexico.

Sanyo	HIP-190BA3	HIP-195BA3	HIP-200BA3
Rated power	190W	195W	200W
Voltage @ max. pwr.	54.8V	55.3V	55.8V
Current @ max. pwr.	3.47A	3.53A	3.59A
Open-circuit voltage	67.5V	68.1V	68.7V
Short-circuit current	3.75A	3.79A	3.83A
Series fuse	15A	15A	15A
LxWxD (inches)	51.9x35.2x1.4	51.9x35.2x1.4	51.9x35.2x1.4
Weight	30.9 lb./14kg	30.9 lb./14kg	30.9 lb./14kg
Price per watt	$6.29	$6.39	$6.48
Item #	**41-0207**	**41-0301**	**41-0209**
Price	**$1,195**	**$1,245**	**$1,295**

Cosmetic Blems or Seconds in PV

From time to time Real Goods comes across PV modules with full output, but with slight cosmetic defects. Typically these have minimal or no warranty, but we've never had a power output problem with them after selling thousands of them. Prices can be as low as 40%-50% below PV's at retail. Call our techs at 800-919-2400 or the Hopland Store at 707-744-2100 ext 2106 for availability.

KYOCERA PHOTOVOLTAIC MODULES

Kyocera 40- and 50-Watt PV Modules

Kyocera's advanced cell-processing technology and automated production produces very highly efficient polycrystalline modules. These conventional 36-cell modules feature tempered glass covers, EVA encapsulant, polyvinyl backsheets, with a rugged full-perimeter anodized aluminum frame, and a weathertight junction box with three ½-inch knockouts. They will intermix with other 36-cell modules for battery charging. These mid-size 40- and 50-watt modules fill a need for modest power requirements at reasonable prices and are an excellent choice for high-volume water features and high-performance attic and greenhouse fans. 25-yr. mfr. warranty. Japan.

Kyocera	KC40T	KC50T
Rated power	40W	50W
Voltage @ max. pwr.	17.4V	17.4V
Current @ max. pwr.	2.48A	3.11A
Open-circuit voltage	21.7V	21.7V
Short-circuit current	2.65A	3.31A
Series fuse	6A	6A
LxWxD (inches)	20.7x25.7x2.1	25.2x25.7x2.1
Weight	9.9 lb./4.5kg	11.0 lb./5.0kg
Warranty	Within 20% of rated output for 25 yr.	Within 20% of rated output for 25 yr.
Price per watt	$6.92	$6.90
Item #	**41-0128**	**41-0129**
Price	**$277**	**$345**

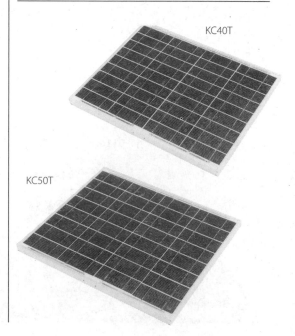

KC40T

KC50T

Uni-Solar Triple Junction PV Technology

Uni-Solar delivers a nice breakthrough in lower-cost, higher-output amorphous film technology. For higher efficiency and lower production costs, each cell is composed of three spectrum-sensitive semiconductor junctions stacked on top of each other: blue light-sensitive on top, green in the middle, red on the bottom. Like all Uni-Solar products, these flexible, unbreakable modules are deposited on a stainless steel backing with DuPont's Tefzel glazing and, due to their unique construction, have the ability to continue useful output during partial shading. The unbreakable construction, good resistance to heat fading, and shade tolerance make these modules an excellent choice for RVs or any site where vandalism is a possibility. The new triple-junction technology brings the price down to a level equal to or better than single-crystal technology.

PV ES-62T

Uni-Solar PV Modules

The venerable US-64 is designed for battery-based systems and has a conventional J-box with terminal strip and a clear anodized frame. The ES-62 is designed for grid-tie with MC connectors and a black anodized frame. Like all amorphous modules, power output will fade approximately 15% over the first year or two. Power was rated after fading. Warranted within 20% of output for 20 years. UL- and CUL-listed. USA.

PV US-64T

Model	ES-62	US-64
Rated watts	62.0W @ 25°C	64.0W @ 25°C
Rated power	15V	16.5V @ 3.88A
Open-circuit voltage	21V	23.8V @ 25°C
LxWxD (inches)	49.5x31.2x1.25	53.8x29.1x1.25
Construction	Anodized aluminum Tefzel glazing	Triple-junction Tefzel glazing
Warranty	20 yr.	20 yr.
Weight	24 lb. 20.2 lb.	
Item #	**41-0220**	**11-228**
Price	**$399**	**$419**

Small PV Panels

GSE Flexible Modules

These weatherproof solar solutions fold to about the size of a paperback book to create a lightweight, portable power station for campers, hikers, boaters, and others on the go. Just open under full Sun and charge lanterns, cell phones, GPS units, iPods®, cameras, and any other device that has a 12-volt car adapter. With grommets to attach them to all kinds of gear, they're trail-tough and sized to pack anywhere. 6.5-, 12-, and 25-watt models have a built-in voltage cap to permit a direct charge to small devices and prevent damage from low-light reverse power flows. These three modules also include a connectivity kit with 12-volt receptacle, 12-volt vehicle power outlet, battery clamps, 4-inch barrel plug, and 8-foot extension cable.

Model	6.5	12	25	55
Rated power	6.5W	12W	25W	55W
HxWxD (inches folded)	11x9x1	9x5x⁷⁄₁₀	11x8¼x⁷⁄₁₀	11x9x1
Weight	0.45 lb.	0.7 lb.	1.8 lb.	3.5 lb.
Item #	41-0249	41-0249	41-0249	41-0250
Price	$99	$195	$399	$959

41-0251	**Extra Connectivity Kit**	**$19**
41-0252	**7A Charge Controller**	**$39**

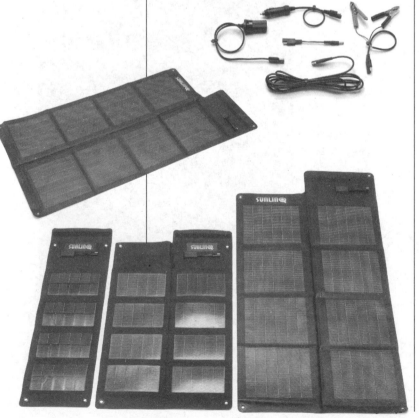

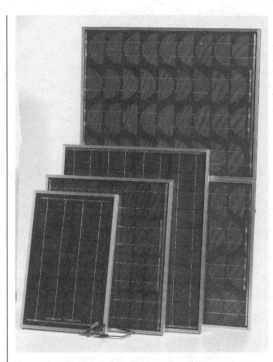

Ever Step Small Solar Modules

Small standalone solar panels have always been hard to find, so we've secured inventory with one of our manufacturing partners to bring you these single-crystal, tempered-glass modules at an affordable price. Each features weathertight construction, sturdy anodized aluminum frames with four adjustable mounting tabs, built-in reverse-current diode, and 7-foot wire pigtail output with color-coded battery clips. Great for trickle battery recharging on RVs, farm equipment, and boats, and perfect for household battery charging or backyard fountains. See chart for specs. 90-day mfr. warranty. China.

Watts	9W	13W	18W	27W	36W
Current rated power	500mA	710mA	1,000mA	1,500mA	2,000mA
Voltage	18V	18V	18V	18V	18V
LxWxD (inches)	16x 10.2x1	15.1x 14.4x1	15.1x 17.67x1	21.2x 17.67x1	35.4x 21.2x1
Weight	3.08 lb.	3.5 lb.	4.4 lb.	6.6 lb.	13.2 lb.
Item #	41-0216	17-0170	41-0216	41-0216	41-0216
Price	$99	$135	$179	$199	$255

PowerFilm® Rollable Solar Modules

Who says you can't take it with you? These PowerFilm lightweight, water-resistant panels make it easy to pack a solar power system on land or sea. Based on NASA technology, the PowerFilm flexible solar panels unroll anywhere to charge and maintain 12-volt batteries and portable electronics, including GPSes, PDAs, cell phones, and laptops (with optional female cigarette-lighter adapter). Flex them to conform to a surface and use the corner grommets to keep them there. Link multiple panels for even greater output. Includes 15-foot extension wire with O-ring connectors. Panels roll up to about a 4-inch diameter and are as durable as they are flexible. USA.

41-0198 Female Cigarette-Lighter Adapter $10

Model	5W	10W	20W
Operating voltage	15.4W	15.4W	15.4W
Operating current	0.3A	0.6A	1.2A
Weight	0.6 lb.	1 lb.	1.9 lb.
LxWxD rolled (inches)	11.5x4x3.75	11.5x4.25x4.25	12x4.25x4.5
LxWxD unrolled (inches)	11.5x21	11.5x38	12x73
Item #	41-0197	41-0197	41-0197
Price	$139	$229	$399

Solar Car Battery Chargers

No matter how long your vehicle sits between starts, keep your battery ready with a Sun-powered trickle charge. Model 135 connects to car and truck batteries via a 2-volt cigarette lighter plug to provide a generous 135mA, 15V charge—even through windshields and clouds. The 400 provides a 400mA, 15V flow to trucks, tractors, RVs, and medium-size boats. For larger applications, two or more Model 400 panels connect in seconds. Both models are weather/rust/shock/UV-resistant and include cigarette lighter adapters, battery clamps, and suction cups for "plug-and-play" use. Model 400 has a charge indicator and also includes terminal rings and connector wires. Not intended for unprotected outdoor use. The 135 is 0.75"x5"x15.5" (HxWxL). The 400 is 1.25"x15"x19" (HxWxL). U.K./China.

17-0339 Model 135 Solar Car Charger $39
17-0339 Model 400 Solar Car Charger $99

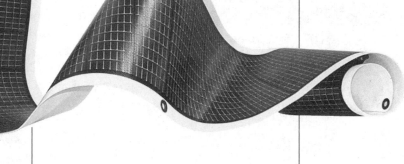

Off-Grid Solar Electric Packages for Home, Cabins/RVs, and Sign-Lighting Applications

(For grid-tie kits, see Chapter 4, Utility Intertie section)

Sol-Provider PV System for Off-Grid Homes

This full-size household system is configured to be expandable and code compliant. The basic system consists of 6 pole-mounted Sharp 180W PV modules, combiner box, OutBack VFX3648 sine-wave inverter, MidNite Solar E-Panel with AC and DC breakers, MX60 MPPT charge controller, OutBack Mate and Tri-Metric Monitors, and enough reserve battery storage for 3 days. Based on the U.S. average annual solar insolation of 5 hours per day, the Starter Kit's estimated output is about 3.1kWh per day. The battery bank consists of 16 L16 batteries and all the required cables.

The system can be easily expanded to triple the output by adding up to 12 more Sharp 180s and 2 more pole mounts. Expansion beyond that requires additional inverters, E-Panels, and MX60's as well. The 1.1kW Expansion Kit includes 6 Sharp 180s, 1 pole-top mount, 2 MC cables, and 2 10A combiner box circuit breakers. Be sure to include an optional autotransformer for 240-volt AC loads such as a well pump or if additional inverters are used. Up to 2 Battery Bank Expansion Kits can be added to the system. It's best to add these initially or at least within 6 months of the Starter Kit battery bank.

While this kit includes the major equipment and parts needed, you will still need to supply certain small parts related to the installation, such as wire, conduit, pipe for the pole mounts, and enclosures/shelving for the batteries. Most jurisdictions will also require a building permit for electrical systems of this nature. See schematic of typical wiring hookup.

53-0500	**1.1kW Sol-Provider Starter Kit**	**$13,995**
53-0501	**1.1kW Sol-Provider Expansion Kit**	**$6,495**
53-0502	**8 L16 Battery Bank Expansion Kit w/Cables**	**$1,795**
49-0232	**OutBack Autotransformer w/Enclosure**	**$390**

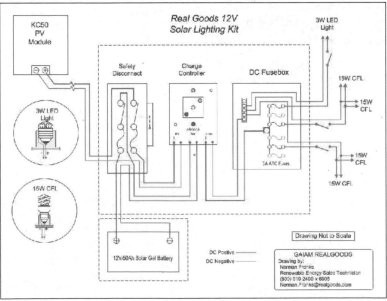

Solar Electric Cabin/RV Kits

For less money than you think, you can have dependable solar power at your remote cabin or vacation camp. Our kits provide everything you need for producing, regulating, and storing DC power safely. We have over 28 years' experience designing solar systems, so we've already done the hard work for you.

Run lights, stereos, computers, water-pressure pumps, and other modest appliances without having to start—and listen to—that noisy, polluting generator. You provide wire and small parts as required by your particular installation, plus DC appliances such as lights, pumps, or entertainment gear. An optional inverter to provide conventional 120 volts AC can be added to any of these kits. If you don't see exactly what you need, our experienced staff can help you create the perfect system at a reasonable price.

Weekend Getaway Kit for Cabins or RVs

Our smallest kit provides power for limited energy needs. The PV array can be doubled in the future without upgrading the controller. The no-maintenance sealed battery is optional (not needed for RVs). Produces about 400+ watt-hours per day of 12VDC output.

53-0106	**Getaway Kit**	**$949**
53-0107	**Getaway Kit w/Battery**	**$1,099**

Includes:
1 120W 12V PV module
1 flush-mount rack
1 BZ 250W MPPT charge controller
1 30A 2-pole safety disconnect w/fuses
1 30A fuse block w/fuse & spare
1 ATC-type DC fuse box w/20A fuses

Battery Option:
1 74Ah sealed gel storage battery

Extended Weekend Kit for Cabins or RVs

Twice the size of the Getaway Kit, the Extended Kit can handle families for longer periods than weekends. The PV array can be doubled in the future without upgrading the charge controller, which also provides digital monitoring of battery voltage, PV amps, and load amps. A combiner box and two breakers and mounting hardware are also required for the additional PV modules. The no-maintenance sealed battery pack is optional (not needed for RVs). Produces about 800+ watt-hours per day of 12VDC output. See schematic of typical wiring hookup.

53-0108	**Extended Weekend Kit**	**$1,899**
53-0109	**Extended Weekend Kit w/Battery**	**$2,499**

Includes:
2 120W PV modules
1 flush-mount rack for two modules
1 BZ 500W MPPT charge controller
1 30A 2-pole safety disconnect w/fuses & spare
1 6-circuit DC load center
4 circuit breakers (one 30A, three 15A)
1 110A Class T fuse & cover

Battery Option:
3 98Ah gel storage batteries w/interconnect cables

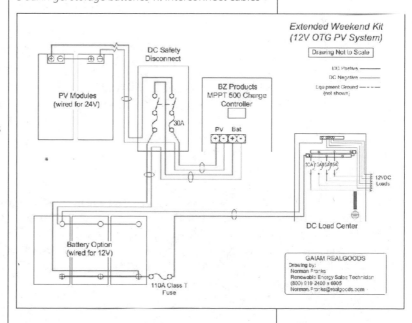

Deluxe Full-Time Cabin Kit

For medium-size weekend places or modest full-time cabins. It will power lights, stereos, TV, laptops, and, with an optional inverter, other small appliances like a small microwave or a desktop computer. The controller delivers maximum power-point tracking for peak PV output under all conditions. It also provides system monitoring with battery voltage and PV amps. You provide a 4-inch steel pipe for the PV array mount. The battery pack is optional. Can be setup initially as a 12- or 24-volt system. Produces about 1,600+ watt-hours per day of 12VDC output.

53-0110 Full-Time Cabin Kit $3,599

53-0111 Full-Time Cabin Kit w/Battery $4,999

Includes:
4 120W PV modules
1 PV combiner box w/fuses
1 30A 2-pole safety disconnect w/fuses & spare
1 pole-top fixed rack
1 BZ 500W MPPT charge controller
1 6-circuit DC load center
4 circuit breakers (one 30A, three 15A)
1 110A Class T fuse & cover

Battery Option:
4 183Ah gel storage batteries w/interconnect cables

Marine Solar Kit for Boats or RVs

With solar power, your marine life can be truly independent. Solar power can keep your galley appliances, communications, navigation gear, running lights, and pumps all running happily. The marine kit employs an unbreakable, shade-tolerant Uni-Solar module with anodized aluminum frame. A 12-amp waterproof charge controller attaches directly to the PV module J-box for a simplified installation. The Universal Battery Monitor shows system state of charge at a glance with red/yellow/green LEDs. The PV array can be doubled in the future without upgrading the controller. Mounting structure, battery, and DC fuse box are not included with this kit. 12 volts DC output.

Includes:
1 Uni-Solar US64 PV Module
1 Morningstar Sunkeeper 12A Charge Controller
1 Universal Battery Monitor
1 fuse holder and 5 20A fuses

53-0114 Marine Solar Kit $550

Solar Floodlights for Signs and More

Our solar-powered sign lighting kits share a core group of proven off-the-shelf components that are combined to deliver the run time you need. These are maintenance-free systems—set them up, turn them on, and they'll take care of themselves.

We provide outdoor floodlight(s), solar module(s), pole-top mount, charge/light controller, sealed battery, and fused safety disconnect. You provide mounting for the floodlight, box or shelter for the battery and control, 2-inch steel pipe for the PV module mount, and wiring/small parts as needed for your particular installation.

Kits state run time for 13-watt floodlight(s) based on 3.5 hours of sunlight, about the national wintertime average.

The Morningstar Sunlight 10 or 20 controller provides daytime charge control, turns on the light at dusk for a selected time period, and shuts it off if the battery gets too low. The sealed, maintenance-free battery provides 3 days of backup power with no Sun and will last 5-10 years. The solar module(s) and pole-top mount are sized to provide sufficient power to run the specified number of hours (actual performance may vary in your location).

24-Hour Solar Sign-Lighting Kit

We're obviously not running 24 hours a day. This kit runs a pair of 13-watt floodlights for 12 hours a day. Provides a pair of Sharp 80-watt modules, a pole-top mount, a pair of 98-ampere-hour sealed batteries, Morningstar Sunlight 20 controller, fused safety disconnect, and a pair of 13-watt floodlights.

**12254   24-Hour Solar
         Sign-Lighting Kit $1,695**

12-Hour Solar Sign-Lighting Kit

Our basic all-nighter lighting kit. Provides a Sharp 80-watt module with pole-top mount, 98-ampere-hour sealed battery, Morningstar Sunlight 10 controller, fused safety disconnect, and 13-watt floodlight.

12238   12-Hour Solar Sign-Lighting Kit $995

8-Hour Solar Sign-Lighting Kit

Provides a Kyocera 50-watt module with pole-top mount, 74-ampere-hour sealed battery, Morningstar Sunlight 10 controller, fused safety disconnect, and 13-watt floodlight.

12253   8-Hour Solar Sign-Lighting Kit $795

4-Hour Solar Sign-Lighting Kit

Provides a Kyocera 40-watt module with side-of-pole mount, 51-ampere-hour sealed battery, Morningstar Sunlight 10 controller, fused safety disconnect, and 13-watt floodlight.

12252   4-Hour Solar Sign-Lighting Kit $645

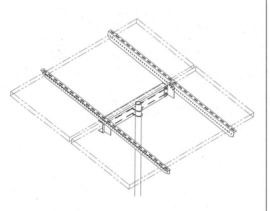

Real Goods 12V Solar Lighting Kit

This DC lighting kit was put together originally for a do-it-yourself article by Associate Editor Charles Higginson in the April/May 2007 issue of the *Mother Earth News*. It is similar to the Weekend Getaway Kit for Cabins described earlier but with a smaller PV module and battery to run 3 12-volt DC 15-watt CFLs and a 12-volt DC 3-watt LED. This kit is intended to get you started with solar without breaking the bank, while providing some useful back-up lighting during a power outage (for example) as well.

It consists of a flush-mounted Kyocera 50-watt PV module, a 20-amp Phocos charge controller, an automotive-type DC fuse box, a 50-ampere-hour gel battery, and the 12-volt DC light bulbs. Not included are the small parts needed such as wire, light fixtures, and battery enclosure. The solar module is sized to run the lights for 2 or 3 hours a day with enough battery backup for twice that (actual performance may vary depending on your location and the time of the year). See schematic of typical wiring hookup.

53-0123 Real Goods 12V Solar Lighting Kit $699

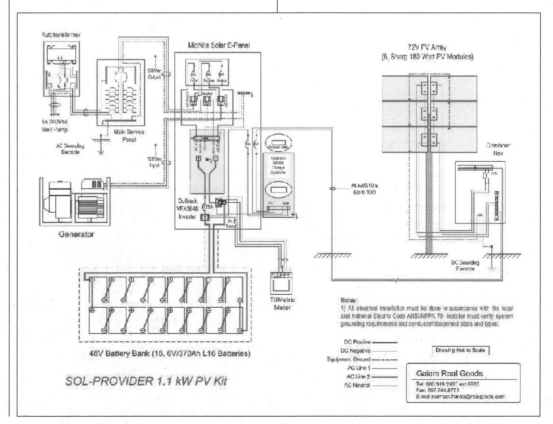

SOL-PROVIDER 1.1 kW PV Kit

Portable Power Packages

Small PV/Battery Packages for Portable Electronics

These small PV/battery packages are intended to provide power and/or recharging for cell phones, GPSes, laptops, or small battery-powered appliances with modest power uses. For larger systems, see the various portable battery packs covered under the Large Storage Batteries section in Chapter 4.

PowerDock Power Station for Mobile Gadgets

The PowerDock has everything you need for running laptops, cell phones, GPSes, and other modest-power gear in remote sites. This sturdy power station has a zip-out, unbreakable 15-watt PV panel and 9.2 ampere-hours of sealed 12-volt batteries. Also included are a pair of fused lighter socket outlets, a power meter to show state of charge, and a water-resistant canvas carrying case with five compartments for converters and accessories. Accommodates laptop computers up to 9"x15"x2". Add the 15-watt Solar Boost to double charging power. Plugs into one of the lighter sockets and folds into the PowerDock for travel. Robust construction; easy access to the batteries (lots of Velcro®). Has four batteries that should be replaced in 2-4 years. Weighs 14.5 lb., yet is comfortable to carry. This product is quality all the way! Comes with an AC wall-watt charger for when the Sun doesn't shine. 1-yr. mfr. warranty; 10 yr. on PV array. USA/Mexico.

53-0112 PowerDock   **$349**

41-0127 PowerDock 15W Boost   **$179**

Portable Solar Electronics Charger

This solar kit charges and runs iPods®, cell phones, digital cameras, GPSes, and PDAs wherever there's Sun. A rugged, hardcover-book-sized plastic case stores a 20-inch USB cable, various adapter jacks, and a 4.4-watt solar module. Open it to the Sun, plug in your gear, and power up miles from the nearest outlet. Includes a 12-volt car-lighter outlet for your gear's own adapters and an AA/AAA battery charger. 19 oz. 9³⁄₁₀"x6"x1½" (HxWxD). Philippines.

17-0340 Portable Solar Electronics Charger   **$139**

Solio Portable Solar Charger

Meet the next generation of solar chargers and our new top pick for portable power. Though its solar cells are rated at just 1 watt, the new Solio's high-capacity battery and onboard firmware provide a variable output as high as 8 watts—nearly double the needs of most gear. Three solar blades retract neatly to keep it pocket-sized, and it holds a charge for up to a year. Includes Nokia, Samsung, Motorola, and Mini USB tips (many others available); cable; AC adapter with international plug set; 12-volt female adapter; and suction cup. Optional urban-hip case made of recycled innertubes. 1"x4¾"x2½" (HxWxD). China.

17-0341 Solio Portable Solar Charger   **$109**

17-0343 Recycled Tube Solio Case   **$34**

Universal Solar Charger

It may be a wired world, but you don't have to be tethered to it! Our compact Universal Solar Charger is the lightest, fastest solar charger on the market because it delivers a straight charge to your iPod®, BlackBerry, or Nokia, Samsung, Sony, Ericsson, Siemens, or Motorola cell phone. It eliminates the need for internal battery storage. Fold it open, connect the mini-USB cord with the proper adapter tip (included), and you'll have a full charge in 2 or 3 hours of direct Sun. Maximum output is 6.58 volts-320mA in full Sun. Weighs only 3 oz. 5½"x3¼"x½" (LxWxD; 6"W open). Netherlands.

17-0336 Universal Solar Charger $119

Mobile DC Power Adapter

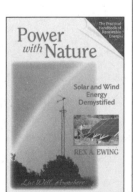

Use a 12-volt source to run and recharge any battery-powered device using less than 12 volts DC. Our adapter delivers stable DC power at 3, 4.5, 6, 7.5, 9, or 12 volts at up to 2 amps. Voltage varies less than 0.5 volt in our tests. Perfect for your camera, DVD or MP3 player, camcorder, Game Boy®, cell phone, or whatever. The 6-foot fused input cord with LED power indicator plugs into any lighter socket; the 1-foot output cord has six adapter plugs. A storage compartment in the power adapter holds your spare plugs to prevent loss. Slide switch selects voltage. China.

53-0105 Mobile DC Power Adapter $24

12-Volt DC PC Power Supplies

Run and recharge your PC from any 12-volt DC source. Available in 65- or 100-watt output with stabilized voltage adjustable from 15 to 21 volts. Output voltage varies less than 0.5 volt from zero to full load in our tests. External voltage adjustment is recessed for safety.

Two-foot fused input cord with LED power indicator plugs into any lighter socket. Four-foot output cord has six different adapters to fit almost any PC laptop with a round power socket. Will not recharge newer Macintosh products requiring 24-volt input. 65-watt unit has maximum 4-amp output. 100-watt unit has maximum 7-amp output. Weighs 8 oz. 3.8"x2.3"x1.5". Taiwan.

53-0103 65W $59
53-0104 100W $75

BOOKS

Power with Nature: Solar and Wind Energy Demystified

Author Rex Ewing is right up to date, explaining all the latest offerings like utility intertie systems and maximum power point tracking charge controls—all with extensive drawings, pictures, sidebars, and witticisms. You'll be enlightened by the wealth of practical, hands-on information for solar, wind, and micro-hydro power, plus commonsense explanations of the necessary components such as batteries, inverters, charge controllers, and more. 255 pages, softcover. .

21-0339 Power with Nature $25

There Are No Electrons

Do you know how electricity works? Want to? Read this book and learn about volts, amps, watts, and resistance. So much fun to read, you won't even notice you're learning. And if you already "know" about all this, you'll enjoy it even more—just be prepared to rethink everything. 322 pages, hardcover.

82646   There Are No Electrons $16

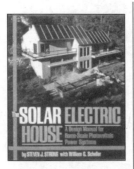

The Solar Electric House

By Steven Strong. The author has designed more than 75 PV systems. This fine book covers all aspects of PV, from the history and economics of solar power to the nuts and bolts of panels, balance of systems equipment, system sizing, installation, utility intertie, standalone PV systems, and wiring instruction. A great book for the beginner. 276 pages, paperback.

80800   The Solar Electric House $21.95

The Solar Electric Independent Home Book

By Paul Jeffrey Fowler. A good, basic primer for getting started with PV, written for the layperson. Lots of good charts, clear graphics, a solid glossary, and appendix. Includes 75 detailed diagrams, and the text includes recent changes in PV technology. This is one of the best all-round books on wiring your PV system. 174 pages, paperback.

**80102   The Solar Electric
             Independent Home Book $20**

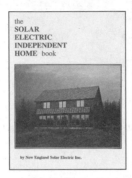

Wiring 12 Volts for Ample Power

The most comprehensive book on DC wiring to date, out of print until recently updated, and self-published by the authors. This book presents system schematics, wiring details, and troubleshooting information not found in other publications. Leans slightly toward marine applications. Chapters cover the history of electricity, fundamentals of DC and AC, battery loads, electric sources, wiring practices, system components, tools, and troubleshooting. By David Smead and Ruth Ishihara. 100 pages, softcover.

**80111   Wiring 12 Volts for
             Ample Power $19.95**

Photovoltaics

Used as a textbook in Solar Energy International workshops, *Photovoltaics* delivers the critical information to successfully design, install, and maintain PV systems. It covers solar electricity basics, PV applications, system components, site analysis, and mounting; how to size standalone and PV/generator hybrid systems; utility-interactive PV systems; system costs; safety issues; and much more. By Solar Energy International staff. 317 pages, softcover.

21-0366 Photovoltaics $59

National Electrical Code® 2005 Handbook

Produced by the NFPA, the popular *National Electrical Code® 2005 Handbook* contains the complete text of the 2005 edition of the NEC® supplemented by helpful facts and figures, full-color illustrations, real-world examples, and expert commentary. An essential reference for students and professionals, this handbook is the equivalent of an annotated edition of the 2005 NEC® that offers insights into new and more-difficult articles to guide users to success in interpreting and applying current code requirements to all types of electrical installations. A valuable information resource for anyone involved in electrical design, installation, and inspection, the *NEC® 2005 Handbook* is updated every 3 years and provides 100% of the information needed to "meet code" and avoid costly errors. 1,250 pages, hardcover.

**21-0525 National Electrical Code®
             2005 Handbook $125**

For the full NEC, please visit www.realgoods.com/nec.pdf.

Photovoltaic Mounting Hardware

A PV mounting structure will secure your modules from wind damage and lift them slightly to allow some cooling air behind them. Mounts can be as small as one module for an RV or big enough to carry hundreds of modules for a utility intertie system. PV systems are most productive if the modules are approximately perpendicular to the Sun at solar noon, the most energy-rich time of the day for a PV module. If you live in the Northern Hemisphere, you need to point your modules roughly south. The best year-round angle for your modules is approximately equal to your latitude. For better winter performance, raise that angle about 10°; for better summer performance, lower that angle about 10°.

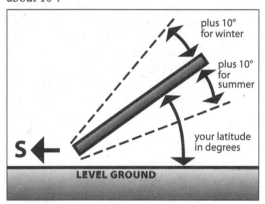

Proper PV mounting angle

Off-the-grid systems should probably have the modules oriented for best wintertime performance, as this is typically when they are most challenged for power delivery. On-grid, utility-intertied systems are usually set up for best summer performance. Most utilities allow credits to be rolled over from one month to the next. We'll make the most of those long summer days to deliver the maximum kilowatt-hours for the year.

You can change the tilt angle of your array seasonally as the Sun angle changes, but on a practical level, many residential systems will have power to burn in the summer. Most folks have found seasonal adjustments to be unnecessary.

The perfect mounting structure would aim the PV array directly at the Sun and follow it across the sky every day. Tracking mounts do this, and in years past, Real Goods has enthusiastically promoted trackers. But times change. The electrical and/or mechanical complexity of tracking mounts assures you of ongoing maintenance chores (ask us how we know . . .), and

the falling cost of PV modules makes trackers less attractive. For most systems now, an extra module or two on a simple fixed mount is a much better investment in the long run.

In the following pages, you'll find numerous mounting structures, each with its own particular niche in an independent energy system. We'll try to explain the advantages and disadvantages of each style to help you decide if a particular mount belongs in your system.

In ascending order of complexity, your choices are:
- RV Mounts
- Home-Built Mounts
- Fixed Roof or Ground Mounts
- Top-of-Pole Fixed Mounts
- Passive Trackers
- Active Trackers

RV MOUNTS

Because of wind resistance and never knowing which direction the RV will be facing next, most RV owners simply attach the module(s) flat on the roof. RV mounts raise the module an inch or two off the roof for cooling. They can be used for small home systems as well. Simple and inexpensive, most of them are made of aluminum for corrosion resistance. Obviously, they're built to survive high wind speeds. These are a good choice for systems with a module or two. For larger systems, the fixed or pole-top racks are usually more cost effective.

HOME-BUILT MOUNTS

Want to do it yourself? No problem. Small fixed racks are pretty easy to put together. Anodized aluminum or galvanized steel are the preferred materials due to corrosion resistance, but mild steel can be used just as well, so long as you're willing to touch up the paint occasionally. Slotted steel angle stock is available in galvanized form at most hardware and home-supply stores and is exceptionally easy to work with. Wood is not recommended, because your PV modules will last longer than any exposed wood. Even treated wood won't hold up well when exposed to the weather for over 40 years. Make sure that no mounting parts will cast shadows on the modules. Adjustable tilt is nice for seasonal angle adjustments, but most residential systems have power to spare in the summer, and seasonal adjustments are usually abandoned after a few years.

The best year-round angle for your modules is approximately equal to your latitude. For better winter performance, raise that angle about 10°; for better summer performance, lower that angle about 10°.

FIXED ROOF OR GROUND MOUNTS

This is easily the most popular mounting structure style we sell. These all use the highly adaptable SolarMount extruded aluminum rails as their base. This à la carte mounting system offers the basic rails in various lengths and strengths (Lite or Standard). Buy enough rails to fit your particular array, then add optional roof standoffs and/or telescoping back legs for seasonal tilt and variable roof pitches. This mounting style can be used for flush roof arrays, low-profile roof arrays with a modest amount of tilt, high-profile arrays that stand way up, ground mounts in either low or high profile, or even flipped over on south-facing walls. Use concrete footings or a concrete pad for ground mounts. The racks are designed to withstand wind speeds up to 100 miles per hour or more. They don't track the Sun, so there's nothing to wear out or otherwise need attention. Getting

snow off of them is sometimes troublesome; it can pile up at the base. Ground mounting can leave the modules vulnerable to grass growing up in front, or to rocks kicked up by mowers. A foot or so of elevation or a concrete pad is a good idea for ground mounts. Note that attaching these racks to the roof will require roof penetrations every 4-6 feet. For example, a 3kW PV array might need 20-24 support points, so the skill of the installer is important here to ensure a leak-free roof.

For larger, flat roof-mounted systems (usually on commercial buildings), several ballast-type racks are available these days. If roof penetrations are undesirable, then securing the mounting hardware with heavy weights (such as concrete blocks or containers of sand) or the weight of the hardware—ballasts—is the answer. The disadvantage of this type of mounting is usually the high cost, even though installation is not too difficult. High-wind or seismic areas will probably also require some sort of mechanical attachment to the roof anyway.

Finally, a mention of BIPV (Building Integrated PV) products should be included here, even though these roof PV modules don't require mounting hardware at all. Intended primarily for new construction, these modules actually are part of the roof covering. They provide an architecturally pleasing look that blends in seamlessly with the rest of the home or building. One type is the frameless roof-integrated module, such as the UniSolar Shingle, which looks much like a typical composition shingle. Another type is more like a standard framed polycrystalline module but is sized and installed like a roof tile. The main advantage of BIPV products is that they are unobtrusive and usually look pretty good if aesthetics are important. Disadvantages include their higher cost and low efficiency due to higher operating temperatures (lack of cooling air circulation). For more information about BIPV and ballast mounts, be sure to call one of our Solar Technicians.

TOP-OF-POLE FIXED MOUNTS

A popular and cost-effective choice, pole-top mounts are designed to withstand winds up to 80 mph and in some cases up to 120 mph. The UniRac rails for the larger pole-top arrays are a heavier-duty version of the standard SolarMount rails. This mounting style is a good choice for snowy climates. With nothing underneath it, snow tends to slide right off.

For small or remote systems, pole-top mounts are the least expensive and simplest choice. We

Do Tilt Angle or Orientation Matter?

Much has been made in years past of PV tracking and tilt angle. You would think that it's nearly a life-or-death matter to point your PV modules *directly* at the Sun during all times of the day. We're here to shake up this belief. Tilt angle and orientation make a lot less difference than most folks believe.

From the Sandia National Labs comes this very interesting chart, which details tilt angle vs. compass orientation, and the resulting effect on yearly power production. South-facing, at a 30° angle (7:12 roof pitch) delivered the most energy. They labeled that point 100%. All other orientations and tilt angles are expressed as a percentage of that number. Note that we can face SE or SW, a full 45° off due south, and lose only 4%. Our tilt angle can be 15° off, and we lose only 3%. In reality, we can face due east or west and lose only 12%! A couple of bird droppings could cost you more energy.

ROOF SLOPE AND ORIENTATION (Northern California Data)

	Flat (0°)	4:12 (18.4°)	7:12 (30°)	12:12 (45°)	21:12 (60°)	Vertical (90°)
South	0.89	0.97	1.00	0.97	0.89	0.58
SSE, SSW	0.89	0.97	0.99	0.96	0.88	0.59
SE, SW	0.89	0.95	0.96	0.93	0.85	0.60
ESE, WSW	0.89	0.92	0.91	0.87	0.79	0.57
East, West	0.89	0.88	0.84	0.78	0.70	0.52

The moral of this story? Shadows, bird droppings, leaves, and dirt will have far more effect on your PV output than orientation. Keep your modules clean and shade free, and don't worry if they aren't perfectly perpendicular to the Sun at noon every day. You'll do fine.

To Track or Not to Track

Photovoltaic modules produce the most energy when situated perpendicular to the Sun. A tracker is a mounting device that follows the Sun from east to west and keeps the modules in the optimum position for maximum power output. At the right time of year, and in the right location, tracking can increase daily output by more than 30%. But beware of the qualifiers: Trackers are often *not* a good investment.

Trackers work best during the height of summer, when the Sun is making a high overhead arc. They add very little in winter unless you live in the extreme southern U.S., or even further south. Trackers need clear access to the Sun from early in the morning until late in the afternoon. A solar window from 9 a.m. to 4 p.m. is workable; if you have greater access, more power to you (literally).

Tracking mounts are expensive, and PV power is getting cheaper. If you have a project that peaks in power use during the summer, such as water pumping or residential cooling, then tracking may be a very good choice. For many water-pumping projects, the most cost-effective way to increase daily production is to simply add a tracking mount.

If your projects peaks in power use during winter, such as powering a typical house, then tracking doesn't offer much. In most of North America, winter tracking will add less than 15%. One of the new generation of MPPT charge controls is a much better investment in this situation. They add 15%-30% and do their best work in the winter. See the Controllers section of Chapter 4 for more info.

almost always use these for one- or two-module pumping systems. Tilt and direction can be easily adjusted. Site preparation is easy, just get your steel pipe cemented in straight. The pole is common schedule 40 steel pipe, which is not included (pick it up locally to save on freight).

Make sure that your pole is tall enough to allow about one-third burial depth and still clear livestock, snow, or weeds. Ten feet total for pole length is usually sufficient. Taller poles are sometimes used for theft deterrence. Pole diameter depends on the specific mount and array

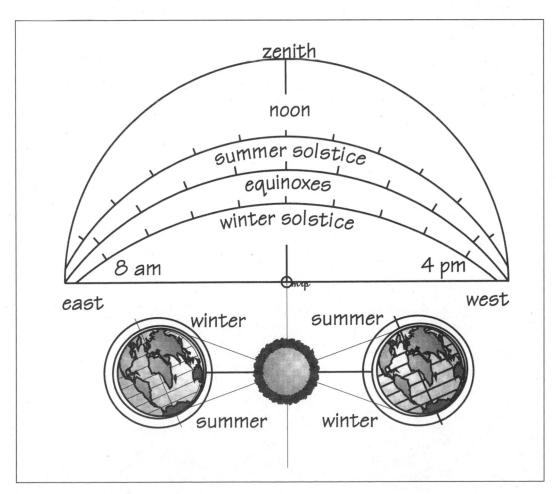

Based on our experience, the experience of hundreds of customers, and the dropping price of PV power, we no longer recommend active trackers. Their high initial cost and continuing maintenance problems just aren't worth the investment any longer. Want more power? Add some more PV. It's hard to beat the reliability of *no moving parts.*

size. Pole sizes listed are for "nominal pipe size." For instance, what the plumbing industry calls "4-inch" is actually 4½-inch outside diameter. When a mount says it fits "4-inch," it's actually expecting a 4½-inch-diameter pipe.

PASSIVE TRACKERS

Tracking mounts will follow the Sun from east to west across the sky, increasing the daily power output of the modules, particularly in summer and in southern latitudes. Trackers are most often used on water-pumping systems with peak demands in summer. See the sidebar "To Track or Not to Track" for a discussion on when tracking mounts are appropriate.

Passive trackers follow the Sun from east to west using just the heat of the Sun and gravity. No source of electricity is needed—a simple, effective, and brilliant design solution. The north-south tilt axis is seasonally adjustable manually. Maintenance consists of two squirts with a grease gun once every year.

Tracking will boost daily output by about 30% in the summer and 10%-15% in the winter. The two major problems with passive technology are wind disturbances and slow "wake-up" when cold. The tracker will go to "sleep" facing west. On a cold morning, it may take more than an hour for the tracker to warm up and roll over toward the east. In winds over 15 mph, the passive tracker may be blown off course. These trackers can withstand winds of up to 85 mph (provided you follow the manufacturer's recommendations for burying the pipe mount) but will not track at high wind speeds. If you have routine high winds, you should have a wind turbine to take advantage of those times, but that's a different subject.

ACTIVE TRACKERS

Active trackers use photocells, electronics, and linear actuators like those on giant old-fashioned satellite TV dishes to track the Sun very accurately. A small controller bolted to the array is programmed to keep equal illumination on the photocells at the base of an obelisk. Power use is minuscule. Active trackers average slightly more energy collection per day than a passive tracker in the same location, but historically they have also averaged more mechanical and electrical problems, too. Based on our experience, the experience of hundreds of customers, and the dropping price of PV power, we no longer recommend active trackers. Their high initial cost and continuing maintenance problems just aren't worth the investment any longer. Want more power? Add some more PV. It's hard to beat the reliability of *no moving parts.*

PV MOUNTING PRODUCTS

ZOMEWORKS UNIVERSAL TRACK RACKS

Passive Solar Tracker for Photovoltaic Modules
Incorporating over two decades of experience with tracker design and more than three decades of innovation of new products, Zomeworks has introduced the F-Series Track Rack™ Passive Solar Tracker to their line of UTR Universal Trackers. This tracker features an integral early-morning rapid-return system. It is shipped partially assembled, is easy to install, and is module specific. Utilizing the Sun's heat to move liquid from side to side allows gravity to turn the Track Rack and follow the Sun—no motors, no gears, and no controls to fail. There are six standard UTR and UTR-F Track Racks, which fit all common photovoltaic modules. They come "knocked down" and packaged in one to four (depending on the size of the rack) easy-to-handle boxes that fit conveniently into a pickup truck.

How the Zomeworks Track Racks Follow the Sun

1. The Track Rack begins the day facing west. As the Sun rises in the east, it heats the unshaded west-side canister, forcing liquid into the shaded east-side canister. As liquid moves through a copper tube to the east-side canister, the tracker rotates so that it faces east.

2. The heating of the liquid is controlled by the aluminum shadow plates. When one canister is exposed to the Sun more than the other is, its vapor pressure increases, forcing liquid to the cooler, shaded side. The shifting weight of the liquid causes the rack to rotate until the canisters are equally shaded.

3. As the Sun moves, the rack follows (at approximately 15 degrees per hour), continually seeking equilibrium as liquid moves from one side of the tracker to the other.

4. The rack completes its daily cycle facing west. It remains in this position overnight until it is "awakened" by the rising Sun the following morning.

Six Sizes Fit All

The Zomeworks Universal Rack System allows for almost limitless adjustment in both the east-west and north-south directions. Available in five standard sizes for holding 2 to 32 modules, Universal Track Racks are designed to fit all common photovoltaic modules, including BP/Solarex, Evergreen, Kyocera, Sharp, and most other popular modules. This flexibility translates to faster delivery, better quality, and overall economy.

Universal Track Rack Features:

- UTR K-040 accommodates up to two shock absorbers and early-morning wake-up fin for fast response. Lighter and easier to install. Ships via UPS.
- UTR F-168 comes standard with four shock absorbers for improved stability (equivalent to a high-wind kit).
- Large Early-Morning Wake-Up Fin—An additional 3 feet of wake-up fin to speed up morning rapid return for UTR F-168. Standard on all UTR F-168 trackers.
- ZW-2003 Shock Absorber—Exclusively from Zomeworks. Now standard on all F-Series Track Racks.
- ZW2006 Shock Absorber—Replacement shock absorber for UTR-series and TRPM-series trackers.
- Stainless steel and zinc-plated hardware.
- Both module-specific and J-Clip designs available.
- 10-yr.= standard warranty with extended warranty available.
- F-Series Track Racks™ ship partially assembled for easy installation.
- 6 Standard Universal Track Racks™ fit all common photovoltaic modules.

Universal* Track Rack™ Sizes:

13-950	UTR 020	20 Sq. Ft. Module Space	$606
42-0386	UTR K-040	40 Sq. Ft. Module Space	$1,177
13-952	UTR F-64	64 Sq. Ft. Module Space	$1,632
42-0288	UTR F-90	90 Sq. Ft. Module Space	$1,846
14336	UTR F-120	120 Sq. Ft. Module Space	$2,085
42-0424	UTR F-168	168 Sq. Ft. Module Space	$3,583

*Although these Universal racks are designed to hold a maximum number of square feet of PV module, we need to know how many of which brand of module are to be placed on the rack so that we can be sure your modules will fit. We supply the correct amount of hardware, and in some cases, we adjust the rail length to customize the fit. The UTR 020 is shipped with J-Clips for mounting the modules. The F-Series Track Racks are either drilled for specific modules, or J-Clips are supplied.

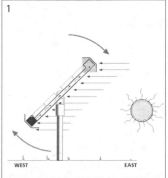

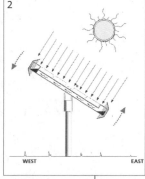

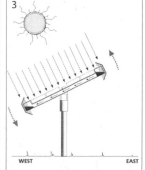

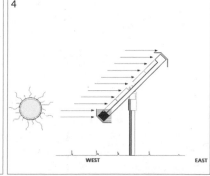

Flush roof mounting

Solar mount rail
cross section

UNIRAC MOUNTING SYSTEMS

UniRac makes a wide variety of high-quality, easy-to-install mounting options. Most are based on their incredibly adaptable extruded SolarMount™ aluminum rail (now available in two strengths—standard and light). The clever bolt slots allow module and footing bolts to be placed at any points needed along the rail. All UniRacs feature the highest quality, noncorrosive components. Constructed of aluminum and galvanized steel with stainless steel fasteners, there are no painted or zinc-plated pieces. Durability is excellent, assembly is fast, and instructions are clear and complete. UniRacs ship directly from the manufacturer via UPS, usually within one week, thanks to standardized construction. There's no waiting for racks and no exorbitant shipping charges since everything packs small and tight. Engineered to sustain wind loads of 50 lb. per sq. ft. (about 120 mph). USA.

UniRac SolarMount™ Roof Mounts

These universal mounting systems are sold à la carte. SolarMount rails are sold in pairs. Buy enough length to carry all your modules (plus 1 inch between modules and 1.5 inches at each end), buy mounting clamp sets as required, then add optional tilt legs or optional standoff mounting points as needed.

The standard rails are supplied with L-bracket feet, which are acceptable for most asphalt shingle roofs. Standoff mounts are a must for tile, ceramic, shake, or flat roofs. Standoffs are a more leak-secure, roofer-friendly option for all roofs. SolarMounts are installed flush with your roof, unless you use the optional tilt legs, which are also used for ground mounting. Both high-profile and low-profile style legs are offered, depending on your module orientation. See example drawings. Stainless steel fastener hardware is supplied for everything. 10-yr. mfr. warranty. All components made in USA.

4-18′ L $99 to $325 (call for details)

UniRac SolarMount™ Light Rails

Lower-cost SolarMount Light Rails use 38% less aluminum than standard rails, yet they're strong enough for flush applications. Use SolarMount standard top-mounting clamps and footing components. Note that they do require a minimum of 48-inch L-feet spacing (the standard heavier rails can be supported every 72 inches). Kits include a pair of rails, L-feet, and hardware to join L-feet to rails. Just add clips to secure PV modules to the rails, and hardware to secure L-feet to the roof.

4′-18′ L $83-$237 (call for details)

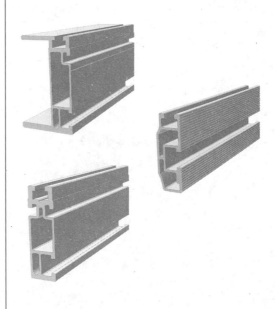

UniRac Top-Mount Rail Clamp Sets

These easy-to-install clamps secure the PV modules from the top using aluminum clamps and stainless nuts and bolts. They require 1 inch between modules and 1.5 inches at row ends. Sold in sets with all the clamps and hardware to secure from 2 to 8 modules. Module frame depths vary. USA.

Clamp sets   **$20-$50 (call for details)**

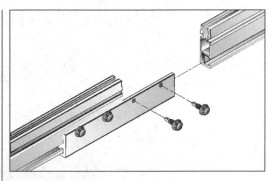

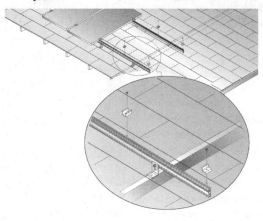

Standoffs

Standoffs are the connection between your roof framing and the PV racking. Standoffs penetrate your roof and use a leakproof flashing just like a plumbing vent.

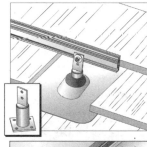

They're even the same size as a typical 1.25-inch vent pipe. This is familiar territory for your roofer. The flat base is lag-bolted to a framing member. Two ⁵⁄₁₆"x3.5" stainless lag bolts are included, then a standard plumbing vent flashing slips over the top. Standoffs are chrome-plated steel for strength with corrosion resistance. Standoff mounts are required for tile, ceramic, shake, and flat roofs. They're a more leak-secure, roofer-friendly option for all roofs.

Raised flange-type standoffs replace the L-feet and fasten directly to the SolarMount rails. These are the most common type used. Flat-top standoffs support common uni-strut stringers that fasten to the standard L-feet, a common technique on flat roofs. All standoffs are available in 3-, 4-, 6-, or 7-inch heights. Order as needed to raise rails above roofing material. Flashings are NOT included. USA.

3"-7" L Standoff   **$18-$21 (call for details)**

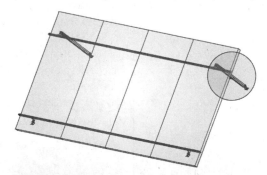

UniRac Rail Splice Kits

For very long flush or low-profile installations, splice kits can be used to join rails end to end. Due to thermal expansion and contraction, no more than two rail sections should be spliced, and splice kits are never used with high-profile tilt legs. Each splice kit contains two splices plus stainless steel hardware. USA.

13666   UniRac SP-2 Splice Kit   **$19.50**

UniRac Low-Profile Tilt Legs

Low-profile orientation minimizes the vertical height of your array. Useful to hide an array behind a parapet on a flat roof, or just to keep the visual impact to a minimum, these can be used for rooftop or ground mounting. The aluminum rear legs are telescoping and adjustable on-site. Longer rail sets, 192"-216", require two sets of legs. USA.

Low-Profile Tilt Legs $32-$100 (call for details)

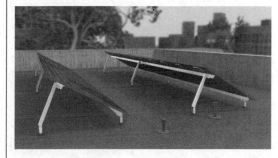

Low-profile tilt mounts

UniRac High-Profile Tilt Legs

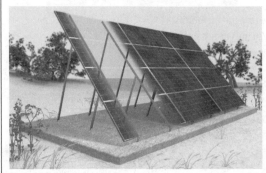
Highprofile tilt mounts

UniRac 5001 Series Pole-Top

High-profile orientation maximizes module density on a given footprint. High-profile legs are telescoping and adjust on-site. Sizing is based on the length and desired angle of your rails. Two legs are provided for rail sets up to 106". For 120"-180" rail sets, a short and a long leg are provided for each rail (four legs total). Rails longer than 180" are not recommended for high-profile mounts. Do NOT use rail splices with any high-profile mounting. USA.

High-Profile Tilt Legs $46-$188 (call for details)

UniRac Pole-Top Mounts

UniRac 5002 Series Pole-Top

Pole-top mounts are the easiest for most homeowners to install and will shed snow better than any other mounting style. Tilt and direction adjust easily. You supply the Schedule 40 or 80 steel pipe locally. Approximately one third of the total pipe length should be buried and backfilled with concrete. Pole heights more than about 10 feet above ground are okay but should have guy wires or be firmly attached to the side of a building. Uni-Rac mounts use only noncorrosive aluminum with stainless steel hardware. All necessary nuts and bolts for modules are included. Modules mount with long dimension perpendicular to the rails, as shown in photo. 10-yr. mfr. warranty. USA.

Pole-Top Mounts $110-$850 (call for details)

UniRac 5003 Series Pole-Top

UniRac 5000 Series Pole-Top

UniRac 5004 Series Pole-Top

UNIRAC SIDE-OF-POLE RACKS

PoleSides Fixed Tilt

Operate a series of remote monitoring stations or other low-power applications. Mount a single PV module quickly, easily, and economically. Racks include module mounting hardware and two U-bolts for 2-inch schedule 40 steel pole. Brackets are noncorrosive aluminum, and hardware is zinc-plated. Wind pressure load is 50 pounds per square foot. 10-yr. warranty. USA.

PoleSides Fixed Tilt Rack   **$29-$112**
(call for details)

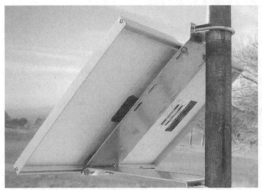

Series 4011: 45 degrees

Series 4012: 45 or 60 degrees

PoleSides Adjustable

Optimize Tilt Angle by Latitude, by Season

Series 4000

Series 4001

These racks include aluminum structural components and stainless steel hardware and can be used for up to four PV modules. Assembly procedure is flexible to suit installer preference for mounting on a pole or the side of a building, for example. They utilize installation-friendly SolarMount® rails, saving time on the job site. Design wind pressure load is 30-50 pounds per square foot depending on rack model, tilt angle, and PV module. 10 yr. warranty. USA.

PoleSides Adjustable Rack   **$67-$390**
(call for details)

Series 4002

RV Mounting Structures

For RVs, we've got simple mounts, and even simpler mounts. The really, really simple solution is RV Mounting Feet. This set of four rustproof aluminum feet fits any single PV module and holds it just above the roof for junction box clearance and cooling. Stainless hardware for attachment to the module is included. You provide hardware to attach to the RV roof as needed. One set of feet mounts one module. The slightly more elegant solution is the RV flush mount, which provides a pair of aluminum channels bolting down to four L-brackets. This provides more cooling clearance, and depending on assembly, mostly will hide under the module. Depending on your module size, these flush mounts can take one or two modules. Choose your size. You provide roof attachment hardware as needed. Both mounts have a 10-yr. mfr. warranty. USA.

42-0169	**RV Mounting Feet**	**$19**
42-0170	**RV Flush Mount 16"**	**$49**
42-0171	**RV Flush Mount 20"**	**$54**
42-0172	**RV Flush Mount 24"**	**$59**
42-0173	**RV Flush Mount 28"**	**$64**
42-0174	**RV Flush Mount 32"**	**$69**
42-0175	**RV Flush Mount 40"**	**$74**
42-0176	**RV Flush Mount 44"**	**$79**

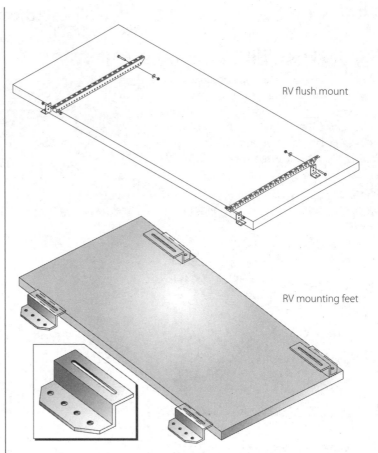

RV flush mount

RV mounting feet

Wind and Hydro Power Sources

Sunlight, wind, and falling water are the renewable energy big three. These are energy sources that are commonly available at a reasonable cost. Solar, or sunlight, the single most common and most accessible renewable energy source, is well covered at the beginning of this chapter. We've found in our years of experience that wind and hydro energy sources are most often developed as a booster or bad-weather helper for a solar-based system. These hybrid systems have the advantage of being better able to cover power needs throughout the year and are less expensive than a similar capacity system using only one power source. When a storm blows through, the solar input is lost, but a wind generator more than makes up for it. The short, rainy days of winter may limit solar gain, but the hydro system picks up from the rain and delivers steady power 24 hours a day. This is not to say that you shouldn't develop an excellent single-source power system if you've got it—like a year-round stream dropping 200 feet across your property, for instance. But for most of us, we'll be further ahead if we don't put all our eggs in one basket. Diversify!

Our experienced technical staff is well versed in supplying the energy needs of anything from a small weekend getaway cabin all the way up to an upscale state-of-the-art resort. We'll be glad to help put a system together for you. There is usually no charge for our friendly and personalized services.

Wind Systems

We generally advise that a good, year-round wind turbine site isn't a place that you'd want to live. It takes average wind speeds of 8-9 mph and up to make a really good site. That's honestly more wind than most folks are comfortable living with. But this is where the beauty of hybrid systems comes in. Many very livable sites *do* produce 8-mph and greater wind speeds during certain times of the year or when storms are passing through. Tower height and location also make a big difference. Wind speeds average 50%-60% higher at 100 feet compared with ground level (see chart in the Wind section). Wind systems these days are almost always designed as wind/solar hybrids for year-round reliability. The only common exceptions are systems designed for utility intertie; they feed excess power back into the utility and turn the meter backward.

Hydroelectric Systems

For those who are lucky enough to have a good site, hydro is really the renewable energy of choice. System component costs are much lower, and watts per dollar return is much greater for hydro than for any other renewable source. The key element for a good site is the vertical distance the water drops. A small amount of water dropping a large distance will produce as much energy as a large amount of water dropping a small distance. The turbine for the small amount of water is going to be smaller, lighter, easier to install, and vastly cheaper. We offer several turbine styles for differing resources. The small Pelton wheel Harris systems are well suited for mountainous territory that can deliver some drop and high pressure to the turbine. The propeller-driven Low Head Stream Engine is for flatter sites with less drop but more volume, and the Stream Engine, with a turgo-type runner, falls in between. It can handle larger water volumes and make useful power from shorter vertical water-drop distances.

Read on for detailed explanations of wind generators and hydro turbines. If you need a little help and guidance putting a system together or simply upgrading, our technical staff, with decades of hands-on experience in renewable energy, will be glad to help. Call us toll-free at 800-919-2400.

> Wind and hydro energy sources are most often developed as a booster or bad-weather helper for a solar-based system. These hybrid systems have the advantage of being better able to cover power needs throughout the year and are less expensive than a similar capacity system using only one power source.

Hydroelectricity

Hydropower, given the right site, can cost as little as a tenth of a PV system of comparable output.

If you could choose any renewable energy source you wanted, hydro is the one. If you don't want to worry about a conservation-based lifestyle—always nagging your kids to turn off the lights, watching the voltmeter, basing every appliance decision on energy efficiency—then you had better settle next to a nice year-round mountain stream! Hydropower, given the right site, can cost as little as a tenth of a PV system of comparable output. Hydropower users are often able to run energy-hog appliances that would

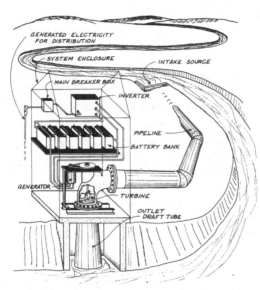

Low-head installation

bankrupt a PV system owner, like large side-by-side refrigerators and electric space heaters. Hydropower will probably require more effort on-site to install, but even a modest hydro output over 24 hours a day, rain or shine, will add up to a large cumulative total. Hydro systems get by with smaller battery banks because they need to cover only the occasional heavy power surge rather than four days of cloudy weather.

Hydro turbines can be used in conjunction with any other renewable energy source, such as PV or wind, to charge a common battery bank. This is especially true in the West, where seasonal creeks with substantial drops flow only in the winter. This is when power needs are at their highest and PV input is at its lowest. Small hydro systems are well worth developing, even if used only a few months out of the year, if those months coincide with your highest power needs. So, what makes a good hydro site, and what else do you need to know?

Small hydro systems are well worth developing, even if used only a few months out of the year, if those months coincide with your highest power needs.

What Is a "Good" Hydro Site?

The Columbia River in the Pacific Northwest has some really great hydro sites, but they're not exactly homestead scale (or low cost). Within the hydro industry, the kind of home-scale sites and systems we deal with are called micro-hydro. The most cost-effective hydro sites are located in the mountains. Hydropower output is determined by water's volume times its fall or drop (jargon for the fall is "head"). You can get approximately the same power output by running 1,000 gallons per minute through a 2-foot drop as by running 2 gallons per minute through a 1,000-foot drop. In the former scenario, where lots of water flows over a little drop, we are dealing with a low-head/high-flow situation, which is not truly a micro-hydro site. Turbines that can efficiently handle thousands of gallons are usually large, bulky, expensive, and site specific. But if you don't need to squeeze every last available watt out of your low-head source, the Low-Head Stream Engine generator will produce very useful amounts of power from low-head/high-flow sites, or the Turgo runner used on the Stream Engine is good at high-volume sites with 15 feet or more of head.

High-head installation

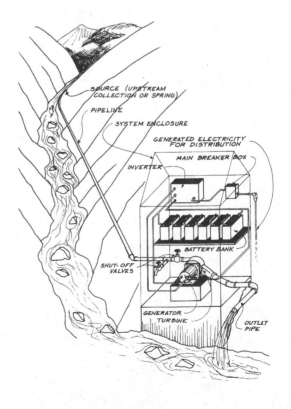

For hilly sites that can deliver a minimum of 50 feet of head, Pelton wheel turbines offer the lowest-cost generating solution. The Pelton-equipped Harris turbine is perfect for low-flow/high-head systems. It can handle a maximum of 200 gallons per minute and requires a minimum 50-foot fall to make useful amounts of power. In general, any site with more than 100 feet of fall will make an excellent micro-hydro site, but many sites with less fall can be very productive also. The more head, the less volume will be necessary to produce a given amount of power. Check the output charts in the Products section for a rough estimate of what your site can deliver.

A hydro system's fall doesn't need to happen all in one place. You can build a small collection dam at one end of your property and pipe the water to a lower point, collecting fall as you go. It's not unusual to use several thousand feet of pipe to collect a hundred feet of head (vertical fall).

Our Hydro Site Evaluation service will estimate output for any site, plus it will size the piping and wiring, and factor in any losses from pipe friction and wire resistance. See the example at the end of this editorial section.

What If I Have a High-Flow/Low-Head Site or Want AC Output?

Typically, high-flow/low-head or AC-output hydro sites will involve engineering, custom metalwork, formed concrete, permits, and a fair amount of initial investment cash. None of this is meant to imply that there won't be a good payback, but it isn't an undertaking for the faint-of-heart or thin-of-wallet. AC generators are typically used on larger commercial systems, or on utility intertie systems. DC generators are typically used on smaller residential systems.

DC generation systems offer several advantages for small hydro. Control is easy and cheap. The batteries we'll use to store energy allow power output surges way over what the turbine is delivering. The DC-to-AC inverters now available will deliver far cleaner and more tightly regulated AC power than a small AC hydro turbine can manage, and the inverter will cost less than a small AC control system.

For the homestead with a good creek but little significant fall, the Low-Head Stream Engine turbine is a far less costly alternative. With just 3 or 4 feet of fall, and some site develop-

ment work, this simple turbine can provide for a modest homestead.

If you'd rather look into the typical low-head scenario, contact the DOE's Renewable Energy Clearinghouse at 800-363-3732, or use Internet access at www1.eere.energy.gov/windandhydro/hydro_engineer_analysis.html or www.microhydropower.net for more free information on low-head hydro than you ever thought was possible.

How Do Micro-Hydro Systems Work?

The basic parts of micro-hydro systems are the pipeline (called the penstock in the trade), which delivers the water; the turbine, which transforms the energy of the flowing water into rotational energy; the alternator or generator, which transforms the rotational energy into electricity; the regulator, which controls the generator or dumps excess energy, depending on regulator style; and the wiring, which de-

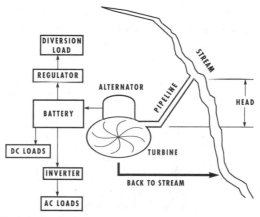

Typical micro-hydro system

livers the electricity. Our micro-hydro systems also use batteries, which store the low-voltage DC electricity, and usually an inverter, which converts the low-voltage DC electricity into 120 or 240 volts AC electricity.

Most micro-turbine systems use a small DC alternator or generator to deliver a small but steady energy flow that accumulates in a battery bank. This provides a few important advantages. The battery system allows the user to store energy and expend it, if needed, in short powerful bursts (like a washing machine starting the spin cycle). The batteries will allow us to deliver substantially more energy for short periods than a turbine is producing, as long as the battery and inverter are designed to handle the load. DC charging means that precise control of

For the homestead with a good creek but little significant fall, the Low-Head Stream Engine turbine is a far less costly alternative.

The bottom line is that a DC-based system will cost far less than an AC system for most residential users, and will perform better.

alternator speed is not needed, as is required for 60-Hz AC output. This saves thousands of dollars on control equipment. And finally, with the quality of the DC-to-AC inverters now available, you'll enjoy cleaner, more tightly controlled AC power through an inverter than through a small AC turbine. The bottom line is that a DC-based system will cost far less than an AC system for most residential users, and will perform better.

DC Turbines

Several micro-hydro turbines are available with simple DC output. We currently offer the Harris and the Stream Engine.

HARRIS TURBINE

The Harris turbine uses a hardened cast-silicone bronze Pelton turbine wheel mated with a low-voltage DC alternator. Pelton wheel turbines work best at higher pressures and lower volumes. Life begins at about 50 feet of head for these turbines and has no practical upper limit. Harris offers a couple choices for the alternator. The standard Harris is based on common 1970s Motorcraft alternators with windings that are customized for each individual application. Bearings and brushes will require replacement at intervals from one to five years, depending on how hard the unit is working. These parts are commonly available at any auto parts store. Harris now comes with a standard permanent-magnet alternator that is custom made. The PM alternators deliver more power under almost all conditions, have no brushes to wear out, and are mounted on larger, more robust bearings with two to three times the life expectancy.

Depending on the volume and fall supplied, Harris turbines can produce from 1kWh (1,000 watt-hours) to 35kWh per day. Maximum alternator instantaneous output is about 2,500 watts in a 48-volt system with cooling options

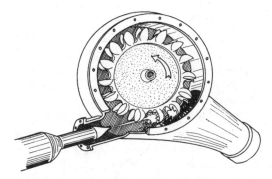

An 1880s vintage single-nozzle Pelton wheel. Only the generator technology has changed.

required. The typical American home consumes 15-25kWh per day with no particular energy conservation, so with a good hydro site, it is fairly easy to live a conventional lifestyle.

The Harris turbine can be supplied with one, two, or four nozzles. The maximum flow rate for any single nozzle can be from 20 to 60 gallons per minute (gpm), depending on the head pressure. The turbine can handle flow rates to about 120 gpm before the sheer volume of water starts getting in its own way. Many users buy two- or four-nozzle turbines with different-sized nozzles, so that individual nozzles can be turned on and off to meet variable power needs or water availability. The brass nozzles are easily replaceable because they eventually wear out, especially if there is grit in the water. They are available in 1/16-inch increments, from 1/16 inch through 1/2 inch. The first nozzle doesn't have a shutoff valve, while all nozzles beyond the first one are supplied with ball valves for easy, visible operation.

STREAM ENGINE

The Stream Engine turbines use a unique brushless, permanent-magnet alternator with three large sealed-shaft bearings. Permanent magnets mean there are no field brushes to wear out. Magnetic field strength is adjusted by varying the air gap between the magnet disk and the stationary alternator windings. Once the unit is set up, there is almost no routine maintenance required. Setup requires some time with this universal design, however, and involves trial and adjustment: selecting one of four alternator wiring setups and then adjusting the permanent-magnet rotor air gap until peak output is achieved. A precision shunt and digital multimeter are supplied to expedite this setup process. Output voltage can be user selected at 12, 24, or 48 volts. Maximum instantaneous output is 800 watts for this alternator.

This low-maintenance alternator is employed on two very different turbines. The original

Two- and four-nozzle Harris turbines. The four-nozzle is upside down to show the Pelton wheel and nozzles.

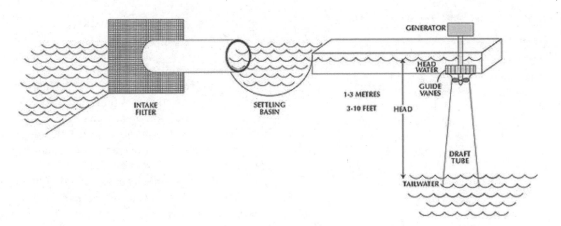

Low-Head Stream Engine installation

Stream Engine uses a cast bronze Turgo-type runner wheel. This Turgo wheel can handle a bit more water volume than the Harris pelton wheel—up to about 200 gpm before it starts choking—and starts to deliver useful output at lower 15- to 20-foot head. Nozzles are cone-shaped plastic casings; you cut them at the size desired, from 1/8 inch up to 1 inch. Two- or four-nozzle turbines are available, with replacement nozzles being a readily available bolt-in.

The Low-Head Stream Engine uses the same alternator but is packaged quite differently. It works on 2-10 feet of fall, which happens on the downstream end of the turbine for a change. Flows of 200-1,000 gpm can be accommodated through the 5-inch propeller turbine. The large draft tube on the output must be immersed in the tail water. (See the illustration.) This turbine requires more site preparation but almost no maintenance or attention once it's installed and tuned.

Power Transmission

One disadvantage of lower-voltage DC hydro systems is the difficulty of transmitting power from the turbine to the batteries, particularly with high-output sites. A typical installation places the batteries at the house on top of the hill, where the good view is, and the turbine at the bottom of the hill, where the water ends its maximum drop. Low-voltage power is difficult to transmit if large quantities or long distances are involved. The batteries should be as close to the turbine as is practical, but if there's more than 100 feet of distance involved, things will work better if the system voltage is 24 or even 48 volts. Transmission distances of more than 500 feet often require expensive large-gauge wire or technical tricks. Don Harris has been working with Outback Power Systems to develop a hydroelectric version of their maximum power point tracking MX60 controller, which allows

inputs up to 140 volts, while feeding the batteries whatever it is that makes them happy. At the time of this writing, the Hydro MX60 isn't ready for release, but beta units are in trial. Please consult with the Real Goods technical staff about this or other transmission options.

Controllers

Hydro generators require special controllers or regulators. Controllers designed for photovoltaics may damage the hydro generator and will very likely become crispy critters themselves if used with one. You can't simply open the circuit when the batteries get full like you can with PV. So long as the generator is spinning, there needs to be a place for the energy to go. Controllers for hydro systems take any power beyond what is needed to keep the batteries charged and divert it to a secondary load, usually a water- or space-heating element. So extra energy heats either domestic hot water or the house itself. These diversion controllers are also used with some wind generators and can be used for PV control as well if this is a hybrid system.

Site Evaluation

Okay, you have a fair amount of drop across your property and/or enough water flow for one of the low-head turbines, so you think micro-hydro is a definite possibility. What happens next? Time to go outside and take some measurements, then fill in the necessary information on the Hydro Site Evaluation form. With the info on your completed form, the Real Goods technicians can calculate which turbine and options will best fill your needs, as well as what size pipe and wire and which balance-of-system components you require. Then we can quote specific power output and system costs so you can decide if hydro is worth the installation effort.

A typical installation places the batteries at the house on top of the hill, where the good view is, and the turbine at the bottom of the hill, where the water ends its maximum drop. Low-voltage power is difficult to transmit if large quantities or long distances are involved.

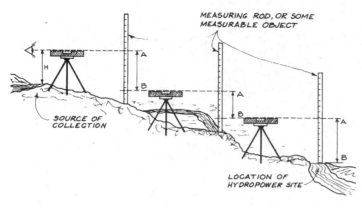

Measuring fall

DISTANCE MEASUREMENTS

Keep the turbine and the batteries as close together as practical. As discussed earlier, longer transmission distances will get expensive. The more power you are trying to move, the more important distance becomes.

You'll need to know the distance from the proposed turbine site to the batteries (how many feet of wire) and the distance from the turbine site to the water collection point (how many feet of pipe). These distances are fairly easy to determine; just pace them off or use a tape measure.

FALL (DROP OR HEAD) MEASUREMENT

Next, you'll need to know the fall from the collection point to the turbine site. This measurement is a little tougher. If there is a pipeline in place already, or if you can run one temporarily and fill it with water, this part is easy. Simply install a pressure gauge at the turbine site, make sure the pipe is full of water, and turn off the water at the bottom. Read the static pressure (which means no water movement in the pipe), and multiply your reading in pounds per square inch (psi) by 2.31 to obtain the drop in feet. If the water pipe method isn't practical, you'll have to survey the drop or use a fairly accurate altimeter or GPS device. A number of relatively inexpensive sports watches come with a built-in altimeter now. If the altimeter can read ±10 feet, that's close enough. Strap it on, take a hike, and record the difference.

The following instructions represent the classic method of surveying. You've seen survey parties doing this, and if you've always wanted to attend a survey party, this is your big chance to get in on the action. You'll need a carpenter's level (or a pocket sight level), a straight sturdy stick about eye-level tall, a brightly colored target that you'll be able to see a few hundred feet away, and a friend to carry the target and to make the procedure go faster and more accurately. (It's really hard to party alone.)

Stand the stick upright and mark it at eye level. (Five feet even is a handy mark that simplifies the mathematics, if that's close to eye level for you.) Measure and note the length of your stick from ground level to your mark. Starting at the turbine site, stand the stick upright, hold the carpenter's level at your mark, make sure it is level, then sight across it uphill toward the water source. With hand motions and body English, guide your friend until the target is placed on the ground at the same level as your sightline, then have your friend wait for you to catch up. Repeat the process, carefully keeping track of how many times you repeat. It is a good idea to draw a map to remind you of landmarks and important details along the way. If you have a target and your friend has a stick (marked at the same height, please) and level, you can leapfrog each other, which makes for a shorter party. Multiply the number of repeats between the turbine site and the water source by the length of your stick(s) and you have the vertical fall. People actually get paid to have this much fun!

FLOW MEASUREMENT

Finally, you'll need to know the flow rate. If you can, block the stream and use a length of pipe to collect all the flow. Time how long it takes to fill a 5-gallon bucket. Divide 5 gallons by your fill time in seconds. Multiply by 60 to get gallons per minute. Example: The 5-gallon bucket takes 20 seconds to fill. So 5 divided by 20 = 0.25 times 60 = 15 gpm. If the flow is more than you can dam up or get into a 4-inch pipe, or if the force of the water sweeps the bucket out of your hands, forget measuring: You've got plenty!

Conclusion

Now you have all the information needed to guesstimate how much electricity your proposed hydro system will generate based on the manufacturer's Output Charts on the next pages. This will give you an indication of whether or not your hydro site is worth developing, and if so, which turbine option is best. If you think you have a real site, fill out the Real Goods Hydro Site Evaluation Form and send it to the Technical Department at Real Goods, or just give us a call. We will run your figures through our computer sizing program, which allows us to size plumbing and wiring for the least power loss at the lowest cost, and a myriad of other calculations necessary to design a working system. You'll find an example of our Hydro Survey Report on the next page, followed by the form for the info we need from you.

Now you have all the information needed to guesstimate how much electricity your proposed hydro system will generate based on the manufacturer's Output Charts on the next pages.

CALCULATION OF HYDROELECTRIC POWER POTENTIAL

Copyright © 1988 by Ross Burkhardt. All rights reserved.

ENTER HYDRO SYSTEM DATA HERE: **Customer: Meg A. Power**

Pipeline Length:	1,300 feet
Pipe Diameter:	4 inches
Available Water Flow:	100 gpm
Vertical Fall:	200 feet
Hydro to Battery Distance:	50 feet (one way)
Transmission Wire Size:	2 AWG
House Battery Voltage:	24 volts
Hydro Generation Voltage:	29 volts

Power produced at hydro:	*Power delivered to house:*
49.78 amps	49.78 amps
29 volts	28.20 volts
1,443.53 watts	1,403.59 watts

4-nozzle, 24V, high-output with cooling turbine required

Pipe Calculations

Head Lost to Pipe Friction:	7.61 feet
Pressure Lost to Pipe Friction:	3.29 psi
Static Water Pressure:	86.62 psi
Dynamic Water Pressure:	83.33 psi
Static Head:	200.01 feet
Dynamic Head:	192.40 feet

Hydropower Calculations

Operating Pressure:	83.33 psi
Available Flow:	100 gpm
Watts Produced:	1,443.53 watts
Amperage Produced:	49.78 amps
Amp-Hours per Day:	1,194.65 amp-hours
Watt-Hours per Day:	34,644.83 watt-hours
Watts per Year:	12,645,362.71 watt-hours

Line Loss (using copper)

Transmission Line One-Way Length:	50 feet
Voltage:	29 volts
Amperage:	49.78 amps
Wire Size #:	2 AWG
Voltage Drop:	0.8 volts
Power Lost:	39.95 watts
Transmission Efficiency:	97.23 percent
Pelton Wheel rpm Will Be:	2,969.85 at optimum wheel efficiency

This is an estimate only! Due to factors beyond our control (construction, installation, incorrect data, etc.), we cannot guarantee that your output will match this estimate. We have been conservative with the formulas used here, and most customers call to report more output than estimated. However, be forewarned! We've done our best to estimate conservatively and accurately, but there is no guarantee that your unit will actually produce as estimated.

Real Goods
Hydroelectric Site Evaluation Form

Name: _____

Address: _____

_____

Phone: _____ Date: _____

Pipe Length: _____ (from water intake to turbine site)

Pipe Diameter: _____ (only if using existing pipe)

Available Water Flow: _____ (in gallons per minute)

Fall: _____ (from water intake to turbine site)

Turbine to Battery Distance: _____ (one way, in feet)

Transmission Wire Size: _____ (only if existing wire)

House Battery Voltage: _____ (12, 24, etc.)

Alternate estimate (if you want to try different variables)

Pipe Length: _____ (from water intake to turbine site)

Pipe Diameter: _____ (only if using existing pipe)

Available Water Flow: _____ (in gallons per minute)

Fall: _____ (from water intake to turbine site)

Turbine to Battery Distance: _____ (one way, in feet)

Transmission Wire Size: _____ (only if existing wire)

House Battery Voltage: _____ (12, 24, etc.)

For a complete computer printout of your hydroelectric potential, including sizing for wiring and piping, please fill in the above information and send it to Real Goods.

HYDROELECTRIC PRODUCTS

Hydro Generation Products

Harris Hydroelectric:
Bringing hydropower home

When Don Harris moved to California's Santa Cruz Mountains in the mid-1970s, he was 7 miles from the nearest power line—and direct sunshine wasn't plentiful enough to make power from the Sun. Hydropower was his best bet, but the only hydro plants in operation were AC and way too big for residential use. Then he visited a hydro museum and saw a way to make micro-hydro work.

In a machine shop from his old drag-racing days, Don created 50 Pelton wheels and launched Harris Hydroelectric, whose very first customer was Real Goods in 1981. Harris has manufactured over 3,000 micro-hydro plants since then—including the one that powers Real Goods President John Schaeffer's home all winter, for an original outlay of just $1,500, providing him with 36kWh per day.

Harris Hydroelectric Turbines

The generating component of the standard Harris turbines is an automotive alternator equipped with custom-wound coils appropriate for each installation. Field strength, and therefore output, is adjustable with a large rheostat. The built-in ammeter shows the immediate effects of tweaking the field.

The permanent-magnet alternator eliminates brushes and annual brush replacement. It has bigger bearings that last longer than the standard automotive bearings, so overall maintenance is greatly reduced. Under most operating conditions, the P.M. alternator delivers more wattage (see output chart). This is now standard with all units. (Also available for older machines in service. Please call.)

The rugged turbine wheel is a one-piece Harris casting made of tough silicon bronze. There are hundreds of these turbine wheels that have been in service for over a decade. Even with dirty, gritty water, it takes at least 10 years to wear out a Harris wheel. The cast aluminum wheel housing serves as a mounting for the alternator and up to four nozzle holders. It also acts as a spray shield, redirecting the expelled water down into the collection/drain box. Expelled water depends on gravity to drain.

Continued next page

Single-nozzle turbine with permanent magnet option.

Harris Turbine Max. Flow Rates in gpm				
	Number of Nozzles			
Feet of Head	1	2	3	4
25	17	35	52	70
50	25	50	75	100
75	30	60	90	120
100	35	70	105	140
200	50	100	150	200
300	60	120	–	–

HARRIS TURBINE OUTPUT IN WATTS									
Feet of Head	Gallons per Minute Flow (permanent-magnet specs in bold)								
	3	6	10	15	20	30	50	100	200
25	–	–	–	20 **25**	30 **40**	50 **65**	115 **130**	200 **230**	–
50	–	–	35 **40**	60 **75**	80 **100**	125 **150**	230 **265**	425 **500**	520 **580**
75	–	25 **30**	60 **75**	95 **110**	130 **160**	210 **250**	350 **420**	625 **750**	850 **900**
100	–	35 **45**	80 **95**	130 **150**	200 **240**	290 **350**	500 **600**	850 **1,100**	1,300 **1,300**
200	30 **45**	100 **130**	180 **210**	260 **320**	400 **480**	580 **650**	950 **1,100**	1,500 **1,500**	–
300	70 **80**	150 **180**	275 **300**	400 **450**	550 **600**	850 **940**	1,400 **1,500**	–	–

Maximum wattage @ voltage: 12V=750, 24V=1,500, 48V=2,500

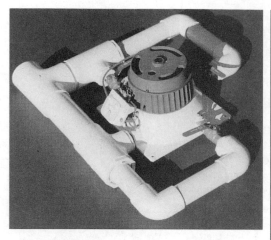

Harris Hydroelectric turbines are available with one, two, or four nozzles to maximize the output of the unit. Having an extra nozzle or two beyond what is absolutely required makes it easy to change flow rates or power output for seasonal or household variations. Just turn different-size nozzles on and off. The individual brass nozzles are screwed in from the underside of the turbine. They do wear with time and will need eventual replacement. Any turbine delivering 20 amps or more (at whatever system voltage) needs the cooling fan option. USA.

17101	1-Nozzle Turbine	$1,800
17102	2-Nozzle Turbine	$1,900
17103	4-Nozzle Turbine	$2,150
17132	24-Volt Option	No charge
17136	48-Volt Option	$200
17134	Extra Nozzles	$8
17135	Optional Cooling Fan Kit	$89

All Harris turbines are custom assembled to match your site. We need to know head, flow rate, pipeline size and length, system voltage, and distance from turbine to batteries. Allow 2-4 weeks delivery time.

Harris Pelton Wheels

For do-it-yourselfers interested in a small hydroelectric system, we offer the same tough and reliable Pelton wheel used on the complete turbines above. Harris silicon bronze Pelton wheels resist abrasion and corrosion far longer than polyurethane or cast-aluminum wheels. These are 5-inch-diameter quality castings that can accommodate nozzle sizes of ¹⁄₁₆-½ inch. Designed with threads for Delco or Motorcraft alternators. USA.

17202	Silicon Bronze Pelton Wheel	$300

The Stream Engine

The highly adjustable Stream Engine uses a Turgo-type runner wheel that can produce useful power from lower heads. The permanent-magnet generator has no brushes to wear out and uses a clever combination of alternator wiring changes and adjustable air gap to compensate for differing flow rates. Setup requires some time with this universal design and involves trial and adjustment, selecting one of four alternator wiring setups and then adjusting the permanent-magnet rotor air gap until peak output is achieved. A precision shunt and digital multimeter are supplied to expedite this setup process. Output voltage is user-settable from 12 to 48 volts. Available in two- or four-nozzle turbines, the Stream Engine is supplied with universal cut-to-size nozzles of ⅛-1 inch. See the Nozzle Chart to determine flow rates at various heads. Head can be 5-400 feet, and flows of 5-300 gpm can be used (note that flow rates above 150 gpm have diminishing returns, as the sheer volume of water starts to get in its own way). Maximum output is approximately 800 watts. See performance chart. Canada.

17109	Stream Engine, 2-nozzle	$1,995
17110	Stream Engine, 4-nozzle	$2,045
17111	Turgo Wheel only (bronze)	$620
17112	Extra Universal Nozzle	$40

Stream Engine Nozzle Flow Chart (in gpm)

Head Pressure		Nozzle Diameter in Inches								Turbine			
Feet	PSI	1/8	3/16	1/4	5/16	3/8	7/16	1/2	5/8	3/4	7/8	1.0	RPM
5	2.2					6.18	8.40	11.0	17.1	24.7	33.6	43.9	460
10	4.3			3.88	6.05	8.75	11.6	15.6	24.2	35.0	47.6	62.1	650
15	6.5		2.68	4.76	7.40	10.7	14.6	19.0	29.7	42.8	58.2	76.0*	800
20	8.7	1.37	3.09	5.49	8.56	12.4	16.8	22.0	34.3	49.4	67.3	87.8*	925
30	13.0	1.68	3.78	6.72	10.5	15.1	20.6	26.9	42.0	60.5	82.4*	107*	1,140
40	17.3	1.94	4.37	7.76	12.1	17.5	23.8	31.1	48.5	69.9	95.1*	124*	1,310
50	21.7	2.17	4.88	8.68	13.6	19.5	26.6	34.7	54.3	78.1*	106*	139*	1,470
60	26.0	2.38	5.35	9.51	14.8	21.4	29.1	38.0	59.4	85.6*	117*	152*	1,600
80	34.6	2.75	6.18	11.0	17.1	24.7	33.6	43.9	68.6	98.8*	135*	176*	1,850
100	43.3	3.07	6.91	12.3	19.2	27.6	37.6	49.1	76.7	111*	150*	196*	2,070
120	52.0	3.36	7.56	13.4	21.0	30.3	41.2	53.8	84.1*	121*	165*	215*	2270
150	65.0	3.76	8.95	15.0	23.5	33.8	46.0	60.1	93.9*	135*	184*	241*	2540
200	86.6	4.34	9.77	17.4	27.1	39.1	53.2	69.4	109*	156*	213*	278*	2930
250	106	4.86	10.9	19.9	30.3	43.6	59.4	77.6*	121*	175*	238*	311*	3270
300	130	5.32	12.0	21.3	33.2	47.8	65.1	85.1*	133*	191*	261*	340*	3590
400	173	6.14	13.8	24.5	38.3	55.2	75.2	98.2*	154*	221*	301*	393*	4140

Stream Engine Power Output (in Watts)

Net Head	Flow Rate in U.S. Gallons per Minute											
Feet	5	10	15	20	30	40	50	75	100	150	200*	300*
5				5	8	10	15	20	30	40		
10			7	12	18	23	30	45	60	80	100	
15	5	10	15	20	30	40	50	75	100	125	150	200
20	8	16	25	32	50	65	85	125	170	210	275	350
30	12	30	45	60	90	120	150	225	300	400	500	700
40	16	40	60	80	120	160	200	300	400	500	600	
50	20	50	75	100	150	200	250	375	500	600		
75	30	75	110	150	225	300	375	560	700			
100	40	100	150	200	300	400	500	650				
150	60	150	225	300	400	550	650					
200	80	200	300	400	550	700						
300	120	240	360	480	720							
400	160	320	480	640								

*Flow rates above 150 gpm are possible but will suffer from diminishing returns.

Low-Head Stream Engine

Using the same brushless, permanent-magnet generator as the regular Stream Engine, the Low-Head model works with drops of 2-10 feet and flows of 200-1,000 gallons per minute. Peak output is 1,000 watts. Wattage output at your site will be determined by available flow rate and drop. The generator is driven by a 5-inch low-head propeller turbine and includes the large 10-foot-long tapered draft tube, which must be immersed in the tailwater. 12- to 48-volt output. Canada.

17328   Low-Head Stream Engine $2,300

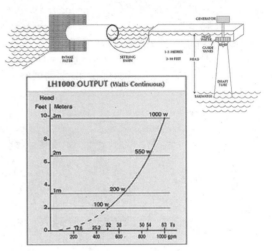

Hydro System Controls and Accessories

Hydroscreen Water Intakes

Maintenance-free clean water

Complete assembly

Crud in your hydro intake is nothing but trouble, reducing output, shortening life expectancy, and requiring regular attention. Self-cleaning Hydroscreens are robust, noncorrosive stainless steel wedge-wire water intake screens with a stainless frame. They're installed at about a 45° angle, so the water washes over them. The wires are flat on top and wedge-shaped. Anything that can't drop through the precise 1.25mm (0.05-inch) gap between wires is flushed off the surface. No maintenance! No regular screen cleaning. High-water floods pass right over. Available as either a flat screen with stainless ramp

Flat screen

Edge-on close-up of wedge wire

and tail (you build your own collection box under it) or as a complete assembly with screen, collection box, and 3- or 4-inch output pipe. Screen area is 12"x12". These models are sized to work with any Harris or Stream Engine turbine installation up to 200 gpm. 1-yr. mfr. warranty. USA.

47-0100 Hydroscreen Flat Screen   **$279**

47-0101 Hydroscreen Complete Assembly   **$679**

TriStar PV or Load Controllers

Full-featured 45- or 60-amp controls

The reliable, UL-listed MorningStar TriStar controller can provide either solar charging, load control, or diversion regulation for 12-, 24-, or 48-volt systems. This is a full-featured, modern, solid-state PWM controller. Features include reverse polarity protection, short-circuit protection, lightning protection, high-

temperature current reduction, four-stage battery charging, simple DIP-switch-controlled voltage set-point setup, and large 1-/1.25-inch knockouts on bottom, sides, and back, with extra wire-bending room and terminals that accept up to #2 AWG wire. Circuit boards are fully conformal coated, the aluminum heat sink is anodized, and the enclosure is powder-coated with stainless fasteners to laugh at high-humidity tropical environments.

This controller is particularly good for hydro diversion because the controller will allow up to 300-amp inrush currents to start motors or warm up cold resistive loads without the electronic short-circuit protection interfering.

The optional 2-line, 16-character digital meter may be mounted to the controller in place of the cover plate, or remotely in a standard double-gang box using standard RJ-11 connectors (Ethernet cables). It displays all system information, self-test results, and set-points with intuitive up/down and left/right scrolling buttons. The remote temperature sensor option with 10-meter cable will automatically adjust voltage set-points and is recommended if your batteries will routinely be exposed to temperatures under 40°F or over 80°F. 10.1"x4.9"x2.3" (HxWxD), weighs 4 lb. CE and UL listed. 5-yr. mfr. warranty. USA.

48-0010 TriStar 45-Amp Controller   **$169**

48-0011 TriStar 60-Amp Controller   **$218**

48-0012 TriStar Digital Meter   **$99**

48-0013 TriStar Remote Temperature Sensor   **$45**

Air Heater Diversion Loads

These resistive loads are enclosed in vented aluminum boxes for safety. They can be used on any DC system between 12 and 48 volts. The box needs at least a 12-inch clearance to combustibles. Both units are shipped in the highest resistance mode and can be easily reconfigured for lower resistance by changing connections in the terminal block. The HL-100 unit can be configured for 30 or 60 amps in nominal 12-volt mode. 2-yr. mfr. warranty. USA.

Item #	Price	Model	Diversion Amps @ Voltage Below			Resistance Setting
			15v	30v	60v	
48-0081	$299	HL-100	30/60	–	–	0.5/0.25 ohm
			15	30	–	1 ohm
			3.8	7.5	15	4 ohm
48-0082	$235	HL-75	20	40	–	0.75 ohm
			5	10	20	3 ohm

Diversion Water Heater Elements

Put your spare energy to work!

These industrial-grade DC heater elements give you someplace to dump extra wind or hydro power. Use with a diversion controller such as the MorningStar Tristar. The water heater elements fit a standard 1-inch NPT fitting. To use with a 240-volt AC water heater, just replace the AC element with one of these. The 12/24V model has a pair of 25-amp/12-volt elements. It can be wired for 25 amps or 50 amps @ 12 volts, or in series for 25 amps @ 24 volts. The 24/48V model has a pair of 30-amp/24-volt elements. It can be wired for 30 amps or 60 amps @ 24 volts, or in series for 30 amps @ 48 volts. USA.

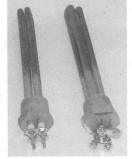

25078   12/24V Water Heating Element $95

25155   48V/30A Water Heating Element $116

Books

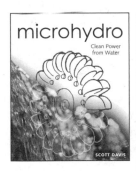

Microhydro

Hydroelectricity is the world's largest source of renewable energy. But until now, there were no books that covered residential-scale hydropower, or micro-hydropower. This excellent guide covers the principles of micro-hydro, design and site considerations, and equipment options, plus legal, environmental, and economic factors. Author Scott Davis covers all you need to know to assess your site, develop a workable system, and get the power transmitted to where you can use it. 157 pages, softcover. Canada.

21-0340 Microhydro $19.95

Microhydro Power Video

This is a program we highly recommend to anyone considering residential hydro. It covers basic technology and hands-on applications, with easy-to-follow explanations of micro-hydro power. It also covers performance and safety issues. With micro-hydro pioneer Don Harris of Harris Hydroelectric. VHS, 42 minutes. USA.

80362   Microhydro Power Video $39

Wind Energy

■ This section was adapted from *Wind Energy Basics*, a Real Goods Solar Living Book by Paul Gipe. He is also the author of *Wind Power: Renewable Energy for Home, Farm, and Business* (2004). Gipe has written and lectured extensively about wind energy for more than two decades.

Small Wind Turbines Come of Age

The debut of micro wind turbines has revolutionized living off the grid. These inexpensive machines have brought wind technology within reach of almost everyone. And their increasing popularity has opened up new applications for wind energy previously considered off-limits, such as electric fence charging and powering remote telephone call boxes, once the sole domain of photovoltaics.

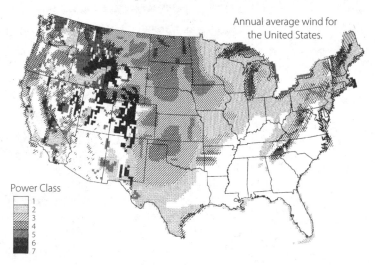

Annual average wind for the United States.

Power Class
1
2
3
4
5
6
7

Micro wind turbines have been around for decades for use on sailboats, but they gained prominence in the 1990s as their broader potential for off-the-grid applications on land became more widely known. While micro wind turbines have yet to reach the status of widely available consumer commodities such as personal computers, the day may not be far off.

The use of wind power is "exploding," say Karen and Richard Perez, the editors of *Home Power* magazine. "There are currently over 150,000 small-scale RE (renewable energy) systems in America and [the numbers] are growing by 30% yearly. [But] [t]he small-scale use of wind power is growing at twice that amount—over 60% per year." And a large part of that growth is due to Southwest Windpower of Flagstaff, Arizona.

Southwest Windpower awakened latent consumer interest in micro wind turbines with the introduction of its sleek Air series. Since launching the line in 1995, they've shipped thousands of the popular machines.

"What Americans, and folks all over the world, are finding out," the Perezes say, "is that wind power is an excellent and cost-effective alternative" to extending electric utility lines and fossil-fueled back-up generators.

Hybrid Wind and Solar Systems

You might say that joining wind and solar together is a marriage made in heaven. The two resources are complementary: In many areas, wind is abundant in the winter when photovoltaics are least productive, and sunshine is abundant in the summer when winds are often weakest. The Sun and wind together not only improve the reliability of an off-the-grid system, but also are more cost effective than using either source alone.

Hybrid systems include a DC source center (for DC circuit breakers), batteries, inverters, and often an AC load center. These components are necessary whether you're using just wind or solar. So it's best to spread the fixed cost of these components over more kilowatt-hours by using PV panels in addition to a wind turbine.

Engineers have found that these hybrids perform even better when coupled with small back-up generators to reduce the battery storage needed. Many of those living off the grid reach the same conclusion by trial and error.

Both wind and PV can happily feed a common battery.

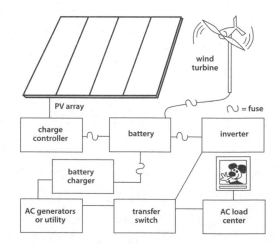

Typically, a micro turbine, such as Southwest Windpower's Air series, or a small wind turbine, such as the Bergey XL1, a modest array of PV panels, batteries, and a small back-up generator will suffice for most domestic uses. Though Pacific Gas & Electric Co. found that most Californians living off the grid had back-up generators, they seldom used them. In a well-designed hybrid system, the backup provides peace of mind but little electricity.

Size Matters

In wind energy, size, especially rotor diameter, matters. Nothing tells you more about a wind turbine's potential than its diameter—the shorthand for the area swept by the rotor. The wind turbine with the bigger rotor will intercept more of the wind stream and will almost invariably generate more electricity than a turbine with a smaller rotor, regardless of their generator ratings.

The area swept by a wind turbine rotor is equivalent to the surface area of a photovoltaic array. When you need more power in a PV array, you increase the surface area of the panels exposed to the Sun by adding more panels. In the case of wind, you find a wind turbine that sweeps more area of the wind—a turbine with a greater rotor diameter.

Micro turbines range in size from 3 to 5 feet in diameter. The lower end of the range is represented by the Southwest Windpower's Air series. These machines are suitable for recreational vehicles, sailboats, fence charging, and other low-power uses. Micro turbines will generate about 300 kilowatt-hours (kWh) per year at sites with average wind speeds typical of the Great Plains (about 12 mph at the height of the turbine).

Mini wind turbines are slightly larger than micro turbines and are well suited for vacation cabins. They span the range from 5 to 9 feet in diameter and include Southwest Windpower's Whisper 200 as well as the Bergey Windpower XL1. Wind turbines in this class can produce 1,000-2,000kWh per year at sites with an average annual wind speed of 12 mph.

Household-size turbines, as the name implies, are suitable for homes, farms, ranches, small businesses, and telecommunications. They span an even broader range than the other size classes and encompass turbines from 10 to 23 feet in diameter. This class includes Bergey's Excel model and African Windpower's 3.6 meter- (12-foot-) diameter turbine. Household-size wind turbines can generate from 2,000kWh to 20,000kWh per year at 12-mph sites.

Generators

Most small wind turbines use permanent-magnet alternators. This is the simplest generator configuration and is nearly ideal for micro and mini wind turbines. There is more diversity in household-size turbines, but again, nearly all use permanent-magnet alternators. Some manufacturers use Ferrite magnets, others use rare-earth magnets. The latter have a higher flux density than ferrite magnets. Both types do the job.

POWER CURVES

The power curve indicates how much power (in watts or kilowatts) a wind turbine's generator will produce at any given wind speed. Power is presented on the vertical axis, wind speed on the horizontal axis. In the advertising wars between wind turbine manufacturers, often the focus is the point at which the wind turbine reaches its "rated," or nominal, power. Though rated power is just one point on a wind turbine's power curve, many consumers mistakenly rely on it when comparison shopping. But not all power curves are created equal. Some power curves are, to be diplomatic, more "aggressive" than others.

Manufacturers may pick any speed they choose to "rate" their turbine. In the 1970s, it was easy for unscrupulous manufacturers to manipulate this system to make it appear that their turbines were a better buy than competing products. By pushing "rated power" higher, they were able to show lower relative costs (turbine cost/rated power) or they were able to increase their price—and profits—proportionally.

Unlike wind farm turbines, the performance of many small wind turbines has not been tested to international standards. As a rule, don't place much faith in power curves, unless the manufacturer has clearly stated the conditions under which the curve was measured (usually in a detailed footnote), or in "rated power." Some power curves can be off by 40% in low winds and as much as 20% at "rated" power. And it's the performance in lower winds that matters most.

Most homeowners seldom see their turbines operating in winds at "rated" speeds of, say, 28 mph. Small wind turbines operate most of the time in winds at much lower speeds, often from 10 to 20 mph. For size or price comparisons, stick with rotor diameter or swept area. Both are more reliable indicators of performance than power curves are.

Installing a 1,500-watt wind turbine on a tilt-up tower using an electric winch.

Bob and Ginger Morgan's Bergey Excel being installed by a crane near Tehachapi, California.
© Paul Gipe

The sleek Air 403
Industrial Turbine

Robustness

Wind turbines work in a far more rugged environment than photovoltaic panels that sit quietly on your roof. You quickly appreciate this when you watch a small wind turbine struggling through a gale. There's no foolproof way to evaluate the robustness of small wind turbine designs.

In general, heavier small wind turbines have proven more rugged and dependable than lightweight machines. Wisconsin's wind guru Mick Sagrillo is a proponent of what he calls the "heavy metal school" of small wind turbine design. Heavier, more massive turbines, he says, typically last longer. *Heavier* in this sense is the weight or mass of the turbine relative to the area swept by the rotor. By this criteria, a turbine that has a relative mass of $10kg/m^2$ may be more robust—and rugged—than a turbine with a specific mass of $5kg/m^2$.

Siting

To get the most out of your investment, site your wind turbine to best advantage: well away from buildings, trees, and other wind obstructions. Install the turbine on as tall a tower as you're comfortable working with. Jason Edworthy's experience at NorEnergy Systems in Canada convinces him that the old 30-foot rule still applies. This classic rule from the 1930s dictates that for best performance, your wind turbine should be at least 30 feet above any obstruction

---

Small Wind Turbines, *Cuisinart for birds or red herring?*

Avian mortality is an issue that always seems to come up with wind turbines. Do birds die from running into wind turbine blades, towers, and guy wires? Yes, they do. In large numbers? Hardly. Let's look at the facts. Birds die all the time, in great numbers, due to human endeavors. The number-one bird killer is glass-faced office buildings. These account for 100 million to 900 million bird deaths per year[1, 2]. Some skyscrapers have been monitored and shown to kill as many as 200 birds per day. Cars and trucks kill another 50 to 100 million birds per year[3, 4]. Cell phone, TV, radio, and other communications towers kill 4 million to 10 million birds per year[5]. And finally, Muffy the housecat and all her friends take at least 100 million birds a year, possibly many more[6, 7].

In comparison, the vast majority of large wind plants report zero bird kills[8], and even the famous Altamont Pass area, in which wind turbine bird kills were first reported, has about 1,000 bird deaths per year. This sounds like a lot, until we note that there are 5,400 turbines on this site. That gives an average of 0.19 bird kills per turbine per year[9]. The problem seems to be the abundance of prey in the tall grasses around the towers, and the fact that this is a heavily used migratory route and early tower designs provided a wealth of good perches.

That's the true story for large power-production turbines. Bird kills are a nonissue. In comparison, residential-size turbines and towers simply do not pose any significant risk to birds. The towers are too short, and the blade swept area is too small. The chance of your residential turbine ever hitting a bird are vanishingly small. If you or your neighbors are truly concerned about bird deaths, getting rid of Muffy would have much more real effect.

—Doug Pratt

1. Curry & Kerlinger, Dr. Daniel Klem of Muhlenberg College has done studies over a period of 20 years, looking at bird collisions with windows. www.currykerlinger.com/birds.htm.
2. Defenders of Wildlife, "Bird mortality from wind turbines should be put into perspective." www.defenders.org/habitat/renew/wind.html.
3. Curry & Kerlinger, "Those statistics were cited in reports published by the National Institute for Urban Wildlife and U.S. Fish and Wildlife Service." www.currykerlinger.com/birds.htm.
4. Defenders of Wildlife, "Bird mortality from wind turbines should be put into perspective." www.defenders.org/habitat/renew/wind.html.
5. Curry & Kerlinger, "U.S. Fish and Wildlife Service estimates that bird collisions with tall, lighted communications towers, and their guy wires result in 4 to 10 million bird deaths a year." www.currykerlinger.com/birds.htm.
6. Curry & Kerlinger, "The National Audubon Society says 100 million birds a year fall prey to cats." www.currykerlinger.com/birds.htm.
7. Defenders of Wildlife, the American Bird Conservancy estimates that feral and domestic outdoor cats probably kill on the order of hundreds of millions of birds per year (Case 2000). One study estimated that in Wisconsin alone, annual bird kill by rural cats might range from 7.8 million to 217 million birds per year (Colemen & Temple, 1995). www.defenders.org/habitat/renew/wind.html.
8. www.currykerlinger.com/studies.htm
9. www.nrel.gov/docs/fy04osti/33829.pdf

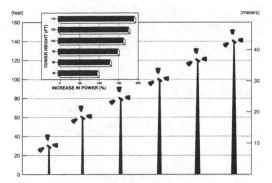

Increase in power with height above 30 ft (10m).

Chart adapted from *Wind Power for Home & Business.*

within 200 feet of the tower. Under the best of conditions, a tower height of 30 feet is the absolute minimum, says Edworthy.

Putting a turbine on the roof is no alternative. Seldom can you get the turbine high enough to clear the turbulence caused by the building itself. Imagine trying to mount a 30-foot-tall tower on a steeply pitched roof—it's a recipe for disaster. Even if you could, turbine-induced vibrations will quickly convince you otherwise. (It's like putting a noisy lawn mower on your roof.) While many small wind turbines are relatively quiet, some are not—another good reason to put them out in the open, well away from buildings.

Towers

Most small wind turbines are installed on guyed, tubular masts that are hinged at the base. With an accompanying gin pole, these towers can be raised and lowered, simplifying installation and service. Some tilt-up towers use thin-walled steel tubing, others use thick-walled steel pipe. Household-size turbines also use guyed masts of steel lattice as well as freestanding truss towers. With the advent of pre-engineered tubular mast kits, there's now less excuse than ever for installing micro and mini wind turbines on inadequate towers.

INSTALLATION

Those with good tool skills—who work safely—can install a micro turbine themselves using a pre-engineered tilt-up tower kit. Installing mini wind turbines, because of the greater forces involved, requires considerably more skill. Household-size turbines should be left only to professionals—and even the pros have made tragic mistakes. If you like doing the work yourself, start with a micro turbine and a tilt-up tower. Once you're satisfied you know what you're do-

ing and you're comfortable with the technology, you can try your hand with a larger turbine.

Typical Costs

The cost of a wind system includes the cost of the turbine, tower, ancillary equipment (disconnect switches, cabling, etc.), and installation. The total cost of a micro turbine can be as little as $1,000, depending upon the tower used and its height, while that of a household-size machine can exceed $50,000. When comparing prices, remember that bigger turbines cost more but often are more cost effective. For an off-the-grid power system, the addition of a wind turbine will almost always make economic sense by reducing the number of photovoltaic panels or batteries needed.

For grid-intertie systems, the economics depend upon the winds at your site and a host of other factors, including the average wind speed, the cost of the wind system, the cost of utility power, and whether your utility provides net billing.

Some Do's and Don'ts

Do plenty of research. It can save a lot of trouble—and expense—later.

Do visit the library. Books remain amazing repositories of information. (We *do* have a bias about books, since we write them. But we've always firmly believed that "you get what you pay for" when it comes to free information, whether it's from the Internet, manufacturers, or their trade associations.)

Do talk to others who use wind energy. They've been there. You can learn from them what they did right and what they'd never do again.

Do read and, equally important, follow directions.

Do ask for help when you're not sure about something. The folks at Real Goods Renewables are there to help.

The Bergey Excel Turbine

ESTIMATED ANNUAL ENERGY OUTPUT at Hub height in thousand kWh/y						
Avg Wind Speed (mph)	**Rotor Diameter, m (ft.)**					
	1 (3.3)	1.5 (4.9)	3 (9.8)	7 (23)	18 (60)	40 (130)
	thousands of kWh per year					
9	0.15	0.33	1.3	7	40	210
10	0.20	0.45	1.8	10	60	290
11	0.24	0.54	2.2	13	90	450

Do build to code. In the end, it makes for a tidier, safer, and easier-to-service system.

Do take your time. Remember, there's no rush. The wind will always be there.

Do be careful. Small wind turbines may look harmless, but they're not.

Don't skimp and **don't** cut corners. Taking shortcuts is always a surefire way to ruin an otherwise good installation.

Don't design your own tower—unless you're a licensed professional engineer.

Don't install your turbine on the roof despite what some manufacturers may say!

And, of course, **don't** believe everything you read in sales brochures.

In general, doing it right the first time may take longer and cost slightly more, but you'll be a lot happier in the long run. ■ ■

SOURCES OF INFORMATION

For more on micro turbines, see *Wind Energy Basics* (Chelsea Green Publishing, 1999), available from Real Goods (see page 533). It fully describes the new class of small wind turbines, dubbed *micro turbines*. These inexpensive machines, when coupled with readily available photovoltaic panels, have revolutionized living off the grid.

For more on small wind turbine technology, see *Wind Power: Renewable Energy for Home, Farm, and Business* (Chelsea Green Publishing, 1993), also available from Real Goods (see page 534). This book explains how modern, integrated wind turbines work and how to use them most effectively.

For information about the commercial wind power industry, see *Wind Energy Comes of Age* (John Wiley & Sons, 1995), a chronicle of wind energy's progress from its rebirth during the oil crises of the 1970s to its maturation on the

Bergey Excel and photovoltaic panels at the home of Dave Bittersdorf, founder of NRG Systems.

Real Goods Solar Living Center's Whisper 3000 wind turbine atop a hinged, tilt-up tower in Hopland, California.

plains of northern Europe in the 1990s. Selected as one of the outstanding academic books published in 1995.

To determine your local wind resources, visit http://rredc.nrel.gov/wind/pubs/atlas/ or search for the National Renewable Energy Laboratory's *Wind Energy Resource Atlas of the United States*.

For tips on installing micro and mini wind turbines using a Griphoist-brand winch, see "Get a Grip!" in *Homepower Magazine* #68 (December 1998/January 1999), or visit www.wind-works.org.

The Gipe Family Do-It-Yourself Wind Generator Slide Show

All photos: © Paul Gipe

1. Taking delivery of a new Air turbine and 45-foot tower.

2. Checking the packing list against parts delivered.

3. Securing the tower's base plate.

4. Aligning the guy anchors.

5. Driving the screw anchors.

6. Unspooling the guy cable.

7. Assembling the mast.

8. Clevis and gated fitting hook.

9. Gin pole and lifting cables.

10. Strain relief for supporting power cables.

11. Final assembly of the turbine.

12. Disconnect switch and junction box.

13. Slowly raising the turbine with a Griphoist-brand hand winch.

14. Air turbine safely installed with a Griphoist-brand hand winch.

WIND PRODUCTS

Site Evaluation Tools

NRG WindWatcher

The next generation of wind loggers. Collects and records wind speed and power; calculates average speed and average power density for accurate power predictions; delivers monthly averages from constant two-second data samples; and saves data at the end of each month. Runs for one year on a single D-cell battery. Includes anemometer with weather boot, stub mast, 100 feet of sensor wire, ground wire, battery, and Wind-Watcher with surface mounting box. 4.7 feet square. Designed for indoor or weather-protected mounting. Will track wind direction and deliver monthly average with optional direction vane kit (includes small separate mounting mast and 100-foot sensor wire). 2-yr. mfr. warranty. USA.

43-0154 NRG WindWatcher Wind Meter $395

43-0159 NRG Wind Direction Vane Kit $229

Wind Speed Indicator

Test how much power you'll get from a wind generator before buying one. Our accurate indicator includes a pickup vane and 50 feet of connecting tube.

16285   Wind Speed Indicator $79

Handheld Wind Speed Meter

This is an inexpensive and accurate wind speed indicator. It features two ranges, 2-10 mph and 4-66 mph. A chart makes easy conversions to knots. Speed is indicated by a floating ball viewed through a clear tube. Many of our customers are curious about the wind potential of various locations and elevations on their property but are reluctant to spend big money for an anemometer just to find out. Here is an economical solution. Try taping this meter to the side of a long pipe (use a piece of tape over the finger hole for high range reading) and have a friend hold it up while you stand back and read the meter with binoculars. This is helpful in determining wind speeds at elevations above ground level. Includes a protective carrying case and cleaning kit.

63205   Handheld Wind Speed Meter $20

Lightning Protection

Do you need lightning protection on a wind system? Without question, you need serious industrial-grade protection here, and it's surprisingly simple and inexpensive.

The first level of protection depends on you to ground the tower and the guy wires and to make sure ALL the grounds in your wind and household system are connected together. Lightning wants to go to ground. Make it easy, and give it a path outside your system wiring. The second level of protection is needed when lightning decides to take a trip through your system wiring. Delta arrestors will clamp any high-voltage

spike and direct most of the energy to ground. They will withstand multiple strikes until capacity is reached and then will rupture, indicating the need for replacement. Delta arrestors can be installed at any convenient point, or points, between the turbine and the battery. Multiple arrestors are a good idea in lightning-prone areas. Lightning does strike twice!

The three-wire DC model is for turbines like the Air X with two-wire DC output. Connect one wire to positive, one to negative, and one to ground. It can take up to 60,000 amps current, or 2,000 joules per pole. Response time to clamp is 5 to 25 nanoseconds depending on amp load.

The four-wire model is for turbines like the Whisper or Bergey with three-wire wild AC output. Connect one wire to each of the output wires and one to ground. It can take up to 100,000 amps current, or 3,000 joules per pole. Same clamp time as above. USA.

25194   DC 3-wire Lightning Protector $45

25749   AC 4-wire Lightning Protector $60

Wind Generators

Southwest Windpower Air-X

The Third Generation of North America's Most Popular Turbine

Southwest Windpower has introduced their third generation of the Air Turbine series, and things just keep getting better for this simple, attractive wind generator.

Proven features from previous models include a sleek noncorrosive cast-aluminum alloy body, three flexible carbon-reinforced blades, a neodymium permanent magnet alternator, and built-in regulation. What's improved is noise control and charge control. The sophisticated new microprocessor-based speed and charge control delivers better battery charging by optimizing the alternator output at all points of the power curve; it won't overcharge smaller battery packs; and it eliminates the blade "flutter" that was a noise problem at high wind speeds by actually controlling and limiting the blade speed.

Both land and marine models are rated at 400 watts output. Land models are unpainted; marine models are powder-coated white. Both feature a large LED in the underbelly to indicate output. Available in 12-, 24-, or 48-volt versions, the voltage cannot be changed in the field. See chart below for full specifications. 3-yr. mfr. warranty. USA.

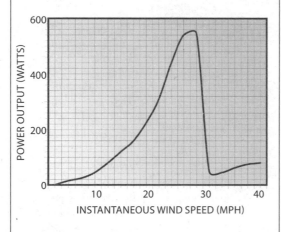

AIR-X TURBINE SPECIFICATIONS

	Land	Marine
Watts output	400	400
Available DC voltages	12/24/48	12/24/48
Rated wattage output	28 mph	28 mph
Max. wind speed	110 mph	110 mph
Cut-in wind speed	8 mph	8 mph
Number of blades	3	3
Rotor diameter	46 in.	46 in.
Tower top weight	13 lb.	13 lb.
Tower type	1.5-in. pipe*	1.5-in. pipe*

*Pipe size shown is for nominal steel pipe.

16277	Air-X Land Wind Turbine 12V	$725
16278	Air-X Land Wind Turbine 24V	$725
16280	Air-X Land Wind Turbine 48V	$725
16281	Air-X Marine Wind Turbine 12V	$915
16282	Air-X Marine Wind Turbine 24V	$915
16284	Air-X Marine Wind Turbine 48V	$915
16129	Air-X Land Wind Turbine Blade Set	$110

Installation Accessories for Air-X

Between your battery and turbine, you need overcurrent protection. If your DC center isn't providing it, Southwest recommends the simple surface-mounted breakers below. Use the larger 50-amp breaker for the 12-volt.

The stop switch disconnects the turbine from the battery and then connects the positive and negative to each other to give maximum braking. It's highly recommended for marine applications, especially if you plan to sleep on the boat. Turbines *do* make noise, and this will help a lot.

16153	Air-X 50-Amp Stop Switch	$32
48-0210	30-AMP Surface-Mount Breaker (for 24V/48V Air-X)	$50
48-0210	50 AMP Surface-Mount Breaker (for 12V Air-X)	$50

Southwest Windpower Air Industrial

The 400-watt industrial model is specially designed for survival in the most extreme environments: on remote mountaintops and off-shore platforms. It will survive winds in excess of 100 mph and temperatures down to –60°F for extended periods without attention. Features of this fully powder-coated model include anodized cooling fins for continuous high-power output, larger internal wiring, and special carbon fiber blades that are twice the strength of standard Air blades. These wind turbines require use of an external diversion control and an air-heater dump load. They're designed for hybrid solar/wind systems in telecommunication stations, offshore plat-forms, monitoring stations, lighthouses, and cathodic protection systems. 3-year mfr. warranty. Industrial models are made to order. Allow 8 weeks minimum delivery time. USA.

16-276	**Air Industrial Wind Turbine 12V**	**$1,187**
43-0222	**Air Industrial Wind Turbine 24V**	**$1,187**
43-0223	**Air Industrial Wind Turbine 48V**	**$1,187**
25-017	**Xantrex C-40 Diversion Load Controller**	**$169**
48-0081	**Air-Heater Diversion Load HL-100**	**$299**

SOUTHWEST WINDPOWER WHISPER WIND TURBINES

The Whisper wind turbines were upgraded and renamed in 2005. They feature a side-furling Angle-Governor to protect the turbine in high winds by turning the alternator and blades out of the wind, reducing turbine exposure. Unlike other wind turbines that lose as much as 80% of their output when furled, the Angle-Governor allows the Whisper to achieve maximum output in any wind.

Included with every Whisper is the new Whisper Controller, which includes a diversion load to ensure quiet, safe operation of your wind turbine when the batteries are charged. Add the optional LCD display and you will instantly receive real-time data on the performance of your Whisper wind turbine. The display can be mounted on the controller or a 1000' (300 meters) away letting you know what is happening. Now you can even add an anemometer to your controller, which lets you measure wind speed and compare it with the output to make sure you are getting the most from your Whisper.

Other features include field-adjustable voltage, a four-bearing spindle for efficiency, an upgraded yaw shaft, and a new bushing for smoother operation. The marine versions, designed for coastal and offshore applications, feature powder coating for corrosion protection, stainless steel hardware, marine-grade wire, and watertight housings. All Whisper models carry a 5-yr. warranty.

Whisper 100

Formerly the Whisper H40, the Whisper 100 was updated in 2005 with new features and added flexibility for your renewable energy system. Now any Whisper 100 can have its voltage changed, 12-48 volts DC, within a few minutes in the field. With its three-blade, 38-square-foot swept area, the Whisper 100 is designed to operate at a site with medium to high wind speed averages of 12 mph (5.4m/s) and greater. The 100 provides 100+kWh per month, or 3.4kWh per day, in a 12-mph average wind. Worried about noise? The Whisper 100 is one of the quietest wind turbines ever tested by the National Renewable Energy Labs.

The Whisper 100 is designed for even the novice. The Whisper can be installed in just a few hours with no welding, no concrete, and no cranes. The SWWP engineers designed the work out of installing wind turbines. Now wind is as easy to use as photovoltaics.

Whisper 200

Formerly the Whisper H80, the Whisper 200 was also updated for 2005. Now any Whisper 200 can have its voltage changed, 12-48 volts DC, within a few minutes in the field. With its three-blade, 64-square-foot swept area, the Whisper 200 is designed to operate at a site with low to medium wind speed averages of 8 mph (3.6m/s) and greater. The 200 provides 200+kWh per month, or 6.8kWh per day, in a 12-mph average wind. The Whisper 200 produces more noise than its 100 counterpart, but it produces almost twice the power.

Like the 100, the Whisper 200 is designed for the novice. It too can be installed in just a few hours with no welding, no concrete, and no cranes.

Whisper 500

Serious power from a medium-size small wind turbine. The Whisper 500 can produce enough energy to power an entire home. Formerly the Whisper 175, the Whisper 500 was completely redesigned in 2004 to work in harsh high-wind environments. The Whisper 500 is a two-bladed fiberglass-reinforced blade (177-square-foot swept area) and incorporates the patented side-furling design that optimizes output at any wind speed. Assuming a 12-mph (5.4m/s) wind, a Whisper 500 will produce as much as 500kWh per month. That is enough energy to power the average California home.

The Whisper 500 is NOT for the novice. Installation of the turbine requires concrete foundations and at times a crane to lift the wind turbine into place. Tower kits and anchors are offered below in Wind Accessories. All Whisper turbines feature a 5-yr. mfr. warranty. USA.

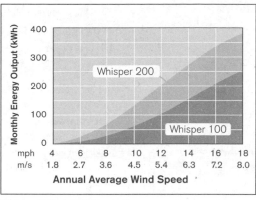

Whisper 100 and 200 Power Curve

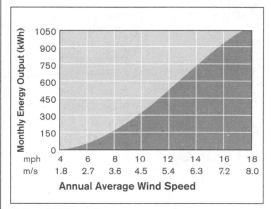

Whisper 500 Power Curve

Whisper Specifications

Model	Whisper 100	Whisper 200	Whisper 500
Swept area ft.^{2}	38	77	177
Rotor diameter	7 ft.	9 ft.	15 ft.
Tower top weight	47 lb.	65 lb.	155 lb.
Tower type	2.5-in. steel pipe	2.5-in. steel pipe	5-in. steel pipe
Max. design wind speed	120 mph	120 mph	120 mph
Cut-in wind speed	7.5 mph	7.0 mph	7.5 mph
Number of blades	3	3	2
Rated wattage (@ mph)	900 @ 28	1,000 @ 26	3,000 @ 24
Generator type	Brushless permanent-magnet alternator		
Available DC voltages	12-48	12-48	24-48

16239	Whisper 100 w/Controller	$2,475
16241	Whisper 200 w/Controller	$2,995
16242	Whisper 500 24V w/Controller	$7,675
63-0200	Whisper 500 48V w/Controller	$7,675
63-0201	Whisper 500 Grid-Tie w/ Controller and Inverter	$12, 125
43-0220	Whisper 100 Marine w/Controller	$2,725
43-0221	Whisper 200 Marine w/Controller	$3,275
43-0155	Whisper 200 Blade Set	$200
48-0084	Digital Display for 100/200 Controller	$109

Southwest Windpower Skystream 3.7 Grid-Tie Wind Turbine

The Skystream 3.7 is a new-generation residential wind generator that hooks up to your home to reduce or eliminate your monthly electric bill. It's the first all-inclusive wind generator (with controls and inverter built in—no batteries needed) designed to provide quiet, clean electricity in very low winds. It reaches rated

power of 1,800 watts at just over 22 mph (10m/s) with a maximum rotor speed of 325 rpm. Because of the exceptionally low rpm, the machine operates nearly sound-free. This makes Skystream ideal for residential homes and small businesses. In areas with an average 12-mph wind speed, the Skystream can produce about 400kWh per month. Assuming an installed cost (before any rebates or tax credits) of $9,000, the cost of the power it produces at $0.09/kWh is less than the retail rate of most utilities. The Skystream 3.7 can be installed on a range of tower heights from 35 to 110 feet. The optional 35-foot freestanding (no guy wire) tower looks much like a standard light pole. Other options include a wireless remote display and USB converter, which allows the display to connect to a PC. 5-yr. warranty. USA.

Specifications

Rated capacity	1.8kW
Weight	170 lb.
3-rotor diameter	12 ft. (3.72m)
Blade material	Fiberglass-reinforced composite
Cut-in wind speed	8 mph (3.5m/s)
Rated wind speed	22 mph (10m/s)
Survival wind speed	140 mph (63m/s)

43-0216 Skystream 3.7 Land 220V Complete GT Wind Turbine **$5,400**

43-0224 Skystream 3.7 Wireless Display Kit **$335**

43-0225 Skystream 3.7 USB Converter Kit **$99**

Kestrel Wind Turbines
Full Axial Flux Discoid Technology

The Kestrel line of small wind turbines are designed to survive and excel in the toughest of environments. They are built to the highest standards of material and workmanship by Eveready Diversified Products in South Africa, manufacturers of dry-cell batteries since the 1930s. Kestrel turbines represent a clear departure from the typical lightweight, mass-produced small turbines on the market today, in that each one is precision made of heavy-gauge steel, hand-laminated fiberglass, and stainless steel fasteners. Every Kestrel machine is electroplated, heavily primed, and then painted with aircraft-grade epoxy finish to produce a true "marine-grade" product at no extra cost.

The Kestrel 600 has proved itself to be an excellent choice for use in the harshest situations where high dependability is a must, such as remote telecommunications sites and farmsteads. Another remarkable feature of these turbines is their near-silent operation in all wind speeds. The 600 is best sited in moderate- to high-wind sites but will still work reasonably well in lower-wind areas, with many successfully employed in dense urban situations.

The Kestrel 800 is the low-wind version of the 600. It has a large 7-foot-diameter fiberglass rotor with a robust variable-pitch hub assembly, which varies the blade angle as a direct result of the actual speed of the rotor. This means that the turbine is not forced to turn away from the wind to protect itself but instead can remain producing full output when the storm winds blow. The 600 can be upgraded to an 800 by replacing the blades, hub, and shroud.

Kestrels are sold primarily as battery-charging units. Various charge controllers can be used with the Kestrels, but at this time we are recommending the TriStar as a diversion-load controller. Kestrel supplies its own dump-load resistors at the prices listed.

Kestrel Specifications		
Model	600	800
Swept area ft.²	19	41
Rotor diameter	5 ft.	7.2 ft.
Tower top weight	46 lb.	55 lb.
Max. design wind speed	90 mph	90 mph
Cut-in wind speed	5.5 mph	7 mph
Number of blades	6	3
Rated wattage (@ mph)	600 @ 30	800 @ 26
Generator type	Brushless permanent-magnet alternator	
Available DC voltages	12, 24, 48	12, 24, 48
Warranty	3 years	2 years

48-0500	**Kestrel 600 Wind Turbine (specify voltage)**	**$1,395**
48-0501	**Kestrel 600 Dump Load**	**$125**
48-0502	**Kestrel 800 Wind Turbine (specify voltage)**	**$1,995**
48-0503	**Kestrel 800 Dump Load**	**$218**
48-0010	**TriStar 45A Diversion Load Controller**	**$169**
48-0011	**TriStar 60A Diversion Load Controller**	**$218**

BERGEY WIND GENERATORS

Bergey wind generators are wonderfully simple: no brakes, pitch-changing mechanism, gearbox, or brushes. Bergey pioneered the automatic side-furling design that practically all wind turbines use now. This furling design forces the generator and blades partially out of the wind at high wind speeds but still maintains maximum rotor speed. AutoFurl uses only aerodynamics and gravity—there are no brakes, springs, or electromechanical devices to reduce reliability. These turbines are designed to operate unattended at wind speeds up to 120 mph.

Bergey XL.1 Wind Generator

The 1,000-watt XL.1 is Bergey's newest turbine. With 24-volt output, this wind charger is intended for off-the-grid or remote-location battery charging. This 2.5-meter (8.2-foot) turbine is designed for high reliability, low maintenance, and automatic operation under adverse conditions. The XL.1 features Bergey's proven AutoFurl storm protection, a direct-drive permanent-magnet alternator, excellent low-wind performance, and nearly silent operation thanks to the state-of-the-art, patent-pending, blade airfoils combined with the oversized, low-speed alternator. Low-wind-speed performance is enhanced by low-end boost circuitry that increases output at wind speeds down to 5.6 mph (2.5m/s).

The BWC PowerCenter provides battery-friendly constant voltage charging and will control up to 30 amps of PV input also. The simple controller shows battery voltage and which charging sources are active with at-a-glance LED indicators. Excellent 5-yr. mfr. warranty. China & USA.

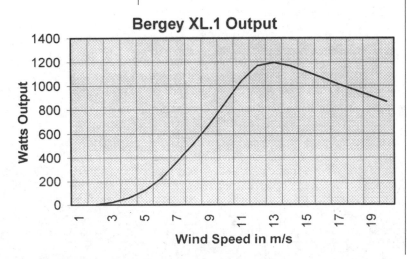

Bergey XL.1 Output — Watts Output vs. Wind Speed in m/s

Bergey XL.1 Wind Turbine Specifications	
Model	24VDC
Swept area ft.²	52.8
Rated wattage (@ mph)	1,000 @ 25 (11m/s)
Cut-in wind speed	6 mph (2.5m/s)
Max. design wind speed	120 mph (54 m/s)
Generator type	Brushless Permanent Magnet, 3-phase
Rotor diameter	8.2 ft. (2.5 m)
Tower top weight	75 lb. (34kg)
Tower stub	Proprietary

16243	**Bergey XL.1 w/Controller (24V only)**	**$2,590**

Proven Wind Turbines

The toughest and quietest turbine for residential use

Proven turbines are built in Scotland, where they have an intimate familiarity with some of the world's harshest weather and most brutal wind turbine conditions. Developed for the ferocious wind climates of northern Scotland, and built with marine quality using all noncorrosive components, Proven turbines have been put to the test in some of the world's worst weather locations and will last a lifetime. When wind turbulence is unavoidable, this is the turbine you want to fly! These turbines belong to the "heavy metal" school of wind machines. They are heavy, tough, slow survivors. Proven turbines are all very slow speed, very quiet turbines. Most owners report they can't hear, or can just barely hear, the turbine even in the strongest winds. Proven turbines use an extremely low rpm brushless permanent-magnet alternator. Low rpm means a low blade tip speed and almost silent operation. In an extremely strong 40-mph wind, they make 55-65 decibels, about the same as a normal conversation. Lower wind speeds are even quieter. The smallest 600-watt turbine spins faster than the larger machines and will be noisier.

Assembling a WT 2500

These unique turbines are all "downwind machines." There's no tail; the blades run on the downwind end. As wind speeds increase, the very clever and robust polypropylene blades hinge inward, forming a progressively smaller cone area for the wind to push against. There is no wind speed at which these turbines stop producing, up to 145 mph!

Three models are available at 600, 2,500, and 6,000 watts. We have put together packages for each model that include the turbine, an appropriate charging/rectifier regulator with diversion control, and the required shipping box. Metering is standard on the 2,500- and 6,000-watt turbines; on the 600-watt turbine, metering is available as an option at 48 volts. Voltage must be specified at time of purchase. Attractive self-supporting

tilt-up monopole masts are available as well as conventional guyed towers. Please see Wind Accessories. 2-yr. mfr. warranty. Scotland.

Pricing subject to fluctuation with international currency. Delivery is 5-6 weeks air freight, 10-12 weeks surface.

Proven Turbine Specifications

Model	WT600	WT2500	WT6000
Swept area ft.^{2}	61.5	104	256.0
Available DC voltages	12, 24, 48	24, 48,	48
Rated wattage	600 @	2,500 @	6,000 @
(@ mph)	22.5	26.8	24.6
Cut-in wind speed	6 mph	6 mph	6 mph
Max. design wind speed	145 mph	145 mph	145 mph
Generator type	Brushless permanent magnet		
Rotor diameter	8.4 ft.	11.5 ft.	18
Tower top weight	154 lb.	419 lb.	1,102 lb.
Tower stub	Use Proven tower or adapter		

43-0176 Proven WT600/12V Turbine Package $5,250

43-0177 Proven WT600/24 or 48V No-Meter Turbine Package (specify voltage) $5,185

43-0178 Proven WT600/48V w/Meter Turbine Package $5,540

43-0179 Proven WT2500/24 or 48V w/m Meter Turbine Package (specify voltage) $11,500

43-0181 Proven WT6000/48V w/Meter Turbine Package $20,360

WT 6000 on mono-pole mast

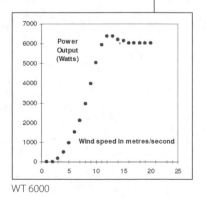

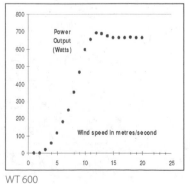

WT 600

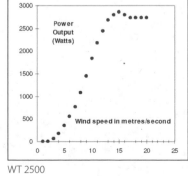

WT 2500

WT 6000

Wind Accessories

Griphoist Hand Winches

The Griphoist is an extraordinarily safe portable hand winch that can be used in any position, with any length of cable, going in or out. This is THE tool for safe, controlled raising and lowering of small residential wind plants with tilt-up towers, such as we offer below. The

Griphoist holds in position the instant the hand lever is released. It cannot get away from you like traditional drum winches. The cable is special and must be ordered separately. You will need a lifting cable approximately twice as long as your tower height. The Pull-All model is acceptable for the lightest turbines, the Air-X for instance. The Super Pull-All model is acceptable for turbines up to about 250 pounds with shorter towers. The Tirfor 508D is a safe choice for turbines 200 pounds and up, or turbines on towers 100 feet or taller.

	Pull-All	Super Pull-All	Tirfor 508D
Lifting capacity	700 lb.	1,500 lb.	2,000 lb.
Weight	3.85 lb.	8.25 lb.	17 lb.
Supplied with	32 ft. of ³⁄₁₆-in. wire rope	32 ft. of ¼-in. wire rope	30 ft. of ⁵⁄₁₆-in. wire rope
Item #	**43-0196**	**43-0202**	**43-0208**
Price	**$165**	**$550**	**$595**
Add'l Cable, 75 Ft.	43-0197 $110	43-0203 $80	43-0209 $110
Add'l Cable, 100 Ft.	43-0198 $145	43-0204 $105	43-0210 $145
Add'l Cable, 150 Ft.	43-0199 $220	43-0205 $160	43-0211 $225
Add'l Cable, 200 Ft.	43-0200 $290	43-0206 $210	43-0212 $290
Add'l Cable, 250 Ft.	43-0201 $365	43-0207 $260	43-0213 $310

TOWER KITS FOR WIND TURBINES

The basic rule of thumb when installing any wind turbine is: Be 30 feet higher than any trees, buildings, or other obstructions within a 200- to 300-foot radius. Anything that ruffles up the wind before it hits your turbine is going to reduce the power available, and it's going to beat up your turbine. You really want to be up in the clean, smooth airflow. Plus, there's more power available as you go up. At 100 feet, there's about 40% more energy than at ground level. "The taller the tower, the greater the power." Don't skimp on your tower!

Towers come in a variety of styles. The most basic division is between self-supporting towers, like the old water pumpers used, and guy-wire supported towers. Guyed towers are significantly less expensive and are usually the tower of choice for smaller homestead turbines. Tubular fold-over guyed towers that use steel pipe or lighter-weight steel tubing for the actual tower, with guy-wire supports, have become the most common do-it-yourself tower type. This style completely assembles the tower, guy-wire system, and turbine while safely on the ground, then stands it all up. Rigging and setup takes awhile during initial assembly, but standing up or folding down the tower takes only a few minutes, and everything happens with your feet firmly planted on the ground.

Most of the tower kits offered below are tilt-up tubular guyed towers.

Roof Mounting Kit for Air-X

We don't recommend this mounting option for any building that people plan to sleep (or lose sleep) in. Wind turbines make noise, and it transmits down the tower very nicely. This kit includes vibration isolators, which helps, but they won't entirely cure the noise problem. This kit also includes clamps, straps, safety leashes, bolts, and roof seal. Mounting pipe and lag bolts are not included. USA.

16235   Air-X Roof Mount w/Seal $143

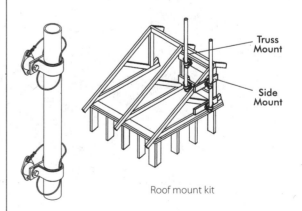

Truss Mount

Side Mount

Roof mount kit

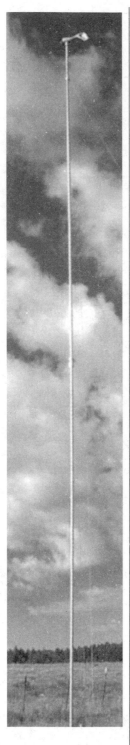

Tower Kits for Air-X

These low-cost, guyed, tubular tower kits provide all the hardware, cabling, and instructions for assembling a standup tower. Pipe and anchors are not included. 1.5-inch nominal steel pipe can be purchased locally; guy anchors are offered below. 36-inch anchors recommended with a 25-foot tower; 48-inch anchors with a 47-foot tower. USA.

16223   **Air-X 27′ Guyed Tower Kit**   **$189**

16234   **Air-X 45′ Guyed Tower Kit**   **$280**

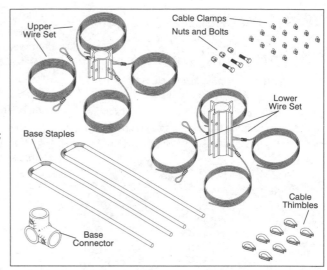

Mounting Kits for Air-X Marine

Here's a quality 9-foot mast mount that's built to withstand hurricane-force winds and looks good, too. Includes white powder-coated aluminum poles, all stainless steel hardware, self-locking nuts, and vibration dampening mounts. Main mast is 1.9 inches o.d. (nominal 1.5-inch pipe); stays are 1 inch o.d. Masts and hardware kits are sold separately for those who want to supply their own masts. USA.

16148   **Air-X Marine Tower Hardware Kit** **$199**

16149   **Air-X Marine Tower 9′ Mast Kit**   **$220**

Our powder-coated tower kits make it easy to install the Air marine

Tower Kits for Whisper 100 or 200 Turbines

These guyed, tubular tower kits provide all the hardware, couplings, hinged base plate, cabling, and instructions for assembling a tilt-up tower. Pipe and anchors are not included. Steel pipe can be purchased locally; guy anchors are offered below. See chart below for anchor recommendations. See Tower Sizing graphic for layout and amount of ground space required. All these towers will fit either the 100 or the 200 turbine and require nominal 2.5-inch pipe (which is actually 2.875 inches o.d.). USA.

All tower kits drop-shipped from manufacturer.

Item #	Tower Height	Anchor Hole Dimensions	Guy Wire Radius	Auger Size	Price
16256	24'	1.5' dia., 3' deep, 12" concrete fill	15'	36"	**$290**
16257	30'	2' dia., 3' deep, 9" concrete fill	20'	48"	**$495**
16258	50'	2' dia., 3' deep, 10" concrete fill	25'	48"	**$619**
16259	65'	2' dia., 3' deep, 12" concrete fill	33'	60"	**$795**
16260	80'	2' dia., 4' deep, 12" concrete fill	40'	60"	**$949**

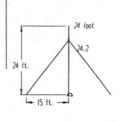

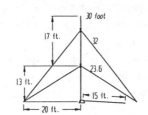

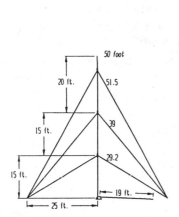

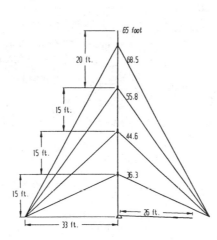

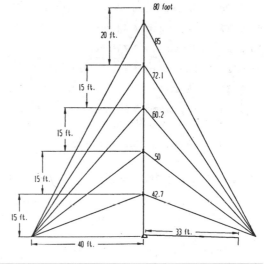

Anchors for Wind Towers

These anchor augers are sold in sets of four. With smaller turbines, you can simply screw them into place. No digging! With larger turbines, you need to dig anchor holes, per the recommendations above, and fill the bottom with concrete. Not for use with the Whisper 500 or Skystream 3.7. USA.

16-253 36" Galvanized Anchor Augers for 24' Wind Towers (Air/100/200) $199

16-254 48" Galvanized Anchor Augers for 30', 50' Wind Towers (Air/100/200) $225

16-255 60" Galvanized Anchor Augers for 65', 80' Wind Towers (Air/100/200) $270

*Means shipped directly from manufacturer.

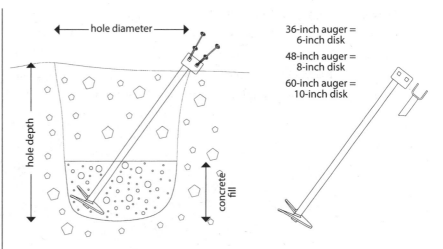

36-inch auger = 6-inch disk

48-inch auger = 8-inch disk

60-inch auger = 10-inch disk

Tower Kits for Whisper 500 and Skystream 3.7 Turbines

These guyed, tubular tower kits provide all the hardware, couplings, hinged base plate, cabling, and instructions for assembling a standup tower. Pipe and anchors are not included. Steel pipe can be purchased locally. See Tower Sizing graphic for layout and amount of ground space required. These towers will fit only the 500 turbine and require nominal 5-inch pipe (which is actually 5.563 inches o.d.) and concrete foundations. The Skystream Monopole tower does not require guy wires. USA.

All tower kits drop-shipped from manufacturer.

16249	30' Guyed Wind Tower Kit for WHI500/SS 3.7 (Guy wire radius = 20')	$980
16250	42' Guyed Wind Tower Kit for WHI500/SS 3.7 (Guy wire radius = 25')	$1,060
16251	70' Guyed Wind Tower Kit for WHI500/SS 3.7 (Guy wire radius = 35')	$1,499
43-0218	33' Monopole Wind Tower Kit for Skystream 3.7	$1,625
43-0227	Skystream 3.7 Tower Bolt Kit (for pier foundation)	$450
43-0228	Skystream 3.7 Tower Bolt Kit (for mat foundation)	$290
43-0229	Skystream 3.7 Tower Gin Pole Kit	$350
43-0230	Skystream 3.7 Tower Hinge Plate Kit	$299
43-0231	Skystream 3.7 Tower Install Kit (Includes Gin Pole and Hinge Plate Kits)	$599
43-0233	Skystream 3.7 Tower Adapter Kit (5")	$139

Towers for Bergey XL.1 Turbines

These are complete tilt-up, guyed tower kits that include galvanized tubular tower sections with associated hardware, guy anchors, guy wires, and gin pole. Only the tower electrical wiring is needed to complete your installation.

The 18-, 24-, and 30-meter towers have 3-meter long, 4.5-inch diameter tower sections, and are shipped via common carrier (semi truck).

16263	Bergey XL.1 Tower Kit, 18m (59')	$1,595
16264	Bergey XL.1 Tower Kit, 24m (78')	$2,050
16265	Bergey XL.1 Tower Kit, 30m (98')	$2.999
16245	Bergey XL.1 Tower Raising Kit	$355
16286	Bergey XL.1 Marine Anticorrosion Protection	$390

Towers for Proven Wind Turbines

Proven provides aesthetically pleasing self-supporting monopole towers. These attractive, uncluttered pole mounts are equipped with a hinged base and come with a gin pole for easy raising and lowering.

Conventional guyed tilt-up pipe tower kits are also offered. These include guy cables, galvanized turnbuckles, anchors, couplings, pivot assembly for base, hardware, and installation manual. You supply the 5-inch schedule 40 steel pipe. Also included with these kits is a Pulley Kit that contains the pulleys, cable, and hardware required to raise the tower. No other wind turbine manufacturer is offering such a complete package. Equip yourself with the appropriate griphoist and you'll be prepared to raise or lower your turbine anytime.

If you're providing your own tower, rather than one of Proven's kits, you'll need the appropriate tower adapter for connecting your turbine.

TILT-UP MONOPOLE TOWERS FOR PROVEN TURBINES

43-0183	5.5m (18') Monopole for WT600	$2,315
43-0184	6.5m (21') Monopole for WT2500	$3,600
43-0185	11m (36') Monopole for WT2500	$6,725
43-0186	9m (30') Monopole for WT6000	$6,500
43-0187	15m (49') Monopole for WT6000	$8,800

TILT-UP GUYED PIPE TOWER KITS FOR PROVEN TURBINES

43-0188	13m (42') Guyed Tower Kit for WT2500	$3,100
43-0189	20m (63') Guyed Tower Kit for WT2500	$3,960
43-0190	27.5m (84') Guyed Tower Kit for WT2500	$4,985
43-0191	34m (105') Guyed Tower Kit for WT2500	$5,790
43-0192	41m (126') Guyed Tower Kit for WT2500	$6,865

Please call for towers for other turbines.

PROVEN TOWER ADAPTERS

43-0193	Proven WT600 Tower Adapter	$245
43-0194	Proven WT2500 Tower Adapter	$345
43-0195	Proven WT6000 Tower Adapter	$465

Books

Wind Power: Renewable Energy for Home, Farm, and Business

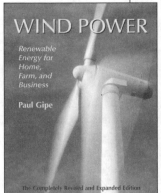

Wind energy technology has truly come of age—with better, more reliable machinery and a greater understanding of how and where wind power makes sense. In addition to expanded sections on gauging resources and turbine siting, this thoroughly updated edition contains a wealth of examples, case studies, and human stories. Covers everything from small off-grid to large grid-tie, plus water pumping. This is the wind energy book for the industry. Everything you ever wanted to know about wind turbines. Has an abundance of pictures, charts, drawings, graphs, and other eye candy, with extensive appendixes. By Paul Gipe. 504 pages, softcover.

21-0343 Wind Power: Renewable Energy for Home, Farm, and Business $45

Wind Energy Basics

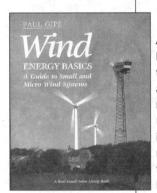

A Guide to Small and Micro Wind Systems

By Paul Gipe. A step-by-step manual for reaping the advantages of a windy ridgetop as painlessly as possible. Chapters include wind energy fundamentals, estimating performance, understanding turbine technology, off-the-grid applications, utility intertie, siting, buying, and installing, plus a thorough resource list with manufacturer contacts and international wind energy associations. A Real Goods Solar Living Book. 222 pages, paperback.

80928   Wind Energy Basics $19.95

Residential Wind Power with Mick Sagrillo

Mick Sagrillo, the former owner of Lake Michigan Wind and Sun, has been in the wind industry for over 16 years, with over 1,000 turbine installation/repairs under his belt. This professionally produced video includes basic technology introductions, hands-on applications, and performance and safety issues. The action moves into the field to review actual working systems, including assembling a real system. Tape runs 63 minutes. Standard VHS format.

80363   Renewable Experts Tape—Wind $39

Hydrogen Fuel Cells

Power Source of Our Bright and Shining Future?

Fuel cells seem to be the darlings of the energy world lately. If we believe all the hype, they'll bring clean, quiet, reliable, cheap energy to the masses, allowing us to continue an energy-intensive lifestyle with no penalties or roadblocks. Are fuel cells going to save our energy-hog butts? Maybe. They sure will help. Can you buy one yet? Certainly . . . for a price. It helps if you're the Department of Defense or have a fat government grant or contract in your back pocket. This is still a very young, very rapidly developing technology. The first small commercial units started appearing in 2003 for relatively outrageous prices. Residential units that will run off natural gas or propane to provide backup power probably won't show up until at least 2010, and they're liable to be fairly expensive. Want a bit more background info before you order one? Good plan, read on.

Traditional energy production, for either electricity or heat, depends on burning a fuel source like gasoline, fuel oil, natural gas, or coal to either spin an internal combustion engine or to heat water, then piping the resulting steam or hot water to warm our buildings or run the turbines to make electricity. Burning fuels produce some byproducts we'd be better off not letting loose into the environment, and every energy conversion costs some efficiency. The increasing industrialization of the world is requiring more and more fuel, creating our current unsustainable run on the world's energy reserves. The ultimate effect of all this fossil fuel burning is to create global climate change and global warming. Most scientists predict that if we fail to curb our fossil fuel consumption, the average world temperature will rise 7-18 degrees Fahrenheit (4-10 degrees Celsius) in the next 50-100 years, and the oceans will rise 3-8 feet in the same time. Other, even less savory results of global warming are still speculative and debatable. What is known is that we're messing with things we don't understand, can't fix quickly, and almost surely won't enjoy. For a sobering perspective, feel free to review Climate Change and the Need to Eliminate Fossil Fuels in Chapter 1.

Fuel cells are chemical devices that go from a source, like hydrogen or natural gas, straight to heat and electrical output, without the combustion step in the middle. Fuel cells increase efficiency by two- or three-fold and dramatically reduce the unintended byproducts. A fuel cell is an electrochemical device, similar to a battery, except fuel cells operate like a continuous battery that never needs recharging. So long as fuel is supplied to the negative electrode (anode) and oxygen, or free air, is supplied to the positive electrode (cathode), they will provide continuous electrical power and heat. Fuel cells can reach 80% efficiency when both the heat and electric power outputs are utilized.

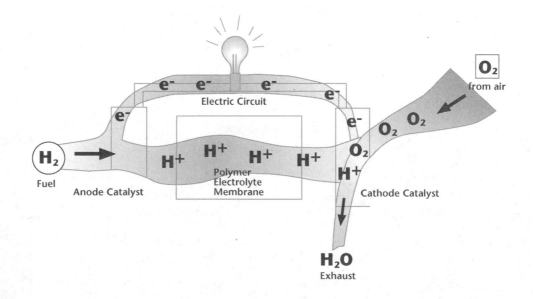

HOW FUEL CELLS OPERATE

A fuel cell is composed of two electrodes sandwiched around an electrolyte material. Hydrogen fuel is fed into the anode. Oxygen (or free air) is fed into the cathode. Encouraged by a catalyst, the hydrogen atom splits into a proton and an electron. The proton passes through the electrolyte to reach the cathode. The electron takes a separate, outside path to reach the cathode. Since electrons flowing through a wire is commonly known as "electricity," we'll make those free electrons do some work for us on the way. At the cathode, the electrons, protons, and oxygen all meet, react, and form water. Fuel cells are actually built up in "stacks" with multiple layers of electrodes and electrolyte. Depending on the cell type, electrolyte material may be either liquid or solid.

Like other electrochemical devices such as batteries, fuel cells eventually wear out and the stacks have to be replaced. Stack replacement cost is typically 20%-25% of the initial fuel cell cost. Current designs have run times of 40,000-60,000 hours. Under continuous operation that's 4.5-6.8 years.

WHAT'S "FUEL" TO A FUEL CELL?

In purest form, a fuel cell takes hydrogen, the most abundant element in the universe, and combines it with oxygen. The output is electricity, pure water (H_2O), and a bit of heat. Period. Very clean technology! No nasty byproducts and no waste products left over. There are some real obvious advantages to using the most abundant element in the universe. We've got plenty of it on hand, and as most anyone can clearly see, a hydrogen-based economy is far more clean and sustainable than a petroleum-based economy.

On the downside, hydrogen doesn't usually exist in pure form on Earth. It's bound up with oxygen to make water, or with other fuels like natural gas or petroleum. If you run down to your friendly local gas supply to buy a tank of hydrogen, what you get will be a byproduct of petroleum refining.

Since hydrogen probably isn't going to be supplied in pure form, most commercial fuel cells have a fuel-processing component as part of the package. Fuel processors, or "reformers," do a bit of chemical reformulation to boost the hydrogen content of the fuel. This makes a fuel cell that can run on any hydrocarbon fuel. Hydrocarbon fuels include natural gas, propane, gasoline, fuel oils, diesel oil, methane, ethanol, methanol, and a number of others. Natural gas and propane are favored for stationary genera-

tors, and methanol and even gasoline have been used in experimental automotive fuel cells. But if you feed the fuel cell something other than pure hydrogen, you're going to get something more than pure water in the output, carbon dioxide usually being the biggest component. Fuel cells emit 40%-60% less carbon dioxide than conventional power generation systems using the same hydrocarbon fuel. Other air pollutants such as sulfur oxides, nitrogen oxides, carbon monoxide, and unburned hydrocarbons are nearly absent, although you'll still get some trace byproducts we'd be better off without.

The dream is to build a fuel cell that accepts water input. But it takes more energy to split the water into hydrogen and oxygen than we'll get back from the fuel cell. One potential future scenario has large banks of PV cells splitting water. The hydrogen is collected and used to run cars and light trucks. Efficiency isn't great, but nasty byproducts are zero.

Competing Technologies

Just as there's more than one "right" chemistry with which to build a battery, there are quite a number of ways to put together a functional fuel cell. As an emerging technology, several fuel cell chemical combinations are receiving experimental interest, substantial funding, and absolutely furious development. At the risk of boring you, or scaring you all away, we're going to list the major chemical contenders for fuel cell power and their strong or weak points.

PHOSPHORIC ACID

This is the most mature fuel cell technology. Quite a number of 200-kilowatt demonstration/experimental units are in everyday operation at hospitals, nursing homes, hotels, schools, office

The typical phosphoric acid fuel cell is about the size of a freight car. This one's in New York's Central Park.

buildings, and utility power plants. The Department of Defense currently runs about 30 of these with 150- to 250-kilowatt output. If you simply must have a fuel cell in your life right now, this is your baby. They are available for sale . . . for a price. Currently that price is about $3,000 per kilowatt. These are large, stationary power plants, usually running on natural gas. Phosphoric acid cells do not lend themselves well to small-scale generators. Locomotives and buses are probably about as small as they will go. Operating temperatures are 375°-400°F, so they need thermal shielding. Efficiency for electric production alone is about 37%; if you can utilize both heat and power, efficiency hits about 73%. Stack life expectancy is about 60,000 hours.

PROTON EXCHANGE MEMBRANE

PEM is the fuel cell technology that is probably seeing the most intense development and is the type of fuel cell you are most likely to see in your lifetime. Because it offers more power output from a smaller package, low operating temperatures, and fast output response, this is the favored cell technology for automotive and residential power use. Thanks to their low noise level, low weight, quick start-up, and simple support systems, experimental PEM fuel cells have even been produced for cell phones and video cameras. Recent advances in performance and design raise the tantalizing promise of the lowest cost of any fuel cell system. Current prices are about $3,000 per kilowatt, but they will drop as mass-production economies begin to kick in. Operating temperature is a very low 175°-200°F, so minimal thermal shielding is needed. Efficiency for electric production alone is about 36%, and if you can utilize both heat and power, efficiency hits about 70%. PEM cells can start delivering up to 50% of their rated power at room temperature, so start-up time is minimal. Stack life expectancy is about 40,000 hours.

A few small demonstration PEM fuel cells are available for sale currently, but for something that's going to run your house, you'll need some patience. A great many manufacturers have residential units under development, and many are in beta testing; and if you want it bad enough, you can find one, although this technology isn't expected to fully mature till sometime after 2010. For current information, visit the nonprofit Fuel Cells 2000 Web site at www.fuelcells.org.

MOLTEN CARBONATE AND SOLID OXIDE

These two fuel cell technologies aren't related, except that they are both large base-load type cells that utility companies can use to supply grid power. These technologies have barely delivered their first prototypes, so operational plants are still far in the future. Very high operating temperatures, 1,200°-1,800°F, are the norm for these cells. Electrical generation efficiencies are over 50%, and pollution output is way, way down, so these technologies are going to be very attractive to utility companies. The high operating temperatures mean that these cells don't turn on or off casually. Start-up can take over 10 hours. These aren't fuel cells for off-the-grid homes.

Cruisin' the City on Fuel Cells?

One of civilization's biggest fossil fuel uses, and the one we seem to have the hardest time weaning ourselves from, is automobiles. Do fuel cells offer any hope for hopeless driving addicts? Yes, they do. Every major auto manufacturer has a well-funded, active fuel cell development program. Several manufacturers, Daimler-Chrysler, Toyota, and Honda, in particular, have several generations of prototypes behind them already. The stakes are high. The company that develops a workable fuel cell-powered auto holds the keys to the future. Several manufacturers are aiming to make the first commercial models to be available. These will probably be methanol-fueled PEM cells with a modest battery or capacitor pack to help meet sudden acceleration surges.

One of the primary problems has been the size of the fuel cell stack required. Mercedes' first prototype was a small bus; its second prototype was a van; and its latest, fifth prototype is in the new compact A-series sedan. Each

PEM cells are smaller, run at lower temperatures, and are more consumer friendly.

prototype has room for just a driver and passenger. So cell stacks are getting smaller and more efficient, but we could still stand some improvement (or an expectation adjustment).

Crash-safe hydrogen storage in a vehicle also continues to be a trying problem. We demand much safer vehicles than we used to. The first generations of fuel cell vehicles will be using on-board reformers with methanol or other liquid fuels, so this isn't a problem that requires an immediate solution.

All in all, we're skeptical about the near-term potential of hydrogen fuel cells to reduce greenhouse gas emissions in the transportation sector. For the foreseeable future, hybrid electric and biofuel-powered vehicles are a much better, more realistic bet. You'll find a fuller explanation of this subject in Chapter 12, "Sustainable Transportation."

The Mazda Premacy FC-EV, a protorype of a fuel cell car.

How Do Fuel Cells Compare with Solar or Wind Energy?

As an electrical supplier, a fuel cell is closer kin to an internal combustion engine than to any renewable energy source. Think of a fuel cell as a vastly improved generator. It burns less fuel, it makes less pollution, there are no moving parts, and it makes hardly any noise. But unless you have a free supply of hydrogen, it will still use nonrenewable fossil fuels for power, which will cost something.

Photovoltaic modules, wind generators, hydro generators, and solar hot-water panels all use free, renewable energy sources. No matter how much you extract today, it doesn't impact how much you can extract tomorrow, the next day, and so on. This is the major difference between technologies that harvest renewable energy, and fuel cells, which are primarily going to continue using nonrenewable energy sources.

In the long run, we think that fast-starting PEM fuel cells will take the place of back-up generators in residential energy systems. They'll be in place to pick up the occasional shortfall in renewable-powered off-the-grid systems or to provide back-up AC power for grid-supplied homes when the utility fails. If your home is off the grid, then solar, wind, or hydro is still going to be the most cost-effective and environmentally responsible primary power source. In an ideal world, every back-up power fuel cell would be supplied by a small PV-powered hydrogen extractor that cracks water all day and stores the resulting hydrogen for later use. Guess we'd better get busy designing renewable energy-powered hydrogen extractors.

CHAPTER 4

From Panel to Plug

How to Get Your Renewable Energy Source Powering Your Appliances

A Renewable Energy System Primer: How All the Pieces Fit Together

IN OTHER CHAPTERS AND SECTIONS of the *Sourcebook*, we provide details about how individual components such as PV modules or inverters work, and why they're needed. This is the place where we explain how and why all the bits and pieces fit together. Read this Renewable Energy System Primer first to get an overall idea of how everything works, and then move on to details as needed.

> The battery is your energy reservoir; it stores up electrons until they're needed to run something.

Standalone Systems (Battery Backup)

Explanations in this section will revolve around systems that use batteries and are capable of standalone operation. If your interest lies in simple utility intertie systems with no battery backup, please see the separate section dealing with utility intertie, which immediately follows on pages 177-196.

IT ALL REVOLVES AROUND THE BATTERY

The battery, or battery pack, is the center of every standalone renewable energy system. Everything comes and goes from the battery, and much of the safety and control equipment is designed either to protect the battery or to protect

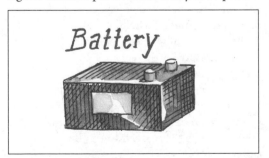

other equipment from the battery. The battery is your energy reservoir; it stores up electrons until they're needed to run something.

The size of your reservoir determines how many total electrons you can store but has minimal effect on how fast electrons can be poured in or out, at least in the short term. A smaller battery will limit how long your system will run

without recharging. A larger battery will allow more days of run time without recharging. However, there are practical limits in each direction. Batteries are like the muscles of your body: They need some exercise to stay healthy but not too much or they get worn out at an early age. We usually aim for three to five days' worth of battery storage. Less than three and your system will be cycling the battery deeply on a daily basis, which is bad for the battery. More than five and your battery starts getting more expensive than a back-up generator or other recharging source.

Once you start removing electrons from your battery, you'd better have a plan to replace them or you'll soon have an empty reservoir—what we call a dead battery—which brings us to . . .

THE CHARGING SOURCE

Batteries aren't the least bit fussy about where their recharging electrons come from. So long as the voltage from the charging source is a bit

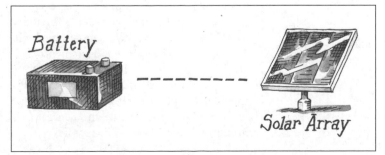

higher than the battery voltage, the electrons will flow "downhill." You can haul your battery into town to get it charged, or you can connect it to your vehicle charging system and charge it on the way to town and back. But batteries are heavy, and the sulfuric acid in them just loves to eat your clothes, so hauling batteries around is no fun. You'll want an on-site charging source. This can be a fossil-fueled (gasoline or propane) generator, a wind turbine, a hydroelectric generator, solar modules (PVs), a stationary bicycle generator, or combinations of these.

As long as incoming energy keeps up with outgoing energy, your battery will remain happy. Up to about 80% of full charge, a battery will accept very large current inputs with no problem. As it gets closer to full, we have to start applying some charge control to prevent overcharging. Batteries will be damaged if they are severely or regularly overcharged.

BATTERIES NEED CONTROL

Someplace in your system, there needs to be a charge controller. Charge controllers use various strategies, but they all have a common goal: to keep the battery from overcharging and becoming toasted. A battery that is regularly overcharged will use excessive amounts of water, may run dry, and sheds flakes of active lead from the battery plates, reducing future capacity. A charge controller watches the battery voltage, and if it gets too high, the controller will either open the charging circuit or dump excess electrons into a heater element.

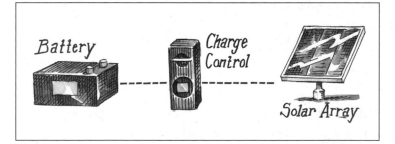

A charge controller watches the battery voltage, and if it gets too high, the controller will either open the charging circuit or dump excess electrons into a heater element.

The best monitors stand alone and get mounted in the kitchen or living room where they're more likely to get some regular attention.

We use different strategies depending on the charging source. Solar modules can use simple controls between the solar array and the battery that just open the circuit if voltage gets too high. Rotary generators, such as wind and hydroelectric turbines, use a controller that diverts excess energy into a heater element. (Really bad things happen if you open the circuit between a generator and the battery while the generator is spinning!)

Now you have a battery, a way to recharge it, a controller to prevent overcharging, but no clue how full or empty it is. You need a system monitor.

MONITORING YOUR SYSTEM

System monitors all have a common goal: to keep you informed about your system's state

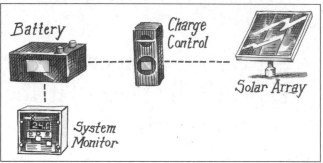

of charge and help you prevent the exceptionally deep discharges that sap the life out of your batteries. Monitors can be as simple as a red or green light, or as complicated as a PC-linked system that logs dozens of system readings every few minutes. At the simplest level, displaying the battery voltage gives an approximate indication of how full or empty your battery is and what's going on at the moment. More complete monitors may show how many amps are flowing in or out at the moment, and the best monitors keep tabs on all this information, then give it to you as a simple "% of full charge" reading. Many charge controllers offer some simple monitoring abilities, but the best monitors stand alone and get mounted in the kitchen or living room where they're more likely to get some regular attention.

You can run a renewable energy system without a monitor, but it's not a smart idea. You could run your car without any gauges or meters too, but you'd be much more likely to exceed speed limits, run out of gas, or do accidental damage. System monitors are a real bargain compared to batteries.

Now you have a battery, a way to recharge it, a controller to prevent overcharging, and a system monitor to tell how full or empty the battery is. Next, you need to be able to use the stored energy.

USING YOUR POWER

Batteries deliver direct current, or DC, when you ask them for power. You can tap this energy to run any DC lights or appliances directly. This is fine if you're running an RV or a boat; vehicles pretty much run on DC, thanks to their

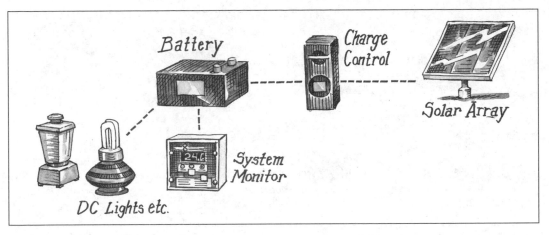

automotive heritage. With a DC breaker or fuse panel and a wiring circuit or two, you're all set to run your DC light or pump.

But what if you want to run an entire household, or a microwave in your RV? Household appliances run on 120-volt AC, or alternating current. AC power is easier to transmit over long distances, so it has become the world standard. AC appliances won't run on DC from a battery. You could convert all your household lights and appliances to DC, but joining the mainstream offers some definite advantages. AC appliances are mass produced, making them less expensive, universally available, and generally of better quality when compared with "RV appliances."

But AC power can't be stored—it has to be generated on demand. To meet this highly variable demand, you need a device called an inverter. The inverter converts low-voltage DC from the battery into high-voltage AC as fast as your household lights and appliances demand it.

Inverters come in a variety of outputs from 50 watts up through hundreds of thousands of watts, and range in cost from under $40 to more than you want to know. You want one that will deliver enough wattage to start and run all the AC appliances you might turn on at the same time, but no bigger than necessary. Bigger inverters cost more and use a bit more power on standby, waiting for something to turn on. However, lower-cost inverters usually leave out some protection equipment, which means a shorter life expectancy. (We've seen really cheap inverters blow up the instant a compact fluorescent lamp was turned on.)

Many of the larger, household-size inverters come with built-in battery chargers that come on automatically anytime outside AC power becomes available by starting a generator or plugging into an RV park outlet. This provides a simple back-up charging source for times when the primary solar or wind charger isn't keeping up.

Now you have a battery, a way to recharge it, a controller to prevent overcharging, a system monitor to tell how full or empty it is, and a way to use your stored energy. Only one component is missing, and it's the most important one: Safety!

HOW TO AVOID BURNING DOWN YOUR HOUSE

For the most part, low-voltage DC systems and batteries are a lot safer than jittery high-voltage AC systems. Just keep your tongue out of the

> The inverter converts low-voltage DC from the battery into high-voltage AC as fast as your household lights and appliances demand it.

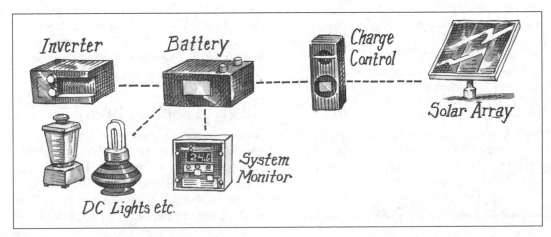

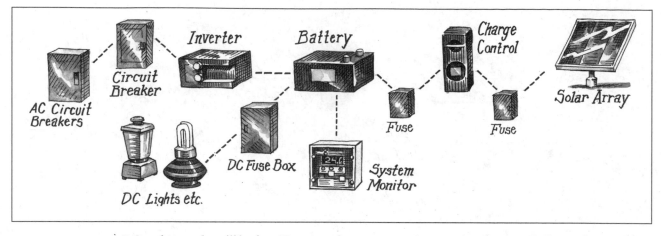

DC sockets and you'll be fine. However, if something does short-circuit a battery, look out! Because they're close by, batteries can deliver more amps into a short circuit than an AC system can, because the AC is strung out over many miles. So common sense, and the Electric Code, says you *must* put some overcurrent protection on any wire attached to a power source. "Overcurrent protection" is code-speak for a fuse or circuit breaker. "Power source" means batteries or any charging source such as PV arrays or a wind turbine. These fuses or circuit breakers must be rated for DC use and appropriately sized for the wires they're protecting. A DC short circuit is harder to interrupt than an AC short, so DC-rated equipment usually contains extra stuff and is therefore more expensive. DC-rated fuses

and circuit breakers are built tougher, and you really, really want them that way. You need to see wires glowing red with burning insulation dripping off them only once to appreciate the serious importance of fuses.

Now you have a battery, a way to recharge it, a controller to prevent overcharging, a system monitor to tell how full or empty it is, a way to use your stored energy, and the safety equipment to prevent burning down your house. That's a complete system!

WHAT SYSTEM VOLTAGE SHOULD YOU USE?

Renewable energy systems can be put together with the battery bank running at 12, 24, or 48 volts. The more energy your system processes per day, the higher your system voltage is likely to be. As systems get bigger, it gets easier to push the DC power around if the voltage is higher. If you double the voltage, you need to push only half the amperage through the wires to perform a given job. There are no hard and fast rules here, but in general, systems processing up to 2,000 watt-hours per day are fine at 12 volts. From 2,000 to about 7,000 watt-hours per day, a system will work better at 24 volts, and from 7,000 watt-hours on up, we prefer to run at 48 volts. These are guidelines, not rules. We can make a system run at any of these three common voltages, but you'll buy less hardware and big wires if system voltage goes up as the watt-hours increase.

Ready ... Set ... Build!

Are you ready to put together a renewable energy system? Of course not. You probably still have a lot of questions about how big, how many, what brand, and more. These questions can be answered by consulting other sections in this chapter: Safety and Fusing, Controllers,

---

The Importance of Conservation

Conservation always comes first, and in the case of off-grid solar, it pays for itself by a factor of 5. That is, every dollar you spend to conserve energy will save you $5 in solar system costs.

The cost to purchase an off-grid solar system is $3,500 per kilowatt-hour used on a daily basis. To outfit an average American home (which uses 30kWh) with solar would cost you $105,000 with no rebates, incentives, or tax credits.

Now consider the importance of conservation. Look at your light bulbs. If you replace 30 of the 60-watt incandescent bulbs in your home with more-efficient compact fluorescent bulbs that run on average five hours per day (national average) and use only 15 watts each, you'll save 6,750 watts per day (45 watts each x 30 bulbs x 5 hours), which equates to a $23,625 savings on your solar system.

The point is: Always implement conservation measures first before you build your renewable energy system. The money you save will more than pay for those more expensive, longer-lasting, energy-efficient appliances!

Monitors, Batteries, Inverters, Wiring, and the System Sizing Worksheet. Each of these sections explains a particular renewable energy component in more detail and helps you make the best choice for your individual needs.

And finally, if you aren't the technical type, and don't get joy from figuring this stuff out yourself—or even if you do but would appreciate a little guidance—there's our highly experienced technical staff, which we think is the best in the business. We're available for consultation, system design, and troubleshooting Monday through Saturday, 7:30 a.m. to 5:00 p.m. Pacific time.

System Sizing Worksheets

Off-the-Grid Systems

The following System Sizing Worksheet provides a simple and convenient way to determine approximate total household electrical needs for off-the-grid systems. Once the worksheet is completed, our Gaiam Real Goods technicians can size the photovoltaic and battery system to meet your needs. Do you want to go with utility intertie? That's simpler—see the next paragraph. Ninety percent of the renewable energy systems we design are PV-based, so these worksheets deal primarily with PV. If you are fortunate enough to have a viable wind or hydro power source, you'll find output information for these sources in their respective sections of Chapter 3 of this *Sourcebook*. Our technical staff has considerable experience with these alternate sources and will be glad to help you size a system. Give us a call.

Utility Intertie Systems

These are easy. Whatever your renewable energy system doesn't cover, your existing utility company will. So we don't need to account for every watt-hour beforehand. Tell us either how much you want to spend, or how many kilowatt-hours of utility power you'd like to displace on an average day. For direct-intertie systems without batteries, you'll invest about $2,500 for every kilowatt-hour per day your solar system delivers. For battery-based systems that can provide limited emergency back-up power, you invest about $3,500-$4,500 for every kilowatt-hour per day. These are very general ballpark figures for initial system costs. Remember, state and other rebates can reduce these costs by up to 50%.

General Information

The primary principle for off-the-grid systems is: Conserve, conserve, conserve! As a rule of thumb, it will cost about $3-$4 worth of equipment for every watt-hour per day you must supply. Trim your wattage to the bone! Don't use incandescent light bulbs or older standard refrigerators.

If you're intimidated by this whole process, don't worry. Our tech staff is here to hold your hand and help you through the tough parts. We do need you to come up with an estimate of total household watt-hours per day, which this worksheet will help you do. We can pick up the design process from there. You are in the best position to make lifestyle decisions: How late do you stay up at night? Are you religious about always turning the light off when leaving a room? Are you running a home business, or is the house empty five days a week? Do you hammer on your computer for 12 hours a day? Does

---

What's It Going to Cost Me to Go Solar?

Three easy steps to get a ballpark calculation for utility-intertie systems.

1. Find your daily utility usage by dividing the kilowatt-hours (kWh) used on an average month's utility bill by 30.

2. Divide that number by 5 (the average number of peak Sun hours in the U.S.), and multiply by 1.43 to account for system losses. This is the size of the solar system, in kilowatts, that you will need for taking care of 100% of your electrical needs.

3. Multiply that number by $9,000 ($9/watt installed) for a good ballpark idea of the gross installed cost.

Can state rebate incentives take a chunk out of that price? Go to www.dsireusa.org to find out what grants or incentives are available in your state. For instance, in California you can multiply your gross installed cost by 0.70 to account for rebates and tax credits. In Colorado, New York, and New Jersey, multiply by 0.5. This Web site of the Database of State Incentives for Renewables & Efficiency also contains handy maps showing where various kinds of incentives are available: www.dsireusa.org/library/includes/topic.cfm?TopicCategoryID=6&CurrentPageID=10&EE=1&RE=1.

What ongoing savings can I expect? Whatever you're paying the utility for electricity now will change to $0 (service charges will still apply).

Call our techs at 800-919-2400 for more information on how solar can work for your home.

---

your pet iguana absolutely require his rock heater 24 hours a day? These are questions we can't answer for you. So figure out your watt-hours to let us know what we're shooting for.

Determine the Total Electrical Load in Watt-Hours per Day

The following form allows you to list every appliance, how much wattage it draws, how many hours per day it runs, and how many days per week. This gives us a daily average for the week, as some appliances, like a washing machine, may be used only occasionally.

Some appliances may give only the amperage and voltage on the nameplate. We need wattage. Multiply the amperage by the voltage to get wattage. Example: A blender nameplate says, "2.5A 120V 60Hz." This tells us the appliance is rated for a maximum of 2.5 amps at 120 volts/60 cycles per second. 2.5 amps times 120 volts equals 300 watts. Be cautious of using nameplate amperage, however. For safety reasons, this must be listed as the highest amperage the appliance is capable of drawing. Actual running amperage is often much less. This is particularly true for refrigerators and entertainment equipment. The Watt Chart for Typical Appliances may help give you a more "real" wattage use. An excellent tool for determining loads is the "Kill-a-Watt" meter (#17-0320 for $30, see p. 225). For 120-volt AC appliances up to 1,500 watts, just plug the appliance into the meter and the meter into your household outlet. The meter gives instant power usage readings (watts) as well as the accumulated energy used over a period of time (watt-hours).

AC device	Device watts	×	Hours of daily use	×	Days of use per week	÷	7	=	Average watt-hours per day
		×		×		÷	7	=	
		×		×		÷	7	=	
		×		×		÷	7	=	
		×		×		÷	7	=	
		×		×		÷	7	=	
		×		×		÷	7	=	
		×		×		÷	7	=	
		×		×		÷	7	=	
		×		×		÷	7	=	
		×		×		÷	7	=	
		×		×		÷	7	=	
		×		×		÷	7	=	
		×		×		÷	7	=	
		×		×		÷	7	=	
		×		×		÷	7	=	
		×		×		÷	7	=	
		×		×		÷	7	=	
		×		×		÷	7	=	
		×		×		÷	7	=	
		×		×		÷	7	=	
		×		×		÷	7	=	

1   Total AC watt-hours/day

2   × 1.1 = Total corrected DC watt-hours/day

DC device	Device watts	×	Hours of daily use	×	Days of use per week	÷	7	=	Average watt-hours per day
		×		×		÷	7	=	
		×		×		÷	7	=	
		×		×		÷	7	=	
		×		×		÷	7	=	
		×		×		÷	7	=	
		×		×		÷	7	=	
		×		×		÷	7	=	
		×		×		÷	7	=	
		×		×		÷	7	=	

3   Total DC watt-hours/day

3 (from previous page)	Total DC watt-hours/day	
4	Total corrected DC watt-hours/day from line 2 +	
5	Total household DC watt-hours/day =	
6	System nominal voltage (usually 12 or 24) ÷	
7	Total DC amp-hours/day =	
8	Battery losses, wiring losses, safety factor × 1.2	
9	Total daily amp-hour requirement =	
10	Estimated design insolation (hours per day of sun, see map on p. 582) ÷	
11	Total PV array current in amps =	
12	Select a photovoltaic module for your system	
13	Module rated power amps ÷	
14	Number of modules required in parallel =	
15	System nominal voltage (from line 6 above)	
16	Module nominal voltage (usually 12) ÷	
17	Number of modules required in series =	
18	Number of modules required in parallel (from line 14 above) ×	
19	Total modules required =	

BATTERY SIZING

20	Total daily amp-hour requirement (from line 9)	
21	Reserve time in days ×	
22	Percent of useable battery capacity ÷	
23	Minimum battery capacity in amp-hours =	
24	Select a battery for your system, enter amp-hour capacity ÷	
25	Number of batteries in parallel =	
26	System nominal voltage (from line 6)	
27	Voltage of your chosen battery (usually 6 or 12) ÷	
28	Number of batteries in series =	
29	Number of batteries in parallel (from line 25 above) ×	
30	Total number of batteries required	

LINE-BY-LINE INSTRUCTIONS

Line 1 Total all average watt-hours/day in the column above.

Line 2 For AC appliances, multiply the watt-hours total by 1.1 to account for inverter inefficiency (typical by 90%). This gives the actual DC watt-hours that will be drawn from the battery.

Line 3 DC appliances are totaled directly, no correction necessary.

Line 4 Insert the total from line 2 above.

Line 5 Add the AC and DC watt-hour totals

to get the total DC watt-hours/day. At this point, you can fax or mail the design forms to us, and after a phone consultation we'll put a system together for you. If you prefer total self-reliance, forge on.

Line 6 Insert the voltage of the battery system; 12-volt and 24-volt are the most common. Talk it over with one of our tech staff before deciding on a higher voltage, as control and monitoring equipment is sometimes hard to find.

Line 7 Divide the total on line 5 by the voltage on line 6.

Line 8 This is our fudge factor that accounts for losses in wiring and batteries, and allows a small safety margin. Multiply line 7 by 1.2.

Line 9 This is the total amount of energy that needs to be supplied to the battery every day on average.

Line 10 This is where guesswork rears its ugly head. How many hours of Sun per day will you see? Our Solar Insolation Maps in the Appendix (page 582) give the average daily Sun hours for the worst month of the year. You probably don't want to design your system for worst possible conditions. Energy conservation during stormy weather, or a back-up power source, can allow use of a higher hours-per-day figure on this line and reduce the initial system cost.

Line 11 Divide line 9 by line 10; this gives the total PV current needed.

Line 12 Decide what PV module you want to use for your system. You may want to try the calculations with several different modules. It all depends on whether you need to round up or down to meet your needs.

Line 13 Insert the amps of output at rated power for your chosen module.

Line 14 Divide line 11 by line 13 to get the number of modules required in parallel. You will almost certainly get a fraction left over. Since we don't sell fractional PV modules, you'll need to round up or down to a whole number. We conservatively recommend that any fraction from 0.3 and up be rounded upward.

If yours is a 12-volt nominal system, you can stop here and transfer your line 14 answer to line 19. If your nominal system voltage is something higher than 12 volts, then forge on.

Line 15 Enter the system battery voltage. Usually this will be either 12 or 24.

Line 16 Enter the module nominal voltage. This will be 12 except for unusual special-order modules.

Line 17 Divide line 15 by line 16. This will be how many modules you must wire in series to charge your batteries.

Line 18 Insert the figure from line 14 and multiply by line 17.

Line 19 This is the total number of PV modules needed to satisfy your electrical needs. Too high? Reduce your electrical consumption, or add a secondary charging source such as wind or hydro if possible, or a stinking, noisy, troublesome, fossil fuel-gobbling generator.

BATTERY SIZING WORKSHEET

Line 20 Enter your total daily amp-hours from line 9.

Line 21 Reserve battery capacity in days. We usually recommend about three to seven days of back-up capacity. Less reserve will have you cycling the battery excessively on a daily basis, which results in lower life expectancy. More than seven days' capacity starts getting so expensive that a back-up power source should be considered.

Line 22 You can't use 100% of the battery capacity (unless you like buying new batteries). The maximum is 80%, and we usually recommend to size at 50% or 60%. This makes your batteries last longer and leaves a little emergency reserve. Enter a figure from 0.5 to 0.8 on this line.

Line 23 Multiply line 20 by line 21, and divide by line 22. This is the minimum battery capacity you need.

Line 24 Select a battery type. The most common for household systems are golf carts at 220 amp-hours, or L-16s at 350 amp-hours. See the Battery section for more details. Enter the amp-hour capacity of your chosen battery on this line.

Line 25 Divide line 23 by line 24; this is how many batteries you need in parallel.

Line 26 Enter your system nominal voltage from line 6.

Line 27 Enter the voltage of your chosen battery type.

Line 28 Divide line 26 by line 27; this gives you how many batteries you must wire in series for the desired system voltage.

Line 29 Enter the number of batteries in parallel from line 25.

Line 30 Multiply line 28 by line 29. This is the total number of batteries required for your system.

Power Consumption Table

Appliance	Watts	Appliance	Watts	Appliance	Watts
Coffeepot	200	Electric blanket	2,000	Compact fluorescent	
Coffee maker	800	Blow dryer	1,000–1,500	Incandescent equivalents	
Toaster	800–1,500	Shaver	15	40 watt equiv.	11
Popcorn popper	250	WaterPik	100	60 watt equiv.	16
Blender	300	Computer		75 watt equiv.	20
Microwave	600–1,700	Laptop	50–75	100 watt equiv.	30
Waffle iron	1,200	PC	200–600	Ceiling fan	10–50
Hot plate	1,200	Printer	100–500	Table fan	10–25
Frying pan	1,200	System (CPU, monitor, laser printer)	up to 1,500	Electric mower	1,500
Dishwasher	1,200–1,500	Fax	35	Hedge trimmer	450
Sink disposal	450	Typewriter	80–200	Weed eater	450
Washing machine		DVD Player	25	¼″ drill	250
Automatic	500	TV 25″ color	150+	½″ drill	750
Manual	300	19″ color	70	1″ drill	1,000
Vacuum cleaner		12″ b&w	20	9″ disc sander	1,200
Upright	200–700	VCR	40–100	3″ belt sander	1,000
Hand	100	CD player	35–100	12″ chain saw	1,100
Sewing machine	100	Stereo	10–100	14″ band saw	1,100
Iron	1,000	Clock radio	1	7¼″ circular saw	900
Clothes dryer		AM/FM car tape	8	8¼″ circular saw	1,400
Electric NA	4,000	Satellite dish/Internet	30–65	Refrigerator/freezer— Conventional Energy Star	
Gas heated	300–400				
		CB radio	5	23 cu. ft.	540 kWh/yr.
Heater		Electric clock	3	20 cu. ft.	390 kWh/yr.
Engine block NA	150–1,000	Radiotelephone		16 cu. ft.	370 kWh/yr.
Portable NA	1,500	Receive	5	Sun Frost	
Waterbed NA	400	Transmit	40–150	16 cu. ft. DC (7)	112
Stock tank NA	100	Lights		12 cu. ft. DC (7)	70
Furnace blower	300–1,000	100W incandescent	100	Freezer—Conventional	
Air conditioner NA		25W compact fluorescent	28	14 cu. ft. (15)	440
Room	1,500	50W DC incandescent	50	14 cu. ft. (14)	350
Central	2,000–5,000	40W DC halogen	40	Sun Frost freezer	
Garage door opener	350	20W DC compact fluorescent	22	19 cu. ft. (10)	112

Utility Intertie Systems

HOW TO SELL POWER TO YOUR ELECTRIC UTILITY COMPANY (AND GET YOURS FOR FREE!)

Utility electric costs are rising, blackouts are increasing, and the power grid we once took for granted is becoming increasingly unreliable. Many of our customers are turning to the reliability and consistency of solar power and utility intertie systems. Solar is clean, reliable, non-polluting energy that is quickly becoming cost effective, with payback periods approaching six years where utility rates are high and rebate programs are in place. State and federal government rebate programs may reduce initial costs by as much as 50%. The time for solar is now.

A utility intertie system makes it possible to generate your own solar power and sell it back to your utility company. Your utility electric meter will run backward anytime your renewable energy system is making more power than your house needs at the moment. If the meter runs backward, your utility company is buying power from you, and if you live in a state with net metering, they will buy it at the same retail cost that you pay them. It feels great! Even if your renewable energy system isn't making more energy than your house needs at the moment, any watt-hours your system delivers will displace an equal number of utility watt-hours, directly reducing your utility bill and reducing your exposure to the fluctuating and often increasing utility rates that are becoming the norm as utility deregulation proceeds.

THE LEGALITIES OF ELECTRICITY AND NET METERING

The 1978 federal Public Utilities Regulatory Policies Act (PURPA) states that any private renewable energy producer in the United States has a right to sell power to their local utility company. The law doesn't state that utilities have to make this easy, or profitable, however. Under PURPA, the utility usually will pay its "avoided cost," otherwise known as wholesale rates. This law was an outgrowth of the 1973 Arab oil embargo and was intended to encourage renewable energy producers. But 2-4 cents per kilowatt-hour was not sufficiently encouraging to have much of an effect on home utility intertie installations.

The next wave of encouragement started arriving in the mid-1980s with net metering laws. There is no blanket federal net metering law (yet). As of early 2007, 36 states plus Washington, DC, have statewide net metering laws, and an additional four have individual utilities that support net metering. These policies allow small-scale renewable energy producers to sell excess power to their utility company, through the existing electric meter, for standard retail prices. Details of these state laws vary widely. For an up-to-date summary, see the Database of State Incentives for Renewables & Efficiency website: www.dsireusa.org. This site also has some great maps, for those of you who are visually oriented: www.dsireusa.org/library/includes/topic.cfm?TopicCategoryID=6&CurrentPageID=10&EE=1&RE=1. Or call your State Energy Office; there's a complete directory in the Appendix.

HOW UTILITY INTERTIE SYSTEMS WORK

An intertie system uses an inverter that takes any available renewable power generated on-site, turns it into conventional AC, and feeds it into your electrical circuit-breaker panel. Electricity flow is much like water flow; it runs from higher to lower voltage levels. When you turn on an electrical appliance, the voltage level in your house slightly decreases, letting current flow in from the utility. The intertie inverter carefully monitors the utility voltage level and feeds its power into the grid at just slightly higher voltage. As long as the voltage level in your house is being maintained by the intertie inverter, no current flows in from the utility. You are displacing utility power with renewable power (in most cases solar power). If your house demands more current than the intertie system is delivering at the moment, utility power compensates seamlessly. If your house is using less current than the intertie system is delivering, the excess pushes out through your meter, turning it backward, back into the grid. This treats the utility like a big 100% efficient battery. You sell it power during the day; it sells power back to you at night for the same price (assuming you have a PV system that is operating during the daytime). If you live in one of the majority of states that allow net metering, then everything goes in and out through a single residential meter. Very simple, and extremely encouraging to renewable energy producers.

KEEP IT SAFE!

Utility companies are naturally concerned about any source that might be feeding power into

> If the meter runs backward, your utility company is buying power from you.

their network. Is this power clean enough to sell to your neighbors? Are the frequency, voltage, and waveform within acceptable specifications? And most important, what happens if the utility power goes off? Utility companies take a very dim view of any power source that might send power back into the grid in the event of a utility power failure, as this could be a serious threat to power line repair workers.

All intertie inverters have precise output specifications, voltage limits, and multilayered automatic shut-off protection features as required by UL specification 1741. This specification details all the necessary output and safety requirements as agreed upon by utility and industry representatives and is a specific UL listing for this new family of intertie inverters. If the utility fails, the inverter will disconnect in no more than 30 milliseconds! Most utility companies agree this is a faster response than their repair crews usually will manage. Even an "island" situation will be detected and shut off within two minutes. An island happens (theoretically) when your little neighborhood is cut off, isolated, and just happens to require precisely the amount of energy that your intertie inverter is delivering at the moment. Safety has been the top concern during development of intertie technology. These inverters just can't cause any problems for the utility. The worst problem they could cause the homeowner is to shut off the intertie system temporarily because utility power drifts outside specifications.

Because small-scale intertie is a new and evolving development, some smaller utility companies may not yet have worked out their official policies and procedures. Many utilities simply follow the lead of Pacific Gas and Electric, the largest utility company in the United States. If PG&E approves a particular product or procedure, it's probably safe enough for the rest of us. Toward the end of 2006, PG&E made the following announcement:

"In response to your requests and in support of the California Solar Initiative, Pacific Gas and Electric Company (PG&E) has modified its policy regarding the installation of an AC disconnect switch on inverter-based generation systems. This includes photovoltaic (PV), fuel-cell, and inverter-based rotating machine technologies. Effective November 21, 2006, customers installing inverter-based systems will no longer be required to include an AC disconnect switch when the facility has a self-contained electric revenue meter (i.e., 0- to 320-amp socket-based meter or 400 amp K-based meters). This type of

meter is used by 98% of all PG&E customers."

Be aware that not all electric meters will support net metering. Some meters are equipped with a ratchet that allows them only to turn forward. Some of the newer digital remote-reading data collection meters simply count how many times the hash mark on the wheel goes by; they just assume it's moving forward. Also remember that not all states have net metering laws for small-scale renewable energy producers. Your utility company will make sure you're equipped with the right kind of meter. There's usually no charge if a change is required.

DO YOU NEED BATTERIES OR NOT?

Utility intertie inverters come in two basic configurations: those that support back-up batteries, and those that don't. This is a major fork in the road of system design, as the two system types operate very differently, require completely different equipment, and have big price differences. To help you choose wisely, read on.

Battery-based systems are basically large standalone renewable energy systems using a sophisticated inverter that also can intertie with the utility. Several manufacturers offer battery-based intertie inverters. The Xantrex SW is the venerable choice, with almost 10 years of experience. The SW-series inverters are large 2,500- to 5,500-watt units with 24- or 48-volt battery packs. Coming in late 2007 is the Xantrex XW series. OutBack's FX and VFX series are newer offerings with units from 2,000 to 3,600 watts and 24- or 48-volt battery packs. Beacon Power has a 5,000-watt all-in-one-box solution that's easy to install for residential customers. SMA has the Sunny Island battery-based inverter that can work along with their existing series of Sunny Boy direct-intertie inverters. And in the near future, we expect a battery-based unit from Fronius. All these inverters will allow selected circuits in your house to continue running on battery and solar power if utility power fails. They probably won't run your entire house, but they can keep the fridge, furnace, and a few lights going through an extended outage, and they'll do it automatically, without noise or fuss. With inverter, safety, charge control, batteries, and a minimal PV array gear, these systems start out at approximately $15,000. Average-size systems that will comfortably (partially) support a typical suburban home run about $25,000-$30,000.

Direct-intertie systems are much simpler. They consist of PV modules, a mounting structure that holds the modules, the intertie

With inverter, safety, charge control, batteries, and a minimal PV array gear, these systems start out at approximately $15,000.

inverter(s), a couple of safety switches, and a run of wiring that connects everything. Without batteries, less safety and control equipment is necessary and less physical space is required on-site. About 5% more of your PV power ends up as useful AC power, and more of your investment tends to end up buying PV power. But if utility power goes off, so does your intertie system, regardless of whether the Sun is shining. Direct-intertie inverters come in modest to large sizes from 700 watts up to 6,000 watts and are produced by an increasing number of companies, including Xantrex, SMA, and Fronius. Direct-intertie systems start at just under $8,000 for a system that delivers approximately 2 kilowatt-hours per day. The average-size system we've sold over the past few years sells for about $26,000 (before any rebates) and delivers approximately 12 kilowatt-hours per day.

Which type of intertie system is right for you? That primarily depends on the reliability of your utility provider. If you live in an area where storms, poor maintenance, or mangled deregulation schemes tend to knock out power regularly, then a battery-based system is going to greatly increase your security and comfort level. Looking out over the darkened neighborhood from inside your comfortable, well-lit home does wonders for your sense of well-being. On the downside, adding batteries will raise your system cost by $6-$12,000 and adds system components that will need some ongoing maintenance and eventual replacement. If utility power is reliable and well maintained in your area, then you have little incentive for a battery-based system. Those dollars could be better spent on PV wattage. You'll see more benefits and get to enjoy almost zero maintenance. Just be aware that if utility power does fail, your solar system will automatically shut off too for safety reasons.

REBATES CAN REDUCE YOUR TOTAL SYSTEM COST

Money is available out there for doing the right thing. A number of states have programs that will pay you an incentive in the form of a rebate, buydown, or grant to defray your renewable energy system costs. It's difficult to keep tabs on all the rapidly changing programs in all the states. The Database of State Incentives for Renewables & Efficiency does a great job on its website: www.dsireusa.org. We also provide a list of State Energy Offices in the Appendix. Give yours a call to ask about net metering and money available for renewable energy projects.

California has a buydown program for intertie systems that our technicians are fully versed in, and we'll be happy to help you figure out how to navigate it. In the Sacramento, San Francisco/North Bay, and Los Angeles areas, we provide complete installation services. Our fully trained technicians have years of experience sizing these systems (800-919-2400). Real Goods has solar-

Grid-Tie Is a Good Investment

An "enthused but discouraged" customer considering a solar electric system asked us the following questions.

Q: Capital invested in a solar system must come from somewhere. If you take it out of savings, you will give up the interest that the money can earn from savings. If you take it from an investment account, you give up the projected annual return on your investment. If you borrow the money, you give up the interest cost for the investment.

Depreciation is a real cost issue, particularly for batteries. Batteries deteriorate over time; golf cart batteries should be replaced after 3-5 years. Other components such as PV arrays may have a long life, but depreciation over a 30-year period is not immaterial.

Electronic components will age. Controllers and other electronics may be subject to failure and replacement after 10 years.

With a fair, complete consideration of all the financial factors, does solar electric really make economic sense?

A: There is almost no reason to be off grid if utility power is available. Being off grid is usually a choice made out of necessity (no grid available), and return on investment is simply not an issue. On the other hand, solar grid-tie systems can be very cost effective where utility rates are high and rebates are available, such as in California. Here, going solar is a good investment—and batteries are not required. This investment can be a better use of one's capital than a savings account. The return is better and the risk very low, as long as utility rates keep going up.

For example, the simple pretax first-year return on a 4.3kW grid-tie PV system in California is about 10% over a 25-year period. Another way of stating this fact is that the savings generated over 25 years will pay for the investment more than 4 times over. (This figure is based on the following information: System cost after rebate and tax credit = $29,700; avoided electric rate = $0.32/kWh; annual electric rate increase = 5%; inverter is replaced after 15 years.) This calculation doesn't even take into account the increase in value of the home due to the home improvement. (For more on that subject, see the accompanying sidebar on Solar Payback.)

ized over 60,000 homes and businesses, more than any other company in the world.

GOOD FOR YOUR POCKETBOOK, GOOD FOR THE PLANET

The equity value of a PV system pays for itself the day it's installed. Also in states with solar incentives, like California, Colorado, New Jersey, New York, and others (see www.dsireusa.org for your state), your system will normally pay for itself with energy savings in 6-15 years, providing a 7%-16% return on investment (ROI). This ROI is way better than the best bonds and CDs or other safe financial investments today. Most state intertie regulations won't let residential customers deliver more energy than their average electric use. The details vary from state to state, but if you make more power than you use, you'll either give it away or sell it at wholesale rates. Besides the economic investment payback value, intertie is also something you do because it feels good to be independent of the utility and to cover your own electricity needs directly with a clean, nonpolluting, renewable power source. It also feels good, of course, to increase the value of your home and make a healthy return on your investment. Solar modules last for decades, require almost no maintenance, and don't borrow from our kids' future, which is probably the best possible reason to invest in renewable energy.

30th ANNIVERSARY

Converting California Sunshine (or Anywhere Else's) into Electricity

All the Details on How You Can Solarize Your Home, Slash Your Electric Bill, and Spin Your Meter Backward

We've talked a lot about how to solarize your home using clean, safe, reliable, and cost-effective power from the Sun. It's easier than you think, and Real Goods is here to take care of all the details. Rebates and tax incentives currently available in California and some other states have made solar power affordable and cost effective, providing a 10% return on investment. There has never been a better time than now to go solar. In most cases, we can provide you with a system that will cost the same or less than what you currently pay for utility power—and in many cases, solar will pay for itself almost instantly.

Real Goods has provided solar energy to over 60,000 homes and businesses in America since 1978. We are the world's oldest and largest supplier of renewable energy. Our customers are our first priority, and we will take care of every detail, from the planning stages to the installation and deployment of your new solar energy system.

Depending upon the circumstances of individual installation, system payback times are as brief as 5-10 years, with annual rates of return as high as 10%-20%. The simple fact is that installing a solar electric system on your home is now cost effective for the first time. With your personal Real Goods Solar Expert walking you through every step of the process, it couldn't be easier. And you'll have peace of mind knowing that you've become part of the energy solution for future generations as we see the twin problems of climate change and dwindling fossil fuel supplies make our world vulnerable to significant or catastrophic disruption.

Your Real Goods Solar Expert will come to your home (if you're in our California or Colorado service territories) to review your electrical needs and your last 12 months of electric bills. He or she will then evaluate your home for solar potential, provide you with a financial analysis and payback period, and give you both a price quotation and an installation time, should you decide to proceed. We'll even take care of the paperwork needed for rebates and tax credits.

TODAY'S SOLAR ECONOMICS—YOUR SYSTEM PAYS FOR ITSELF THE DAY IT'S INSTALLED!

At Real Goods, we've been selling solar for 30 years. When we started in 1978, the price was well over $300 per watt, or more than $6 per kilowatt-hour. Today, with state of California incentives, that $6 per kWh cost has come down to around $0.14/kWh, a 97% drop! This means that you'll be guaranteeing yourself an electric rate of $0.14/kWh for the next 30 years—and we all know that over that time, utility rates will be sharply increasing.

What do these lower prices and governmental incentives mean for the average homeowner? We frequently see 15% annual returns on investment (ROI)—far better than the stock market, bond market, money markets, and long-term CDs.

According to the National Appraisal Institute (*Appraisal Journal*, October 1999), your home's value increases $20 for every $1 reduction in annual utility bills. This means our 3kW system increases your home's value by $21,900 while costing you only $17,500. You're $4,400 ahead from day one!

STATE AND FEDERAL INCENTIVES

In 1996, the state of California, through its California Energy Commission (CEC), decided to level the playing field for solar energy by providing cash rebates directly to customers who installed solar systems on their homes, provided the homes were connected to the utility company grid (because the money came from the utilities . . .).

As of January 1, 2007, the rebate provided to customers of Southern California Edison (SCE), Pacific Gas and Electric Company (PG&E), and

Fetzer Vineyards, Hopland, California

> "The installation of photovoltaic panels on the Fetzer Vineyards Administration Building marks another milestone in our continuing quest toward sustainability of our winery. We are especially pleased to have our neighbors and friends at Real Goods oversee the project."
>
> —PAUL DOLAN,
> PRESIDENT, FETZER VINEYARDS

San Diego Gas & Electric Company (SDG&E) went to $2.50 per delivered watt, or about 30% of the total system cost. The rebate program was assured to be around for 10 more years as the California legislature passed the California Solar Initiative, providing $3.1 billion in far-reaching solar incentives.

In addition to this generous rebate, there is a further federal tax credit of 30%, or $2,000 per year (whichever is less), of the cost of your renewable energy system (solar or wind) after rebates. This comes directly off of the amount of federal income tax that you owe.

Other states, for example Colorado, New Jersey, and New York, have similar rebate and tax credit programs for renewable energy systems. The information we provide here is for California, but it serves as an example of how similar incentives in your state may make the choice to go solar financially attractive. Check with your state energy office or a local renewable energy contractor to learn the details of your state's program (or check out www.dsireusa.org). Then call Real Goods for further consultation.

Businesses can also get a 10% federal income tax credit with no cap. The "MACRS" five-year accelerated depreciation schedule is also applicable. With these incentives, we have often seen paybacks as low as three years and internal rates of return in excess of 25%.

TIME-OF-USE METERING

California's net metering laws (over 40 states have them now) mandate that your utility company cannot charge you more for your electricity than they pay you for the solar power you generate.

The physics of solar power make it the perfect energy source—it puts out the most power in mid-afternoon, which is exactly when California utilities have the greatest demand due to summer air conditioning. To decrease the summer midafternoon power load, the utilities have instituted "time-of-use" metering to encourage homeowners to conserve in the afternoon. This is great news for solar system owners!

Time-of-use metering is an excellent choice for PV systems that meet a minimum of 50% of your electrical needs. This means you can sell your solar power to the utility between noon and 6 p.m. for as much as $0.46/kWh, and then you can buy that same power back at a rate as low as $0.095/kWh. This gives you a huge financial benefit in the long run and also reduces the initial size of your solar system and consequently your ultimate payback time.

YOUR PROPERTY TAX WON'T GO UP

California law prohibits county property tax assessors from increasing your tax assessment because of value increases from your installation of a solar system. This is great news for your cash flow, as you'll be saving hundreds or even thousands of dollars every year from your new solar system, increasing your property's resale value by $21,900 due to your utility bill savings and not adding even one penny in property tax!

FINANCING YOUR SOLAR SYSTEM

Real Goods has researched numerous quality home-equity lenders and other loan sources in California. If you choose not to self-fund your solar system, there are many cost-effective alternatives. Many lenders are open to financing solar systems for less than prime rate. Call Real Goods at 888-507-2561 to get our latest recommendations on the best financing options. For many homeowners, this means instant positive cash flow. For a more detailed explanation of the payback opportunities, read the following article by Andy Black, page 184.

HOW MUCH POWER CAN YOUR SOLAR SYSTEM PRODUCE?

If you've checked with any of our competitors, you'll see widely divergent opinions on how much real electricity a solar system can produce. Our industry is relatively new, so consistent standards of measurement haven't yet been adopted to level the playing field for everyone. In the meantime, all we can do is be completely honest with our customers about the true output they can expect from their solar systems.

STC or "Nameplate" Ratings: STC stands for Standard Test Conditions. It is the rated output in watts that the manufacturer puts on its photovoltaic (PV) modules under laboratory-perfect conditions. STC ratings are generally used by the solar industry, installers, our competitors, and the general public.

PTC Ratings: PTC stands for Practical Test Conditions, or the ratings under the PVUSA Test Conditions. This is the standard used by the California Energy Commission and in general runs about 6%-12% less than STC. PTC ratings generally are used to calculate the California rebate.

Real-Life Expectations: Many industry professionals have studied the issue of photovoltaic ratings and are uncomfortable with both STC and PTC ratings because they seem overrated to real-world conditions. A couple of years ago, Real Goods attended an excellent seminar at the national solar conference in Portland, Oregon, and the consensus was that to be conservative in your expectations, you should expect your solar system to yield in AC output to your electrical panel about 75% of STC (manufacturer's nameplate) ratings (that is, multiply STC by 0.75).

In Summary: Throughout our literature, like all in the solar industry, we will use STC ratings to describe our systems by size. However, to calculate performance for actual "real life" expectations, we use the STC rating times 75%. What this means is well summarized by the example at the right featuring a solar panel we commonly use for 3kW systems.

This is how a 3kW system at the outset becomes a 2kW system in actual electricity yield.

STC rating	208W STC x 14 = 2,912 watts STC
PTC rating	183.3W PTC x 14 = 2,566 watts PTC
Real-life rating	208W STC x 14 x 75% = 2,184 watts actual

HOW REAL GOODS SOLAR PROFESSIONALS WORK TO SOLARIZE YOUR HOME

1. Feasibility and Cost/Benefit Analysis: After you contact one of our solar experts, we go to work immediately to determine if solar is right for you. We ask you to provide us with:

- copies of your last 12 months of utility bills
- a description or rough drawing of your home's directional orientation
- a description of your roofing material, condition, and age
- a description of your current electrical service and installation
- an indication of your desire to be 100% energy independent, or what fraction thereof

Next, our technician will discuss with you the possibility of making a personal visit to your home to assess your location, access to solar gain, and your electrical service. This service is provided for a $50 assessment fee to cover our expenses, refundable upon purchase. We'll get back to you with:

- a comprehensive recommendation of which solar system is right for your situation
- a price quotation that includes all equipment, delivery to your home, building permit fees, installation, building inspections, and amount of rebate available from the state or your utility
- a recommendation on whether time-of-use metering will work for you
- a financial analysis of return on investment for your system, years to payback, and internal rate of return, including any state or federal tax credits or incentives available to you
- a variety of financing options if you choose not to purchase the solar system outright

2. Contract Signing, Procurement, and Installation Scheduling: When you have made a decision to proceed with a solar installation, after personal consultation in your home, the next steps are:

PROJECTED RATES OF RETURN ON YOUR 3KW RESIDENTIAL SOLAR SYSTEM

The following figures are calculated based on an electric bill "baseline" of 383kWh per 30 days AND time-of-use (TOU) metering as previously described. Note: TOU metering may not be the best solar solution for all homes. Savings vary slightly from these numbers depending upon your utility company. These are averages for California.

If your monthly electric bill is usually:	A 3kW solar electric system will reduce it to:
$50	$5
$75	$20
$100	$30
$150	$60

If you choose to pay cash for your system, the payback period will range from 6 to 13 years, depending upon how you value the discounted pre-tax cash flow and the anticipated inflation rate. Your return on investment will typically be 8%-16%, far better than just about any investment available today . . . or you can choose to finance your solar system with a home equity loan at 7.75%-7.875% (current rate from www.solarhomefinancing.com).

For electric bills in excess of $150 per month, contact one of our solar technicians at 888-212-5640 for a thorough financial payback analysis. Solar systems for homes with large utility bills are always cost effective and have shorter payback times and higher returns on investment due to the higher base electricity rates.

THE BOTTOM LINE

Total installed cost of 3kW system (incl. sales tax) before rebate and tax credit	$25,567
CEC rebate ($2.50/W PTC)	($6,031)
Amount paid to Real Goods after installation completed	$19,536
Federal tax credit (30% of post-rebate installed cost up to $2,000)	($2,000)
Total installed cost (incl. sales tax) after rebate and tax credit	$17,536

Estimated monthly cost based on projected production ratings:

Electricity cost per month before solar @ E1 rate using 840kWh/mo.	$150
Less solar electricity cost savings per month (avoided electric cost of $.25/kWh @ E6 rate) (2,184kW x $0.25/kWh x 5.5 hr./day x 30.4 days/mo.)	($90)
Plus monthly payment on financing $17,536 @ 7.5%, 30-year loan (incl. interest tax deduction)	$87
Total monthly cost for electricity after installing solar system	$147

The bottom line is that, in certain situations, your cash flow can actually increase by going solar!

3KW RESIDENTIAL SOLAR SYSTEM COMPONENTS

Component	Quantity
2,912 watts of PV modules (14, Sharp 208W)	2.9kW
Fronius IG 3000 DC-to-AC inverter with metering	1
Solar panel mounting kits and roof standoffs	2
Utility-required safety AC disconnect switch w/fuses	1
High-voltage solar array DC disconnect switch	1
Delta lightning arrestor	1
Weatherproof output cable (50 ft., male-female)	2

- sign our installation contract and provide us with a $1,000 deposit
- submit your application (this is handled by our solar experts) to the California Energy Commission or other relevant agency, qualifying you for the maximum rebate
- procure a building permit; you will have approximately seven months to complete your installation before the CEC rebate runs out
- provide any additional information

needed by our solar technician if you are building a new house

3. Installation of Your Solar System:
Our professionals will typically install your complete system in one to three days.

- 50% of the payment is due the day we begin installation. The balance is due upon completion of the installation. We will collect the rebate on your behalf so you won't

be responsible for any income tax. You pay us only the after-rebate price, without waiting for the rebate.
- Upon completion and payment, we'll schedule an inspection with the local building inspector and utility for final approval.
- You'll be making your own clean solar power and contributing to the energy solution!

What's the Payback?

How to calculate the return on your solar electric system investment before you buy.

For years, questions about returns on the expensive investment in a solar electric system were dismissed with the analogy, "What's the payback on your swimming pool?" That sentiment might speak to the converted, but for most people considering a solar energy system, finances are a major deciding factor.

Fortunately, photovoltaic technology has matured enough that the payback question can now be given a serious answer, backed by solid math and accounting. The answers vary significantly by local climate, utility rates, and incentives. In the best cases—California and New Jersey—

the compound annual rate of return is well over 10%, the cash flow is positive, and the increase in property resale value more than covers the cost of the PV system. In other parts of the country where electric rates are low and incentives may be less, a grid-tied system may barely cover its maintenance costs.

This article focuses on residential analyses. Similar calculations can be done for commercial situations, but significant differences in the tax and accounting rules exist. See the resources mentioned at the end to learn more about commercial analyses.

WHAT FACTORS IMPROVE PAYBACK?

The most important factors that make solar an attractive investment are high electric rates, financial incentives, net-metering policies, and good sunlight (available in most of the continental U.S.).

Electric rates vary by region, state, and utility. California, Hawaii, and New York have the highest average rates, well above $0.15 per kilowatt-hour. California's tiered pricing system also penalizes large residential users with prices as high as $0.33 per kilowatt-hour (see Figure 1).

Under most net-metering laws (these also vary by state or utility), solar energy offsets the retail cost of the electricity generated. Even better, in California, solar systems are allowed to operate on a time-of-use rate

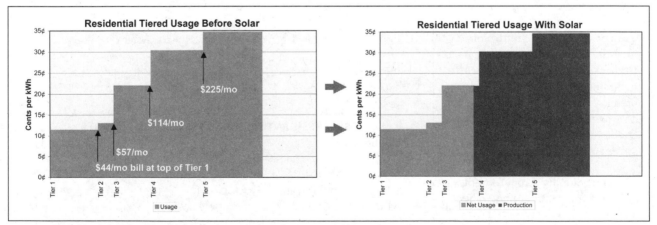

Fig. 1. Effect of Tiered Electric Rates on Large Users, and How Solar Helps
Tiered-rate pricing penalizes large users most with high marginal electricity rates. Solar energy offsets highest-tier usage first, making the solar customer look like a small user, with a lower marginal cost.

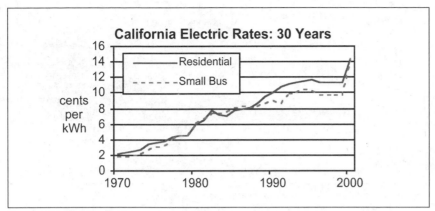

Fig. 2. California Electric Rates: 30 Years
Inflation is a major factor in photovoltaic system returns. In California, rates increased an average 6.7% per year from 1970 to 2000. This article assumes 5% electric-rates inflation going forward. Source: California Public Utility Commission "Electric Rate Compendium," November 2001.

will be positive, and the increase in resale value will exceed the system cost. These scenarios are common in certain markets.

Compound annual rate of return (CARR) on an investment is another term for interest-rate yield, which is a way of comparing one investment to another. For example, a savings account might pay 1% interest and the long-term stock market has paid about 10.5%. In California, New Jersey, and a few other locations solar electric systems can see a pretax CARR of 10% or more.

Several examples are shown in the table "Sample Scenarios for Residential System Payback." For more detailed information on these calculations, download www.ongrid.net/papers/PaybackOnSolarSERG.pdf.

The cash flow will be positive, either immediately or within the first few years, for many homeowners who finance their solar systems using home equity loans. Cash-flow calculation compares the estimated savings on the electric bill with the cost of the loan. Monthly cost is the principal plus interest payment required to pay off the loan, less any tax savings. In the case of "deductible" loans, such as home equity-based loans, the interest is usually tax deductible and thus the loan effectively costs less. Home equity loans are also excellent sources of funds because interest rates on real estate-secured loans are relatively low and payment terms can be long.

schedule, which enables users to sell back electricity to the utility at peak rates, which can be even more valuable. The higher the price your PV power can fetch, the better payback scenario you have.

Direct incentives can include tax benefits such as credits or depreciation. The most celebrated recent incentive is the federal tax credit for solar systems that went into effect January 1, 2006. This credit is for 30% of the system cost up to $2,000 for residential systems (no cap on commercial credits). For PV systems, that typically means a $2,000 credit on the purchaser's tax return for the year the system was installed. The federal credit can be coupled with state incentives such as rebates, which can discount up to 60% of the system cost. Some states also have a state tax credit, which can further reduce the up-front cost of a system. Consult a certified tax advisor to check the applicability of such incentives to your situation.

New forms of direct incentives are PBIs (performance-based incentives) and RECs (renewable energy credits, or "green tags"). Both are paid on a per-kilowatt-hour basis. They don't reduce the up-front cost, but they do increase the cash payments received after commissioning the system. Payments can be as much as $0.39 per kilowatt-hour for five years for the PBI, and more than $0.17 per

kilowatt-hour for a 10⁺-year contract on RECs. Because these payments often can be combined with net-metering value, the PV system is capable of garnering substantial revenue per kilowatt-hour generated.

Inflation in electric rates is another factor that can improve payback (see Figure 2). Solar is an inflation-protected investment because it offsets electricity costs at the current prevailing retail rate. As rates rise, you save even more money.

DETERMINING THE PAYBACK

The economic value of a solar system can be measured in several ways: *compound annual rate of return, cash flow,* and *increase in property resale value.* In strong cases, the returns will be over 10%, the cash flow

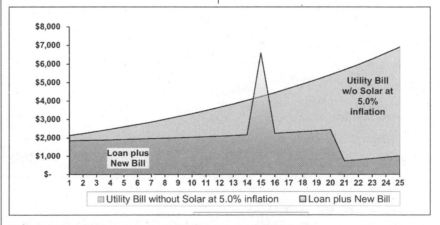

Fig. 3. Inflation's Effect on Loan Costs vs. Electric Costs Without Solar
In this case, the photovoltaic system loan term is 20 years, with inverter replacement occurring at year 15 (the large spike).

Inflation plays an important part in the cash-flow equation because as electric rates rise, savings from a solar system will increase. But inflation doesn't affect loan rates, particularly for fixed-rate loans, so your savings grows while the cost of the loan stays relatively constant (it rises a little over time as the interest portion and thus the tax deductibility of the payment declines). See Figure 3 for an example.

Figure 4 highlights the difference in the curves in Figure 3. It shows the net annual savings—old bill minus new bill, loan, maintenance, and inverter-replacement costs—and the effects of inflation over time.

Figure 5 shows the accumulation of net annual savings. This accumulation is free and clear with no initial outlay of cash, because that was covered by the loan. In the example, the system will save the owner about $90,000 over 25 years, with no up-front cost. The savings are small though significant in the first years but really jump when the loan payments stop after year 20. Of course, you can select any loan term that suits your needs. Note that these are

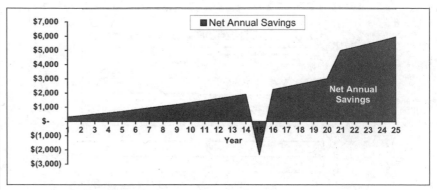

Fig. 4. Net Annual Savings with a Photovoltaic System
Net annual savings consists of the old electric bill minus the new bill, loan principal payments, loan net interest (after taxes), maintenance, and inverter-replacement costs. This example system is cash-positive from the first day at no "out of pocket" costs to the purchaser.

Payback Calculators and Resources

Database of State Incentives for Renewable Energy:
www.dsireusa.org

The Clean Power Estimator:
www.consumerenergycenter.org/renewable/estimator

FindSolar.com:
www.findsolar.com

OnGrid Solar Financial Analysis Calculator:
www.ongrid.net/payback

PV Watts: **http://rredc.nrel.gov/solar/codes_algs/PVWATTS**

RETScreen: **www.retscreen.net**

Solar Energy Industries **Association Guide to Federal Tax Incentives: www.seia.org/manualdownload.php**

examples of ideal cases in the states with the best economics.

An Increase in Property Resale Value occurs in homes with solar electric systems because these systems decrease utility operating costs. According to a 1998 *Appraisal Journal* article, a home's value increases $20,000 for every $1,000 reduction in annual operating costs from energy efficiency. (See www.icfi.com/Markets/Community_Development/doc_files/apj1098.pdf.)

The rationale is that the money from the reduction in operating costs can be spent on a larger mortgage with no net change in monthly cost of ownership. Historically, mort-

gage costs have an after-tax effective interest rate of about 5%. If $1,000 of reduced operating costs is put toward debt service at 5%, it can support an additional $20,000 of debt. To the borrower, total monthly cost of home ownership is identical. Instead of paying the utility, the homeowner pays the bank, but the total cost is unchanged and you have free electric power.

The column labeled "Appraisal Equity/Increase" in the Sample Scenarios table shows the increase in home value that you can expect. This increase can effectively reduce the payback period to zero years if you chose to sell the property immedi-

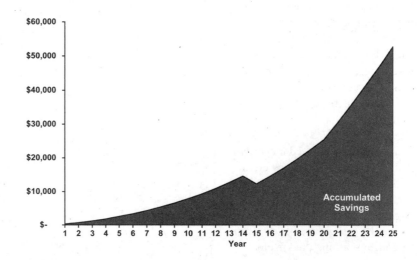

Fig. 5. Accumulation of Net Annual Savings
The accumulation of the net annual savings shows total lifetime electric bill savings experienced with a photovoltaic system, net of all costs. All initial costs are included in the loan.

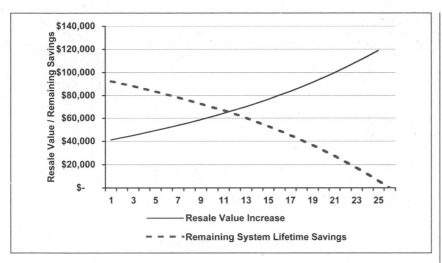

Fig. 6. House Resale Value over Time

The rising curve represents the increasing property value based on 20 times annual savings. The declining curve is the remaining savings of the photovoltaic system before the end of 25 years of conservatively estimated life. The resale value estimate is based on the lower of these two curves. This estimate can be compared with the accumulated savings in Figure 5. In later years, resale value declines, but greater accumulated savings have been enjoyed.

the end of 25 years—a conservative estimate since the panels are warranted at 25 years to work at 80% of their original capability. Figure 6 shows both the increasing property resale value due to increasing annual savings and the remaining-value limitation that takes over at approximately year 11. As the NREL resale study suggests, however, actual resale could be much higher depending on the market mood for solar.

CREATING MARKETS THAT REWARD INVESTMENT

For more information on calculating photovoltaic system payback, access www.ongrid.net/papers. Additional resources are listed below. Be aware that some tools don't account for tiered or time-of-use electric rates,

ately, and it removes the purchase risk. It could even lead to a profit on resale in some cases. However, note that this increase is largely hypothetical at present because a high fraction of grid-tied solar electric systems have been installed since 2001. Most of these homes have not been sold, so no broad studies of comparable resale values are available. But emerging evidence suggests that some sellers of solar homes are realizing significant jumps in resale value. For example, a 2004 National Renewable Energy Laboratory (NREL) study demonstrated that San Diego zero-energy homes with solar features increased in value faster than comparable conventional homes in a nearby community.

On average, value increased $40,000 more than the conventional homes, at a higher rate of appreciation and with a shorter length of ownership. This boost outstrips the estimates shown in the table.

PV systems will appreciate over time, rather than depreciate as they age. The appreciation comes from the increasing annual savings the system will yield as electric rates and bill savings rise. Such appreciation cannot continue forever, being limited by the system's remaining life. Here the system is assumed to be worthless at

How Does Solar Compare with Remodeling Investments?

Why should a homeowner pay more for a house with a solar system when she could buy a nearby nonsolar house and install a solar system for less money? Yet other remodeling upgrades increase resale value in this way.

Nationally, decks return an average 104% of their cost on resale (Remodeling Online, www.remodeling.hw.net). Kitchen and bathroom remodeling have similar results, depending on geography. So it makes sense that in certain regions where the Sun shines brightly and electric rates are high, solar would return more than its installed cost, with a national average comparable to other types of house upgrade. The table below lists projected resale value of various solar systems, compared with nationwide averages for other home improvements.

Resale Value Comparisons of Various Home Improvements

Home Improvement Type	Investment Amount/ Net System Cost	Resale Value Increase	% Return
4.3-kilowatt Photovoltaic System, California	$23,000	$22,000	95%
9.4-kilowatt Photovoltaic System, California	$48,000	$77,000	160%
Deck Addition	$6,300	$6,700	104%
Bathroom Remodel	$10,100	$9,100	89%
Window Replacement	$9,600	$8,200	85%
Kitchen Remodel	$44,000	$33,000	75%

Analysis based on data from Remodeling Online's 2003 survey.

SAMPLE SCENARIOS FOR RESIDENTIAL SYSTEM PAYBACK

Example	Electric Bill Before Solar	Usage Before Solar per Month (kWh)	Solar System Size (Standard Test Conditions)	Final Net Cost with Tax Credits and Rebate	Pre-Tax Annual Return	Appraisal Equity/ Resale Increase in First Year	Net Monthly Cash Flow Compared with 8% 30-Year Loan in First Year
California: Medium System	$100	675	4.0kW	$23,000	10.1%	$21,000	$(28)/mo.
California: Large System	$375	1,500	9.5kW	$53,000	16.6%	$84,000	+$87/mo.
New Jersey	$139	1,000	9.9kW	$38,000	13.2%	$29,000	+$144/mo.
Hawaii	$167	600	4.0kW	$28,000	11.9%	$31,000	$(18)/mo.

and as such, their results may over- or underestimate the value of a PV system in a particular situation.

In addition to avoiding oversimplified estimates, be cautious of aggressive sales pitches and "optimistic" financial analyses. Going solar, like any other major purchase, is a "buyer beware" situation. For 10 tips on how to detect unrealistic analyses, see the last page of the "Payback" article posted at www.ongrid.net/papers/PaybackOnSolarSERG.pdf.

Solar has finally come into its own in certain markets. To encourage widespread adoption of solar energy, we need to empower everyone with knowledge of the financial benefits of renewable electricity and expand the programs that make it possible—tiered rates, time-of-use net metering, RPSes with solar requirements, and RECs. Market forces will take it from there.

Adapted from an article by Andy Black in *Solar Today* (May-June 2006) and printed here with permission. Andy Black is the owner of OnGrid Solar, providing solar financial analysis tools and consultation. He serves as a board member of the American Solar Energy Society and its San Francisco Bay Area chapter, NorCal Solar. Contact him at 408-428-0808 or www.ongrid.net. Black regularly teaches and consults on "Payback on Solar Electricity" for many audiences, including the Solar Liviing Institute.

UTILITY INTERTIE INVERTERS
Straight from the Sun to Your Utility Meter

PV modules that produce conventional AC power have long been a dream of the renewable energy industry. Small individual inverters that delivered AC power output from each PV module were introduced in the mid-1990s, turned out to be stunningly expensive, and have since disappeared in the United States. It costs a lot more to build many small inverters rather than a single large one. Using some reasonable economies of scale, we can offer several cost-effective ways to connect PV modules directly to your household AC system.

Trace/Xantrex provided the first generation of utility intertie inverters with their large multi-tasking sine wave series in the mid-1990s. The original full-featured SW-series can intertie and support emergency back-up batteries. The ability to intertie and provide battery backup is now shared by the new generation of highly adaptable OutBack GFX- and GVFX-series inverters, by the new Beacon M5 inverter, the Sunny Island from SMA America, and other offerings due to debut in the near future. This type of fairly large inverter is the proper choice if you absolutely must have backup batteries due to unreliable utility power, but they're expensive for smaller arrays, and the batteries with their associated control and safety equipment add to the cost.

Direct-intertie inverters don't use batteries at all. They use the utility grid as if it was a 100% efficient battery. These are the best choice for systems that don't need back-up power. Incoming PV power is converted to conventional AC power and fed to your household breaker panel, displacing utility use. When there's excess power, your meter runs backwards, giving you retail credit; when there's a deficit, power is purchased, running the meter forward again. Xantrex, SMA, Fronius, and an increasing number of others make direct-intertie inverters.

An additional advantage of direct-intertie inverters is the high energy efficiency of the system. PV modules are controlled with power point tracking, which allows the modules to run at their maximum power output, rather than a lower battery-charging voltage. Batteries are eliminated, which typically costs 5%-20% of charging power; and all transmission is done at high voltage, allowing smaller wire sizes, and less loss.

A multiple number of intertie inverters can be paralleled for system expansion. Direct-intertie inverters will not operate without an AC power source to feed back into, and will

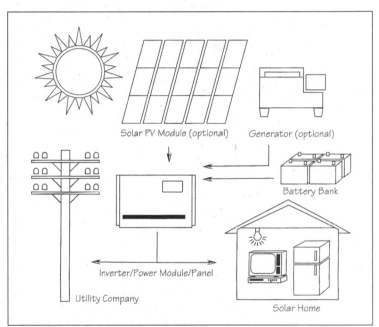

Solar PV Module (optional) Generator (optional)

Battery Bank

Inverter/Power Module/Panel

Utility Company

Solar Home

A typical grid interactive system with back-up power.

not charge batteries. All inverters designed for utility intertie, whether they're battery-base or direct-intertie, are UL-1741 listed.

Some states and utility companies are offering rebate or buy-down programs for intertie systems. Please call your local utility, www.dsireusa.org, or our technical staff for details.

Utility Direct-Intertie Inverters (Without Batteries)

FRONIUS INVERTERS

One of the world's most experienced utility-intertie inverter manufacturers

For over 55 years, Fronius has been developing power-conversion technologies. A time dominated by persistent research, diligence, and personal commitment. Today Fronius is a world market leader in high-frequency welding technology and one of the leading PV inverter manufacturers, with more than 60,000 inverters installed worldwide. The company is considered an innovator in all its fields of activity. The performance and ease of installing these inverters is borne out by the many that Real Goods has installed over the past couple of years. In addition, because of its reliability (very few service calls), it is fast becoming our inverter of choice.

Fronius Intertie Inverters

Fronius is one of the largest suppliers of direct-intertie inverters in Europe for good reason. They feature advanced high-frequency technology that delivers high efficiency, precision maximum power point tracking, and a lighter-weight design. Active cooling allows full power from -4° to 122°F, and a wide voltage input range of 150-450 volts permits highly flexible array sizing. Though we strongly encourage system designers and installers to avoid shading as much as possible, the net effects of using the Fronius IG inverter with a shaded array versus two separate units or a unit with separate MPP tracking functions is minimal—overall, under 1%. The Fronius IG also outperforms other inverters in situations that involve similar strings at different angles. The backlit LCD display is standard, and shows instant AC output, daily output, output since installation, CO_2 emissions avoided, and much more, including fault displays for quick repairs. The integrated DC and AC disconnects mean a clean installation with less hardware on the wall. The NEMA 3R enclosure can be outside or inside mounted. Expansion slots allow future upgrades or remote displays. There are data collection options for those who want to connect to their PC. The IG2000 model delivers a maximum 2,000 watts of output; the IG3000, 2,700 watts; the IG4000, 4,000 watts; and the IG5100, 5,100 watts. They are CSA listed to UL 1741. IG 2000 and IG3000 dimensions and weight are 18.5" x 16.5" x 8.8" and 26 lb. The IG4000 and IG 5100 dimensions and weight are 28.5" x 16.5" x 8.8" and 42 lb. 10-yr. mfr. warranty. Germany.

49-0150 Fronius IG2000 Inverter		**$2,170**
49-0149 Fronius IG3000 Inverter		**$2,460**
49-0197 Fronius IG4000 Inverter		**$3,690**
49-0200 Fronius IG5100 Inverter		**$3,995**
49-0249 Extended 10-Yr. Warranty		**$69**

Fronius Monitoring Options

The Fronius data communication features are easily added on to your PV system. With the three included expansion slots, you can add external sensors and remote displays. To connect to a PC (free software) simply add a Com Card and the Datalogger Easy Box.

41-0210 Fronius IG Personal Remote Display	**$345**
41-0211 Fronius Wireless Card	**$175**
41-0230 Fronius Datalogger pro Box (1-10 inverters)	**$695**
41-0231 Fronius COM Card, Retofit	**$135**
41-0232 Fronius Datalogger easy Box (1 inverter)	**$440**

GRID-CONNECTED SOLAR ELECTRIC SYSTEM

electrical meter

fuse box

inverter

Inverter conditions electricity for use.

SMA Sunny Boy Intertie Inverters

Direct-intertie with German engineering

Sunny Boy now offers several closely related models at 700, 1,100, 1,800, 2,500, 3,000, 3,300, and 6,000 watts. All models feature the best of high-quality German engineering. High efficiency, redundant processors, easy data communications, maximum power point tracking, an excellent reliability record, and powder-coated, stainless steel, outdoor-rated boxes are only a few of the quality design features. Outdoor installation in a shady location is preferred.

Sunny Boy inverters require high-voltage series-strings of PV modules. The number of modules varies with inverter model, PV modules used, climate in which they'll be installed, and the owner's budget and/or output requirements. Smaller inverters require shorter, lower-voltage PV strings. Bigger inverters require longer strings. Call for details.

Output is delivered to your household main breaker panel by feeding through a dedicated circuit breaker. Conversion efficiency is an excellent 93.0%-95.6%. Internal power use is less than 7 watts during operation, less than 0.1 watts during standby, and the inverter will disconnect from the utility at night to eliminate any phantom load.

The display option is one of the slickest in the inverter industry. The 2-line, 40-character backlit display is acoustically activated by knocking on the lid and scrolls through: Instantaneous AC Watts; DC Volts; Daily kWh; Total kWh (since the inverter was turned on); Operating Hours; and Inverter Status. Several optional remote monitoring and professional data acquisition options are also available; please call for details.

All Sunny Boy installations will need a high-voltage, DC-rated raintight disconnect. Up to three Sunny Boy inverters can be connected to the high-voltage DC disconnect listed below. The lightning protection is optional, but highly recommended. All Sunny Boy inverters have a 5-yr. mfr. warranty. Germany.

49-0500	Sunny Boy 7000U w/Display	$4,395
49-0168	Sunny Boy 6000U w/Display	$4,095
49-0501	Sunny Boy 5000U w/Display	$3,995
49-0502	Sunny Boy 4000U w/Display	$2,995
49-0503	Sunny Boy 3000U w/Display	$2,395
27904	Sunny Boy 1800U w/Display	$1,895
49-0117	Sunny Boy 700U w/Display	$1,295
24648	Hi-Volt DC Disconnect	$208
25775	Hi-Volt Lightning Protector	$40

Sunny Boy Installation and Monitoring Options

The Sunny Breeze fan clips to the top of the heat-sink fins, provides thermostatic-controlled cooling as needed, and includes an AC power supply. It's needed with indoor installations, and some heavily loaded outdoor installations.

Please call for help with remote monitoring. There are many options.

49-0107	Sunny Breeze Fan Option	$150
24687	Sunny Boy GFI Fuse, 1A/600V	$19
27902	Sunny Boy RS-232 Comm. Module	$179
49-0118	Sunny Boy RS-485 Comm. Module	$187
27858	Sunny Boy Control	$660
24710	On-Board Digital Display (after purch) (specify inverter model & color)	$317

Xantrex Intertie Inverters

These direct-intertie inverters deliver superior PV energy harvesting, high reliability, easy installation, and an attractive cost. They also offer some of the highest efficiencies in the industry—94.8%-96.5%. An input voltage range of 195-550 volts allows more flexible system sizing and helps to make use of marginal sunlight. The maximum power point tracking software features rapid response to make the best of cloudy weather. A large aluminum heat sink ensures excellent thermal performance. They feature an easy-access wiring box with an NEC-compliant integrated PV/utility disconnect, so only the NEMA 3R outdoor rated box is needed.

The Built-in display cycles through instant, daily, and lifetime energy production, utility voltage/frequency, online selling time, any fault messages, and other customized screens. Also included is an isolated RS232 port and two RJ45 (Ethernet) communications ports. No additional cards or other gizmos are needed to connect to a PC, and The Viewer software program can be downloaded for free. Size is 28.5" x 15.9" x 5.7" (HxWxD). The GT 5.0 weighs 58 lb. while the others weigh 49-51 lb. CSA listed to UL 1741, FCC Class B. 5-yr. mfr. warranty. China.

49-5001	Xantrex GT 2.8 inverter	$2,775
49-0234	Xantrex GT 3.3 inverter	$2,795
49-5002	Xantrex GT 4.0 inverter	$2,995
49-5000	Xantrex GT 5.0 inverter	$3,795

Direct-Intertie Installation Accessories

MC Cable Extensions

Most PV modules used for intertie now are being supplied with the quick-connect MC connectors for fast series connections. Sometimes you need some extra length, without switching to standard wiring and back.

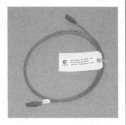

Our MC extension cables make it easy, with a male connector on one end and a female on the opposite end. We almost always use one to get from the far end of the array back to the wiring J-box. USA.

26662	PV MC-Type Output Cable 6' M/F	$18
26666	PV MC-Type Output Cable 15' M/F	$20
26667	PV MC-Type Output Cable 30' M/F	$30
26664	PV MC-Type Output Cable 50' M/F	$38
26665	PV MC-Type Output Cable 100' M/F	$74
48-0273	Waterproof Strain Relief for MC Cables (1/2", 2 hole)	$5

Distribution Block

Our 3-pole distribution block is handy for joining multiple PV strings before making the long run to the DC disconnect and inverter. 3-pole for positive, negative, and ground. Accepts one #14 to #2/0

wire on the primary side, and up to four #14 to #4 wires on the secondary side. Up to two of these distribution blocks will fit nicely in the raintight 10" x 8" x 4" J-box listed below.

| 24705 | 3-Pole Distribution Block | $39 |
| 24213 | J-Box 10x8x4 R/T | $49 |

AC Combiner Panel

The AC Combiner Panel is necessary when you have more than one direct-intertie inverter on your house. We need to combine all the power outputs before passing through the AC disconnect. The utility requires there only be one disconnect. Combiner is raintight for exposed outdoor use. Use a 15A/240V breaker for each

Sunny Boy 2500, a 20A/120V for each Sunny Boy 1800. A pair of Sunny Boy 1100 or 700 models can land on each 15A/240V breaker. Combiner will accept up to four 240V, or up to eight 120V breakers (or combinations).

48-0018	AC Rain Tight Combiner Load Center (SqD #HOM48L125GRB)	$87
24707	15A/240V Circuit Breaker	$16.50
24709	20A/120V Circuit Breaker	$8.50

AC and DC Disconnects

Some utilities require a lockable disconnect between your intertie inverter(s) and the main panel. This is not an electric code requirement; but it allows your local utility lineman-an extra measure of safety by positively locking your inverter off during nearby repairs. These disconnects are both raintight with visible blades and a handle that's lockable in the off position. The 240VAC inverters require two fuses; 120VAC inverters, one. PV array DC disconnects are usually unfused; however, with some multiple string arrays, a fused disconnect may be required. USA.

24644	30A/2-Pole Fused Disconnect, Raintight (Square D# D221NRB)	$66
24645	60A/2-Pole Fused Disconnect, Raintight (Square D D222NRB)	$109
48-0054	60A/2-Pole Fused Disconnect, Raintight (GE TG3222R)	$85
48-0068	15A FRN-R15 Class RK5 Fuse	$5
24516	20A FRN-R20 Class RK5 Fuse	$5
24501	30A FRN-R30 Class RK5 Fuse	$6
48-0209	40A FRN-R40 Class RK5 Fuse	$9
24502	45A FRN-R45 Class RK5 Fuse	$11
24503	60A FRN-R60 Class RK5 Fuse	$10
24505	100A FRN-R100 Class RK5 Fuse	$20
24648	30A/600V/3-pole Unfused Disconnect, Raintight (Square D HU361RB)	$208
48-0055	60A/600V/2-pole Unfused Disconnect, Raintight (GE THN2262RDC)	$325
24229	60A/240V/3-pole Unfused Disconnect, Raintight (GE TGN3322R)	$159
24-5003	30A 2-P RT Fused 250VDC/ 240 VAC Disconnect (Square D H221NRB)	$320
49-0517	60A 2-P RT Fused 240VAC/ 250 VDC Disconnect (Square D H222NRB)	$320

UTILITY INTERTIE INVERTERS WITH BACK-UP ABILITY

If you want back-up ability from your solar electric system, you'll need batteries. How big those batteries are depends on how much you want to run, and for how long. For some folks, a few hours would be just fine. Three days is probably about the practical limit. For longer outages, a back-up generator is the answer. Run a couple days. Start the generator. Recharge. Run a couple more days . . . repeat as needed. Talk with one of our tech staff to size your battery pack.

In all cases, we strongly recommend sealed batteries for intertie systems, partly because they're likely to be forgotten and won't get regular watering, and partly because sealed batteries are going to be a lot happier about floating along for weeks, months, and years between discharge cycles. Sealed batteries have their chemistry tweaked for "set it and forget it" service, and you'll get longer, more dependable life from them. And you'll never need to kick yourself for forgetting to water them.

There are at least three manufacturers with an inverter that's smart enough to manage utility intertie and a back-up battery pack. (And maybe more by the time you read this.) Let's give a quick introduction to the known prospects, with an honest review of their good and bad points. In chronological order . . .

XANTREX

Trace started the whole business of small-scale everyday folks selling excess renewable power back to the utility in the mid-1990s with the SW inverter series. In the late 1990s, Trace was purchased by Xantrex and before long much of the engineering staff left to form Out-Back Engineering. This wasn't a happy time for Xantrex products. Now Xantrex is fully staffed and has been rolling out a nice upgrade to the well-proven SW series, called the XW.

The Xantrex XW Series inverters for battery-based systems are for both off-grid and on-grid applications (since they are field

selectable). These inverters offer an innovative, integrated design that minimizes external balance-of-system components allowing for much quicker and easier installation. The XW PDP power distribution panel includes all AC/DC disconnects and wiring. The XW Series offers pure sine-wave capability as well as split-phase operation right out of the box for 12VAC and/or 240VAC solutions (no need for auto-transformer or stacking of inverters). There will initially be three sizes available: 4,000W/24V, 4,500/48V, and 6,000/48V.

Xantrex inverters and options are fully covered in the Inverters section of this chapter.

OUTBACK ENGINEERING

OutBack formed when Xantrex bought Trace and much of the engineering staff decided they didn't enjoy working for a big company. These folks probably have more experience with residential inverters than any engineering team on Earth. The OutBack inverter is a clean-sheet-of-paper next-generation product. As such, it has many good and a few bad points. On the plus side, it delivers an absolutely perfect sine wave, and does it quietly, from a package that's about half the size and weight of the Xantrex SWP. Access for repairs in the field is easy on this inverter, and components are bundled into just a few major pieces. Much of the software is outside the inverter on the small, light,  easily returned and upgraded Mate. If you have a problem, a call to the manufacturer likely will put you in touch with the engineer who built it. The level of service from these folks is really terrific. On the negative side OutBack is an engineer-owned and operated company, and engineers will *never* stop developing their product until someone takes it away. So there will always be new capabilities and functions on next year's model. These will probably always be developing products, and although OutBack has the most extensive beta-testing programs our industry has ever seen, we still find occasional bugs.

OutBack systems are also à la carte packages, with several separate boxes for inverters, DC, and AC centers, and the like, and are available as preassembled Power Centers. Hey, it's a techie product, after all.

OutBack inverters and options are fully covered in the Inverters section of this chapter.

BEACON POWER

The Beacon M5 inverter is a fully integrated one-box package for utility intertie with battery backup. This is the nontechie solution when you don't want to play with it, or modify it, or have a raft of options to sort through. You just want it to work well, and take a minimum of space. The inverter, charge controller, DC and AC overcurrent protection, and all switchgear is housed in a single outdoor-rated enclosure. On the plus side the M5 is tidy, compact, relatively quick to install, and delivers a big healthy

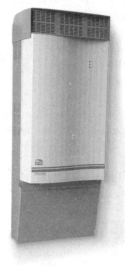

Beacon M5 Inverter

5,000 watts. On the negative side, it's 120-volt output only; it won't run 240-volt appliances. But then you shouldn't be running any 240-volt appliances during a utility outage anyway. On first glance, the M5 seems expensive. Actually, because all the bits and pieces are included, the M5 tends to cost less for a complete system in the 5,000-watt range.

SMA AMERICA

The Sunny Island sine-wave inverter from the makers of the Sunny Boy line of grid-tie inverters adds battery back-up functionality to new or existing grid-tie-only systems. The Sunny Island 4248U provides a continuous power output of 4,200 watts. It includes a built-in transfer relay rated for 60 amps at 120 volts AC and an internal battery charger supplying up to 100 amps. For an existing grid-tie setup, you need to install an essential load subpanel, a 240V/120V auto transformer, the Sunny Island, and a bank of deep-cycle gel batteries (in that order) and watch the nuisance of power outages seamlessly and quietly disappear.

Battery-Based Intertie Products

Beacon Power M5 Inverter

The single box intertie solution

In 2004, Beacon Power introduced a new concept to the large household inverter market. The M5 is a single, compact outdoor-rated unit that handles all the necessary functions for solar-electric intertie with battery backup. Rated at a big healthy 5,000 watts, the M5 converts any incoming PV energy to conventional AC power and uses it to displace utility power. If more PV energy is available than your house can use, the extra is pushed back through your meter, giving you credit toward later use. The M5 provides three ground-fault protected 50-amp PV input breakers with state-of-the-art Maximum Power Point Tracking control of the PV array to extract the maximum amount of energy at any moment. If the utility fails, the M5 instantly switches to battery power. It will easily keep the fridge, furnace, computer, small kitchen appliances, household lights, and entertainment equipment running during utility outages.

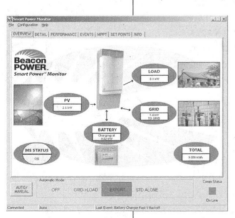

The M5 has an easy hang-on-the-wall mount with intelligent wiring interfaces that speed and simplify installation. The 50A, 120V output will run a separate electrical subpanel for all the circuits you need to back up.

For true set-it-and-forget-it operation, the M5 shows it's working with a simple set of LED indicators. For those who want to play with it a bit, the optional PC-based Smart Power Monitor lets you track a range of performance parameters plus total PV power generated, using common Ethernet plug and cable. Field reprogrammable setpoints provide complete system adjustment and control.

A complete system will require a 48-volt PV array with string combiners, a 48-volt sealed battery pack, 2/0 battery cables, and a 175A battery breaker. The M5 inverter is UL 1741 certified, measures 42" x 16" x 10" (HxWxD), weighs 120 lb., and has a 5-yr. mfr. warranty. USA.

Beacon Power M5 Specifications
Continuous output: 5,000 VA
Surge amps: 62.5 amps
DC voltage: 48 VDC
Charging rate: No AC charger
Preferred circuit breaker: 175 amp
DC cable size: 2/0 cu
Weight: 120 lb.

Item	Description	Price
49-0151	Beacon Power M5 Inverter	$6,995
27098	2/0, 10-ft Inverter Cables, 2-lug pair	$140
48-0186	Beacon 175A Breaker Assembly	$250
49-0152	Beacon Smart Power Monitor software	$249

Sealed Batteries for Utility Intertie Systems

Solar Gel Batteries

These true deep-cycle gel batteries come highly recommended by the National Renewable Energy Labs. They are completely maintenance-free, with no spills or fumes, and can be operated in any position. Because the electrolyte is gelled, no stratification occurs, and no equalization charging is required. Self-discharge is under 2% per month. Because of thinner plate construction, gel batteries can be charged or discharged more rapidly than other deep-cycle batteries. All Solar Gel batteries now come with secure, bolt-up, flag-type terminals.

To avoid gassing, gel batteries must be charged at lower peak voltages; typically, 13.8-14.2 volts maximum is recommended. Higher-charging voltages will drastically shorten life expectancy. Life expectancy depends on battery size and use in service, but no sealed battery will perform better or longer than these. Typical life expectancy is five years or more. Amp-hour rating at standard 20-hour rate. USA.

15210	**8GU1 12V 32Ah (24.2 lb., 7¾" x 5⅕" x 7¼")**	**$76**
15211	**8G22NF 12V 50Ah (37.6 lb., 9½" x 5½" x 9¼")**	**$122**
15212	**8G24 12V 74Ah (53.6 lb., 10⅘" x 6¾" x 9⅘")**	**$149**
15213	**8G27 12V 86Ah (63.2 lb., 12¾" x 6¾" x 9⅘")**	**$185**
15214	**8G31 12V 98Ah (71.7 lb., 12 9/10" x 6¾" x 9⅖")**	**$210**
15215	**8G4D 12V 183Ah (129.8 lb., 20¾" x 8½" x 10")**	**$379***
15216	**8G8D 12V 225Ah (160.8 lb., 20¾" x 11" x 10")**	**$453***
15217	**8GGC2 6V 180Ah (68.4 lb., 10¼" x 7⅕" x 10⅘")**	**$205**

**Shipped via truck.*

Hawker Envirolink Sealed Industrial Battery Packs

The best choice for emergency power back-up systems

The Hawker Envirolink series offers true gel construction, not the cheaper, lower life-expectancy AGM construction. These batteries are the top-quality choice for long-life, low-maintenance emergency power back-up systems. No routine maintenance is required, and this battery type thrives on float service. Life expectancy is 10-15 years with proper care and a charger/controller that keeps voltage below 2.35V/cell. Delivered in pre-assembled 12- or 24-volt packs with smooth steel cases and no exposed electrical contacts. Cycle life expectancy is 1,250 cycles to 80% depth of discharge. Interconnect cables are only needed between 12- or 24-volt trays; cell interconnects are soldered in place at the factory. 5-yr. mfr. warranty. USA.

Hawker Envirolink Sealed Industrial 12-Volt Battery Packs

Item #	Ah @ 20 hr.	Weight lb.	Size (") LxWxH	Price
17-0296	369	390	26.3 x 6.5 x 23	**$2,470**
15819	553	564	31 x 7.75 x 23	**$3,315**
15820	925	885	31.13 x 13 x 23	**$5,293**
15821	1,110	1,050	29.3 x 13 x 23	**$6,250**
15822	1,480	1,374	25.7 x 19.6 x 23	**$8,086**

Hawker Envirolink Sealed Industrial 24-Volt Battery Packs

Item #	Ah @ 20 hr.	Weight lb.	Size (") LxWxH	Price
17-0297	369	780	25.7 x 11 x 23	**$4,680**
15823	553	1,128	31 x 13 x 23	**$4,680**
15825	925	1,770	38.5 x 16.6 x 23	**$9,985**
15826	1,110	2,100	38.7 x 19.7 x 23	**$9,272**
15827	1,480	2,748	38.7 x 25.6 x 23	**$15,230**

Complete Intertie Systems

Typical 3,000-Watt Residential Intertie System

Almost every intertie system is site specific. Available mounting space and style vary, power needs and budgets vary, and we're happy to put together just what you need. But for information purposes, here's what a typical system looks like: (18) Sharp 208 watt PV modules, roof mounted racking, a Fronius IG 3000 intertie inverter, and the required DC and AC disconnect switches. Installation and permit costs are estimates based on Northern California averages. Wiring is supplied locally as needed to get from roof to inverter to main breaker panel.

Based on the North American yearly average of 5.5 hours noon-equivalent sunlight per day, this system will deliver approximately 340 kWh/month. Rebates and tax credits vary. In California (one of the states where our residential techs design, provide, and install several hundred turn-key solar systems every year), the current (1-1-07) rebate is $2.50/watt. A 30% federal tax credit (capped at $2,000) is available through the end of 2008 (unless it is extended). So, together, these solar incentives reduce the bottom line below by around 30% off the top. Check out www.dsireusa.org for any applicable incentives for your state.

Item #	Qty.	Description	Retail	Extended Price
41-0195	14	**SHARP 208-WATT MODULE**	$1,149	$16,086
41-0149	1	**FRONIUS IG 3000 INTERTIE INVERTER**	$2,460	$2,460
42-0282	2	**UNIRAC SOLARMOUNT 4-RAIL KIT**	$533	$1,066
14375 F	2	**UNIRAC SOLARMOUNT TOP-MOUNT CLAMP SET**	$39	$78
24644	1	**2-POLE SAFETY DISCONNECT**	$66	$66
24501	2	**30-AMP FUSE**	$6	$12
24648	1	**30-AMP SAFETY DISCONNECT**	$208	$208
25775	1	**DELTA LIGHTNING ARRESTOR**	$40	$40
26664	2	**SHARP 50' PV OUTPUT CABLE**	$38	$76
		System component subtotal		$20,092
		Installation including permit fees and cost to pull permit		$5,468
		Total installed system price		$25,560*

Shipping and sales tax (where applicable) are not included.

Go Solar— Declare Independence from Skyrocketing Energy Costs!

Starting at $9,995 for the 1.2kW Real Goods Grid-Tie Starter Kit, you can produce your own electricity from solar energy and sell about 140 kWh per month back to your utility company. Depending on your state or utility company, rebates and a federal tax credit are available to reduce this cost. See www.dsireusa.org for details. And by adding PV modules, you can expand the 1.2kW Starter Kit to 3.5kW. For $26,995, the 4kW Mid-Size Kit can produce about 450kWh per month.

The PV modules have a 25-yr. warranty; the other equipment, 10 yr.; with an expected system life of 30-40 yr. The Starter Kit includes the following equipment: (15) Sharp 80W PV modules (expandable to 44), roof mounting hardware by UniRac, and a Xantrex GT 3.0 sine wave grid-tie inverter. The Mid-size Kit includes (21) Evergreen 190W PV modules, roof mounting hardware by UniRac, and a Fronius IG 4000 sine wave grid-tie inverter. We also include the PV array interconnect cables, a DC disconnect switch and an AC disconnect switch. Your installer supplies wire, conduit and necessary small parts. If your home or business is located in California or Colorado, we can install the system for an additional charge (see www.realgoods.com/calsolar or www.realgoods.com/cosolar). If not, the Real Goods technical staff can provide expert training and assistance to ensure that your chosen installer (a licensed general, electrical or solar contractor in your area) gets the system up and running properly. (Note that for best results you need a shade-free south-to-southwest facing roof to mount the PV modules.)

53-0121	**1.2kW PV Grid-Tie Starter System Kit**	**$9,995**
53-0122	**4.0kW PV Grid-Tie Mid-Size System Kit**	**$26,995**

Fully Integrated Controls or Preassembled Power Centers

Fully integrated controls, otherwise known as power centers, are how most renewable energy systems go out the door now. A power center puts the inverter(s), DC protection/distribution, charge controller(s), monitoring, and AC protection/distribution hardware all in one tidy, pre-wired, UL-approved package. You get *ONE* component to hang on the wall. The alternative is to buy all the bits and pieces a la carte, hang them on the wall, and run all the wiring between them. If you're good (and/or lucky), it will only take half a dozen trips to the hardware store for small parts, but you will save yourself several hundred dollars. Most of the remote home power systems we've sold in the past few years have been built around one of these integrated control packages.

Intelligent manufacturers have taken the safety, charge control, and monitoring functions and combined them onto one neat, fully assembled and wired, UL-approved board, or enclosed box. The technical staff at Real Goods endorses this concept wholeheartedly, as it generally results in safer, faster, better-looking installations that rarely, if ever, cause problems with building inspectors. Most integrated controls are easily adaptable for diverse system demands, and expandable in the future. So now, instead of dealing with lots of bits and pieces that you (or your paid electrician) have to wire together on site, you can have it all delivered prewired.

Power centers are partially custom assembled to match the particular needs of a renewable energy system. Everybody needs theirs set up a little differently, so these are usually built to order. You must communicate with one of our technical staff to order an integrated control. If there's anything better, cheaper, or safer for your application, we'll let you know during this interview period.

At this time, Xantrex and OutBack both are offering various power center assemblies with a wealth of options.

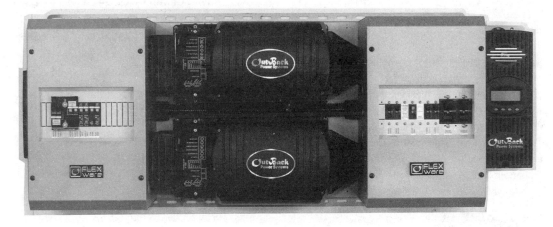

A typical residential power center using OutBack products.

PREASSEMBLED POWER CENTERS

OUTBACK FLEXWARE POWER CENTERS

Preassembled, prewired, UL-approved, and ready to hang on the wall
This is how most residential and commercial systems are installed now, even by the pros. Let a specialist with immediate access to all the small bits and pieces do the major component assembly. It costs a little more initially but saves big-time in on-site labor and trips to the hardware store.

Using quality OutBack components, each power center is assembled in an ETL-approved shop. Each custom-built panel is prewired, tested, and ready to go. You simply add batteries, charging source(s), and output wiring on site. All power centers come with DC and AC enclosures, the inverter(s), controllers, and other options of your choice, on a sturdy OutBack mounting plate.

The FLEXware Power Centers come in three basic sizes: two mounting plates with full-size AC and DC enclosures and space for up to four OutBack inverters—the FLEXware 1000; the half-size rack with smaller AC and DC enclosures and space for up to two OutBack inverters—the FLEXware 500; and the new single-inverter system—the FLEXware 250. The half-size racks are usually more than adequate for most residential uses and naturally cost less. In addition to half- and full-size racks, OutBack inverters come in sealed or vented models. Vented models are for indoor installations and are the right choice for most installations. Sealed models are for marine, jungle, bug-infested, or other challenging sites. So four power center styles are listed below.

In addition to the FLEXware equipment, the PS1 Power Panel is designed exclusively for the grid-tie market. It is a high-performance, easy to install, and low-maintenance battery back-up solution to grid-tie systems. OutBack has bundled their top-performing GFX sine-wave inverter and their MX60 MPPT controller in a factory-wired ETL-listed, weatherproof assembly. Maximum solar input is 3,200 watts at up to 72 volts nominal. The PS1 enclosure includes the AC bypass, DC inverter breaker, 50A house feed for your essential loads, MX60 battery disconnect, and the code-required GFP. Separate battery enclosure includes cable interconnects, hold-downs, and the rain-tight hub conduit.

Options? We have lots of options! All the controls, breakers, transformers, GFIs, and more that OutBack makes, plus a few things they don't make, are available assembled on power centers. We've listed the more common basic racks and options below; this is not a complete listing. Please work with one of our experienced technicians to specify a panel that fits your current and future needs. Prices run from $3,200 to $12,000 depending on size and options. Proudly made in the USA. *(See the Inverter section for a complete introduction to OutBack inverters.)*

OutBack PS1 Grid-Tie Solar with Battery Backup

49-0193	**OutBack PS1-3000 (GVFX3648)**	**$4,869**
49-0194	**OutBack PS1-2500 (GVFX2548)**	**$4,869**
49-0195	**OutBack PS1-BE Battery Enclosure**	**$599**
15-214	**98Ah/12V 8G31 Gel Battery (min. of 4 required per PS1-BE)**	**$210**

Standard OutBack Vented FLEXware 500 Power Centers

Base price for standard centers
All 20"x44"x12.75" (HxWxD). Weight 145 lb. w/one inverter, 215 lb. w/two inverters (2).

53-2000	**OutBack FLEXware 500 Vented Power Center VFX2812**	**$4,250**
53-2001	**OutBack FLEXware 500 Vented Power Center VFX2812 (2)**	**$7,395**
53-2009	**OutBack FLEXware 500 Vented Power Center VFX3524**	**$4,250**
53-2010	**OutBack FLEXware 500 Vented Power Center VFX3524 (2)**	**$7,395**
53-2019	**OutBack FLEXware 500 Vented Power Center VFX3648**	**$4,250**
53-2020	**OutBack FLEXware 500 Vented Power Center VFX3648 (2)**	**$7,395**

Standard OutBack Sealed FLEXware 500 Power Centers

Base price for standard centers
All 20"x44"x12.75" (HxWxD). Weight 145 lb. w/one inverter, 215 lb. w/two inverters (2).

53-2004	**OutBack FLEXware 500 Sealed Power Center FX2524**	**$3,850**
53-2005	**OutBack FLEXware 500 Sealed Power Center FX2524 (2)**	**$6,595**
53-2035	**OutBack FLEXware 500 Sealed Power Center FX2548**	**$3,850**
53-2036	**OutBack FLEXware 500 Sealed Power Center FX2548 (2)**	**$6,595**
53-2014	**OutBack FLEXware 500 Sealed Power Center FX3048**	**$3,850**
53-2015	**OutBack FLEXware 500 Sealed Power Center FX3048 (2)**	**$6,595**

Standard OutBack Vented FLEXware 1000 Power Centers

Base price for standard centers
All 36"x50"x12.75" (HxWxD). Weight 260 lb. w/two inverters (2); add additional 70 lb. per inverter.

53-1001 OutBack FLEXware 1000 Vented Power Center VFX2812 (2) $8,450

53-2002 OutBack FLEXware 1000 Vented Power Center VFX2812 (3) $11,350

53-2003 OutBack FLEXware 1000 Vented Power Center VFX2812 (4) $14,695

53-2011 OutBack FLEXware 1000 Vented Power Center VFX3524 (2) $8,450

53-2012 OutBack FLEXware 1000 Vented Power Center VFX3524 (3) $11,350

53-2013 OutBack FLEXware 1000 Vented Power Center VFX3524 (4) $14,695

53-2021 OutBack FLEXware 1000 Vented Power Center VFX3648 (2) $8,450

53-2022 OutBack FLEXware 1000 Vented Power Center VFX3648 (3) $11,350

53-2023 OutBack FLEXware 1000 Vented Power Center VFX3648 (4) $14,695

Standard OutBack Sealed FLEXware 1000 Power Centers

Base price for standard centers
All 36"x50"x12.75" (HxWxD). Weight 260 lb. w/two inverters (2); add additional 70 lb. per inverter.

53-2006 OutBack FLEXware 1000 Sealed Power Center FX2524 (2) $7,650

53-2007 OutBack FLEXware 1000 Sealed Power Center FX2524 (3) $10,150

53-2008 OutBack FLEXware 1000 Sealed Power Center FX2524 (4) $13,095

53-2016 OutBack FLEXware 1000 Sealed Power Center FX3048 (2) $8,195

53-2017 OutBack FLEXware 1000 Sealed Power Center FX3048 (3) $9,995

53-2018 OutBack FLEXware 1000 Sealed Power Center FX3048 (4) $12,895

Please call us for quotes on larger FLEXware 1000 systems with up to four inverters.

OUTBACK FLEXWARE FOR ON- OR OFF-GRID

OutBack inverters burst onto the renewable energy stage only a few years ago. Thanks to their small, light, quiet, highly dependable, user-friendly design, they've rapidly become the inverter of choice for both pure off-grid systems and for grid-tied systems that want the security of automatic back-up power. The software is slightly different for on- or off-grid. So if you're thinking grid-tied, be sure to order one of the models that starts with "G." Earlier models can be returned to the factory for upgrading if needed.

Most OutBack inverters go out the door preassembled into a Power Center for faster installation on-site and include the laboratory listing that inspectors love.

Standard OutBack Grid-Tie Vented FLEXware 500 Power Centers

Base price for standard centers
All 20"x44"x12.75" (HxWxD). Weight 145 lb. w/one inverter, 215 lb. w/two inverters (2).

53-2024 OutBack FLEXware 500 Vented Power Center GVFX2524 $3,850

53-2025 OutBack FLEXware 500 Vented Power Center GVFX2524 (2) $6,595

53-2028 OutBack FLEXware 500 Vented Power Center GVFX3048 $3,850

53-2029 OutBack FLEXware 500 Vented Power Center GVFX3048 (2) $6,595

53-2026 OutBack FLEXware 500 Vented Power Center GVFX3524 $4,250

53-2027 OutBack FLEXware 500 Vented Power Center GVFX3524 (2) $7,395

53-2030 OutBack FLEXware 500 Vented Power Center GVFX3648 $4,250

53-2031 OutBack FLEXware 500 Vented Power Center GVFX3648 (2) $7,395

OUTBACK FLEXWARE POWER CENTER OPTIONS

A wide variety of options are available installed on your OutBack Power Center.

Ground Fault Protection

If your PV array is installed on a residence, this protection is code-required. Uses three of the four breaker spaces; provides two 60-amp breakers.

27931   GFP Installed $195

OutBack MX60 Controller

The best charge control on the market, with MPPT, accepts up to 140-volt input, delivers up to 60 amps at 12-60 volts. Installed price includes two 60-amp breakers.

27930   MX60 Controller Installed $899

DC Panel-Mount Circuit Breakers

Extra input or output breakers for wind, hydro, or any DC appliance.

27915	1A DC Breaker Installed	$39
27915	10A DC Breaker Installed	$39
27915	15A DC Breaker Installed	$39
27915	30A DC Breaker Installed	$39
27915	40A DC Breaker Installed	$45
27915	60A DC Breaker Installed	$45
27915	100A DC Breaker Installed	$75

AC DIN Mount Circuit Breakers

Extra input or output breakers for generators, AC appliances, or subpanels.

27922	Single 15A AC Breaker Installed	$29
27922	Single 20A AC Breaker Installed	$29
27922	Single 50A AC Breaker Installed	$39
27922	Dual 15A AC Breaker Installed	$45
27922	Dual 20A AC Breaker Installed	$55
27922	Dual 50A AC Breaker Installed	$69

Flexware X240 Transformer

Used for a variety of step-up, step-down, or balancing chores. Can output 240 volts AC from a single inverter, or balance two 120-volt AC inputs from a generator for battery charging, or allow two OutBack inverters to share a heavy 120-volt AC load, while still wired for 240-volt AC output. (Note: Only OutBack inverters can do this trick.) Wiring varies with intended use; please let us know your intention at time of order.

27929   X240 Transformer Installed $554

OutBack DCA Conduit Adapter

An aluminum conduit adapter required when mounting an OutBack inverter to a Power Center or to a 2-inch conduit. Required for a NEC-compliant installation.

27908   DCA Conduit Adapter $45

OutBack ACA Conduit Adapter

A cast-polycarbonate AC wiring-compartment extension. Provides a 2-inch conduit adapter, two 1-inch conduit knockouts, and an AC outlet knockout. Required when mounting an FX inverter to an OutBack Power Center.

27909   ACA Conduit Adapter $65

XANTREX XW POWER DISTRIBUTION PANELS

Xantrex is offering a new line of components for the new XW Inverter/Chargers. All components will ship by UPS or other courier to your installation site. No expensive truck service, and with smaller, lighter components, now systems can be installed without a crew or expensive lifting equipment.

The basic power distribution panel requires 30"x36" (HxW) of wall space for an XW inverter with DC and AC boxes. All pieces are powder-coated for continued good looks. No backing plate is needed. The new taller AC and DC conduit boxes have space for up to two XW-series inverters plus ground fault protectors, terminal and distribution strips, and all the incoming, outgoing, or bypass circuit breakers you'll want, including the big DC ones for the inverter(s).

Basic Power Distribution Panel Packages

Xantrex offers basic Power Distribution Panel Packages with conduit box, factory wired, and labeled to support one inverter in a code-compliant manner. There is room for additional inverters, charge controllers, or other equipment so the system can be expanded to support larger and/or three phase systems. The prewired breaker panel for one inverter has wiring space and conduit and breaker knockouts to add up to three inverters and/or four charge controllers. The wall mounting bracket and multiple cable strap points facilitate clean and profes-

Item #	Model	AC Output	DC Input	Price
49-0506	XW PDP 4024 (1)	4.0kW	24VDC	$5,500
49-0507	XW PDP 4024 (2)	8.0kW	24VDC	$8,750
49-0508	XW PDP 4024 (4)	16.0kW	24VDC	$15,250
49-0509	XW PDP 4548 (1)	4.5kW	48VDC	$5,750
49-0510	XW PDP 4548 (2)	9.0kW	48VDC	$9,350
49-0511	XW PDP 4548 (4)	18.0kW	48VDC	$16,550
49-0512	XW PDP 6048 (1)	6.0kW	48VDC	$6,650
49-0513	XW PDP 6048 (2)	12.0kW	48VDC	$11,150
49-0514	XW PDP 6048 (4)	24.0kW	48VDC	$20,150

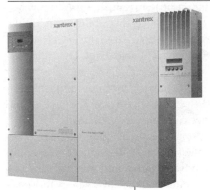

sional looking installations. The basic panel comes with DC inverter input breakers, and AC input/output bypass breakers and one inverter/charger. Certified to UL1741 and CSA for utility interactive applications.

A one-inverter power panel with one charge controller.

XW AC/DC Conduit Box

This is a bare conduit box (no wires) that can be used to create systems with more than two inverters or to retrofit XW inverters into existing systems which may already have AC/DC disconnects. It interfaces with XW4024, XW4548, and XW6048 inverter/chargers and the XW DPD. Includes conduit box/raceway with barriers to ensure proper separation between low-voltage communication cables and AC and DC wires.

49-0515 XW Conduit Box   **$175**

PVGFP PV Ground Fault Breakers

If you mount your PV array on a residence, you need ground fault protection. These bolt into the DC conduit box and take 2, 3, 4, or 5 small breaker slots (one more slot than the number of poles). Each pole can handle one charge controller, up to 100 amps. USA.

48-0191 PVGFP-CF-1 1-Pole Grd. Fault Protector   **$250**

48-0192 PVGFP-CF-2 2-Pole Grd. Fault Protector   **$275**

48-0193 PVGFP-CF-3 3-Pole Grd. Fault Protector   **$300**

48-0194 PVGFP-CF-4 4-Pole Grd. Fault Protector   **$325**

CF60 DC Circuit Breaker

A 60-amp DC circuit breaker for input or output power as needed. Takes one small breaker slot. This CF-type breaker is rated for up to 160vdc open circuit. USA.

48-0195 CF60 Circuit Breaker   **$40**

XW Automatic Generator Start

If your generator supports two- or three-wire remote start ability, this module will provide automated start and stop functions based on battery voltage. Programmable thru the SWP inverter programming. USA.

49-0518 XW Automatic Generator Start   **$199**

ALM Auxiliary Load Module

Provides three relays that can turn stuff off or on based on battery voltage. Want to automatically turn some appliance(s) off if the battery gets too low, or start some other appliance once the battery gets mostly charged? Have it either or both ways, here's your answer. A nerd's dream! Programmed thru the inverter. USA.

25865   Auxiliary Load Module   **$159**

Inverter Cables

Our inverter and battery cables are all flexible, cable with color coding for polarity and shrink-wrapped lugs for 3⁄8" bolt studs. These two-lug versions (lugs on both ends) are great for Xantrex power centers and disconnects. Just cut them at the perfect point to be both the inverter cables as well as the battery cables. USA.

15851	2/0, 3' Inverter Cables, 2 lug, single	$30
15852	4/0, 3' Inverter Cables, 2 lug, single	$40
27311	4/0, 5' Inverter Cables, 2 lug, Red or Black	$60
27312	4/0, 10' Inverter Cables, 2 lug, Red or Black	$105
27099	2/0, 5' Inverter Cables, 2 lug, Red or Black	$50
27098	2/0, 10' Inverter Cables, 2 lug, Red or Black	$70

Safety and Fusing

Any wire that attaches to a battery must be fused!

Any time we have a power source and wiring, we need to ensure that the current flow will be kept within the capability of the wire. Too much current through too small a wire leads to melted insulation or even fires. Renewable energy systems have at least two power sources. The solar, wind, or hydro generator is one (or more), and the battery pack is the other. The battery requires particular concern. Even small battery packs store up what can be an awesome amount of energy for future use. So we'll cover battery safety first.

In many ways, batteries are safer than traditional AC power. It's not impossible, but it's pretty difficult to shock yourself with the low-voltage direct current (DC) in the 12- or 24-volt battery systems we commonly use. On the other hand, because batteries are right there in your house, they can deliver many more amperes into an accidental short circuit than a long-distance AC transmission system can. These high-amperage flows from battery systems can turn wires red hot almost instantly, burning off the insulation and easily starting a fire. (The National Electrical Code says the same thing, although it spends a lot more words saying it.)

And if red-hot wires and melting insulation aren't enough to teach caution, it's difficult to stop a DC short circuit. In a really severe short, popping open a circuit breaker or fuse may not stop the electrical flow. The current will simply arc across the gap and keep on cooking! Once a DC arc is struck, it has much less tendency to self-snuff than an AC arc. Arc-welders greatly

An Outback DC Enclosure with large breaker for the inverter, ground fault protection for the PV array, smaller breakers for the charge controller, and lots of space left over.

prefer using DC for this very reason, but impromptu welding is not the sort of thing you want to introduce into your house. Fuses that are rated for DC use a special snuffing powder inside to prevent an arc after the fuse blows. AC-rated fuses rarely use this extra level of protection.

Many DC circuit breakers have limited current interruption ability, and require an upstream main fuse that's designed to stop very large (20,000-amp) current flows. All the Class T fuses we offer have this ability and rating. When large household-sized inverters started becoming widely available, large DC-rated circuit breakers, DC disconnects, and DC power centers started showing up in the industry. For household systems, these are the safest, and most aesthetically pleasing, solution.

The National Electrical Code requires "overcurrent protection and a disconnect means" between any power source and any appliance. A DC-rated circuit breaker covers both those requirements, and has become the first choice. For small systems, a breaker box and circuit breakers may get expensive and folks are tempted to cut corners. Do so at your own peril.

Large-inverter safety requires some specialized equipment. Full-size 3,000- to 5,000-watt inverters can safely handle up to 500 amps input at times. Systems of this size will use either an OutBack or Xantrex DC power center that provides all the fusing, safety disconnect, ground fault protection, and interconnection space for everything DC in one compact, prewired, engineered, UL-approved package.

For midsize, 400- to 1,000-watt inverters a 110-amp Class T fuse assembly with an appropriate Blue Sea circuit breaker makes a cost-effective and legal safety disconnect. The small inverters of 100 to 300 watts simply plug or hardwire into a fused outlet.

Fuses are sized to protect the wires in a circuit. See the chart near the fuse listings for standard sizing/ampacity ratings. These ratings are for the most common wire types. Some types of wire can handle slightly more current, and all types of wires can safely handle much more current for a few seconds. If a particular appliance, rather than the wire, wants protection at a lower amperage, then usually a separate or in-line fuse is used just for that appliance.

Lightning Protection and Grounding

Let's preface this subject with a warning: *Lightning is capricious.* That means it does whatever it wants to do, whenever it wants to. This does not mean you're defenseless, however. You can do plenty to avoid lightning damage 99% of the time.

The most important thing to keep in mind about lightning is that it *wants* to go to ground. Your job is to give it a clear, easy path, and then stay out of the way. Good, thorough grounding is your first, and most effective, line of defense. Solar arrays have a nice conductive aluminum frame around each module. But frames are usually anodized, which makes them act as an insulator. There will not be a good electrical path between the frames and the mounting structure. You need to run bare copper wire, no smaller gauge than your PV interconnect wiring, across the backs of all the modules, with a good mechanical connection to each module frame. Stainless steel self-tapping screws with stainless star washers are the best choice. Run this ground wire to your household ground rod, if it's within about 50 feet. If your array is more remote, install a ground rod at the array and ground to it. It's fine to have multiple ground rods in a single system . . . however . . . *all ground points must be connected to each other with #6-gauge copper wire.* This is probably the most important point about lightning

safety. Connect all the grounds together!

Why? Let's say your array just took a hit, but the lightning took the easy path to ground you provided. Now the earth around your array is at several thousand volts potential, and that energy is looking for a way to spread out, the quicker and easier the better. It sees a nice fat ground rod over at your house, and if it just jumps into this highly conductive copper PV-negative wire that someone nicely provided, it can get over to the house ground rod real easy...so it does. Goodbye charge controller, and whatever else gets in the way. If your ground rods are all connected together with bare copper wire, however, the lightning is going to take the easy route, and

pretty much stay out of your system wiring. If your bare ground wire is buried in direct contact with the ground, not inside a conduit, the lightning will like it even better, because it has more ground contact to disperse into.

For wind turbine towers, which obviously are just begging for lightning strikes, you need a ground rod at the base of the tower, and a ground rod at each guy wire anchor. Connect them all together, and then connect back to the house ground.

Good grounding like this will save your system from damage through better than 90% of strikes. For that occasional capricious strike that decides to take an excursion through your wiring, you need to have lightning protectors installed. These $45 gizmos short out internally if voltage goes above 300, and route as much energy as possible to ground. You should have at least one on each set of input wires. If you live in a high-lightning zone, a couple of protectors on each input is cheap insurance.

Good grounding and lightning protectors are sufficient protection for residential systems. If you're installing in a really severe lightning climate, like a mountaintop (or Florida, according to some folks), seek professional lightning protection assistance! Much more can be done for extreme sites.

SAFETY EQUIPMENT

OutBack PV Combiner Box

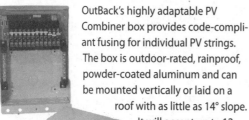

OutBack's highly adaptable PV Combiner box provides code-compliant fusing for individual PV strings. The box is outdoor-rated, rainproof, powder-coated aluminum and can be mounted vertically or laid on a roof with as little as 14° slope. It will accept up to 12 DC breakers for circuits up to 150 VDC, or it can be set up with up to 7 touch-safe midget fuses for high-voltage systems up to 600 VDC. Breakers or fuse holders are sold separately as needed. Can be configured for one, or two separate PV output circuits. You heard right. Run two PV controller circuits from a single combiner box! Set-screw output terminals will accept up to 1/0 gauge bare cable. Plenty of space for current sensors or lightning protection. The TBB (insulated terminal bus bar) option is required when running two RV Power Products Solar Boost controls, to provide a second isolated negative circuit. PV combiner box measures 13.1"x9.5"x3.5" (HxWxD) and weighs 6 lb. (plus options). ETL-listed, made in USA.

24653	PV Combiner Box	$139
24660	OBB-6-125VDC-DIN, 6-amp Breaker	$13
24661	OBB-9-125VDC-DIN, 9-amp Breaker	$13
24662	OBB-10-125VDC-DIN, 10-amp Breaker	$13
24663	OBB-15-125VDC-DIN, 15-amp Breaker	$13
460517	OBB-20-125VDC-DIN, 20-amp Breaker	$13
24664	OBB-30-125VDC-DIN, 30-amp Breaker	$13
24665	OBFH-30-600VDC-DIN Hi-Volt Fuseholder	$18
24667	OBF-6-600VDC Hi-Volt Fuse, 6-amp	$18
24668	OBF-10-600VDC Hi-Volt Fuse, 10-amp	$18
24669	OBF-15-600VDC Hi-Volt Fuse, 15-amp	$18
24654	TBB-White Optional 2nd Neg. Bus Bar	$19

OUTBACK POWER GEAR

All the safety and connection pieces for a professional installation

OutBack supplies a full range of mounting, safety, control, and enclosure equipment to make your system safe and good looking (at least to techie types). If you're going the pre-assembled power center route, these necessary parts are already covered for you. If you're assembling on site, forge on (and don't hesitate to call us for help).

Power gear is available for full-size (FLEXware 1000) racks that will accept up to four OutBack inverters, or half-size (FLEXware 500) racks that accept up to two OutBack inverters. All OutBack metal work is highest quality, fits perfectly, has a wealth of prepunched knockouts and mounting points, and is powdercoated for great corrosion resistance and continued good looks.

OUTBACK AC POWER GEAR

OutBack AC Enclosures

Provides space for AC bypass switches, the incredibly useful transformer option, and for smaller systems it can function as the complete household AC breaker panel. The AC input/output/bypass breaker come in 30A or 60A sizes. USA.

48-0254	FW250 AC &/or DC Breaker Enclosure	$99
48-0256	FW500 AC Breaker Enclosure	$289
48-0258	FW1000 AC Breaker Enclosure	$489
27909	ACA AC Conduit Adapter	$65

OutBack AC Autotransformer

Can be used for step-up, step-down, or load balancing. Unique to OutBack, two FX inverters can share a heavy 120-VAC load, while wired for 240-VAC output with this transformer. Rated to 4kVA. Includes a 25A two-pole breaker. Mounts inside FW500 or FW1000 AC enclosures. USA.

24690	AC Autotransformer (FW-X240)	$390

OutBack AC Input-Output-Bypass Assemblies

These are field installable kits with color coded wire, all required terminal bus bars and sliding interlock plate. The FW250 breakers are single pole PANEL mount; all others are single pole DIN mount.

48-0259 Single inverter I-O-B for FW250 only
(120VAC, 60A output) **$99**

48-0260 Single inverter I-O-B for FW250 only
(230VAC, 30A output) **$99**

48-0261 Dual inverter I-O-B for FW500
(120/240VAC, 60A output) **$169**

48-0262 Dual inverter I-O-B for FW500
(120VAC, 120A output) **$169**

48-0263 Dual inverter I-O-B for FW500
(230VAC, 60A output) **$169**

48-0264 Quad inverter I-O-B for FW1000
(120/240VAC, 120A output) **$339**

48-0265 Quad inverter I-O-B for FW1000
(120VAC, 240A output) **$339**

48-0266 Quad inverter I-O-B for FW1000
(230VAC, 120A output) **$339**

48-0267 Triple inverter I-O-B for FW1000
(120/208VAC, 60A output) **$299**

48-0268 Triple inverter I-O-B for FW1000
(230/400VAC, 30A output) **$299**

OutBack AC Circuit Breakers

Extra input or output breakers for generators, AC appliances, or subpanels. These are DIN mount. Single pole are .5" wide; dual pole 1" wide.

24700 Single 15A OutBack AC Circuit Breaker
(15A 120VAC single pole) **$39**

24700 Dual 15A OutBack AC Circuit Breaker
(15A 120/240VAC dual pole) **$39**

24700 Single 20A OutBack AC Circuit Breaker
(20A 120VAC single pole) **$15**

24700 Dual 20A OutBack AC Circuit Breaker
15A 120/240VAC dual pole) **$39**

24700 Dual 25A OutBack AC Circuit Breaker
(25A 120/240VAC dual pole) **$35**

Terminal Bus Bars

The AC enclosures come with all the bus bars most folks will need, but for the creative types, here you go. Isolated bus bars, with 12 holes. Choose your color. USA.

**24670 TBB-Black (use as AC L1 hot
or DC negative** **$19**

**24702 TBB-Red (use as AC L2 hot or
DC positive** **$19**

**24654 TBB-White (use as AC L1 hot
or DC negative)** **$19**

**46-0514 TBB-Blue (use as phase C on
three phase systems)** **$19**

**46-0515 TBB-Brown (use as AC hot on
European systems)** **$19**

**46-0516 TBB-GROUND (ground/neutral w/
mounting screw-no insulators)** **$19**

OUTBACK DC POWER GEAR

OutBack FLEXware Mounting Plates

FLEXware MP is a one piece powder-coated aluminum mounting plate for FLEXware 500 and 1000. Utilizing stainless steel mounting hardware, the integrated locating bolts make installation quick and easy by providing guides to line up enclosures and inverter/chargers. A single MP is designed to accommodate the FLEXware 500 while two are required for the FLEXware 1000 configuration. Includes enclosure and inverter mounting screws..

48-0072 FW-MP Mounting Plate **$129**

FW1000 DC enclosure

FW500 DC enclosure

OutBack DC Enclosures

Provides space for large DC inverter breakers, shunt, controller breakers, GFP breaker, and bus bars. The full-size FW1000 enclosure supports up to four inverters, and accepts eleven ¾" breakers, nine 1" breakers, and six 1.5" breakers. The half-size FW500 enclosure supports up to two inverters, and accepts eight ¾" breakers and two 1.5" or three 1" breakers. The single inverter FW250 AC/DC enclosure supports four ¾" breakers and one 1.5" breaker.

48-0254	**FW250 AC &/or DC Breaker Enclosure**	**$99**
48-0256	**FW500 AC Breaker Enclosure**	**$289**
48-0258	**FW1000 AC Breaker Enclosure**	**$489**
27908	**DCA DC Conduit Adapter for FW500/1000**	**$45**
46-0519	**Single side mounted MX Charge Controller bracket for FW500/1000**	**$39**
46-0520	**Dual side mounted MX Charge Controller bracket for FW500/1000**	**$59**
46-0521	**Dual top mounted MX Charge Controller bracket for FW500/1000**	**$59**

OutBack DC Circuit Breakers

OutBack DC breakers can be used for additional inverters, input from DC charging sources, or output to DC appliances. They come in three widths and are panel-mounted

24674	**250A DC Breaker, 1.5"**	**$129**
24675	**175A DC Breaker, 1.5"**	**$129**
46-0522	**125A DC Breaker, 1.0"**	**$59**
24676	**100A DC Breaker, 1.0"**	**$59**
46-0523	**80A DC Breaker, ¾"**	**$25**
24677	**60A DC Breaker, ¾"**	**$25**
46-0524	**50A DC Breaker, ¾"**	**$25**
24678	**40A DC Breaker, ¾"**	**$25**
24679	**30A DC Breaker, ¾"**	**$25**
46-0525	**20A DC Breaker, ¾"**	**$25**
24680	**15A DC Breaker, ¾"**	**$25**
24681	**10A DC Breaker, ¾"**	**$25**
46-0526	**5A DC Breaker, ¾"**	**$25**
46-0527	**1A DC Breaker, ¾"**	**$25**

OutBack Ground Fault Protection

Required for PV arrays mounted on a residence. For one or two MX charge controllers. This simple kit includes an 80A 125VDC dual pole panel-mount breaker with an isolated negative bus bar and a ground bus bar. Takes three ¾" mounting spaces.

24673	**Outback GFP (OBB-GFP-80D-125VDC-PNL)**	**$129**

Xantrex DC Disconnects

To provide DC system overload/disconnect equipment for everything passing in or out of the battery bank, here's an alternative that meets all the code require-

ments. Best used with systems that only use DC power for one or two loads, the Xantrex DC Disconnects are available in two basic versions, 175 amps and 250 amps, with a variety of options.

No fumbling for expensive replacement fuses in the dark— these units feature easy to reset circuit breakers. Breakers are rated for 25,000-amps interrupting capacity at 65 volts, and are UL-listed for DC systems up to 125 volts. Main breaker lugs accept up to #4/0 AWG fine strand cable, no ring terminals required. 0.75", 2", and 2.5" knockouts are provided. Mates up with the conduit box option for DR and SW series Xantrex inverters. (Matching color scheme, too.) Space is available inside the disconnect for the DC current shunt(s) required by most monitors.

Options, all of which can be installed in the field, include a second main breaker for dual inverter systems, a DC bonding block for negative and ground wires, and up to four smaller DC breakers. 15-, 20-, and 60-amp circuit breakers are offered for PV or other charging input, or DC power output to loads. The negative/ground bonding block provides a necessary connection point for any negative wires. It accepts up to four 4/0, two #1/0, two #2, and four #4 cables. USA.

25-010	Xantrex DC 175A Disconnect	$345
25-011	Xantrex DC 250A Disconnect	$345
25-012	Xantrex 2nd 175ABreaker	$195
25-013	Xantrex 2nd 250A Breaker	$195
25-014	Xantrex 15A Auxiliary Breaker	$29
25-008	Xantrex 20A Auxiliary Breaker	$29
25-015	Xantrex 60A Auxiliary Breaker	$42
25-016	Xantrex Neg/Grd Bonding Block	$140

Code-Approved DC Load Centers

We find that a simple six-circuit center will handle the DC fusing needs for 99% of systems. This six-circuit center meets all safety standards and utilizes only UL-listed components. All current-handling devices have been UL approved for 12-volt through 48-volt DC applications.

These DC circuit breakers require a main fuse in-line before the breaker box to provide catastrophic overload protection. If you already have a large Class T fuse or circuit breaker for your inverter, it can be used for this protection. If not, you must use the 110-amp class T fuse block on the input. USA.

Breakers must be ordered separately.

23119	6-Circuit DC Load Center (Square D #QO6-12L100D)	$49
24507	Fuse Block, 110A Fuse and Cover	$53
23131	10A Circuit Breaker	$19
23132	15A Circuit Breaker	$15
23133	20A Circuit Breaker	$15
23134	30A Circuit Breaker	$15
23140	60A Circuit Breaker	$19

Class T DC Fuse Blocks with Covers

For those who want to protect their midsized inverter or other DC systems, we offer the excellent DC-rated, slow-blowing Class T fuses with block, cover, and #2/0 lugs. Dimensions are 2" x 2.5" x 5.5", including the cover. Rated for DC voltage up to 125 volts.

For quick and easy basic catastrophic protection, just bolt the bare replacement fuse between your battery + terminal and your output cable. Has 5/16" bolt holes. USA.

Use with surface-mount DC breakers below for a disconnect means and NEC compliance.

24511	Fuse Block, 400A Fuse & Cover	$75
24518	Fuse Block, 300A Fuse & Cover	$75
24212	Fuse Block, 200A Fuse & Cover	$55
24507	Fuse Block, 110A Fuse & Cover	$53
48-0272	400A Class T Replacement Fuse	$40
24217	300A Class T Replacement Fuse	$54
24211	200A Class T Replacement Fuse	$18
24508	110A Class T Replacement Fuse	$18

Surface-Mount DC Circuit Breakers

High amperage protection, small, inexpensive package

Here is a compact DC circuit breaker at an affordable price. These sealed, surface-mounting breakers are water- and vaporproof. They are thermally activated, and manually resettable, with the reset handle providing a positive trip identification. The red trip button can also be activated manually, allowing the breaker to function as a switch. They are "trip-free" and cannot be held closed against an overload current. Thermal tripping allows high starting loads, without tripping. They have ¼" studs for easy electrical connection.

These simple-to-install, surface-mount breakers are an excellent low-cost choice for small to midsize inverters, wind generators, control and power diversion equipment, or any other application with current flows of 50-150 amps. For use with DC power systems up to 30 volts. USA.

Rated at minimal 3,000-amp interrupting capacity. Requires a class T fuse inline before the circuit breaker for NEC compliance. See above.

48-0210	**30A** Circuit Breaker	**$50**
48-0210	**50A** Circuit Breaker	**$50**
48-0210	**70A** Circuit Breaker	**$50**
48-0210	**100A** Circuit Breaker	**$50**
48-0210	**150A** Circuit Breaker	**$50**

Safety Disconnects

Safety disconnects provide overcurrent protection and a quick disconnect means between any power source and appliance. Both the batteries and the renewable generator (PV, wind, hydro turbine) are considered a power source. The 2-pole units have two separate paths for power, 3-pole units have three paths, so these can provide protection for different circuits. We carry several different sizes. Fuses should be sized according to the wire size you are using. Raintight boxes should be used if the box will be exposed to the weather. USA.

AC ONLY

24201	**30A 2-Pole Fused Disconnect,** Indoor (Eaton #DG221NGB)	**$45**
24644	**30A/2-Pole Fused Disconnect,** Raintight (Square D #D221NRB)	**$66**
24404	**30A 3-Pole Fused Disconnect,** Indoor (Square D #D321N)	**$99**
24-4045	**30A 3-Pole Fused Disconnect,** Raintight (Square D #D321NRB)	**$124**

24202	**60A 2-Pole Fused Disconnect,** Indoor (Square D #D222N)	**$65**
48-0054	**60A 2-Pole Fused Disconnect,** Raintight (GE #TG3222RB)	**$85**
24645	**60A 2-Pole Fused Disconnect,** Raintight (Square D #D222NRB)	**$109**
24203	**60A 3-Pole Fused Disconnect,** Indoor (Square D #D322N)	**$114**
24-4046	**60A 3-Pole Fused Disconnect,** Raintight (Square D #D322NRB)	**$189**
24-4047	**100A 2-Pole Fused Disconnect,** Indoor (Square D #D323N)	**$150**
24204	**100A 2-Pole Fused Disconnect,** Raintight (Square D #D323NRB)	**$199**

DC RATED

24-5003	**30A 2-Pole Fused Disconnect,** Raintight (Square D #H221NRB)	**$320**
49-0517	**60A 2-Pole Fused Disconnect,** Raintight (Square D #H222NRB)	**$320**

Fuses sold separately below.

DC-Rated Fuses

These fuses fit the Safety Disconnects and the Minimum Protection Fuseblocks. Class RK5 fuses are rated for DC service up to 125 volts. They are time delay types that will allow motor starting or other surges. The 45A fits a 60A holder. Buy extras! You won't find these DC-rated fuses at your local hardware store (when you need one bad). USA.

48-0068	**15A FRN-R15 Class RK5 Fuse**	**$5**
24516	**20A FRN-R20 Class RK5 Fuse**	**$5**
24501	**30A FRN-R30 Class RK5 Fuse**	**$6**
48-0209	**40A FRN-R40 Class RK5 Fuse**	**$9**
24502	**45A FRN-R45 Class RK5 Fuse**	**$11**
24503	**60A FRN-R60 Class RK5 Fuse**	**$10**
24505	**100A FRN-R100 Class RK5 Fuse**	**$20**

Minimum Protection Fuseblocks

For simple, inexpensive protection, we offer larger diameter, lower resistance fuseblocks that offer more contact surface area than automotive-type fuses. Single fuseblocks with spring clips. 30A block takes up to 10-gauge wire; 60A block takes up to 2-gauge wire. Order fuses separately. USA.

| 24401 | 30A Fuseblock | $13 |
| 24402 | 60A Fuseblock | $18 |

ATC-Type DC Fusebox

Using the newer, safer, ATC fuse style, this covered plastic box has wire entry ports on the sides. Six individually fused circuits with a pair of main lugs for battery positive, and a negative bus bar. Will accept up to 10-gauge wire. Requires a Torx T-15 screwdriver. 8.75" x 5.5" x 2". USA.

| 24-418 | ATC-Type DC Fusebox | $39 |

Inline ATC Fuse Holder

A simple inline holder for any ATC-type fuse. With 14-gauge pigtails. Not for exposed outdoor use.

| 25540 | ATC Inline Fuse Holder | $2.95 |

ATC-Type DC Fuses

These are the newer plastic body DC fuses that most automotive manufacturers have been using lately. Compared to old-style round glass fuses, they're safer, as energized metal parts aren't exposed, they have more metal to metal surface contact, the amp rating is easier to read and color-coded, and with the different colors they look cool! These come in little boxes of five fuses. USA.

24219	ATC 2A Fuse	$2.95
48-0227	ATC 3A Fuse	$2.95
48-0228	ATC 5A Fuse	$2.95
48-0229	ATC 10A Fuse	$2.95
48-0230	ATC 15A Fuse	$2.95
48-0231	ATC 20A Fuse	$2.95
24225	ATC 25A Fuse	$2.95
48-0232	ATC 30A Fuse	$2.95

DC Lightning Protector

Here is simple, effective protection that can be used on any PV, wind, or hydro system up to 300 volts DC. These high-capacity silicon oxide arrestors will absorb multiple strikes until their capacity is reached and then rupture, indicating need for replacement. Three-wire connection, positive, negative, and ground, can be done at any convenient point in the system. The AC unit is identical in operation and capacity; only the wire colors vary. The 4-wire unit is for wind turbines with 3-phase output (African, Proven, Whisper, Bergey turbines). USA.

25194	DC Lightning Protector LA302 DC	$45
25277	AC Lightning Protector LA302 R	$45
25749	4-Wire Lightning Protector LA603	$60

Raintight Junction Boxes and Power Distribution Block

Most battery-based PV systems will need a fused combiner box where the small array wiring joins the larger lead-in wiring, but small systems and utility intertie systems, which rarely require a fused combiner box, can use a distribution block and raintight box. This is the easy, secure way to join your lead-in cables to the smaller interconnect wires from the PV array. Insert cables and tighten set screws. The 2-pole block accepts positive and negative, the 3-pole also accept ground. For every pole of both blocks, the primary side accepts one large cable, #14 to 2/0, the secondary side accepts four smaller cables, #14 to #4. For use with copper or aluminum conductors.

Raintight boxes are 16-gauge zinc-coated steel with gray finish. Removable cover is fastened at bottom by a screw. For noncorrosive environments. USA.

48-0014	2-Pole Power Distribution Block	$27
24705	3-Pole Power Distribution Block	$39
24213	Junction Box 10"x8"x4"	$49
24214	Junction Box 12"x10"x4"	$65

Toggle Circuit Breakers– AC/DC

Toggle circuit breakers combine switching and circuit protection needs for 12/24/48VDC systems. Typically used for DC charge or load circuits rated 65 VDC/277VAC max. All are single pole magnetic action in a phenolic case with screw terminals Compact size allows for less area consumed on DC panel assemblies 0.75"x2", toggle is red color. UL, CE recognized.

Model	Current Rating Amp	Weight lb.	MSRP
42-0525	10	0.17	$17.50
42-0525	15	0.17	$17.50
42-0525	20	0.17	$17.50
42-0525	30	0.17	$17.50

Mounting Panel–Toggle Breakers

Attractive dark aluminum ⅛" panels hold one or two toggle type breakers or panel switches.

45-0526 ONE for one breaker
2.63 x 3.75 0.08   **$18.00**

45-0526 TWO for two breakers
2.63 x 3.75 0.08   **$22.50**

45-0526 SCREWS circuit breaker screws (6 pk) lat head, black to match panel   **$2.25**

Automatic Transfer Switch

Transfer switches are designed as safety devices to prevent two different sources of voltage from traveling down the same line to the same appliances. This transfer switch will safely connect an inverter and an

AC generator to the same AC house wiring. If the generator is not running, the inverter is connected to the house wiring.

When the generator is started, the house wiring is automatically disconnected from the inverter and connected to the generator. A time-delay feature allows the generator to start under a no-load condition and warm up for approximately one minute. Available in both 110-volt and 240-volt models. Each will handle up to a maximum of 30 amps. These switches are great for applications where utility power may be available only a few hours per day, or if frequent power outages are experienced. Housed in a metal junction box with hinged cover. Wires are clearly marked and installation schematic is included inside cover. Two-yr. warranty. USA.

23121	**Transfer Switch, 30A, 110V**	**$99**
23122	**Transfer Switch, 30A, 240V**	**$149**

Charge Controllers

Almost any time batteries are used in a renewable energy system, a charge controller is needed to prolong battery life. The most basic function of a controller is to prevent battery overcharging. If batteries routinely are allowed to overcharge, their life expectancy will be reduced dramatically. A controller will sense the battery voltage, and reduce or stop the charging current when the voltage gets high enough. This is especially critical with sealed batteries that do not allow replacement of the water that is lost during overcharging.

The only exception to the need for a controller is when the charging source is very small and the battery is very large in comparison. If a PV module produces 1.5% of the battery's ampacity or less, then no charge control will be needed. For instance, a PV module that produces 1.5 amps charging into a battery of 100 amp-hours capacity won't require a controller, as the module will never have enough power to push the battery into overcharge.

PV systems generally use a different type of controller than a wind or hydro system requires. PV controllers can simply open the circuit when the batteries are full without any harm to the modules. Do this with a rapidly spinning wind or hydro generator and you will quickly have a toasted controller, and possibly a damaged generator. Rotating generators make electricity whenever they are turning. With no place to go, the voltage will escalate rapidly until it can jump the gap to some lower-voltage point. These mini lightning bolts can do damage. With rotating generators, we generally use diversion controllers that take some power and divert it to other uses. Both controller types are explained below.

PV Controllers

Most PV controllers simply open or restrict the circuit between the battery and PV array when the voltage rises to a set point. Then, as the battery absorbs the excess electrons and voltage begins to drop, the controller will turn back on. With some controllers, these voltage points are factory preset and nonadjustable, while others can be adjusted in the field. Early PV controllers used a relay, a mechanically controlled set of contacts, to accomplish this objective. Newer solid-state controllers use power transistors and PWM (Pulse Width Modulation) technology to turn the circuit rapidly on and off, effectively floating the battery at a set voltage. PWM controllers have the advantage of no mechanical contacts to burn or corrode, but the disadvantage of greater electronic complexity. Both types are in common use, with most new designs favoring PWM, as electronics have been gaining greater reliability with experience.

The latest generation of PV controls employs an electronic trick called Maximum Power Point Tracking, or MPPT. This allows the controller to run the PV array at whatever voltage delivers the highest wattage. This is often at a higher voltage level than the batteries would tolerate. The excess voltage is converted to amperage that the batteries can digest happily before it leaves the controller. So MPPT controllers usually push more amps out than they're taking in. On average, an MPPT controller will deliver about 15% more energy per year than a standard controller. And they do their best work in winter, when most off-grid homes need all the charging help they can get. Cold temperatures tend to elevate PV voltages, and long hours of darkness tend to lower battery voltages. We expect MPPT controls to gradually take over most of the PV controller market.

Controllers are rated by how much amperage they can handle. National Electric Code regulations require controllers to be capable of withstanding 25% overamperage for a limited time. This allows your controller to survive the occasional edge-of-cloud effect, when sunlight availability can increase dramatically. Regularly or intentionally exceeding the amperage ratings of your controller is the surest way to turn your controller into a crispy critter. It's perfectly okay to use a controller with more amperage capacity than you are generating. In fact, with this piece of hardware, buying larger to allow future expansion is often smart planning, and usually doesn't cost much.

A PV controller usually has the additional job of preventing reverse-current flow at night. Reverse-current flow is the tiny amount of electricity that can flow backward through PV modules at night, discharging the battery. With smaller one- or two-module systems, the amount of power lost to reverse current is negligible. A dirty battery top will cost you far more power loss. Only with larger systems does reverse-current flow become anything to be concerned about. Much has been made of this

> If batteries are routinely allowed to overcharge, their life expectancy will be reduced dramatically.

> On average, a Maximum Power Point Tracking controller will deliver about 15% more energy per year than a standard controller. And they do their best work in winter, when most off-grid homes need all the charging help they can get.

If a hydro or wind turbine is disconnected from the battery while still spinning, it will continue to generate power; but with no place to go, the voltage rises dramatically until something gives.

problem in the past, and almost all charge controllers now deal with it automatically. Most of them do this by sensing that voltage is no longer available from the modules, when the Sun has set, and then opening the relay or power transistor. A few older or simpler designs still use diodes—a one-way valve for electricity—but relays or power transistors have become the preferred methods (see sidebar on Dinosaur Diodes).

Hydroelectric and Wind Controllers

Hydroelectric and wind controllers have to use a different strategy to control battery voltage. A PV controller can simply open the circuit to stop the charging and no harm will come to the modules. If a hydro or wind turbine is disconnected from the battery while still spinning, it will continue to generate power; but with no place to go, the voltage rises dramatically until something gives. With these types of rotary generators, we typically use a diverting charge controller. Examples of this technology are the Trace C-series, or the MorningStar TriStar series of controllers. A diverter-type control will

monitor battery voltage, and when it reaches the adjustable set point, will turn on some kind of electrical load. A heater element is the most common diversion load, but DC incandescent lights can be used also. (Incandescent lights are nothing but heater elements that give off a little incidental light, anyway.) Both air and water heater elements are commonly used, with water heating being the most popular. It's nice to know that any power beyond what is needed to keep your batteries fully charged is going to heat your household water or hot tub. Diversion controllers can be used for PV regulation as well, in addition to hydro or wind duties in a hybrid system with multiple charging sources. This controller type is rarely used for PV regulation alone, due to its higher cost.

Controllers for PV-Direct Systems

A PV-direct system connects the PV module directly to the appliance we wish to run. They are used most commonly for water pumping and ventilation. By taking the battery out of the system, initial costs are lower, control is simpler, and maintenance is virtually eliminated.

Although they don't use batteries, and therefore don't need a controller to prevent overcharging, PV-direct systems often do use a device to boost pump output in low light conditions. While not actually a controller, these booster devices are closer kin to controllers than anything else, so we'll cover them here. The common name for these booster devices is LCB, short for Linear Current Booster.

An LCB is a solid-state device that helps motors start and keep running under low light conditions. The LCB accomplishes this by taking advantage of some PV module and DC motor operating characteristics. When a PV module is exposed to light, even very low levels, the voltage jumps way up immediately, though the amperage produced at low light is very low. A DC motor, on the other hand, wants just the opposite conditions to start. It wants lots of amps but doesn't care much if the voltage is low. The LCB provides what the motor wants by downconverting some of the high voltage into amperage. Once the motor starts, the LCB will automatically raise the voltage back up as much as power production conditions allow without stalling the motor. Meanwhile, on the PV module side of the LCB, the module is being allowed to operate at its maximum power point, which

Dinosaur Diodes

When PV modules first came on the market a number of years ago, it was common practice to use a diode, preferably a special ultra-low forward-resistance Schottky diode, to prevent the dreaded reverse-current flow at night. Early primitive charge controllers didn't deal with the problem, so installers did, and gradually diodes achieved a mystical must-have status. Over the years, the equipment has improved and so has our understanding of PV operation. Even the best Schottky diodes have 0.5- to 0.75-volt forward voltage drop. This means the module is operating at a slightly higher voltage than the battery. The higher the module voltage, the more electrons that can leak through the boundary layer between the positive and negative silicon layers in the module. These are electrons that are lost to us; they'll never come down the wire to charge the battery.

The module's 0.5-volt higher operating voltage usually results in more power being lost during the day to leakage than the minuscule reverse-current flow at night we're trying to cure. Modern charge controllers use a relay or power transistor, which has virtually zero voltage drop, to connect the module and battery. The relay opens at night to prevent reverse flow. Diodes have thus become dinosaurs in the PV industry. Larger, multimodule systems may still need blocking diodes if partial shading is possible. Call our tech staff for help.

Tracking Mounts or Maximum Power Point Tracking?

Which delivers more power?

Until recently, tracking mounts have been the only way to increase the daily output from your PV array. By following the sun from east to west, trackers can increase output as much as 30%. Tracking works best in the summer, when the Sun makes a high arc across the sky. Tracking is the best choice for power needs that peak in the summer, such as water pumping and cooling equipment. In the winter, tracking gains are much less, usually about 10%-15%.

What if you're looking for more winter output? The new MPPT charge controls can increase your cold-weather output up to 30%. MPPT controls are a refinement of linear current boosting, which allows the PV module to run at its maximum power point, while converting the excess voltage into amps. MPPT controls perform best with cold modules and hungry batteries, a wintertime natural! Some MPPT controllers also allow wiring and transmitting at higher voltage. A 72-volt array can charge a 24-volt battery, for instance. This saves additional costs in wire and reduces transmission losses.

Bottom line: For most residential power systems, an MPPT control will do more for your power well-being than a tracking mount.

is usually a higher voltage point than the motor is operating at. It's a constant balancing act until the PV module gets up closer to full output, when, if everything has been sized correctly, the LCB will check out of the circuit, and the module will be connected directly to the motor.

LCBs cost us something in efficiency, which is why we like to get them out of the circuit as the modules approach full power. All that fancy conversion comes at a price in power loss—usually about 10%-15%—but the gains in system performance more than make up for this. An LCB can boost pump output as much as 40% by allowing the pump to start earlier in the day and run later. Under partly cloudy conditions, an LCB can mean the difference between running or not. We usually recommend them strongly with most PV-direct pumping systems. With PV-direct fan systems, where start-up isn't such a bear, LCBs are less important.

PV, HYDRO, AND WIND CHARGE CONTROLLERS

Morningstar SunGuard

A small controller at the right price!

For systems with a single module of 75 watts or smaller, you can't beat this controller. Featuring a 5-yr. warranty, temperature compensation, 25% overload rating, extreme heat (and cold) tolerance, reverse-current leakage less than 0.01mA, solid-state series type 0 to 100% duty cycle PWM, epoxy encapsulated circuitry, 1500W transorb lightning protection, and more! For 12-volt systems only, it is rated for up to 4.5-amp PV and regulates at 14.1 volts. Okay for sealed batteries. USA.

**25739   Morningstar SunGuard
4.5A/12V Controller $30**

SunKeeper Junction Box-Mounted Solar Controller

Morningstar's SunKeeper solar controller maximizes battery life in small solar power applications. Epoxy encapsulated and rated for outdoor use. Delivers weatherproof connection when mounted directly to the module junction box and wired through the junction box knockout.

6-amp or 12-amp version available (both at 12VDC). Controller is rated to 70°C and is certified for use in Class 1, Division 2 hazardous locations, making it ideal for solar powered oil/gas applications. Dimensions are 3⁹⁄₁₀"x2"x½" (HxWxD). 5-yr. mfr. warranty. USA.

**48-0241   SunKeeper SK-12 Charge
Controller $89**

**48-0252   SunKeeper SK-6 Charge
Controller $63**

Phocos CML-Series PV Charge Controllers

Sophisticated German-engineered controllers at affordable prices, in 5-, 10-, 15-, and 20-amp sizes. These are 12- or 24-volt controls with automatic selection when you connect the battery. Packed with features, including PWM 3-stage charging, a 3-LED graphic battery charge indicator, LED charging and overload indicators, electronic circuit and load protection, and automatic low voltage load disconnect with acoustic warning (25 beeps). Charging voltage temperature compensation is built in. Choose between sealed and wet-cell battery charging programs. Will accept up to #10 AWG wire. 3.9"x3.2"x1.4" (HxWxD). 2-yr. mfr. warranty. German-engineered, made in China.

48-0069	Phocos CML5	$25
48-0070	Phocos CML10	$35
48-0071	Phocos CML15	$45
48-0086	Phocos CML20	$55

SunSaver 10- and 20-Amp Charge Controllers

The SunSaver charge controllers provide high-reliability, low-cost PWM charge control for small 12- or 24-volt systems. Featuring an easy-to-access, graphic terminal strip for connections, a weather-resistant anodized aluminum case, fully encapsulated potting, reverse-current protection, low voltage disconnect, and field selection of battery type for flooded or sealed batteries. Has an LED to show charging activity, and features automatic temperature compensation, based on the temperature of the controller. Accepts up to #10 gauge wire. 6"x2.2"x1.3" (HxWxD), 8 oz. 5-yr. mfr. warranty. USA.

48-0275	Morningstar SunSaver SS-10L-12V	$70
48-0019	Morningstar SunSaver SS-10L-24V	$85
48-0506	Morningstar SunSaver SS-20L-12V	$95
48-0507	Morningstar SunSaver SS-20L-24V	$101

Sunlight 10- and 20-Amp PV and Light Controllers

A simple PV and light control

The Sunlight 12-volt solid-state PWM charge controller has automatic light control, and low-voltage discon-

nect for battery protection. Connect a battery, a PV panel, and a 12V gizmo, and you've got a system that automatically turns on your sign lighting, stairway light, or other PV-powered task at dusk. If voltage falls below 11.7V, a low-voltage disconnect occurs and will not let the light run until battery voltage recovers to 12.8V. A "test" button overrides LVD for five minutes. Load control has 10 settings from off through dusk to dawn, including three unique settings that run a few hours after dusk, turn off, then run an hour or two before dawn. Charge control can be set for either sealed batteries at 14.1V or wet-cell batteries at 14.4V with a simple jumper, and provides reverse-current protection at night. Automatic charge voltage temperature compensation is built in. Reverse polarity protected. Available in 10- or 20-amp models for 12-volt applications. 5-yr. mfr. warranty. USA.

25742	Morningstar Sunlight SL-10L-12V	$109
25743	Morningstar Sunlight SL-10L-24V	$139
48-0224	Morningstar Sunlight SL-20L-12V	$141

Xantrex C-12

Charge/load/lighting controller

This 12-volt controller can do multiple functions. For starters, it's an electronic, 3-stage 12-amp PV charge controller with easily adjustable bulk and float voltages, nighttime disconnect, and automatic monthly equalization cycles.

It's also a 12-amp automatic load controller with 15-amp surge and electronic overcurrent protection, automatically reconnecting at 12-second and then 1-minute intervals. Voltage set points for connect and disconnect are also easily adjustable. Loads will blink five minutes before disconnect, allowing the user to reduce power use. Even after disconnect, it allows one 10-minute grace period by pushing an override button.

Plus, it's an automatic lighting controller that will turn on at dusk for sign or road lighting. Run time is user adjustable from two to eight hours, or dusk to dawn. If voltage gets too low, low-voltage disconnect will override run time to protect the battery.

The LED mode indicator will show the approximate state of charge and indicate if low-voltage disconnect, over-load, or equalization has occurred.

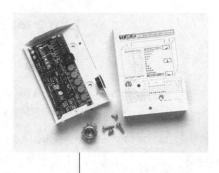

The C-12 is compatible with any 12V battery type including sealed or nicads. Interior terminal strip will accept up to #10 AWG wire. The box is rain-tight, and the electronics are conformal coated. Certified by ETL to UL specs. 2-yr. mfr. warranty. USA.

27315	Xantrex C-12 Multi Controller	$110
27319	Battery Temperature Probe	$29

BZ MPPT 250 and 500 Charge Controllers

MPPT technology is available for smaller 12V systems at a reasonable price. Run your PV modules at their max power point (usually 16V to 19V on a 12V nominal panel) while down-converting that extra voltage to amperage to get up to 15%-30% more power from your array, especially in winter. This can mean a lot of savings in the number of solar panels you need to meet your demand. Digital volt and amp meter displays battery voltage and charging current.

The 250 unit even has an extra AUX connection available to keep a generator starter battery topped off with a 100mA 13.8 trickle charge. The BZ 250 accepts up to 250 watts of solar. The BZ 500 will accept up to 500 watts of solar. The PWM float charge is set at 14.1 volts (perfect for lead-acid batteries) and is adjustable from 13 to 15 volts. ATC-type automotive fuses supplied on the input and output. LVD (low voltage disconnect) is also included on the 250 model to automatically shut off a load if the batteries' charge drops below 12V and reconnects at 12.6V.

Max direct load via LVD is 15A. Will accept up to 12 AWG wire. Comes ready for flush mounting to save space in your RV or cabin. Best results occur with 24V array input and 12V battery output. Battery temp sensor included. BZ 250, 5⅛"x7¾"x2½" (HxWxD); BZ 500, 10"x8½"x2½" (HxWxD). 5-yr. mfr. warranty. USA.

48-0226 250	BZ 250 MPPT Charge Controller	$140
48-0226 500	BZ 500 MPPT Charge Controller	$250

ProStar PV Charge Controllers

Control and monitoring in a single package

MorningStar is one of the largest and most reliable PV controller manufacturers in the world. The ProStar line, with models at 15 or 30 amps, features solid-state, constant-voltage PWM control, automatic selection of 12- or 24-volt operation, simple manual selection of sealed or flooded battery charging profile, nighttime disconnect, built-in temperature compensation, automatic equalize charging, reverse-polarity protection, electronic short circuit protection, moderate lightning protection, and roomy terminal strips that accept up to #6 AWG wire sizes. A built-in LCD display scrolls through battery voltage, charging current, and load current (if any loads are connected directly to the controller). An automatic low-voltage disconnect will shut off connected loads if batteries get dangerously low for more than 55 seconds. Maximum internal power consumption is a minimal 22 milliamps. Dimensions are 6.01"x4.14"x2.17" (HxWxD). 5-yr. mfr. warranty. USA.

25771	ProStar 15-amp PV Controller w/ meter	$179
25729	ProStar 30-amp PV Controller w/ meter	$216

TriStar PV or Load Controllers

Full-featured 45- or 60-amp controls

The reliable, UL-listed MorningStar TriStar controller can provide either solar charging, load control, or diversion regulation for 12-, 24-, or 48-volt systems. This is a full-featured, modern, solid-state PWM controller. Features include reverse-polarity protection, short circuit protection, lightning protection, high temperature current reduction, 4-stage battery charging, simple DIP switch controlled voltage setpoint setup, large 1"/1.25" knockouts on bottom, sides, and back, with extra wire bending room and terminals that accept up to #2 AWG wire. Circuit boards are fully conformal coated, the aluminum heat sink is anodized, and the enclosure is powder-coated with stainless fasteners to laugh at high humidity tropical environments.

This controller is particularly good for hydro diversion because the controller will allow up to 300-amp inrush currents to start motors or warm up cold resistive loads without the electronic short circuit protection interfering.

The optional 2-line, 16-character digital meter may be mounted to the controller in place of the cover plate, or remotely in a standard double-gang box using standard RJ-11 connectors (ethernet cables). Displays all system information, self-test results, and setpoints with intuitive up/down and left/right scrolling buttons. The

remote temperature sensor option with 10-meter cable will adjust voltage setpoints automatically, and is recommended if your batteries routinely will be exposed to temperatures under 40°F or over 80°F. 10.1"x4.9"x2.3" (HxWxD), weighs 4 lb. CE and UL listed. 5-yr. mfr. warranty. USA.

48-0010	TriStar 45-Amp Controller	$169
48-0011	TriStar 60-Amp Controller	$218
48-0012	TriStar Digital Meter	$99
48-0013	TriStar Remote Temperature Sensor	$45

Xantrex C-Series Multi-Function DC Controllers

The versatile C-series of solid-state controllers can be used for PV charge control, DC load control or DC diversion. They only operate in one mode, so PV charge control and DC load control requires two controllers. The 40-amp controller can be selected manually for 12, 24 or 48 volts. The 35- and 60-amp controllers can do 12 or 24 volts. All setpoints are field adjustable, with removable knobs to prevent tampering. They are protected

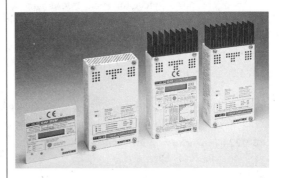

electronically against short circuits, overload, overtemp, and reverse polarity with auto-reset. No fuses to blow! Two-stage lightning and surge protection is included.

The three-stage PV charge control uses solid-state pulse width modulation control for the most effective battery charging. Has an automatic "equalize" mode every 30 days, which can be turned off for sealed batteries. A battery temperature sensor is optional but recommended if your batteries will routinely see temps below 40°F or above 90°F. The diversion control mode will divert excess power to a dummy heater load and offers the same adjustments and features as the PV control. Your dummy load must have a smaller amperage capability than the controller. The DC load controller has adjustable disconnect and auto-reconnect voltages with a time delay for heavy surge loads.

The optional LCD display with backlighting replaces the standard front panel, or can be mounted remotely. It continuously displays battery voltage, DC amperage, cumulative amp-hours, and a separate resettable "trip" amp-hour meter. Note that this meter can sense and display only current flows passing through the charge controller.

9"x5"x2" (HxWxD), 4 lb. Certified by ETL to UL standards. 2-yr. mfr. warranty. USA.

25-027	C-35	$119
25-017	C-40	$169
25-028	C-60	$199
27319	Battery Temperature Sensor	$29
25018	Xantrex C-Series LCD Display	$99
25774	Remote C-Series LCD Display w/50" cord	$135

OutBack MX60 MPPT Charge Controller

Get the most from your PV array

This is the absolute queen of PV charge controls. The MX60 charge controller will deliver up to 70 amps continuous output to any battery system between 12 and 60 volts. The Maximum Power Point Tracking will run your PV array at its greatest power point, and down-convert the excess voltage to something your batteries can digest. Power conversion efficiency is an incredible 99.1% at 40 amps, or 97.3% at the full 70 amps output. MPPT technology will deliver about 15% more power to your batteries on a yearly average, and delivers up to 30% boost in the winter, when your batteries probably need all the help they can get. Any PV array with less than 140 volts open circuit output (usually 72-volt nominal) can be used to charge a lower battery voltage, reducing wire size and power loss. Set points are fully adjustable for any battery type, chemistry, and charging profile. Periodic equalization cycles may be programmed (for wet cells), or not (for sealed cells).

The MX60 comes with a four-line, 80-character backlit LCD display for monitoring and programming. It's easy to read, easy to use, and easy to understand. Our photo here is showing the home screen. From the top, it's displaying:

PV volts in, Bat volts out

Amps in, Amps out

Watts at the moment

Aux state

kWh delivered today, functional state

Performance data for the past 60 days is saved in the easily accessible logging file. The Auxiliary output is a 12-volt signal that can be programmed to turn on or off at specific voltages. It's always 12V regardless of system voltage, so it can run an external relay or warning light. The optional OutBack Mate will plug into the MX60 and will allow control and monitoring of up to eight MX60 controls from up to 300 feet away. The Mate can also output to a PC computer for data logging or remote monitoring.

Standby power use is less than 1 watt. For installation in indoor or protected locations. Accepts 14- to 4-guage wire. Size: 14.5"H x 5.75"W x 5.75" D. Weighs 10 lb. Two-year manufacturer's warranty with extended option. Made in USA.

25812   OutBack MX60 Charge Controller$649

Apollo T80 Charge Controller

The T80 integrates battery charge management, charge data reporting and communications into a single

device. Breakthrough, high-efficiency technologies lower your total system costs by reducing the number of PV panels, eliminating the need for heavy gauge wiring and extending battery life. Enjoy 80 Amps of continuous output at up to 40°C ambient temperatures. Process up to 5300 Watts of PV power—30% more than the next largest capacity in the industry — with algorithms that start early and lock into peak power during rapidly changing insolation (usuable sunlight) and temperature. A built-in battery energy monitor tracks power production and use while calculating remaining battery supplies and storing 90 days of historical data. You can monitor system performance and upload software updates via your PC. Wire your modules in series up to 72 VDC nominal and charge 12-, 24-, 36-, or 48-volt batteries. Supports Flooded Lead Acid (FLA), GEL, and Absorbed Glass Mat (AGM) batteries. USA.

41-0253	Apollo T80 Charge Controller	$849
41-0254	Remote Monitor	$249

Controller Accessories and Specialty Controllers

Air Heater Diversion Loads

These resistive loads are enclosed in vented aluminum boxes for safety. They can be used on any DC system between 12 and 48 volts. Box needs at least 12" clearance to combustibles. Both units are shipped in the highest resistance mode, and can be reconfigured easily for lower resistance by changing connections in the terminal block. The HL-100 unit can be configured for 30 or 60 amps in nominal 12-volt mode. Two-year manufacturer's warranty. USA.

Item #	Price	Model	Diversion Amps @ Voltage Below			Resistance Setting
			15V	30V	60V	
48-0081	$299	HL-100	30/60	—	—	0.5/0.25 ohm
			15	30	—	1 ohm
			3.8	7.5	15	4 ohm
48-0082	$235	HL-75	20	40	—	0.75 ohm
			5	10	20	3 ohm

Diversion Water Heater Elements

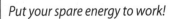

Put your spare energy to work!

These industrial-grade DC heater elements give you someplace to dump extra wind or hydro power. Use with a diversion controller such as the MorningStar Tristar.

The water heater elements fit a standard 1" NPT fitting. The 12/24V model has a pair of 25A/12V elements. Can be wired for 25A or 50A @ 12V, or in series for 25A @ 24V. The 24/48V model has a pair of 30A/24V elements. Can be wired for 30A or 60A @ 24V, or in series for 30A @ 48V. USA.

25078	12/24V Water Heating Element	$95
25155	48V/30A Water Heating Element	$116

Voltage-Controlled Switches

Got something you want to turn on, or off, based on your battery voltage? This is your ticket. Maybe you'd like to divert power to pumping or fans once the battery is charged? Or you'd like to shut off some appliances if the battery gets dangerously low? This 30-amp relay switch is adjustable for any DC voltage 10-63 volts. You set the high and low trip points. There's a normally open and a normally closed connection, so you can turn things on or off at your trip point. The "active high" unit turns ON the relay at the high trip point and holds it until the voltage drops below the low trip point. The

"active low" unit is the opposite. Quiescent power use is 17ma. The relay coil draws 75/47/32ma at 12/24/48V. 1-yr. mfr. warranty. Canada.

26657	Voltage Controlled Sw. Active High	$89
26658	Voltage Controlled Sw. Active Low	$89

30-Amp Power Relays

These general purpose DPDT (double-pole double-throw) power relays have either a 12-volt/169-milliamp, a 24-volt/ 85-milliamp, or a 120VAC/85-milliamp pull-in coil, depending on model chosen. Other controls like timers, load disconnects, or float switch, can be limited in their amperage capability. These relays can be used remotely to switch larger

loads, up to 30 amps per pole. They can also be used to switch an appliance that's a very different voltage. Turn on an AC gizmo with a DC control, or vice-versa. Because these are highly adaptable DPDT relays, they can control multiple switching tasks simultaneously. Screw terminal connections. Approx. 3.5"x2.5"x 2.5"(LxWxH). Choose your system voltage. USA.

25112	12V/30A Relay	$40
25151	24V/30A Relay	$40
25160	120vac/30A Relay	$40

Isolation Diodes

There used to be controls that needed this big diode, but they're rare now. If you need it, you'll know it. USA.

48-0201	100-amp Diode with Heat Sink	$59
48-0274	10-amp Diode	$10

Spring-Wound Timer

Our timers are the ultimate in energy conservation; they use absolutely no electricity to operate. Turning the knob to the desired timing interval winds the timer. Timing duration can be from 1 to 12 hours (or 1 to 15 minutes), with a hold feature that allows for

continuous operation. This is the perfect solution for automatic shutoff of fans, lights, pumps, stereos, VCRs, and Saturday morning cartoons! It's a single-pole timer, good for 10 amps at any voltage, and it mounts in a standard single gang switch box. A brushed aluminum faceplate is included. 1-yr. warranty.

25399 Spring-Wound Timer, 15-minute, no hold $35

12-Volt Digital Timer

This accurate quartz clock will control lights, pumps, fans, or other appliances from any 12VDC power source. The switched connections are isolated, so the switched voltage can be up to 36 volts DC or 240 volts AC, and up to 8 amps. Normally open and normally closed contacts are provided, as well as a manual on/off/auto button. The 7-day timer can accept up to eight on/off events per day. The

internal battery will retain programming for one year after loss of 12V power (but won't switch on/off events). 4"x4"x2.5". 1-yr. mfr. warranty. USA.

25064 12V Digital Timer $85

Battery Selector Switch

Few renewable energy homes use dual battery banks, but most boats and RVs do. This high-current switch permits selection between Battery 1 or Battery 2, or both in parallel. The off position acts as a battery disconnect. Wires

connect to ⅜" lugs. Capacity is 250 amps continuous, 360 amps intermittent. Okay for 12- or 24-volt systems. 5.25" diameter x 2.6" deep. Marine UL-listed. USA.

25715 Battery Switch $41

Dual Battery Isolator

For RVs or boats, this heavy-duty relay automatically disconnects the household battery from the starting battery when the engine is shut off. This prevents inadvertently running down the starting battery. When the engine is started, the isolator connects the batteries, so both will be recharged by the alternator. The relay is large enough to handle heavy, sustained charging amperage. Easy hookup: Connect a positive wire from each battery to the large terminals. Connect ignition power to one small terminal, ground the other small terminal. USA.

25716 Dual Battery Isolator $27

About Diodes

Diodes are one-way valves for electricity. In PV systems, they help the electrons get where they're supposed to go, and stay out of places they aren't supposed to go. But even the best valves have some restriction to flow in the forward direction, so diodes are used for strictly limited jobs.

The most common diode is a blocking diode. These are usually already installed inside the J-box of most larger modules. Installed in the middle of the series string of PV cells that make up the individual module, a blocking diode prevents a shaded module from stealing power from its neighbors. Smaller modules can have blocking diodes added in-line. But unless you have a multiple-module array that is so poorly positioned that it will get hard shade during prime midday sun hours, don't worry about blocking diodes. Large RV or sailboat arrays need blocking diodes (if they aren't already installed by the module manufacturer).

Blocking or bypass diodes

Another common use is the bypass diode, used in series strings.

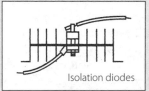
Isolation diodes

If one module in the string has limited output due to shading or other problems, a bypass diode installed across the output terminals will allow the other modules in the string to continue output. They are usually only necessary on long, high-voltage series strings. You must have an amperage rating higher than the module string.

Isolation diodes (such as Schottky diodes) are used to prevent backflow from batteries, or when combining parallel strings. Reverse-current flow from batteries is usually better handled by the charge controller, but a few controllers don't do nighttime disconnect.

DC-to-DC Converters

Got a DC appliance you want to run directly, but your battery is the wrong voltage? Here's your answer from Solar Converters. These bi-directional DC-to-DC converters work at 96% efficiency in either direction. For instance, the 12/24 unit will provide 12V @ 20A from a 24V input, or can be connected backwards to provide 24V @ 10A from a 12V input. Current is electronically limited and has fuse back-up. Four units are offered: The 12/24 unit described above; a larger 12/24 unit that delivers 25A or 50A depending on direction, a 24/48 unit that delivers 5A or 10A depending on direction, and a 12/48 unit that delivers 2.5A or 10A depending on direction. The housing is a weatherproof NEMA 4 enclosure measuring 4.5"x2.5"x2", good for interior or exterior installation (keep out of direct sunlight to keep the electronics happy). Rated for –40°F to 140°F (–40°C to 60°C). 1-yr .mfr. warranty. Canada.

25709   12/24 DC-to-DC Converter, 20A $199

25710   24/48 DC-to-DC Converter, 10A $210

PV Direct Pump Controllers

These current boosters from Solar Converters will start your pump earlier in the morning, keep it going longer in the afternoon, and give you pumping under lower light conditions when the pump would otherwise stall. Features include Maximum Power Point Tracking, to pull the maximum wattage from your modules; a switchable 12- or 24-volt design in a single package; a float or remote switch input; a user-replaceable ATC-type fuse for protection; and a weatherproof box. Like all pump controllers, a closed float switch will turn the pump off, an open float switch turns the pump on.

Four amperage sizes are available in switchable 12/24 volt. Input and output voltages are selected by simply connecting or not connecting a pair of wires. Amperage ratings are surge power. Don't exceed 70% of amp rating under normal operation. The 7-amp model is the right choice for most pumping systems, and is the controller included in our submersible and surface pumping kits. 7- and 10-amp models are in a 4.5"x2.25" x2" (LxWxD) plastic box. 15-amp model is a 5.5"x2.9"x 2.9" (LxWxD) metal box. 30-amp model is supplied in a NEMA 3R rain-tight 10"x8"x4" metal box. 1-yr .mfr. warranty. Canada.

25002   7A, 12/24V Pump Controller $119

25003   10A, 12/24V Pump Controller $159

25004   15A, 12/24V Pump Controller $239

25005   30A, 12/24V Pump Controller $395

Ballast Resistor

Made of automotive-grade porcelain this ceramic resistor lowers potentially damaging high voltage produced by PV modules to the 13.6 volts maximum comfortably handled by most pumps. This is not a charge controller but is an effective low cost way to protects sensitive electronics and fan and pump motors from over-voltage in small solar-direct applications. 3"x½"x⅝" (LxWxD). ⅓ lb. Mexico.

43-0235 Ballast Resistor $10

Universal Generator Starter

Automatic start and stop based on voltage

The Solar Converters Genset starter will automatically control the starting and stopping of a backup generator based on the voltage of any 12- to 48-volt battery. Start and stop voltage points are user adjustable. Supports 2,

3, or 4 wire, and advanced start systems. There's a time delay to prevent false starts, and a 20-second precrank relay closure for diesel preheaters. A 30-second delay for engine warm up is provided before closing the load relay. (Optional external power relays will be required for some functions.) LEDs indicate operation. A manual start/stop switch is on the board, and an external one can be added. All connections land on a clearly labeled screw terminal strip that will accept up to #12-gauge wire. On-board relays rated at 5 amps. Available as either a bare, open frame, as pictured, measuring 5"x5"x1", or mounted in a NEMA 3R outdoor box, measuring 10"x 8"x4". Quiescent power use less than 20ma. 1-yr .mfr. warranty. Canada.

48-0017 Univ. Gen. Starter, bare $265

48-0083 Univ. Gen. Starter w/Enclosure $359

Monitors

If you're going to own and operate a renewable energy system, you might as well get good at it. Your reward will be increased system reliability, longer component life, and lower operating costs. The system monitor is the gizmo that allows you to peer into the electrical workings of your system and keep track of what's going on.

The most basic, indispensable, minimal piece of monitoring gear is the voltmeter. A voltmeter measures the battery voltage, which in a lead-acid battery system can be used as a rough indicator of system activity and battery state of charge. By monitoring battery voltage, you can avoid the battery-killing over- and under-voltages that come naturally with ignorance of system voltage. No battery, not even the special deep-discharge types we use in RE systems, likes to be discharged beyond 80% of its capacity. This drastically reduces its life expectancy. Since a decent voltmeter is a tiny percentage of battery cost, there is no excuse for not equipping your system with this minimal monitoring capability.

Beyond the voltmeter, the next most common monitoring device we use is the ammeter. Ammeters measure current flow (amps) in a circuit. They can tell us how much energy is flowing in or out of a system. Small ammeters are installed in the wire. Larger ammeters use a remote sensing device called a shunt. A shunt is a carefully calibrated resistance that will show a certain number of millivolts drop at a specific number of amps current flow. For instance, our 100-amp shunt is rated at 100 millivolts. For every millivolt difference the meter sees between one side of the shunt and the other, it knows that one amp is flowing. By noticing which side is higher, it knows if the flow is into or out of the battery, and will display a + or − sign. A single shunt gives us the net amperage in or out for the entire system, but won't tell us, for example, that 20 amps are coming down from the PV array, but 15 amps are going straight into the inverter, and only 5 amps are passing through the shunt.

The most popular monitoring device is the accumulating amp-hour meter. This is an ammeter with a built-in clock and simple computing ability that will give a running cumulative total of amp-hours in or out of the system. This allows the system owner to monitor easily and accurately how much energy has been taken from the battery bank. Our most popular meter, the TriMetric, can display remaining energy as an easily understood percentage. When the meter says your battery system is 80% full, that's a number anyone can understand. Not every system or every person needs this kind of high-powered monitor/controller, but for those with an in-depth interest or an aversion to technology, it's available at modest cost.

INSTALL YOUR MONITOR WHERE IT WILL GET NOTICED

The best, most feature-laden monitor won't do you a bit of good if nobody ever looks at it. We usually recommend installing your monitor in the kitchen or living room so it's easy for all family members to notice and learn from. It's surprising how fast kids can learn given the proper incentive. A "No computer games or Saturday morning cartoons unless the batteries are charged" rule works wonders to teach the basics of battery management. Sometimes this even works on adults, too.

Elastic Voltmeters!

A voltmeter is a basic battery monitoring tool. It can give you a rough indication of how full or empty a lead-acid battery is, because the at-rest voltage will rise or fall slightly with state of charge. HOWEVER . . . the most important thing to know about a voltmeter is . . . this is a very *e l a s t i c* sort of indicator. Voltage stretches up or down when charging or discharging. When charging, a 12-volt battery can read over 14 volts. Almost the instant you stop charging, it will drop under 13 volts. Did you lose power when the voltage dropped? Not at all. It was just the voltage snapping back down to normal. Now put that battery under a heavy load like a toaster or microwave. The voltage will drop well under 12 volts, maybe even under 11 volts depending on how big the battery is, how well charged it is, and how heavy the load. But it'll bounce right back up as soon as the microwave goes off. The lesson here is that you need to know what else is happening when you read your voltmeter. A reading of 11.8 volts isn't a problem while the microwave is running, but if nothing is on, and the batteries have been sitting awhile to stabilize, then 11.8 volts is a very nearly dead battery. You'll learn with practice.

SYSTEM METERS AND MONITORS

Fat Spaniel Technologies Data Tracker

Wonder how much energy your PV system is producing or how much you are using at any moment? Now you can find out—in real time. The Fat Spaniel Data Tracker

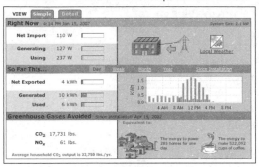

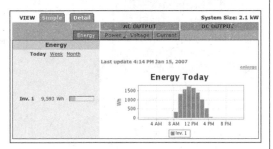

Performance Monitor sends the hard data you need to any web-enabled device anywhere in the world. Connected to your inverter, it feeds live and historical data to an interactive dash-board interface. Log on at any time to see how much solar, wind, or other alternative energy your system is producing, you are consuming, and how much is being sent to the utility grid. And because you can log on from anywhere, you don't have to wonder about your renewable energy system when you are off-site. Five years of monitoring service included. Broadband internet connection required. USA.

57-0112 Basic Residential Monitoring-
Inverter Direct-Simple View $899
57-0114 Detail View and Alerts Option $339

Analog Voltmeters and Ammeters

Use these low-cost ammeters to monitor either charging or discharging current flow. Meter reads in one

direction only, so full monitoring may require two meters. Voltmeter size is 2.5" square, designed for panel mounting in a 2" round hole. USA.

25313	0–30A Ammeter	$24
25778	11–16 Volt Analog Voltmeter	$27
25304	22–32 Volt Analog Voltmeter	$30

Universal Battery Monitor

Here's simple at-a-glance system monitoring that anyone can understand. Ten LEDs are labeled in 10% increments and range from green to yellow to red. Draws only 1/4 watt. Can be set for 12-, 24-, or 48-volt operation. Easy mounting tabs, 2-foot cable that can be spliced and extended with small 18-gauge wire as needed. 5.2"x2.75"x1.1". 2- amp fuse recommended, listed below. 5-yr .mfr. warranty. USA.

57-0100	Universal Battery Monitor	$65
25540	ATC In-line Fuseholder	$2.95
24219	ATC 2A Fuse (5)	$2.95

Digital Meters

Our easy-to-read, surface-mount digital voltmeter monitors battery voltage up to 40 volts (for 12- and 24-volt systems), and reads three digits (tenths of a volt). Draws only 8 milliamps to operate. Our Digital Ammeter works on 12- or 24-volt systems, reads four digits up to 199.9 amps, shows reverse flow with a negative sign, and includes a 100A shunt. 4"x2"x1.75".1-yr .mfr. warranty. USA.

25297	Voltmeter	$58
25298	Ammeter	$89

TriMetric 2020 Monitor

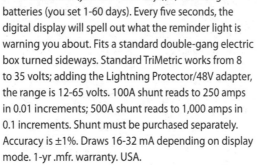

The TriMetric is a user-friendly 12- to 48-volt digital monitor that displays volts, amps or percentage of full charge, and tracks a number of battery management items. The bright 3-digit LED display is now 33% larger, and the face and label are larger, the deeper battery management data is easier to access and is listed right on the label. A "charging" light shows when positive amp flow is happening or flashes to show that "charged" criteria were met during the last day. A new feature is the "battery reminder" light, which can be programmed to flash a warning for any of three conditions: 1) low battery voltage (you set the voltage); 2) equalize batteries (you set 1-250 days); 3) full charge batteries (you set 1-60 days). Every five seconds, the digital display will spell out what the reminder light is warning you about. Fits a standard double-gang electric box turned sideways. Standard TriMetric works from 8 to 35 volts; adding the Lightning Protector/48V adapter, the range is 12-65 volts. 100A shunt reads to 250 amps in 0.01 increments; 500A shunt reads to 1,000 amps in 0.1 increments. Shunt must be purchased separately. Accuracy is ±1%. Draws 16-32 mA depending on display mode. 1-yr .mfr. warranty. USA.

25371	**TriMetric 2020 Monitor**	**$165**
25303	**Lightning Protector/48V Adapter**	**$26**
23145	**TriMetric Surface-Mount Box**	**$14**
25351	**100-Amp Shunt**	**$26**
25364	**500-Amp Shunt**	**$35**
48-0188	**Meter Wire, 18 ga. 3 twisted pairs**	**$1.20/ft.**

Please specify length.

PentaMetric Monitor

New from Bogart Engineering is the PentaMetric Monitor. It offers additional capabilities beyond the TriMetric monitor. It includes input unit, display unit, and/or computer interface unit. Monitor up to three shunts—to measure total solar input and wind input, to monitor battery state of charge. Access data using LCD display unit and buttons up to 1,000 feet from the batteries. Relay output to control generator or external alarm. Audible and visual alarms for high and low battery conditions. 1-yr .mfr. warranty. USA.

57-0104	**PentaMetric Meter with Computer Interface**	**$299**
57-0105	**LCD display for PentaMetric Meter**	**$199**

Protect Your Monitor Investment

Monitors are expensive, protection is cheap. What more do we need to say? Inline fuse holder works with all monitors.

25540	**Inline ATC Fuse Holder**	**$2.95**
24219	**2A ATC Fuse 5/box**	**$2.95**

TM500 System Monitor

Batteries are a significant and expensive part of any independent energy system. They are also the part of your system that is most vulnerable to mistreatment. The Xantrex TM500 meter keeps track of the energy your system has available as well as energy consumed, ensuring adequate reserve power and proper treatment of your bat-

teries. It measures DC system voltage (needs the TM48 adapter for 48-volt systems), net amperage, cumulative amp-hours, days since full, peak voltage, minimum voltage, and more! Measuring 4.55 inches square by 1.725 inches deep, it may be surface or flush mounted. A 500- amp shunt is required (Xantrex and OutBack power centers, modules, and panels already have a shunt wired in). We strongly recommend purchasing the version with the shunt included. Their shunt is prewired for simply plugging in. Otherwise some tedious wiring is required. USA.

25740	**TM500 Meter with Prewired Shunt**	**$245**
25724	**TM500 Meter without Shunt**	**$195**
25725	**TM48 48-Volt Adapter for TM500**	**$40**

PAK TRAKR BATTERY MONITORS

These new battery monitors "give your batteries the attention they've been looking for" by monitoring and then detecting battery problems before they occur. The Pat Trakr automatically and continuously monitors up to 24 individual 6V, 8V, or 12V batteries in: electric vehicles, golf carts or home renewable energy systems. It has the capability of detecting individual battery problems before they destroy your entire pack. The easy-to-read two-line LC graphic can display 1) pack state of charge; 2) individual battery voltages; 3) 30-day logged data. The digital display scrolls between battery voltages, amps/watts (with optional sensor) and battery bay temperature for seamless monitoring. An added option not common in battery monitors is the Pak Trakrs ability to do audible text alerts to signify failing batteries, if low water is likely, a pack imbalance, damaged batteries, low state of charge or overcharging.

The PakTrakr 600EV package monitors 6 batteries. Three additional 6-battery Remotes may be connected, to monitor up to 24 total batteries during charging, under load, and at idle. It is self-calibrating for 6V, 8V, or 12V lead-acid batteries. The Pak Trakr 800HEM package monitors up to 8 batteries; with one 8-battery remote for monitoring 6V or 8V batterires. The PakTrakr includes a serial interface for logging data to a PC by sending out a CSV data stream every second, containing all the battery data. By attaching a PC to the PakTrakr via the optional serial interface (coming soon), the data may be easily read into a spreadsheet for analysis.

The PakTrakr 600EV and 800HEM systems includes: PakTrakr LCD Display Unit—mounts in 2⅟₁₆" dash cutout or on any flat surface; PakTrakr Remote (max. 4 per system) mounts in battery bay and connects to display; unit with 8' cable and 6 batteries with 3' and battery connect wires with ⅜" ring terminals. An optional current sensor may be added at any time. The current sensor is a hall-effect device (no shunt required). Current draw 10mA. 2.35"x2.35"x0.9", 8 oz. 1-yr. limited warranty. USA.

PakTrakr 600EV

PakTrakr Display and one 6-battery remote, for monitoring six 6V, 8V, or 12V batteries. Monitor additional batteries by adding one or more 6-battery remotes.

48-0001 PakTrakr 600EV	$159

PakTrakr 800HEM

PakTrakr Display and one 8-battery Remote for monitoring up to eight 6V or 8V batteries.

48-0002 PakTrakr 800HEM	$169
48-0003 Additional Remote	$79
48-0004 Optional Current Sensor	$59

AC/DC Monitor

With quad display panel

Because it displays both AC and DC values, this is a great monitor for renewable energy systems, boats, RVs, and battery-based intertie systems. The large 4"x 2.5" LCD display with switchable backlighting shows four measurements simultaneously, and can quickly cycle through everything it's monitoring with the clearly

labeled front panel buttons. It measures two DC battery voltages, DC net current, DC net wattage, AC voltage, AC frequency, AC current (with optional sensor), plus time and temperature. Min/Max alarms can be set for all inputs. There is an audible alarm and an external alarm terminal to turn on a remote alarm. Pushing any button stops the audible alarm, and there's a master button to turn all alarms on or off. We are offering this meter with the optional AC current sensor and larger 250-amp shunt as a package. For 12- or 24-volt systems. Can be used on 48-volt systems so long as a 12V power source is connected to BAT 1 terminals. Will display up to 60 volts on BAT 2. Draws only 5.5ma with backlight off. For panel or wall mounting, wiring terminal strip protrudes out back. 6"x4"x0.6". Has shunt on positive side, requires three inline fuses, below. 1-yr. mfr. warranty. USA.

57-0108	AC/DC Monitor	$365
25540	Inline ATC Fuse Holder (needs 3)	$2.95
24219	2A ATC Fuse 5/box	$2.95

Brand Electronics ONE Meter

Multichannel digital monitoring for utility intertie systems

Ever wonder how many watt-hours your battery-based intertie system is really contributing? Here's the ultimate meter to answer all your questions, and a few you hadn't even thought of yet. In the three-channel basic package we're offering below, the ONE Meter monitors DC power input from your PV, wind, or hydro system, plus AC power output through the inverter output terminals, plus AC power input/output through the inverter AC1 terminals.

Results are displayed on the 4x20 backlit LCD display. Two displays are available, the flush mounted oak-trimmed one we've shown, which mounts in a standard 3-gang electric box, or a fully enclosed stand-alone display. We'll supply the oak-trimmed flush display unless ordered otherwise. Additional displays (up to 3) can be run from a single base.

The ONE Meter is computer-ready with a common RS-232 serial-port connection for downloading to a PC, and comes with a Windows-compatible BASIC terminal program for logging, compiling, and display. The ONE Meter also has standalone logging capability. AC current sensors are simple snap-on types, DC current sensors are toroid hall effect types, cabling is standard 8-conductor CAT-5 LAN type with RJ-45 connectors, widely used for computer networking. Cabling is supplied with the meter. Additional data channels (up to 9) can be monitored with a single base display. Optional extra channels must be specified for AC or DC. Our basic package is set up for a single Xantrex SW-series inverter. Each additional SW inverter in your system will require two more AC channels for complete monitoring. Made in USA. 1-yr. mfr. warranty.

25766	Brand ONE Meter Basic Package	$799
25767	Brand Optional DC Channel	$150
25768	Brand Optional AC Channel	$150
25769	Brand Optional Oak Display	$150
25770	Brand Optional Stand-Alone Display	$150

Homeowner's Digital Multi-meter

If you're going to be the owner/operator of a renewable energy system, you need one of these. A multi-meter is your first line of help when things go wrong. This is a nice one with all the features you'll ever need, and a great price. LCD 1" display reads DC volts, AC volts, amps up to 20, plus continuity and auto negative polarity indication. Has good quality, flexible test leads. Powered by an included 9-volt battery. 7½"x3½"x1½". China.

57-0109 Digital Multi-meter $39

Kill-a-Watt Electricity Usage Monitor

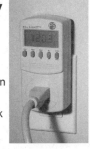

Ever wondered what that old fridge, freezer, or other appliance costs to run? Just plug the Kill-a-Watt between your appliance and outlet, and it will show current wattage plus track peak wattage and cumulative power use for as long as it's plugged in. Enter your power cost per kilowatt-hour, and it shows what your appliance costs per day, month, and year based on actual use. The internal battery stores all data until reset, so power failures won't erase your history. Simple, intuitive operation. Maximum load 1,875 watts. Accuracy is +/- 0.2%. Indoor, 120VAC/60Hz use only. China.

17-0320 Kill-a-Watt Meter $30

Watt's Up AC Power Measurement Tool

Determine power consumption of any AC appliance (up to 15 amps) in your home. Simply plug into this accumulating watt-hour meter for hours or months to see how much power it's using. Displays: watts, within 0.1; kWh; time; average monthly cost; voltage now; current now; power factor; max/min voltage; current; duty cycle (% of time the appliance runs). Plus memory storage and the ability

to download the data to a PC (serial cable and software included). One thousand data points are stored, so years of usage can be accurately recorded. 1-yr. mfr. warranty. USA.

25818	Watt's Up	$119
25819	Watt's Up Pro	$149

AC Kilowatt-Hour Meter

Keep track of your intertie system's output with this simple, familiar, rebuilt utility meter. Equipped with the easy-read dial, this analog digital meter displays kilowatt-hours, and will monitor both 120- or 240-volt circuits. Our round meter socket has 1" threaded outlets top and bottom. Both meter and socket are required for operation, and can be mounted inside or outside. USA.

25780	Meter	$49
25781	Socket	$25

Large Storage Batteries

Lead-acid batteries offer the best balance of capacity per dollar, and are far and away the most common type of battery storage used in standalone power systems.

Batteries provide energy storage, and are required for any remote, standalone, or back-up renewable energy system. Batteries accumulate energy as it is generated by various renewable energy devices such as PV modules, wind, or hydro plants. This stored energy runs the household at night or during periods when energy output exceeds energy input. Batteries can be discharged rapidly to yield more power than the charging source can produce by itself, so pumps or motors can be run intermittently. For personal safety and good battery life expectancy, batteries need to be treated with some care, and get properly recycled at the end of their life. If commonsense caution is not used, batteries can easily provide enough power for impromptu welding and even explosions.

Batteries 101

A wide variety of differing chemicals can be combined to make a functioning battery. Some combinations are very low cost, but they have very low power potential; others, like the lithium-ion batteries used in better laptops, can store astounding amounts of power but have astounding costs to go with them. Lead-acid batteries offer the best balance of capacity per dollar, and are far and away the most common type of battery storage used in standalone power systems.

This battery type is also well known as the common automotive battery. Although storage batteries and starting batteries have very different internal construction, they're both lead-acid types, and that's what we're going to cover in this section of the *Sourcebook*.

Simplified Lead-Acid Battery Operation

The lead-acid battery cell consists of positive and negative lead plates made of slightly different alloys, suspended in a diluted sulfuric acid solution called an electrolyte. This is all contained in a chemically and electrically inert case. If we lower the voltage on the battery terminals by turning on a load such as a light bulb, the cell will discharge. Sulfur molecules from the electrolyte bond with the lead plates, releasing electrons, which flow out the negative battery terminal, through the light, and back into the positive terminal. This flow of electrons is what we call electricity. If we raise the voltage on the battery terminals by applying a charging source, the cell recharges. Electrons push back into the battery, bond with the sulfur compounds, and force the sulfur molecules back into the electrolyte solution. Electrons will always flow from higher voltage to lower voltage if a path is available.

This back-and-forth energy flow isn't perfectly efficient. Lead-acid batteries average about 20% loss. For every 100 watt-hours you put into the battery, you can pull about 80 watt-hours back out. Efficiency is better with new batteries, and drops gradually as batteries age. Still, this is the best energy storage medium for the price.

A single lead-acid cell produces approximately two volts, regardless of size. Each individual cell has its own cap. A battery is simply a collection of individual cells. A 12-volt automotive starting battery consists of six cells, each producing 2 volts, connected in series. Larger cells provide more storage capacity; we can run more electrons in and out, but the voltage out-

Why Batteries?

It's a reasonable question. Why don't renewable energy systems simply produce standard 120-volt AC power like the utility companies? It's simple. No technology exists to store AC power; it has to be produced as needed. For a utility company, with a power grid spread over half a state or more, considerable averaging of power consumption takes place. So it can (usually) deliver AC power as demanded. A single household or remote homestead doesn't have the advantage of power averaging. Batteries give us the ability to store energy when an excess is coming in, and dole it out when there's a deficit. Batteries are essential for remote or back-up systems. Utility intertie systems make do without batteries, by using the utility grid like a battery to make up shortfalls or overages. But this requires the presence of grid power and doesn't provide any back-up ability in case of a power outage caused by grid failure.

put stays in the 2.0- to 2.5-volt potential of the chemical reaction that drives the cell. Cells are connected in series and parallel to achieve the needed voltage and storage capacity.

Battery Capacity

Think of a battery as a bucket. It will hold a specific amount of energy, and no amount of shoving, compressing, or wishing is going to make it hold any more. For more capacity, you need bigger buckets or more buckets.

A storage battery's capacity is rated in amp-hours: the number of amps it will deliver, times how many hours. How fast or slow we pull the amps out will affect how much energy we get. A slower discharge will yield more total amps. So battery capacity figures need to specify how many hours the test was run. The 20-hour rate is the usual standard for storage batteries, and is the standard we use for all the storage batteries in our publications. A 220 amp-hour golf cart battery, for example, will deliver 11 amps for 20 hours. This rating is designed only as a means to compare different batteries to the same standard and is not to be taken as a performance guarantee. Batteries are electrochemical devices, and are sensitive to temperature, charge/discharge cycle history, and age. The performance you get from your batteries will vary with location, climate, and usage patterns. But in the end, a battery rated at 200 amp-hours will provide you with twice the storage capability of one rated at 100 amp-hours.

What if the battery you're looking at only has a rating for Cold Cranking Amps? That's a battery designed for engine-starting service, not storage. Stay away from it. Starting batteries suffer short, ugly lives when put into storage service.

Batteries are less-than-perfect storage containers. For every 1.0 amp-hour you remove from a battery, it is necessary to pump about 1.2 amp-hours back in, to bring the battery back to the same state of charge. This figure varies with temperature, battery type, and age, but is a good rule of thumb for approximate battery efficiency.

Advantages and Disadvantages of Lead-Acid Batteries

Lead-acid batteries are the most common battery type. Thanks to the automotive industry, they are well understood, and suppliers for purchasing, servicing, and recycling them are practically everywhere. Lead-acid batteries are the most recycled item in U.S. society, averaging better than a 95% return rate in most states. Of all the energy-storage mediums available to us, lead-acid batteries offer the most bang for the buck by a very wide margin. All the renewable energy equipment on the market is designed to work within the typical voltage range of lead-acid batteries. That's the good part.

Now the bad part: The active ingredients, lead and sulfuric acid, are toxins in the environment, and need to be handled with great respect. (See "Real Goods' Battery Care Class," pp. 232-234, for more info.) Acid can cause burns, and just loves to eat holes in your blue jeans. Lead is a danger if it enters the water cycle, is a strategic metal, and has salvage value, so it should be recycled. Lead is also incredibly heavy. More lead equals more storage capacity, but also more weight. Work carefully with batteries to avoid strains and accidents.

Lead-acid batteries produce hydrogen gas during charging, which poses a fire or explosion risk if allowed to accumulate. The hydrogen must be vented to the outside. See our drawings of ideal battery enclosures for both indoor and outdoor installations at the end of this section. Lead-acid batteries will sustain considerable damage if they are allowed to freeze. This is harder to do than you may imagine. A fully charged battery can survive temperatures as low as −40°F without freezing, but as the battery is discharged, the liquid electrolyte becomes closer to plain water. Electrolyte also tends to stratify, with a lower concentration of sulfur molecules near the top. If a battery gets cold enough and/or discharged enough, it will freeze. At 50% charge level, a battery will freeze at approximately 15°F. This is the lowest level you should intentionally let your batteries reach. If freezing is a possibility, and the house is heated more or less full time, the batteries should be kept indoors, with a proper enclosure and vent. For an occasional-use cabin, the batteries may be buried in the ground within an insulated box.

Batteries will slowly self-discharge. Usually the rate is less than 5% a month, but with dirty battery tops to help leak a bit of current, it can be 5% a week. Batteries don't fare well sitting around in a discharged state. Given some time, the sulfur molecules on the surface of the discharged battery plates tend to crystallize into a form that resists recharging, and will even block off large areas of the plates from doing active service. This is called sulfation, and it kills batteries

Batteries are less-than-perfect storage containers. For every 1.0 amp-hour you remove from a battery, it is necessary to pump about 1.2 amp-hours back in, to bring the battery back to the same state of charge.

Lead-acid batteries produce hydrogen gas during charging, which poses a fire or explosion risk if allowed to accumulate.

Unit of Electrical Measurement

Most electrical appliances are rated by the watts they use. One watt consumed over one hour equals one watt-hour of energy. Wattage is the product of current (amps) times voltage. This means that one amp delivered at 120 volts is the same amount of wattage as 10 amps delivered at 12 volts. Wattage is independent of voltage. A watt at 120 volts is the same amount of power as a watt at 12 volts. To convert a battery's amp-hour capacity to watt-hours, simply multiply the amp-hours times the voltage. The product is watt-hours. To figure how much battery capacity is required to run an appliance for a given time, multiply the appliance wattage times the number of hours it will run to yield the total watt-hours. Then divide by the battery voltage to get the amp-hours. For example, running a 100-watt lightbulb for one hour uses 100 watt-hours. If a 12-volt battery is running the light, it will need 8.33 amp-hours (100 watt-hours divided by 12 volts equals 8.33 amp-hours).

far before their time. The cure is regular charge and discharge exercise, or lacking that, trickle charging that will keep the battery fully charged. Special pulse chargers that deliver high-voltage spikes of charging power have proven very effective at reducing and preventing sulfation in batteries that don't get much exercise.

Lead-acid batteries age in service. Once a bank of batteries has been in service for six months to a year, it generally is not a good idea to add more batteries to the bank. A battery bank performs like a chain, pulling only as well as the weakest link. New batteries will perform no better than the oldest cell in the bank. All lead-acid batteries in a bank should be of the same capacity, age, and manufacturer as much as possible.

Much like the muscles of your body, wet-cell batteries need regular exercise to maintain good performance. This makes them less than ideal for emergency and back-up power systems.

Why Can't I Use Car Batteries?

Batteries are built and rated for the type of "cycle" service they are likely to encounter. Cycles can be "shallow," reaching 10%-15% of the battery's total capacity, or "deep," reaching 50%-80% of total capacity. No battery can withstand 100% cycling without damage, often severe.

Automotive starting batteries and deep-cycle storage batteries are both lead-acid types, but they have important construction and even materials differences. Starting batteries are designed for many, many shallow cycles of 15% discharge or less. They deliver several hundred amperes for a few seconds, and then the alternator takes

over and the battery is quickly recharged. Deep discharges cause the lead plates to shed flakes and sag, reducing life expectancy. Deep-cycle batteries are designed to deliver a few amperes for hundreds of hours between recharges. Their lead plates are thicker, and use different alloys that don't soften with deep discharges. Neither battery type is well suited to doing the other's job, and will deteriorate quickly and not last very long if forced to do the wrong service.

Automotive engine-starting batteries are rated for how many amps they can deliver at a low temperature, aka cold cranking amps (CCA). This rating is not relevant for storage batteries. Beware of any battery that claims to be a deep-cycle storage battery and has a CCA rating.

Sealed or Wet?

Lead-acid batteries come in two basic flavors, traditional "wet cells" or the newer "sealed cells." Wet cells have caps that can be removed. In fact, you have to remove them every couple of months to add distilled water. Wet-cell batteries cost less, their problems are more easily diagnosed, and they're usually the right choice for remote homesteads. On the downside, they do require routine maintenance, they produce hydrogen gas during charging that needs to be vented outside, and they have to ship as hazardous freight, which makes them expensive to move. Much like the muscles of your body, wet-cell batteries need regular exercise to maintain good performance. Without regular cycling, they become "stiff" chemically, and won't perform well until they're limbered up. This makes them less than ideal for emergency and back-up power systems.

This is where sealed batteries come in. Sealed cells do better at sitting around for long periods waiting for activity. Plus, nobody has to remember to water them periodically, and they can be installed in places where hydrogen gassing couldn't be tolerated. Sealed batteries come in two basic styles. Gel batteries have the liquid electrolyte in a jellied form. AGM (absorbed glass mat) batteries use a fiberglass sponge-like mat to hold liquid electrolyte between the plates. There is intense debate and rivalry as to which technology is better. AGM is less expensive and fussy to build, and the gels tend to be a bit more robust and longer lived. We offer small sealed batteries as AGM types, and larger sealed batteries as gel types.

Sealed batteries require special charge controls. To prevent gassing and the irreplaceable loss of water, charging voltage on sealed cells needs to be held to a maximum of 14.1 volts (or 2.35 volts per cell). Higher charging voltages, as commonly used for wet cells, will seriously reduce the life expectancy of sealed batteries. Almost all the charge controls we offer are either voltage adjustable, or have a selection switch for sealed or wet-cell batteries.

How Big a Battery Pack Do I Need?

We usually size remote household battery banks to provide stored power for three to five days of autonomy during cloudy weather. For most folks, this is a comfortable compromise between cost and function. If your battery bank is sized to provide a typical three to five days of back-up power, then it will also be large enough to handle any surge loads that the inverter is called upon to start. A battery bank smaller than three days' capacity will get cycled fairly deeply on a regular basis. This isn't good for battery life. A larger battery bank cycled less deeply will cost less in the long run. Banks larger than five days' worth start getting more expensive than a back-up power source (such as a modest-sized generator).

As a general rule of thumb, you'll be better off building your battery pack with the fewest

How Long Will Batteries Last?

Life expectancy depends mostly on what battery type you purchase initially, and then how well you treat it. Any fool can destroy even the best battery pack within six months, while a few saints manage to make their batteries last twice the average expectancy. We've learned from experience approximately what to expect from different batteries on average.

From low to high:

RV Marine types: 1.5-2.5 years

Golf Cart types: 3-5 years

L-16 types: 6-8 years

Solar Gel (sealed) types: 5-10 years

Industrial-quality traction cells (IBE and Hawker [formerly Yuasa] brands): 15-25 years

Triple the Life of Your Batteries

With regular cleanings, water top-off, and equalizations, you can triple the life of your batteries. Here's how:

Lesson 1: Equalize every time the batteries drop below 50% state of charge, sometimes two to four times a month.

Lesson 2: Every chance you get, fully recharge your batteries. "Mostly" charging most of the time doesn't cut it.

Lesson 3: Check your batteries often. Look for crud on top—it can build up and actually short between the posts, causing a constant discharge. And be sure to give your batteries a good cleaning every two to three months. Get on your grubby clothes, eye protection, and gloves, and ensure you have plenty of baking soda next to your batteries. Then mix up a mild soap and water solution and start cleaning. (Use an old paintbrush and rinse all connections with clean water while checking for tightness and corrosion.) Treat any posts and connections as required. Equalize the battery bank, top off the water, and check each cell's voltage and log it.

Lesson 4: Don't overcharge a gel cell battery. There's no way to replace the expelled electrolyte, and you'll shorten the battery's lifespan.

number of cells. Larger cells with more ampacity means fewer interconnects, fewer cells to water, less maintenance time, and a smaller chance that one cell will fail early, causing the entire pack to be replaced before its time. Bigger batteries tend to be higher quality, and also have longer lives.

Occasionally we run into situations with ¾-horsepower or larger submersible well pumps or stationary power tools requiring a larger battery bank simply to meet the surge load when starting. Call the Real Goods technical staff for help if you are anticipating large loads of this type. Our System Sizing Worksheet at the beginning of Chapter 4 or the Appendix has a quick battery sizing section to help you out.

New Technologies?

Compared to the electronic marvels in the typical renewable energy package, the battery is a very simple, proven technology. Tremendous amounts of research have been directed lately into energy-storage technology. Auto manufacturers are searching desperately for a lightweight battery with high energy density and low cost—the Magic Battery. Several dozen battery technologies are currently under intense development in the laboratory. Several of these are bearing fruit now for cell phones, laptops, and hybrid vehicles, but they are still far too expensive for the amount

If your battery bank is sized to provide a typical three to five days of back-up power, then it will also be large enough to handle any surge loads that the inverter is called upon to start.

of energy storage required in a renewable energy system. The nickel-metal hydride battery pack in the Toyota RAV4 EV is reportedly worth $30,000, and it would only make a very modest-sized remote home battery pack.

Lead-acid batteries are also in the laboratory. The possibilities of lead-acid technology are far from tapped out. This old dog is still capable of learning some new tricks. For now we must coexist with traditional battery technology—a technology that is nearly one hundred years old, but is tried and true and requires surprisingly little maintenance. The care, feeding, cautions, and dangers of lead-acid batteries are well understood. Safe manufacturing, distribution, and recycling systems for this technology are in place and work well. Could we say the same for a sulfur-bromine battery?

Real Goods' Battery Troubleshooting Guide

What Could Go Wrong?

Famous last words . . . Judging by the phone calls we get, plenty can go wrong. Folks have more trouble with batteries than with any other single component of their systems. Here's a troubleshooting guide.

GOT WATER?

Check the fluid level in all the cells. It should be above the plates, but not above the split ring about ½ inch below the top of the cell. It's bad for the plates to be exposed to air. Add distilled water as needed. In an emergency, clean tap water is better than no water, but not by much. Distilled water is really the right stuff.

Auto parts stores sell special battery-filler bottles with a nifty spring-loaded automatic shut-off nozzle. Just stick it in the cell, push down, and wait until it stops gurgling. The fluid level will be perfect and there will be no drips or mess. It's a tool worth having. Or consider automatic battery-watering systems, which make refills on large industrial batteries a snap. (Also, see the PRO-FILL 6-Volt Battery Watering System on page 239).

> Voltage can be tricky, however. This is a very elastic sort of indicator. Voltage will stretch upward when a battery is being charged, and it will contract downward when a battery is being discharged.

CONNECTIONS TIGHT AND CLEAN?

Check battery connections by wiggling all the cables, snug up the hardware, and make sure no corrosion is growing.

Got corrosion? This is like cancer—you've got to clean it all out, or it will return. Disassemble the connection, wire brush and scrape off all the crud you can, then attack what's left with a baking soda and water solution until all traces of corrosion are gone. Clean the bare metal with sandpaper or a wire brush, assemble, and then coat all exposed metal with grease (such as NO-OX-ID grease on page 239) or petroleum jelly to prevent future corrosion.

IS IT CHARGED?

Every system needs some way to monitor the battery state of charge. A voltmeter is the basic minimum. Voltage can be tricky, however. This is a very elastic sort of indicator. Voltage will stretch upward when a battery is being charged, and it will contract downward when a battery is being discharged. Only when the battery has been sitting for a couple hours with no activity can you get a really accurate sense of the battery's state of charge with a voltage

Volt Readings for your Deep-Cycle Battery

Get the longest life from deep-cycle batteries by keeping the state of charge at 60% or more for lead-acid and 70% for gel cells. For accurate volt readings, don't charge or discharge batteries for 2 hours.

Voltage Reading	State-of-Charge
12.75	100
12.70	95
12.65	90
12.60	85
12.55	80
12.50	75
12.45	70
12.40	65
12.30	55
12.25	50
12.25	45

reading. First thing in the morning is usually a good time.

A fully charged 12-volt battery will read 12.6 volts, or a bit higher. (Those with 24- or 48-volt packs multiply accordingly.) Lower voltage readings mean a lower state of charge, or that something is turned on. At 12.0 volts, you're in the caution zone, and at 11.6 volts any further electrical use is doing damage that will reduce the battery life expectancy. When charging, the voltage needs to get up to 14.0 volts or higher periodically. If your battery doesn't climb above 13.5 volts, it isn't getting close to being fully charged. You don't need to achieve a full charge every day, but at least once a week is good.

IS IT HEALTHY?

The very best tool for checking your battery's state of health is a hydrometer. This looks like an oversized turkey baster with a graduated float inside it. You must wait at least 48 hours after adding water before you run a hydrometer test. A sample of electrolyte is drawn up from a cell until the float rises. Note at what level it's floating. It measures the specific gravity of the electrolyte to three decimal places. A typical reading would be 1.220. The decimal point is universally ignored, so our example reads as "twelve twenty." Under normal conditions, all the cells in a battery pack should read within 10 points of each other. That would be 1.215-1.225 in our example. As a general rule, it doesn't matter how high or low the readings are, so long as they're all about the same. We're looking only for differences between cells.

When a cell starts to go bad, it will read much higher or lower—usually lower—than its neighbor cells. At 20-30 points difference, you may only need a good equalizing charge to bring all the cells into line. At 40-50 points difference, you probably have a failing cell. Plan for replacement within two or three months. At 60 points or more difference, you have a failed cell that's sucking the life out of all the surrounding cells, and needs to be replaced as soon as possible.

Common Battery Questions and Answers

What voltage should my system run? I notice that components for 12-, 24-, and 48-volt battery banks are available.

Voltage is selected for ease of transmission on the collection and storage side of the system. Low voltage doesn't transmit well, so as home power systems get bigger and are producing greater amounts of energy, it's easier to do at higher voltages. Generally, systems under 2,000 watt-hours per day are fine at 12 volts, systems up to 7,000 watt-hours per day are good at 24 volts, and larger systems should be running at 48 volts. These aren't ironclad guidelines by any means. If you need to run hundreds of feet from a wind or hydro turbine, then higher voltage will help keep wire costs within reason. Or if you're heavily invested in 12-volt with a large inverter and 12-volt lights, you can still grow your system without upgrading to a higher voltage.

I'm just getting started on my power system. Should I go big on the battery bank assuming I'll grow into it?

The answer is yes and no. Yes, you should start with a somewhat larger battery bank than you absolutely need, perhaps sizing for four to five days of autonomy instead of the bare minimum three days. Over time, most folks find more and more things to use power for once it's available.

But if this is your first venture into remote power systems and battery banks, then we usually recommend that you start with some "trainer batteries." You're bound to make some mistakes and do some regrettable things with your first set of batteries. You might as well make mistakes with inexpensive ones. So no, don't invest too heavily in batteries your first couple of years. The golf cart-type deep-cycle batteries make excellent trainers. They are modestly priced, will accept moderate abuse without harm, and are commonly available. In three to five years, when the golf cart-type trainers wear out, you'll be much more knowledgeable about what you need and what quality you're willing to pay for.

The battery bank I started with two years ago just doesn't have enough capacity for us anymore. Is it okay to add some more batteries to the bank?

Lead-acid batteries age in service. The new batteries will be dragged down to the performance level and life expectancy of the old batteries. Your new batteries will be giving up two years of life expectancy right off the bat. Different battery types have different life expectancies, so we really need to consider how long the bank should last. For instance, it would be acceptable to add more cells to a large set of forklift

> Generally, systems under 2,000 watt-hours per day are fine at 12 volts, systems up to 7,000 watt-hours per day are good at 24 volts, and larger systems should be running at 48 volts.

batteries at two years of age because this set is only at 10% of life expectancy. But an RV/marine battery at two years of age is at 100% of life expectancy.

I keep hearing rumors about some great new battery technology like "flywheels" that will make lead-acid batteries obsolete in the near future. Is there any truth to this, and should I wait to invest in batteries?

The truth is that several dozen battery technologies are currently under intense development. Some of them, like nickel-metal hydride, look very promising, but none of them are going to give lead-acid a run for your money within the foreseeable future. Lead-acid is also in the laboratory. Lead-acid technology is going to be around, and is going to continue to give the best performance per dollar for a long time to come.

Real Goods' Battery Care Class

Basic Battery Safety

1. Protect your eyes with goggles and hands with rubber gloves. Battery acid is a slightly diluted sulfuric acid. It will burn your skin after a few minutes of exposure, and your eyes almost immediately. Keep a box or two of baking soda and at least a quart of clean water in the battery area. Flush any battery acid contact with plenty of water. If you get acid in your eyes, flush with clear water for 15 minutes and then seek medical attention.

2. Tape the handles of your battery tools or treat them with Plastic-Dip so they can't possibly short out between battery terminals. Even small batteries are capable of awesome energy discharges when short circuited. The larger batteries we commonly use in RE systems can easily turn a 10-inch crescent wrench red hot in seconds while melting the battery terminal into a useless puddle, and for the grand finale possibly explode and start a fire at the same time. This is more excitement than most of us need in our lives.

Simple Lead-Acid Battery Care

1. Take frequent voltage readings. The voltmeter is the simplest way to monitor battery activity and approximate state of charge.

2. Baking soda neutralizes battery acid. Keep at least a couple boxes on hand, inside the battery enclosure, in case of spills or accidents.

3. Batteries should be enclosed and covered to prevent casual access to terminals. Wet-cell batteries must be vented to prevent accumulation of hydrogen and other harmful gasses. A 2″ PVC vent at the highest point in the battery box is sufficient.

4. Do not locate any electrical equipment inside a battery compartment. It will corrode and fail.

5. Check the water level of your batteries once a month until you know your typical usage pattern.

Use distilled water only. The trace minerals in tap water kill battery capacity. Get a battery-filler bottle from the auto parts store to make this job easy.

6. Be extra careful with any metal tools around batteries. Tape handles or treat with Plastic-Dip so tools can't possibly short between terminals if dropped.

7. Protect battery terminals and any exposed metal from corrosion. After assembly, coat them with grease or Vaseline, or use one of the professional sprays to cover all exposed metal.

8. Never smoke or carry an exposed flame around batteries, particularly when charging.

9. Wear old clothes when working with batteries. Electrolyte loves to eat holes.

10. Chemical processes slow down at colder temperatures. Your battery will act as if it's smaller. At 0°F, batteries lose about 50% of their capacity. In cold climates,

install your batteries in the warmest practical location.

11. All batteries self-discharge slowly, and will sulfate if left unattended for long periods with no trickle charging. Clean, dry battery tops reduce self-discharging.

12. Batteries tend to gain a memory for typical use and will resist wider discharge cycles initially. Like the muscles of your body, periodic stretching exercises increase strength and flexibility. It's good to do an occasional deep discharge and equalizing charge to maintain full battery capacity.

13. A few large cells are better than many small cells when building a battery pack. Larger cells tend to have thicker plates and longer life expectancies. Fewer interconnections and cells mean less maintenance time, and less chance of one cell failing early and requiring the entire battery pack to be replaced early.

3. Wear old clothes! No matter how careful you are around batteries, you'll probably still end up with holes in your jeans. Wear something you can afford to lose, or at least have holes in.

4. Now stop thinking, "Oh, none of that will happen to me!" Even the most experienced professionals can make mistakes if they aren't careful. Melted terminals, fires, explosions, and personal injury are not worth the risk of cutting corners. Don't take chances: Safety measures are easy, and the potential harm is permanent.

Regular Maintenance for Wet-Cell Batteries

1. Check the water level after charging and fill with distilled water. Batteries use more water when they're being fully charged every day, and they'll also use more water as they get older. Check the water level every couple of months. Fluid level should be above the plates, but not above the split ring about ½″ below the top of the cell. It's bad for the plates to be exposed to air. Add distilled water as needed. In an emergency, clean tap water is better than no water, but not by much. Distilled water is really the right stuff.

Don't fill before charging; the little gas bubbles will cause the level to rise and spill electrolyte. Only pure water goes back into the battery. The acid doesn't leave the cell, so you never need to replenish it.

Auto parts stores sell special battery-filler bottles with a nifty spring-loaded automatic shut-off nozzle. Just stick it in the cell, push down, and wait until it stops gurgling. The fluid level will be perfect and there will be no drips or mess. It's a tool worth having. Consider an automatic battery watering device if you have a large industrial battery bank.

2. Clean the battery tops. The condensed fumes and dust on the tops of the batteries start to make a pretty fair conductor after a few months. Batteries will discharge significantly across the dirt between the terminals. That power is lost to you forever! Sponge off the tops with a baking soda and water solution, or use the battery cleaner sprays you can get at the auto parts store. Follow the baking soda cleaner with a clear water rinse. Make sure that cell caps are tight and that none of the cleaning solution gets into the battery cell! This stuff is deadly poison to the battery chemistry. Clean your battery tops once or twice a year.

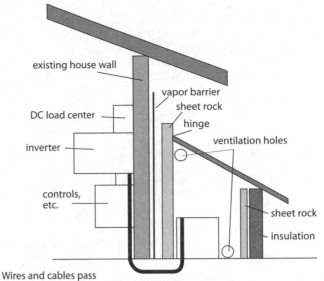

Wires and cables pass through wall at bottom of battery box to prevent hydrogen from entering house. (Hydrogen rises.)

Ideal exterior battery enclosure

3. Clean and/or tighten the battery terminals. Lead is a soft metal and will gradually "creep" away from the bolts. If you were smart and coated all the exposed metal parts around your battery terminals with grease or Vaseline when you installed them, they'll still be corrosion-free. If not, then take them apart, scrub or brush as much of the corrosion off as possible, then dip or brush with a baking soda and water solution until all fizzing stops and then scrape some more. Keep this up until you get all the blue/green crud off. (It's like cancer; if you don't get it all, it will return.) Then carefully cover all the exposed metal parts with grease when you put it back together. If your terminals are already clean, then just gently snug up the bolts.

4. Run a hydrometer test on all the cells. Wait at least 48 hours after adding water before you run this test. Your voltmeter is great for gauging battery state of charge; the hydrometer is for gauging state of health. Use the good kind of hydrometer with a graduated float (not the cheapo floating-ball type). We sell one for $8 (#15-702). A sample of electrolyte is drawn up from a cell until the float rises. Note at what level it's floating. It measures the specific gravity of the electrolyte to three decimal places. A

BATTERY STATE-OF-CHARGE	
Voltage Reading	**Percent of Full Charge**
12.6	100%
12.4–12.6	75–100%
12.2–12.4	50–75%
12.0–12.2	25–50%
11.7–12.0	0–25%

Ideal interior battery enclosure

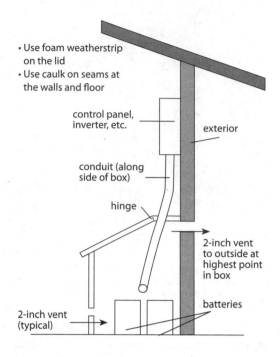

- Use foam weatherstrip on the lid
- Use caulk on seams at the walls and floor

control panel, inverter, etc.

exterior

conduit (along side of box)

hinge

2-inch vent to outside at highest point in box

batteries

2-inch vent (typical)

typical reading would be 1.220. The decimal point is universally ignored, so our example reads as "twelve twenty." Under normal conditions, all the cells in a battery pack will read within 10 points of each other. That would be 1.215-1.225 in our example. What we're looking for in this test is not the state of charge, but the difference in points between cells. In a healthy battery bank, all cells will read within 10 points of each other. Any cell that reads 20-25 points lower is probably starting to fail, although you should run an equalizing charge before casting judgment. You may have three to six months to round up a replacement set. At 60 points difference or more, the bad cell is sucking the life out of your battery bank and needs to get out now! Don't pay any attention to the color-coded good-fair-recharge markings on the float. These only pertain to automotive batteries, which use a slightly hotter acid.

5. Run an equalizing charge (wet-cell batteries only). An equalizing charge is a controlled slight overcharge. This will even out any small differences among cells that have developed over time, and help push any sulfation back into the electrolyte. It's like a minor tune-up for your batteries. It's a good idea to do an equalization every one to six months. Equalizing is more important in the winter when batteries tend to run at lower charge levels than in the summer. *Do NOT equalize sealed batteries! Permanent damage will occur!*

Run your batteries up to about 15.0 volts (or 2.50 volts per cell) and hold them at between 15.0 and 15.5 volts (2.5-2.6 volts per cell) for two to four hours. You'll get fairly vigorous gassing and bubbling on all cells. Do not take the caps off or loosen them during the equalizing charge, you'll just lose more water and spatter acid over the top of the battery. Do NOT top up the water just before equalizing. All that gas production can raise the electrolyte level above the cell top and get messy. Top up the water a day or two later.

Annual Maintenance for Sealed Batteries

1. Clean the battery tops. The crud and dust on the tops of the batteries start to make a pretty fair conductor eventually, even on sealed batteries. Batteries can discharge significantly across the dirt between the terminals. That power is lost to you forever! Sponge off the tops with a baking soda/water solution, or use the battery cleaner sprays you can get at the auto parts store. Follow the baking soda cleaner with a clear water rinse. Cleaning the tops of your sealed batteries once a year is sufficient.

2. Clean and/or tighten the battery terminals. Lead is a soft metal and will gradually "creep" away from the bolts. If your terminals are already clean, then just gently snug up the bolts. Terminal corrosion is rare, but not impossible on sealed batteries. If needed, take them apart, scrub or brush as much of the corrosion off as possible, then dip or brush with a baking soda and water solution until all fizzing stops and then scrape some more. Keep this up until you get all the blue/green crud off. (It's like cancer; if you don't get it all, it will return.) Then carefully cover all the exposed metal parts with grease when you put it back together.

LARGE BATTERIES, CHARGERS, AND ACCESSORIES

Portable Power Supplies

Wagan Power Dome

Stash this portable all-in-one power station in your car, boat, RV, or basement, and you'll be ready for anything from an emergency to an outdoor party. It restores dead car batteries with 600 amps of current and inflates flat tires with a 275 psi air compressor. Two AC outlets send

400W of power with a 1,000W peak surge to everything from tools to TVs while two 12VDC outlets run or charge small gadgets and electronics. There's even an on board 5 LED lamp.

Overload and other safety features protect both you and your gear, and the entire system recharges via AC or DC current to keep you prepared. The 18Ah battery is sealed, spill-proof, airline approved, and replaceable, with a three- to four-year life expectancy. LED indicators and gauges show battery state of charge. #4 AWG jumper cables and an 18" air hose with side storage pockets are included. 20.0 lb. 11"x10.5"x8" (LxWxD). 1-yr. mfr. warranty. China.

17-0342 Power Dome **$149**

XPower 1500 Powerpack

Our largest portable powerpack, the XPower 1500 delivers 120VAC and 12VDC. It packs a large 60Ah sealed battery and a 1,500-watt inverter with dual outlets in a nice, rugged wheeled cart with a removable 38"-high waist handle for ease of movement. Includes a 5-amp battery charger for when AC power is available from a wall plug or generator, and a heavy-duty DC charging cord for lighter socket or a solar panel of your choice. Has a battery charge indicator. Weighs 60 lbs. Measures 14.8"x15.6"x 12.3" (LxWxD) without handle. 1-yr. mfr. warranty. Canada.

17-0283 XPower 1500 Powerpack **$420**

Harvester Solar Electric Generator

When you need backup-power fast, Real Goods' portable Harvester Solar Electric Generator delivers it. Keep radios and cell phones charged during lengthy power outages, or load it up and power up the sound system for a backcountry gathering.

Mounted on a two-wheeled cart, the Harvester includes a power center equipped with a deep-cycle, sealed, FAA-approved, 105Ah battery and True Sine inverter, and a maintenance-free, 80-watt solar panel that tilts to give you maximum solar exposure. The unit can accept up to three panels for extra charging power, and assembly takes less than 30 minutes. Power is fed through an inverter and converted to AC.

Power Center: Heavy-duty yellow and white plastic; 20"x10"x20" (LxWxH); 70 lb. with 105Ah battery and 600W inverter with 1200W surge capacity. Solar Module: 80W; 17V; 4.7A; 21⅖"x1⅕"x47³⁄₁₀" (LxWxH); 21 lbs. Charge controller: 15A PWM with LCD display and electronic overcurrent protection. 10-yr. limited warranty for solar module, 6-mo. warranty for battery, 5 yr. warranty for controller. USA.

41-0218 Harvester	**$2,475**
41-0228 Additional 80W Module	**$785**
41-0229 Additional Battery	**$295**

Wet-Cell Batteries

"Golf Cart"-Type Deep-Cycle Batteries

These batteries are excellent "trainer" batteries for folks new to battery storage systems. First-time users are bound to make a few mistakes as they learn the capabilities and limitations of their systems. These are true deep cycle batteries and will tolerate many 80% cycles without suffering unduly. Even with all the millions of dollars spent on battery research, conventional wisdom still indicates that these lead-acid "golf cart" batteries are the most cost-effective battery storage solution for smaller to medium-sized systems. Typical life expectancy is approximately three to five years. Cycle life expectancy is typically about 225 cycles to 80% depth of discharge. USA.

Battery Voltage: 6 volts
Rated Capacity: 220 amp-hours
Lx WxD: 10.25"x7.25"x10.25"
Weight: 63 lb./28.6kg

15101   Deep-Cycle Battery, 6V/220Ah $129
Free shipping in the continental U.S. for 10 or more.

L-16 Series Deep Cycle Batteries

For larger systems or folks who are upgrading from the "golf cart" batteries, these L-16s have long been the workhorse of the alternative energy industry. The larger cell sizes offer increased ampacity and lower maintenance due to larger water reserves. L-16s typically last from six to eight years. Cycle life expectancy is typically about 300 cycles to 80% depth of discharge. USA.
Battery Voltage: 6 volts
Rated Capacity: 370 amp-hours
LxWxD: 11.25"x7"x16"
Weight: 128 lb./58kg

15102   Deep Cycle Battery, 6V/370Ah $220
Free shipping in the continental U.S. for 10 or more.

Sealed AGM-Type Batteries

Do not buy until your battery is ready for replacement! Batteries will die sitting on the shelf. We buy in small batches, or ship directly from the manufacturer to ensure fresh batteries These small, sealed batteries may be

just what your next expedition needs to keep those laptops running and lights on after dark. The 6V/4Ah size is the replacement for several brands of solar sensor lights. The 12V/18Ah size is a replacement for the Power Dome and is a useful 12-volt storage battery for small power systems. These smaller batteries use an absorbed glass mat construction. AGM costs a bit less than gel batteries but is slightly less durable. Usual life expectancy is three to four years. USA.

15198	6V/4Ah Sealed Battery (2.8"L x 1.9"W x4.25"H , 2 lb.)	$14.50
15200	12V/7Ah (6"L x 2.5"W x 3.9"H, 5.5 lb.)	$28
15202	12V/12Ah (6"L x 3.9"W x 3.9"H, 8.7 lb.)	$50
15209	12V/17.5Ah (7.2"L x 3"W x 6.6"H, 12.8 lb.)	$59

30th ANNIVERSARY

Industrial-Quality IBE Batteries

These Industrial Battery Engineering, Inc. "forklift" batteries are some of the best available. With proper maintenance, they will easily last fifteen to twenty years or more. Each cell is individually packaged in a steel or plastic case with two lifting handles and coated with acid-resistant paint. These individual cells can be carried by two people, making them the best choice for sites without a forklift, or with difficult access. The industrial 2-volt cells provide increased performance, greater reliability, and less frequent maintenance. Every performance and life-enhancing trick known to the battery trade has been incorporated into these cells. Cycle life expectancy is 1,500 cycles to 80% depth of discharge, or 5,000 cycles to 20% depth of discharge.

Priced as 6-cell, 12-volt batteries. For 24-volt systems you must purchase two batteries. These batteries are totally recyclable. Interconnects and input/output leads are sold separately. 7-yr. warranty (1st 5 years free replacement; then 2 years prorated) USA.

15600	**IBE Battery, 12V/1015Ah**
15601	**IBE Battery, 12V/1107Ah**
15602	**IBE Battery, 12V/1199Ah**
15603	**IBE Battery, 12V/1292Ah**
15605	**IBE Battery, 12V/1464Ah**
15606	**IBE Battery, 12V/1568Ah**
15607	**IBE Battery, 12V/1673Ah**

Note: Call for pricing.
Specify system voltage and plastic or steel case when ordering. Allow 4–6 weeks for delivery. Plastic cases may require an additional 3–4 weeks.
Shipped freight collect.

Preassembled HUP Solar-One Batteries

Absolutely the Best Battery Made for Renewable Energy Systems

From EnerSys, Inc., the world's largest battery manufacturer, these industrial strength wet cell lead acid batteries consist of six propylene cells in an epoxy coated steel tray. Features include heat-sealed cell covers, lead-plated copper intercell connectors, and free freight in the continental U.S. to sites accessible by truck. No need to have heavy equipment on site for installation. By using copper buss bar between each cell within the tray, you can break down the 12-volt battery into individual 2-volt cells for removal and reinstallation after the Tray is in its permanent location. If heavy equipment is on site, you can install the battery as is. Sold with six cells or 12 volts per tray. Purchase multiple trays for higher system voltages. With proper maintenance these cells will last over 20 years. HUP stands for High Utilization Positive. These use a patented process that reduces positive plate flaking and shedding, the primary cause of eventual battery failure. No battery offers better performance, a longer life expectancy, or a better manufacturer's warranty. Cycle life expectancy is 2,100 cycles to 80% depth of discharge, or 4,000 cycles to 50% depth of discharge. Interconnect cables are only needed between 12-volt trays; no cell interconnects required. 10-yr. warranty (1st 7 years free replacement; then 3 years prorated). USA.

HUP SOLAR-ONE 12-VOLT BATTERY PACKS				
Item#	**Ah Capacity (@20hrs)**	**Weight (lbs)**	**Size (inches) L x W x H**	**Price**
15810	845	742	40.0 x 7.75 x 25.0	**$2,306**
15811	950	808	40.0 x 8.25 x 25.0	**$2,470**
15812	1,055	880	40.0 x 8.75 x 25.0	**$2,555**
15813	1,160	959	40.0 x 9.0 x 25.0	**$2,778**
15814	1,270	1036	40.0 x 10.25 x 25.0	**$2,910**
15815	1,375	1102	40.0 x 11.25 x 25.0	**$3,075**
15817	1,585	1252	40.0 x 12.75 x 25.0	**$3,569**
15818	1,690	1336	40.0 x 13.5 x 25.0	**$4,820**

Most batteries are in stock for immediate shipment.
Means shipped by truck.

Sealed Batteries

Real Goods Solar Gel Batteries

These true deep cycle gel batteries come highly recommended by the National Renewable Energy Labs. They are completely maintenance-free, with no spills or fumes. Because the electrolyte is gelled, no stratification occurs, and no equalization charging is required. Self-discharge is under 2% per month. May be airline transported as approved by DOT (Department of Transportation), ICAO (International Commercial Airline Organization), and IATA (International Airline Transport Association).

To avoid gassing, gel batteries must be charged at lower peak voltages; typically, 13.8-14.2 volts maximum is recommended. Higher charging voltages will drastically shorten life expectancy. Life expectancy depends on battery size and use in service, but no sealed battery will perform better or longer than these. Typical life expectancy is five to ten years. Smaller batteries will ship by UPS; larger sizes are shipped freight. Free shipping on any order of 10 or more batteries! Batteries will be shipped from the nearest warehouse in AZ, CA, CO, GA, IL, MD, MO, NJ, NV, TX, UT, or WA. Amp-hour rating at standard 20-hour rate. These batteries are all supplied with flag-type terminals for secure bolt up connections. Dimensions are L x W x H including terminals. USA.

15210 32AH/12V SU1 Solar Gel Battery,
23.4 lb. 7.75" x 5.2" x 7.25" $69

15211 51AH/12V S22NF Solar Gel Battery,
37.0 lb. 9.4" x 5.5" x 9.25" $119

15212 74AH/12V S24 Solar Gel Battery,
52.0 lb. 10.25" x 6.75" x 9.8" $149

15213 86AH/12V S27 Solar Gel Battery,
62.7 lb. 12.75" x 6.75" x 9.25" $159

15214 98AH/12V S31 Solar Gel Battery,
69.5 lb. 12.9" x 6.75" x 9.75" $210

15215 183AH/12V S4D Solar Gel Battery,
127.0 lb. 20.75" x 8.5" x 10.6" $379

15216 225AH/12V S8D Solar Gel Battery,
157.0 lb. 20.75" x 11" x 10.6" $459

15217 180AH/6V S6VGC Solar Gel Battery,
68.4 lb. 10.25" x 7.2" x 10.8" $205

Hawker Envirolink Sealed Industrial Battery Packs

The best choice for emergency power back-up systems

The Hawker Envirolink series offers true gel construction, not the cheaper, lower life-expectancy AGM construction. These batteries are the top-quality choice for long-life, low-maintenance emergency power back-up systems. No routine maintenance is required, and this battery type thrives on float service. Life expectancy is 10-15 years with proper care and a charger/controller that keeps voltage below 2.35V/cell. Delivered in preassembled 12- or 24-volt packs with smooth steel cases and no exposed electrical contacts. Cycle life expectancy is 1,250 cycles to 80% depth of discharge. Interconnect cables are only needed between 12- or 24-volt trays; cell interconnects are soldered in place at the factory. 5 yr. mfr. warranty. USA.

Hawker Envirolink Sealed Industrial 12-Volt Battery Packs

Item #	Ah @ 20hr.	Weight (lb.)	Size (in.) L x W x H	Price
17-0296	369	390	26.3 x 6.5 x 23	**$2,470**
15819	555	564	31 x 7.75 x 23	**$3,315**
15820	925	885	31.13 x 13 x 23	**$5,295**
15821	1,110	1,050	29.3 x 13 x 23	**$6,250**
15822	1,480	1,374	25.7 x 19.6 x 23	**$8,089**

Hawker Envirolink Sealed Industrial 24-Volt Battery Packs

Item #	Ah @ 20hr.	Weight (lb.)	Size (in.) L x W x H	Price
17-0297	369	780	25.7 x 11 x 23	**$4,680**
15823	555	1,128	31 x 13 x 23	**$4,680**
15825	925	1,770	38.5 x 16.6 x 23	**$9,985**
15826	1,110	2,100	38.7 x 19.7 x 23	**$10,500**
15827	1,480	2,748	38.7 x 25.6 x 23	**$15,230**

Iota DLS-Series Battery Chargers

Efficient, reliable, and affordable

Iota chargers are solid-state, high-efficiency chargers, very similar to the now-out-of-business Todd charger we've offered for years. Iota features include light, compact construction; automatic fan cooling; reverse-polarity protection; easy terminal access; external, replaceable fuses; rugged reliability; and UL listing. The Iota is a safe constant-voltage charger, so as the battery voltage gradually rises with charging, the charger will gradually reduce amperage flow.

Chargers are available in 12-volt models at 30, 55, 75, and 90 amps; in 24-volt models at 25 and 40 amps; and in a 48-volt model at 10 amps. Standard charger models are safe to use with any battery type, or with any battery pack that is constantly float-charged. Each charger has switchable dual voltage output, with about a 0.6-volt difference. Use low voltage for float charging, or high voltage for the most rapid battery charging. Voltage range is selected by inserting a phone-type plug; a customer-supplied switch can be inserted easily for manual control, or Iota offers their optional IQ switch that delivers automatic 3-stage smart charging. We offer three custom high-volt 12-volt models for wet-cell battery charging designed to keep generator run time to a minimum. The hi-volt models only are set at 14.8/15.4-volt. 2-yr. mfr. warranty. USA.

15791	12V/30A Std. Charger	$135
15793	12V/55A Std. Charger	$170
15795	12V/75A Std. Charger	$335
17-0277	12V/90A Std. Charger	$380
15792	12V/30A Hi-Volt Charger	$150
15794	12V/55A Hi-Volt Charger	$185
15796	12V/75A Hi-Volt Charger	$350
17-0278	24V/25A Std. Charger	$260
17-0279	24V/40A Std. Charger	$390
17-0280	48V/10A Std. Charger	$315
15853	Iota IQ3 Switch	$35

Std. Charger Model	12V/30A	12V/55A	12V/75A	12V/90A	24V/25A	24V/40A	48V/10A
Max. Amp Draw (AC):	8A	15A	18A	21A	12A	20A	15A
Size: 6.5" x 3.5" x	7"	7"	10"	10"	7"	10"	10"
Weight:	5.5 lb.	7 lb.	10 lb.	8 lb.	5.5 lb.	8 lb.	8 lb.
DC Output Voltages:	13.6–14.2	13.6–14.2	13.6–14.2	13.6–14.2	27.2–27.7	27.2–27.7	54.0–55.0

Battery Accessories

Electrical Grease

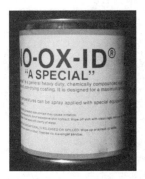

Electrically conductive grease prevents corrosion and lubricates your battery connections. Prevents the formation of corrosive deposits on copper, lead, and aluminum conductors and on steel connectors. Recommended when conductors are exposed to battery acid, salt, moisture, and other corrosive environments. Mandatory when dissimilar metals such as aluminum and copper or lead and steel (galvanized or stainless) are in contact with each other in a corrosive environment (e.g., battery terminals). USA.

48-0276 NO-OX-ID, 1 pint can   **$14.00**

PRO-FILL 6V Battery Watering Systems

Fill up to eight 6V batteries from a remote position using "PRO-FILL" the Flow-Rite 6V battery-watering system for solar systems, boats, golf carts and utility vehicles. Automatic shutoff valves replace standard vent caps. Water flows into each cell to the proper level and pressure buildup tells the operator when all the cells have been topped. Use with any distilled water container. Kits include valves, manifolds, tubing, couplings, caps, pump and instructions. For more than eight batteries, order multiple kits. USA.

48-9248	PRO-FILL 4-Battery Kit	$149
48-9249	PRO-FILL6-Battery Kit	$199
48-9250	PRO-FILL8-Battery Kit	$210
48-9251	PRO-FILL 8-Battery Kit for Non-Std. 3.3" cell spacing	$210
48-0247	Qwik Fill Kit A for 1-2 Batteries	$69
48-0246	Extra Clampless Tubing, 20'	$30

DC-to-DC Converters

Got a DC appliance you want to run directly, but your battery is the wrong voltage? Here's your answer from Solar Converters. These bi-directional DC-to-DC converters work at 96% efficiency in either direction. For instance, the 12/24 unit will provide 12V @ 20A from a 24V input, or can be connected backward to provide 24V @ 10A from a 12V input. Current is electronically limited and has fuse backup.

Four units are offered: The 12/24 unit described above; a larger 12/24 unit that delivers 25A or 50A depending on direction, a 24/48 unit that delivers 5A or 10A depending on direction, and a 12/48 unit that delivers 2.5A or 10A depending on direction. The housing is a weatherproof NEMA 4 enclosure measuring 4.5"x2.5"x2", good for interior or exterior installation (keep out of direct sunlight to keep the electronics happy). Rated for –40°F to 140°F (–40°C to 60°C). 1-yr. mfr. warranty. Canada.

25709	12/24 DC-to-DC Converter, 20A	$199
25782	12/24 DC-to-DC Converter, 50A	$399
25710	24/48 DC-to-DC Converter, 10A	$210
25711	12/48 DC-to-DC Converter, 10A	$210

Full-Size Battery Hydrometer

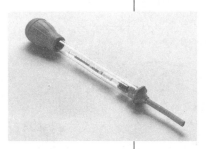

This full-size graduated-float-style specific gravity tester is accurate and easy to use in all temperatures. Specific gravity levels are printed on the tough, see-through plastic body. It has a one-piece rubber bulb with a neoprene tip. USA. *Note: Ignore the color-coded green/yellow/ red zones on the float. These are calibrated for automotive starting batteries, which use a denser acid solution. Storage batteries will read usually in the red or yellow zones. Just watch for differences among individual cells.*

15702	Full-Size Hydrometer	$11

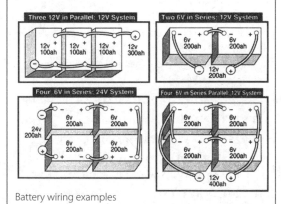

Battery wiring examples

Inverter Cables

Our inverter and battery cables are all flexible cable with color coding for polarity and shrink-wrapped lugs for ⅜" bolt studs. These are two-lug versions (lugs on both ends). USA.

15851	2/0, 3' Inverter Cables, 2 lug, single	$30
15852	4/0, 3' Inverter Cables, 2 lug, single	$40
27311	4/0, 5' Inverter Cables, 2 lug, single red or black	$60
27312	4/0, 10' Inverter Cables, 2 lug, single red or black	$105
27099	2/0, 5' Inverter Cables, 2 lug, single red or black	$50
27098	2/0, 10' Inverter Cables, 2 lug, single red or black	$79

Heavy-Duty Insulated Battery Cables

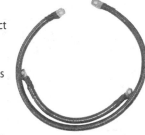

Serious, beefy, 2/0 or 4/0 gauge battery interconnect cables for systems using 1,000-watt and larger inverters. These single cables have ⅜" lug terminals on both ends, and are shrink-wrapped for corrosion resistance. Pick your gauge, length, and color. USA.

15841	4/0, 20" Black	$32
15840	4/0, 20" Red	$32
15839	4/0, 13" Black	$27
15850	2/0, 20" Black	$24
15849	2/0, 20" Red	$24
15848	2/0, 12" Black	$16
15847	2/0, 9" Black	$14

Battery Selector Switch

Few renewable energy homes use dual battery banks, but most boats and RVs do. This high-current switch permits selection between Battery 1 or Battery 2, or both in parallel. The off position acts as a battery disconnect. Wires connect to ⅜" lugs. Capacity is 250 amps continuous, 360 amps intermittent. Okay for 12- or 24-volt systems. 5.25" diameter x 2.6" deep. Marine UL-listed. USA.

25715   Battery Switch   **$41**

Dual Battery Isolator

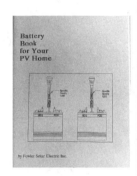

For RVs or boats, this heavy-duty relay automatically disconnects the household battery from the starting battery when the engine is shut off. This prevents inadvertently running down the starting battery. When the engine is started, the isolator connects the batteries, so both will be recharged by the alternator. The relay is large enough to handle heavy, sustained charging amperage. Easy hookup: Connect a positive wire from each battery to the large terminals. Connect ignition power to one small terminal, ground the other small terminal. USA.

41-0242 Dual Battery Isolator   **$27**

Battery Book for Your PV Home

This booklet by Fowler Electric Inc. gives concise information on lead-acid batteries. Topics covered include battery theory, maintenance, specific gravity, voltage, wiring, and equalizing. Very easy to read and provides the essential information to understand and get the most from your batteries. Highly recommended by Dr. Doug for every battery-powered household. 22 pages, softcover.

80104   Battery Book for Your PV Home   **$8**

Battery Care Kit

Our kit includes the informative *Battery Book for Your PV Home,* and a full-size graduated battery hydrometer. The *Battery Book* explains how batteries work and shows how to care for your batteries and get the best performance. Every battery-powered household needs a copy. The hydrometer is your most important tool for determining battery state of health. Detect and fix small problems before they require complete battery bank replacement.

Use Tip: The color-coded red, yellow, and green zones on the hydrometer are calibrated for automotive starting batteries: ignore them. Just look for differences among cells.

15754   Hydrometer/Book Kit   **$14**

Rechargeable Batteries and Chargers

Americans toss over 3 billion small consumer batteries into the landfill every year. The rest of the world adds another few billion to the total.

If you've waded this deeply into the *Sourcebook,* we probably don't need to bore you with why you ought to be using rechargeable batteries. They're far less expensive, they don't add toxins to the landfill, and they don't perpetuate a throw-away society. You know all that already, right? Well, just in case you want a review, we'll provide a brief one. Are you already feeling secure in your knowledge and just want to choose the best rechargeable battery or charger for your purposes? Then skip ahead to Choosing the Right Battery.

Why We Need to Use Rechargeable Batteries

Americans toss over three billion small consumer batteries into the landfill every year. The rest of the world adds another few billion to the total. The vast majority of these are alkaline batteries. Alkaline batteries are the common Duracell, Energizer, and Eveready brands you find in the grocery store checkout lane. While domestic battery manufacturers have refined their formulas in the past few years to eliminate small amounts of mercury, alkaline batteries are still low-level toxic waste. Casual disposal in the landfill may be acceptable to your local health officials, but what a waste of good resources! Our grandchildren, maybe even our children, will be part of the next gold rush, when we start mining our 20th-century landfills for all the refined metals and other depleted resources that are waiting there.

So throw-away batteries are wasteful and a health hazard. They're also surprisingly expensive. Alkaline cells cost around $0.90-$2.50 per battery, depending on size and brand. Use it once, and then throw it away. In comparison, rechargeable nickel-metal hydride cells—including the initial cost of the battery, a charger, and one or two cents worth of electricity—cost $0.04 to $0.10 per cycle, assuming a very conservative lifetime of 400 cycles. Chargers outlive batteries, so these rechargeable cost figures err on the high side. Typically, throw-away batteries cost $0.10/hour to operate, while rechargeable batteries cost only 0.1¢/($\frac{1}{10}$ of one cent) per hour.

ALKALINE BATTERIES VERSUS RECHARGEABLES

What's the smart choice?

2,000 Alkaline Batteries @ $2.₈₅ / 4 pack
4,000 Ah for $1,425.⁰⁰

=

4 Nimh Batteries	$12.⁰⁰
1 Charger	$32.⁰⁰
2,000 Charges @ 1¢ ea.	$20.⁰⁰
4,000 Ah for	$64.⁰⁰

When *Not* to Use Rechargeable Batteries

Rechargeable batteries cost far less, they reduce the waste stream, and they usually present fewer disposal health hazards. What's not to love about them? It isn't a perfect world; rechargeable technology does have deficiencies. Specifically, lower capacity, lower operating voltage, self-discharge, and, in the case of NiCad cells only, toxic waste. Let's look at each issue individually.

Lower capacity. This disadvantage is rapidly disappearing. The latest generation of NiMH (nickel-metal hydride) AA-size batteries actually has *more* storage capacity than the premium name-brand alkalines! As this trend continues and spreads to the larger C- and D-size cells, capacity simply won't be an issue. NiMH batteries will support high-drain appliances such as digital cameras or remote-control toys longer than any other battery chemistry, including throwaway batteries. And they will do it hundreds of times before hitting the trash.

Lower operating voltage. NiCad and NiMH batteries operate at 1.25 volts per cell. The alkaline cells that many battery-powered appliances are expecting to use operate at 1.50 volts per cell. This quarter of a volt is sometimes a problem for voltage-sensitive appliances, or devices such as large boom boxes that use many cells in series. The cumulative voltage difference over eight cells in series can add up to a problem if the device isn't designed for rechargeable battery use.

Self-discharge. All battery chemistries discharge slowly when left sitting. Some chemistries, like the alkaline and lithium cells used for throw-away batteries, have shelf lives of five to ten years. Rechargeable chemistry will self-discharge much faster, with a shelf life of only two to four months. These batteries aren't the ones you want in your glove box emergency flashlight, and probably aren't the best possible choice for battery-powered clocks.

Toxic NiCad cells. NiCad cells contain the element cadmium. Human beings do not fare well when ingesting even tiny amounts of cadmium. Real Goods stopped selling NiCad batteries several years ago as soon as a better alternative was available. These cells absolutely must be properly recycled. We accept all NiCad batteries returned to us for recycling. Radio Shack stores also accept NiCads for recycling. Do not dispose of NiCad cells casually! They are toxic waste! Now, that said, let us point out that the newer, higher-capacity nickel-metal hydride batteries are not toxic at all, and will recharge just fine in any NiCad battery charger you might have already.

> Rechargeable batteries aren't the ones you want in your glove box emergency flashlight, and probably aren't the best possible choice for battery-powered clocks.

RECHARGEABLE BATTERY COMPARISONS			
	Standard Alkaline	**Rechargeable Alkaline**	**NiMH**
Volts	1.5	1.5	1.25
Capacity (compared to alkaline)	100%	90% initially Diminishes each cycle	70%-110% No loss with cycles
Capacity in mAh AAA AA C D	750mAh 2,000mAh 5,000mAh 10,000mAh	750mAh 1,800mAh 3,000mAh 7,200mAh	750mAh 2,200mAh 3,500mAh 7,000mAh
Avg. Recharge Cycles	1	12–25	400–600
Self-Discharge Rate	Negligible; 5-year shelf life	Negligible; 5-year shelf life	Modest; 15% loss @ 60 days
Strengths	1.5-volt standard Long shelf life High capacity	1.5-volt standard Long shelf life Good capacity	Sustains high draw No voltage fade Many recharges Nontoxic
Weaknesses	Highest cost Throw-away	Loses capacity each cycle Special charger required	1.2-volt Self-discharges
Best Uses	Clocks Emergency stuff	Clocks Emergency stuff Toys, games, radios, flashlights	Digital cameras! Remote-control toys Flashlights High-drain appliance

Choosing the Right Battery

There are three basic rechargeable battery chemistries to choose from: rechargeable alkaline, nickel-metal hydride (NiMH), and nickel cadmium (NiCad). Each has strengths and weaknesses, although NiCads compare so poorly now that we aren't even going to dignify them by charting. We've summarized everything and made the comparisons with standard throwaway alkaline batteries in the accompanying chart.

Regardless of your chemistry choice, buy several sets of batteries so you can always have one set in the charger while another set is working for you.

Choosing the Right Battery Charger(s)

Now that you've seen the choices in rechargeable batteries, you'll also need to decide which charger(s) to use. The three types are AC charger (plugs into a wall socket), solar charger (charges directly from the Sun), and 12-volt charger (plugs into your car's cigarette lighter or your home's renewable energy system).

SOLAR CHARGERS

The great benefit of solar battery chargers is that you don't need an outlet to plug them into, you just need the Sun! They can be used anywhere the Sun shines. However, most solar chargers work more slowly than plug-in types, and it may take several days to charge up a set of batteries. This needn't be a problem. We strongly recommend you buy an extra set of batteries with your solar charger, so you can charge one set while using the other. Given time and patience (prerequisites for a prudent, sustainable lifestyle in any case), solar chargers are the best way to go.

DC CHARGERS

The 12-volt charger will recharge your batteries from a car, boat, or from your home's 12-volt renewable energy system (if you're so blessed). It provides a variable charge rate depending on battery size. In contrast, most AC chargers put the same amount of power into every battery, regardless of size. Even though the 12-volt charger is a fast charger, it will not drain your car battery unless you forget and leave it charging for several days. Some 12-volt chargers can overcharge batteries, however, so pay close attention to how long it takes to charge, and don't leave them in there much longer. Leaving them in the 12-volt charger for one-and-a-half times the recommended duration will not harm the batteries.

AC CHARGERS

The AC (plug in the wall) chargers are by far the most convenient for the conventional utility-powered house. Most of us are surrounded by AC outlets all day long and all we have to do is plug in the charger and let it go to work. All these chargers turn off when charging is complete, and go into a trickle charge maintenance mode. Your batteries will always be fully topped up and waiting for you. Most AC chargers automatically figure out what chemistry and size each battery is, then charges it accordingly.

Frequently Asked Questions About Rechargeable Batteries

Is it true that rechargeable batteries develop a "memory"?

This was a NiCad problem many years ago. If you use only a little of the battery's energy every time, and then charge it up, the battery will develop a "memory" after several dozen of these short cycles. It won't be able to store as much energy as when new. This was true only with NiCads, not with NiMHs or alkalines. Improved chemistry has practically eliminated any memory problems. Short cycles have never been a problem for NiMH or alkaline batteries.

How do I tell when a NiMH (or NiCad) battery is charged?

Because these batteries always show the same voltage from 10%-90% of capacity, it's tricky to know when they're charged. A voltmeter tells little or nothing about the battery's state of charge. Either use one of the smart chargers like the AccuManager, or time the charge cycle. (Calculate the recommended charging time by dividing the battery capacity by the charger capacity and add 25% for charging inefficiency. For instance: A battery of 1,100mAh capacity in a charger that puts out 100 milliamps will need 11 hours, plus another 3 hours for charging inefficiency, for a total of 14 hours to recharge completely.) A cruder method to use is the touch method. When the battery gets warm, it's finished charging. (This method is only applicable with the plug-in chargers.)

My new NiMH (or NiCad) batteries don't seem to take a charge in the solar-powered charger.

All batteries are chemically "stiff" when new. NiMH batteries are shipped uncharged from the factory, and are difficult to charge initially. A small solar charger may not have enough power to overcome the battery's internal resistance; hence, no charge. The solution is to use a plug-in charger for the first few cycles to break in the battery. Another solution is to install only one battery at a time in the solar charger. Once the batteries have been broken in, the problem disappears. These batteries will give you many, many years of use for no additional cost with solar charging.

My tape player (or other device) doesn't work when I put NiMHs in. What's wrong?

Some battery-powered equipment will not function on the lower-voltage NiMH batteries, which produce 1.25 volts (versus 1.5 volts produced by fresh alkalines). Some manufacturers use a low-voltage cutoff that won't allow using rechargeables. If the cutoff won't let you use NiMHs, it's also forcing you to buy new alkalines when they're only 50% depleted! Sometimes with devices that require a large number of batteries (six or more), the G-volt difference between rechargeables and alkalines gets magnified to a serious problem. Even a fully charged set of NiMHs will be seen as a depleted battery pack by the appliance. You need a new appliance that's designed for rechargeable batteries.

Do I have to run the rechargeable battery completely dead every time?

No. For greatest life expectancy, it's best to recharge NiMH batteries when you first sense the voltage is dropping. You'll be down to 10% or 15% capacity at this point. Further draining is not needed or recommended.

Alkaline rechargeable batteries really hate deep cycles. They'll do best when they're run only down about 50%.

RECHARGEABLE PRODUCTS

Batteries That Really Just Keep Going and Going!

Americans use nearly 3 billion disposable batteries every year; the rest of the world adds another few billion to that total. Beyond being expensive and a colossal waste of resources, this presents a huge low-level toxic disposal problem. Rechargeable batteries are the obvious, easy, money-saving solution. Buy an extra set. That way you have one set in your charger at all times and one set in the appliance. When the appliance gets weak, just swap batteries with the fresh ones. If you change batteries once a week, they will last for 10-15 years.

Rechargeable Batteries with MORE Energy than Throw-Away Batteries!

The capacity of NiMH batteries keeps increasing as the technology develops. Our high-capacity AA-size NiMH batteries are now equal to or better than throw-away alkaline cells. Our tests show name-brand alkaline AA batteries deliver approximately 1,800-2,100mAh, then you throw them away. Our AA NiMH cells deliver 2,200mAh and will do that about 500 times before you throw them away. In addition, NiMH chemistry delivers at a steady voltage all the way down to 10% of charge. No voltage fade like standard alkaline cells! This means you get to actually *use* more of the available capacity with environmentally friendly rechargeable batteries.

ALKALINE BATTERIES VERSUS RECHARGEABLES
What's the smart choice?

2,000 Alkaline Batteries @ $2.85 / 4 pack	4,000 Ah for $1,425.00

=

4 Nimh Batteries	$12.00
1 Charger	$32.00
2,000 Charges @ 1¢ ea.	$20.00
4,000 Ah for	$64.00

Fast Charger & NiMH Batteries

Nickel-metal hydride batteries just keep getting better and better. Ours excel at hard use and deliver steady voltage with no sag all the way down to 10% of charge. This makes them a better choice for heavy-discharge uses like digital cameras. There is no memory effect from short cycling and no toxic content. Because they self-discharge over three to six months, they aren't a good choice for emergency lights. Our speedy, affordable battery charger comes with both AC and DC power cords, charges AA or AAA size NiMH or NiCad batteries in pairs. Charges most batteries in four to five hours or less. Taiwan.

17-9175 C 2-pack (3,500mAh)	**$11**
17-9176 D 2-pack (7,000mAh)	**$20**
17-0284 9V single (160mAh)	**$9**
17-0285 9V 2-pack (160mAh)	**$16**
AA (2,200mAh)	
17-0324 4-pack	**$14**
17-9324 8-pack	**$24**
AAA (750mAh)	
17-0325 4-pack	**$14**
17-9325 8-pack	**$24**
17-0323 Fast Charger	**$29**

AccuManager 20

Using a pulse-charging technology that rechargeable batteries just love, the AccuManager 20 charger will recharge your NiMH or NiCad batteries faster, and better, than any other charger we've ever seen. You can mix different battery types, sizes and charging times. After charging, each cell is stored at the peak of readiness with a gentle trickle charge, until you're ready for it. There are four bays that accept AAA, AA, C or D cells, plus two bays that accept 9-volt cells. Charge times will run from 20 minutes to 12 hours depending on type and size. Supplied with both AC and DC charging cords and will run off standard 120V AC, automotive 12 volts, or a 10- to 15-watt solar panel with appropriate plug, such as our solar option sold at right. 3-yr. mfr. warranty. For rechargeable batteries only. Germany/China.

50269   AccuManager 20 $69

Hi-Speed Digital Charger

Our speedy, affordable battery charger comes with both AC and DC power cords, charges AA- or AAA-size NiMH or nicad batteries in pairs, and stops charging automatically when batteries are full, then switches to genlty trickle-charge your batteries until you're ready for them. Charges most batteries in 5 hours or less. LEDs show red when charging, green when finished, and flashes red/green if a battery is defective. Taiwan.

17-0281 Hi-Speed Digital Charger $29

Battery Xtender

Extend the life of almost any battery using this Battery Xtender. Its smart technology easily accommodates alkaline, carbon-zinc, AccuManager 20, titanium, nickel-metal hydride and 1.5-volt batteries, sensing battery chemistry and automatically adapting its charging parameters for optimum recharging of each and every cell. Simultaneously accepts AAA, AA, C and D cells in any combination of sizes and chemistries. A built-in indicator gauges the status of each battery. Costs less than 1 cent in electricity per battery. 11"H x 7½"W x 2"D. 1.5 lb. China.

17-0181 Battery Xtender $49

Portable Solar Electronics Charger

This solar kit charges and runs iPods®, cell phones, digital cameras, GPSs and PDAs wherever there's sun. A rugged, hardcover-book-sized plastic case stores a 20" USB cable, various adapter jacks and a 4.4W solar module. Open it to the Sun, plug in your gear, and power up miles from the nearest outlet. Includes 12V car-lighter outlet for your gear's own adapters and an AA/AAA battery charger. 19 oz. 9³⁄₁₀"H x 6"W x 1½"D. Philippines.

**17-0340 Portable Solar Electronics
              Charger $139**

10-Year Smoke Detector Battery

While NOT a rechargeable, our 9V lithium battery offers more than a 10-yr. warranty—it gives you the peace of mind of knowing your smoke detector battery isn't going to fail if you really need it. 10-yr. mfr. warranty when used with ionization-type UL-listed smoke detectors. USA.

**17-0158 10-Year Smoke Detector
              Battery $12.95**

Inverters

If you want to run conventional household appliances with your renewable energy system, you need a device to produce AC house current on demand. That device is an inverter.

What is an inverter, and why would you want one in your renewable energy system? This is a reasonable question. Some small renewable energy systems don't need an inverter.

The inverter is an electronic device that converts direct current (DC) into alternating current (AC). Renewable energy sources such as PV modules and wind turbines make DC power, and batteries will store DC power. Batteries are the least expensive and most universally applicable energy storage method available. Batteries store energy as low-voltage DC, which is acceptable, in fact preferable, for some applications. A remote sign-lighting system, or a small cabin that only needs three or four lights, will get by just fine running everything on 12-volt DC. But most of the world operates on higher-voltage AC. AC transmits more efficiently than DC and so has become the world standard. No technology exists to store AC, however. It must be produced as needed. If you want to run conventional household appliances with your renewable energy system, you need a device to produce AC house current on demand. That device is an inverter.

Inverters are a relatively new technology. Until the early 1990s, the highly efficient, long-lived, relatively inexpensive inverters that we have now were still a pipe dream. The world of solid-state equipment has advanced by extraordinary leaps and bounds. More than 95% of the household power systems we put together now include an inverter.

Benefits of Inverter Use

The world runs on AC power. In North America, it's 120-volt, 60-cycle AC. Other countries may have slightly different standards, but it's all AC. Joining the mainstream allows the use of mass-produced components, wiring hardware, and appliances. Appliances may be chosen from a wider, cheaper, and more readily available selection. Electricity transmits more efficiently at higher voltages, so power distribution through the house can be done with conventional 12- and 14-gauge Romex wiring using standard hardware, which electricians appreciate and inspectors understand. Anyone who's wrestled with the heavy 10-gauge wire that a low-voltage DC system requires will see the immediate benefit.

Modern brand-name inverters are extremely reliable. The household-size models we sell have failure rates well under 1%. Efficiency averages about 85%-90% for most models, with peaks at up to 95%. In short, inverters make life simpler, do a better job of running your household, and ultimately save you money on appliances and lights.

An inverter/battery system is the ultimate clean, uninterruptible power supply for your computer, too. In fact, an expensive UPS (uninterruptible power supply) system is simply a battery and a small inverter with an expensive enclosure and a few bells and whistles. Just don't run the power saw or washing machine off the same inverter that runs your computer. We often recommend a small secondary inverter just for the computer system. This ensures that the water pump or some other large, unexpected load can't possibly cause problems for other sensitive equipment.

Electrical Terminology and Mechanics

Electricity can be supplied in a variety of voltages and waveforms. As delivered to and stored by our renewable energy battery system, we've got low-voltage DC. As supplied by the utility network, we've got house current: nominally 120 volts AC. What's the difference? DC electricity flows in one direction only, hence the name direct current. It flows directly from one battery terminal to the other battery terminal. AC current alternates, switching the direction of flow periodically. The U.S. standard is 60 cycles per second. Other countries have settled on other standards, but it's usually either 50 or 60 cycles per second. The electrical term for the number of cycles per second is Hertz, named after an early electricity pioneer. So AC power is defined as 50 Hz or 60 Hz.

Our countrywide standard also defines the voltage. Voltage is similar to pressure in a water line. The greater the voltage, the higher the pressure. When voltage is high, it's easier to transmit a given amount of energy, but it's more difficult to contain and potentially more dangerous. House current in this country is delivered at about 120 volts for most of our household appliances, with the occasional high-consumption appliance using 240 volts. So our power is defined as 120/240 volts/60 Hz. We usually use the short form, 120 VAC, to denote this particular voltage/cycle combination. This voltage is by no means the world standard, but is the most common one we deal with. Inverters are available in international voltages by special order, but not all models from all manufacturers.

Description of Inverter Operation

If we took a pair of switching transistors and set them to reversing the DC polarity (direction of current flow) 60 times per second, we would have low-voltage alternating power of 60 cycles or Hertz (Hz). If this power was then passed through a transformer, which can transform AC power to higher or lower voltages depending on design, we could end up with crude 120-volt/60 Hz power. In fact, this was about all that early inverter designs of the 1950s and 1960s did. As you might expect, the waveform was square and very crude (more about that in a moment). If the battery voltage went up or down, so did the output AC voltage, only ten times as much. Inverter design has come a long, long way from the noisy, 50% efficient, crude inverters of the 1950s. Modern inverters hold a steady voltage output regardless of battery voltage fluctuations, and efficiency is typically in the 90%-95% range. The waveform of the power delivered has also been improved dramatically.

Waveforms

The AC electricity supplied by your local utility is created by spinning a bundle of wires through a series of magnetic fields. As the wire moves into, through, and out of the magnetic field, the voltage gradually builds to a peak and then gradually diminishes. The next magnetic field the wire encounters has the opposite polarity, so current flow is induced in the opposite direction. If this alternating electrical action is plotted against time, as an oscilloscope does, we get a picture of a sine wave, as shown in our wave-form gallery above. Notice how smooth the curves are. Transistors, as used in all inverters, turn on or off abruptly. They have a hard time reproducing curvy sine waves.

Early inverters produced a square-wave alternating current (see the wave-form gallery again). While a square wave does alternate, it is considerably different in shape and peak voltage. This causes problems for many appliances. Heaters or incandescent lights are fine, motors usually get by with just a bit of heating and noise, but solid-state equipment has a really hard time with it, resulting in loud humming, overheating, or failure. It's hard to find new square-wave inverters anymore, and nobody misses them.

Most modern inverters produce a hybrid waveform called a quasi-, synthesized, or modi-

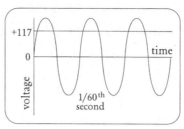

AC sine-wave power.

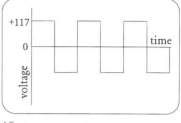

AC square-wave power.

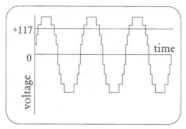

AC modified sine-wave power.

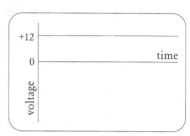

DC 12-volt power.

fied sine wave. In truth, this could just as well be called a modified square wave, but manufacturers are optimists. (Is the glass half full or half empty?) Modified sine-wave output cures many of the problems associated with the square wave. Most appliances will accept it and hardly know the difference. There are some notable exceptions to this rosy picture, however, which we'll cover in detail below in the Inverter Problems section.

Full sine-wave inverters have been available since the mid-1990s, but because of their higher initial cost and lower efficiency, they were only used for running very specific loads. This is changing rapidly. A variety of high-efficiency, moderate-cost pure sine-wave inverters is available now. Xantrex, OutBack, and other manufacturers have unveiled a whole series of sine-wave units, and it's obviously the wave of the future. True sine-wave inverters have very nearly become the standard for larger household power systems already. With very rare exceptions, sine-wave inverters will happily run any appliance that can plug into utility power. Motors run cooler and more quietly on sine wave, and solid-state equipment has no trouble. True sine-wave inverters deliver top-quality AC power and are almost always a better choice for household use.

Inverter Output Ratings

Inverters are sized according to how many watts they can deliver. More wattage capability will cost more money initially. Asking a brand-name

True sine-wave inverters deliver top-quality AC power and are almost always a better choice for household use.

Most households end up with one of the full-size inverters because household loads tend to grow, and larger inverters are often equipped with very powerful battery chargers.

Reading Modified Sine-Wave Output with a Conventional Voltmeter

Most voltmeters that sell for under $100 will give you weird voltage readings if you use them to check the output of a modified sine-wave inverter. Readings of 80 to 105 volts are the norm. This is because when you switch to "AC Volts" the meter is expecting to see conventional utility sine-wave power. What we commonly call 110- to 120-volt power actually varies from 0 to about 175 volts through the sine-wave curve. 120 volts is an average called the Root Mean Square, or RMS for short, that's arrived at mathematically. More expensive meters are RMS corrected; that is, they can measure the average voltage for a complex waveform that isn't a sine wave. Less expensive meters simply assume if you switch to "AC Volts" it's going to be a sine wave. So don't panic when your new expensive inverter checks out at 85 volts: The inverter is fine, but your meter is being fooled. Modern inverters will hold their specified voltage output, usually about 117 volts, to plus or minus about 2%. Most utilities figure they're doing well by holding variation to 5%.

inverter for more power than it can deliver will result in the inverter shutting down. Asking the same of a cheapo inverter may result in a crispy critter. All modern inverters are capable of briefly sustaining much higher loads than they can run continuously, because some electric loads, like motors, require a surge to get started. This momentary overcapacity has led some manufacturers to fudge on their output ratings. A manufacturer might, for instance, call its unit a 200-watt inverter based on the instantaneous rating, when the continuous output is only 140 watts. Happily, this practice is fading into the past. Most manufacturers are taking a more honest approach as they introduce new models and are labeling inverters with their continuous wattage output rather than some fanciful number. In any case, we've been careful to list the continuous power output of all the inverters we carry. Just be aware that the manufacturer calling the inverter a 200-watt unit does not nec-

essarily mean it will do 200 watts continuously. Read the fine print.

How Big an Inverter Do You Need?

The bigger the inverter, the more expensive initially. So you don't want to buy a bigger one than you need. On the other hand, an inverter that's too small is going to frustrate you because you'll need to limit usage.

A small cabin with an appliance or two to run doesn't need much of an inverter. For example, if you want to run a 19-inch TV, a DVD player, and a light all at once, total up all the wattages (about 80 for the TV, 25 for the DVD, and 20 for a compact fluorescent light, a total of 125 watts), pick an inverter that can supply at least 125 watts continuously, and you're all set.

To power a whole house full of appliances and lights might take more planning. Obviously not every appliance and light will be on at the same time. Mid-size inverters of 600-1,000 watts do a good job of running lights, entertainment equipment, and small kitchen appliances, in other words, most common household loads. What a 1,000-watt inverter will not do is run a mid- to full-size microwave, a washing machine, or larger handheld power tools. For those loads, you need a full-size 2,000-plus-watt inverter. In truth, most households end up with one of the full-size inverters because household loads tend to grow, and larger inverters are often equipped with very powerful battery chargers. This is the most convenient way to add battery charging capability to a system (and the cheapest, too, if you're already buying the inverter).

Chargers, Lights, Bells, and Whistles

As you go from small plug-in-the-lighter-socket inverters to larger household-size units, you'll find increasing numbers of bells and whistles, most of which are actually pretty handy. At the very least, all inverters have an LED light showing that it's turned on. Midsize units may feature graphic volt and amp meters; better units may have an LCD display and be able to signal generators and other remote devices to turn on or off.

Battery chargers are the most useful inverter option, and for most folks, the cheapest and easiest way to add a powerful, automatic battery charger to their system. Chargers are pretty much standard equipment on all larger household-size inverters. With a built-in charger, the inverter will have a pair of "AC input" terminals. If the inverter detects AC voltage at these terminals, because you just started the back-up generator, for instance, it will automatically connect the AC input terminals to the AC output terminals, and then go to work charging the batteries. Inverter-based chargers tend to be extremely robust and adjustable, and will treat your batteries nicely.

Many financially challenged off-the-grid systems start out with just a generator, an inverter/charger, and some batteries. You run the generator every few days to recharge the batteries. Add PV charging as money allows, and the generator gradually has to run less and less.

How to Cripple Your Inverter

Do you want to reduce your inverter's ability to provide maximum output? Here's how: Keep your battery in a low state of charge; use long, undersized cables to connect the battery and inverter; and keep your inverter in a small airtight enclosure at high temperatures. You'll succeed in crippling, if not outright destroying, your inverter. Low battery voltage will severely limit any inverter's ability to meet a surge load. Low voltage occurs when the battery is undercharged, or undersized. If the battery bank isn't large enough to supply the energy demand, then volt-

age will drop, and the inverter probably won't be able to start the load. For instance, a Xantrex DR2412 will easily start most any washing machine, but not if you've only got a couple of golf cart batteries. The starting surge will demand more electrons than the batteries can supply, voltage will plummet, and the inverter will shut off to protect itself (and your washing-machine motor). Now suppose you have eight golf cart batteries on the DR2412, which is plenty to start the washer with ease, but decided to save some money on the hookup cables, and used a set of $16.95 automotive jumper cables. The battery bank is capable, the inverter is capable, but not enough electrons can get through the undersized jumper cables. The result is the same: low voltage at the inverter, which shuts off without starting the washer. Do not skimp on inverter cables or on battery interconnect cables. These items are inexpensive compared to the cost of a high-quality inverter and good batteries. Don't cripple your system by scrimping on the electron-delivery parts.

All inverters produce a small amount of waste heat. The harder they work, the more heat they produce. If they get too hot they will shut off or limit their output to protect themselves. Give the inverter plenty of ventilation. Treat it like a piece of stereo gear: dry, protected, and well ventilated.

How Big a Battery Do You Need?

We usually size household battery banks to provide stored power for three to five days of autonomy during cloudy weather. For most folks, this is a comfortable compromise between cost and function. With three- to five-day sizing, the battery will be large enough to handle any surge loads that the inverter is called upon to start. A battery bank smaller than three-days' capacity will cycle fairly deeply on a regular basis. This isn't good for battery life. A larger battery bank cycled less deeply will cost less in the long run, because it lasts longer. Banks larger than five-days' capacity are more expensive than a back-up power source (such as a modest-sized generator). However, we occasionally run into situations with one-horsepower or larger submersible well pumps or stationary power tools requiring a larger battery bank simply to meet the surge load when starting. Call the Gaiam Real Goods technical staff for help if you are anticipating large loads of this type.

Treat your inverter like a piece of stereo gear: dry, protected, and well ventilated.

Happy Inverters

1. Keep the inverter as close to the battery as possible, but not in the battery compartment. Five to 10 feet and separated by a fireproof wall is optimal. The high-voltage output of the inverter is easy to transmit. The low-voltage input transmits poorly.
2. Keep the inverter dry and as cool as possible. They don't mind living outdoors if protected.
3. Don't strangle the inverter with undersized supply cables. Most manufacturers have recommendations in the owner's manual. If in doubt, give us a call.
4. Fuse your inverter cables and any other circuits that connect to a battery!

Inverter Safety Equipment and Power Supply

In some ways batteries are safer than conventional AC power. It's fairly difficult to shock yourself at low voltage. But in other ways batteries are more dangerous: They can supply many more amps into a short circuit, and once a DC arc is struck, it has little tendency to self-extinguish. One of the very sensible things the National Electrical Code requires is fusing and a safety disconnect for any appliance connected to a battery bank. Fusing is extremely important for any circuit connected to a battery! Without fusing, you risk burning down your house.

Several products exist to cover DC fusing and disconnect needs for inverter-based systems. For full-size inverter fusing, you can use a properly sized OutBack or Xantrex or MidNite Solar DC power center. Our technical staff can answer your questions about which one is better for you. Midsize inverters can use a Class T fuse with a small surface-mounted circuit breaker to provide safe and compliant connection. All this safety and connection equipment is covered in the Safety section of this chapter, beginning on page 202. We've recommended fusing or breaker sizing, and required cable sizing with all larger inverters.

The size of the cables providing power to the inverter is as important as the fusing, as we noted earlier. Do not restrict the inverter's ability to meet surge loads by choking it down with undersized or lengthy cables. Ten feet is the longest practical run between the battery and inverter. This is true even for small 100- to 200-watt inverters. Put the extension cord on the AC side of the inverter! Larger models will include cable and circuit breaker requirement charts. Batteries should either be the sealed type, or live in their own enclosure. Don't put your inverter in with the batteries!

Potential Problems with Modified Sine-Wave Inverter Use—
and How to Correct Them

We have been painting a rosy picture up to this point, but now it is time for a little brutal honesty to balance things out. If you know about these problems beforehand, it is usually possible to work around them when selecting appliances.

Wave-Form Problems

Wave-form problems are probably the biggest category of inverter problems we encounter. We talked about sine wave versus modified sine wave in the technical description above. The problems discussed here are caused only by modified sine-wave inverters. If you are planning to use a pure-sine-wave unit, you can skip this section.

AUDIO EQUIPMENT

Some audio gear will pick up a 60-cycle buzz through the speakers. It doesn't hurt the equipment, but it's annoying to the listener. There are too many models and brands to say specifically which have a problem and which do not. We've had better luck with new equipment recently. Manufacturers are starting to put better power supplies into their gear. We can only recommend that you try it and see.

Some top-of-the-line audio gear is protected by SCRs or Triacs. These devices are installed to guard against powerline spikes, surges, and trash (nasties that don't happen on inverter systems). However, they see the sharp corners on modified sine wave as trash and will sometimes commit electrical hara-kiri to prevent that nasty power from reaching the delicate innards. Some are even smart enough to refuse to eat any of that ill-shaped power, and will not power up. The only sure cure for this (other than more tolerant equipment) is a sine-wave inverter. If you can afford top-of-the-line audio gear, chances are you've already decided that sine-wave power is the way to go.

COMPUTERS

Computers run happily on a modified sine wave. In fact, most of the uninterruptible power supplies on the market have modified sine-wave or even square-wave output. The first thing the computer does with the incoming AC power is to run it through an internal power supply. We've had a few reports of the power supply being just a bit noisier on a modified sine wave, but no real problems. Running your prized family computer off an inverter will not be a problem. What can be a problem is large start-up power surges. If your computer is running off the same household inverter as the water pump, power tools, and microwave, you're going to have trouble. When a large motor, such as a skilsaw, is starting, it will pull the AC system voltage way down momentarily. This can cause computer crashes. The fix is a small, separate inverter that only runs your computer system. It can be connected to the same household battery pack, and have a dedicated outlet or two.

LASER PRINTERS

Many laser printers are equipped with SCRs, which cause the problems detailed above. Laser printers are a very poor choice for renewable energy systems anyway, due to their high standby power use keeping the heater warm. Lower-cost inkjet printers can do almost anything a laser printer can do while using only 25-30 watts instead of 400-900 watts.

CEILING FANS

Many variable-speed ceiling fans will buzz on modified sine-wave current. We've also had some reports of fan remote controls not working on modified sine wave. The fans spin up fine, but the noise is annoying, and you can't change speeds.

Potential Problems with ALL Inverters—

and How to Correct Them

You sine wave buyers can stop smirking here, the following problems apply to all inverters.

RADIO AND TV INTERFERENCE

All inverters broadcast radio noise when operating. The FCC put out new guidelines concerning interference, and as manufacturers come out with new models like the OutBack FX series or the Xantrex XW series, this is much less of a problem. Most of this interference is on the AM radio band. Do not plug your radio into the inverter and expect to listen to the ball game; you'll have to use a battery-powered radio and be some distance away from the inverter. This is occasionally a problem with TV interference when inexpensive TVs and smaller inexpensive inverters are used together. Distance helps. Put the TV (and the antenna) at least 15 feet from the inverter. Twisting the inverter input cables may also limit their broadcast power (strange as it sounds, it works).

PHANTOM LOADS AND VAMPIRES

A phantom load isn't something that lurks in your basement with a half-mask, but it's close kin. Many modern appliances remain partially on when they appear to be turned off. That's a phantom load. Any appliance that can be powered up with a button on a remote control must remain partially on and listening to receive the "on" signal. Most TVs and audio gear these days are phantom loads. Anything with a clock—VCRs, coffee makers, microwave ovens, or bedside radio-clocks—uses a small amount of power all the time.

Vampires suck the juice out of your system. These are the black power cubes that plug into an AC socket to deliver lower-voltage power for your answering machine, electric toothbrush, power-tool charging stand, or any of the other huge variety of appliances that use a power cube on the AC socket. These villainous wastrels usually run a horrible 60%-80% inefficiency (which means that for every dime's worth of electricity consumed, they throw away 6 or 8 cents' worth). Most of these nasties always draw power, even if no battery/tooth brush/ razor/cordless phone is present and charging. It would cost their manufacturers less than 25 cents per unit to build a power-saving standby mode into the power cube, but since you, the consumer, are paying for the inefficiency, what do they care? The appliance might be turned off, but the vampire keeps sucking a few watts. Have you ever noticed that power cubes are usually warm? That's wasted power being converted to heat. By the way, cute and appropriate as it is, we can't take credit for the vampire name. That's official electric industry terminology (bless their honest souls).

So what's the big deal? It's only a few watts, isn't it? The problem isn't the power consumed by vampires and phantom loads, it's what is required to deliver those few watts 24 hours per day. When there's no demand for AC power, a full-size inverter will drop into a sleep mode. Sleep mode keeps checking to see if anything is asking for power, but it takes only a tiny amount of energy, usually 2-3 watts. If the cumulative phantom and vampire loads are enough to

Do not plug your radio into the inverter and expect to listen to the ball game; you'll have to use a battery-powered radio and be some distance away from the inverter.

Many modern appliances remain partially on when they appear to be turned off. That's a phantom load.

The solution for phantoms and vampires is outlets or plug strips that are switched off when not in use

awaken the inverter, it consumes its own load plus the inverter's overhead, which is typically 15-20 watts. The inverter overhead is the real problem. An extra 15-20 watts might sound like grasping at straws, but over 24 hours every day, this much power lost to inefficiency can easily add a couple $450 modules to the size of the PV array and maybe another battery or two to the system requirements. You want to make sure your inverter can drop into sleep mode whenever there's no real demand for AC power.

KEEPING THE VAMPIRES AT BAY

With some minimal attention to details, you can keep the vampires and phantoms under tight control and save yourself hundreds or thousands of dollars on system costs.

The solution for clocks is battery power. A wall-mounted clock runs for nearly a year on a single AA rechargeable battery. We have found a wide selection of good-quality battery-powered alarm clocks, and most of the other timekeepers anyone could possibly require, just by looking around. Clocks on house current are ridiculously wasteful.

The solution for phantoms and vampires is outlets or plug strips that are switched off when not in use (such as our Smart Power Strip, see page 308) The outlets only get switched on when the appliance is actually being used or recharged. As a side benefit, this cures PMS—perpetual midnight syndrome—on your VCR, too!

If you're aware of the major energy-wasting gizmos we all take for granted, it's easy to avoid or work around them. When you consider the minor inconvenience of having to flip a wall switch before turning on the TV against the $400-$500 cost of another PV module to make up the energy waste, it all comes into perspective. A few extra switched outlets during construction looks like a very good investment. If your house is already built, then use switched plug strips.

Inverter Conclusions

Inverters allow the use of conventional, mass-produced AC appliances in renewable energy-powered homes. They have greatly simplified and improved life off the grid. Modern solid-state electronics have made inverters efficient, inexpensive, and long lived. Batteries and inverters need to live within 10 feet of each other, but not in the same enclosure. Modified sine-wave inverters will not be perfectly problem-free with every appliance. Expect a few rough edges. The best inverters produce pure sine-wave AC power, which is acceptable to all appliances. Sine-wave units tend to cost a bit more, but are strongly recommended for all full-size household systems. Keep the power-wasting phantom loads and power cube vampires under tight control, or be prepared to spend enough to overcome them.

INVERTERS AND RELATED PRODUCTS

Modified Sine-Wave Inverters

Modified sine-wave inverters represent the lower-cost end of residential inverters. It's easier to build and control this type of inverter, and they tend to be slightly more efficient than pure sine-wave inverters. There are two basic design types. High-speed electronic switchers are the most common type of modified sine-wave inverter. These are all the small to medium-sized, largely overseas-produced units. The Go Power! inverters are an example. Electronic switchers are relatively lightweight. Cost is usually a pretty fair indicator of how long you can expect them to last. To make them cheaper, manufacturers will cut corners on protection circuitry. Buyer beware! The other basic design type is transformer based. These units tend to come in larger wattages, they are much more robust, and will withstand a lot more abuse. The Xantrex DR-series is an example.

Small Mobile Inverters for Battery-Powered Gizmos

This series of incredibly handy Wagan inverters are great for recharging laptops, PDAs, cameras, cell phones, or any small AC appliance up to the wattage rating of the inverter. Plugs into any standard 12-volt lighter socket. Overload, short-circuit, and low-battery protected. Shuts off at 10 volts. Surges to about twice the rated power. Single 3-prong outlet. Weighs 6.4, 9.6, and 14.5 oz. respectively. 1 yr. mfr. warranty. China.

49-0130 75-Watt	$29
49-0131 110-Watt	$35
49-0132 150-Watt	$45

Go Power! Modified Sine-Wave Inverters

These Go Power! modified sine-wave inverters let you run standard AC appliances wherever and whenever you need. Silent, lightweight and simple to use, these inverters bring electricity to your RV, boat, car or even your home. Simply plug the item you wish to run into one of the AC receptacles, and you can run your AC load right off your batteries day or night without a genera-

tor. All six models have an on/off switch, cooling fan, thermal and overload protection, and dual AC outlets. 1 yr. mfr. warranty.

The 300-watt model is the easy solution for small power needs. Compact, lightweight, and portable! Cooling fan runs continuously.

49-0204 300-Watt, 12V	**$36**

The 600-watt model includes the same features as the 300-watt and is the perfect inverter for your middle-range power needs—powerful yet economical. Great for lights, drills and blenders. 860-watt surge.

49-0204 600-Watt, 12V	**$76**

Run your standard AC appliances wherever you travel with these silent, lightweight and simple-to-use inverters. Designed for RVs, campers, long-haul trucks, boats, cars, mobile offices, service vehicles, and emergency back-up power. The Go Power! 1,000-watt inverter has a digital display. The 1,750-watt is a heavy-duty inverter designed to run high-wattage appliances. 2,500-watt is capable of being hard wired with optional accessory kit.

49-0204 1,000-Watt, 12V	**$158**
49-0204 1,750-Watt, 12V	**$310**
49-0204 2,500-Watt, 12V	**$599**
49-0205 500-Watt with 10A Battery Charger	**$358**

Go Power! Modified Sine-Wave Specifications					
	300-Watt	**600-Watt**	**1,000-Watt**	**1,750-Watt**	**2,500-Watt**
Continuous Output	300 watts	600 watts	1,000 watts	1,500 watts	2,500 watts
Surge Output	400 watts	860 watts	1,300 watts	2,100 watts	3,600 watts
No-Load Current	0.2 amps	0.2 amps	0.2 amps	1.1 amps	77 amps
Dimensions (inches)	8.5 x 4.8 x 2	9 x 3.4 x 2	10.4 x 5.7 x 2.8	19 x 9.4 x 3.25	18.7 x 9.5 x 3.5
Weight	1.8 lb.	2.0 lb.	4.2 lb.	12.1 lb.	13½ lb.

Go Power! Install Kits

These complete, easy-to-use kits are recommended for every Go Power! inverter installation. Each kit includes a properly sized fuse and mounting block, 10-foot cables, and fasteners to protect the inverter from damage by overloads and short circuits. Go Power! Install Kits are also required by the Electrical Codes (NEC/CEC) and CSA. Choose the correct wattage and voltage for your inverter from the chart at right to determine which kit to use. Canada.

Go Power! Install Kits		
Model	12-Volt	24-Volt
GP-DC-Kit 2	1,000 W	1,000-1,800 W
GP-DC-Kit 3	1,500-1,750 W	2,000-3,000 W
GP-DC-Kit 4	2,000-2,500 W	3,100-4,000 W
GP-DC-Kit 5	2,600-3,000 W	4,100-6,000 W

49-0276 GP-DC-Kit 2	**$139**
49-0277 GP-DC-Kit 3	**$139**
49-0278 GP-DC-Kit 4	**$239**
49-0279 GP-DC-Kit 5	**$249**

The Xantrex DR-Series Inverters

All DR inverters come equipped with high-performance battery chargers as standard equipment. They are available for 12- or 24-volt systems in a variety of wattages.

DR-series inverters are conventional, high-efficiency, modified sine wave inverters. All models are Electrical Testing Lab approved to UL specifications. These inverters will protect themselves automatically from overload, high temperature, high- or low-battery voltages; all the things we've come to expect from modern inverters. You just can't hurt them externally. Fan cooling is temperature regulated; low battery cut-out is standard and is adjustable for battery bank size. All models come with a powerful three-stage adjustable battery charger and all models have a 30-amp transfer switch built in. Charge amperage is adjustable to suit your batteries or generator. Charging voltages are tailored by a front panel knob. Pick your battery type, eight common choices are given plus two equalization selections, and you've automatically picked the correct bulk and float voltage for your batteries.

The optional battery temperature probe will allow the inverter to factor all voltage set points by actual battery temperature. This is a good feature to have when your batteries live outside and go through temperature extremes. Comes with a 15-foot cord. The conduit box option allows the battery cables to be run in 2-inch conduit to meet NEC.

The DR-series is stackable in series (must be identical models) for three-wire splitable 240-volt output. Stacking is now easier and less expensive, too; just plug in the inexpensive stacking cord, bolt up the battery cable jumpers, and it's done.

All models are designed for wall mounting with conventional 16-inch stud spacing. All Xantrex inverters carry a 2-yr. mfr. warranty. USA.

Model	DR1512	DR2412	DR1524	DR2424	DR3624
Input Voltage:	10.8 to 15.5 volts DC	10.8 to 15.5 volts DC	21.6 to 31.0 volts DC	21.6 to 31.0 volts DC	21.6 to 31.0 volts DC
Output Voltage:	120 volts AC true RMS ± 5%	120 volts AC true RMS ± 5%	120 volts AC true RMS ± 5%	120 volts AC true RMS ± 5%	120 volts AC true RMS ± 5%
Wave Form:	Modified sine wave	Modified sine wave	Modified sine wave	Modified sine wave	Modified sine wave
Continuous Output:	1,500 watts	2,400 watts	1,500 watts	2,400 watts	3,600 watts
Surge Capacity:	3,200 watts	6,000 watts	4,200 watts	7,000 watts	12,000 watts
Idle Current:	0.045A @ 12V (0.54 watt)	0.045A @ 12V (0.54 watt)	0.030A @ 24V (0.72 watt)	0.030A @ 24V (0.72 watt)	0.030A @ 24V (0.72 watt)
Charging Rate:	0 to 70 amps	0 to 120 amps	0 to 35 amps	0 to 70 amps	0 to 70 amps
World Voltages Available:	yes, call for details	no	yes, call for details	yes, call for details	no
Average Efficiency:	94% at half-rated power	94% at half-rated power	94% at half-rated power	95% at half-rated power	95% at half-rated power
Recommended Fusing	200A Class T or Trace DC-175	300A Class T or Trace DC-175	110A Class T or Trace DC-175	200A Class T or Trace DC-175	300A Class T or Trace DC-250
Dimensions:	8.5" x 7.25" x 21"	8.5" x 7.25" x 21"	8.5" x 7.25" x 21"	8.5" x 7.25" x 21"	8.5" x 7.25" x 21"
Weight:	35 lb.	45 lb.	35 lb.	40 lb.	45 lb.
Item #	**27237**	**27245**	**27238**	**27242**	**27246**
Price	**$1,029**	**$1,395**	**$979**	**$1,395**	**$1,595**

Conduit Box Option

The optional conduit box allows the very heavy input wiring from the batteries to be run inside conduit to comply with the NEC. Has three (one on each side) ½", ¾", and 2" knockouts and plenty of elbow room inside for bending large wire. USA.

27-324   Conduit Box option-DR Series $69

DR Stacking Kit

The DR stacking kit includes the plug-in comm-port cable, like the SW, plus a pair of 2/0 battery cables to jumper from the primary to the secondary inverter. USA.

27-313   DR Stacking Intertie Kit $85

Battery Temperature Probe

Comes standard with the SW-series; recommended for the DR-series if your batteries are subject to temperatures below 40°F or above 90°F. Will adjust the battery charger automatically to compensate for temperature. 15-foot length, phone plug on inverter end, sticky pad on battery end. This same temp probe also works with all C-Series Trace controllers. USA.

27-319   Battery Temperature Probe, 15' $29

Remote Control for DR, UX, & TS Series Inverters

A 50-foot ON/OFF remote control with led indicator. Plugs into all Trace DR, UX, and TS series inverters. USA.

27-304   Trace RC8/50 Remote $69

Pure Sine Wave Inverters

Pure sine-wave inverters increasingly represent the future of residential inverters. They tend to be more expensive, but higher quality and longer lasting than modified sine units. Efficiency is slightly lower on pure sine-wave inverters, but there are hardly ever compatibility problems with appliances. Just like modified sine units, we have high-speed switchers in smaller sizes for lower prices, like the GoPower! or Morningstar units, and transformer-based units in larger wattages at higher prices, like the OutBack or Xantrex units.

Morningstar Pure Sine-Wave Inverter

When you need a pure, clean source of AC power, install Morningstar's new SureSine™ Pure Sine Wave Inverter. This 300-watt (12VDC input) toroidol transformer features outstanding surge capability, high efficiency and extensive electronic protections. It was designed specifically to meet the rural PV electrification needs that require AC power as well as to power small PV systems for telecom, remote cabins, weekend homes, and RV/caravans and boats. Highly reliable—having no internal cooling fan or other moving parts prone to failure—the SureSine was developed using Morningstar's power electronics expertise and 25 years of experience with remote off-grid PV systems. 8⅖"x6"x4¹⁄₁₀" (LxWxD). 10 lb. 2-yr. mfr. warranty. Taiwan.

Specs:

AC Output Voltage (RMS)	115V ± 10%
Nominal Power Rating	300 Watts
Peak Power Rating	600 Watts
DC Input Voltage	10.0V-15.5V
Self Consumption (off)	25 mA
Self Consumption (on)	450 mA
Peak Efficiency	92%

49-0218 Morningstar Pure Sine-Wave Inverter $599

Go Power! Sine-Wave Inverters

Some appliances just aren't happy with anything less than pure, clean sine-wave power. Video and audio equipment are often a problem when using modified sine-wave inverters. Here's a selection of high-quality, pure sine-wave inverters that all have dual 3-prong outlets for output, on/off switches and power-on indicator lights, and they feature automatic overload, overheat, high- or low-battery shutdown and thermostatically controlled cooling fans. Pure sine-wave inverters are perfect for appliances that demand a clean, pure source of AC power. With sine-wave power, your audio, video and electronic equipment will run without buzzing or humming, and appliances like microwaves will run more efficiently. Available in 12V or 24V. 2-yr. mfr. warranty on Go Power!; 1-yr. mfr. warranty on Exeltech.

Go Power! Sine-Wave Inverters

49-0215 150-Watt, 12 or 24V (8" x 5³⁄₁₀" x 2⁹⁄₁₀", 5.9 lb.) **$195**

49-0216 300-Watt, 12 or 24V (9³⁄₁₀" x 6¹⁄₁₀" x 2⁴⁄₅", 7.7 lb.) **$240**

49-0217 600-Watt, 12 or 24V (11³⁄₅" x 7¹⁄₁₀" x 2⁴⁄₅", 5.4 lb.) **$399**

49-0218 1,000-Watt, 12 or 24V (15¹⁄₁₀" x 7¹⁄₅" x 3½", 8.8 lb.) **$599**

49-0219 1,500-Watt, 12 or 24V (16½" x 10⁹⁄₁₀" x 4¹⁄₁₀", 17 lb.) **$950**

49-0220 2,000-Watt, 12 or 24V (16³⁄₅" x 8½" x 6³⁄₁₀", 21 lb.) **$1,320**

Exeltech Pure Sine-Wave Inverters

The purest sine wave on the market

The Exeltech 1,100-watt unit is our most reliable midsized pure sine-wave inverter. When your video, audio, or other sensitive electronic components demand clean power, Exeltech has

proven to be durable and reliable. Delivers 1,100 watts continuously, or up to 2,200 watts surge. Will automatically protect itself against overload, overheating, or battery voltage too high or low. Has a thermostatically controlled cooling fan, and runs almost silently. Features high operating efficiency (for sine-wave inverters) of 87%-90%. Dual AC outlets. Measures 7.7"x3.6"x13.8" and weighs 10 lb. Available in 12- or 24-volt versions, or 32, 36, 48, 66, or 108 volts by special order. 1-yr. warranty. USA.

27147 Exeltech XP 1100/12V Sine-Wave Inverter **$850**

27148 Exeltech XP 1100/24V Sine-Wave Inverter **$1020**

27882 Exeltech XP 1100 (specify volts) Sine-Wave Inverter **$1099**

OUTBACK FX-SERIES INVERTERS

The most advanced and adaptable inverters available

OutBack is an engineer-owned, customer-focused power equipment manufacturer. That means they make stuff that *works* and they actively support their customers. OutBack has introduced several new models and increased the power output of their incredibly clean, efficient, sine wave FX series inverters in the past year. The FX is available either in vented models for protected, indoor installations, or in sealed models for installation in marine, jungle, bug-infested, or other challenging sites. The vented models cost a bit less per watt, and are going to be the choice for most folks. But if corrosion, moisture, or critters are a problem where you live, make your life problem-free with a sealed unit.

Grid-tie units are available in both sealed and vented cases for 24- and 48-volt units. The grid-tie software package has some substantial differences, so these are being offered as separate models, at the same pricing. No additional accessories or extra-cost options are required for code-approved grid-tie with OutBack FX-series inverters.

OutBack inverters deliver a very clean sine wave power, with strong surge abilities up to 50 amps on all models. A built-in 60-amp transfer switch with automatic battery charging takes over whenever an outside power source such as a generator or utility power is available. The optional Mate allows adjustment

Fighting for Your Independence

OutBack Power Systems

of battery charging and other operations, but is not required for daily operation. Any custom programming is retained even without battery power. (Yea! No more lost programming!) Multiple inverters can be connected for 120-, 240-, or even 3-phase 208-volt AC output. Multiple inverter stacks *do* require the Mate and Hub. All OutBack inverter installations will require additional parts for mounting and overcurrent protection. Call for assistance. Inverters listed below are North American standard 120V/60Hz output. Export models with 230V/50Hz output are available by special order, same pricing. All inverters are 16.25"x8.25"x11.5" (LxWxH) and weigh 60-62 lb. USA of course. 2-yr. mfr. warranty with optional extended warranty.

In the table below, FX models are sealed, VFX are vented, a G or GT indicates grid-tie. First two digits are rated wattage, second two digits are battery voltage. A GTFX2524 is a grid-tie, sealed unit, 2,500 watts output, 24-volt input, for instance.

OUTBACK INVERTER OPTIONS

OutBack FX-Series Adapter Kit

The easy way to order what every inverter installation needs. The FX Adapter Kit contains one each of the DC Conduit Adapter and AC Adapter described below. Order one Adapter Kit for each FX-series inverter, and you're covered. USA.

49-9909 FX Adapter Kit   **$75**

OutBack FX DC Conduit Adapter

An aluminum conduit adapter required when mounting an OutBack inverter to a Power Center, or to a 2" conduit. Required for a NEC-compliant installation. USA.

27908   DC Conduit Adapter (DCA)   **$45**

OutBack FX AC Adapter

A cast-polycarbonate AC wiring-compartment extension. Provides a 2" conduit adapter, two 1" conduit knockouts, and an AC outlet knockout. Required when mounting an FX to an OutBack Power Center. USA.

27909   AC Adapter (ACA)   **$35**

OutBack Remote Temperature Sensor

The Remote Temperature Sensor lets you take full advantage of the temperature compensation feature of your OutBack inverter/chargers. The RTS will allow the system to know the precise ambient temperature of the batteries and provide optimum charging voltage. USA.

48-0021 Remote Temperature Sensor   **$29**

Model	FX2012T*	VFX2812	GTFX2524*	VFX3524 and GVFX3524	FX3048T	GTFX3048*	VFX3648 and GVFX3648
Battery Voltage:	12 volts	12 volts	24 volts	24 volts	48 volts	48 volts	48 volts
Cooling:	Sealed	Vented	Sealed	Vented	Sealed	Sealed	Vented
Continuous Output:	2,000W	2,800W	2,500W	3,500W	2,500W	3,000W	3,600W
Wave Form:	Pure sine	Pure sine	Pure sine	Pure sine	Pure sine	Pure sine	Pure sine
Surge Capacity:	4,800VA	4,800VA	4,800VA	5,000VA	4,800VA	4,800VA	5,000VA
Battery Charging Rate Max.:	80 amps	125 amps	55 amps	85 amps	35 amps	35 amps	45 amps
Recommended DC Breaker:	OBDC-250	OBDC-250	OBDC-175	OBDC-250	OBDC-100	OBDC-100	OBDC-175
Recommended DC Cable:	4/0 Ga.	4/0 Ga.	2/0 Ga.	4/0 Ga.	2/0 Ga.	2/0 Ga.	2/0 Ga.
Item #:	**49-0134**	**49-0125**	**49-0143**	**49-0145**	**49-0246**	**49-0144**	**49-0146**
Price:	**$1,995**	**$2,345**	**$1,995**	**$2,345**	**$1,995**	**$2,345**	**$2,345**

Equipped with Turbo cooling fan

Classic black Mate

Classic white Mate

Mate 2

OutBack Mate

The brains of your OutBack system

Rather than put all the complex and often upgraded software into the big, heavy inverter, the clever engineers at OutBack have put it into the lightweight, hand-size Mate. Upgrades are simple now! Single inverter systems will run just fine without the Mate, but for larger multi-inverter systems, or folks who like to fine-tune, the Mate is a necessity. The classic oval Mate is available in white or black; the new square Mate2 is black. The 4-

line, 80-character backlit display is also a highly detailed system monitor, and can be 1,000' from the system. Runs on common Ethernet cable (Cat 5E cable with RJ45 connectors). An opto-isolated RS232 port will talk to PC computers. The Mate is supplied with a 50' cable and will talk to one appliance. Hubs let the Mate communicate with multiple inverters and charge controls. Four- and 10-appliance Hubs available. 5.75"x4.25"x2" (WxHxD), 1 lb. 2-yr. mfr. warranty. USA.

25813	**OutBack Mate, White** w/50' cable	**$295**
48-0184	**OutBack Mate, Black** w/50' cable	**$295**
48-0185	**OutBack Mate2, Black w** /50' cable	**$295**
25814	**OutBack Hub-4 w/2-3' &** 2-6' cables	**$195**
48-0039	**OutBack Hub-10 w/2-3' &** 4-6' cables	**$375**

XANTREX XW INVERTER SERIES

Next-generation hybrid inverter/charger for off-grid pure sine-wave power

Xantrex introduces a new, improved version of their well-proven true sine-wave inverter technology. Available in 24- or 48-volt versions, the XW-series offers con-

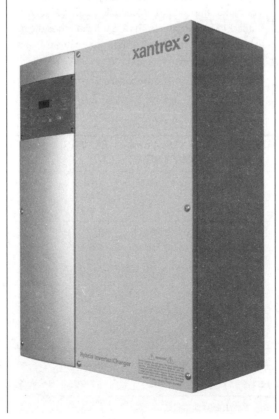

tinuous AC power output at 4.0, 4.5, and 6.0kW. The XW features an inverter/charger that incorporates a DC to AC inverter, a battery charger, and an AC auto-transfer switch. The superior integration of the XW Series design minimizes external balance-of-system components allowing for much quicker and easier installation. All XW system components communicate together over the Xanbus™ network. Common system settings are shared between devices on the network to facilitate setup.

This highly capable inverter/charger can be used for independent households, or to provide automatic back-up power when utility power fails. Peak efficiency is 95%, and search mode power is under 8 watts! The XW-series is still using the basic SW operating system that was first introduced in the mid-1990s, and is now well known and respected. One of the most welcome improvements with the XW is that all inverter programming is held in non-volatile memory so customized charge or control settings won't be lost if DC power is interrupted. The XW-series is also FCC Part B compliant, which means less potential interference with radio and telephone equipment.

Capable of being grid-intertied or off-grid, the XW can operate with generators and renewable energy sources to provide full-time or backup power. Many of the standard equipment bells and whistles from the former SW-series are now options. Options include an AC/DC conduit box for code-compliant installations. With the optional Automatic Generator Start, it will automatically start and stop any remote-start capable generator based on battery voltage and customer needs. A System Control Panel w/25' cable can be installed for remote operation, monitoring, or connection

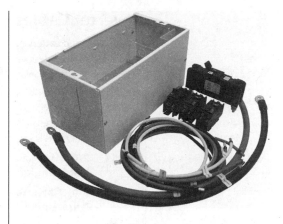
to a PC. Powder-coated steel enclosure rated for indoor use. 120/240 -volt AC split-phase operation. 23"x16"x9" (WxHxD). Certified to UL1741 and CSA for grid-tie applications. 5-yr.mfr. warranty. USA.

XANTREX XW INVERTER OPTIONS

XW Automatic Generator Start

An add-on module that plugs into the XW inverter for automatic generator control. Automatically activating generator to recharge depleted battery bank or assist with heavy loads to provide intelligent power management.

49-0518 Xantrex XW Automatic Generator Start $200

XW System Control Panel

Offers a complete LCD control panel with all controls, indicators, and display. Includes 25 feet of cable. Plugs into Xanbus network and provides a central user interface to configure and monitor all components in the system. Please specify voltage.

49-0519 Xantrex XW System Control Panel $300

XW Solar Charge Controller

60-amp solar charge controller with integrated PV ground fault protection. Accepts PV arrays with open circuit voltage up to 150 volts DC. Employs dynamic Maximum Power Point Tracking (no mini or full sweeps). Convection cooled—no fan. 5-yr. standard warranty. Xanbus network enabled.

49-0520 Xantrex XW Solar Charge Controller $650

XW Connection Kit for Second Inverter

Includes conduit box and all AC/DC breakers and wires needed to connect a second inverter into the XW Power Distribution Panel. All wires are precut and labeled to facilitate easy installation.

49-0521 Xantrex XW Connection Kit for Second Inverter $750

XW Conduit Box

Bare conduit box (no wires) for systems with more than two inverters or to retrofit XW inverters into existing systems that may already have AC/DC disconnects.

49-0515 Xantrex XW Conduit Box $175

Xantrex XW Series Comparison Chart			
Model	XW6048	XW4548	XW4024
Continuous Output:	6,000W	4,500W	4,000W
Wave Form:	Pure Sine	Pure sine	Pure sine
Surge Current:	105Arms (7 sec.)	52.5Arms (7 sec.)	75Arms (20 sec.)
DC Input Voltage Range:	44 – 64VDC	44 – 64VDC	22 – 32VDC
Charging Rate:	0 – 100A	0 – 85A	0 – 150A
Preferred Circuit Breaker:	250A	250A	250A
DC Cable Size:	4/0	4/0	4/0
Weight:	125 lb.	115 lb.	115 lb.
Item #:	49-0273	49-0273	49-0273
Price:	$4,500	$3,600	$3,250

Inverter Cables

Our inverter and battery cables are all flexible cable with color coding for polarity and shrink-wrapped lugs for ⅜" bolt studs. These are two-lug versions (lugs on both ends). USA.

15851	**2/0, 3' Inverter Cables, 2 lug, single**	**$30**
15852	**4/0, 3' Inverter Cables, 2 lug, single**	**$40**
27311	**4/0, 5' Inverter Cables, 2 lug, red or black**	**$60**
27312	**4/0, 10' Inverter Cables, 2 lug, red or black**	**$105**
27099	**2/0, 5' Inverter Cables, 2 lug, red or black**	**$50**
27098	**2/0, 10' Inverter Cables, 2 lug, red or black**	**$70**

Heavy-Duty Insulated Battery Cables

Serious, beefy, 2/0 or 4/0 gauge battery interconnect cables for systems using 1,000-watt and larger inverters. These single cables have ⅜" lug terminals on both ends and are shrink-wrapped for corrosion resistance. Pick your gauge, length, and color. USA.

15841	**4/0, 20" Black**	**$32**
15840	**4/0, 20" Red**	**$32**
15839	**4/0, 13" Black**	**$27**
15850	**2/0, 20" Black**	**$24**
15849	**2/0, 20" Red**	**$24**
15848	**2/0, 12" Black**	**$16**
15847	**2/0, 9" Black**	**$14**

30th ANNIVERSARY

Wire, Adapters, and Outlets

Wires are freeways for electrical power. If we do a poor job designing and installing our wires, we get the same results as with poorly designed roads: traffic jams, accidents, and frustration. Big, wide roads can handle more traffic with ease, but costs go up with increased size, so we're looking for a reasonable compromise. Properly choosing wire and wiring methods can be confusing, but with sufficient planning and thought, we can create safe and durable paths for energy flow.

The National Electrical Code (NEC) provides broad guidelines for safe electrical practices. Local codes may expand upon or supersede this code. It is important to use common sense when dealing with electricity, and this might be best done by acknowledging ignorance on a particular subject and requesting advice and help from experts when you are unsure of something. The Gaiam Real Goods technical staff is eager to help with particular wiring issues, although you should keep in mind that local inspectors will have the final say on what they consider the most appropriate means to the end of a properly installed electrical system. Advice from your local inspector or local electrician, who will be familiar with local code requirements and renewable energy systems, is probably your best advice.

Wire

Wire comes in a tremendous variety of styles that differ in size, number, material, and type of conductors, as well as the type and temperature rating of insulation protecting the conductor. One of the basic ideas behind all wiring codes is that current-carrying conductors must have at least two layers of protection. Permanent wiring should always be run within electrical enclosures, conduit, or inside walls. Wires are prone to all sorts of threats, including but not limited to abrasion, falling objects, gnawing critters, and children using them as a jungle gym. The plastic insulation around the metal conductor offers minimal protection, based on the assumption that the conductors will be otherwise protected. Poor installation practices that forsake the use of conduit and strain-relief fittings can lead to a breach of the insulation, which could cause a short circuit and fire or electrocution. Electricity, even in its most apparently benign manifestations, is not a force to be managed carelessly.

A particular type and gauge (thickness) of wire is rated to carry a maximum electrical current. The NEC requires that we not exceed 80% of this current rating for continuous-duty applications. With the low-voltage conditions that we run across in independent energy systems, we also need to pay careful attention to voltage drop. As voltage decreases, amperage must increase in order to perform any particular job. So lower-voltage DC systems are likely to be pushing higher amperages. This problem becomes more acute with smaller conductors over longer runs. Some voltage drop is unavoidable when moving electrical energy from one point to another, but we can limit this to reasonable levels if we choose the proper wire size. Acceptable voltage drop for 120-volt circuits can be up to 5%, but for 12- or 24-volt circuits, 2-3% is the most we want to see. We provide a good all-purpose chart and formula on page 266 for figuring wire size at any voltage drop, any distance, any voltage, and any current flow. Use it, or give us a call and we'll help you figure it out.

DC Adapters and Outlets

With the rising popularity and reliability of inverters that let folks run conventional AC appliances, demand for DC fixtures and outlets continues to diminish, which is mostly a good thing. The roots of the independent-energy movement grew from the automobile and recreational-vehicle industries, so plugs and outlets for low-voltage applications are usually based on that somewhat flimsy creation, the cigarette lighter socket. This is an unfortunate choice. The limited metal-to-metal contact in this plug configuration seriously limits the current-carrying capacity. Cigarette lighter plugs and outlets are usually rated for 15 amps of surge, but 7-8 amps continuously is absolutely the maximum you can expect without meltdown. Lighter sockets are not approved by the National Electric Code. There is no plug-and-socket combo that is approved officially for 12-volt use. A large, open receptacle accessible to small children presents an obvious safety hazard. If you must use lighter socket outlets, be sure they're fused!

The conventional rules for DC outlets and plugs says the center contact of the receptacle (female outlet) and the tip of the plug (male adapter) should be positive, and the outer shell of the receptacle and side contact of the plug

It is important to use common sense when dealing with electricity, and this might be best done by acknowledging ignorance on a particular subject and requesting advice and help from experts when you are unsure of something.

should be negative. However, the polarity can be reversed in a variety of ways during installation. For some appliances, such as incandescent or halogen lights, it doesn't make a bit of difference. But for those expensive 12-volt compact fluorescent lamps, or that costly DC television you just bought, it matters greatly. It may even be a matter of life or death. Don't assume. Check the polarity with a handheld voltmeter before plugging in.

Another convention is that the power source is presented by the female outlet. This way, it is more difficult to accidentally short out against something or electrocute yourself. Some small solar modules have DC male adapters on them, even though they are a power source, so they

can plug into the dashboard lighter socket and trickle charge the battery. These are such a small power source that they present little to no hazard. It will be difficult to get a visible spark from these trickle-charger panels.

Some people would rather not use lighter sockets and plugs in their houses because they look dangerous . . . and they can be. Little fingers and tools fit conveniently within the supposedly "protected" live socket. Probably the safest option is to use an oddball, but still locally available, AC plug-and-socket configuration that has the prongs perpendicular, slanted, or in some other configuration that makes it impossible to plug in to a standard 120-volt AC plug. These have lots of metal-to-metal contact, and

For Frey Vineyards, Harvesting Grapes and the Sun Go Hand in Hand

"Even if the payoff for the investment amounted to little over many years—viewed in plain dollars—it would still make ecological sense."
—Nathan Frey

Certain wines are known for having good years.

And that means 2006 is a very good vintage for Frey organic and biodynamic wine. It was last April that the Redwood Valley, Calif., winery installed a 17.3kW grid-tied photovoltaic system consisting of 91 panels. Now, in addition to watching their 61 acres of vines ripen, they're watching their electric bills wither on the vine.

"The array sits right next to our vineyards, and it's nice to see both leaves and solar panels soaking up the rays side by side," says Nathan Frey, who works in marketing at the family-owned and -operated Frey Vineyards.

As America's oldest and largest purely organic vineyard, Frey Vineyards has always been a lot greener. They power their tractors with biodiesel and make compost from leftover grape skins.

"We try to adopt sustainable practices whenever we can, and going solar fits right in," Frey says. "We enjoy how clean it is—just pure energy from the Sun helping run our electrical machinery."

That machinery includes a crusher, press, refrigeration system, pumps, air compressors, bottling machine, and electrical forklift. On their sunniest summer days, their solar array produces about half of their total energy needs.

Frey estimates that the system will pay for itself within 6-10 years, depending on future utility rates. But, he adds, "even if the payoff for the investment amounted to little over many years—viewed in plain dollars—it would still make ecological sense and, ultimately, economical sense, considering that during that time we will have put much less demand on conventional sources of power that damage our planet.

Nathan Frey explains the finer points of biodynamic apple and grape growing at the solar-powered Frey Vineyards.

We can't sustain our economy on a failing ecology."

We can all drink to that.

Tech specs:

Solar system size:	17.3 kilowatts
Solar panels:	91 Sanyo HIP-190BA3 190-watt modules
Inverter:	1 Xantrex PV15208 15kW-rated inverter

will be rated for at least 15 amps continuously. Higher-amperage sockets are readily available if you need them. Although it is tempting to use standard inexpensive 120-volt AC outlets, we've seen this lead to disaster too often, and strongly advise against it. Murphy's Law finds enough ways to trip us up without leaving the welcome mat out. We know from personal experience that any savings are quickly lost when a wrong-voltage appliance is plugged in and destroyed.

Do NOT mix AC and DC wiring within a single electrical box! That's a big-time Code (and commonsense) violation. It's fine to put your AC and DC outlets right next to each other—but they need to be in separate boxes and should be labeled.

Wire Sizing Chart/Formula

This chart is useful for finding the correct wire size for any voltage, length, or amperage flow in any AC or DC circuit. For most DC circuits, particularly between the PV modules and the batteries, we try to keep the voltage drop to 3% or less. There's no sense using your expensive PV wattage to heat wires. You want that power in your batteries!

Note that this formula doesn't directly yield a wire gauge size, but rather a "VDI" number, which is then compared to the nearest number in the VDI column, and then read across to the wire gauge size column.

1. Calculate the Voltage Drop Index (VDI) using the following formula:

   VDI = AMPS x FEET ÷ (% VOLT DROP x VOLTAGE)

   Amps = Watts divided by Volts

   Feet = One-way wire distance

   % Volt Drop = Percentage of voltage drop acceptable for this circuit (typically 2%-5%)

2. Determine the appropriate wire size from the chart below.

   A. Take the VDI number you just calculated and find the nearest number in the VDI column, then read to the left for AWG wire gauge size.

   B. Be sure that your circuit amperage does not exceed the figure in the Ampacity column for that wire size. (This is not usually a problem in low-voltage circuits.)

Example: Your PV array consisting of four Sharp 80-watt modules is 60 feet from your 12-volt battery. This is actual wiring distance, up pole mounts, around obstacles, etc. These modules are rated at 4.63 amps x 4 modules = 18.5 amps maximum. We'll shoot for a 3% voltage drop. So our formula looks like:

   VDI = (18.5A x 60 ft.) ÷ (3% x 12V) = 30.8

Looking at our chart, a VDI of 31 means we'd better use #2 wire in copper, or #0 wire in aluminum. Hmmm. Pretty big wire.

What if this system was 24-volt? The modules would be wired in series, so each pair of modules would produce 4.4 amps. Two pairs x 4.63 amps = 9.3 amps max.

   VDI = (9.3A x 60 ft.) ÷ (3% x 24V) = 7.8

Wow! What a difference! At 24-volt input you could wire your array with little ol' #8 copper wire.

Wire Size	Copper Wire		Aluminum Wire	
AWG	VDI	Ampacity	VDI	Ampacity
0000	99	260	62	205
000	78	225	49	175
00	62	195	39	150
0	49	170	31	135
2	31	130	20a	100
4	20	95	12	75
6	12	75	•	•
8	8	55	•	•
10	5	30	•	•
12	3	20	•	•
14	2	15	•	•
16	1	•	•	•

Chart developed by John Davey and Windy Dankoff. Used with permission.

WIRE AND ADAPTER PRODUCTS

Specialty Wire

Meter Wire

Special 18-gauge wire for meters. Has four twisted pairs, a total of eight conductors. Each twisted pair is shielded for an ultraclean signal. Poly jacketed bundle. Custom cut and sold by the foot.

26659   18 AWG 4-Pair Meter Wire $1.50/ft.

Sub Pump Cable

This flat two-conductor, 10-gauge jacketed submersible cable is required for the SHURflo Solar Sub splice kit and is sometimes difficult to find locally. Cut to length. Specify how many feet you need.

26540   10-2 Sub Pump Cable $1.50/ft.

DC Adapters

12-Volt Fused Replacement Plug

This fused plug has a unique polarity-reversing feature. It includes a 5-amp fuse and four sizes of snap-in strain reliefs, which accommodate wire gauges from 24 to 16 AWG. It is rated at 8 amps continuous duty and can be fused to 15 amps for protecting any electric device. USA.

26103   Fused Replacement Plug $3.95

12-Volt Plug Adapter

This simple, durable plastic adapter provides an elegant conversion from any standard 110-volt fixture. Simply insert your 110-volt plug into the connector end of this adapter (no cutting!), insert a 12-volt bulb into the light socket of your lamp, and you're in 12-volt heaven for cheap! USA.

26104   Plug Adapter $5

DC Power Adapter

Use a 12-volt source to run and recharge any battery-powered device using less than 12 volts DC. Our adapter delivers stable DC power at 3, 4.5, 6, 7.5, 9, or 12 volts at up to 2 amps. Voltage varies less than 0.5 volt in our tests. Perfect for your camera, DVD or MP3 player, camcorder, Game Boy®, cell phone, or whatever. The 6-foot fused input cord with LED power indicator plugs into any lighter socket, the 1-foot output cord has six adapter plugs. A storage compartment in the power adapter holds your spare plugs to prevent loss. Slide switch selects voltage. China.

53-0105 Mobile DC Power Adapter $24

12-Volt Extension Cord Receptacle

This receptacle, made of all brass parts in a break-resistant plastic housing, connects with 12-volt cigarette lighter plugs. Includes 6-inch lead wire. USA.

26108   Extension Cord Receptacle $6.95

12-Volt Extension Cord with Battery Clips

These unique cords have spring-loaded clips to attach directly to battery terminals. They increase the use and value of 12-volt products like lights, TV sets, radios, or appliances. The maximum amperage that can be put through the 10-foot cord is 4 amps. USA.

26112   Extension Cord and Clips, 10 ft. $14.95

Emergency Preparedness

Be Ready When the Grid Goes Down

YOU DON'T HAVE TO LOOK VERY HARD to find plenty of news out there that demonstrates dramatically that the world can be a hazardous place even in the best of times. Power grid failures, disastrous storms and fires, fuel price spikes, terrorist attacks: We all know in our deepest fears that our society has become vulnerable to disruption of the various infrastructures that most of us depend on in our daily lives. Who can doubt that factors as diverse as growing population pressures, grinding poverty, economic globalization, political conflict, climate change, and other natural disasters will continue to place enormous strains on the networks that supply us with food, water, power, and other necessities?

Short of living completely off the grid—several of them, really—most of us are likely to experience these kinds of disruptions and breakdowns in the foreseeable future. Won't you rest easier knowing that you've taken steps to protect yourself and your family from events beyond your control that threaten to undermine your ability to meet your basic needs? Gaiam Real Goods offers a variety of products and resources to get you up to speed on emergency preparedness.

> Won't you rest easier knowing that you've taken steps to protect or insulate yourself and your family from events beyond your control that threaten to undermine your ability to meet your basic needs?

Want Some Protection, or at Least a Little Backup?

The place to start, of course, is with electric power. For one thing, that's what we do best. But the key point is that, while not always apparent, many domestic systems, appliances, and operations that are not fueled by electricity are nevertheless dependent on it to function. When the grid goes down, you're going to lose a lot more than your lights and your computer.

The renewable energy industry has spent its maturing years learning how to provide highly reliable home energy systems under adverse and remote conditions. No utility power? No problem! No generator fuel? No problem! No

easy access to the site? No problem! Gaiam Real Goods is ready to provide off-the-shelf solutions for short-term, long-term, or lifetime energy problems. We've collected and discussed a few of them in this chapter. Like individual lifestyles, energy needs and the necessary equipment to meet them are individual matters. If you don't see what you need, give our experienced technical staff a call. We're masters at putting together custom systems to meet individual needs, climates, and sites. Reliable power for your peace of mind is our business.

Steps to a Practical Emergency Power System

The renewable energy industry has spent its maturing years learning how to provide highly reliable home energy systems under adverse and remote conditions. No utility power? No problem! No generator fuel? No problem! No easy access to the site? No problem!

Following are some practical solutions to utility disruption. We've arranged them in order of cost with the least expensive options first. Be aware that lowest-cost options will involve some sacrifices and will be acceptable solutions only for brief time periods—emergencies, in other words. As back-up systems get more robust, they impose fewer restrictions and are more pleasant to live with.

1. INSTALL A GOOD-QUALITY BACK-UP GENERATOR

A generator will be useful for short power outages, or for longer ones if you have fuel and if you, and the neighbors, are willing to endure the noise. Higher-quality generators tend to be much quieter. Propane fuel is a good choice. It's delivered in bulk to your site (in all but the

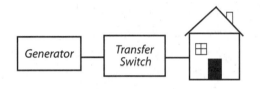

most remote locations), it doesn't deteriorate or evaporate with age, and it's piped to the generator, so you never need to handle it. Generators live longer on propane because it produces less carbon buildup. Your electrician will need to install a transfer switch between the utility line and the generator so your generator doesn't try to run the neighborhood and doesn't threaten any utility workers.

Simple back-up generators use manual starting and a manual transfer switch. When power fails, somebody has to start the generator, then throw the switch. Fancier systems will do all this automatically. With any generator-based back-up system, even automated ones, there will be some dead time between when the power goes out and when the back-up systems come online.

Most generator-based systems will require you to practice some conservation. Don't plan on running big watt-sucker appliances like air conditioning, electric room or water heaters, or electric clothes dryers unless you're willing to support a huge generator of 12,000 watts or larger. Generators run most efficiently at 50%-80% of full load. A big generator is going to suck almost as much fuel to run a few lights as it is to run all the lights and the air conditioning. Generators are not a case of "bigger is better."

2. ADD BATTERIES AND AN INVERTER/CHARGER

The generator supplies power only when it's running, and it isn't practical or economical to run it all the time. Running a load of just 200 watts is particularly expensive with a generator. They're happiest running at about 80% of full-rated wattage, which keeps carbon buildup to a minimum and is the most efficient in terms of watts delivered per quantity of fuel consumed. Adding batteries and an inverter/charger allows the best use of your generator. Whenever the generator runs, it will automatically charge the batteries. This means the generator is working harder, building up less carbon, and deliver-

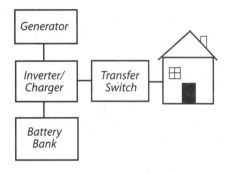

ing more energy for the fuel consumed. Energy stored in the batteries can be used later to run lights, entertainment equipment, and smaller loads without starting the generator.

Many medium- to larger-size inverters come with built-in, automatic battery chargers. So long as the inverter senses that utility or generator power is available, that line power is passed right through. If outside AC power fails, the inverter will start using stored battery power automatically to supply any AC demands. When outside AC power returns, the batteries will be recharged automatically. For a power failure that lasts more than a couple of days, this means you'll need to run the generator only for a few hours per day, yet you'll have AC power full time.

Putting an inverter and battery pack into the system means that any AC power interruptions will be measured in seconds, or less, depending on the equipment chosen.

3. ADD SOLAR, HYDRO, OR WIND POWER FOR BATTERY CHARGING

A generator has the lowest initial cost for power backup, but it's cheap only if you don't run it.

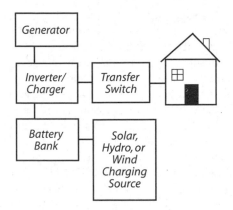

Generators are expensive and noisy to operate, and fuel supplies have to be replenished. If you anticipate power outages lasting longer than a few weeks, then solar, hydro, or wind power will save you money in the long run and deliver a reliable long-term energy supply.

Once you have the battery pack, it can accept recharging from any source. Electrons all taste the same to your battery. Whether they come from a belching, cheapo gas generator or a clean, silent PV module, the effect is the same on the battery. Renewable energy sources have a higher initial cost but very low costs over time for both your wallet and the environment.

Emergency and Back-Up Power System Design

The average American home uses around 20kWh of electricity per day. While it's certainly possible to put together an emergency power system to supply this level of use, most folks would find it prohibitively expensive. Most of us are willing to modify our lifestyle in the face of an emergency. Energy conservation is the cornerstone of reduced consumption, which will make getting by in case of disruptions that much more manageable.

What Do You Really Need in an Emergency?

Refrigeration, heat, drinking water, lights, cooking equipment, and communications are on the typical short list of what you really need in an emergency. Let's take a quick look at each of these necessities and see how they can best be supplied when disaster strikes and the grids go down.

REFRIGERATORS AND FREEZERS

Refrigeration is a major concern for most emergency or back-up systems. The fridge will probably be your major power consumer, typically using 3-5kWh of energy per day. It's going to require a 1,500-watt generator or a robust 1,000-watt inverter, at the very least, to start and run a home refrigerator or freezer. If your refrigerator is 1993 vintage or older, consider getting a new one. Refrigerators have made enormous advances in energy efficiency over the past several years. The average new 22-cubic-foot fridge will use half the power of one produced before 1993. If you're shopping, watch for the EnergyStar label and those yellow EnergyGuide tags. They really level the playing field! All models are compared to the same standard, which happens to be a pessimistic 90°F ambient temperature. Most of us will see slightly lower operating costs most of the time. Any fridge model sporting a tag for under 500kWh per year will serve you well and drop energy use to near 1.5kWh per day. The very best mass-produced fridges will

have ratings for 400kWh/year or less. Top-and-bottom units have a slight energy advantage over side-by-side units.

If you expect power to be out for more than a few weeks, then a propane-powered fridge might be a better choice. Although small by usual American standards (about 7-8 cubic feet), they'll handle the real necessities for about 1.5 gallons of propane per week. Propane freezers are also available.

HEAT

If a natural disaster happens in winter, keeping warm suddenly may become overwhelmingly important, while keeping the food from spoiling is of no concern at all. Anyone with a passive solar house, a wood stove, and a supply of firewood will be sitting pretty when the next ice storm hits. Obviously, any heating solution that doesn't involve electricity will keep you warm more effectively following a natural disaster. Beyond wood stoves, other nonelectric solutions may include gas-fired wall heaters or portable gas or kerosene heaters. Make sure you get one with a standing pilot instead of electric ignition or you'll be out of luck. Also be aware that if your thermostat is electric, it won't work in an extended power outage and you'll need to use manual controls.

Wall heaters are preferable, because they're vented to the outside. Use extreme caution with

Obviously, any heating solution that doesn't involve electricity will keep you warm more effectively following a natural disaster.

Disease carried by polluted drinking water does far more human harm than actual natural disasters.

If you have your own well or other private water supply that requires a pump for water delivery, you'll need some kind of power system that can run your water pump.

If you use compact fluorescent lamps instead of common incandescent lamps, you'll get four to five times as much light for your power expenditure.

portable heaters or any other "ventless" heater that uses room oxygen and puts its waste products into the air you're breathing! Combustion waste products combined with depleted oxygen levels are a dangerous and potentially deadly combination—neither brain damage nor freezing are acceptable choices. Those in earthquake country shouldn't count on a natural gas supply to continue following a large earthquake.

If a central furnace or boiler is your only heat choice, you'll need some electricity to run it. If your generator- or inverter/battery-based emergency power system is already sized to run a fridge, it probably will be able to pick up the heating chores instead of the fridge during cold weather. (Put the food outside; Mother Nature will keep it cold.) Furnaces, which use an air blower to distribute heat, will generally use a bit more power than the average refrigerator. Boilers, which use a water pump or pumps to distribute heat, will generally use a bit less power than a fridge. Depending on the capabilities of your emergency power system, you might need to limit other uses while the heater is running.

Pellet stoves are a very poor choice for emergency heating. Although the fuel is compact and stores nicely, pellet stoves require a steady supply of power in order to operate. These machines have blowers for combustion air, other blowers for circulation of room air, and pellet-feed motors.

WATER

If you're hooked up to a city-supplied water system, you may have a problem that an emergency power system can't solve. Most cities use a water tower or other elevated storage to provide water pressure. They can continue to supply water for only a few days, or maybe just a few hours, when the power fails. Other than filling the bathtub or draining the water heater tank, which can get you by for a few days, you may want to provide yourself with some emergency water storage. Water storage tanks are available from any farm supply store and some larger building and home supply stores. There are "utility" and "drinking water" grade tanks. Make sure you buy the drinking water grade with an FDA-approved nonleaching coating.

If you have your own well or other private water supply that requires a pump for water delivery, you'll need some kind of power system that can run your water pump. AC water pumps require a large surge current to start, the size of the surge depending on pump type and

horsepower. Submersible pumps require more starting power than surface-mounted pumps of similar horsepower.

For short-term backup, the easiest and cheapest solution is to use a generator to run the pump once or twice a day to fill a storage or large pressure tank. While the generator is running, you can also catch a little battery charging for later use to run more-modest power needs that can be supplied by an inverter/battery system.

Longer-term solutions will require either a bigger inverter and battery bank to run the water pump on demand, or a smaller pump that can run more easily on solar or battery power. Our technical staff will by happy to discuss your situation and make the best recommendations for your needs and budget.

WATER QUALITY (IS IT SAFE TO DRINK?)

Natural disasters often leave municipal water sources polluted and unsafe to drink, sometimes for weeks. So even though you may have water, you'd better think twice before drinking it. Disease carried by polluted drinking water does far more human harm than natural disasters themselves.

With just a little preparation, and at surprisingly low cost, you can be ready to treat your drinking water. And considering how truly miserable, long lasting, and even life threatening many water-borne diseases are, an ounce of prevention is worth many tons of cure. You can boil, which is time and energy intensive; you can treat with iodine, which is cheap and effective but distasteful; or you can filter, which is quick and effective with the right equipment. The "right equipment" usually means either ultrafine ceramic filtration or a reverse osmosis system. We offer several models of both types in the product section that follows.

LIGHTING

If you use compact fluorescent lamps (CFLs) instead of common incandescent ones, you'll get four to five times as much light for your power expenditure. Because power is valuable and expensive in an emergency or back-up situation, these efficient lights are a must. If CFLs happen to be part of your normal household lighting anyway—and they should be, if you're concerned about global warming and saving money—you'll enjoy some energy savings while waiting for The Big One. Since CFLs will outlast 10-13 normal incandescents and will use 75% less energy, they'll end up paying for themselves by putting hundreds of dollars back in your pocket over their lifetime. CFLs are a great investment, both for everyday living and for energy-efficient emergency backup.

Smaller battery-powered, solar-recharged lanterns and flashlights can be extremely helpful during short-term outages and can easily be carried to wherever light is needed. We offer several quality models that can be charged with their onboard solar cells, or simply have fresh batteries slipped in when time and conditions won't allow solar recharging.

COOKING

Cooking is one of the easier problems to tackle in a power failure. Let's start by looking at what you're using to cook and bake with now.

Electric Stoves and Ovens. These use massive amounts of electricity. It's not going to be practical to run your electric burners or oven with a back-up system, or even with a generator. Break out the camping gear! Camp stoves running on white gas or propane are a good alternative and widely available at modest cost. Charcoal barbecues can be used as well, but *only outdoors!* They produce carbon monoxide, which is odorless and deadly if allowed to accumulate indoors. Want a longer-term solution? Consider a gas stove.

Gas Stoves and Ovens. These can be fired by either propane or natural gas. Propane, which depends only on a small, locally installed tank, may be more dependable in a major emergency, particularly an earthquake. Major earthquakes break buried natural gas lines, though these systems won't be affected by electrical outages. In either case, a power failure will put your modern spark-ignition burner lighters out of commission. Just use a match or camping-type stove igniter for the burners. Older stoves use a pilot light for ignition, which will work as normal with or without electricity.

Your gas oven probably won't work without electricity. Older stoves, using a standing pilot light, will be unaffected; all others will use either a "glow bulb" or a "spark igniter." Spark igniters, like those used on stovetop burners, will allow you to light the oven with a match. This is good. On the other hand, glow bulbs require 200-300 watts of electrical power all the time the oven is on and will not allow match lighting. This is a problem. You can tell which type of igniter you have by opening the bottom broiler door and then turning on the oven. You'll hear or see the spark igniter, and the unit will light up quickly.

Glow bulbs take 30 seconds or longer to light up, and you'll probably see the glowing orange-red bulb once it gets warmed up.

Solar Ovens. If the weather allows, solar ovens do a great job at zero operational cost. They do require full sunlight and are excellent for anything, including a cup of rice, a batch of muffins, or a pot full of stew. We offer several ovens, kits, and plans in the product section.

COMMUNICATIONS

For simple communication needs, we offer several models of radios with wind-up and/or solar power sources. Then it gets more complicated.

The Internet is increasingly becoming a basic mode of communication, and that means some minimum power use for a computer. If you have the choice, laptop computers use a fraction of the power that desktop units require. The average desktop with 17-inch color monitor will use approximately 120 watts; the average laptop with color-matrix screen uses only 20-30 watts. If you plan to spend more than a couple hours per day on the computer without grid support, then a good laptop is going to pay for itself in energy system savings. Laptops, and cell phones as well, require frequent recharging. These kinds of loads are easily manageable with a small generator back-up system or a renewable energy system. Most other personal communications devices, including cell phones and CB, ham, and two-way radios, are designed for, or easily adapted to, 12-volt DC power use. This means the energy in your battery storage can be utilized directly. And if you want to be totally free of the telephone grid, DirecWay or Star-Band two-way satellite devices provide faster-than-DSL Internet service for net surfing.

Conclusion

An ancient Chinese curse says, "May you live in interesting times." Like it or not, we have been thoroughly cursed with interesting times. We can make the best of it with some modest preparation, or we can dally, do nothing, and wait to see what gets dealt to us and how much it hurts. In the product section that follows, we've put together a few items and kits that can ease the pain a bit, or even a whole lot. Choose the level of protection that feels comfortable to you. After you're prepared, you can sit back feeling secure and smug, knowing that you're independent from the grid—at least for a while. And if you're really ambitious, talk to our technicians about a full-time off-the-grid system.

It's not going to be practical to run your electric burners or oven with a back-up system, or even with a generator. Break out the camping gear!

If the weather allows, solar ovens do a great job at zero operational cost.

EMERGENCY PREPAREDNESS PRODUCTS

Preparedness Now!

By Aton Edwards, Executive Director of the NYC-based nonprofit organization International Preparedness Network (IPN). IPN has worked with the Red Cross, Center for Disease Control, New York City Police Department, and other organizations to train thousands domestically and overseas to prevent and respond to emergencies and disasters. Preparedness Now! provides information and techniques that can help mitigate the destructive effects of disasters, whatever the cause. With illustrations, photographs, and step-by-step instructions, this manual delivers practical advice. 336 pages, paperback.

21-0526 Preparedness Now!　　　　**$14.95**

Multi-Function Survival Tool

Shovel, hammer, saw, hatchet, bottle opener, nail puller, and wrench—all in one. Unscrew the calibrated compass and find matches, fishing hooks and line, nails, and more in the waterproof bag stashed in the handle. Comes with a protective snap-on carrying case. No car trunk should be without one. 12"x4½" (LxW). China.

14-0194 Multi-Function Survival Tool　　**$20**

Survival Kit in a Sardine Can

This lightweight, waterproof tin puts an entire survival kit in your pocket. We still can't believe all it holds: compass, whistle, matches, first-aid instructions, razor blade, pencil, nonaspirin pain reliever, fire starter cube, adhesive bandage, energy nugget, reflective signal surface, fish hook and line, duct tape, wire clip, note paper, tea, sugar, salt, gum, watertight bag, safety pin, antibiotic ointment, and alcohol prep pad. All this gear is stashed in the smallest possible space to keep bikers, hikers, campers, boaters, and adventurers prepared. And if that isn't enough, it even floats. 3"x4¼"x⅞". USA.

14-0399 Survival Kit in Can　　　　**$14**

The Extractor

In the backcountry or the backyard, insect and snake bites range from unpleasant to life threatening. The Extractor treats them all. A simple plunger push generates 11 psi of powerful suction to instantly pull toxic venom of all kinds out of the body. Pain and danger are alleviated without scalpel cuts or delay. Pocket case includes four cups for different bite sizes, safety razor to shave body hair if needed, alcohol and pain-relief pads, bandages, and a mini-guidebook for treating wounds, including those from snakes, bees, wasps, hornets, yellow jackets, fire ants, ticks, scorpions, mosquitoes, spiders, flies, and marine life. USA/China.

14-0401 The Extractor　　　　**$19**

Marie's Poison Oak Soap

This truly amazing product prevents poison oak or poison ivy outbreaks even after exposure and helps to prevent and heal many types of common rashes when used on a daily basis. Even after an irritation has started, healing is greatly accelerated. The soap's natural blend of oils, herbs, extracts, clay, oatmeal, and glycerin pulls poison oak and poison ivy oils off the skin, promotes healing, and helps stop itching and redness due to the soothing antihistamine properties of the herbs. The Oregon Highway Department buys more than 2,000 bars every year for its employees. Contains no dyes, artificial ingredients, or scents. USA.

08-0346 Marie's Soap, 3 bars　　　　**$15**
　　　　　Marie's Soap, 6 bars　　　　**$28**

EMERGENCY LIGHTING PRODUCTS

Flashlights

Shake Light

This is an incredibly innovative flashlight that never wears out or runs down. Just shake the light to charge it up. Ten seconds of shaking delivers about two minutes of light. The internal magnetic generator turns the shaking motion into electricity, which is stored in a large capacitor to run the bright LED lamp. Clear construction lets you see the internal workings. This is one flashlight that will never let you down. Great for emergencies and car trunks. Taiwan.

17-0326 Shake Light $29

Solar Flashlight

As little as four hours of direct Sun on this flashlight's built-in solar panel provides all the power its three ultrabright lifetime LEDs need for shining more than eight hours. The shock- and water-resistant housing has a momentary on/off function that can be used instead of the steady-on setting whenever you need to preserve your night vision. Built-in battery lasts 500 cycles. 7"x1" (L x diameter). China.

06-0624 Solar Flashlight $19

Xenon Flashlights

Designed for extreme conditions where intense, failsafe illumination is vital, these "tactical" flashlights are the brightest you can carry. (Remember that the next time you have to change a tire at night.) A high-power, focusable xenon beam combines with a shock-resistant aluminum case and a sealed O-ring design for 100% waterproof performance. Includes two CR123A batteries. Runs for about 1¾ hours per battery set. Bulbs last 50 hours. Small has twist on/off operation. 4.4"x1.2" (L x diameter). Large features power switch and a built-in compass. 5.1"x1" (L x diameter). China.

06-0647 Small Xenon Flashlight $16

06-0646 Large Xenon Flashlight $20

06-9648 CR123A Batteries, set of 2 $10

06-0649 Xenon Bulb $5

Dynamo Flashlight

The hand-cranked Dynamo Flashlight produces a steady beam of light with just an easy squeeze of the handle. Translucent casing holds an assortment of colored gears to captivate and intrigue young minds. Includes a shockproof and shatterproof lens, extra replacement bulb, and peg to attach a cord. Buy preassembled or as a great build-it-yourself gift that combines science and utility. 5"x3"x3½" (LxWxH). Canada.

06-9402 Preassembled Dynamo Flashlight $15

06-0456 Make-Your-Own Dynamo Kit $19
[pick-up image]

TerraLux LED Flashlight Conversion

Our astonishing LED conversion for common Mini Maglite® 2AA flashlights delivers two to three times more bright blue-white light, runs two to four times longer on a set of batteries, and has a lamp that will likely outlive you, with up to 100,000 hours' life expectancy. Our kit supplies the MicroStar2 lamp and a custom reflector. You supply the MiniMag PowerPush™ technology, which keeps the bulb at full-light output even with dropping battery voltage. When batteries are exhausted, it shifts to emergency Moon mode for reduced, but continuing, light output. USA.

06-0524 MicroStar2 Flashlight Kit $25

The MicroStar2 kit is not made or endorsed by Mag Instruments, Inc.

Lanterns and Spotlights

Solar Lantern

Keep the sunshine working for you long after darkness has fallen with this great-to-take- anywhere solar-powered lantern. It's made of durable materials, including an acrylic housing, so you'll never arrive at the campsite only to discover a broken lantern. The built-in solar panel is fully adjustable to make best use of the sunshine for charging and will give off 3½ hours of light on a full charge. Choose between two modes: bright lantern light or front-mounted flashlight. The lantern mode offers 360° of adjustable-angle fluorescent light. A DC adapter is included for car or other 12-volt charging. 9"x4½"x4" (HxWxD). China.

14-0354 Solar Lantern $59

Freeplay Indigo Table Lantern

Keep this versatile hand-powered lantern in your car, tent, or home, and it won't matter how far off the grid you get. Just one minute of cranking provides three hours of low-level illumination or six minutes of intense 360° white light from the dimmer-controlled cluster of permanent LEDs. Or use the built-in, single-LED side lamp for directional light. Industrial-grade components stand up to rugged use, and there's even an adapter for AC charging. 8"x3½" (H x diameter). China.

14-0431 Freeplay Indigo Table Lantern $69

LED Lantern

Whether you pack it on trips or keep it for emergencies, this new member of our lantern family puts 12 lifetime LEDs on a dimmer switch that lets you adjust its output from a gentle glow to blazing white light. Weather-resistant construction makes it the one to have when the going gets wet. Runs for an incredible 40 hours on a single battery set. Uses four D batteries (not included). 10.8"x6" (H x diameter). China.

14-0444 LED Lantern $49

EverLite Solar Spotlight

For camping, backyard, or emergency use, our high-tech all-weather solar spotlight runs for more than 24 hours per charge. Just six hours of full Sun delivers 12 hours of ultrabright LED run time from the built-in NiMH battery pack. Even 6 hours of solid overcast will deliver 3-4 hours of light. The bright blue-white light has eight LEDs, delivering a total of 50 lumens. The lightweight lamp and base stows on the back of the solar panel along with the 10-foot cord. The lamp assembly can be unplugged for full portability. If the lamp arm is left in the ON position (up), our spotlight will automatically turn on at dusk and off at dawn. Comes with a convenient nylon mesh bag that allows daytime charging right through the bag. Options include plug-in chargers for 12 or 120 volts that allow for quick recharging where the Sun doesn't shine, and a nifty 12-volt DC output cable to deliver power from the lamp base to your cell phone or other small 12-volt appliance. 7.08"x6.6"x2.5". 1.3 lb. total, 0.5 lb. for LED assembly only. 2.7 lb. China.

14-0299 EverLite Solar Spotlight with Carrying Case $119

17-0290 EverLite 12VDC Charger $12

17-0289 EverLite 120VAC Charger $12

17-0291 EverLite 12VDC Output Adapter $25

Mini Everlite

Our Mini Everlite gets its energy from the Sun via a portable solar panel, but this model is so tiny you can slip it into your pocket or pack without a thought. Three high-intensity LEDs deliver 20 lumens of light for 20 hours on a six-hour, full-Sun charge (4-5 hours on a 6-hour solid overcast charge). A light-sensing auto-on/off function—with optional manual shutoff—keeps it shining brightly from dusk to dawn if needed, while protecting against accidental daytime operation. Includes Velcro® pads to attach it anywhere from a tent roof to a car dash or hat. Panel is 4.7"x2.7"x0.5" (HxWxD). Lamp is 2.9"x2.25"x1.4" (HxWxD). China.

14-0376 Mini Everlite $59

Water Collection

Rainwater Collection for the Mechanically Challenged

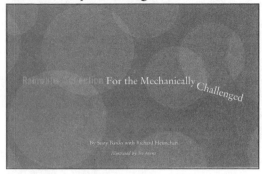

Revised and expanded in 2003. Laugh your way to a home-built rainwater collection and storage system. This delightful paperback is not only the best book on the subject of rainwater collection we've ever found, it's funny enough for recreational reading and comprehensive enough to lead a rank amateur painlessly through the process. Technical information is presented in layman's terms and accompanied with plenty of illustrations and witty cartoons. Topics include types of storage tanks, siting, how to collect and filter water, water purification, plumbing, sizing system components, freezeproofing, and wiring. Includes a resources list and a small catalog. 108 pages, softcover.

80-704   Rainwater Challenged Book $20

Rainwater Collection for the Mechanically Challenged Video

You say you're *really* mechanically challenged and want more than a few pictures? Here's your salvation. From the same irreverent, fun-loving crew that wrote the book above. See how all the pieces actually go together as they assemble a typical rainwater collection system and discuss your options. This is as close as you can get to having someone else put your system together. 37 minutes and lots of laughs.

90-0168 Rainwater Challenged DVD $19.95

Water Storage

This definitive work on water storage includes all you need to know to design, build, and maintain water tanks and ponds and to sustainably manage groundwater storage. It also delves into fire protection and disaster preparedness and includes the best building instructions we've ever seen for making your own ferrocement storage tanks. Like Art Ludwig's other books, the value-density of this book is exceptional. As the National Drinking Water Clearinghouse put it, "On the average water system, this book will pay for itself a hundred times over in errors avoided and maintenance savings." 125 pages, softcover.

21-0374 Water Storage $20

Rain Barrel and Diverter

Capture irrigation water at no cost with our virtually indestructible, 0.1875-inch-thick, recycled food-grade polyethylene Rain Barrel. Multiple barrels can be linked with an ordinary garden hose. Features overflow fitting, drain plug, screw-on cover, and threaded spigot. 39"x24" (H x diameter). 60-gallon capacity. 20 lb. Optional galvanized steel Rain Diverter (shown) fits any downspout (metal or plastic). When full, flip the diverter to a closed position to let downspout function as usual. USA.

14-9201 Rain Barrel and Diverter $149 ($50)

14-0238 Rain Barrel only $134 ($50)

46209   Diverter only $22

Allow 4-6 weeks for delivery. Can be shipped to customers in the contiguous U.S. only. Cannot be shipped to P.O. boxes. Sorry, express delivery not available.

Kolaps-a-Tank

These handy and durable nylon tanks fold into a small package or expand into a very large storage tank. They are approved for drinking water, withstand temperatures to 140°F, and fit into the beds of several truck sizes. They will hold up under very rugged conditions, are self-supported, and can be tied down with D-rings. All tanks have a 1.5-inch plastic valve with standard plumbing threads for input/output, and a 6-inch-diameter by 2-foot-long filler sleeve at the top for filling. Our most popular size is the 525-gallon model, which fits into a full-size long-bed (5'x8') pickup truck. USA.

47-401	75-gal. tank (40"x50"x12")	$369
47-402	275-gal. tank (80"x73"x16")	$539
47403	525-gal. tank (65"x98"x18")	$649
47404	800-gal. tank (6'x10'x2')	$749
47405	1,140-gal. tank (7'x12'x2')	$1,029
47406	1,340-gal. tank (7'x14'x2')	$1,159

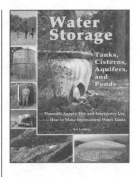

Water Purification

Handcrafted Terra-Cotta Crock

Now you can keep cool filtered water right on your countertop with our handcrafted Water Crock. Simply pour in a gallon of tap water, then let the natural force of gravity go to work. Water passes through a ceramic housing that contains a granular activated-charcoal filter made from coconut shells and a KDF media, then drips into the lower chamber where it's ready for drinking. Removes 95% of chlorine, pesticides, iron, aluminum, and lead and 99% of cryptosporidium, giardia, and sediment. Filter lasts six months or 300 gallons. 17"x8" (H x diameter). 13 lb. Brazil.

01-0465 Water Crock   **$169**

01-0466 Replacement Filter   **$59**
Not available in California

Nonelectric Distiller

An economical, high-output stovetop water distiller that operates on a variety of heat sources for daily and emergency use. Distillate capacity, based on a 2,600-watt electric burner: 3.2 quarts in 1.2 hours; up to 16 gallons per day. Stainless steel with no moving parts or fan; digital timer with alarm; 12"x12" (H x diameter). 9 lb. 3-yr. limited warranty. China/USA.

How It Works:
- The stainless steel boiler kills microbes in water by heating to 212°F (100°C).
- Steam rises, leaving behind dead microbes, dissolved solids, salts, heavy metals, and other substances.
- Low-boiling light gases are discharged through the gaseous vent.
- Steam is condensed in the three-level stainless steel tray.
- Purified/distilled water is captured in the collection cup assembly.
- The 100% steam-distilled water flows through the collection tube into the 3-liter stainless steel storage container.

01-0481 Nonelectric Distiller   **$369**

The Katadyn Drip Filter

With no moving parts to break down, superior filtration, and a phenomenal filter life, there is simply no safer choice for potentially pathogen-contaminated water. There are no better filters than Katadyn for removing bacteria, parasites, and cysts. Three 0.2-micron ceramic filters process one gallon per hour. Clean filters by brushing the surface. Ideal for remote homes, RVs, campsites, and home emergency use. Food-grade plastic canisters stack to 25"x11" (H x diameter). 10 lb. 1-yr mfr. warranty. Switzerland.

55-0083 Katadyn Drip Filter   **$289**

55-0082 Replacement Filter (needs 3)   **$75 ea.**

Base Camp Water Filter

Ideal for emergencies and backcountry trips alike, this portable water filter serves whole families with ease. Just fill its 2.6-gallon water bag, hang it anywhere, and watch as a pumpless gravity-feed design effortlessly produces 16 ounces of safe drinking water per minute. A state-of-the-art glass fiber/carbon combination filter removes odors, tastes, and impurities down to 0.3 microns in size to make virtually any water source potable. Includes a 48-inch outlet hose with valve. Filters last for approximately 200 gallons. Switzerland.

01-0517 Water Filter   **$79**

01-0518 Replacement Filter   **$39**

Filtering Water Bottles

These polycarbonate, flip-top water bottles have an advanced fast-flow in-line filter so you can do more yet pack less. Our Filtering Bottle exceeds EPA standards to remove 99.9999% of all bacteria and cysts like giardia and cryptosporidium for 2,000 fill-ups. Our Purifying Bottle removes 99.9997% of all viruses as well, lasts for 12,000 fill-ups, and includes an adapter for use with a tap. Just fill and sip normally. It's that simple and that safe. In-line filters can also be used with CamelBak or gravity bags. 32-oz. capacity with filter. USA/China.

14-0407	Filtering Water Bottle	$49
14-0408	Purifying Water Bottle	$129

Stainless Steel Bottles

These containers won't impart unpleasant tastes and are nonleaching. Stainless steel with a white plastic cap, 10"x3½" (H x diameter). China.

14-0436 40 oz. Stainless Steel Bottle $29

Water Heating and Washing

4-Gallon Solar Shower

No more cold showers when you're camping. This low-tech invention uses solar energy to heat water for all your washing needs. The large 4-gallon capacity provides ample hot water for at least two hot showers. On a 70°F day, the Solar Shower will heat 60°F water to 108°F in only three hours. Great for camping, car trips, or emergency use. This has been one of our catalog's all-time bestsellers. Taiwan.

17-0169 Solar Shower $19

James Washer

The James hand-washing machine is made of high-grade stainless steel with a galvanized lid. A pendulum agitator swings on nylon bearings, sweeps in an arc around the bottom of the tub, and prevents clothes from lodging in the corner or floating on the surface. This ensures that hot suds are mixed thoroughly with the clothes. The James is sturdily built. The corners are electrically spot-welded. All moving parts slide on nylon surfaces, reducing wear. The faucet at the bottom permits easy drainage. Capacity is about 17 gallons. 30 lb. USA.

63411 James Washer $449
Hand-constructed to order. Please allow up to 12-16 weeks for delivery.

Hand Wringer

The hand wringer will remove 90% of the water, while automatic washers remove only 45%. It has a rustproof, all-steel frame and a very strong handle. Hard maple bearings never need oil. Pressure is balanced over the entire length of the roller by a single adjustable screw. We've sold these wringers without a problem for more than 17 years. USA.

63412 Deluxe Hand Wringer $169

50 Feet of Drying Space

This is the most versatile clothes dryer we've seen yet. Don Reese and his family are still making them by hand, from 100% sustainably harvested New England pine and birch. Don's sturdy dryer can be positioned to create either a peaked top or a flat top ideal for sweaters. A generous 50 feet of drying space—and you can fold it up to carry back in the house (or to the next sunny spot) still loaded with laundry. Measures 53"x30"x33" (HxWxD) in peaked position, 46"x30"x40" (HxWxD) in flat-top position, and holds an average washer load. Folds to 58"x4½" (L x diameter). USA. *Requires FedEx ground delivery.*

10-8002 Wooden Dryer $89.

(See Chapter 11 for more laundry products . . .)

Solar Cooking Products

Hand-Cranked Blender

The human-powered Vortex blender is perfect for picnics, tailgate parties, and camping. It features a 48-ounce (1.5-liter) graduated Lexan pitcher with an O-ring sealing top, a removable pour spout that's a 1-ounce shot glass, a brushed stainless steel finish, soft stable rubber feet, an ergonomic handle, two-speed operation, and a C-clamp for stable operation. The base fits inside the pitcher for easy packing. High-speed operation is noisy so your camping neighbors will know you're making margaritas! 4.5 lb. China.

63867   Hand-Cranked Blender $88

Sun Oven

A great portable solar cooker weighing only 21 pounds, our Sun Oven is ruggedly built with a strong, insulated fiberglass case and tempered-glass door. The reflector folds up and secures for easy portability on picnics, etc. It's completely adjustable and comes with a built-in thermometer. The interior oven dimensions are 9"x14"x14" (HxWxD) and temperatures range from 350°F-400°F. This is a very easy oven to use and it will cook most anything! Includes a handy 16-page recipe booklet. USA.

63421   Sun Oven $249

HotPot Simple Solar Cooker

This is the most convenient and portable solar cooker we sell. It heats up fast and can produce cooking temperatures within an hour and be used for up to six hours, perfect for baked goods or slow-cooking stews. The HotPot's features include a 5.3-quart enameled steel pot. Respected institutions such as Florida Solar Energy Center and Energy Laboratories, Inc. tested and refined the product's design over a six-year period. Sales of the HotPot help subsidize the distribution of this device in developing countries to reduce deforestation and respiratory diseases caused by traditional cooking methods. 12½"x12½"x15" (HxWxD). USA/Mexico.

01-0478 HotPot Solar Cooker $99

Sport Solar Oven

We love the look on friends' faces when we present fresh bread and other home-baked treats while camping. So will you when you take our lightweight solar oven on your next trip. It concentrates the renewable heat of the Sun to effortlessly (and deliciously!) roast meats, steam vegetables, bake breads and cookies, and prepare rice, soups, and stews. Ideal for backyard use, camping, boating, and picnicking. Requires only minimal Sun aiming to cook most foods in two to four hours. Our complete kit includes two pots, oven thermometer, water pasteurization indicator, and recipe book. 12"x27"x16" (HxWxD). 11 lb.

01-0454 Sport Solar Oven $150

Cooking with the Sun:
How to Build and Use Solar Cookers

By Beth and Dan Halacy. Solar cookers don't pollute, they use no wood or any other fuel resource, they operate for free, and they run when the power is out. And a simple and highly effective solar oven can be built for less than $20. The first half of this book, generously filled with pictures and drawings, presents detailed instructions and plans for building a solar oven that will reach 400°F or a solar hot plate that will reach 600°F. The second half has 100 tested solar recipes. These simple-to-prepare dishes range from everyday Solar Stew and Texas Biscuits to exotica like Enchilada Casserole. 116 pages, softcover.

21-0273 Cooking with the Sun $9.95

(See Chapter 11 for more cooking products . . .)

Communications Products

Solar Radio

A built-in solar panel, hand crank, and AC adapter give you charging versatility with our Solar Radio.

Only 60 seconds of cranking powers the internal NiMH battery for 15 minutes of AM and FM. Runs 12 hours on a full charge and has an LED charge indicator. Features a see-through body so you can watch it work, an LED flashlight, and an earphone socket. 4½"x7¼"x2" (HxWxD). 3.5 lb. China.

17-0330 Solar Radio $50

Freeplay Solar-Powered Summit Radio

For a trip to the beach or a global trek, this stylish, compact radio works anywhere in the world! Covers four radio bands: FM, AM, SW, and LW for local and world radio. It's powered by a rechargeable NiMH battery pack. Charge with the built-in solar panel, with the built-in dynamo crank (30 seconds of cranking gives about 30 minutes of listening), or with the included AC/DC power adapters. An indicator light shows optimum charging. Earphone socket included for private listening, or use the built-in speaker. Thirty station presets are available: 10 each for AM and FM, five each for SW and LW, with precise digital tuning. Includes an international shortwave guide. 6.8"x3.1"x3.5" (LxWxH). 1.3 lb. 2-yr. mfr. warranty. Designed in UK, made in China.

53-0101 Freeplay Summit Radio $99

World's Smallest Weatherband Radio

Up-to-the-minute weather reports aren't just a convenience, they're often a necessity. This tiny weatherband radio puts always-on NOAA forecasts and warnings in a rugged palm-size case so you can take the news you need anywhere and access it during emergencies, vacations, day trips, and everyday life. Includes AM/FM bands and built-in 180-degree swiveling LED mini-lamp. Two AA batteries not included. 3"x4"x1" (HxWxD). China.

53-0120 Weatherband Radio $19

SideWinder™ Cell Phone Charger

The SideWinder is the world's smallest, lightest, and most powerful portable cell phone charger ever made. Two minutes of cranking delivers five to six minutes of talk time, or up to 30 minutes standby. Our universal kit has cords and connectors that fit more than 95% of phones, including Motorola V series, Nokia, Samsung/Kyocera, Ericsson, and Audiovox. Also has a bright-white LED light that provides more than five minutes of light, with only 30 seconds of cranking. Includes zippered cordura bag. 2½"x1¾"x1¼" (HxWxD). 2.5 oz. USA/China.

17-0327 SideWinder Charger $29

Universal Solar Charger

It may be a wired world, but you don't have to be tethered to it! Our compact Universal Solar Charger is the lightest, fastest solar charger on the market because it delivers a straight charge to your iPod®; BlackBerry; or Nokia, Samsung, Sony, Ericsson, Siemens, or Motorola cell phone. It eliminates the need for internal battery storage. Fold it open, connect the mini-USB cord with the proper adapter tip (included), and you'll have a full charge in two or three hours of direct Sun. Maximum output is 6.58 volts-320mA in full Sun. 5½"x3¼"x½" (LxWxD). 6"W open. 3 oz. Netherlands.

17-0336 Universal Solar Charger $119

Battery Chargers and Power Supplies

Solio Portable Solar Charger

Meet the next generation of solar chargers and our new top pick for portable power. Though its solar cells are rated at just 1 watt, the new Solio's high-capacity battery and onboard firmware provide a variable output as high as 8 watts—nearly double the needs of most gear. Three solar blades retract neatly to keep it pocket-size, and it holds a charge for up to a year. Includes Nokia, Samsung, Motorola, and Mini USB tips (many others available); cable; AC adapter with international plug set; 12-volt female adapter; and suction cup. Optional urban-hip case made of recycled innertubes. 1"x4¾"x2½" (HxWxD). China.

17-0341	**Solio Portable Solar Charger**	**$109**
17-0343	**Recycled Tube Solio Case**	**$29**

Rechargeable Batteries with More Energy than Throwaway Batteries!

Our tests show that name-brand alkaline AA batteries deliver approximately 1,800-2,100mAh—then you throw them away. Our AA NiMH cells deliver 2,200mAh and will do that about 500 times before you throw them away. All without the voltage fade of standard alkaline cells!

Fast Charger and NiMH Batteries

Nickel-metal hydride batteries just keep getting better and better. Ours excel at hard use and deliver steady voltage with no sag all the way down to 10% of charge. This makes them a better choice for heavy-discharge uses like digital cameras. There is no memory effect

from short cycling and no toxic content. Because they self-discharge over three to six months, they aren't a good choice for emergency lights. Our speedy, affordable battery charger comes with both AC and DC power cords, charges AA- or AAA-size NiMH or NiCad batteries in pairs. Charges most batteries in four to five hours or less. Taiwan.

17-9175	**C 2-pack (3,500mAh)**	**$11**
17-9176	**D 2-pack (7,000mAh)**	**$20**
17-0284	**9V single (160mAh)**	**$9**
17-0285	**9V 2-pack (160mAh)**	**$16**

more next column

AA (2,200mAh):

17-0324 4-pack	**Reg. $14 SALE $12**
17-9324 8-pack	**Reg. $24 SALE $20**

AAA (750mAh):

17-0325 4-pack	**Reg. $14 SALE $12**
17-9325 8-pack	**Reg. $24 SALE $20**
17-0323 Fast Charger	**$29**

AccuManager 20

Using a pulse-charging technology that rechargeable batteries just love, the AccuManager 20 charger will recharge your NiMH or NiCad batteries faster, and better, than any other charger we've ever seen. You can mix different battery types, sizes, and charging times. After charging, each cell is stored

at the peak of readiness with a gentle trickle charge until you're ready for it. There are four bays that accept AAA, AA, C, or D cells, plus two bays that accept 9-volt cells. Charge times will run from 20 minutes to 12 hours depending on type and size. Supplied with both AC and DC charging cords, it will run off standard 120-volt AC, automotive 12 volts, or a 10- to 15-watt solar panel with appropriate plug, such as our solar option shown at left. 3-yr. mfr. warranty. For rechargeable batteries only. Germany/China.

50269	**AccuManager 20**	**$69**

Battery Xtender

Extend the life of almost any battery using this Battery Xtender. Its smart technology easily accommodates alkaline, carbon-zinc, AccuManager 20, titanium, nickel-metal hydride, and 1.5-volt batteries, sensing battery chemistry and automatically adapting its charging parameters for optimum recharging of each and every cell. Simultaneously accepts AAA, AA, C, and D cells in any combination of sizes and chemistries. A built-in indicator gauges the status of each battery. Costs less than one cent in electricity per battery. 11"x7½"x2" (HxWxD). 1.5 lb. China.

17-0181	**Battery Xtender**	**$49**

Portable Solar Electronics Charger

This solar kit charges and runs iPods®, cell phones, digital cameras, GPSes, and PDAs wherever there's Sun. A rugged, hardcover-book-size plastic case stores a 20-inch USB cable, various adapter jacks, and a 4.4-watt solar module. Open it to the Sun, plug in your gear, and power up miles from the nearest outlet. Includes 12-volt car-lighter outlet for your gear's own adapters and an AA/AAA battery charger. 9.3"x6"x1.5" (HxWxD). 19 oz. Philippines.

**17-0340 Portable Solar Electronics
Charger $139**

Mobile DC Power Adapter

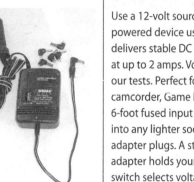

Use a 12-volt source to run and recharge any battery-powered device using less than 12 volts DC. Our adapter delivers stable DC power at 3, 4.5, 6, 7.5, 9, or 12 volts at up to 2 amps. Voltage varies less than 0.5 volt in our tests. Perfect for your camera, DVD or MP3 player, camcorder, Game Boy®, cell phone, or whatever. The 6-foot fused input cord with LED power indicator plugs into any lighter socket; the 1-foot output cord has six adapter plugs. A storage compartment in the power adapter holds your spare plugs to prevent loss. Slide switch selects voltage. China.

53-0105 Mobile DC Power Adapter $24

12-Volt DC PC Power Supplies

Run and recharge your PC from any 12-volt DC source. Available in 65- or 100 watt output with stabilized voltage adjustable from 15 to 21 volts. Output voltage varies less than 0.5 volt from zero to full load in our tests. External voltage adjustment is recessed for safety. Two-foot fused input cord with LED power indicator plugs into any lighter socket. Four-foot output cord has six different adapters to fit almost any PC laptop with a round power socket. Will not recharge newer Macintosh products requiring 24-volt input. 65-watt unit has maximum 4-amp output. 100-watt unit has maximum 7-amp output. 3.8"x2.3"x1.5" (HxWxD). 8 oz. Taiwan.

53-0103 65W Power Supply $59

53-0104 100W Power Supply $75

Power Dome

Stash this portable all-in-one power station in your car, boat, RV, or basement, and you'll be ready for anything from an emergency to an outdoor party. It restores dead car batteries with 600 amps of current and inflates flat tires with a 275-psi air hose. Two AC outlets send 400 watts of power with a 1,000-watt peak surge to everything from tools to TVs, while two 12-volt DC outlets run or charge small gadgets and electronics. There's even an onboard 5-LED lamp. Overload and other safety features protect both you and your gear, and the entire system recharges via AC or DC current to keep you permanently prepared. 11"x10½"x8" (HxWxD). China.

17-0342 Power Dome $149

XPower 1500 Powerpack

Our largest portable powerpack, the XPower 1500 delivers 120 volts AC and 12 volts DC. It packs a large 60Ah sealed battery and a 1,500-watt inverter with dual outlets in a nice, rugged wheeled cart with a removable 38-inch-high waist handle for ease of movement. Includes a 5-amp battery charger for when AC power is available (from a wall plug or generator) and a heavy-duty DC charging cord for a lighter socket or solar panel and charge controller. Has a battery charge indicator. 14.8"x15.6"x12.3" (HxWxD) without handle. 60 lb. 1-yr. mfr. warranty. Canada.

17-0283 XPower 1500 Powerpack $420

Small PV Panels

Small standalone solar panels have always been hard to find, so we've secured inventory with one of our manufacturing partners to bring you these single-crystal, tempered-glass modules at an affordable price. Each features weathertight construction, sturdy anodized aluminum frames with four adjustable mounting tabs, a built-in reverse-current diode, and a 7-foot wire pigtail output with color-coded battery clips. Great for trickle battery recharging on RVs, farm equipment, and boats, and perfect for household battery charging or backyard fountains. See chart for specs. 90-day mfr. warranty. China.

Watts	9W	13W	18W	27W	36W
Current Rated Power	500mA	710mA	1,000mA	1,500mA	2,000mA
Voltage	18V	18V	18V	18V	18V
Size (HxW xD)	16"x10.2" x1"	15.1"x14.4" x1"	15.1"x17.67" x1"	21.2"x17.67" x1"	35.4"x21.2" x1"
Weight	3.08 lb.	3.5 lb.	4.4 lb.	6.6 lb.	13.2 lb.
Item #	41-0216	17-0170	41-0216	41-0216	41-0216
Price	$99	$135	$179	$259	$329

GSE Flexible Modules

These weatherproof solar solutions fold to about the size of a paperback book to create a lightweight, portable power station for campers, hikers, boaters, and others on the go. Just open under full Sun and charge lanterns, cell phones, GPS units, iPods®, cameras, and any other device that has a 12-volt car adapter. With grommets to attach them to all kinds of gear, they're trail-tough and sized to pack anywhere. 6.5-, 12-, and 25-watt models have a built-in voltage cap to permit a direct charge to small devices and prevent damage from low-light reverse power flows. These three

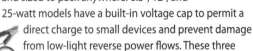

modules also include a connectivity kit with 12-volt receptacle, 12-volt vehicle power outlet, battery clamps, 4-inch barrel plug, and 8-foot extension cable. The 6.5-watt module is 11"x9"x1" (HxWxD) folded, 0.45 lb. The 12-watt module is 9"x5"x0.7" (HxWxD), 0.7 lb. The 25-watt module is 11"x8.25"x0.7" (HxWxD), 1.8 lb. The 55-watt module is 11"x9"x1" (HxWxD), 3.5 lb.

41-0249	6.5W Flexible Module	$99
41-0249	12W Flexible Module	$195
41-0249	25W Flexible Module	$399
41-0250	55W Flexible Module	$959
41-0251	Extra Connectivity Kit	$19
41-0252	7A Charge Controller	$39

PowerDock Power Station for Mobile Gadgets

The PowerDock has everything you need to run laptops, cell phones, GPSes, and other modest-power gear in remote sites. This sturdy power station has a zip-out, unbreakable 15-watt PV panel and 9.2Ah of sealed 12-volt batteries. Also included is a pair of fused lighter socket outlets, a power meter to show state of charge, and a water-resistant canvas carrying case with five compartments for converters and accessories. Accommodates laptop computers up to 9"x15"x2". Add the 15-watt Solar Boost to double charging power.

Plugs into one of the lighter sockets and folds into the PowerDock for travel. Robust construction; easy access to the batteries (lots of Velcro®). Has four batteries that should be replaced in two to four years. Weighs 14.5 pounds yet is comfortable to carry. This product is quality all the way! Comes with an AC wall-watt charger for when the Sun doesn't shine. 1-yr. mfr. warranty; 10 years on PV array. USA/Mexico.

53-0112	PowerDock	$349
41-0127	PowerDock 15W Boost	$179

CHAPTER 6

Energy Conservation

Superefficient Lighting, Heating, and Cooling Is Step 1

WE LIVE IN A DYNAMIC SOCIETY with a constantly growing economy. All economic activities require energy inputs, so economic growth demands increased energy consumption. To fuel this process, according to the conventional wisdom, we need to pump more oil out of the ground and build more power plants to generate more electricity. Right?

Wrong. It is a well-documented fact that conservation is the cheapest, most cost-effective way to "produce" energy: Renewable energy guru Amory Lovins calls it "negawatts." You know the drill: Raised on a steady diet of cheap fossil fuel energy during the 20th century, especially since World War II, our society is enormously wasteful of energy. Not only are those fossil fuel supplies finite and dwindling, but we now recognize the dire ecological impact of fossil fuel energy on Earth's climate. Energy conservation is the fundamental starting point for a sustainable future.

Did we forget to mention that energy conservation saves us all money, too? A lot of money.

Here are a couple of quick examples to illustrate the glories of energy conservation. We have calculated, using data from the California Energy Commission, that if every household in the state replaced four 100-watt incandescent light bulbs with four 27-watt compact fluorescent bulbs (the CF equivalent of a 100-watt light), burning on average for five hours a day, the state would save 22 gigawatt-hours per day (a gigawatt is 1,000 megawatts)—enough energy to *shut down* 17 power plants. Similarly, if each of those households replaced one average-flow showerhead with a low-flow, energy-saving showerhead, California would save an additional 19.2gWh per day—enough to shut down another 15 power plants. Conservation is a very powerful tool. Or try this: For every dollar you spend on energy conservation, you save $5 on the cost of your solar electric system!

So it's really a no-brainer. Conserving energy reduces greenhouse gas emissions, slows down the depletion of natural resources, decreases environmental pollution, takes strain off the planet's organic life-support systems, and takes a smaller bite out of your wallet. None of us can afford *not* to conserve energy. The further beauty of conservation is that regardless of what not-so-enlightened pundits and critics say, conservation does not mean sacrifice. Many European and Scandinavian societies enjoy a comparable standard of living to those of us in the United States, but they do it on a much tighter energy budget. As the authors of *Limits to Growth: The 30-Year Update* (Chelsea Green, 2004) note, "It seems certain that the U.S. economy could do everything it now does, with currently available technologies and at current or lower costs, using half as much energy" (Chelsea Green, 2004; p. 96), which would bring it to the efficiency levels of Western Europe. It's all about awareness, habits, and using energy efficiently. The tools, techniques, and technologies are readily available and well tested. We just have to start implementing them on a much wider basis.

> It is a well-documented fact that the cheapest, most cost-effective way to increase energy supplies under contemporary conditions is through simple acts of conservation.

> For every dollar you spend on energy conservation, you save $5 on the cost of your solar electric system!

The lower your household energy consumption, particularly for heating and electricity, the more easily renewable resources can contribute to meeting your needs with clean, reliable, sustainable energy.

Energy conservation is especially important if you want to achieve some degree of self-sufficiency, independence from conventional utility networks, or off-the-grid living. The lower your household energy consumption, particularly for heating and electricity, the more easily renewable resources can contribute to meeting your needs with clean, reliable, sustainable energy. But regardless of your mode of living, the same few conservation principles apply to everyone: Build smaller rather than larger. Make sure your building envelope is tight. Use passive solar strategies to minimize your heating and cooling loads. Install low-flow showerheads, low-flush or composting toilets, and compact fluorescent lights. Buy the most efficient appliances possible for your needs and budget. Strive to use renewable energy resources.

Gaiam Real Goods supplies many types of products, appliances, systems, and informational resources to help families conserve energy and consume it efficiently. The material in this chapter focuses on domestic space heating and cooling, household appliances, and lighting. Energy conservation as applied to water heating and transportation is addressed in later chapters.

We'd like to mention one other idea. If you want to take the concept of energy conservation to its furthest reaches, consider the notion of *embodied energy*. Every commodity or product that you consume "embodies" all the energy it took to produce it and get it to you. From the big-picture perspective, then, you can also conserve by assessing the comparative energy required to produce and transport various goods that you choose to buy. If you live on the East Coast, for example, a locally grown organic tomato in season embodies much less energy than a California organic tomato that's calling out to your taste buds in January. The scenarios and possible comparisons are endless, of course, and it's not always easy to figure out how much energy is embodied in any given thing, or to measure the actual impact of one choice over another. Thinking about this issue will probably take you places you may not want to go, and drive you crazy in the process. Nonetheless, the concept of embodied energy has real environmental import. We suggest that cultivating the habit of thinking about embodied energy is a conscientious act of global citizenship in service to sustainability.

Living Well but Inexpensively

Imagine a pair of similar suburban homes on a quiet residential street. Both house a family of four living the American suburban lifestyle. The homes appear to be identical, yet one spends under $50 per month on utilities, and the other over $400. How can that be? This huge cost difference demonstrates the dramatic savings that careful building design, landscaping, and selection of energy-efficient appliances make possible without affecting a family's basic lifestyle.

Our example isn't based on some bizarre construction technique, or appliances that can be operated only by rocket scientists, but on simple, commonsense building enhancements and off-the-shelf appliances, all of which you can find information about in this *Sourcebook*. Studies have shown that no investment pays as well as conservation. Banks, mutual funds, real estate investments . . . none of these options will bring the 100%-300% returns that are achievable through simple, inexpensive conservation measures. Not even our own dearly beloved re-

newable energy systems will repay your investment as quickly as conservation.

At Gaiam Real Goods, we are experts in energy conservation. We have to be! After 30 years designing renewable electrical systems for remote locations, we have learned to squeeze the maximum work out of every precious watt. Although our focus is on energy generation through solar modules or wind and hydroelectric generators, we often have to backtrack a bit to basic building design, or retrofitting of existing buildings, before we start selecting appliances.

We'll tackle the broad, multifaceted subject of energy conservation first by looking at building design, followed by retrofitting an existing building, and then by offering tips for selecting appliances. After that, we'll delve more into space heating, cooling, energy-efficient appliances, and superefficient lighting. Feel free to skip sections that do not apply to your current needs.

It All Starts with Good Design

The single most important factor that affects energy consumption in your house is design. All the intelligent appliance selection or retrofitting in the world won't keep an Atlanta, Georgia, family room with a 4x6-foot skylight and a west-facing 8-foot sliding glass door from overheating every summer afternoon and consuming massive amounts of air-conditioning energy. Intelligent solar design is the best place to start. Passive solar design works compatibly with your local climate conditions, using the seasonal Sun angles at your latitude to create an interior environment that is warm in winter and cool in summer, without the addition of large amounts of energy for heating and cooling.

Passive solar buildings cost no more to design or build than energy-hog buildings yet cost only a fraction as much to live in. You don't have to build a totally solar-powered house to take advantage of some passive solar savings. Just a few simple measures, mostly invisible to your neighbors, can substantially improve your home's energy performance!

For an in-depth examination of passive solar design strategies, from the simple to the complex, see Chapter 2, "Land and Shelter."

House orientation affects heating and cooling costs.

Retrofitting: Making the Best of What You've Got

If you're like many of us, you already have a house, have no intention of building a new one, and just want to make it as comfortable and economical as possible. By retrofitting your home with energy savings in mind, you can lower operating costs, boost efficiency, improve comfort levels, save money, and feel good about what you've accomplished.

WEATHERIZATION AND INSULATION

Increasing insulation levels and plugging air leaks are favorite retrofitting pastimes, which produce well-documented, rapid paybacks in comfort and savings on your space-heating bills. Short of printing your own money, weatherization and insulation are the best bets for putting cash in your wallet—and they're a lot safer in the long run than counterfeiting.

Weatherization

Weatherization, the plugging and sealing of air leaks, can save 25%-40% of heating and cooling bills. The average unweatherized house in the U.S. leaks air at a rate equivalent to a 4-foot-square hole in the wall. Weatherization is the first place for the average homeowner to concentrate his or her efforts. You'll get the most benefit for the least effort and expense.

WHERE TO WEATHERIZE

■ The following suggestions are adapted from *Homemade Money,* by Richard Heede and the Rocky Mountain Institute. Used with permission.

Here's a basic checklist to help you get started. Weatherization points are keyed in parentheses to the illustration on the next page.

1. In the attic

- Weather-strip and insulate the attic access door (6).
- Seal around the outside of the chimney with metal flashing and high-temperature sealant such as flue caulk or muffler cement (3).
- Seal around plumbing vents, both in the attic floor and in the roof. Check roof flashings (where the plumbing vent pipes pass through the roof) for signs of water leakage while you're peering at the underside of the roof (8).
- Seal the top of interior walls in pre-1950s houses anywhere you can peer down into the wall cavity. Use strips of rigid insulation, and seal the edges with silicone caulk (7).
- Stuff fiberglass insulation around electrical wire penetrations at the top of interior walls and where wires enter ceiling fixtures (but not around recessed light fixtures unless the fixtures are rated IC [for insulation contact]). Fluorescent fixtures usually are safe to insulate

Each year in the U.S. about $13 billion worth of energy, in the form of heated or cooled air—or about $150 per household —escapes through holes and cracks in residential buildings.

—American Council for an Energy-Efficient Economy

Attic:
1 Dropped ceiling
2 Recessed light
3 Chimney chase
4 Electric wires and box
5 Balloon wall
6 Attic entrance
7 Partition wall top plate
8 Plumbing vent chase
9 Exhaust fan

Basement and crawlspace:
10 Dryer vent
11 Plumbing/ utility penetrations
12 Sill plate
13 Rim joist
14 Bathtub drain penetration
15 Basement windows and doors
16 Block wall cavities
17 Water heater and furnace flues
18 Warm-air ducts
19 Plumbing chase
20 Basement/crawlspace framing
21 Floorboards

Living area:
22 Window sashes and doors
23 Laundry chute
24 Stairwell
25 Kneewall/framing intersection
26 Built-in dresser
27 Chimney penetration
28 Built-in cabinet
29 Cracks in drywall
30 Warm-air register
31 Window and door frames
32 Baseboards, coves, interior trim
33 Plumbing access panel
34 Sink drain penetration
35 Dropped soffit

36 Electrical outlets and switches
37 Light fixture

Exterior:
38 Porch framing intersection
39 Missing siding and trim
40 Additions, dormers, overhangs
41 Unused chimney
42 Floor joist

around; they don't produce a lot of waste heat. Incandescent fixtures should be upgraded to compact fluorescent lamps, but that's another story; see below, pages 327-332 (2, 4).

- Seal all other holes between the heated space and the attic (9).

2. In the basement or crawlspace

- Seal and insulate around any accessible heating or A/C ducts. This applies to both the basement and the attic (18).
- Seal any holes that allow air to rise from the basement or crawlspace directly into the living space above. Check around plumbing, chimney, and electrical penetrations (11, 14, 19, 21).
- Caulk around basement window frames (15).
- Seal holes in the foundation wall as well as gaps between the concrete foundation and the wood structure (at the sill plate and rim joist). Use caulk or foam sealant (12, 13).

3. Around windows and doors

- Replace broken glass and reputty loose panes. See the Window section, page 289, about upgrading to better windows or retrofitting yours (22).
- Install new sash locks, or adjust existing ones on double-hung and slider windows (22).

- Caulk on the inside around window and door trim, sealing where the frame meets the wall and all other window woodwork joints (31).
- Weather-strip exterior doors, including those to garages and porches (31).
- For windows that will be opened, use weather stripping or temporary flexible rope caulk.

4. In living areas

- Install foam-rubber gaskets behind electrical outlet and switch trim plates on exterior walls (36).
- Use paintable or colored caulk around bath and kitchen cabinets on exterior walls (26, 35).
- Caulk any cracks where the floor meets exterior walls. Such cracks are often hidden behind the edge of the carpet (32).
- Got a fireplace? If you don't use it, plug the flue with an inflatable plug, or install a rigid insulation plug. If you do use it, make sure the damper closes tightly when a fire isn't burning. See the section on Heating and Cooling (pages 293-302) for more tips on fireplace efficiency (41).

5. On the exterior

- Caulk around all penetrations where electrical, telephone, cable, gas, dryer vents, and water lines enter the house. You may want to stuff some fiberglass insulation in the larger gaps first (9, 10, 11).

- Caulk around all sides of window and door frames to keep out the rain and reduce air infiltration.
- Check your dryer exhaust vent hood. If it's missing the flapper, or it doesn't close by itself, replace it with a tight-fitting model (10).
- Remove window air conditioners in winter; or at least cover them tightly, and make rigid insulation covers for the flimsy side panels.
- Caulk cracks in overhangs of cantilevered bays and chimney chases (27, 40). ■ ■

Insulation

Many existing homes are woefully underinsulated. Exterior wall cavities or underfloors are often completely ignored, and attic insulation levels are sometimes more in tune with the 1950s, when heating oil was 12¢ per gallon, than they are with the early 21st century, when oil prices are more around 15-20 times higher and promising to continue to escalate. Adequate insulation rewards you with a building that is warmer in the winter, cooler in the summer, and much less expensive to operate year-round.

Recommended insulation levels vary according to climate and geography. Your friendly local building department can give you the locally mandated standards, but for most of North

R-Values of Loose-Fill Insulation

MATERIAL	R-VALUE PER INCH	USE
Fiberglass (low density)	2.2	Walls and ceilings
Fiberglass (medium density)	2.6	Walls and ceilings
Fiberglass (high density)	3.0	Walls and ceilings
Cellulose (dry)	3.2	Walls and ceilings
Wet-spray cellulose	3.5	Walls and ceilings
Rock wool	3.1	Walls and ceilings
Cotton	3.2	Walls and ceilings

R-Values of Rigid Foam and Liquid Foam Insulation

MATERIAL	R-VALUE PER INCH	USE
Expanded polystyrene	3.8-4.4	Foundations, walls, ceilings, and roofs
Extruded polystyrene	5.0	Same
Polyisocyanurate	6.5-8.0	Same
Roxul (mineral wool)*	4.3	Foundation exteriors
Icynene	3.6	Walls and ceilings
Air Krete	3.9	Walls and ceilings

*Rigid board insulation made from mineral wool.

—From Dan Chiras, *The Solar House: Passive Heating and Cooling*

Insulation

Since warm air rises, the best place to add insulation is in the attic. This will help keep your upper floors warm. Recommended insulation levels depend on where you live and the type of heating system you use. For most climates, a minimum of R-30 will do, but colder areas, such as the northern tier and mountain states, may need as much as R-49. Two useful booklets, BOE/CE-0180 (Aug. 1997) and DOE/GO-10097-431 (Sept. 1997), provide information about insulating materials and insulation levels needed in various locations by zip code. Write to the U.S. Department of Energy, Office of Technical Information, P.O. Box 62, Oak Ridge, TN 37831.

There are four types of insulating materials: batts, rolls, loose fill, and rigid foam boards. Each is suitable for various parts of your house. Batts, designed to fit between the studs in the walls or the joists in the ceilings, are usually made of fiberglass or rock wool. Rolls are also made of fiberglass and can be laid on the attic floor. Loose-fill insulation, which is made of cellulose, fiberglass, or rock wool, is blown into attics and walls. Rigid board insulation, designed for confined spaces such as basements, foundations, and exterior walls, provides additional structural support. It is often used on the exterior of exposed cathedral ceilings.

America, you want a minimum of R-11 in floors, R-19 in walls, and R-30 in ceilings. These levels will be higher in more northern climes and lower in more southern climes.

INSULATE IT YOURSELF?

Some retrofit insulation jobs, like hard-to-reach wall cavities, are best done by professionals who have the spiffy special tools and experience to do the job quickly, correctly, and with minimal disruption. There are insulation and weatherproofing companies that will inspect, advise, and give an estimate to bring your home up to an acceptable standard.

Easier-to-reach spaces, like attics, are improvable by the do-it-yourselfer with either fiberglass batt insulation or easier- and quicker-to-install blown-in cellulose insulation. Many lumber and building supply centers will rent or loan the blowing equipment (and instructions!) when you buy the insulation. Some folks may choose to leave this potentially messy job to the pros.

Underfloor areas are usually easy to access, but working overhead with fiberglass insulation is no picnic. You might consider a radiant barrier product for underfloor use because of the ease of installation. A radiant barrier stapled on the bottom of floor joists performs equally to R-14 insulation in the winter. Performance

Adequate insulation rewards you with a building that is warmer in the winter, cooler in the summer, and much less expensive to operate year-round.

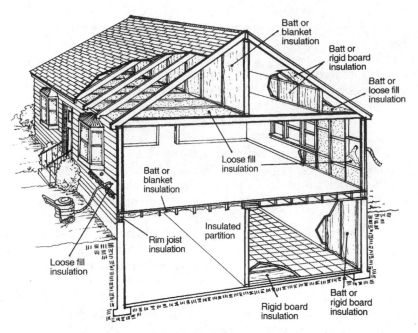

Where to Insulate.
Illustration adapted from Reader's Digest (1982), *Home Improvements Manual*, p. 360, Pleasantville, NY.

Two of the best applications of radiant barriers are under rafters to keep out unwanted summer heat, and under floors to keep desirable winter heat inside your house.

will be less in the summer keeping heat out (R-8 equivalent), but this isn't a big consideration in most climates.

RADIANT BARRIERS—THE NEWEST WRINKLE IN INSULATION

Over the past few years, a new type of insulation called "radiant barrier" has become popular. Radiant barriers work differently from traditional "dead-air space" insulation. See the sidebar for a detailed explanation.

A radiant barrier will stop about 97% of radiant heat transfer. Multiple layers of barrier provide no cumulative advantage; the heat practically all stops at the first layer. Radiant barriers do not affect conducted or convected heat transfers, so they are usually best employed in conjunction with conventional dead-air-space types of insulation, such as fiberglass.

R-values can't be assigned to radiant barriers because they're only effective with the most common radiant transfer of heat. But for the sake of having a common understanding, the performance of a radiant barrier is often expressed in R-value equivalents. For instance, when heat is trying to transfer downward, like from the underside of your roof into your attic in the summer, a radiant barrier tacked to the underside of the rafters will perform equivalently to R-14 insulation. When that same heat is trying to transfer back out of your attic at night, however, the radiant barrier only performs equivalently to R-8, since it has no effect on the convection transfer that is in the upward direction. Therefore, conventional insulation is best for keeping warm in the winter when all

the heat is trying to go up through convection. Radiant barrier material is most effective when it's reflecting heat back upward (the direction of convection). So, two of the best applications of this product are under rafters to keep out unwanted summer heat, and under floors to keep desirable winter heat inside your house.

The independent, nonprofit Florida Solar Energy Center has tested radiant barrier products extensively and has concluded that installation on the underside of the roof deck will reduce summer air-conditioning costs by 10%-20%. And that's in a climate where humidity is more of a problem than temperature is. This means rapid payback for a product as inexpensive, clean, and easy to install as a radiant barrier. FSEC also paid close attention to roof temperatures, as there's a logical concern about where all that reflected heat is going. They found that shingle temperatures increased by 2°-10°F. To a roof shingle, that's beneath notice and will have no effect on roof life span.

We offer only perforated radiant barriers now, due to the moisture-barrier properties of unperforated barriers causing occasional problems when misapplied.

AIR-TO-AIR HEAT EXCHANGERS (AKA HEAT-RECOVERY VENTILATORS)

Building and appliance technologies have made tremendous gains in the past 25 years. Improvements in insulation, infiltration rates, and weather stripping do much to keep our expensive warmed or cooled air inside the building. In fact, we've gotten so good at making tight buildings and reducing infiltration of outside air that indoor air pollution has become a problem.

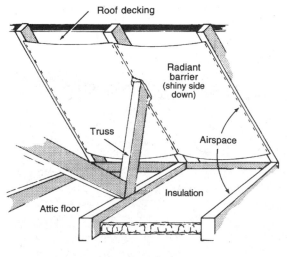

Typical radiant barrier attic installation.

How Radiant Barriers Work

Heat is transmitted three ways:

Conduction is heat flow through a solid object. A frying pan uses heat conduction.

Convection is heat flow in a liquid or gas. The warmer liquid or gas rises and the cooler liquid or gas comes down. Floor furnaces heat the air, which then rises in the room.

Radiation is the transmission of heat without the use of matter. It follows line of sight and you can feel it from the Sun and from a wood stove. Radiant heat flow is the most common method of heat transmission.

Radiant barriers use the reflective property of a thin aluminum film to stop (i.e., reflect) radiant heat flow. Gold works even better, but its use is limited mainly to spacecraft, for obvious reasons. A thin aluminum foil will reflect 97% of radiant heat. If the foil is heated, it will emit only 5% of the heat a dull black object of the same temperature would emit. For the foil to work as an effective radiant barrier, there must be an air space on at least one side of the foil. A vacuum is even better and is used in glass thermos bottles, where the double glass walls are aluminized and the space between them is evacuated. Without an air space, ordinary conduction takes place and there is no blocking effect.

A radiant barrier blocks only heat flow. It has no effect on conduction or convection. But then, so long as there's a small air space, the only practical way that heat can be transmitted across it is by radiant transfer.

See our product section for a selection of perforated radiant barriers in various widths (page 309).

A whole new class of household appliances has been developed to deal with the byproducts of normal living. Air-to-air heat exchangers, also known as heat-recovery ventilators, take the moisture, radon, and chemically saturated indoor air and exchange it for outdoor air. They strip up to 75% of the heat from the outbound air in the winter or the inbound air in the summer, thereby increasing your home's energy efficiency. These heat exchangers use minimal power, and controls can be triggered by time, humidity, or usage. If you use gas for cooking, live in a moist climate, have radon or allergy problems, or simply live in a good, tight house, then air-to-air heat exchangers are worth considering.

The air-to-air heat exchanger.
A heat recovery ventilator's heat exchanger transfers 50%–70% of the heat from the exhaust air to the intake air. Illustration courtesy of Montana Department of Natural Resources and Conservation.

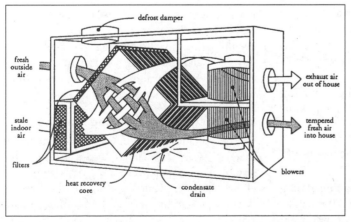

Windows

Window technology is one of the fastest developing fields of building technology. And it's about time! Windows are the weakest link in any building's thermal barrier. Until a few years ago, a window was basically a hole in the wall that let light in and heat out. The R-value of a single-pane window is a miserable 1. Nearly half of all residential windows in the United States provide only this negligible insulation value. An "energy-saving" thermopane (or double-pane) window is worth only about R-2. When you consider that the lower-cost wall around the window is insulated to at least R-12, and probably to over R-16, a window needs a solid justification for being there. In cold climates, windows are responsible for up to 25% of a home's winter heat loss. In warmer climates, solar radiation, entering through improperly placed or shaded windows, can boost air-conditioning bills similarly.

WORTH RETROFITTING?

Major innovations in window construction have occurred in the past several years. The most advanced superwindows now boast R-values up to R-10. While superwindows cost only 20%-50% more than conventional double-pane windows, the payback period for such an upgrade in an existing house is 15-20 years due to installation costs. That's too long

> The R-value of a single-pane window is a miserable 1. Nearly half of all residential windows in the United States provide only this negligible insulation value. An "energy-saving" thermopane (or double-pane) window is worth only about R-2.

for most folks. But if you're planning to replace windows anyway due to remodeling or new construction, the new high-tech units are well worth the slightly higher initial cost and will pay for themselves in just a few years.

SNEAKY R-VALUES

When shopping for windows, be forewarned that until recently, most windows were rated on their "center of glass" R-value. Such ratings ignore the significant heat loss through the edge of the glass and the frame. Inexpensive aluminum-framed windows, which transmit large amounts of heat through the frame, benefit disproportionately from this dubious standard. A more meaningful measure is the newer "whole-unit" value. Make sure you're comparing apples to apples. Wood- or vinyl-framed units outperform similar metal-framed units, unless the metal frames are using "thermal break" construction, which insulates the outer frame from the inner frame. The *Consumer Guide to Home Energy Savings*, published by the American Council for an Energy-Efficient Economy, provides an in-depth overview of how to understand various window ratings.

START WITH LOW-COST OPTIONS

With windows, as with all household energy-saving measures, start with jobs that cost the least and yield the most. A commonsense combination of three or four of the ideas in the next section will result in substantial savings on your heating/cooling bill, minimized drafts, more constant temperatures, and enhanced comfort, especially in areas near windows.

WINDOW SOLUTIONS FOR EFFICIENT HEATING

■ The following suggestions (cold- and warm-weather window solutions) are adapted from *Homemade Money,* by Richard Heede and the Rocky Mountain Institute. Used with permission.

First, stop the wind from blowing in and around your windows and frames by caulking and weather stripping. After you've cut infiltration around the windows, the main challenge is to increase the insulating value of the window itself while continuing to admit solar radiation. Here are some suggestions for beefing up your existing windows in winter.

Install clear plastic barriers on the inside of windows

Such barriers work by creating an insulating dead-air space inside the window. After caulking, this is the least expensive temporary option

for cutting window heat loss. Such barriers can cut heat loss by 25%-40%.

Repair and weatherize exterior storm windows

If you already own storm windows, just replace any broken glass, reputty loose panes, install them each fall, and seal around the edges with rope caulk.

Add new exterior or interior storm windows

Storm windows are more expensive than temporary plastic options but have the advantages of permanence, reusability, and better performance. Storm windows cost about $7.50-$12.50 per square foot and can reduce heat loss by 25%-50%, depending on how well they seal around the edges. Exterior storm windows will increase the temperature of the inside window by as much as 30°F on a cold day, keeping you more comfortable.

Apply low-e films

Low-e films substantially reduce the amount of heat that passes through a window, with minimal effect on the amount of visible light passing through. The accompanying sidebar explains how these films work. Don't use low-e films on, or built into, south-facing windows if you want solar gain! They can't tell the difference between winter and summer. See our product section for films that can be applied to existing windows. Note that low-e films will retard either heat loss or heat gain, depending on where they are applied.

Exotic infills

The other new technology commonly found in new windows is exotic infills. Instead of filling the space between panes with air, many windows are now available with argon or krypton gas infills that have lower conductivity than air and that boost R-values. Krypton has a higher R-value but costs more. These inert gases occur naturally in the atmosphere and are harmless even if the window breaks. (Krypton-filled windows are also safe from Superman breakage.) All this technology adds up to windows with R-values in the 6-10 range.

Install tight-fitting insulating shades

These shades incorporate layers of insulating material, a radiant barrier, and a moisture-resistant layer to help prevent condensation. Several designs are available. One popular favorite is Window Quilts. This quilted-looking material consists of several layers of spun polyester

> With windows, as with all household energy-saving measures, start with jobs that cost the least and yield the most.

Low-E Films

The biggest news in window technology is "low-e" films, for low-emissivity. These thin metal coatings allow the shortwave radiation of solar energy to pass in but block most of the long-wave thermal energy trying to get back out. A low-e treated window has less heat leakage in either direction. A low-e coating is virtually invisible from the inside, but most brands tend to give windows a semimirror appearance from the outside.

Low-e windows are available ready-made from the factory, where the thin plastic film with the metal coating is suspended between the glass panes; or low-e films can be applied to existing windows. We offer a do-it-yourself product that is applied permanently with soap and a squeegee, or it can be applied professionally. These films offer the same heat-reflecting performance as the factory-applied coatings and are relatively modest in cost (especially compared with new windows!). They do require a bit of care in cleaning, as the plastic film can be scratched. Some utility companies offer rebates for after-market low-e films.

and radiant barriers with a cloth outer cover. Depending on style, they fold or roll down over your windows at night, providing a tight seal on all four sides, high R-value insulation, privacy, and soft quilted good looks. This allows your windows to have all the daytime advantages of daylighting and passive heat gain, while still enjoying the nighttime comfort of high R-values and no cold drafts. Because there is little standardization of window sizing, Window Quilts are generally custom cut to size, making them unsuitable for a mail-order retail operation like ours. You can order directly from the manufacturer, or they can direct you to a local retailer who can supply installation services. Contact them at 877-966-3678 or on the Internet at www.1windowquilts.com.

Construct insulated pop-in panels or shutters

Rigid insulation can be cut to fit snugly into window openings, and a lightweight, decorative fabric can be glued to the inside. Pop-in panels aren't ideal, as they require storage whenever you want to look out the window, but they are cheap, simple, and highly effective. They are especially good for windows you wouldn't mind covering for the duration of the winter. Make sure they fit tightly so moisture doesn't enter the dead-air space and condense on the window.

Close your curtains or shades at night

The extra layers increase R-value, and you'll feel more comfortable not being exposed to the cold glass.

Open your curtains during the day

South-facing windows let in heat and light when the Sun is shining. Removing outside screens for the winter on south windows can increase solar gain by 40%.

Clean solar-gain windows

Keep those south-facing windows clean for better light and a lot more free heat. Be sure to keep those same windows dirty in the summer. (Just kidding!)

WINDOW SOLUTIONS FOR EFFICIENT COOLING

The main source of heat gain through windows is solar gain—sunlight streaming in through single or dual glazing. Here are some tips for staying cool:

Install white window shades or miniblinds

Using shades or blinds is a simple, old-fashioned practice. Since our grandparents didn't have air conditioners, they knew how to keep the heat out. Miniblinds can reduce solar heat gain by 40%-50%.

Close south- and west-facing curtains

Do this during the day for any window that lets in direct sunlight. Keep these windows closed too.

Install awnings

Awnings are another good, old-fashioned solution. Awnings work best on south-facing windows where there's insufficient roof overhang to provide shade. Canvas awnings are more expensive than shades, but they're more pleasing to the eye, they stop the heat on the outside of your building, and they don't obstruct the view.

Hang shades outside the window

Hang tightly woven screens or bamboo shades outside the window during the summer. Such shades will reduce your view, but they are

> Pop-in panels aren't ideal, as they require storage whenever you want to look out the window, but they are cheap, simple, and highly effective.

Appropriate landscaping on the west and south sides of your house provides valuable summertime shading that will reduce unwanted heating by as much as 50%. Those are better results than we get from more expensive projects like window and insulation upgrades!

inexpensive and stop 60%-80% of the Sun's heat from getting to the window.

Plant trees or build a trellis

Deciduous (leaf-bearing) trees planted to the south or, particularly, to the west of your building provide valuable shade. One mature shade tree can provide as much cooling as five air conditioners (although they're a bit difficult to transplant at that stage, so the sooner you plant the better). Deciduous trees block summer Sun but drop their leaves to allow half or more of the winter Sun's energy into your home to warm you on clear winter days.

Low-e films and exotic gas infills

Both low-e films and inert-gas infill windows will improve cooling efficiency as well as heating efficiency. See above, page 290, and the sidebar on page 291 for further details. ■ ■

Landscaping

When retrofitting existing buildings, most folks don't think about the impact of landscaping on energy use. Appropriate landscaping on the west and south sides of your house provides valuable summertime shading that will reduce unwanted heating by as much as 50%. Those are better results than we get from more expensive projects like window and insulation upgrades! If your landscaping is deciduous, losing its leaves in winter to let the warming winter Sun through, you have the best of both worlds. Landscaping that makes outdoor patio and yard spaces cooler, more livable, and inviting in the summer can also block cold winter winds that push through the little cracks and crevices of a typical house. For more information, see Home Cooling on page 302.

Conserving Electricity

Appliance selection is one area we generally have control over. If you're renting, you may not have control over your house design, window selection and orientation, or heating plant. But you can select the light bulbs in your lamps and the showerhead in your bathroom. And you can determine whether appliances get turned off (really off!) when not in use. Electricity generation is one of the largest contributors of greenhouse gas emissions, along with the internal combustion engine. So in addition to saving you money, conserving electricity contributes meaningfully to saving the planet.

Many household appliances are important enough to justify entire chapters in the *Sourcebook*. Lighting, heating and cooling (with refrigeration), water pumping, water heating, composting toilets, and water purification are big topics that have significant impact on total home energy use. Please see the individual chapters or sections on the above products. Here we provide general information that applies to all appliances.

DON'T USE ELECTRICITY TO MAKE HEAT

Avoid products that use electricity to produce heat when you have a choice. Making heat from electricity is like using bottled water for your lawn. It gets the job done just fine, but it's terribly expensive and wasteful. Electric space heaters are not the only electrical appliances that produce heat; in fact, most of them do. This includes electric water heaters, electric ranges

and ovens, hot plates and skillets, waffle irons, waterbed heaters, and the most common household electric heater . . . the incandescent light bulb. Most of these appliances can be replaced by other appliances that cost far less to operate. Incandescent bulbs are one of the most dramatic examples. Standard light bulbs return only 10% of the energy you feed them as visible light. The other 90% disappears as heat. Compact fluorescent lamps return better than 80% of their energy as visible light, and they last more than 10 times longer per lamp.

If you have a choice about using gas or electricity for residential heating, water heating, or cooking, use gas, even if this means buying new appliances. If you live in an area that has natural gas service, your monthly bills will decline by 50%-60%. Even if your only option is bottled propane, gas will be 30%-40% cheaper than electric. The same goes for clothes drying. Better yet, use the zero-energy-cost option: the clothesline, which has the side benefit of making your clothes last longer and smelling wonderfully like fresh air instead of perfumed detergent or dryer sheets. Dryer lint is your clothes wearing out by tumbling.

PHANTOMS AND VAMPIRES!

Now what's that odd quote in the margin about televisions using power when "off"? Appliances that use power even when they're off create what are called phantom loads. Any device that uses a remote control is a phantom load, because part of the circuitry must remain on in order to receive the "on" signal from the remote. For most TVs, this power use is 15-25 watts. VCRs are typically 5-10 watts. Together that'll cost you over $20 per year. Either plug these little watt-burners into a switched outlet, or use a switched plug strip so they really can be turned off when not in use.

The other increasingly common villains to watch out for are the little transformer cubes that live on the end of many small-appliance power cords. In the electric industry, these are officially known as "vampires," because they constantly suck juice out of your system, even when there's no electrical demand at the appliance. Ever noticed how those little cubes are usually warm to the touch? That's wasted wattage being converted to heat. Most vampire cubes draw a few watts continuously. Vampires need to be unplugged when not in use, or plugged into switched outlets.

Go Low-Flow, Too

Showers typically account for 32% of home water use. A standard showerhead uses about 2-5 gallons of water per minute, so even a five-minute shower can consume 25 gallons. According to the U.S. Department of Energy, heating water is the second largest residential energy user. With a low-flow showerhead, energy use and costs for heating water for showers may drop as much as 50%. This is particularly important if you heat your water with electricity, which is the most expensive and energy-inefficient way. A study a few years ago showed that changing to a low-flow showerhead saved 27¢ worth of water and 51¢ of electricity per day for a family of four. So, besides being good for the Earth, a low-flow showerhead will pay for itself in about two months!

WATER-SAVING SHOWERHEADS

As noted above, showers account for one-third of the average family's home water use, and heating water is the second-largest residential energy user. Our low-flow showerheads can easily cut shower water consumption by 50%. Add one of our instantaneous water heaters for even greater savings. See Chapter 8 for details.

> All the remote control televisions in the U.S., when turned to the "off" position, still use as much energy as the output of one Chernobyl-sized plant.
>
> —Amory Lovins, The Rocky Mountain Institute

House Heating and Cooling

Heating and cooling a home typically accounts for more than 40% of a family's energy bill (as much as two-thirds in colder regions) and costs on average well over $800 a year. That's a good bit of money, and any of us could find more interesting ways to spend it. Fortunately, it isn't difficult to cut our heating and cooling bills dramatically with a judicious mixture of weatherization, additional insulation, window upgrades, landscaping, improved heating and cooling systems, and careful appliance selection.

Making improvements to your building envelope with weatherization, insulation, window treatments, and landscaping for energy conservation is covered in the preceding section. In this section, we'll cover home heating and cooling systems, looking at sustainability issues, selecting the best options for your home, and improvements and fine-tuning to keep those systems working efficiently.

Sustainability and Renewables

Whether you're concerned about costs, environmental impact, or both, it makes sense to consider home heating and cooling from a sustainability perspective. We know that the fossil fuels are going to run out eventually; we know that fossil fuel prices are very likely to increase, probably dramatically, as supplies become used up during the 21st century; we know that our current dependence on oil from the Middle East, which will continue into the foreseeable future, carries various risks that portend the real possibility of price hikes; and we know that our enormous consumption of fossil fuels produces tremendous, potentially catastrophic, environmental damage. The solutions to all these problems depend upon greater use of renewable energy resources: Sun, wind, water, biomass, geothermal heat.

Home heating and cooling options can be more or less sustainable, depending on your choice of strategies, systems, and fuels. As discussed above and in Chapter 2, maintaining a tight building envelope and incorporating passive solar strategies to whatever extent possible are the places to start.

> If a house isn't resource-efficient, it isn't beautiful.
>
> —Amory Lovins, The Rocky Mountain Institute

Passive Solar Heat

Solar energy is the most environmentally friendly form of heating you can find. The amount of solar energy we take today in no way diminishes

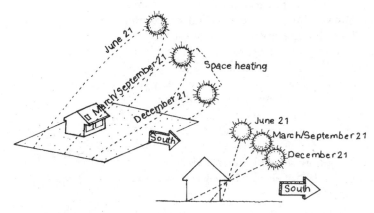

Sun path diagrams.
Passive solar heating is practicable in every climate. This simplified illustration shows the importance of a calculated roof overhang to allow solar heating in the winter but prevent unwanted solar heat gain in the summer. Adapted from an illustration by E SOURCE (1993), Space Heating Technology Atlas.

> A carefully designed passive solar building can rely on the Sun for half or more of its heating needs in virtually any climate in the United States.

the amount we can take tomorrow, or next decade, or next millennium. Free fuel is always the cheapest fuel. Passive solar heating is nonpolluting and does not produce greenhouse gases. It uses no moving parts, just sunshine through south-facing insulated windows and thermal mass in the building structure to store the heat. (Those south of the equator, please make the usual direction adjustments.)

A carefully designed passive solar building can rely on the Sun for half or more of its heating needs in virtually any climate in the United States. Many successful designs have cut conventional heating loads by 80% or better. Some buildings, such as the Real Goods Solar Living Center in Hopland, California, or the Rocky Mountain Institute's headquarters high in the Rocky Mountains in Snowmass, Colorado, have no need for a central heating system. RMI even grows banana trees indoors! By the same token, in hot climates, good design, shading, and passive cooling strategies can eliminate the need for mechanical cooling. Our Solar Living Center can survive over four weeks of daily high temperatures exceeding 100°F—and a steady flow of overheated visitors—while the interior temperature stays below 78°F without using air conditioning or mechanical cooling in any way.

Adding south-facing windows or a greenhouse is a way to retrofit your house to take advantage of free solar gain. See Chapter 2 for more-detailed information and books about passive solar design.

ACTIVE SOLAR HEATING: BEST FOR HOT WATER

In many climates around the U.S., flat-plate collector systems can provide enough solar heat to make a central heating system unnecessary, but in most regions, it is prudent to add a small back-up system for cloudy periods. Active solar systems can be added at any time to supplement space- and water-heating needs. However, active solar space heating has a high initial cost, is complex, and has potential for high maintenance. We don't recommend it; use passive solar! The beauty of passive systems is in their simplicity and lack of moving parts. Remember the KISS rule: Keep It Simple, Stupid.

The best and most common use of active systems is for domestic water heating. Real Goods offers several solar hot-water systems for differing budgets. See Chapter 8 for a detailed discussion of solar hot-water systems.

Buying a New Heating System

If your old heating system is about to die, you're probably in the market for a new one. Or, if you're currently spending over $1,000 per year on heating, it's likely that the $800-$4,500 you'll spend on a more efficient heating system will reduce your bills enough to pay for your investment in several years. Your new system also will be more reliable, and it will increase the value of your house. As you shop, keep in mind the following factors that will save you money and increase your comfort:

If you've weatherized and insulated your home—and we'll remind you again that these are by far the most cost effective things you can do—then you can downsize the furnace or boiler without compromising the capability of your heating system to keep you warm. An oversized heater will cost more to buy up front and more to run every year, and the frequent short-cycle on and off of an oversized system reduces efficiency. Ask your heating contractor to explain any sizing calculations and make sure he or she understands that you have a tight, well-insulated house, to verify that you don't get stuck with an oversized model (which the contractor gets to charge more money for).

Wood Stoves, Masonry Stoves, and Pellet Stoves

Wood heat is a mixed blessing. If harvested and used in a responsible manner, firewood can be a sustainable resource. As long as the wood you burn is part of a managed cycle that includes replanting trees, heating with wood is a carbon-neutral process that does not make a net contribution to global warming. That's because the

CO_2 released when the wood burns is equal to the CO_2 the tree absorbed while it was growing. Burning it does emit CO_2, but a tree newly planted to replace it will absorb an equivalent amount over its lifetime.

But, to be honest, most wood gathering is not done sustainably. If you have your own woodlot, you can create a sustainable forest management plan with the help of a professional forester, if necessary. If you buy wood, you can try to ascertain if your wood supplier cuts in a responsible manner or buys logs from someone who does—and you can encourage him to do so. You can also participate in reforestation programs run by organizations such as American Forests; it costs very little to offset your annual fuel wood consumption by funding the planting of new trees.

Pollution is still a problem, though progress is being made. There are 27 million wood-burning stoves and fireplaces in the United States, the majority of them older pre-EPA designs, that contribute millions of tons of pollutants to the air we breathe. In Washington State, for example, wood heating contributes 90% of the particulates in two of the counties with the worst air pollution problems. Modern wood stoves have catalytic combustors and other features that boost their efficiencies into the 55%-75% range while lowering emissions by two-thirds. Older wood stoves without air controls have efficiencies of only 20%-30%. Wood burns at a higher temperature and the gases go through a secondary combustion in a catalytic or a newfangled noncatalytic wood stove, thereby minimizing pollution and creosote buildup.

If you're buying a new wood stove, make sure it's not too large for your heating needs. This will allow you to keep your stove stoked with the damper open, resulting in a cleaner and more efficient burn. Some stoves incorporate more cast iron or soapstone as thermal mass to moderate temperature swings and prolong fire life. If you live in a climate that requires heat continuously for several months, you definitely want the heaviest stove you can afford. Masonry heaters take the thermal mass idea one very comfortable step further.

MASONRY STOVES

Masonry stoves, also called Russian stoves or fireplaces, developed by a people who know heating, have long been popular in Europe. These large, freestanding masonry fireplaces work by circulating the heat and smoke from the combustion process through a long twist-

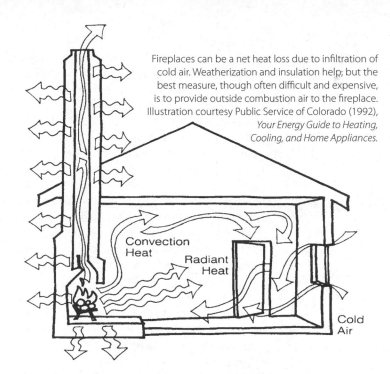

Fireplaces can be a net heat loss due to infiltration of cold air. Weatherization and insulation help, but the best measure, though often difficult and expensive, is to provide outside combustion air to the fireplace. Illustration courtesy Public Service of Colorado (1992), *Your Energy Guide to Heating, Cooling, and Home Appliances.*

ing labyrinth masonry chimney. The mass of the heater slowly warms up and radiates heat into the room for hours. Masonry stoves are best suited for very cold climates that need heat most of the day. They are much more efficient than fireplaces. They have very low emissions because airflow is restricted only minimally during burning, giving a clean, high-temperature burn. How many wood stoves have seats built into them so that you can snuggle in? Russian fireplaces commonly do. As Mark Twain observed, "One firing is enough for the day . . . the heat produced is the same all day, instead of too hot and too cold by turns."

Masonry stoves are expensive initially but will last for the life of the home and are one of the cleanest, most satisfying ways to heat with wood. They're easiest to install during initial construction and are best suited to well-insulated homes with open floor plans that allow the heat to radiate freely.

PELLET STOVES

If you can't burn wood but are looking for a similar way to wean yourself from oil, a pellet stove may be right for you. As renewable energy expert Greg Pahl observes, pellets have clear environmental benefits: "Pellets are a renewable resource, so burning pellets does not add any net carbon dioxide to the atmosphere. Because of the extremely hot combustion in most pellet-fired appliances, pellets burn cleanly with extremely low emissions and produce virtually no creosote. Compared with other fuels, burning 1 ton of pellets instead of heating with electricity will save 3,323 pounds of carbon emissions.

Masonry stoves are the best wood-burning solution for extreme climates that need 24-hour heating.

You'll save 943 pounds of emissions per ton by replacing oil with pellets and 549 pounds per ton if you're replacing natural gas."

Pellet stoves have the advantage that the fuel pellets of compressed sawdust, cardboard, grass, or agricultural waste are fed automatically into the stove by an electric auger. The feed rate is dialed up or down with a rheostat, so they'll happily run all night if you can afford the pellets. You fill the reservoir on the stove once every day or so. Pellet stoves are cleaner burning than wood stoves are because combustion air is force-fed into the burning chamber. Unlimited access to oxygen and drier fuel than firewood mean a more efficient combustion. But the pellets may be slightly more expensive than firewood in most parts of the country and simply unavailable in some areas. Pellets must be kept bone dry in order to auger and burn.

Pellet stoves have disadvantages, too. They have more mechanical parts than wood stoves do, so they're subject to maintenance and breakdown, and they require electricity. No power, no fire—the fire will go out without the combustion blower running. Pellet stoves usually can-

not be used with renewable electricity systems, unless you've got considerable extra power in the winter, like from a hydro system.

Displace Oil with Biodiesel

If you're stuck with an oil-burning furnace or boiler, don't just sit there and grimace about all the greenhouse gases you're putting into the atmosphere—you *can* do something about it (besides making sure your house is tight and your system efficiency is high): biodiesel. The original diesel engine exhibited at the Paris World's Fair in 1900 ran on vegetable oil, and the concept remains more than viable. Biodiesel is a renewable fuel that can be made easily from a variety of biomass feedstocks, even from recycled cooking oil. Biodiesel has been promoted primarily as a vehicle fuel, but it also works as a home heating fuel additive. Increasing numbers of people are burning B100, or pure biodiesel, while others burn a blend of #2 heating oil mixed with 10% or 20% biodiesel; no conversion is required, and the extra maintenance needed is minimal. As of this writing, B20 blend is just slightly more ex-

> Biodiesel is a renewable fuel that can be made easily from a variety of biomass feedstocks, even from recycled cooking oil. Biodiesel has been promoted primarily as a vehicle fuel, but it also works as a home heating fuel additive.

To Burn or Not to Burn

I'm totally confused. The subject is wood, a material I have used to heat my home for the past 25 years. I know it's more work to burn wood, but I like the exercise. It makes more sense to burn calories splitting and hauling hardwood than running in front of a television set on a treadmill at the health spa. I enjoy the fresh air. I like the warmth. I like the flicker and glow of embers.

Sometimes, when I am stacking wood so that a summer's worth of sunshine can make it a cleaner, hotter fuel, I think of those supertankers carrying immense cargos of black gold from the ancient forests of the Middle East. I think of the wells that Saddam Hussein set afire in Kuwait, and the noxious roar of the jet engines of the warplanes we used to ensure our country's access to cheap oil.

My wood comes from hills that I can see. It's delivered by a guy named Paul, who cuts it with his chainsaw, then loads it in his one-ton pickup. We talk about the weather, the conditions in the woods, and the burning characteristics of different species. We finish by bantering about whether he's delivered a large cord, a medium, or a small. I can't imagine having the same conversation with the man who delivers oil or propane.

Unfortunately (and this is where I begin to get confused), there is another side to the wood-burning issue. Along comes evidence that the emissions from airtight stoves contain insidious carcinogens, and further indications that a home's internal environment can be more adversely affected by burning wood than by passive cigarette smoke. So now I burn my wood in a "clean" stove. But my environmentally active friends say that the only good smoke is no smoke and that even

my supposedly high-tech stove is smogging up their skies.

Theoretically, I should be able to go back to feeling good about wood, but life is never that simple. On the positive side of the ledger comes the information that wood-burning, in combination with responsible reforestation, actually helps the environment by reversing the greenhouse effect. The oxidation of biomass, whether on the forest floor or in your stove, releases the same amount of carbon into the atmosphere. It makes more sense for wood to be heating my home than contributing to the brown skies over Yellowstone.

This is obviously a simplistic analysis, but wood should be a simple subject. In the meantime, I have to keep warm, so I'm choosing to burn wood. I keep coming back to the fact that humans have been burning oil for more than 50 years and wood for 5 million.

—Stephen Morris

---

pensive than heating oil, and the price is bound to become more competitive over time. Biodiesel manufacturing facilities are springing up everywhere, and fuel distributors are beginning to supply it to consumers. Seek and ye shall find.

Any biodiesel you use to displace oil from your heating system lowers carbon dioxide emissions and reduces particulate pollution. Call a local fuel distributor or your state energy office for more information about obtaining biodiesel for heating purposes. See Chapter 12, "Sustainable Transportation," for details on biodiesel as an alternative motor vehicle fuel.

Fuel Switching: Natural Gas Furnaces and Boilers

When available, natural gas is the lowest-cost heating fuel for most homes. If you have a choice, switching fuels may be cost effective but only if this does not also require substantial changes to your existing heat distribution system. For example, replacing the old hot-water boiler with a high-efficiency forced-air furnace would require installation of all new forced-air ducts.

It is always cost effective to pay a little more up front in return for more efficiency. But it may not be cost effective to pay a lot more for the most efficient gas unit, since your reduced heating needs mean a longer payback. The differences between various models of gas furnaces and boilers can be significant, and it's worth your while to examine efficiency ratings before making a decision.

If you are currently using electricity for heating, switching to any other fuel will be cost effective. See the next subject.

Electric Resistance Heating

Electric baseboard heating is by far the most expensive way to warm one's home; it's cheap to buy and install, but it costs two to three times more to heat with electricity as it does to heat with gas. The life-cycle cost of electric heating is, without exception, far higher than that of gas-fired furnaces and boilers, even taking into account the higher installation cost of the latter, and even in regions with exceptionally low electricity prices. Electric resistance is an energy-inefficient way to produce heat, and because grid-generated electric power is a key greenhouse gas culprit, this is a terrible environmental choice. (Electric heating may be excused if you have a plentiful, seasonal supply of renewably generated power, for example from a year-round micro-hydro system.)

If you must heat with electricity, making sure that your home is well insulated and weatherized, along with installing a programmable thermostat, are no-brainer investments that you should do immediately. If you also use air conditioning and live in a mild winter climate, consider switching to a heat pump.

Geothermal Energy: Heat Pumps

Geothermal heating is a simple and elegant concept. Earth has been absorbing and storing solar energy for millions of years, and it's a massive object: Voilà, the epitome of the phrase "thermal mass." A heat pump is a mechanical device that makes use of this fact, and the fact that the temperature just below the surface remains constant year-round.

Heat pumps are considered a renewable energy system because they consume no fuel to generate heat. In fact, they don't generate heat at all; they operate by moving heat around from a warmer place to a cooler place. However, a heat pump uses electricity to perform this sleight of hand. Heat pumps, therefore, can be considered the most efficient form of electric heat. And they are very efficient.

There are three types of heat pumps: air-to-air, ground-source, and water-source. They collect heat from the air, water, or ground depending on type, concentrate the warmth, and

ENERGY CONSERVATION

The life-cycle cost of electric heating is, without exception, far higher than that of gas-fired furnaces and boilers.

Ground-source heat pump.
Ground-source heat pumps can be highly cost effective in new construction over the life cycle of heating and cooling equipment compared with more conventional options. Several designs are on the market; this illustration shows the Slinky™ design. Adapted from an illustration by E SOURCE (1993), *Space Heating Technology Atlas*.

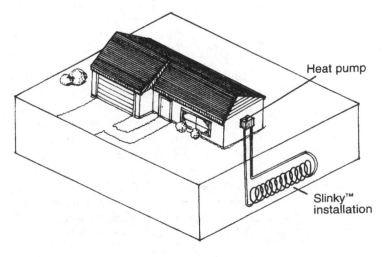

Heat pump

Slinky™ installation

distribute it through the home. Heat pumps typically deliver three times more energy in heat than they consume in electric power. Heat pumps also can be used to cool homes by reversing the process—collecting indoor heat and transferring it outside the building. Some newer models are designed to provide an inexpensive source of household hot water as well.

Not all heat pumps are created equal, but fortunately the Air Conditioning and Refrigeration Institute rates all the ones on the market. Look at the efficiency ratings and purchase a system designed for a colder climate. Some electric utilities offer rebates and other incentives to help finance the higher capital costs of these more efficient systems.

Air-to-air heat pumps are the most popular, the least expensive initially, and the most expensive to run in colder weather. Air-to-air units must rely on inefficient back-up heating mechanisms (usually electric resistance heat) when outside temperatures drop below a certain point. That point varies from model to model, so make sure you buy one that's designed for your climate—some models can cope with much colder weather before resorting to back-up heat. A heat pump running on backup is no better than standard electric resistance heating. Some newer pumps have more efficient gas-fired back-up mechanisms, but if gas is available, why not use it directly?

Ground- or water-source heat pumps are among the most efficient heating systems around and are substantially more cost effective in colder climates. Because of the extensive burial of plastic pipe required, these systems are best installed during new construction. Tens of thousands of ground-source heat pumps have been installed in Canada, New England, and other frost-belt areas. See the annualized heating/cooling system cost chart on the next page for a comparison.

Fireplace Inserts

As noted by the authors of the *Consumer Guide to Home Energy Savings*, a fireplace is essentially part of a room's décor, not a source of heat. A customary fireplace installation generally will lose more heat than it provides. However, if outfitted with a well-sealed insert, a fireplace can provide some useful heat.

A fireplace insert fits inside the opening of a fireplace, operates like a wood stove, and offers improved heat performance. An insert is difficult to install because it must be lifted into the fireplace opening and the space around it covered with sheet metal and sealed with a cement grout to reduce air leaks. Such inserts are only 30%-50% efficient, which is a good improvement over a fireplace but less efficient than a good freestanding wood stove, which is free to radiate heat off all sides.

A fireplace can be retrofitted with an airtight wood stove installed in, or better yet in front of, the fireplace. This is the most efficient choice, since all six heat-emitting sides are inside the living space, on the hearth. The stove's chimney is linked into the existing flue, and good installations line the chimney all the way to the top with stainless pipe to limit creosote buildup and provide easy flue cleaning. Simply shoving a few feet of wood-stove pipe into the existing chimney is begging for a chimney fire of epic proportions. Any creosote in the smoke stream will condense on the large, cool chimney walls, providing a spectacular amount of high-temperature fuel in a few years. This has been well proven to be an excellent way to burn a house down—further demonstrations and experiments aren't necessary. Line your chimney when retrofitting a wood stove.

Gas Fireplaces

Gas fireplaces have combustion efficiencies up to 80%, whereas the initially less expensive gas logs are only 20%-30% efficient. However, in case of gas leaks, many local codes require dampers to be welded open. This reduces their overall efficiency unless they have tight-fitting doors to cut down the amount of warmed interior air allowed to float out freely.

People Heaters

It often makes sense to use a gas or electric radiant or convection heater to warm only certain areas of the home or, as in the case of radiant heaters, to keep individual people warm and comfortable. Radiant systems keep you and objects around you warm, just like the Sun warms your skin. This allows you to set the thermostat for the main heating system lower by 6-8 degrees F.

Even if you heat with gas, electric convective or radiant spot heaters can save you money, depending on how the system is used. Radiant systems are designed and used more like task lights; you turn them on only when and where heat is needed, rather than heating the whole house. Most central heating systems give you little control in this respect, since they are designed to heat the entire home.

Heating Costs Comparison

While operating costs will vary by climate and region, and fuel and electricity prices differ across the country, the following chart represents estimated installation and operating costs for selected heating and cooling systems in a typical single-family house. It assumes that heat pumps will provide air conditioning in the summer as well as space heating in the winter.

Fine-Tuning Your Heating System

Let's assume that your home is already well insulated and weather tight. You have reduced the amount of heat escaping in the most cost effective way. It now makes sense to look at how efficiently your heating system produces and delivers heat to your living space. This section gives tips for making your existing heating system run more efficiently, whether it is powered by gas, electric, oil, wood, or solar energy. The following suggestions are either relatively inexpensive or free. All are highly cost effective.

Remember, if your existing system appears close to a natural death, or its maintenance costs are high, these fine-tuning tips may not be particularly cost effective. Modern heating systems have seen dramatic efficiency gains over the past few years, so replacing an old one could well save you money.

For safety reasons, adjustments, tune-ups, and modifications to your heating system itself are best done by a heating system professional. Home owners and renters can improve system performance by insulating ducts and pipes, cleaning registers, replacing filters, and installing programmable thermostats.

FURNACES AND BOILERS
Heating System Tune-Ups

Gas furnaces and boilers should be tuned every two years, oil units once a year. Your fuel supplier can recommend or provide a qualified technician. Expect to pay $60-$150—money well spent. Also have the technician do a safety test to make sure the vent does not leak combustion products into the home. You can do a simple test by extinguishing a match a couple of inches from the spillover vent: The smoke should be drawn up the chimney.

Comparing Heating Fuel Costs*	
IN DOLLARS PER MILLION BTU	
ELECTRICITY	
Ground-source heat pump (COP** = 3.0)	$11
Air-source heat pump (COP = 2.0)	$16
Baseboard resistance heater (COP = 1.0)	$32
NATURAL GAS	
Central furnace (AFUE*** = 85%)	$16
PROPANE	
Central furnace (AFUE = 85%)	$26
FUEL OIL	
Central furnace (AFUE = 80%)	$22
CORDWOOD	
Air-tight stove (50% efficient)	$19
WOOD PELLETS	
Pellet stove (80% efficient)	$15

*Fuel costs are 2006 national averages from www.eia.doe.gov: electricity = $0.11/kWh; natural gas = $1.40/therm.; propane = $2/gal.; and fuel oil = $2.50/gal. Other costs are based on cordwood @ $200/cord and pellets @ $0.10/lb.
**COP = Coefficient of Performance
***AFUE = Annual Fuel Utilization Efficiency

During a furnace tune-up, the technician should clean the furnace fan and its blades, correct the drive-belt tension, oil the fan and the motor bearings, clean or replace the filter (make sure you know how to perform this monthly routine maintenance), and help you seal ducts if necessary.

Efficiency Modifications

While the heating contractor is there to tune up your system, he or she may be able to recommend some modifications, such as reducing the nozzle (oil) or orifice (gas) size, installing a new burner and motorized flue damper, or replacing the pilot light with an electronic spark ignition. If you have an older oil burner, installing a flame-retention burner head (which vaporizes the fuel and allows more-complete combustion) typically will pay back your investment in two to five years.

Turn Off the Pilot Light During the Summer

This simple act will save you about $2-$4 per month. Do this only if you can safely light it again yourself, so you don't have to pay someone to do it. Federal regulations now require that new natural gas-fired boilers and furnaces be equipped with electronic ignition, saving $30-$40 per year in gas bills. Propane-fired units cannot be sold without a pilot light and cannot be retrofitted safely with a spark ignitor unless you install expensive propane-sniffing equipment. Propane is heavier than air, and any

If you have an older oil burner, installing a flame-retention burner head (which vaporizes the fuel and allows more-complete combustion) typically will pay back your investment in two to five years.

A clogged furnace filter impedes airflow, makes the fan work harder, makes the furnace run longer, and cuts overall efficiency. Filters are designed for easy service, so the hardest part of this maneuver is turning into the hardware store parking lot with a note in hand of the correct size filter.

leaking propane can pool around the unit, creating an explosion potential when the unit tries to start. Natural gas is lighter than air.

Insulate the Supply and Return Pipes on Steam and Hot-Water Boilers

Use a high-temperature pipe insulation such as fiberglass wrap for steam pipes. The lower temperatures of hydronic or hot-water systems may allow you to use a foam insulation (make sure it's rated for at least 220°F).

Clean or Change Your Filter Monthly

A clogged furnace filter impedes airflow, makes the fan work harder, makes the furnace run longer, and cuts overall efficiency. Filters are designed for easy service, so the hardest part of this maneuver is turning into the hardware store parking lot with a note in hand of the correct size filter. While you're there, pick up several filters, since you'll want to replace the filter every month during the heating season. Most full-house air-conditioning systems use the same fan, ducts, and filter. So monthly changes during A/C season are a must, too. They'll set you back a buck or two apiece. For five bucks, you can buy a reusable filter that will need washing or vacuuming every month but will last for a year or two.

Seal and Insulate Air Ducts

Losses from leaky, uninsulated ducts—especially those in unheated attics and basements—can reduce the efficiency of your heating system by as much as 30%. Don't blow your expensive heated air into unheated spaces! Seal ducts thoroughly with mastic, caulking, or duct tape, and then insulate with fiberglass wrap.

RADIATORS AND HEAT REGISTERS

Vacuum the cobwebs out of your registers. Anything that impedes airflow makes the fan work harder and the furnace run longer.

Reflect the Heat from Behind Your Radiator

You can make foil reflectors by taping aluminum foil to cardboard, or use radiant barrier material (left over from your attic or basement). Place them behind any radiator on an external wall with the shiny side facing the room. Foil-faced rigid insulation also works well.

Vacuum the Fins on Baseboard Heaters

Vacuuming up dust improves airflow and efficiency. Keeping furniture and drapes out of the way also improves airflow from the radiator.

Bleed the Air out of Hot-Water Radiators

Trapped air in radiators keeps them from filling with hot water and thus reduces their heating capacity. Doing this should also quiet down clanging radiators. You can buy a radiator key at the hardware store. Hold a cup or a pan under the valve as you slowly open it with the key. Close the valve when all the air has escaped and only water comes out. If it turns out that you have to do this more than once a month, have your system inspected by a heating contractor.

THERMOSTATS
Install a Programmable Thermostat

One-half of homeowners already turn down their heat at night, saving themselves 6%-16% of heating energy. A programmable or clock thermostat can do this for you automatically. In addition to lowering temperatures while you sleep, it also will raise temperatures again before you get out from under the covers. It not only sounds good, it's cost effective too. Such thermostats can be used to drop the house temperature during the day as well if the house is unoccupied. *Home Energy* magazine reports savings in excess of 20% with two eight-hour setbacks of 10 degrees F each.

Electronic setback thermostats are readily available for $25-$150. If you have a heat pump or central air conditioning, make sure the thermostat is designed for heat and A/C. Instal-

Thermostat Myths

Myth: Turning down your thermostat at night or when you're gone means you'll use more energy than you saved when you have to warm up the house again.

Fact: You always save by turning down your thermostat no matter how long you're gone. (The only exception is older electric heat pumps with electric resistance heaters for backup.)

Myth: The house warms up faster if you turn the thermostat up to 90°F initially.

Fact: Heating systems have only two "speeds": ON or OFF. The house will warm up at the same rate no matter what temperature you set the thermostat at, as long as it is high enough to be "ON." Setting it higher only wastes energy by overshooting the desired temperature.

lation is simple, but make sure to turn off the power to the heating plant before you begin. If connecting wires makes you jittery, have an electrician or heating technician do it for you. It's well worth the investment.

HEAT PUMPS

Change or Clean Filters Once a Month

Dirty filters are the most common reason for heat pump failures.

Keep Leaves and Debris Away from the Outside Unit

Good airflow is essential to heat pump efficiency. Clear a 2-foot radius. This includes shrubbery and landscaping.

Listen Periodically to the Compressor

If the compressor cycles on for less than a minute, something is wrong. Have a technician adjust it immediately.

Dust Off the Indoor Coils of the Heat Exchanger at Least Once a Year

This dust-off should be part of the yearly tune-up if you pay a technician for this service.

Have the Compressor Tuned Up

Every year or two, a technician should check refrigerant charge, controls, and filters, and oil the blower motors.

WOOD STOVES

Use a Wood Stove Only if You're Willing to Make Sure It Burns Cleanly

Wood stoves aren't like furnaces or refrigerators; you can't just turn them on and forget them. "Banking" a wood stove for an overnight burn only makes the wood smolder and emit lots of pollution. Sure, we've all done it, but we didn't know any better. Now we do. A well-insulated house will hold enough heat that you shouldn't need to keep the stove going all night. Heavier masonry stoves are a good idea in colder climates. Once the large mass is warmed up, it emits slow, steady heat overnight without a smoky fire.

Burn Well-Seasoned Wood and No Trash

Wood should be split at least six months before you burn it. Never burn garbage, plastics, plywood, or treated lumber.

Hotter Fires Are Cleaner

Another good case for masonry stoves, which burn hot, clean, short fires with no air re-

striction. The heat is soaked up by the masonry and emitted slowly over many hours.

If a Wood Stove Is Your Main Source of Heat, Use an EPA-Approved Model

EPA regulations now require wood stoves to emit less than 10% of the pollution that older models produced.

Check Your Chimney Regularly

Blue or gray smoke means your wood is not burning completely.

FIREPLACES

Fireplaces offer a good light show but are not an effective source of heat. In fact, fireplaces are usually a net heat loser because of the huge volume of warm interior air sucked up the chimney, which has to be replaced with cold air leaking into the house. Fireplaces average minus 10% to plus 10% efficiency. If you've done a good job of weatherizing your house and sealing up air leaks, you may lack adequate fresh air for combustion, or the chimney won't draw properly, filling the house with smoke. The two best things to do with a fireplace are (1) don't use it and plug the flue, or (2) install a modern wood stove and line the chimney.

Improve the Seal of the Flue Damper

To test the damper seal, close the flue, light a small piece of paper, and watch the smoke. If the smoke goes up the flue, there's an air leak. Seal around the damper assembly with refractory cement. (Don't seal the damper closed.) If the damper has warped from heat over the years, get a sheet-metal shop to make a new one, or consider a wood stove conversion.

Install Tight-Fitting Glass Doors

Controlling airflow improves combustion efficiency by 10%-20% and may help reduce air leakage up a poorly sealed damper.

Use a C-Shaped Tubular or Blower-Equipped Grate

Tube grates draw cooler air into the bottom and expel heated air out the top. Blower-equipped grates do the same job with a bit more strength.

Caulk Around the Fireplace and Hearth

Do this where they meet the structure of the house, using a butyl rubber caulk.

> Fireplaces offer a good light show but are not an effective source of heat. In fact, fireplaces are usually a net heat loser because of the huge volume of warm interior air sucked up the chimney, which has to be replaced with cold air leaking into the house.

Use a Cast-Iron Fireback

Firebacks are available in a variety of patterns and sizes. Fireplace efficiency is improved by reflecting more radiant heat into the room.

Locate the Fire Screen Slightly Away from the Opening

Moving the screen forward into the room a bit allows more heated air to flow over the top of the screen. While fire screens do prevent sparks from flying into the room, they also prevent as much as 30% of the heat from entering the room.

If You Never Use Your Fireplace, Put a Plug in the Flue to Stop Heat Loss

Seal the plug to the chimney walls with good-quality caulk and tell this to anyone who might build a fire (or leave a sign). Temporary plugs can be made with rigid board insulation, plywood with pipe insulation around the edges, or an inflatable plug.

Home Cooling Systems

Cooling your home, like heating it, can be accomplished in sensible, inexpensive, energy-efficient ways, or in the wasteful way that Americans have favored for the past several decades in the era of cheap fossil fuels: Install a big air conditioner, crank it up, and pay the bills. It's much better for your pocketbook, and for the planet, to pursue the common-sense approach: Reduce your cooling load, take advantage of passive solar cooling strategies, utilize low-energy methods like fans, and increase the efficiency of your existing air conditioner. Finally, if you determine that it's cost effective, you should explore the prospect of buying new, efficient cooling equipment.

Mature trees shading your house will keep it much cooler by lowering roof and attic temperatures, blocking unwanted direct solar heat radiation from entering the windows, and providing a cooler microclimate outdoors. Adapted from an illustration by Saturn Resource Management, Helena, Montana.

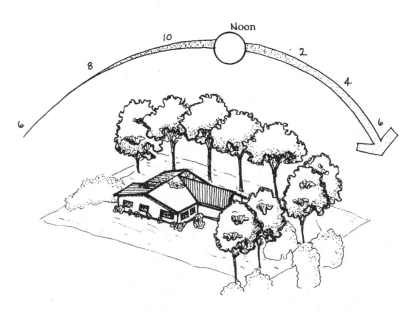

Reduce the Cooling Load Through Smart Design and Passive Solar Strategies

The best strategy for keeping a dwelling cool is to keep it from getting hot in the first place. This means preventing outside heat from getting inside, and reducing the amount of heat generated inside by inefficient appliances such as incandescent lights, unwrapped water heaters, or older refrigerators.

In a hot, humid climate, 25% of the cooling load is the result of infiltration of moisture; 25% is from outdoor heat penetrating windows, walls, and roof; 20% is from unwanted solar gain through windows; and the remaining 30% is from heat and moisture generated within the home. It makes a lot of economic sense to cut these loads before getting that new monster air conditioner. You'll be able to get by with a smaller, less-expensive unit that costs less to operate for the rest of its life. And by cutting your power consumption, you'll significantly reduce your personal emissions of greenhouse gases and other environmental pollutants.

The same kinds of smart design choices that reduce your heating load and make your home energy efficient also serve to reduce the cooling load. Weatherizing, insulating, radiant barriers, good-quality windows, environmental landscaping, effective ventilation, and roof whitening each has the capacity to reduce your cooling costs substantially, especially in combination. Funny how we keep hammering those points. These tricks for keeping outside heat from getting inside should be considered, and implemented, before you move on to even low-energy methods that consume electricity—especially if you rely on renewable power systems.

Each of these methods, plus others, is recapped briefly here. For more-detailed discussions of these aspects of energy-efficient building design, see Chapter 2.

WEATHERIZATION

Weatherization, usually done in northern climates to keep heat in, can also effectively keep heat and humidity out. Cutting air infiltration by half, which is easily done, can cut air-conditioning bills by 15% and save $50-$100 per year in the average house in the southern United States. Much of the work done by an air conditioner in humid climates goes into removing moisture, which increases comfort. Reducing air infiltration is more important in humid climates than in dry ones.

INSULATION

Insulation slows heat transfer from outside to inside your home, or vice versa in the winter. Attic insulation levels should be R-19 or higher. If you also need to heat your house in the winter, insulation is even more important because of the greater temperature differential between inside and outside. In that case, you'll want to have insulation levels of R-30 or higher. In hot climates, if you already have some significant amount of insulation, it will probably be more cost effective to install a radiant barrier instead of adding more insulation.

RADIANT BARRIERS

Radiant barriers are thin metal films on a plastic or paper sheet to give them strength. They will reflect or stop approximately 97% of long-wave infrared heat radiation. Such barriers typically are stapled to the underside of attic rafters to lower summertime attic temperatures. Laying these barriers on the attic floor is also an option, but research indicates that the effectiveness of the barrier is reduced by dust buildup. Lowering attic temperatures—which easily can reach 160°F on a sunny day—reduces heat penetration into the living space below and is frequently a cost-effective measure to lower air-conditioning bills. A study by the Florida Solar Energy Center found that properly installed radiant barriers reduced cooling loads by 7%-21%. The study also found that temperatures of the roofing materials were increased by only 2-10 degrees F, which will have little or no impact on the life expectancy of a shingle. A radiant barrier will reduce, but not eliminate, high attic temperatures, so insulating to at least R-19 is still advisable to slow down heat penetration.

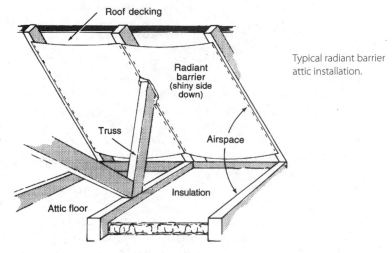

Typical radiant barrier attic installation.

The cost for radiant barrier material is very moderate at approximately 20¢-25¢ per square foot. See the product section that follows.

WINDOWS

For too long, a window has simply been a hole in the wall that lets light and heat through. Windows, even so-called "thermopane" (dual-glazed) windows, have very low R-values. Single-pane units have an R-value of 1. Thermopane windows have an R-value of 2. Both are pretty miserable.

Conventional windows conduct heat through the glass and frame, permit warm, moist air to leak in around the edges, and let in lots of unwanted heat in the form of solar radiation. South-facing windows with properly sized roof overhangs or awnings won't take in the high summer Sun but will accept the lower winter Sun when it usually is wanted. East- and west-facing windows cannot be protected so simply. See Chapter 2, "Land and Shelter," for a complete discussion of window treatments, including newer high-tech windows with R-values up to 8 or 10.

LANDSCAPING FOR SHADE AND PASSIVE COOLING

You can take advantage of so-called "environmental landscaping" to provide shading and/or to channel cooling breezes through your home.

Shading that blocks summer Sun on the east, south, and west sides of your house, but not summer breezes, is one of the most effective ways to keep your home cooler. Planting shade trees, particularly on the west and south sides of your house, can greatly increase comfort and coolness. (Remember that deciduous trees, which lose their leaves in winter but grow a thick canopy of leaves in the summer, should

Shading that blocks summer Sun on the east, south, and west sides of your house, but not summer breezes, is one of the most effective ways to keep your home cooler.

Renewable Energy Systems and Home Cooling

Traditional air conditioners, even small room-size units, and renewable energy systems are generally incompatible, but it can be done. The equipment and knowledge is readily available. The problem is the sheer quantity of electricity required for air conditioning, and the long hours of continuous operation, augmented by the high initial cost of renewable energy systems. As a ballpark estimate, we have found that it will require about $3.50 worth of renewable energy "stuff" for every watt-hour needed.

A 1,500-watt room-size A/C unit running for six hours consumes 9,000 watt-hours and needs over $30,000 worth of renewable energy system to support it.

be used on the south side in colder climates to maximize solar gain in winter and minimize it in summer.) Awnings, porches, or trellises on those same sides of a building will reduce solar gain through the walls as well as through the windows. A home's inside temperature can rise as much as 20 degrees F or more if the east and west windows and walls are not shaded.

Trees and shrubs planted in the proper configurations also can funnel and concentrate breezes to increase airflow through and around a house. Increased airflow translates into increased cooling action, all thanks to Mother Nature.

> Storing "coolth" works better if your house has thermal mass such as concrete, brick, tile, or adobe that will cool down at night and help prevent overheating during the day. These are the same things that make passive solar houses work better at heating in the winter.

MECHANICAL SHADING

In addition to trellises and awnings, various types of mechanical window-shading devices can be cost effective. Some are internal (such as drapes, curtains, and shades); some are external (such as louvers, shutters, and exterior roller shades). In the overall scheme of things, these devices are relatively inexpensive and they operate effectively to block Sun and heat from entering your home. Explore the options and decide which ones will work best in various applications.

ROOF WHITENING

A white roof is an obvious solution. If you're replacing a roof, using a white or reflective roofing surface will reduce your heat load on the house. If you are not reroofing, coating the existing roof with a white elastomeric paint is a cost-effective measure in some climates. Be sure to get a specialty paint specifically designed for roof whitening: Normal exterior latex won't do.

The Florida Solar Energy Center reports savings from white roofs of 25%-43% on air-conditioning bills in two poorly insulated test homes, and 10% savings in a Florida home with good R-25 roof insulation already in place. Costs are relatively high—the coatings cost 30¢-70¢ per square foot of roof surface area—and adding insulation or a radiant barrier may be a more cost-effective measure. Light-colored asphalt or aluminum shingles are not nearly as effective as white paint in lowering roof and attic temperatures.

PASSIVE NIGHTTIME COOLING

Opening windows at night to let cooler air inside is an effective cooling strategy in many regions. This method is far more effective in hot, dry areas than in humid areas, however, because dry air tends to cool down more at night and because less moisture is drawn into the house. The Florida Solar Energy Center found that when apartments in humid areas (and Florida certainly qualifies) opened their windows at night to let the cooler air in, the air conditioners had to work much harder the next day to remove the extra moisture that came along with the night air. Storing "coolth" works better if your house has thermal mass such as concrete, brick, tile, or adobe that will cool down at night and help prevent overheating during the day. These are the same things that make passive solar houses work better at heating in the winter.

ACTIVE SOLAR VENTILATION

Ventilating your attic is important for moisture control and keeping cool. Roofs can reach temperatures of 180°F, and attics can easily exceed 160°F. Unless you get rid of this heat, it will eventually soak through even the best insulation. Solar-powered vent fans are a very cost-effective way to move this unwanted heat out. By using solar power, they'll run only when the Sun shines, and the brighter the Sun, the faster they'll work. We offer complete kits in two different sizes; they include fan, temperature switch—so it will run only when the attic temperature is above 80°F—photovoltaic module(s), and mounting. See the product section.

WHOLE-HOUSE, CEILING, AND PORTABLE FANS

A whole-house fan is an effective means of cooling and is far less expensive to run than an air conditioner. It can reduce indoor temperatures by 3-8 degrees F, depending on the outside temperature. Through prudent use of a whole-

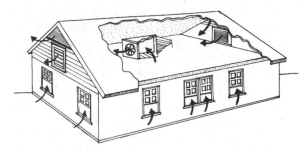

A whole-house fan can cool the house by bringing in cooler air—especially at night—and lowering attic temperatures. Adapted from an illustration by Illinois Department of Energy and Natural Resources (1987), *More for Your Money ... Home Energy Savings.*

house fan, you can cut air-conditioner use by 15%-55%. This is true particularly if you live in a climate that reliably cools off in the evening. A high R-value house can be buttoned up during the heat of the day, heating up inside slowly like an insulated ice cube. In the evening, when it's cooler outside than inside, a whole-house fan blows the hot air out of your house while drawing cooler air in through the windows. It should be centrally located so that it draws air from all rooms in the house. Be sure your attic has sufficient ventilation to get rid of the hot air. To ensure a safe installation, the fan must have an automatic shut-off in case of fire.

Ceiling, paddle, and portable fans produce air motion across your skin that increases evaporative cooling. A moderate breeze of one to two miles per hour can extend your comfort range by several degrees and will save energy by allowing you to set your air conditioner's thermostat higher or eliminate the need for air conditioning altogether. Less frequent use of air conditioning by setting the thermostat higher will cut cooling bills greatly. See our product section for more details.

EVAPORATIVE COOLING

If you live in a dry climate, an evaporative, or "swamp," cooler is an excellent cooling choice, saving 50% of the initial cost and up to 80% of the operating cost of an air conditioner. The lower the relative humidity, the better they work. Evaporative coolers are not effective when your relative humidity is higher than about 40%.

PERSONAL COOLING

Cool drinks, the Real Goods Solar Hat Fan, a minimum of lightweight clothing, siestas, going for a swim, or sitting on a shady porch—all of these personal cooling methods work.

Improving Air-Conditioner Efficiency

Once you've reduced heat penetration and taken advantage of passive solar and low-energy cooling methods, if you still have a need for additional cooling, it makes sense to improve the efficiency of your air conditioner.

CHECK DUCTS FOR LEAKS AND INSULATE

With central A/C systems, inspect the duct system for leaks and seal with duct tape or special duct mastic where necessary. The average home loses more than 20% of its expensive cooled or heated air before it gets to the register. It's well worth your time to find and seal these leaks. Just turn the fan on and feel along the ducts. Distribution joints are the most common leaky points.

If your ducts aren't already insulated in the attic or basement, insulate them with foil-faced fiberglass duct insulation. Batts can be fastened with plastic tie wraps, wire, metal tape, or glue-on pins. Your A/C technician can do this job or recommend someone who can, if you're not up for it.

HAVE YOUR HEAT PUMP OR CENTRAL A/C SERVICED REGULARLY

Regular maintenance will include having the refrigerant charge level checked—both under- and overcharging compromises performance—oiling motors and blowers, removing dirt build-up, cleaning filters and coils, and checking for duct leakage. Proper maintenance will ensure maximum efficiency as well as the equipment's longevity.
• Set your A/C to the recirculate option if your system has this choice—drawing hot and

Personal cooling.
Buying a block of ice isn't cost effective, but feeling cool is important. Cool drinks, small fans, less clothing, siestas, or sitting on a shady porch all work. Illustration courtesy of Saturn Resource Management, Helena, Montana.

Cool drinks, the Real Goods Solar Hat Fan, a minimum of lightweight clothing, siestas, going for a swim, or sitting on a shady porch—all of these personal cooling methods work.

Air conditioners have become much more efficient over the last 20 years, and top-rated models are 50%-70% better than the current average.

humid air from outside takes a lot more energy. Many newer systems only recirculate, so if you can't find this option, it's probably already built in.

- Set your A/C thermostat at 78°F—or higher if you have ceiling fans. For each degree you raise the thermostat, you'll save 3%-5% on cooling costs.
- Don't turn the A/C thermostat lower than the desired setting—cooling only happens at one speed. The house won't cool any faster, and you'll waste energy by overshooting.
- Turn off your A/C when you leave for more than an hour. It saves money.
- Close off unused rooms or, if you have central A/C, close the registers in those rooms and shut the doors.
- Install a programmable thermostat for your central A/C system to regulate cooling automatically. Such a thermostat can be programmed to ensure that your house is cool only when needed.
- Trim bushes and shrubs around outdoor condenser units so they have unimpeded airflow; a clear radius of 2 feet is adequate. Remove leaves and debris regularly.
- Provide shade for your room A/C or for the outside half of your central A/C if at all possible. This will increase the unit's efficiency by 5%-10%.
- Remove your window-mounted A/C each fall. Their flimsy mounting panels and drafty cabinets offer little protection from winter winds and cold. If you are unable to remove the unit, at least close the vents and tightly cover the outside.
- Clean your A/C's air filter every month during cooling season. Normal dust buildup can reduce airflow by 1% per week.
- Clean the entire unit according to the manufacturer's instructions at least once a year. The coils and fins of the outside condenser units should be inspected regularly for dirt and debris that would reduce airflow. This is part of the yearly service if you're paying a technician for the service.

IF YOU LIVE IN A HUMID CLIMATE

An important part of what a conventional air conditioner does is removing moisture from the air. This makes the body's normal cooling—sweating—work better, and you feel cooler. Unfortunately, some of the highest-efficiency models don't dehumidify as well as less-efficient air conditioners. A/C units that are too big for their cooling load have this problem also. High humidity leads to indoor condensation problems and gives us a sweaty mid-August New York City feeling—minus the physical and aural assaults and the great food and nightlife. People tend to set their thermostats lower to compensate for the humidity, using even more energy. Here are some better solutions.

- Reduce the fan speed. This makes the coils run cooler and increases the amount of moisture that will condense in the A/C rather than inside your house.
- Set your A/C to recirculate if this is an option and you haven't already done so.
- Choose an A/C with a "sensible heat fraction" (SHF) less than 0.8. The lower the SHF, the better the dehumidification.
- Size your A/C carefully. Contractors should calculate dehumidification as well as cooling capacity. Many air-conditioning systems are oversized by 50% or more. The old contractor's rule of thumb—1 ton of cooling capacity per 400 square feet of living space—results in greatly oversized A/C units when reasonable weatherization, insulation, and shading measures have been implemented to reduce cooling load. One ton per 800-1,000 square feet may be a more reasonable rule of thumb.
- Investigate using a Dinh heat pipe/pump. These devices cool the air before it runs through the A/C evaporator and warm it as it blows out, and this without energy input. They can help your A/C dehumidify several times more effectively.
- Explore "desiccant dehumidifiers," which, coupled with an efficient A/C, may save substantial energy. Desiccant dehumidifiers use a water-absorbing material to remove moisture from the air. This greatly reduces the amount of work the A/C has to do, thereby reducing your cooling bill.

Buying a New Air Conditioner

Air conditioners have become much more efficient over the last 20 years, and top-rated models are 50%-70% better than the current average. Federal appliance standards have eliminated the least-efficient models from the market, but builders and developers have little incentive to install a model that's more efficient than required, since they won't be paying the higher electric bills.

If you have an older model, it may be cost effective to replace it with a properly sized, efficient unit. The Office of Technology Assessment—a U.S. Congress research organization

—estimates that buying the most efficient room air conditioner, costing $70 more than a standard unit, has a payback of six and a half years or less, if the additional features the better units include are taken into account.

AIR-CONDITIONER EFFICIENCY

Several studies have found that most central air-conditioning systems are oversized by 50% or more. To avoid being sold an oversized model, or one that won't remove enough moisture, ask your contractor to show you the sizing calculations and dehumidification specifications. Many contractors may simply use the "1 ton of cooling capacity per 400 square feet of living space" rule of thumb. This will be far larger than you need, especially if you've effectively weatherized and insulated your home and done some of the other measures we discuss here to reduce heat gain and the need for cooling.

One ton of cooling capacity—an archaic measurement equaling approximately the cooling capacity of one ton of ice—is 12,000 BTUs per hour. Room air conditioners range in size from ½ ton to 1½ tons, while typical central A/Cs range from 2 to 4 tons.

Desuperheaters

It's possible to use the waste heat removed from your house to heat your domestic water. In hot climates, where A/C is used more than five months per year, it makes economic sense to install a desuperheater to use the waste heat from the A/C to heat water. Some utilities give rebates to builders or homeowners who install such equipment.

Evaporative Coolers

If you live in a dry climate, an evaporative, or "swamp," cooler can save 50% of the initial cost and up to 80% of the operating cost of normal A/C. An evaporative cooler works by evaporating water, drawing fresh outside air through wet,

porous pads. This is an excellent choice for most areas in the southwestern Sunbelt. The lower the relative humidity, the better they work. At relative humidity above 30%, performance will be marginal; above 40% humidity, forget it! "Indirect" or "two-stage" models that yield cool, dry air rather than cool, moist air are also available, but the same low humidity climate restrictions apply.

Heat Pumps

If you live in a climate that requires heating in the winter and cooling in the summer, consider installing a heat pump in new construction or if you are replacing the existing cooling systems. The ground- or water-coupled models are the most efficient; see the discussion on page 297 and the comparison chart included there. The air-conditioning efficiency of a heat pump is equivalent to that of a typical air conditioner, as they're basically the same thing, but the heating mode will be much more efficient than the cooling mode, outperforming everything except natural gas-fired furnaces.

Appliances for Cooling

Since our core business is renewable energy, and the high energy consumption of air-conditioning equipment makes renewable energy systems very expensive, we offer a minimal selection of appliances. Apply the weatherization, insulation, shading, and alternative cooling tips we have discussed, and your need for energy-intensive mechanical cooling will be eliminated in all but the most severe climates.

Real Goods sells a selection of DC-powered cooling appliances that can be run from batteries or directly from photovoltaic modules. These include small, highly efficient evaporative coolers, a selection of fans for attic venting or personal cooling, and some ceiling-mounted paddle fans.

Apply the weatherization, insulation, shading, and alternative cooling tips we have discussed, and your need for energy-intensive mechanical cooling will be eliminated in all but the most severe climates. Real Goods sells a selection of DC-powered cooling appliances that can be run from batteries or directly from photovoltaic modules.

CONSERVATION PRODUCTS

Smart Power Strip

Even when they're off, today's electronics continue to draw electricity we pay for but don't use. This revolutionary power strip prevents that waste. Plug your main device (computer, TV, etc.) into the primary outlet and its peripherals (printer/scanner or VCR/cable box, etc.) into the others. High-tech sensors know when you shut down the main device, and they cut off everything else. Saves up to 72% of the energy your systems use and offers state-of-the-art surge protection too. With 6-foot cord, six no-idle outlets, and three always-on outlets. 16"x6.2"x2" (LxWxH). 2.2 lb. China.

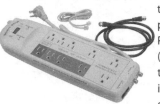

13-0062 Smart Power Strip $39

Hobo Temperature Logger

You won't find a better energy management tool than the cordless Hobo. It records a constant stream of temperature and humidity readings in any room. Upload this data to any Mac or PC, and you'll see exactly when and where cooling or heating is and isn't needed. Use this invaluable (and often surprising!) information to fine-tune thermostat settings and wring every last drop of energy savings from your home. Variable sampling intervals, programmable start times, and extended operation abilities adjust to any situation. Operating range is -4°F to 158°F; 0%-95% relative humidity. Software, USB cable, and CR2032 lithium battery included. Free software is downloadable. 2.4"x1.9"x0.8" (HxWxD). USA.

57-0113 Hobo Temp Logger $79

Insulation, Shading, and Radiant Barrier Products

Insulated Curtains

Fusing fashion with function, our insulated window coverings handsomely complement existing décor while helping preserve precious energy resources. With a heavy-duty sueded liner that keeps cold drafts and sweltering heat out, they're an affordable way to help keep your utility bills from going out of control. Sienna, linen, natural, or sage. Order Insulation Panels to make your existing tab-top curtains energy-smart. Reduces light by 50%. Canada.

01-0498 Tab-Top Curtain Panels (set of 2)
63"x40" (LxW) $54
84"x40" (LxW) $68

01-9497 Insulation Panels for Existing Drapes and Curtains (set of 2)
77"x39" (LxW) $44
77"x45" (LxW) $48

Our Roman-style coverings have the same energy-saving qualities as our drapes but also have a pull cord. Natural only. All are 72"L. Imported.

01-0535 Roman Shade

27"W	$58
30"W	$58
32"W	$64
34"W	$64
36"W	$68
38"W	$68
48"W	$99

Express delivery not available. Cannot be shipped to P.O. boxes. Can be shipped to customers in contiguous U.S. only. Gift wrap not available.

Draft Dodger

Classic weatherbeater locks out drafts so you stay warmer for less

It's a simple idea that's worked without fail for generations. Covered in hardy ThermaFleece, trimmed in satin piping, stuffed with insulating poly fibers, and weighted to stay right where you need it most, the Draft Dodger shuts out cold air to lower energy bills and keep you cozy. 36"x4" (LxH). China.

17-0183 Draft Dodger $20

Water Absorber

Filled with all-natural crushed corncobs that absorb 1,000% of their weight in water, the Water Absorber folds around problem windows, doors, tubs, basement areas, and more to soak up leaks and spills. Just air-dry when saturated and re-use. Fully dries in about 3-4 hours. 36"x5¾" (LxW). China.

01-0501 Water Absorber $16

KeepCool Radiant Barrier

A radiant barrier is your best line of defense for keeping excess heat out of your attic in the summer and keeping your floors warm in the winter. In a barn, chicken coop,

or anywhere it's vital to keep farm animals and livestock warm in winter and cool in summer, a radiant barrier will provide invaluable protection. Stapled under the rafters or joists, it reflects 97% of the radiant heat. The nonprofit Florida Solar Energy Center has shown that an attic radiant barrier can reduce air-conditioning costs by at least 20%. Performance in drier climates is even better. This perforated product won't become an unintentional moisture barrier. When the heat source is above the radiant barrier, it performs equivalently to R-14 insulation. To function, it must have an air space on at least one side. It doesn't matter if it's the hot or cold side. Installation is a snap—just staple the radiant barrier to the bottom of rafters or joists. Perforated foil over a lightweight, tear-resistant paper/poly-fiber backing that passes UL-723 flame spread tests. Supplied in 48-inch-wide rolls to cover 500 or 1,000 square feet. USA.

56730   KeepCool Radiant Barrier 125'L $90

56731   KeepCool Radiant Barrier 250'L $170

All-Weather Shades

Keep your house and outdoor dining areas pleasantly cool when outdoor temperatures rise with the Coolaroo Shade Sails and the easy roll-up Coolaroo Window Shade. Tough, all-weather knit fabric blocks up to 90% of the Sun's ultraviolet rays, allows plenty of air and light to pass through, and resists mold, mildew, and rot. Shade Sails attach easily to existing objects with corner grommets to create a stylish shaded area. Hardware not included. Choose from triangle or square. Sand or green. Window Shades can be installed on exterior windows or used to enclose your porch to make a cool sunroom. Includes installation hardware, tie-downs, and easy-to-follow instructions. Pebble or smoke. All wash clean with a spray of the hose. Australia.

14-0451 Triangle Shade Sail (11' per side) $168

14-0450 Triangle Shade Sail
**   (16'5" per side) $188**

14-0452 Square Shade Sail (11'10"W) $228

14-0453 Square Shade Sail (17'9"W) $329

14-0232 Coolaroo Window Shade
**   6'x4' (LxW) $68 ($15)**
**   6'x6' (LxW) $88 ($15)**
**   6'x8' (LxW) $118 ($28)**

Allow 4-6 weeks for delivery. Can be shipped to customers in the contiguous U.S. only. Cannot be shipped to P.O. boxes. Express delivery not available.

Shadecloth

The Sun Tex Shadecloth blocks up to 80% of the Sun's glare, and the Super Solar Screen blocks up to 90%. A great way to reduce cooling costs and create privacy during the day, these cloths appear virtually opaque from the outside yet allow you to see out clearly. Made from vinyl-coated polyester, they're mildew- and fade-resistant, too. Easy to install and easy to remove. Try on windows, skylights, greenhouses, patios, and doors. Available in 7-foot-long rolls. USA.

06-0349 Sun Tex 80% Shadecloth 36"W $18

06-0350 Sun Tex 80% Shadecloth 48"W $24

06-0351 Super Solar 90% Screen 36"W $22

06-0352 Super Solar 90% Screen 48"W $28

Window Tinting

Our do-it-yourself window-tinting solution reduces UV transmission by 97%. Equally important, window tinting cuts your utility demand and saves you money because it decreases heat penetration into your home by 60%. It applies easily to wet windows or Plexiglas, and static electricity—not adhesive—holds it in place. Lift it off instantly when the seasons change. Reduces glare and protects furniture, drapes, and carpeting from fading. Accommodates a variety of window widths. For

multipane windows, apply to the outer pane. Available in 15-foot-long rolls. USA.

17-0177 Window Tinting

24"W	**$38**
36"W	**$48**
48"W	**$58**

17-0178 Window Tinting Installation Kit (squeegee and knife)   **$4**

HEATING PRODUCTS

Space Heating

EMPIRE DIRECT-VENT HEATERS

Direct venting makes these gas heaters safe and efficient for bedrooms, bathrooms, other closed rooms, or airtight homes. The entire combustion chamber is sealed off from inside air. Outside air is used for combustion, so there's no chance of combustion byproducts in your living space, oxygen depletion, or wasting of already heated air. Most of these heaters must mount on an outside wall because intake and exhaust go out the back.

Empire MV-Series Heaters

- Room-size heaters, with 8,400, 14,000, and 20,000 BTU models
- Thermostat with modulating control
- Matchless piezo igniter with standing pilot
- No electricity
- Natural gas or LP (propane)
- 10-yr. warranty on combustion chamber, 1-yr. on all else

The slim, attractive, tan Euro-styled MV-series seems to float a couple inches off the wall with safe, pleasant radiant heat output. The thermostat modulates the gas flame for steady heat output with no fans or noise. Intake and exhaust are straight out the rear through a single double-wall pipe. Mounts to any wall 4½"-12" thick. Optional extensions to 18" wall thickness are available. Minimum clearances: floor 4¾", sides 4", back 0", top 36". AGA and CGA approved. Specify LP or Natural Gas. Argentina. Warranted by U.S. importer.

Model #	MV-120	MV-130	MV-145
Max. (BTUs/hr.)	8,400	14,000	20,000
Min. (BTUs/hr.)	3,800	6,000	6,000
Width	21.2"	27.2"	27.2"
Height	24.4"	24.4"	24.4"
Depth	6.75"	6.75"	6.75"
Weight	33 lb.	42 lb.	55 lb.
Item #	54-0045	54-0046	54-0047
Price	$460	$545	$620

Empire DV-Series Heaters

- For medium to larger rooms, with 25,000 and 35,000 BTU models
- Matchless piezo ignition with standing pilot
- Includes direct-vent kit and millivolt thermostat
- Optional automatic internal blower package
- Natural gas or LP (propane) models available
- 10-yr. warranty on combustion chamber, 1-yr. on all else

Heating larger areas is convenient, economical, and safe with Empire Direct-Vent models. The millivolt thermostat requires no electricity. An optional automatic blower package delivers 75-cfm airflow to help distribute the heat more evenly for larger spaces. Outside vent cap included. The vinyl siding option offers extra heat protection. Beige color. Mounts to any outside wall 4½"-13" thick. Minimum clearances: floor 4", sides 6", back 0", top 48". AGA and CGA approved. Specify LP or Natural Gas. USA.

54-0050 Automatic Blower Option		**$125**
54-0051 Vinyl Siding Vent Option		**$30**

Model #	DV-25-SG	DV-35-SG
Input (BTUs/hr.)	25,000	35,000
Width	37"	37"
Height	27.75"	27.75"
Depth	11.5"	11.5"
Weight	98 lb.	101 lb.
Item #	**54-0048**	**54-0049**
Price	**$690**	**$765**

Empire Console-Type Heaters

- Standard vented units (not direct-vent)
- For larger rooms, with 65,000 BTU models
- Matchless piezo ignition with standing pilot
- Includes thermostat
- Automatic internal blower
- Natural gas or LP (propane) models available
- 10-yr. warranty on combustion chamber, 1-yr. on all else

These larger standard vented units are available with either a closed front or with flame-view glass. A ceramic log option is available for the glass-front model. Automatic blowers put warm air at floor level. The 5-inch exhaust vent can run vertically or horizontally from heater back. Minimum clearances: floor 0", sides 6", back 2", top 55". AGA and CGA approved. Specify LP or Natural Gas. USA.

54-0065 Ceramic Log Option for RH-65B $175

Model #	RH-65CB	RH-65B (w/glass)
Input (BTUs/hr.)	65,000	65,000
Width	34"	34"
Height	29.6"	29.6"
Depth	20"	20"
Weight	142 lb.	142 lb.
Item #	**54-0064**	**54-0063**
Price	**$899**	**$1,115**

Ecofan Warms Without Electricity

Maximize the warmth achieved from precious timber resources by boosting your wood stove's efficiency up to 30% with the innovative Ecofan. Using simple thermodynamic technology, the hotter your stove gets, the faster it quietly propels heat back into your living space—using zero electricity. Peak performance occurs at surface temperatures of 400°-650°F. A temperature-sensitive, bimetal strip on the unit automatically tilts it slightly to prevent overheating. The new and improved Ecofan Plus has a larger heat-absorbing surface area that multiplies its capacity to transfer the fire's BTUs into your environment. The 9-inch-diameter tri-blade unit moves an impressive 150 cubic feet of air per minute. Made of anodized aluminum, fans won't rust or corrode. Canada.

06-0146 Ecofan	**$109**
17-0180 Ecofan Plus	**$150**

Holy Smokes! Firestarter

Ignitable sticks set fires aglow without combustible chemicals or paper. Crafted from a combination of recycled church candles and wood fiber, Holy Smokes! Firestarters get a fire blazing in minutes. One-quarter stick of starter gets logs going. Habitat for Humanity receives a donation with your purchase. One package starts more than 30 fires. Scented. USA.

07-9294 Holy Smokes! Firestarter (set of 2) $16

Germ-Free Humidifiers

Get ready for winter with the best Warm-Mist Humidifier we've ever found (a cool-mist version is available for summer). Both models create moist air for a healthier home and sterilize the vapor with an internal ultraviolet light that's proven effective through independent laboratory testing to kill 99.99% of germs and spores. Use the Warm-Mist Humidifier any time of year. Both are excellent for asthma and allergy relief, sickroom humidification, infant rooms, and general purpose. Ultraquiet fan propels warm, sterile vapor into the air without white dust. Removable easy-fill tanks offer enough capacity for two to three days of continuous operation on both models, and each includes a refill indicator light, easy-maintenance stainless steel reservoir, and two long-lasting Mineral Absorption Pads to reduce scale (replace every 2-3 months, depending on buildup). Taiwan.

01-0180 Germ-Free Warm-Mist Humidifier $159
    3.5 gal.; 14"x14¼"x13¼" (HxWxD); 9.5 lb.

01-0511 Germ-Free Cool-Mist Humidifier $129
    2 gal.; 13"x13¾"x12½" (HxWxD);
    10.6 lb. (See gaiamliving.com for photo.)

**01-0174 Mineral Absorption Pad
    Refills (set of 6) $12**

Whole Home Humidifier

Relieve dry skin, chapped lips, and other symptoms of unhealthy dry air without maintaining a humidifier in every room. The Whole Home Humidifier produces 8 gallons of warm mist per day to humidify up to 1,700 square feet of space. The compact space-saving design hides an easy-fill water tank with indicator light; programmable digital controls adjust in 5% increments between 30% and 90% relative humidity; a humidistat automatically controls output; and a special mineral-filtering wick produces clean mist. Choose day speed or quiet night speed and enjoy properly moisturized air throughout your home. Wicks last 2-3 months. 18"x12½"x21" (HxWxD). 17 lb. USA.

01-8026 Whole Home Humidifier $189 ($15)
01-8025 Replacement Wick $20

People Heating

Heated Foot Mat

Handy foot mat stops the cold where it starts, before it starts

Cold extremities easily lead to head-to-toe chills. The cure for such discomfort often lies in highly localized warmth that stops the source of the sensation for full body relief. Using less energy than a light bulb, the carpeted Heated Foot Mat employs this efficient strategy. Under desks and tables, or next to sofas and chairs, two convenient remote-switched settings of radiant heat deliver gently soothing foot warmth that never burns and can't overload. On full power, the unit uses 70 watts; low power uses 35 watts. 18"x19" (LxW). USA.

13-0031 Heated Foot Mat $99

12-Volt Heated Blanket

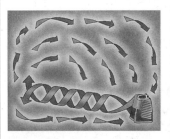

Everyone who travels during the winter should have one of these in the car. Soft, 100% polyester fleece travel blanket has an 8-foot cord and plugs into a 12-volt lighter socket for quick warmth whenever you need it. Draws just 2.9 amps, about as much as your parking lights, so you can stay warm without running the engine continuously. Navy. 57"x41" (LxW). China.

55339   Heated Blanket $39

Digital Vortex Heater

Consider a space heater to directly heat yourself or to heat very small spaces. Our compact Digital Vortex Heater automatically delivers the right amount of heat by continuously monitoring intake air temperature and projecting a stream of warm air around the room. It doesn't cycle on and off like a conventional thermostatically controlled heater, so heat flow is constant and consistent. Easily accessible digital control panel with large numbers is simple to understand and operate. Features a fan-only setting for warm-weather use, lighted power switch, and cool-touch cabinet. Auto shut-off protects against overheating. 11¼"x9¾"x11¾" (LxWxH). UL-listed. USA.

01-8024 Digital Vortex Heater $129

COOLING PRODUCTS

Evaporative Air Cooler

Cool your space for about a dime a day with the energy-efficient Evaporative Air Cooler. Simple yet effective evaporative cooling technology provides effective heat relief without ozone-depleting CFCs. Use alone or to boost older A/C systems. Chills air as far as 7 feet. Includes two-speed fan, air filter, adjustable louvers, and large-capacity water reservoir for 10 hours of continual cooling. Works best when relative humidity is 50% or below. One-year replaceable wick. 21½"x12"x127/8" (LxWxH). China.

01-0112 Evaporative Air Cooler   **$149**

01-0427 Replacement Wick   **$49**

Turbo-Aire™ Fan with Stand

From even 20 feet away, the precision-engineered Turbo-Aire Aerodynamic Fan will leave you feeling fresh and cool. Using the same power required of a 100-watt light bulb, the fan creates a powerful, focused column of air that draws in surrounding air and delivers 100% more cooling power. The result is a fan that's 300% more energy efficient than comparable fans. Includes three speed settings and adjustable-angle ABS housing. Fan arrives fully assembled; stand needs simple assembly. Made from 10% recycled materials. 8-foot cord. 18"H. 12.5 lb. Canada.

01-0526 Turbo-Aire Fan   **$115**

Portable Air Conditioner

Cool only where you need it, when you need it for energy-saving comfort. This Portable Air Conditioner features a 24-setting programmable timer, durable caster wheels and side-carry handles, and a quiet, highly efficient compressor. The dehumidifying mode eliminates wasteful condensation as it filters the air with dripless operation (no bucket to empty). Features remote control, single-exhaust hose, window bracket, and permanent washable filters. 10,000 BTUs. 30½"x19½"x18½" (HxWxD). 77 lb. Italy.

01-0544 Portable Air Conditioner   **$998**

Millennia Evaporative Cooler

For just pennies a day, this state-of-the-art evaporative cooler chills whole rooms to replace or augment energy-gobbling A/C units. Powerful, 1,500-cubic-feet-per-minute airflow removes 3,350 BTUs of heat each hour from up to 200 square feet of space while filtering dust, pollen, and other allergens. Set on wheels that easily roll it anywhere, it has three fan speeds and a huge 3.7-gallon capacity. Works best when relative humidity is 50% or less. 30"x22"x14" (HxWxD). Australia.

01-0541 Millennia Evaporative Cooler   **$369**
    Extra Shipping   **($15)**

01-0542 Pads   **$29**

Dehumidifier

Eliminating growth of mold, mildew, and bacteria in humid climates is one of the most effective ways to reduce allergy symptoms. The Energy-Saving Dehumidifier keeps the humidity in your home at a healthy and comfortable level, while the Energy Star™ rating ensures that you keep your energy bills low. A 24-hour programmable timer and electronic controls with LCD display allow for easy and constant monitoring. Never empty a bucket again with the patented pump system  that removes moisture from the collection tank via the included 16-foot or 3-foot hoses. The dehumidifier function can be operated with or without the pump. For added health, the permanent, washable air filter traps dust particles and extends the life of the unit. Compact size and high-efficiency airflow design contribute to very quiet operation. Will remove up to 40 pints of water in 24 hours. Rolling casters and handles allow full portability. 14½"x15½"x23" (LxWxH). 52 lb. Italy.

01-0456 Energy-Saving Dehumidifier   **$399**
    Extra Shipping   **($80)**

Allow 4-6 weeks for delivery. Can be shipped to customers in the contiguous U.S. only. Cannot be shipped to P.O. boxes. Express delivery not available.

Solar Attic Fans

Keep your attic optimally ventilated using only Sun power. These screened fans employ advanced circuitry and a built-in PV panel to capture sunlight (even on overcast days) and power a 12-inch fan pitched for maximum airflow of 850 to 1,250 cubic feet per minute (cfm). Attics stay cooler and drier, preventing mold, mildew, dry rot, and roof damage and saving on home cooling costs. UV-stabilized materials are designed to blend into most roofs, easily install in about an hour, and resist extreme weather. 21¼"x25½"x7" (LxWxH). USA.

14-0438	**850 cfm Solar Attic Fan**	**$499**
	1,050 cfm Solar Attic Fan	**$599**
	1,250 cfm Solar Attic Fan	**$699**
14-0439	**Fixed Thermostat Option**	**$29**

Solar-Powered Attic Fan Kits

Use solar power to move hot, stagnant air out of your attic, greenhouse, or barn. Our kits include the DC fan and a pair of 13-watt PV modules. The 12-inch fan moves 600-800 cfm in full Sun. The 16-inch fan moves 800-1,000 cfm. Fans have a 3-yr. mfr. warranty; USA. PV modules have a 90-day warranty; China.

64-479	**12" Attic Fan Kit**	**$359**
64-480	**16" Attic Fan Kit**	**$379**
64-216	**12" DC Fan**	**$149**
64-217	**16" DC Fan**	**$169**

Solar-Powered Attic Fan Kits			
	Input	Draw	Output
12" Fan	14 volts	1.47 amps	1,050 cfm
	28 volts	4.20 amps	1,430 cfm
16" Fan	14 volts	1.84 amps	1,100 cfm
	28 volts	5.30 amps	2,275 cfm

12-Volt DC Freedom Fan

High-efficiency DC cooling, ideal for boats, RVs, cabins, or anywhere that 12-volt power is used. The two-speed Freedom Fan is quieter and more efficient and powerful than any other 12-volt fan in its class. Built in the USA by the Amish, this rugged fan uses only 1.25 amps at low speed and 3 amps at high speed. It has a 12-inch fan blade protected by a durable plastic housing. Place it on a table and tilt it to any angle, or mount it to the wall. Comes with either battery clips or a cigarette lighter plug on the power cord.

50-0122 DC Freedom Fan   **$145**

Oscillating 8-Inch DC Table Fan

This 12-volt table fan features a weighted base with rubber feet and two-speed operation. Draws about 1.2 amps on low, and 2 amps on high. Oscillating feature can be switched on or off just like an AC fan.

64241   Oscillating DC Table Fan   **$49**

Compact Desk Fan

Powerful, quiet, battery-operated fan keeps air fresh and cool

This powerful, variable-speed fan will keep you cool and comfortable. The computerized speed control allows for a complete range of fan speeds—a better option over typical three-speed fans. The last setting is conveniently remembered when the fan is turned off and on again, and the tilt of the head is adjustable. Offering an impressive airflow, it runs for more than 300 hours on just one set of four "D" batteries (not included). Or use the included 6-volt AC/DC adapter and plug it in anywhere. 35/8"x5¾"x8¾" (LxWxH). 1-yr. mfr. warranty. Canada.

64255   Compact Desk Fan   **$25**

Vari-Cyclone DC Ceiling Fan

The Vari-Cyclone utilizes revolutionary "Gossamer Wind" blade design technology that delivers 40% more airflow without any increase in power use. Intended for 12- or 24-volt off-grid DC power systems, this 60-inch-diameter fan can be assembled with either three or four blades. Finish of blades and body is white. Can be mounted as either a close flush mount for flat ceilings or with a down rod ball mount for sloped ceilings. Hardware is included for both. The close mount hangs down 9.5 inches; the down rod takes 13 inches. The fan becomes variable speed and reversing with the addition of 12- or 24-volt solid state control listed below. Performance specs (with 3-blade configuration): At 12 volts: 60 rpm, 7 watts, 2,500 cfm. At 24 volts: 120 rpm, 27 watts, 3,400 cfm. We recommend the Voltage Doubler adjustable fan speed control (below) for 12-volt use. Fan weight approximately 14 lb. USA.

53-0100 Vari-Cyclone DC Fan $279

Solar Hat Fan

A practical solar gadget that really cools you down on a hot day! It clamps onto your favorite hat (stiffer-type brims and baseball caps are recommended) and gets to work harvesting enough solar energy to directly power the working fan. Features a small photovoltaic panel with an adjustable mount. We sold these solar hat fans to a roofing company, whose workers can't be without them as they cool down from the 100°F temperatures. 2¾"x21/8"x2¼" (LxWxH). China.

17-0328 Solar Hat Fan $10
3 for $8 each

Voltage Doubler

Delivers 24 volts from a 12-volt source for DC ceiling fans. Two-amp output at 99% efficiency. Installed on an ivory faceplate with on/off switch. Fits in a standard single-gang electric box. Accepts 8V-16V DC input. Output voltage will be double the input. USA.

64505   Voltage Doubler $59

BOOKS

Consumer Guide to Home Energy Savings, 8th Ed.

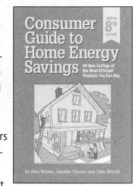

Energy efficiency pays. Written by the nonprofit American Council for an Energy-Efficient Economy, the *Consumer Guide* has helped millions of folks do simple things around their homes that save hundreds of dollars every year. The new 8th edition has up-to-date lists of all the most energy-efficient appliances by brand name and model number, plus an abundance of tips, suggestions, and proven energy-saving advice for homeowners. 264 pages, softcover.

21-0344 Consumer Guide to
Home Energy, 8th Ed. $9.95

The Fuel Savers: A Kit of Solar Ideas for Your Home

What can you do right now, with your funky existing home and a minimum of cash, to save energy and live more comfortably? This book has a wealth of ideas from insulating curtains to solar hot-water heaters. Each idea is examined for materials costs, fuel reduction, advantages, disadvantages, and cost effectiveness. A great step-by-step guide for improving the comfort and efficiency of any building. Updated in 2002. 83 pages, softcover.

21-0357 The Fuel Savers $15

Energy-Efficient Appliances

Why Bother to Shop for Energy Efficiency?

Wasted energy translates into carbon dioxide production, air pollution, acid rain, and lots of money down the drain. The average American household spends more than $1,100 per year on appliances and heating and cooling equipment. You can easily shave off 50%-75% of this expense by making intelligent appliance choices.

For example, simply replacing a 20-year-old refrigerator with a new energy-efficient model will save you about $100 per year in reduced electric bills, while saving 1,000kWh of electricity and reducing your home's CO_2 contribution by about a ton per year. Highly efficient appliances may be slightly more expensive to buy than comparable models with lower or average efficiencies. However, the extra first cost for a more efficient appliance is paid back through reduced energy bills long before the product wears out.

WHAT APPLIANCES CAN I BUY FROM REAL GOODS?

Most appliance manufacturers prefer to sell through established appliance stores. Your mail-order choices are limited to more specialized or high-efficiency appliances. At present, we offer several small washers; the full-size, highly efficient Staber horizontal-axis washing machine; the Spin-X centrifugal dryer; several varieties of incredibly efficient clotheslines; gas and DC refrigerators and freezers specifically for off-the-grid folks; high-efficiency tankless water heaters in both gas and electric; and the most problem-free solar water heaters available. We want to enable our customers to make intelligent appliance purchases. Thanks to the nonprofit American Council for an Energy-Efficient Economy, we can offer the up-to-date 8th edition of the *Consumer Guide to Home Energy Savings*. This is a more-or-less annual listing of the most efficient refrigerators, clothes washers, dishwashers, tank-type water heaters, air conditioners, and home heating appliances available. Fortify yourself with this wonderful book before venturing into any major appliance purchase.

Refrigerators and Freezers

The refrigerator is likely to be the largest single power user in your home aside from air conditioning and water heating. Refrigerator efficiency has made enormous strides in the past 15 years, largely due to insistent prodding from the Feds with tightening energy standards. An average new fridge with top-mounted freezer sold today uses under 500 kilowatt-hours per year, compared with more than 1,800 kilowatt-hours annually for the average 1973 model. The most efficient models available today are under 400 kilowatt-hours per year and still dropping. The typical refrigerator has a life span of 15-20 years. The cost of operation over that time period will easily be two to three times the initial purchase price, so paying somewhat more initially for higher efficiency offers a solid payback. It may not be worth scrapping your 15-year-old clunker to buy a new energy-efficient model. But when it does quit, or it's time to upgrade,

> The average American household spends more than $1,100 per year on appliances and heating and cooling equipment. You can easily shave off 50%–75% of this expense by making intelligent appliance choices.

Refrigerator Shopping Checklist

- Smaller models obviously will use less energy than larger models. Don't buy a fridge that's larger than you need. One large refrigerator is more efficient than two smaller ones, however.
- Models with top- or bottom-mounted freezers average 12% less energy use than side-by-side designs.
- Features like through-the-door ice, chilled water, or automatic icemakers increase the profit margin for the manufacturer, and the purchase price by about $250. So ads and salespeople tend to push them. They also greatly increase energy use and are far more likely to need service and repair. Avoid these costly, troublesome options.
- Make sure that any new refrigerator you buy is CFC-free in both the refrigerant and the foam insulation.
- Be willing to pay a bit more for lower operating costs. A fridge that costs $75 more initially but costs $20 less per year to operate due to better construction and insulation will pay for itself in less than four years.

buy the most efficient model available. A great source for listings of highly efficient appliances is the Environmental Protection Agency's Energy Star® Web site, www.epa.gov/energystar/.

Be sure to have the Freon removed professionally from your clunker before disposal. Most communities now have a shop or portable rig to supply this necessary service at reasonable cost.

Also note that snazzy features, such as side-by-side refrigerator-freezer units and through-the-door ice or cold water, increase energy consumption. If you can do without, it's cost effective to avoid these features.

REFRIGERATION AND RENEWABLE ENERGY SYSTEMS

When your power comes from a renewable energy system, which has a high initial cost, the payback for most ultraefficient appliances is immediate and handsome. You will save the price of your appliances immediately in the lower operating costs of your power-generating system. This is particularly true for refrigerators.

Many owners of smaller or intermittent-use renewable energy systems prefer gas-powered fridges and freezers, especially when the fuel is already on site anyway for water heating and cooking. Gas fridges use an environmentally friendly mixture of ammonia and water, not Freon, as a refrigerant. We offer several gas-powered freezers and refrigerator/freezer combinations.

For many years, Sun Frost has held the enviable status as manufacturer of the "world's most efficient refrigerators," and that has produced a steady flow of renewable energy customers willing to pay the high initial cost in exchange for the savings in operating cost. Real Goods has been an unabashed supporter of this product for many years. But times are changing. The major appliance manufacturers have managed to implement huge energy savings in their mass-produced designs over the past several years. While few of the mass-produced units challenge Sun Frost for king-of-the-efficiency-hill, many are close enough, and less expensive enough, to deserve consideration. The difference between a $900 purchase and a $2,700 purchase will buy quite a lot of photovoltaic wattage. A number of mass-produced fridges have power consumption in the 1.0 to 1.5 kilowatt-hours per day range and can be supported on renewable energy systems. Any reasonably sized fridge with power use under 1.5 kilowatt-hours per day is welcome in a renewable system; that's $45-$50

Refrigerators and CFCs

All refrigerators use some kind of heat-transfer medium, or refrigerant. For conventional electric refrigerators, this has always been Freon, a CFC chemical and prime bad guy in the depletion of the ozone layer. Manufacturers all have switched to non-CFC refrigerants, and by now it should be nearly impossible to find a new fridge using CFC-based refrigerants in an appliance showroom.

What most folks probably didn't realize is that of the 2.5 pounds of CFCs in the typical 1980s fridge, 2 pounds are in the foam insulation. Until very recently, most manufacturers used CFCs as blowing agents for the foamed-in-place insulation. Most have now switched to non-CFC agents, but if you're shopping for new appliances, it's worth asking.

The CFCs in older refrigerators must be recovered before disposal. For the 0.5 pound in the cooling system, this is easily accomplished, and your local recycler, dump, or appliance repair shop should be able either to do it or to recommend a service that will. To recover the 2 pounds in the foam insulation requires shredding and special equipment, which sadly happens only rarely.

per year on the yellow EnergyGuide tag that every new mass-produced appliance wears in the showroom.

In Sun Frost's defense, we need to point out that conventional refrigerators run on 120 volts AC, which must be supplied with an inverter and will cost an extra 10% due to inverter efficiency. Also, if your inverter should fail, you could lose refrigeration until it gets repaired. Sun Frosts (if ordered as DC-powered units) will run on direct battery power, making them immune to inverter failures.

IMPROVING THE PERFORMANCE OF YOUR EXISTING FRIDGE

Here is a checklist of tips that will help any refrigerator do its job more easily and more efficiently.

- Cover liquids and wrap foods stored in the fridge. Uncovered foods release moisture (and get dried out), which makes the compressor work harder.
- Clean the door gasket and sealing surface on the fridge. Replace the gasket if damaged. You can check to see if you are getting a good seal by closing the refrigerator door on a dollar bill. If you can pull it out without resistance, replace the gasket. On new refrigerators with magnetic seals, put a flashlight inside the fridge some evening, turn off the room lights, and check for light leaking through the seal.

- Unplug the extra refrigerator or freezer in the garage. The electricity the fridge is using—typically $130 a year or more—costs you far more than the six-pack or two you've got stashed there. Take the door off, or disable the latch so kids can't possibly get stuck inside!

- Move your fridge out from the wall and vacuum its condenser coils at least once a year. Some models have the coils under the fridge. With clean coils, the waste heat is carried off faster and the fridge runs shorter cycles. Leave a couple inches of space between the coils and the wall for air circulation.

- Check to see if you have a power-saving switch or a summer-winter switch. Many refrigerators have a small heater (yes, a heater!) inside the walls to prevent condensation buildup on the fridge walls. If yours does, switch it to the power-saving (winter) mode.

- Defrost your fridge if significant frost has built up.

- Turn off your automatic icemaker. It's more energy efficient to make ice in ice trays.

- If you can, move the refrigerator away from any stove, dishwasher, or direct sunlight.

- Set your refrigerator's temperature between 38° and 42°F, and your freezer between 10° and 15°F. Use a real thermometer for this, as the temperature dial on the fridge doesn't tell real temperature.

- Keep cold air in. Open the refrigerator door as infrequently and briefly as possible. Know what you're looking for. Label frozen leftovers.

- Keep the fridge full. An empty refrigerator cycles frequently without any mass to hold the cold. Beer makes excellent mass, and you probably always wanted a good excuse to put more of it in the fridge, but it tends to disappear. In all honesty, plain water in old milk jugs works just as well.

A new, more efficient refrigerator can save you $70–$80 a year on electricity and will completely pay for itself in nine years.

BUYING A NEW REFRIGERATOR

A new, more efficient refrigerator can save you $70–$80 a year on electricity and will completely pay for itself in nine years. It also frees you from the guilt of harboring a known ozone killer: Since 1999, all new fridges in the showroom are completely CFC-free.

Shop wisely by carefully reading the yellow EPA EnergyGuide label found on all new appliances. Use it to compare models of similar size.

The American Council for an Energy-Efficient Economy is a nonprofit organization dedicated to advancing energy efficiency as a means of promoting economic development and environmental protection. Their invaluable *Consumer Guide to Home Energy Savings* lists all the most efficient mass-produced appliances by manufacturer, model number, and energy use. We strongly recommend consulting this guide before venturing into any appliance showroom. It has an amazingly calming effect on overzealous sales personnel and allows you to compare energy use against the top-of-the-class models. See the product section.

Clothes Washers and Dryers

Laundry is another domestic activity that consumes considerable power. When it comes to appliances for washing and drying clothes, however, the energy-efficiency focus is decidedly lopsided. Much progress has been made recently on resource-efficient washing machines. Clothes dryers, on the other hand, are not yet required to display EnergyGuide labels, so it is difficult to compare their energy usage and, apparently, manufacturers have little incentive to improve efficiency. The good news about drying, of course, is that it's easy to dry clothes using solar power—by hanging them up on a line or a rack, indoors or outdoors—thereby eliminating or greatly reducing the need to use a power-hungry machine.

Most of the energy use in washing clothes is for heating water. The new resource-efficient washing machines use less water than conventional models do, which cuts their energy consumption by up to two-thirds. Some models

The EPA and Refrigerator Power Use Ratings

We're all familiar with the yellow energy-use tags on appliances now. These level the playing field and make comparisons between competing models and brands simple. With some appliances, like refrigerators, the EPA takes a "worst case" approach. Fridges are all tested and rated in 90°F ambient temperatures. The test fridge is installed in a "hot box" that maintains a very stable 90° interior temperature. This results in higher energy use ratings than will be experienced in most settings. At the mid-70° temperatures that most of us prefer inside our houses, most fridges will actually use about 20%–30% less energy than the EPA tag indicates.

are horizontal-axis front loaders, in which the clothes tumble through the water. Others are redesigned vertical-axis top loaders that spray the clothes or bounce them through the water rather than submerging them. Both types also use higher spin speeds to extract more water from the clean clothes, thereby reducing the energy required for drying (solar or otherwise).

The only energy-saving features available on dryers are sensors that shut off the machine when the clothes are dry. Most better-quality models on the market today have this feature. In addition to using a resource-efficient washer that spins more water out of your clean clothes, and taking advantage of free solar energy as much as possible, here are a few more tips for using a clothes dryer efficiently (courtesy of the American Council for an Energy-Efficient Economy).

- Separate clothes and dry similar types together.
- Dry two or more loads in a row to take advantage of residual heat already in the machine.
- Clean the filter after each use. A clogged filter will reduce airflow.
- Dry full loads when possible. Drying small loads wastes energy because the dryer will heat up to the same extent regardless of the size of the load.
- Make sure the outdoor exhaust vent is clean and that the flapper opens and closes properly. If that flapper stays open, cold air will infiltrate into the house through the dryer, increasing your wintertime heating costs.

See the product section that follows for more information about the efficient washers and dryers Gaiam Real Goods sells.

ENERGY-EFFICIENT HOME APPLIANCES AND PRODUCTS

ENERGY CONSERVATION

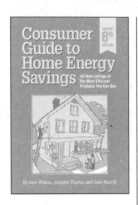

Consumer Guide to Home Energy Savings, 8th Ed.

Energy efficiency pays. Written by the nonprofit American Council for an Energy-Efficient Economy, the *Consumer Guide* has helped millions of folks do simple things around their homes that save hundreds of dollars every year. The new 8th edition has up-to-date lists of all the most energy-efficient appliances by brand name and model number, plus an abundance of tips, suggestions, and proven energy-saving advice for homeowners. 264 pages, softcover.

21-0344 Consumer Guide to Home Energy, 8th Ed.   **$9.95**

Electrically Powered Refrigeration

Sun Frost AC or DC Refrigerators and Freezers

Sun Frost employs a variety of design innovations that result in energy efficiency and ultraquiet operation. These units are a particularly good choice for off-grid customers who can run directly on DC power. The

smaller 12- and 16-cubic-foot Sun Frosts are so efficient they can be powered by as little as 150 watts of PV.

Compressors are top-mounted for thermal efficiency. Without the usual mechanical equipment underneath, a 13-inch stand is recommended to raise bottom shelf level. Sun Frost's optional stand has two drawers. Standard models are off-white Navimar (a Formica-like laminate). Colors are available for $100 extra, and unfinished wood veneers for $150. All models are 34½"x27½" (WxD). Heights vary by model. All models are available in 12- or 24-volt DC, or 120-volt AC, and can be hinged on the right or left (when facing unit). Voltage and hinge side not field switchable, so specify when ordering. Models available in 4, 10, 12, 16, or 19 cubic feet, as fridges and freezers, or in larger sizes as combo units. Call for models not listed. R = refrigerator, F = freezer, followed by cubic footage. USA.

62135-DC	**Sun Frost RF-19, DC** (66"H, 1,020Wh/day)	**$3,299**
62135-AC	**Sun Frost RF-19, 120VAC**	**$3,099**
62125	**Sun Frost R-19, DC**	**$2,899**
62125-AC	**Sun Frost R-19, 120VAC**	**$2,799**
62115	**Sun Frost F-19, DC**	**$3,299**
62115-AC	**Sun Frost F-19, 120VAC**	**$3,199**
62134-DC	**Sun Frost RF-16, DC** (62½"H, 700Wh/day)	**$3,099**
62134-AC	**Sun Frost RF-16, 120VAC**	**$2,899**
62133-DC	**Sun Frost RF-12, DC** (49.3"H, 470Wh/day)	**$2,299**
62133-AC	**Sun Frost RF-12, 120VAC**	**$2,199**
62122	**Sun Frost R-10, DC** (43½"H, 280Wh/day)	**$1,849**
62122-AC	**Sun Frost R-10, 120VAC**	**$1,749**
62112	**Sun Frost F-10, DC**	**$1,949**
62112-AC	**Sun Frost F-10, 120VAC**	**$1,849**
62131	**Sun Frost RF-4, DC**	**$1,599**
62121	**Sun Frost R-4, DC**	**$1,599**
62111	**Sun Frost F-4, DC**	**$1,599**
62116	**Sun Frost 13" Stand, White**	**$259**
62117	**Sun Frost Stand, Color**	**$335**
62118	**Sun Frost Stand, Woodgrain**	**$335**
62120	**Sun Frost Color Option**	**$150**
62119	**Sun Frost Woodgrain Option**	**$150**

Prices include $60 crating charge. Made to order; usually requires 4-6 weeks. Ships by truck from northern California.

Sun Frost Technical Data

Model	Power Usage @ 70°F*	Power Usage @ 90°F*	Weight Crated	Volume Crated	Exterior Dimensions (HxWxD)
RF-19	770Wh/day	1,030Wh/day	320 lb.	46 cu. ft.	66"x34.5"x27.75"
R-19	380Wh/day	630Wh/day	310 lb.	46 cu. ft.	66"x34.5"x27.75"
F-19	1,250Wh/day	1,630Wh/day	320 lb.	46 cu. ft.	66"x34.5"x27.75"
RF-16	560Wh/day	810Wh/day	300 lb.	44 cu. ft.	62.5"x34.5"x27.75"
RF-12	350Wh/day	560Wh/day	230 lb.	36 cu. ft.	49.3"x34.5"x27.75"
R-10	190Wh/day	310Wh/day	215 lb.	32 cu. ft.	43.5"x34.5"x27.75"
F-10	690Wh/day	880Wh/day	215 lb.	32 cu. ft.	43.5"x34.5"x27.75"
R-4	110Wh/day	160Wh/day	160 lb.	23 cu. ft.	31.5"x34.5"x27.75"
F-4	350Wh/day	450Wh/day	160 lb.	23 cu. ft.	31.5"x34.5"x27.75"
RF-4	160Wh/day	240Wh/day	160 lb.	23 cu. ft.	31.5"x34.5"x27.75"
RFV-4	160Wh/day	230Wh/day	160 lb.	23 cu. ft.	31.5"x34.5"x27.75"

*24-hour closed-door test with stated ambient outside temperature.

Interior Dimensions of Sun Frost Refrigerators and Freezers

Model	FREEZER SECTION Height	Depth	Width	Volume	REFRIGERATOR SECTION Height	Depth	Width	Volume
R, F, RF-19	24"	20.75"	28"	8.1 cu. ft.	24"	20.75"	28"	8.1 cu. ft.
RF-16	13"	20"	26"	3.91 cu. ft.	31"	20.75"	28"	10.4 cu. ft.
RF-12	6.5"	21"	26"	2.05 cu. ft.	24"	20.75"	28"	8.1 cu. ft.
R-10	—	—	—	—	28"	20.5"	27.5"	9.13 cu. ft.
F-10	28"	20.5"	27.5"	9.13 cu. ft.	—	—	—	—
R-4	—	—	—	—	13"	20"	26"	3.9 cu. ft.
F-4	13"	20"	26"	3.91 cu. ft.	—	—	—	—
RF-4	2.5"	20"	26"	68 cu. ft.	10.5"	20"	26"	3.2 cu. ft.
RFV-4	4"	20"	26"	1.2 cu. ft.	6"	20"	26"	1.8 cu. ft.

Chest-Style Off-Grid SunDanzer Refrigerators and Freezers

SunDanzer fridges and freezers provide the lowest energy use and save the most money we've ever seen for refrigeration—in fact, the SunDanzer often pays for itself in saved system expenses. Sorry AC users, these units are only for 12- or 24-volt DC use (the unique compressor accepts either voltage without modifica-

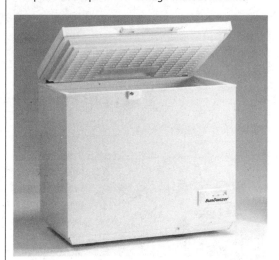

tion). SunDanzer is available in 5.8- or 8.1-cubic-foot units, configured either as a fridge or a freezer. Features include thick 4.33-inch polyurethane insulation, a lockable lid with interior light, baskets for food organization, a patented low-frost system, automatic adjustable thermostat, ozone-safe R-134a refrigerant, powder-coated aluminum interior and galvanized steel exterior, and an easy-to-clean bottom drain. And did we mention the ultralow power use that saves the remote homeowner big dollars on support hardware? SunDanzer units are manufactured by one of the world's leading appliance manufacturers using stringent standards for quality and efficiency. We've heard nothing but rave reviews from folks who have purchased these coolers. See power-use specs and sizing info below. 1-yr. mfr. warranty. USA. *Shipped by truck from Texas.*

	SunDanzer			
	Fridge 5.8 cu. ft.	Fridge 8.1 cu. ft.	Freezer 5.8 cu. ft.	Freezer 8.1 cu. ft.
Watt-hours/ day @ 90°F	220	240	620	800
External width (All are 34.5"x26.2" (HxD)	36.8"	46.9"	36.8"	46.9"
Product weight	120 lb.	140 lb.	120 lb.	140 lb.
Item number	**62581**	**62582**	**62583**	**62584**
Price	**$974**	**$1,074**	**$974**	**$1,074**

Dometic RGE 400 Propane Refrigerator

The state-of-the-art 7.7-cubic-foot RGE 400 model was introduced in the late 1990s. Average energy use per day is 0.27 gallon (1.1 lb.) propane, or 3.9kWh electric. The body is white, the doors are hinged on the right as you face the unit (and are reversible on site), and there are four adjustable stainless shelves in the fridge and two in the freezer. Also includes two vegetable bins, egg and dairy racks, frozen juice rack, ice-cube trays, ice bucket, and battery-powered interior light. Base-mounted control panel with gas/electric selector, thermostat, piezo igniter, and flame safety valve. Operates on either propane or 120-volt AC. 6.0 cu. ft. fridge, 1.7 cu. ft.

freezer. 63½"x23"x26½" (HxWxD). 167 lb.; 195 lb. shipping weight. 1-yr. mfr. warranty. Sweden/USA.

| 62303 | **Dometic RGE 400 Propane Fridge** | **$1,295** |

Shipped freight from manufacturer.

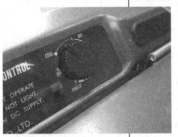

Engel Portable Fridge/Freezers

The most reliable and efficient portable refrigerator/freezers available

Engel portable 12-volt coolers are extremely reliable and are widely used in emergency vehicles. These models represent the latest, most efficient compressor technology out of Japan. They are robust and are built to deliver many years of reliable operation. They are thermostat controlled and can be either a freezer or a refrigerator depending on your needs. The power use is lower than any other portable cooler line thanks to high-efficiency patented "swing compressor" technology. This is a great compressor design for portable coolers because it continues to function perfectly even when jostled and tilted, and has only one moving part. The larger models are ideal for sailboats or other power-challenged locations. All models have a 9-foot detachable power cord and an attractive two-tone gray color scheme. 12-volt only. CFC-free. Japan.

62575	**Engel 15 Portable Cooler**	**$399**
62576	**Engel 35 Portable Cooler**	**$899**
62577	**Engel 45 Portable Cooler**	**$999**
62578	**Engel 65 Portable Cooler**	**$1,399**

Model	Cap.	Weight	Outside Dimensions	Mfr. War.
Engel 15	14 qt.	23.5 lb.	17"x11"x14.5"	1 yr.
Engel 35	34 qt.	46.3 lb.	25.5"x14.3"x16"	2 yr.
Engel 45	43 qt.	52.9 lb.	25.5"x14.3"x20"	2 yr.
Engel 65	64 qt.	68.3 lb.	31.1"x19.3"x17.4"	2 yr.

CRYSTAL COLD PROPANE REFRIGERATORS

Crystal Cold units are the largest and most fuel-efficient Crystal Cold units are the largest and most fuel-efficient gas refrigerators available. Made in the American Midwest, these high-quality units were developed for the Amish with large families and high quality expectations. Styling in '50s retro with heavy-duty casters for easy moving. Interiors are powder-coated white with front-mounted push-button igniter. Gas fridges need 10" to 12" of clearance above them to properly vent waste heat. The 15- and 18-cubic-foot models both use the same absorption cooling unit. In really hot, humid climates, the smaller 15-cubic-foot model will perform better. Available in propane or optional natural gas conversion. Natural gas models are available by special order; add $50 to cost. 1-yr. mfr.. warranty. USA. *All Crystal Cold units ship by truck from Illinois.*

Refrigerator/Freezer Combo

This is the largest gas fridge on the market. Features a pair of vegetable crispers, reversible doors and great retro styling. Burns approximately 1½ lbs. (0.35 gal.) of propane per day at 75°F ambient temperature. The best unit for hot climates, the 15-cubic-foot model, has a single, deep, vegetable crisper across the bottom. Doors are reversible on site. 18-cubic-foot: Refrigerator section is 13 cu. ft.; freezer is 4 cu. ft. Exterior size is 65½"H x 28½"W x 34½"D. Shipping weight is 324 lbs. 15-cubic-foot: Refrigerator section is 10 cu. ft., freezer is 3.3 cu. ft. Burns approximately 1¼ lbs. (0.29 gal.) of propane per day at 75°F ambient temperature. Exterior size is 63½"H x 28½"W x 34½"D. Shipping weight is 289 lbs. USA.

62579	**18-Cubic-Foot Refrigerator/ Freezer**	**$2,499**
62586	**15-Cubic-Foot Refrigerator/ Freezer**	**$2,430**

Upright Freezer

The largest, most dependable gas freezer on the market. Three quick-freeze shelves and four door shelves. Built-in thermometer, magnetic door seals and white exterior. Single door. Has three quick-freeze shelves and four door shelves. Built-in thermometer, magnetic door seals and white exterior. 18-cubic-foot: 62½"H x 43½"W x 31½"D. Shipping weight is 340 lbs. 15-cubic-foot: Single door. 61"H x 40"W x 28½"D. Shipping weight is 324 lbs. USA.

62585	**18-Cubic-Foot Upright Freezer**	**$2,829**
62574	**15-Cubic-Foot Upright Freezer**	**$2,680**

Clothes Washing and Drying

Staber Horizontal-Axis Washing Machine

The most advanced, efficient, and convenient washer available

The Staber System 2000 clothes washer combines top-loading convenience with the superior cleaning and low power use of horizontal-axis design. With the money you'll save on water, electricity, and laundry products, the Staber washer can actually pay for itself in about three years. The Staber is designed and built by a company that first spent 20 years rebuilding commercial washers. This washer is simple, durable, and (if it ever needs it) extremely easy to fix.

The Staber's unique horizontal-axis design that loads from the top is more convenient for the user and allows stronger support on both sides of the six-sided all stainless steel drum. Horizontal-axis machines have proven to deliver superior cleaning with less power and water use due to the tumbling action. Compared with conventional vertical-axis washers, the Staber uses 66% less water, 50% less electricity, and 75% less detergent. The well-supported drum design allows much higher spin speeds, which shortens drying time too.

Other features that set the Staber apart include power use so low that a 1,500-watt inverter will run it

happily; no complex, power-robbing transmission; a self-cleaning, maintenance-free pump; rust and dent-proof ABS resin tops and lids with a 25-year warranty; easy front access to all parts; and a complete, fully illustrated service/owners manual for easy owner maintenance. Average power use per wash cycle is only 150 watt-hours. Average electric draw is 300 watts; peak is 550 watts. Water use: 6 gallon wash, 6 gallon first rinse, 6 gallon second rinse (6 gallon prewash).

Standard features include a large 18-pound capacity; variable water temperature; high/low water level selection; multiple wash cycles (normal, permanent press, delicate, prewash); an automatic dispensing system for detergent, prewash detergent, bleach, and softener; and a safety lid lock when running. EnergyGuide tagged at $7/year for a natural gas water heater, or $26/year for an electric water heater (hot water costs are included in washer comparisons). 42"x27"x26" (HxWxD). Manufacturer's warranties: 25 yr. on outer tub module; 10 yr. on tub and lid; 5 yr. on bearings and suspension; 1 yr. on all else. USA.

63645   Staber HXW-2304 Washer $1,405
Shipped freight collect from Ohio.

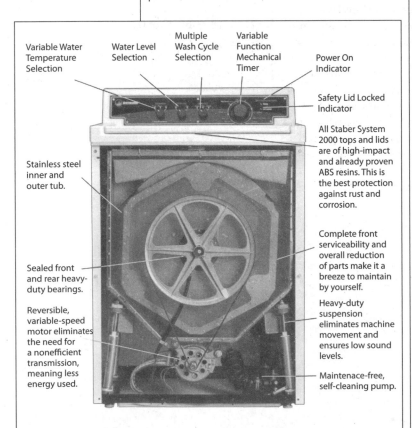

Variable Water Temperature Selection

Water Level Selection

Multiple Wash Cycle Selection

Variable Function Mechanical Timer

Power On Indicator

Safety Lid Locked Indicator

All Staber System 2000 tops and lids are of high-impact and already proven ABS resins. This is the best protection against rust and corrosion.

Stainless steel inner and outer tub.

Complete front serviceability and overall reduction of parts make it a breeze to maintain by yourself.

Sealed front and rear heavy-duty bearings.

Reversible, variable-speed motor eliminates the need for a nonefficient transmission, meaning less energy used.

Heavy-duty suspension eliminates machine movement and ensures low sound levels.

Maintenace-free, self-cleaning pump.

Spin X Spin Dryer

Clothes removed from your washer still contain considerable amounts of water. A tumble dryer takes up to an hour to remove that water, while the tumbling wears out your clothes and the heat costs you money. In three

minutes at 3,300 rpm, Spin X removes 60% of the water with centrifugal force. Clothes come out barely damp, greatly reducing or even eliminating expensive dryer run time. 25½"x13½" (H x diameter). 20 lb. UL-approved. 1-yr. mfr. warranty by the U.S. importer. Germany.

63829   Spin X Spin Dryer $499
Allow 4-6 weeks for delivery.

Filtrol Septic System Protector

Polyester, nylon, and other synthetic fibers leave your washing machine as lint and don't break down as they flush through your septic

system. These fibers clog up the soil of your leach field, causing your septic system to die prematurely and expensively. The simple Filtrol 200 is your solution. Installed on your washer drain and equipped with a reusable 200-micron filter bag, the Filtrol effectively removes most clothing lint before it gets to your drain. Simply dump the accumulated lint out of the filter bag periodically. Filter bags should be replaced every one to three years. Package includes clear filter housing, two 200-micron filter bags, sturdy wall-mounting bracket, 6-foot washer drain hose extension, and hardware. Existing washer drain hose will fit 1-inch barbed fitting at bottom of Filtrol housing. Filter housing is 12"x7" (H x diameter). USA.

10-0043 Filtrol 200 Septic Protector Kit $179

10-0044 Filtrol 200 Replacement Bag $29

James Washer

The James hand-washing machine is made of high-grade stainless steel with a galvanized lid. A pendulum agitator swings on nylon bearings, sweeps in an arc around the bottom of the tub, and prevents clothes from lodging in the corner or floating on the surface. This ensures that hot suds are mixed thoroughly with the clothes. The James is sturdily built. The corners are electrically spot-welded. All moving parts slide on nylon surfaces, reducing wear. The faucet at the bottom permits easy drainage. Capacity is about 17 gallons. 30 lb. USA.

63411   James Washer $499
Hand-constructed to order. Please allow up to 12-16 weeks for delivery.

Hand Wringer

The hand wringer will remove 90% of the water, while automatic washers remove only 45%. It has a rustproof, all-steel frame and a very strong handle. Hard maple bearings never need oil. Pressure is balanced over the entire length of the roller by a single adjustable screw. We've sold these wringers without a problem for more than 17 years. USA.

63412   Deluxe Hand Wringer $169

Dryer Vent Cleaner

Save money on utility bills, spend less time doing laundry, and most important, avoid dryer fires with the Rotary Dryer Vent Cleaner. Damp clothes at the end of a typical drying cycle, longer drying times than normal, or excessive lint in the laundry room are all signs of lint blocking your clothes dryer's exhaust system. With the assistance of your shop vacuum/blower and a cordless drill, this brush cleaner navigates the twists and turns in your venting systems with a powerful spinning action. Ultrasoft bristles reach up to 15 feet to clean behind, underneath, and all around your dryer. Effective on metal, foil, or white vinyl vents. Includes five 36-inch-long flexible rods, 4-inch auger brush, 2¼-inch lint trap brush, vacuum adapter, dryer adapter, blockage removal tool, and instructions. 4 lb. USA.

06-9605 Rotary Dryer Vent Cleaner $64

50 Feet of Drying Space

This is the most versatile clothes dryer we've seen yet. Don Reese and his family are still making them by hand, from 100% sustainably harvested New England pine and birch. Don's sturdy dryer can be positioned to create either a peaked top or a flat top ideal for sweaters. A generous 50 feet of drying space—and you can fold it up to carry back in the house (or to the next sunny spot) still loaded with laundry. Measures 53"x30"x33" (HxWxD) in peaked position, 46"x30"x40" (HxWxD) in flat-top position, and holds an average washer load. Folds to 58"x4½" (L x diameter). USA. *Requires FedEx ground delivery.*

10-8002 Wooden Dryer   **$89**

Umbrella Clothes Dryer

Having beautifully fragrant clothes with the scent of the fresh outdoors is easier than ever! Our easy-to-assemble folding clothes dryer is a classic energy-efficient method for drying. It rotates easily, and has 164-feet of PVC-coated line with 4" spacing to allow easy hanging. The line holes in the arms are protected to prevent line chafing. It can accommodate four to five loads of laundry, yet folds to only 4.5"W x 78.5"H. The diagonal width is an impressive 9'6" when open. The height conveniently adjusts from 4' to 6'. Polished aluminum 1.75" post, galvanized steel arms. Easy opening and closing for stowing between uses. Cover for folded dryer prevents bird spots and UV degradation of line. Netherlands.

06-0229 Umbrella Cothes Dryer   **$159 ($5)**
10-0034 Cover   **$16**

Retractable Clothesline

Our Retractable Clothesline makes air drying laundry easy in any location. Extendable up to 40 feet, the tangle-free line hooks to a bracket you secure at any desired distance and retracts into the impact-resistant cover for out-of-the-way storage. Supports a load's worth of sheets and towels while remaining taut. Easily installs to wood, metal, or drywall for indoor or outdoor use. Stainless steel springs won't rust in outdoor applications. Includes mounting hardware. Wipes clean with a damp cloth. 6" diameter. USA.

10-0037 Retractable Clothesline   **$20**

Superefficient Lighting

A revolution has taken place in commercial and residential lighting. Until recently, we were bound by Thomas Edison's ingenious discoveries. But that was more than a century ago, and Mr. Edison's magic now appears crude, even outdated. Mr. Edison was never concerned with how much energy his incandescent bulb consumed. Today, however, economic, environmental, and resource issues force us to look at energy consumption much more critically. Incandescent literally means "to give off light as a result of being heated." With standard light bulbs, typically only 10% of the energy consumed is converted into light, while 90% is given off as heat. So, in short, incandescent lights are in reality "incandescent heaters"!

New technologies are available today that improve on both the reliability and the energy consumption of Edison's invention. Although the initial price for these new, energy-efficient products is higher than that of standard light bulbs, their cost effectiveness is far superior in the long run. (See the discussion of whole-life cost comparison below.) When people understand just how much money and energy can be saved with compact fluorescents, they often replace regular bulbs that haven't even burned out yet.

Light output of a lamp is measured in lumens; energy input to a lamp is measured in watts. The efficiency of a lamp is expressed as lumens per watt. In the efficiency chart below, this number is listed for various lamps. High-pressure sodium and low-pressure sodium lights are primarily used for security and street lighting.

"Energy-Saving" Incandescent Lights?

Inside an incandescent light is a filament that gives off light when heated. The filament is delicate and eventually burns out. Some incandescents incorporate heavy-duty filaments or introduce special gases into the bulb to increase life. While this does not increase efficiency, it increases longevity by up to four times. Most manufacturers are now offering "energy-saving" incandescent bulbs. These are nothing more than lower-wattage bulbs that are marketed as having "essentially the same light output as . . ." In fact, General Electric lost a multimillion-dollar lawsuit relating to this very issue. A 52-watt incandescent bulb does not put out the same amount of light as a 60-watt incandescent bulb. An "energy-saving" incandescent bulb will use less power, true, but it will also give you less light. This specious and oxymoronic advertising claim is just a megacorporation's excuse to avoid retooling by selling you a wimpier light bulb at a higher price. Don't be hoodwinked; you can do much better.

Quartz-Halogen Lights

Tungsten-halogen (or quartz) lamps are really just "turbocharged" incandescents. They are typically only 10%-15% more efficient than standard incandescents, a step in the right direction, but nothing to write home about. Compared with standard incandescents, halogen fixtures produce a brighter, whiter light and are more energy efficient because they operate their tungsten filaments at higher temperatures than do standard incandescent bulbs. In addition, unlike the standard incandescent light bulb, which loses approximately 25% of its light output before it burns out, a halogen light's output depreciates very little over its life, typically less than 10%.

To make these gains, lamp manufacturers enclose the tungsten filament inside a relatively small quartz-glass envelope filled with halogen gas. During normal operation, the particles that evaporate from the filament combine with the halogen gas and are eventually redeposited back on the filament, minimizing bulb blackening and extending filament life. Where halogen lamps are used on dimmers, they need to be operated occasionally at full output to allow this regenerative process to take place.

The higher operating temperatures used in halogen lamps produce a whiter light, which eliminates the yellow-reddish tinge associated with standard incandescents. This makes them an excellent light source for applications where good color rendition is important or fine-detail work is performed. Because tungsten-halogen lamps are relatively expensive compared with standard incandescents, they are best suited for applications where the optical precision possible with the compact reflector models can be effectively utilized. They are a favorite of high-rent retail stores.

Never touch the quartz-glass envelope of a halogen lamp with your bare hands. The natural

With standard light bulbs, typically only 10% of the energy consumed is converted into light, while 90% is given off as heat.

Never touch the quartz-glass envelope of a halogen lamp with your bare hands. The natural oils in your skin will react with the quartz glass and cause it to fail prematurely.

oils in your skin will react with the quartz glass and cause it to fail prematurely. This applies to automotive halogen headlight bulbs as well. Because of this phenomenon, and for safety reasons, many manufacturers incorporate the halogen lamp capsule (generally about the size of a large flashlight bulb) within a larger outer globe.

Fluorescent Lights

Fluorescent lights are still trying to overcome a bad reputation. For many people, the term "fluorescent" connotes a long tube emitting a blue-white light with an annoying flicker, a death-warmed-over color, and headaches. However, these limitations have been overcome completely with technological improvements.

The fluorescent tube, however it is shaped, contains a special gas at low pressure. When an arc is struck between the lamp's electrodes, electrons collide with atoms in the gas to produce ultraviolet light. This, in turn, excites the white phosphors (the white coating on the inside of the tube), which emit light at visible wavelengths. The quality of the light that fluorescents produce depends largely on the blend of chemical ingredients used in making the phosphors; dozens of different phosphor blends are available. The most common and least expensive are "cool white" and "warm white." These provide a light with relatively poor color-rendering capabilities, making colors appear washed out, lacking luster and richness.

The tube may be long and straight, as with standard fluorescents, or there may be a series of smaller tubes, in a configuration that can be screwed into a common light fixture. This latter configuration is called a compact fluorescent light (or CF light).

Consumer awareness of and interest in compact fluorescents has exploded in recent years, and the market for them is growing rapidly. A good variety of lamp wattages, sizes, shapes, and aesthetic packages is now readily accessible. Because this marketplace is growing and changing so quickly, the more-frequently published Gaiam Real Goods catalogs may feature lamp types not described here.

COLOR QUALITY

The color of the light emitted from a fluorescent bulb is determined by the phosphors that coat the inside surface. The terms "color temperature" and "color rendering" are the technical terms used to describe light. Color temperature, or the color of the light that is emitted, is measured in degrees Kelvin (°K) on the absolute temperature scale, ranging from 9,000°K (which appears blue) down to 1,500°K (which appears orange-red). The color-rendering index (CRI) of a lamp, on a scale of 0 to 100, rates the ability of a light to render an object's true color when compared with sunlight. A CRI of 100 is perfect, 0 is a cave. The accompanying table will demystify the quality-of-light debate.

Phosphor blends are available that not only render colors better, but also produce light more efficiently. Most notable of these are the fluorescent lamps using tri-stimulus phosphors, which have CRIs in the 80s. These incorporate relatively expensive phosphors with peak luminance in the blue, green, and red portions of the visible spectrum (those that the human eye is most sensitive to) and produce about 15% more visible light than standard phosphors do. Wherever people spend much time around fluorescent lighting, specify lamps with higher (80+) CRI.

BALLAST COMPARISONS

All fluorescent lights require a ballast to operate, in addition to the bulb. The ballast regulates the voltage and current delivered to a fluorescent lamp and is essential for proper lamp operation. The electrical input requirements vary for each type of compact fluorescent lamp, and so each type/wattage requires a ballast specifically designed to drive it. There are two types of ballasts that operate on AC: magnetic and electronic. The magnetic ballast, the standard

EFFICIENCY AND LIFETIME OF LAMPS

Lifetime Benchmark	10,000 Hours		Electric Rate		$0.15 per kWh
Lamp Type	Power	Lumens per Watt	Lifetime in Hours	Lamps per 10,000 Hours	Cost per Lumen per 10,000 Hours ($)
Incandescent bulb	25W	8.4	2,500	4	0.18
Incandescent bulb	60W	14.0	1,000	10	0.11
Halogen bulb	50W	14.2	2,000	5	0.11
Incandescent bulb	100W	16.9	750	14	0.09
Halogen bulb	90W	17.6	2,000	5	0.09
Halogen floodlight	300W	19.8	2,000	5	0.08
G-16, compact fluorescent	16W	53.0	10,000	1	0.03
SLS 20, compact fluorescent	20W	60.0	10,000	1	0.03
Metal halide, floodlight	150W	60.0	10,000	14	0.03
D lamp, compact fluorescent	39W	71.0	10,000	1	0.02
T8, 4-ft. fluorescent tube	32W	86.0	20,000	1	0.02
High-pressure sodium light	150W	96.0	24,000	1	0.02
Low-pressure sodium light	135W	142.0	16,000	1	0.01

The types of efficient lighting on the market include new-design incandescents, quartz-halogens, standard fluorescents, and compact fluorescents.

Clean Lights Save Energy

Building owners can save up to 10% of lighting costs simply by keeping light fixtures clean, says NALMCO, the National Association of Lighting Management Companies. Over the past several years, crews working on a U.S. EPA-funded project have collected scientific field measurements at offices, schools, health facilities, and retail stores to determine the benefits—improved lighting as well as reduced costs—of simple periodic cleaning of light fixtures. Recognizing that fixtures produce less light as they get dirty, current design standards compensate by adding more fixtures to maintain the desired light level. But if building owners committed to regular cleanings, designers could recommend fewer fixtures, reducing installation costs while trimming lighting energy use by at least 10%. For more information, call NALMCO at 515-243-2360.

since fluorescent lighting was first developed, uses electromagnetic technology. The electronic ballast, only recently developed, uses solid-state technology. All DC ballasts are electronic devices.

Magnetic ballasts can last up to 50,000 hours and often incorporate replaceable bulbs. Magnetic ballasts flicker when starting and take a few seconds to get going. They also run the lamp at 60 cycles per second, and some people are affected adversely by this flicker.

Electronic ballasts weigh less than magnetic ballasts, operate lamps at a higher frequency (30,000 cycles per second vs. 60 cycles), are silent, generate less heat, and are more energy efficient. However, electronic ballasts cost more, particularly DC units. Much-longer-lived electronic ballasts are possible, but costs go up dramatically.

Electronic ballasts last about 10,000 hours, the same as the bulb, and most do not have replaceable bulbs. Electronic ballasts start almost instantly with no flickering. They run the lamp at about 30,000 cycles per second. For the many people who suffer from the "60-cycle blues," this is the energy-efficient lamp of choice.

All the compact fluorescent lamp assemblies and prewired ballasts reviewed here are equipped with standard medium, screw-in bases (like normal household incandescent lamps). The ballast portion, however, is wider than an incandescent light bulb just above the screw-in base. Fixtures having constricted necks or deeply recessed sockets may require a socket "extender" (to extend the lamp beyond the constrictions). These are readily available either in our catalogs or at most hardware stores.

Note: There is no difference between a fluorescent tube for 120 volts and one for 12 volts. Only the ballasts are different.

SAVINGS

The main justification for buying fluorescent lights is to save energy and money. Compact fluorescents provide opportunities for tremendous savings without any inconveniences. Simple payback calculations prove their cost effectiveness (see the Cost Comparison table). Fluorescent lights typically last 10-13 times longer and use one-fourth the energy of standard incandescent lights. The U.S. Congress's Office of Technology Assessment calls compact fluorescents the best investment in America today, with a 1.2-year payback.

The next question, of course, is, "How much do compact fluorescents save?" The more a light is used, the more energy and money you can save by replacing it with a compact fluorescent. Installing CFs in the fixtures that are used most and have high wattage saves the most. An excellent candidate, for example, is an all-night security light. The calculations in the Cost Comparison table assume that the light is on for an average of six hours per day and that power costs 15¢ per kWh (the approximate national average). The total cost of operation is the cost of the bulb(s) plus the cost of the electricity used. The table shows that using the compact fluorescent saves about $75 and 550kWh of electricity.

HEALTH EFFECTS

Migraine headaches, loss of concentration, and general irritation have all been blamed on fluorescent lights. These problems are caused not

Electronic ballasts start almost instantly with no flickering. They run the lamp at about 30,000 cycles per second. For the many people who suffer from the "60-cycle blues," this is the energy-efficient lamp of choice.

Light, Color Rendering Index, and Color Temp		
Type of Light	CRI	°Kelvin
Incandescent	90-95	2,700
Cool white fluorescent	62	4,100
Warm white fluorescent	51	3,000
Compact fluorescent	82	2,700

CRI = Color Rendition Index. A measure of how closely a light source mimics sunlight. 100 is perfect sunlight, 0 is darkness. Most standard CFLs are around 82 CRI; most full-spectrum lights are 90-92 CRI.

CFs are available that fit into just about any fixture that currently uses a standard incandescent bulb. Not surprisingly, these CFs have quickly become the most popular model in the marketplace.

by the lights themselves, but by the way they operate. Common magnetic ballasts run the lamps at the same 60 cycles per second that is delivered by our electrical grid. This causes the lamps to flicker noticeably 120 times per second, every time the alternating current switches direction. Approximately one-third of the human population is sensitive to this flicker on a subliminal level.

The cure is to use electronic ballasts, which operate at around 30,000 cycles per second. This rapid cycling totally eliminates perceptible flicker and avoids the ensuing health complaints.

Fluorescents and Remote Energy Systems

Fluorescent lights are available for both AC and DC. Most people using an inverter choose AC lights because of the wider selection, a significant quality advantage, and lower price. Some older inverters may have problems running some magnetic-ballasted lights. If you have an older Heart inverter, buy one light to try it first.

Most inverters operate compact fluorescent lamps satisfactorily. However, because all but a few specialized inverters produce an alternating current having a modified sine wave (versus a pure sinusoidal waveform), they will not drive compact fluorescents that use magnetic ballasts as efficiently or "cleanly" as possible, and some lamps may emit an annoying buzz. Electronic-ballasted compact fluorescents, on the other hand, are tolerant of the modified sine-wave

input and will provide better performance, silently.

COMPACT FLUORESCENT APPLICATIONS

Due to the need for a ballast, a compact fluorescent light bulb is shaped differently from an incandescent. This is the biggest obstacle in retrofitting light fixtures. Compact fluorescents are longer, heavier, and sometimes wider. The ballast, the widest part, is located at the base, right above the screw-in adapter. Today, new "mini-spiral" CFs are available that fit into just about any fixture that currently uses a standard incandescent bulb. Not surprisingly, these CFs have quickly become the most popular model in the marketplace.

Compact fluorescents have many household applications—table lamps, floor lamps, recessed cans, desk lamps, bathroom vanities, and more. As manufacturers become attuned to this relatively new market, more light fixtures suited for compact fluorescents are becoming available. Gaiam Real Goods now offers a limited selection of fixtures; watch future catalogs for the latest offerings in this rapidly expanding field.

Table lamps are an excellent application for compact fluorescents. The metal harp that supports the lampshade may not be long enough or wide enough to accommodate these bulbs. We offer an inexpensive replacement harp that can solve this problem. Be aware also that heavier CFs may change a lightweight but stable lamp into a top-heavy one.

Recessed can lights are limited by diameter and sometimes cannot accept the wide ballast

An Energy Tale, or Real Goods Walks Its Talk

Several years ago, Real Goods headquarters moved into a 10,000-square-foot office building. Built in the early 1960s, the building had conventional 4-foot fixtures, each holding four 40-watt cool white tubes: the dreary, glaring "office standard." Light levels were much brighter than recommended when employees are using computers a lot. Plus, a number of our employees were having headache, energy level, and "attitude" problems after moving, probably caused

by the 60-cycle flicker from the old magnetic ballasts. Reducing overall light levels for less eyestrain was needed, as was giving employees a healthier working environment. And if we could save some money too, it wouldn't hurt.

Our solution was to retrofit the existing fixtures with electronic ballasts for no-flicker lighting, install specular aluminum reflectors, and convert from four cool-white 40-watt T-12 lamps per fixture to two warm-colored 32-watt T-8 lamps. Power use per fixture was reduced by over 60%, but desktop light levels, because of the reflectors and more efficient lamps and ballasts,

were reduced by only 30% or less. The entire retrofit cost about $6,000 (not counting the $2,300 rebate from our local utility), yet our electrical savings alone are close to $5,000 per year. Plus our employees enjoy flicker-free, warm-colored, nonglaring light and greatly reduced EMF levels. The entire project paid for itself in less than a year! Improved working conditions came as a freebie!

Editor's Note: Even though this story is now more than 10 years old, it's continuing relevance justifies repeating it again in this 30th-anniversary edition of the *Sourcebook*.

—JS

at the screw-in base. The base depth is adjustable on most recessed cans. There are a number of special bulbs available now just for recessed cans.

Desk lamps can be difficult to retrofit, but complete desk lights that incorporate compact fluorescents are readily available.

Hanging fixtures are one of the easiest applications for compact fluorescent lights. One of the best uses is directly over kitchen and dining room tables. Usually the shades are so wide that they won't get in the way. It may even be possible to use a Y-shaped two-socket adapter (available at hardware stores) and screw in two lights if more light is desired.

Track lighting is one of the most common forms of lighting today. When selecting your track system, choose a fixture that will not interfere with the ballasts located right above the screw-in base. It is best to use one of the reflector lamps, such as the SLS/R series.

Dimming

With limited exceptions, the current generation of compact fluorescent lights should never be dimmed. Using these lights on a dimmer switch may even pose a hazard. If you have a fixture with a dimmer switch that you wish to retrofit, you can either buy a CF lamp that specifically says that it is dimmable, or replace the switch. Replacing the switch is a simple, inexpensive procedure. Remember to turn off the power first!

Three-Way Sockets

Recently manufacturers have begun to make three-way compact fluorescent bulbs. While more efficient than conventional three-way bulbs, they are more expensive than one-way CFs and have a shorter life span.

Start-Up Time

The start-up time for compact fluorescent lamps varies. It is normal for most magnetic CFs to flicker for up to several seconds when first turned on while they attempt to strike an arc. Most electronically ballasted units start their lamps instantly. All fluorescent lamps start at a lower light output; depending on the ambient temperature, it may take anywhere from several seconds to several minutes for the lamp to come up to full brightness. The very brief start-up time, which is apparent only in some of the fluorescent lights, is a small price to pay for the energy savings and your own peace of mind about doing less harm to our fragile environment.

COLD-WEATHER/OUTDOOR APPLICATIONS

Fluorescent lighting systems are sensitive to temperature. Manufacturers rate the ability of lamps and ballasts to start and operate at various temperatures. Light output and system efficiency both fall off significantly when lamps are operated above or below the temperature range at which they were designed to operate. Most fluorescent lamps will reach full brightness at 50°-70°F. In general, the lower the ambient temperature, the greater the difficulty in starting and attaining full brightness. Manufacturers are usually conservative with their temperature ratings. We have found that most lamps will start at temperatures 10-20 degrees Fahrenheit lower than those stated by the manufacturer. Most compact fluorescent lamps are not designed for use in wet applications (for example, in showers or in open outdoor fixtures). In such environments, the lamp should be installed in a fixture rated for wet use.

SERVICE LIFE

All lamp-life estimates are based on a three-hour duty cycle, meaning that the lamps are tested by turning them on and off once every three hours until half of the test batch of lamps burn out. Turning lamps on and off more frequently decreases lamp life. Since the cost of operating a light is a combination of the cost of electricity used and the replacement cost, what is the optimum no-use period before turning off a light? Generally, it is more cost effective to turn off standard fluorescent lights whenever you leave a room for more than 15 minutes. For compact fluorescents, the interval is three minutes, and you should turn incandescents off whenever you leave the room. Since the rated lamp life

> Generally, it is more cost effective to turn off standard fluorescent lights whenever you leave a room for more than 15 minutes. For compact fluorescents, the interval is three minutes, and you should turn incandescents off whenever you leave the room.

COST COMPARISON*		
	20W Evolution CFL	75W Incandescent Light Bulb
Cost of bulb	$9	$0.25
Product life (years)	4.6	0.5
Bulbs used per 10,000 hours	1	10
Energy used annually (kWh)	44	164
Energy used per 10,000 hours (kWh)	200	750
Total cost per 10,000 hours	$39	$113
Savings per 10,000 hours	$74	
Energy saved in 4.5 years (kWh)	550	
Greenhouse gases saved (pounds of CO_2)	847	

*Based on 6 hours of lamp use per day and electricity cost of $0.15/kWh.

Fluorescent lighting is four to five times more efficient than incandescent lighting. This means you can produce the same quantity of light with only 20%-25% of the power use.

represents an average, some lamps will last longer and some shorter. A typical compact fluorescent lamp will last as long as (or longer than) 10 standard incandescent AC light bulbs or five standard incandescent AC floodlights, saving you the cost of numerous bulbs in addition to a lot of electricity due to the greater efficiency. See the Cost Comparison table on page 331.

FULL-SPECTRUM FLUORESCENTS

Literally, "full-spectrum" refers to light that contains all the colors of the rainbow. As the term is used by several manufacturers, "full-spectrum" refers to the similarity of their lamps' light (including ultraviolet light) to the midday Sun. While we do believe that the closer an electric light source matches daylight, the healthier it is, we have several practical reservations. The quality of light produced by "full-spectrum" lighting available today is very "cool" in color tone and, in our opinion, not flattering to people's complexions or most interior environments. Also, the light intensity found indoors is roughly 100 times less than that produced by sunlight. At these low levels, we question whether people receive the full benefits offered by full-spectrum lighting.

Ideally, we encourage you to get outdoors every day for a good dose of sunshine. Indoors, we think it is best to install lighting that is complimentary to you and the ambiance you desire to create. For most applications, people prefer a warmer-toned light source. Most of the compact fluorescent lamps we offer mimic the warm rosy color of 60-watt incandescent lamps, but we do offer a small selection of full-spectrum lamps for those who prefer them.

DC Lighting

When designing an independent power system, it is essential to use the most efficient applianc-

es possible. Fluorescent lighting is four to five times more efficient than incandescent lighting. This means you can produce the same quantity of light with only 20%-25% of the power use. When the power source is expensive, like PV modules, or noisy and expensive, like a generator, this becomes terribly important. Fluorescent lighting is quite simply the best and most efficient way to light your house.

Usually we recommend against using DC lights in remote home applications except in very small systems where there is no inverter. The greater selection, lower prices, and higher quality of AC products have made it hard to justify DC lights. DC wiring and fixture costs are significantly higher than with AC lighting, and lamp choices are limited. However, you might want to consider putting in a few DC lights for load minimization and emergency backup. A small incandescent DC night-light uses less than its AC counterpart plus inverter overhead, and a single AC device may not draw enough power to be detected by the inverter's load-seeking mode.

Since the introduction of efficient, high-quality, long-lasting inverters, the low-voltage DC appliance and lighting industries have almost disappeared except for RV equipment, which tends to be of lower quality. This is because most RV equipment is designed for intermittent use, where short life or lower quality isn't as noticeable.

Tube-type fluorescents, such as the Thin-Lite series, are great for kitchen counters and larger areas. These are RV units, with lower life expectancies compared with conventional AC units. Replacement ballasts for these units are available, and the fluorescent tubes are standard hardware-store stock. We also offer incandescent and quartz-halogen DC lights but suggest caution in using these for renewable energy systems, where every watt is precious.

NATURAL LIGHTING AND LIGHTING PRODUCTS

SUNPIPE-9

17-0303	2-ft. Kit	$262
17-0304	4-ft. Kit	$291
17-0305	6-ft. Kit	$335
63-158	9-in. Pipe Extension, 2-ft.	$43
63-823	9-in. Steep Flashing Option	$99
63-002	9-in. Elbow Option	$66

SUNPIPE-13

17-0306	2-ft. Kit	$334
17-0307	4-ft. Kit	$374
17-0308	6-ft. Kit	$410
63-181	13-in. Pipe Extension, 2-ft.	$55
63-814	13-in. Steep Flashing Option	$179
63-004	13-in. Elbow Option	$78

SunPipe Skylights

Free sunlight is the best light

The SunPipe adds beautiful natural daylight to your house without high expense, heat gain or loss, bleaching or color fading of fabrics, or framing and drywall work. This is how skylights were always meant to be. Sunlight is collected above the roof and transferred through a mirror finish aluminum pipe to a diffusing lens in your ceiling. The SunPipe transmits no direct Sun rays, so there's no solar heat gain. No cutting of joists and rafters is required. Installation takes two to four hours, on average, and can be done easily by any home handyman. The light is uniformly diffused by the interior lens, and the pipe is sealed top and bottom to prevent heat loss or condensation.

Two sizes are available. Approximately speaking, the SunPipe-9 is for up to 100 square feet of floor space; the SunPipe-13 covers up to 240 square feet. At midday, the 13-inch SunPipe delivers more cool, free sunlight than a dozen 100-watt light bulbs. The 9-inch SunPipe deliver about half as much light.

The SunPipe Kits provide all the parts you'll need, including templates and a complete installation guide. Order a pipe length that equals at least your ceiling-to-roof distance through the attic, plus 1 foot. The pipe has to extend up through the flashing. Add an extra 2-foot section if in doubt. Extra pipe length will telescope if needed. The standard galvanized flashing covers 0:12-6:12 roof pitches; the steep roof flashing covers 6:12-12:12 roof pitches and is a substitute for the standard flashing in the kit. The all-metal flashings are heavily galvanized to G-90 standard and have a 60-year life expectancy. The optional elbows adjust from 0° to 45°. 10-yr. mfr. warranty. USA.

AC Lighting
Our energy-efficient light bulbs save you more and conserve more resources!

Bulb	Product Description and Price	Special Features	Usage (Watts)	Incandescent Equivalent (Watts)	Lumens	Required Operating Temp.	Size (HxW)	Color Temp.[1]
	Sunwave Bulb 31772 20W $25 31773 23W $25 31774 26W $25	Most advanced full-spectrum lighting in the world reduces eyestrain and increases visual acuity. Instant-starting electronic ballast. China.	20W 23W 26W	75W 100W 120W	1,300 1,600 2,000	14°F 14°F 14°F	5.7"x2.2" 5.7"x2.2" 6.2"x2.5"	5,550°K 5,550°K 5,550°K
	Compact Fluorescent Bulb 11-0226 9W $6 11-0306 9W (4 pk.) $18 11-0227 13W $6 11-0305 13W(4pk.) $18	Our CFLs, especially our smallest 9-watt Ultra Mini, bring efficient lighting to more fixtures. 9-watt is rated at 6,000 hours. 13-watt is available in both warm glow or natural daylight. China.	9W 13W	40W 60+ W	500 900	14°F 14°F	3.6"x1.2" 4.4"x1.4"	2,700°K 2,700°K
	Compact Fluorescent Bulb 11-0219 PAR 38, 23W $16	Our reflector floodlight shines more brightly than a 60-watt bulb while drawing only 23 watts. Rated at 6,000 hours. China.	23W	60+ W	960	-6°F	6.5"x5"	3,000°K
	Evolution Light Bulb 11-0192 11W $8 20W $9 14W $8 25W $10	Same size and shape as standard incandescents. Use anywhere, even when lampshade attaches to bulb. Available in warm glow or natural daylight. China.	11W 14W 20W 25W	50W 60W 75W 100W	550 850 1,200 1,500	-5°F -5°F -5°F -5°F	3.75"x2.375" 4.5"x2.375" 4.75"x2.375" 5"x2.375"	WG 2,700°K ND 5,500°K
	Mini Globe 11-0194 $12	Replace your vanity lights with a bulb that offers the same brightness but uses only 11 watts.	11W	40W	440	14°F	5.08"x3"	3,000°K
	LED Floodlight 11-0215 PAR 20 $45 11-0215 PAR 30 $55 11-0215 PAR 38 $60	Our LED floodlights fit in standard fixtures, burn safely cool, have a 60,000-hour life, and need almost no electricity. China.	3W 6W 8W	No Equivalent No Equivalent No Equivalent	80 150 200	14°F 14°F 14°F	3.5"x2.5" 3.65"x3.9" 5.25"x4.75"	5,500°K 5,500°K 5,500°K
	LED Bulb 11-0207 18-LED $35 11-0207 36-LED $45	For low-light applications, LEDs are much more efficient than incandescent bulbs. LEDs fit into standard fixtures, use little electricity, generate zero heat, and last for 60,000 hours. China.	2W 3W	No Equivalent No Equivalent	31 60	14°F 14°F	3.0"x1.75" 3.25"x2.25"	5,500°K 5,500°K

Chart is based on information provided by manufacturer. All CFL bulbs have flicker-free electronic ballasts and a 10,000-hour life unless otherwise noted. CFL = Compact Fluorescent Lamp. Compact fluorescents can go anywhere, but use this guide for the best applications.

 Open Indoor: Hanging pendant, ceiling lamps, and wall fixtures

 Shaded Lamps: Table and floor lamps

 Enclosed Fixture: Ceiling and wall fixtures, exterior yard post lights, wall brackets, and hanging fixtures (check minimum operating temperature). Life expectancy decreases in enclosed fixtures.

 Recessed Can: Recessed ceiling downlights

 Track Light: Ceiling and wall fixtures

[1]Color Temperature—Not to be confused with bulb temperature, this is a color spectrum rating; the higher the number, the whiter the light.

ENERGY CONSERVATION

Natural Daylight Fluorescent Tubes

Our Natural Spectrum® Fluorescent Tubes come in three sizes. The 20W (24" length; 800 lumens; 94.5 CRI; 6,280°K) and 40W (48" length; 2,100 lumens; 94.5 CRI; 6,280°K) sizes are T-12 diameter, about 1¼ inches. The 32W (48" length; 2,900 lumens; 85 CRI; 6,000°K) size is T-8 diameter, about 1 inch. The T-8 bulbs (use with electronic ballasts only) provide more light with less power. You must match your existing bulb size! All sizes offer an extra-long life of 26,000 hours. T-12 lamps, France. T-8 lamps, USA.

11-0052 40W (4-pk.)   **$50**

11-0134 32W (4-pk.)   **$50**

11-9001 20W (2-pk.)   **$25**

Socket Extender

Sometimes your compact fluorescent lamps may have a neck that is too wide for your fixture. This socket extender increases the length of the base. The extender is also useful if a bulb sits too deeply inside a track lighting or can fixture, losing its lighting effectiveness. Extends lamp 1.75 inches. USA.

11-0104 Socket Extender   **$2.95**

Dimmable Compact Fluorescent Multi-Paks

Our CFLs, especially our smallest 9W Ultra Mini, bring efficient lighting to more fixtures. 9W is rated at 6,000 hours. 13W available in both warm glow or natural daylight. China.

11-0226 9W (4-pk.)   **$7**

11-9226 13W (4-pk.)   **$16**

11-0227 20W (4-pk.)   **$7**

11-9227 23W (4-pk.)   **$16**

Lamps and Fixtures

Heritage™ Floor and Desk Lamps

Instantly see more clearly with the bright, natural white light and advanced glare protection of Natural Spectrum® light. Patented 27W energy-saving bulb lasts 10,000 hours and delivers the brightness of an ordinary 150W bulb. Optix™ Glare Control Filter virtually eliminates reflective glare. Three-way brightness control adjusts between 75W, 100W, and 150W. Multidirectional head swivels 360° to shine light where you need it. Height-adjustable base. Antiqued brushed brass or antiqued brushed nickel finish. Taiwan/China.

11-0203 Floor, 39", 53" x10½" (H x diameter), 15.2 lb.   **$229**

11-0223 Desk, 11½", 17½" x8½" (H x diameter), 10.8 lb.   **$199**

11-0204 Replacement Bulb, 10,000 hr.   **$30**

Natural Spectrum® Task Lamps

You can't buy sunshine, but these lamps come close. Each style filters out yellow tones to provide a natural, flicker-free light suitable for any task—and has a light grid to reduce eye-tiring glare (see diagram). Adjustable gooseneck and ergonomic swivel design afford easy placement. All include a compact fluorescent bulb that lasts 10,000 hours yet uses only 27 watts to provide the light of a regular 150W bulb. 8-foot cord. Both swivel and floor lamps come with a convenient utility tray for

glasses, books, and more; plus, swivel lamp features an EasyLatch® handle that easily adjusts the height from 38½" to 54". Graphite, putty, or burl wood finish (floor and desk models), or putty (swivel model only). China.

11-0199 Swivel Floor Lamp, 38½", 54" x10" (HxW)   **$169**

11-0204 Replacement Bulb, 10,000 hr.   **$30**

Gooseneck LED Lamp

We don't throw the word "perfect" around lightly, but it's the best way to describe this lamp. Twenty hyper-efficient, cool-to-the-touch, 100-year LEDs use less than 3 watts to provide brilliantly warm, flicker-free illumination while an amazing 36-inch neck adjusts infinitely to aim that light exactly where you need it. As a bedside lamp that won't disturb your partner, a precision task light, a reading lamp, and more, it's unrivaled at saving energy, eyesight, and even relationships! China.

11-0214 Gooseneck LED Lamp   **$99**

Littlite

Our high-intensity 5-watt quartz halogen bulb concentrates light in a tightly controlled area, easily aimed exactly where you need it. Also great for jobs involving accurate, close, or detail work—drawing, computer, or work on a music stand or by a stereo system. The small hooded end won't get in your way; you have just the right amount of light, and you save energy. Includes 18-inch flexible gooseneck mounting base and fully adjustable dimmer. Weighted base converts Littlite into a movable freestanding lamp. USA.

33514	12V Model	$48
50-0112	AC Littlite	$57
33401	Weighted Base	$18

DC Lighting

11-watt

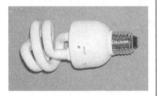
15- and 30-watt

DC Compact Fluorescent Lamps

More choices and better quality! DC users at both 12 and 24 volts can finally have quality compact fluorescent lamps in a variety of sizes and at reasonable prices. Our new German-engineered supplier manufactures in China and actually does a one-hour burn-in on every lamp to eliminate possible failures. Available in 11-, 15-, and 30-watt sizes, which are approximately equivalent to 45-, 60-, and 120-watt incandescent lamps. Warm lamps are 2,700°K; cool lamps are 6,400°K. Will operate at 14°-104°F and at 11-15 volts (22-30 volts for 24V). Standard Edison medium base. 2-yr. mfr. warranty. Choose warm glow or natural daylight. Germany/China.

50-0104 12V, 11W	$14
50-0104 12V, 15W	$19
50-0104 12V, 30W	$39
50-0105 24V, 15W	$20

LED 12-Volt Bulbs

At last! 12-volt DC light bulbs that bring all the energy-saving benefits of the new wave of LED lighting to solar homes and systems. Just like our original LED bulbs (page 000), these technological marvels fit into standard fixtures, use negligible amounts of electricity, generate no heat, and last an astonishing 60,000 hours. But the 12VDC design means solar system owners can now join the revolution, too. The 3-watt model has 36 LEDs that produce an output of 90 lumens. The 2-watt model uses 18 LEDs to produce 40 lumens. China.

50-0116 3W	$45
50-0116 2W	$35

LED Undercabinet Lighting

LED lighting is the perfect answer for high-efficiency DC task-lighting needs. These low-profile lights fit along the underside of cabinets or shelves above your kitchen counter, workbench, or desk. One primary strip (16"L) and up to three secondary strips (14½"L) can be chained together for longer runs. All contain 60 LEDs. They have three brightness levels and consume only 3.8 watts per strip. The on/off/brightness switch on the primary unit controls the chain. The primary unit comes with an AC or DC cord. (AC cord has a transformer; DC cord has bare wire.) Each secondary strip comes with a 10-inch jumper cord to chain it to the "upstream" light strip. Aside from undercabinet lighting, other applications include RV lighting, overcabinet accent lighting, or any application where a light must be left on for long periods of time. Other features include: USA.

- Life expectancy of 60,000 hours. That's 40 years at four hours per day!
- 4,000°K color temperature. That's in between "natural daylight" and "warm glow."
- Durability—no glass or filaments to break. Extruded aluminum housing.
- Convenience—easy to install. No bulbs to replace.
- Sleek, attractive design—and only ¾" deep. Black or nickel finish.

50-9116 Primary AC	$169
50-9117 Primary DC	$159
50-9118 Secondary AC/DC	$149

High-Output Individual White LEDs

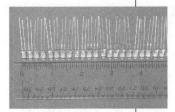

Build your own low-voltage DC lights. Each 5mm LED delivers 11 candelas of blue/white light with 3.3 volts, 20 milliamps input. This is three times the light output for the same power and same price as our previous LED packs! Wire three of these bulbs in series for 12-volt input. Sold in packs of 12 or 24. Resistor needed in series connections. Call for sizing. China.

32323	White 8cd LED, 12-pk.	$20
32324	White 8cd LED, 24-pk.	$35

12-Volt Holiday Lights

It just doesn't seem like Christmas or Solstice without lights on the tree, and these 12-volt lights will dazzle any old fir bush. These lights are actually great year-round for decorating porches, decks, your rolling-art automobile, or your motorcycle. The light strand is 20 feet long and consists of 35 mini, colored, nonblinking lights (draw is 1.2 amps) with a 12-volt cigarette lighter socket on the end. Note: You can't connect them in series as you can with many 120-volt lights. Uses conventional 3-volt mini-lights, same as any 120-volt mini-light string, just wired for 12-volt input. Taiwan.

37-301	12V Holiday Lights	$15

Or for the ultimate efficiency, try our LED lights—18-foot cord with 18 LEDs, 0.1 amp draw. Taiwan. Choose green and red, green, red, or yellow.

37389	12V LED Lights	$39

Solar Icicles

Holiday lighting needn't involve tapping the grid. Our spectacular Solar Icicles are beautifully illuminated from within by bright, high-efficiency LEDs that draw their power from an unobtrusive solar pod placed in a sunny spot. Three small and three large sparkling acrylic icicles spaced 19 inches apart on a sturdy wire strand light up your season with sustainable cheer. Icicles are 10 inches and 7 inches and hang about 11 inches from main wire. 18-foot strand includes 10-foot lead wire. Solar pod is 3"x6"x6" (HxWxD) on a 17½-inch stake. Solar pod takes four AA NiMH batteries, included. China.

11-0208 Solar Icicles	$89

Solar Outdoor Lighting

Solar Sensor Dual Light

Our hands-free Solar Sensor Dual Light automatically turns on when motion is detected within a 35-foot radius and stays lit until all is still for 30 seconds. Ideal for security, safety, and convenience, its twin adjustable 5-LED lamps light instantly in all conditions while sensitivity controls reduce false alarms to conserve power. Mounts easily anywhere; no wiring or electricity needed. Solar module's 15-foot cord allows optimal placement. Fully charges in five Sun hours. Includes mounting kit and replaceable battery. 8.2"x9.75" (HxW). Extends 7 inches from wall. 1-yr. mfr. warranty. China.

14-0433 Solar Sensor Dual Light	$79

Durable and Powerful DC Floodlight

This tough polycarbonate outdoor floodlight is completely watertight with a UV-stabilized, nonyellowing lens. The 13-watt compact fluorescent lamp delivers 900 lumens of focused light, thanks to the ellipsoidal mirror reflector. Standard ½-inch male pipe thread mounting with lamp angle adjustable through 90°. This efficient floodlight is a great choice for automatic PV-powered sign lighting and is included in our sign lighting kits. 12VDC only. USA.

32307	DC Floodlight	$69
50-0108	Replacement Bulb	$9

Mission-Style Solar Lights

Thanks to revolutionary white LED technology and an internal reflector, our Mission Solar Lights disperse light more brightly and evenly than most other solar lights available. Lamp fixture removes from hook for use as a table lantern. Made of rust-proof antiqued metal and durable bubble glass. Set of two light fixtures includes ground stakes and adjustable hanging hooks. Requires only three hours of direct sunlight to charge. No wiring required. Lights switch on automatically at night. Set of two. 8"x6"x6" (HxWxD). China.

14-0266 Mission-Style Solar Lights   **$126**

Candle Pathmarkers

Our new pathmarkers look just like flickering candles set inside stylish, frosted-glass votives. But their moving "flame" actually comes from superefficient LEDs powered by a miniature solar cell via an included rechargeable AA battery. A wireless garden-stake design allows instant installation anywhere, and light-sensing auto-on operation means dependable illumination from dusk to dawn without a thought. 13¾"H. Votlves are 3.9"H. Solar panel is 2¾" diameter. China.

14-0409 Candle Pathmarkers, set of 4   **$79**

Solar-Powered Yard Critters

Each of our captivating critters has a tiny, brilliant, color-changing LED producing dazzling colors that radiate throughout the clear acrylic body. Power comes from an unobtrusive solar pod/battery pack on an 18-inch aluminum stake. Garland style has six same-style critters; garden-stake style has a hummingbird, dragonfly, and butterfly, each on a 30-inch-tall black plastic stake. Stakes are spaced over 10 feet, with an additional 10 feet of wire back to the solar power pod. Solar pod takes four AA NiMH batteries, included. Also available is our new single solar starburst on a 34-inch black metal stake, with single AA battery, included. China.

14-0272 Garland Yard LED 6-Light Kit   **$95**
    (specify hummingbirds or starbursts)

14-0273 Stake Yard LED 3-Light Kit   **$95**

14-0360 Starburst Single Solar Stake   **$24**

FLASHLIGHTS

Shake Light

This is an incredibly innovative flashlight that never wears out or runs down. Just shake the light to charge it up. Ten seconds of shaking delivers about two minutes of light. The internal magnetic generator turns the shaking motion into electricity, which is stored in a large capacitor to run the bright LED lamp. Clear construction lets you see the internal workings. This is one flashlight that will never let you down. Great for emergencies and car trunks. Taiwan.

17-0326 Shake Light   **$29**

Solar Flashlight

As little as four hours of direct Sun on this flashlight's built-in solar panel provides all the power its three ultrabright lifetime LEDs need for shining more than eight hours. The shock- and water-resistant housing has a momentary on/off function that can be used instead of the steady-on setting whenever you need to preserve your night vision. Built-in battery lasts 500 cycles. 7"x1" (L x diameter). China.

06-0624 Solar Flashlight   **$19**

Xenon Flashlights

Designed for extreme conditions where intense, failsafe illumination is vital, these "tactical" flashlights are the brightest you can carry. (Remember that the next time you have to change a tire at night.) A high-power, focusable xenon beam combines with a shock-resistant aluminum case and a sealed O-ring design for 100% waterproof performance. Includes two CR123A batteries. Runs for about 1¾ hours per battery set. Bulbs last 50 hours. Small has twist on/off operation. 4.4"x1.2" (L x diameter). Large features power switch and a built-in compass. 5.1"x1" (L x diameter). China.

06-0647 Small Xenon Flashlight   **$16**

06-0646 Large Xenon Flashlight   **$20**

06-9648 CR123A Batteries, set of 2   **$10**

06-0649 Xenon Bulb   **$5**

Dynamo Flashlight

The hand-cranked Dynamo Flashlight produces a steady beam of light with just an easy squeeze of the handle. Translucent casing holds an assortment of colored gears to captivate and intrigue young minds. Includes a shockproof and shatterproof lens, extra replacement bulb, and peg to attach a cord. Buy preassembled or as a great build-it-yourself gift that combines science and utility. 5"x3"x3½" (LxWxH). Canada.

06-9402 Preassembled Dynamo Flashlight $15

06-0456 Make-Your-Own Dynamo Kit   **$19**

TerraLux LED Flashlight Conversion

Our astonishing LED conversion for common Mini Maglite® 2AA flashlights delivers two to three times more bright blue-white light, runs two to four times longer on a set of batteries, and has a lamp that will likely outlive you, with up to 100,000 hours' life expectancy. Our kit supplies the MicroStar2 lamp and a custom reflector. You supply

the MiniMag PowerPush™ technology, which keeps the bulb at full-light output even with dropping battery voltage. When batteries are exhausted, it shifts to emergency Moon mode for reduced, but continuing, light output. USA.

06-0524 MicroStar2 Flashlight Kit   **$25**

The MicroStar2 kit is not made or endorsed by Mag Instruments, Inc.

Lanterns and Spotlights

Solar Lantern

Keep the sunshine working for you long after darkness has fallen with this great-to-take- anywhere solar-powered lantern. It's made of durable materials, including an acrylic housing, so you'll never arrive at the campsite only to discover a broken lantern. The built-in solar pan-

el is fully adjustable to make best use of the sunshine for charging and will give off 3½ hours of light on a full charge. Choose between two modes: bright lantern light or front-mounted flashlight. The lantern mode offers 360° of adjustable-angle fluorescent light. A DC adapter is included for car or other 12-volt charging. 9"x4½"x4" (HxWxD). China.

14-0354 Solar Lantern   **$59**

Freeplay Indigo Table Lantern

Keep this versatile hand-powered lantern in your car, tent, or home, and it won't matter how far off the grid you get. Just one minute of cranking provides three hours of low-level illumination or six minutes of intense 360° white light from the dimmer-controlled cluster of permanent LEDs. Or use the built-in, single-LED side lamp for directional light. Industrial-grade components stand up to rugged use, and there's even an adapter for AC charging. 8"x3½" (H x diameter). China.

14-0431 Freeplay Indigo Table Lantern   **$69**

LED Lantern

Whether you pack it on trips or keep it for emergencies, this new member of our lantern family puts 12 lifetime LEDs on a dimmer switch that lets you adjust its output from a gentle glow to blazing white light. Weather-resistant construction makes it the one to have when the going gets wet. Runs for an incredible 40 hours on a single battery set. Uses four D batteries (not included). 10.8"x6" (H x diameter). China.

14-0444 LED Lantern   **$49**

LED Lighting

LED development as a light source is going the way of early computer development. Output doubles about every 18 months while prices slowly drift downward. We're starting to see some high-quality white light sources that may be appropriate for many off-the-grid home applications.

Good Things About LEDs . . .

The white LEDs have a bright, slightly bluish light output that is extremely easy to read by. Black print just jumps off a white page with these lamps. The life expectancy of LEDs is exceptional, often exceeding 50,000 hours. They aren't bothered by vibration or modest impacts. Your flashlight won't die just because you dropped it. We're finally starting to see LED manufacturers listing the light output of their devices. Often listed in milli-candelas; 1,000 milli-candelas equals 1 candela, which is also equal to 1 lumen.

Not So Good Things About LEDs . . .

Most LEDs are still fairly modest light sources, and they're expensive. LEDs tend to be highly directional in their output; they don't make good general illumination sources without serious diffusion help. Their lumens-per-watt ratio is improving but currently is slightly less than that of compact fluorescent lamps. In late 2003, LEDs were averaging 25-30 lumens per watt, while CFLs averaged 40-60.

LED Worklight

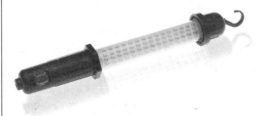

Whether you're at a table or in a crawlspace, this rechargeable cordless light stick uses 60 LEDs to cast a wide swath of brilliant white illumination wherever needed. A 90-minute charge via the included 110V or 12V adapters provides four hours of versatile light during hobby work, repairs, emergencies, camping trips, and more. Or just keep it plugged in for continuous use. 15"x1½" (L x diameter).

14-0432 LED Worklight   **$59**

EverLite Solar Spotlight

For camping, backyard, or emergency use, our high-tech all-weather solar spotlight runs for more than 24 hours per charge. Just six hours of full Sun delivers 12 hours of ultrabright LED run time from the built-in NiMH battery pack. Even 6 hours of solid overcast will deliver 3-4 hours of light. The bright blue-white light has eight LEDs, delivering a total of 50 lumens. The lightweight lamp and base stows on the back of the solar panel  along with the 10-foot cord. The lamp assembly can be unplugged for full portability. If the lamp arm is left in the ON position (up), our spotlight will automatically turn on at dusk and off at dawn. Comes with a convenient nylon mesh bag that allows daytime charging right through the bag. Options include plug-in chargers for 12 or 120 volts that allow for quick recharging where the Sun doesn't shine, and a nifty 12-volt DC output cable to deliver power from the lamp base to your cell phone or other small 12-volt appliance. 7.08"x6.6"x2.5". 1.3 lb. total, 0.5 lb. for LED assembly only. 2.7 lb. China.

14-0299 EverLite Solar Spotlight with Carrying Case   **$119**

17-0290 EverLite 12VDC Charger   **$12**

17-0289 EverLite 120VAC Charger   **$12**

17-0291 EverLite 12VDC Output Adapter   **$25**

Mini EverLite

Our Mini EverLite gets its energy from the Sun via a portable solar panel, but this model is so tiny you can slip it into your pocket or pack without a thought. Three high-intensity LEDs deliver 20 lumens of light for 20 hours on a six-hour, full-Sun charge (4-5 hours on a 6-hour solid overcast charge). A light-sensing auto-on/off function—with  optional manual shutoff—keeps it shining brightly from dusk to dawn if needed, while protecting against accidental daytime operation. Includes Velcro® pads to attach it anywhere from a tent roof to a car dash or hat. Panel is 4.7"x2.7"x0.5" (HxWxD). Lamp is 2.9"x2.25"x1.4" (HxWxD). China.

14-0376 Mini EverLite   **$59**

Water Development

Reckoning with Our Most Precious and Endangered Resource

WATER IS THE SINGLE MOST IMPORTANT INGREDIENT in any homestead. Without a dependable water source, you can't call any place home for very long. An age-old saying goes, "You buy the water and the land comes for free." Today, unfortunately, many people literally *are* buying water, unnecessarily and at outrageous expense, most of whom can least afford it. Some are swayed by multimillion-dollar ad campaigns by corporations selling bottled water; others are victims of the privatization of previously public water supplies. Around the world, *more than a billion* people still lack access to clean water, and as supplies increasingly become polluted, privatized, or gobbled up by for-profit corporate enterprises, that number is projected to steeply rise. See the accompanying sidebars for a shocking picture of the current state of our most precious and endangered resource.

Gaiam Real Goods offers a wide variety of water development solutions for remote, and not so remote, homes. We specialize in products that are made specifically for solar-, battery-, wind-, or water-powered pumping. These pumps are designed for long hours of dependable duty in out-of-the-way places where the utility lines don't reach. For instance, ranchers are finding that small solar-powered submersible pumps are far cheaper and more dependable than the old wind pumps that still dot the Great Plains. Many state and national parks use our renewable-energy-powered pumps for their backcountry campgrounds. The great majority of pumps we sell are solar- or battery-powered electric models, so we'll cover them first.

The initial cost of solar-generated electricity is high. Therefore, most solar pumping equipment is scaled toward modest residential needs, rather than larger commercial or industrial applications. Solar-powered pumps tend to be far more efficient than their conventional AC-powered cousins. By wringing every watt of energy out of a system, we're able to keep the start-up costs reasonable.

> Around the world, *more than a billion* people still lack access to clean water, and as supplies increasingly become polluted, privatized, or gobbled up by for-profit corporate enterprises, that number is projected to steeply rise.

The Three Components of Every Water System

Every rural water system has three easily identified basic components.

A Source. This can be a well, a spring, a pond, a creek, collected rainwater, or the big expensive tanker truck that hauls in a load every month.

A Storage Area. This is sometimes the same as the source, and sometimes it is an elevated or pressurized storage tank.

A Delivery System. This used to be as simple as the bucket at the end of a rope. But given a choice, most of us would prefer to have our water arrive under pressure from a faucet. Hauling water, because it's so heavy, and because we use such surprising amounts of it, gets old fast.

The first two components, source and storage, you need to produce locally; however, we can offer a few pointers based on experience and some of the better, and worse, stories we've heard.

Sources

WELLS

The most common domestic water source is the well, which can be hand dug, driven, or drilled. Wells are less prone than springs, streams, or ponds to picking up surface contamination from animals, pesticides and herbicides, or other runoff. Hand-dug wells are usually about 3 feet or larger in diameter and rarely more than 40 feet deep. These are difficult to build, they need regular cleaning, and it is dangerous to have them lying around unused on your property. Drilled or driven wells are standard practice now. The most common size for a residential driven or drilled well is either 4 or 6 inches. Beware of the "do-it-yourself" well-drilling rigs, which rarely can insert pipe larger than 2 inches. This restricts your pump choices to only the least-efficient jet-type pumps. A 4-inch casing is the smallest you want—that's the minimum for a submersible pump.

SPRINGS

Many folks with country property are lucky enough to have surface springs that can be developed. At the very least, springs need to be securely fenced to keep out wildlife. More commonly, springs are developed with either a backhoe and 2- or 3-foot concrete pipe sections sunk into the ground, or drilled and cased with one of the newer, lightweight horizontal drilling rigs. This helps to ensure that the water supply won't be contaminated by surface runoff or animals. To be respectful to your land and the critters that lived there before you came, prior to developing a spring it's a good idea to give some consideration to any ecosystems it supports. If there are no other springs nearby, try to ensure that some runoff will still continue after development.

PONDS, STREAMS, AND OTHER SURFACE WATER

Some water systems are as simple as tossing the pump intake into the existing lake or stream and turning on the switch. However, surface water usually is used only for livestock or agricultural needs—although for many homesteads with gardens and orchards, this can be 90% of your total water use. If you don't have a surface water source on your property, consider putting in a pond. Ponds are one of the least expensive

> To be respectful to your land and the critters that lived there before you came, prior to developing a spring it's a good idea to give some consideration to any ecosystems it supports.

Global Water Crisis

More than 1 billion people today don't have access to clean drinking water, and millions die every year for lack of it. Water resources around the globe are threatened by climate change, misuse, and pollution. But there are solutions: By using innovative water efficiency and conservation strategies, community-scale projects, smart economics, and new technology, we can provide for people's basic needs while protecting the environment.

That 1.2 billion people lack access to clean water is surely one of the greatest development failures of the modern era. That as many as 5 million people—mainly children—die every year from preventable, water-related disease is surely one of the great tragedies of our time.

Unfortunately, despite a growing recognition that more must be done to help those without clean water or adequate sanitation, almost nothing is being done. The problem is not merely a lack of aid (although more money is needed) or a lack of technology. It is a failure of vision and will. According to many international water experts, hundreds of billions of dollars are needed to bring safe water to everyone who needs it. Since international water aid is so paltry, many of these experts claim that privatization of water services is the only way to help the poor.

But many critics of this approach note that community-scale infrastructure and efficiency and conservation can bring basic water services to the millions who need it without breaking the bank. And many critics of the "gold-plated" approach argue that water privatization, although it can play some productive role, will never be able to bring water to the world's poorest people.

There are solutions to the global water crisis that don't involve massive dams, large-scale infrastructure, and tens or hundreds of billions of dollars. First and foremost, we must use what the Pacific Institute calls "soft path" solutions to the global water crisis. Soft path solutions aim to improve the productivity of water rather than seek endless new supply; soft path solutions complement centrally planned infrastructure with community-scale projects; and soft path solutions involve stakeholders in key decisions so that water deals and projects protect the environment and the public interest.

For more information on global water issues, challenges, and solutions, visit www.pacinst.org and www.worldwater.org.

—Courtesy of Peter Gleick, Pacific Institute

The Real Cost of Bottled Water

San Franciscans and other Bay Area residents enjoy some of the nation's highest-quality drinking water, with pristine Sierra snowmelt from the Hetch Hetchy reservoir as our primary source. Every year, our water is tested more than 100,000 times to ensure that it meets or exceeds every standard for safe drinking water. And yet we still buy bottled water. Why?

Maybe it's because we think bottled water is cleaner and somehow better, but that's not true. The federal standards for tap water are higher than those for bottled water.

The Environmental Law Foundation has sued eight bottlers for using words such as "pure" to market water that contains bacteria, arsenic, and chlorine. Bottled water is no bargain either: It costs 240 to 10,000 times more than tap water. For the price of one bottle of Evian, a San Franciscan can receive 1,000 gallons of tap water. Forty percent of bottled water should be labeled bottled tap water because that is exactly what it is. But even that doesn't dampen the demand.

Clearly, the popularity of bottled water is the result of huge marketing efforts. The global consumption of bottled water reached 41 billion gallons in 2004, up 57% in just five years. Even in areas where tap water is clean and safe to drink, such as in San Francisco, demand for bottled water is increasing—producing unnecessary garbage and consuming vast quantities of energy. So what is the real cost of bottled water?

Most of the price of a bottle of water goes for its bottling, packaging, shipping, marketing, retailing, and profit. Transporting bottled water by boat, truck, and train involves burning massive quantities of fossil fuels. More than 5 trillion gallons of bottled water is shipped internationally each year. Here in San Francisco, we can buy water from Fiji (5,455 miles away) or Norway (5,194 miles away) and many other faraway places to satisfy our demand for the chic and exotic. These are truly the Hummers of our bottled water generation. As further proof that the bottle is worth more than the water in it, starting in 2007, the state of California will give 5¢ for recycling a small water bottle and 10¢ for a large one. The next time you see someone drinking "Fiji" water, do them a favor and let them know they're blowing it!

Just supplying Americans with plastic water bottles for one year consumes more than 47 million gallons of oil, enough to take 100,000 cars off the road and 1 billion pounds of carbon dioxide out of the atmosphere, according to the Container Recycling Institute. In contrast, San Francisco tap water is distributed through an existing zero-carbon infrastructure: plumbing and gravity. Our water generates clean energy on its way to our tap—powering our streetcars, fire stations, the airport, and schools.

More than 1 billion plastic water bottles end up in California's trash each year, taking up valuable landfill space, leaking toxic additives such as phthalates into the groundwater, and taking 1,000 years to biodegrade. That means bottled water may be harming our future water supply.

The rapid growth in the bottled water industry means that water extraction is concentrated in communities where bottling plants are located. This can have a huge strain on the surrounding ecosystem. Near Mount Shasta, the world's largest food company, Nestlé, is proposing to extract billions of gallons of spring water, which could have devastating impacts on the McCloud River.

So it is clear that bottled water directly adds to environmental degradation, global warming, and a large amount of unnecessary waste and litter. All this for a product that is often inferior to San Francisco's tap water. Luckily, there are better, less expensive alternatives:

- In the office, use a water dispenser that taps into tap water. The only difference your company will notice is that you're saving a lot of money.
- At home and in your car, switch to a stainless steel water bottle and use it for the rest of your life, knowing that you are drinking some of the nation's best water and making the planet a better place.

© Jared Blumenfeld and Susan Leal, *San Francisco Chronicle*, February 18, 2007, reprinted with permission. Jared Blumenfeld is the director of the San Francisco Department of the Environment. Susan Leal is the general manager of the San Francisco Public Utilities Commission.

WATER DEVELOPMENT

---

and most delightfully pleasing methods of supplying water, not only for your own needs but also for the wildlife population that your pond will soon support. They often are used to hold winter and spring runoff for summertime use. Don't drink or cook with surface water unless it has been treated or purified first (see Chapter 9 for more help on this topic).

Storage

You might need to provide some means of water storage for any number of reasons. The most common are: To get through long dry periods, to provide pressure through elevated storage or pressurized air, to keep drinking water clean and uncontaminated, and to prevent freezing.

John Schaeffer's pond in Hopland, California.

SURFACE STORAGE

Those with ponds, lakes, streams, or springs may not need any additional storage. Let the livestock find their own way to the water, or pump straight to whatever needs irrigation. However, many systems will need to pump water to an elevated storage site or a large pressure tank in order to develop pressure. You may need sea-

Approximate Daily Water Use for Home and Farm	
Usage	Gallons per Day
Home	
National average residential use	60-70 per person
As above, without flush toilet	25-40 per person
Drinking and cooking water only	5-10 per person
Lawn—Garden—Pool	
Lawn sprinkler, per 1,000 sq. ft., per sprinkling	600 (approx. 1 inch)
Garden sprinkler, per 1,000 sq. ft., per sprinkling	600 (approx. 1 inch)
Standard garden hose	10 gallons per minute
Swimming pool maintenance, per 100 sq. ft. surface area	30 per day
Farm (maximum needs)	
Dairy cows	20 per head
Dry cows or heifers	15 per head
Calves	7 per head
Beef cattle, yearlings (90°F)	20 per head
Beef, brood cows	12 per head
Sheep or goats	2 per head
Horses or mules	12 per head
Swine, finishing	4 per head
Swine, nursing	6 per head
Chickens, laying hens (90°F)	9 per 100 birds
Chickens, broilers (90°F)	6 per 100 birds
Turkeys, broilers (90°F)	25 per 100 birds
Ducks	22 per 100 birds
Dairy sanitation—milk room and milking parlor	500 per day
Flushing barn floors	10 per 100 sq. ft.
Sanitary hog wallow	100 per day

sonal freeze protection, which means protected or underground storage.

POND LINERS

Custom-made polyvinyl liners are available for ponds or leaking tanks. Liners have revolutionized pond construction. They are designed for installation during pond construction and are then buried with 6 inches of dirt around the edges, practically guaranteeing a leak-free pond even when working with gravel, sand, or other problem soils. When buried, the life expectancy of these liners is 50 years or more. They make reliable pond construction possible in locations that normally could not accommodate such inexpensive water storage methods.

TANKS

Covered tanks of one sort or another are the most common and longest-lasting storage solutions. The cover must be screened and tight enough to keep critters such as lizards, mice, and squirrels from drowning in your drinking water (always an exciting discovery). The most common tank materials are polypropylene, fiberglass, and concrete or ferro-cement.

Plastic Tanks. Both fiberglass and polypropylene tanks are commonly available. Your local farm supply store is usually a good source. All plastic tanks will suffer slightly from UV degradation in sunlight. Simply painting the outside of the tank will prevent this problem and probably make the tank nicer to look at. If this tank is for your drinking water, make sure it's internally coated with an FDA-approved material for drinking water. Most farm supply stores will offer both "drinking water grade" and "utility grade" tanks. Utility grade tanks often use recycled plastic materials and probably will leach some polymers into the water, particularly when new. Some plastic tanks can be partially or completely buried for freeze protection. Ask before you buy if this is a consideration for you.

Concrete Tanks. Concrete is one of the best and longest-lasting storage solutions, but these tanks are expensive and/or labor intensive initially. All but the smallest concrete tanks are built on site. They can be concrete block, ferrocement, or monolithic block pours. Although monolithic pours require hiring a contractor with specialized forming equipment, these tanks are usually the most trouble free in the long run. Any concrete tank will need to be coated internally with a special sealer to be watertight. Concrete tanks can be buried for freeze protection and to keep the water cool in hot climates. They

A Brief Glossary of Pump Jargon

Flow: The measure of a pump's capacity to move liquid volume. Given in gallons per hour (gph), gallons per minute (gpm), or liters per minute (Lpm).

Foot Valve: A check valve (one-way valve) with a strainer. Installed at the end of the pump intake line, it prevents loss of prime and keeps large debris from entering the pump.

Friction Loss: The loss in pressure due to friction of the water moving through a pipe. As flow rate increases and pipe diameter decreases, friction loss can result in significant flow and head loss.

Head: Two common uses: 1) the pressure or effective height a pump is capable of raising water; 2) the height a pump is actually raising the water in a particular installation.

Lift: Same as head. Contrary to the way this term sounds, pumps do not suck water, they push it.

Prime: A charge of water that fills the pump and the intake line, allowing pumping action to start. Centrifugal pumps will not self-prime. Positive-displacement pumps will usually self-prime if they have a free discharge—no pressure on the output.

Submersible Pump: A pump with a sealed motor assembly designed to be installed below the water surface. Most commonly used when the water level is more than 15 feet below the surface or when the pump must be protected from freezing.

Suction Lift: The difference between the source water level and the pump. Theoretical limit is 33 feet; practical limit is 10-15 feet. Suction lift capability of a pump decreases 1 foot for every 1,000 feet above sea level.

Surface Pump: Designed for pumping from surface water supplies such as springs, ponds, tanks, or shallow wells. The pump is mounted in a dry, weatherproof location less than 10-15 feet above the water surface. Surface pumps cannot be submerged and be expected to survive.

---

can cost 40¢-60¢ per gallon, but in the long run they're well worth it.

Pressure Tanks. These are used in pumped systems to store pressurized water so that the pump doesn't have to start for every glass of water. They work by squeezing a captive volume of air, since water doesn't compress. Pressure tanks are rated by their total volume. Draw-down volume, the amount of water that actually can be loaded into and withdrawn from the tank under ideal conditions, is typically about 40% of total volume. So a "20-gallon" pressure tank can really deliver only about 8 gallons before starting the pump to refill. Pressure tanks are one of the few things in life where bigger really is better. With a larger pressure tank, your pump doesn't have to start as often, it will use less power and live longer, your water pressure will be more stable, and if power fails you'll have more pressurized water in storage to tide you over. Forty-gallon capacity is the minimum for residential use, and more is better.

Delivery Systems

A few lucky folks are able to collect and store their water high enough above the level of intended use (you need 45 vertical feet for 20 pounds per square inch) that the delivery system will simply be a pipe, and the weight of the water will supply the pressure free of charge. Most of us are going to need a pump, or two, to get the water up from underground and/or to provide pressure. We'll cover electrically driven solar and battery-powered pumping first, then water-powered and wind-powered pumps.

The standard rural utility-powered water system consists of a submersible pump in the well delivering water into a pressure tank in some location that's safe from freezing. A pressure tank extends the time between pumping cycles by saving up some pressurized water for delivery later. This system usually solves any freezing problems by placing the pump deep inside the well, and the pressure tank indoors. The disadvantage is that the pump must produce enough volume to keep up with any potential demand, or the pressure tank will be depleted and the pressure will drop dramatically. This requires a ⅓-hp pump minimally, and usually ½ hp or larger. Well drillers often will sell a much larger than necessary pump because it increases their profit and guarantees that no matter how many sprinklers you add in the future, you'll have sufficient water-delivery capacity. (And they don't have to pay your electric bill!) This is fine when you have large amounts of utility power available to meet heavy surge loads, but it's very costly to power with a renewable energy system because of the large equipment requirements. We try to work smarter, smaller, and use less-expensive resources to get the job done.

WATER CONVERSIONS

1 gal. of water = 8.33 lb.

1 gal. of water = 231 cu. in.

1 cu. ft. of water = 62.4 lb.

1 cu. ft. of water = 7.48 gal.

1 acre ft. of water = 326,700 gal.

1 psi of water pressure = 2.31 ft. of head

Beware of Privatizing the Essence of Life

Just like air, water is precious and sustains all life on earth. Increased demand for water by industry and agriculture is draining away the planet's rivers, lakes, and other freshwater sources. Meanwhile, a profit-driven industry increasingly controls our water supply. How we address the impending water crisis will have tremendous implications for people's health and the environment in the 21st century.

Though only a little less than 1% of Earth's water is available for human use, there is still more than enough fresh water to sustain every person on the planet—all 6 billion of us. Because of its essential, even sacred, role in life, many believe that water is a common resource to be shared by all.

In North America, most people receive water from a public utility. But not everyone in the world has access to the water they and their families need. The United Nations estimates that more than 1 billion people lack access to safe drinking water. Hundreds of millions must walk miles every day to gather enough water to survive. The need to travel great distances to collect water is a leading reason that young girls are not able to attend school in many African countries. And nearly 2 million children die every year of diseases caused by unsafe water. These situations are likely to worsen. According to the UN, by 2025, two-thirds of the world's population—more than 5 billion of us—will lack access to water. The World Bank has predicted that the wars of tomorrow will no longer be fought over oil but over water.

How is it possible that the water crisis could explode within a single generation?

There are many causes: the escalating use and abuse by water-intensive industries such as mining, paper production, and power generation; a growing population and an increasing need for irrigation; a spread of industrial pollution that fouls lakes and rivers, especially in developing countries; and spreading droughts induced by climate change. The World Bank estimates that rising temperatures and decreasing rainfall associated with climate change will reduce the amount of rain-fed farmland by 11% within the lifetimes of today's children.

There's Huge Profit in Thirst

A limited water supply, coupled with the growing demand for water, is seen by corporations as a huge profit-making opportunity. *Fortune Magazine* maintains that "water will be to the 21st century what oil was to the 20th century," and corporations are racing to stake claims to this "blue gold."

In fact, corporations have already been meeting behind closed doors for more than a decade, vying for control of the world's water resources. They have pushed officials at the World Bank and International Monetary Fund to make industry-friendly water policies a condition of debt assistance to developing countries. Water corporations have pushed trade ministers and officials at the World Trade Organization to craft industry-biased trade agreements. In March of 2006, Coca-Cola sponsored the World Water Forum, where giant corporations met with representatives of the United Nations, governments, and the World Bank, to promote profit-oriented water policies around the world.

Nowhere is the corporate water grab more insidious than its escalating control of drinking water. Supplying water is already a $420-billion-a-year business. Throughout the world, powerful corporations are gaining control of public water systems, reducing a shared common resource into simply another opportunity to profit. For example, Suez—the corporation that built the Suez Canal—has recently been snatching up government contracts to take over municipal water systems.

And if controlling our taps were not enough, Coke, Pepsi, and Nestlé are bottling our water and selling it back to us at prices that are hundreds, even thousands of times greater than what tap water costs. Today, water is one of the world's fastest-growing branded beverages.

Bottled Water: Is It Better?

In countries like the U.S., most water services are hidden from public view. We catch a glimpse of them when we turn on the shower or flush the toilet. Then they retreat into the background and we go about our day.

Bottled water is an exception. Bottled water corporations aim to brand the water we drink and turn it into a status symbol. But these

---

Bottled water brands in North America owned by Nestlé:

Perrier	Aberfoyle-Nestlé	Santa Maria	Great Bear
Arrowhead	Pure Life	Aqua Pana	Vitell
Montclair	Calistoga	Deer Park	Ice Mountain
San Pellegrino	Poland Spring	Zephyrhills	Ozarka

Bottled water brands in North America owned by Coca-Cola:

Dasani	Evian	Volvic	Crystal
Alhambra	Sparkletts	Dannon	Ciel (Mexico)
AquaPenn	Pure American	Dasani Nutriwater	

Pepsi's big brand:

Aquafina

---

companies, led by Coke, Nestlé, and Pepsi, have sold us a bill of goods. Misleading advertising is fueling the explosive growth of the bottled water business. In 2005, bottled water corporations spent over $158 million to portray their products as "pure," "safe," "clean," "healthy," and superior to tap water. Today, three out of four Americans drink bottled water and one in five drinks *only* bottled water, even though it is much more expensive than tap water and can sometimes be less safe.

Water bottling is one of the least-regulated industries in the U.S. Tap water and bottled water are subject to similar standards, but tap water is tested far more frequently and is more rigorously monitored and enforced by the EPA. In contrast, there are significant gaps in FDA regulation of bottled water, and the agency largely relies on the corporations to police themselves. A comprehensive study conducted by the Natural Resources Defense Council in 1999 found that samples of various bottled water brands contained elevated levels of arsenic, bacteria, and other contaminants. In 2004, Coke recalled 500,000 bottles of its Dasani water in the United Kingdom after authorities discovered elevated levels of the carcinogen bromate in some of the bottles.

People are paying a high price for this deception, and price gouging is only the beginning. Bottled water corporations use their political and economic clout to secure sweetheart deals, block legislative efforts to protect local water rights, and pursue costly and time-consuming litigation against individuals and governments.

Communities Resist Privatization

Imagine that you live in a town where some of the wells are contaminated with elevated levels of naturally occurring radiation. The contamination is known to the managers of the corporate-controlled water system, who shut down the wells when government inspectors arrive to take water samples. Their decep-

tion discovered, the managers are fired, indicted, and replaced with new managers. Months later, you turn on the tap while entertaining guests for a Memorial Day picnic, only to find it dry. You call the water company, but your call is disconnected after 40 minutes on hold because no one at the private water company could be reached. Welcome to Toms River, New Jersey, where Suez controls the water system.

Imagine residing in a town where the local water has become unfit to drink. The private corporation that supplies tap water refuses to repair

> Water bottling is one of the least-regulated industries in the U.S. Tap water and bottled water are subject to similar standards, but tap water is tested far more frequently and is more rigorously monitored and enforced by the EPA. In contrast, there are significant gaps in FDA regulation of bottled water.

the water system or to connect to another source and instead provides 25 gallons of bottled water a week to each household. Welcome to San Jerardo, California.

Imagine a major U.S. city that spends $1 million on a marketing campaign to boost people's confidence in the public water system, which is widely regarded as one of the highest-quality systems in the country. Then imagine opening the newspaper one day to discover that the city has simultaneously been spending $2 million to provide expensive bottled water for city workers. Welcome to San Francisco.

In Stanwood, Michigan, concerned citizens have been fighting Nestlé

for years. Nestlé has drained tens of millions of gallons of water from local water sources and ecosystems, causing significant damage to the local environment—a stream, two lakes, and rich diverse wetlands have been harmed. A local environmental group, Michigan Citizens for Water Conservation (MCWC), won a major court victory in 2003 that shut down the well field where Nestlé bottled water. But Nestlé retaliated. The corporation used its political and financial leverage to appeal the ruling and won temporary permission to continue to extract and bottle 218 gallons of water per minute. The citizens group is currently appealing the case to the Michigan Supreme Court.

Unfortunately, Nestlé's story is not unique. Communities affected by Coke's and Pepsi's dangerous bottling practices in India—where hundreds of wells have dried up, water tables have dropped precipitously, and plant effluent has contaminated surface water supplies—are also eagerly awaiting a court decision that will determine whether or not these corporations will be held accountable for their actions.

The reality behind this industry's slick public relations and marketing is that bottled water threatens our health and our ecosystems, costs far more than tap water, and undermines local democratic control of a common resource. Bottled water corporations take water from underground springs and municipal sources without regard to scarcity or human rights. Corporations are trying to make a profit-driven commodity out of a public resource that should not be bought or sold.

Courtesy of Corporate Accountability International, whose "Think Outside the Bottle" campaign aims to educate the public about the dangers of corporate water privatization. For more information, visit www.StopCorporateAbuse.org.

Solar-Powered Pumping

Centrifugal pumps are good for moving large volumes of water at relatively low pressure. As pressure rises, however, the water inside the centrifugal pump "slips" increasingly, until finally a pressure is reached at which no water is actually leaving the pump.

Where Efficiency Is Everything

PV modules are expensive, and water is surprisingly heavy. These two facts dominate the solar-pumping industry. At 8.3 pounds per gallon, a lot of energy is needed to move water uphill. Anything we can do to wring a little more work out of every last watt of energy is going to make the system less expensive initially. Because of these economic realities, the solar-pumping industry tends to use the most efficient pumps available. For many applications, that means a positive-displacement type of pump. This class of pumps prevents the possibility of the water slipping from high-pressure areas to lower-pressure areas inside the pump. Positive-displacement pumps also ensure that even when running very slowly—such as when powered by a PV module under partial light conditions—water still will be pumped. As a general rule, positive-displacement pumps manage four to five times the efficiency of centrifugal pumps, particularly when lifts over about 60 feet are involved. Several varieties of positive-displacement pumps are commonly available. Diaphragm pumps, rotary-vane pumps, piston pumps, and the newest darling, helical-rotor pumps, will all be found in our product pages.

Q: Will Solar Pumps Run Only During Direct Sunlight?

A. No. While we often run pumping systems directly off solar electric modules without batteries, battery-based systems can be used for round-the-clock pumping. Direct pumping is usually the better option (approximately 20% more efficient), but this works only during sunny times of day. PV direct is usually the preferred power delivery if you're pumping to a storage tank, doing direct irrigation, or running the backyard fountain. A PV-direct pump controller often is needed to help start the pump earlier in the day, keep it running later in the afternoon, or make any pumping possible on cloudy days.

Positive-displacement pumps have some disadvantages. They tend to be noisier, as the water is expelled in lots of little spurts. They usually pump smaller volumes of water, they must start under full load, most require periodic maintenance, and most won't tolerate running dry. These are reasons that this class of pumps isn't used more extensively in the AC-powered pumping industry.

Most AC-powered pumps are centrifugal types. This type of pump is preferred because of easy starting, low noise, smooth output, and minimal maintenance requirements. Centrifugal pumps are good for moving large volumes of water at relatively low pressure. As pressure rises, however, the water inside the centrifugal pump "slips" increasingly, until finally a pressure is reached at which no water is actually leaving the pump. This is 0% efficiency. Single-stage centrifugal pumps suffer at lifts over 60 feet. To manage higher lifts, as in a submersible well pump, multiple stages of centrifugal pump impellers are stacked up.

In the solar industry, centrifugal pumps are used for pool pumping and for some circulation duties in hot-water systems. But in all applications where pressure exceeds 20 psi, you'll find us recommending the slightly noisier, occasional-maintenance-requiring but vastly more efficient positive-displacement-type pumps. For instance, an AC submersible pump running at 7%-10% efficiency is considered "good." The helical-rotor submersible pumps we promote run at close to 50% efficiency.

For Highest Efficiency, Run PV Direct

We often design solar pumping systems to run PV direct. That is, the pump is connected directly to the photovoltaic (PV) modules with no batteries involved in the system. The electrical-to-chemical conversion in a battery isn't 100% efficient. When we avoid batteries and deliver the energy directly to the pump, 20%-25% more water gets pumped. This kind of system is ideal when the water is being pumped into a large storage tank or is being used immediately for irrigation. It also saves the initial cost of the batteries, the maintenance and periodic replacement they require, and the charge controllers and the fusing/safety equipment that they demand. PV-direct pumping systems, which are designed to run all day long, make the most of your PV investment and help us get around the lower gallon-per-minute output of most positive-displacement pumps.

However (every silver lining has its cloud), we like to use one piece of modern technology on PV-direct systems that isn't often found on battery-powered systems. A linear current booster, or LCB for short, is a solid-state marvel that will help get a PV-direct pump running earlier in the morning, keep it running later in the evening, and sometimes make running a possibility on hazy or cloudy days. An LCB will convert excess PV voltage into extra amperage when the modules aren't producing quite enough current for the pump. The pump will run more slowly than if it had full power, but if a positive-displacement pump runs at all, it delivers water. LCBs will boost water delivery in most PV-direct systems by 20% or more, and we usually recommend a properly sized one with every system.

Direct Current (DC) Motors for Variable Power

Pumps that are designed for solar use DC electric motors. PV modules produce DC electricity, and all battery types store DC power. DC motors have the great advantage of accepting variable voltage input without distress. Common AC motors will overheat if supplied with low voltage. DC motors simply run slower when the voltage drops. This makes them ideal partners for PV modules. Day and night, clouds and shadows; these all affect the PV output, and a DC motor simply "goes with the flow"!

Which Solar-Powered Pump Do You Want?

That depends on what you're doing with it and what your climate is. We'll start with the most common and easiest choices and work our way through to the less common.

PUMPING FROM A WELL

Do you have a well that's cased with a 4-inch or larger pipe, and a static water level that is no more than 750 feet below the surface? Perfect. We carry several brands of proven DC-powered submersible pumps with a range of prices, lift, and volume capabilities. The SHURflo Solar Sub is the lowest-cost system with lift up to 230

feet and sufficient volume for most residential homesteads. The bigger helical-rotor-type sub pumps like the Grundfos and Lorentz pumps are available in over a dozen models, with lifts up to 750 feet or volume over 25 gpm, depending on the model. Performance, prices, and PV requirements are listed in the product section. The SHURflo Solar Sub is a diaphragm-type pump, and unlike almost any other submersible pump, it can tolerate running dry. The manufacturer says just don't let it run dry for more than a month or two! This feature makes this pump ideal for many low-output wells.

Complete submersible pumping systems—PV modules, mounting structure, LCB, and pump—range from $1,900 to $12,000 depending on lift and volume required. Options such as float switches that will automatically turn the pump on and off to keep a distant storage tank full are inexpensive and easy to add when using an LCB with remote control, as we recommend.

Because many solar pumps are designed to work all day at a slow but steady output, they won't keep up directly with average household fixtures, like your typical AC sub pump. This often requires some adjustment in how your water supply system is put together. For household use, we usually recommend the following, in order of cost and desirability:

Option 1. Pumping into a storage tank at least 50 feet higher than the house, if terrain and climate allow.

Option 2. Pumping into a house-level storage tank, if climate allows, and using a booster pump to supply household pressure.

Option 3. Pumping into a storage tank built into the basement for hard-freeze climates, and using a booster pump to supply household pressure.

Option 4. Using battery power from your household renewable energy system to run the submersible pump, and using a big pressure tank (80 gallons minimum).

Option 5. Using a conventional AC-powered submersible pump, large pressure tank(s), and your household renewable energy system with large inverter. There is some loss of efficiency in this setup, but it's the standard way to get the job done in freezing climates, and your plumber won't have any problems understanding the system. We strongly recommend one of the helical-rotor-type Grundfos SQ Flex pumps for this option. Start-up surge will be kind to your inverter, output is 5-9 gpm, and power use is a quarter that of conventional AC pumps.

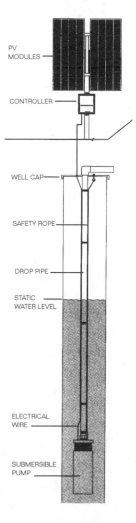

A simple PV-direct solar pumping system.

PV MODULES

CONTROLLER

WELL CAP

SAFETY ROPE

DROP PIPE

STATIC WATER LEVEL

ELECTRICAL WIRE

SUBMERSIBLE PUMP

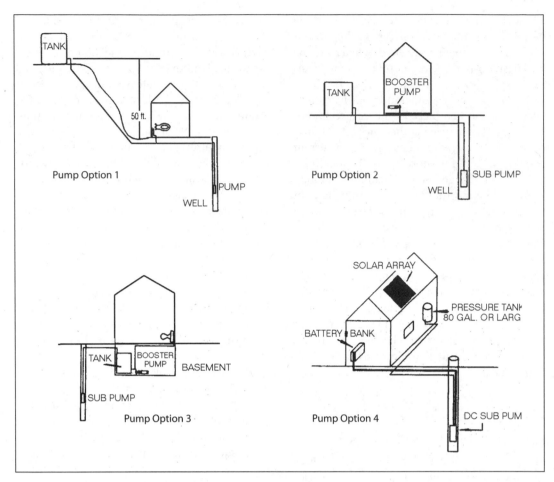

Many folks, for a variety of reasons, already have an AC-powered submersible pump in their well when they come to us but are real tired of having to run the generator to get water. For wells with 6-inch and larger casings, it's usually possible to install both the existing AC pump and a submersible DC pump. If your AC pump is 4 inches in diameter, the DC pump can be installed underneath it. The cabling, safety rope, and ½-inch poly delivery pipe from the DC submersible will slip around the side of the AC pump sitting above it. Just slide both pumps down the hole together. It's often comforting to have emergency backup for those times when you need it, like when it's been cloudy for three weeks straight, or the fire is coming up the hill and you want a lot of water fast!

PUMPING FROM A SPRING, POND, OR OTHER SURFACE SOURCE

Your choices for pumping from ground level are a bit more varied, depending on how high you need to lift the water and how many gallons per minute you want. Surface-mounted pumps are not freeze tolerant. If you live in a freezing climate, make sure that your installation can be (and is!) completely drained before freezing

weather sets in. If you need to pump through the hard-freeze season, we recommend a submersible pump as described above.

PUMPS DON'T SUCK! (THEY PUSH)

Pumps don't like to pull water up from a source. Or, put simply, Real Goods' #1 Principle of Pumping: Pumps don't suck; pumps push. To operate reliably, your surface-mounted pump must be installed as close to the source as practical. In no case should the pump be more than

Need Pump Help?

If you would like the help of our technical staff in selecting an appropriate pump, controller, power source, etc., we have a Solar Water Supply Questionnaire at the end of this chapter. It will give our staff the information we need in order to thoroughly and accurately recommend a water supply system for you. You can mail or fax your completed form. Please give us a daytime phone number if at all possible! Call us at 800-919-2400.

10 feet above the water level. With some positive-displacement pumps, higher suction lifts are possible but not recommended. You're simply begging for trouble. If you can get the pump closer to the source and still keep it dry and safe, do it! You'll be rewarded with more dependable service, longer pump life, more water delivery, and less power consumption.

LOW-COST SOLUTIONS

For modest lifts up to 50 or 60 feet and volumes of 1.5-3 gpm, we have found the SHURflo diaphragm-type pumps to be moderately priced and tolerant of abuses that would kill other pumps. They can tolerate silty water and sand without distress. They'll run dry for hours and hours without damage. But you get what you pay for. Life expectancy is usually two to five years, depending on how hard and how much the pump is working. Repairs in the field are easy, and disassembly is obvious. We carry a full stock of repair parts, but replacement motors don't cost much less than a new pump. Diaphragm pumps will tolerate sand, algae, and debris without damage, but these may stick in the internal check valves and reduce or stop output, necessitating disassembly to clean out the debris. Who needs the hassle? Filter your intake!

LONGER-LIFE SOLUTIONS

For higher lifts, or more volume, we often go to a pump type called a rotary-vane. Examples include the Slowpump and Flowlight Booster pumps. Rotary-vane pumps are capable of lifts up to 440 feet and volumes up to 4.5 gpm, depending on the model. Of all the positive-displacement pumps, they are the quietest and smoothest. But they will not tolerate running dry, or abrasives of any kind in the water. It's very important to filter the input of these pumps with a 10-micron or finer filter in all applications. Rotary-vane pumps are very long-lived but will eventually require a pump head replacement or a rebuild.

HOUSEHOLD WATER PRESSURIZATION

We promote two pumps that are commonly used to pressurize household water: The Chevy and Mercedes models, if you will.

The Chevy model. SHURflo's Medium-Flow pump is our best-selling pressure pump. It comes with a built-in 20- to 40-psi pressure switch. With a 2.5- to 3-gpm flow rate, it will keep up with most household fixtures, garden hoses excluded. The diaphragm pump is reliable and easy to repair, but it's somewhat noisy

and has a limited life expectancy. We recommend 24-inch flexible plumbing connectors in a loop on both sides of this pump, and a pressure tank plumbed in as close as possible to absorb most of the buzz.

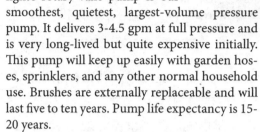

Diaphragm-type pump

The Mercedes model. Flowlight's rotary-vane pump is our smoothest, quietest, largest-volume pressure pump. It delivers 3-4.5 gpm at full pressure and is very long-lived but quite expensive initially. This pump will keep up easily with garden hoses, sprinklers, and any other normal household use. Brushes are externally replaceable and will last five to ten years. Pump life expectancy is 15-20 years.

Any household pressure system requires a pressure tank. A 20-gallon tank is the minimum size we recommend for a small cabin; full-size houses usually have 40-gallon or larger tanks. Pressure tanks are big, bulky, and expensive to ship. Get one at your local hardware or building supply store.

SOLAR HOT-WATER CIRCULATION

Most of the older solar hot-water systems installed during the tax-credit heydays of the early 1980s used AC pumps with complex controllers and multiple temperature sensors at the collector, tank, plumbing, ambient air, etc. This kind of complexity allows too many opportunities for Murphy's Law.

The smarter solar hot-water systems simply use a small PV panel wired directly to a DC pump. When the Sun shines a little bit, producing a small amount of heat, the pump runs slowly. When the Sun shines bright and hot, producing lots of heat, the pump runs fast. Very simple, but absolutely perfect. System control is achieved with an absolute minimum of gadgetry. We carry several hot-water circulation pumps for systems of various sizes. The best choice for most residential systems is the El-Sid pump.

Hot-water circulation pump

This is a solid-state, brushless DC circulation pump that was designed from scratch for PV-direct applications. The El-Sid pump comes in small, medium, and large sizes now, with open-discharge flow rates of 2, 3.3, and 6 gallons per minute. El-Sid pumps require a 5- to 30-watt PV module for drive, and life expectancy is three to four times longer than any other DC circulation pump. Volume and lift are sharply

Ram-type water pumps use the energy of falling water to force a portion of that water up the hill to a storage tank.

limited, however. These are circulation pumps, not lift pumps. They're meant to stir the fluid round and round in a closed system.

For solar hot-water systems with long, convoluted collection loops creating a lot of pipe friction, we can use multiple El-Sid pumps. If some amount of lift is involved, we have to look at more-robust pumps like the Hartell or the SunCentric series, which will require substantially more PV power.

SWIMMING POOL CIRCULATION

Yes, it's possible to live off the grid and still enjoy luxuries like a swimming pool. In fact, pool systems dovetail nicely with household systems in many climates. Houses generally require a minimum of PV energy during the summer because of the long daylight hours, yet the maximum of energy is available. By switching a number of PV modules to pool pumping in the summer, then back to battery charging in the winter, you get better utilization of resources.

DC pumps run somewhat more efficiently than AC pumps, so a slightly smaller DC pump can do the same amount of work as a larger AC pump. We also strongly recommend using a low-back-pressure cartridge-type pool filter. Diatomaceous-earth filters are trouble. They have high back pressure and will greatly slow circulation or increase power use.

Water-Powered Pumps

A few lucky folks have access to an excess supply of falling water. This falling-water energy can be used to pump water. Both the High Lifter and ram-type water pumps use the energy of falling water to force a portion of that water up the hill to a storage tank.

RAM PUMPS

Ram pumps have been around for many decades, providing reliable water pumping at almost no cost. They are more commonly used in the eastern U.S. where modest falls and large flow rates are the norm, but they will work happily almost anyplace their minimum flow rate can be satisfied. Rams will work with a minimum of 1.5 feet to a maximum of about 20 feet of fall feeding the pump. Minimum flow rates depend on the pump size; see the product section for specs.

Here's how ram pumps work: A flow is started down the drive pipe and then shut off suddenly. The momentum of moving water slams to a stop, creating a pressure surge that sends a

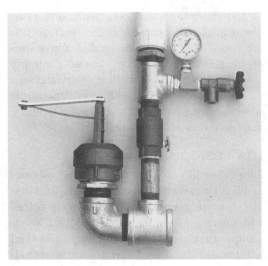

Ram pump

little squirt of water up the hill. How much of a squirt depends on the pump size, the amount of fall, and the amount of lift. Output charts accompany the pumps in the product section. Each ram needs to be tuned carefully for its particular site. Ram pumps are not self-starting. If they run short of water, they will stop pumping and simply dump incoming water, so don't buy too big. Rams make some noise—a lot less than a gasoline-powered pump, but the constant 24-hours-a-day chunk-chunk-chunk is a consideration for some sites. Ram pumps deliver less than 5% of the water that passes through them, and the discharge must be into an unpressurized storage tank or pond. But they work for free and you can expect them to last for decades.

THE HIGH LIFTER PUMP

This pump is unique. It works by simple mechanical advantage. A large piston at low water pressure pushes a smaller piston at higher water pressure. High Lifters recover a much greater percentage of the available water than ram pumps do, but they generally require greater fall into the pump. This makes them better suited for

High Lifters recover a much greater percentage of the available water than ram pumps do, but they generally require greater fall into the pump.

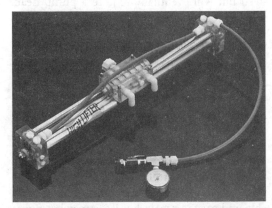

High Lifter pump

more mountainous territory. They are available in two ratios, 4.5:1 and 9:1. Fall-to-lift ratios and waste-water-to-pumped-water ratios are also either 4.5:1 or 9:1. Note, however, that as the lift ratio gets closer to theoretical maximum, the pump is going to slow down and deliver fewer gallons per day. High Lifters are self-starting. If they run out of water, they will simply stop and wait, or slow down to match what water is available. This is a very handy trait for unattended or difficult-to-attend sites. The only serious disadvantage of the High Lifter is wear caused by abrasives. O-rings are used to seal the pistons against the cylinder walls. Any abrasives in the water wear out the O-rings quickly, so filtering of the intake is strongly encouraged. High Lifter pumps can be overhauled fairly easily in the field, but the O-ring kit costs a lot more than filter cartridges. Output charts for this pump are included in the product section.

Wind-Powered Water Pumps

If you've been reading this far because you want to buy a nostalgic old-time jack-pump windmill, we're going to disappoint you. They are still made but are very expensive, typically $7,500 and up. We don't sell them and don't recommend them. They are quite a big deal to set up and install into the well and require routine yearly service at the top of the tower. This is technology that largely has seen its day. Submersible pumps powered by PV panels are a much better choice for remote locations now.

There are a couple of ways to run a pump with wind power that have seen good success.

COMPRESSED-AIR WATER PUMPS

Airlift pumps use compressed air. Three models are available, depending on lift and volume needs. They all use a simple pole-mounted turbine that direct-drives an air compressor with a wind turbine. The air is piped down the well and runs through a carefully engineered air injector. As it rises back up the supply tube, it carries slugs of water in between the bubbles. The lift/submergence ratio of this pump is fairly critical. Lift is the vertical rise between the standing water level in the well and your tank. Approximately 30% of the lift is the recommended distance for the air injector to be submerged below the standing level. As lifts edge over 200 feet, the submergence ratio rises to a maximum of 50%. Too little submergence and the air will separate from the water; too much and the air will not lift the water, though considerable latitude exists be-

Wind-powered pump

tween these performance extremes. This pump isn't bothered by running dry. Output depends on wind speed, naturally, but the largest model is capable of over 20 gpm at lower lifts, or can lift a maximum of 315 feet. The air compressor requires an oil change once a year. See the product section for more details and output specs.

WIND GENERATORS RUNNING SUBMERSIBLE PUMPS

The larger wind generator manufacturers, Bergey and Whisper, both offer options that allow the three-phase wind turbine output to power a three-phase submersible pump directly. These aren't residential-scale systems and are mostly used for large agricultural projects, or village pumping systems in less industrialized countries. They are moderately expensive to buy. Contact our technical staff for more information on these options.

Freeze Protection

In most areas of the country, freezing is a major consideration when installing plumbing and water storage systems. For outside pipe runs, the general rule is to bury the plumbing below frost level. For large storage tanks, burial may not be feasible, unless you go with concrete. In moderate climatic zones, simply burying the bottom of the tank a foot or two along with the input/output piping is sufficient. In some locations, due to climate or lack of soil depth, outdoor storage tanks simply aren't feasible. In these situations, you can use a smaller storage tank built into a corner of the basement with a separate pressure-boosting pump, or you can pump directly into a large pressure-tank system.

The air is piped down the well and runs through a carefully engineered air injector. As it rises back up the supply tube, it carries slugs of water in between the bubbles.

Other Considerations and Common Questions

The usual rule for sizing PV-direct arrays is to add about 20% to the pump wattage in a mild climate, or add 30% in a hot climate.

We hope that by this point, you've zeroed in on a pump or pumps that seem to be applicable to your situation. If not, our technical staff will be happy to discuss your needs and recommend an appropriate pumping system—which may or may not be renewable-energy powered. Filling out the Solar Water Supply Questionnaire that follows will supply all the answers we're likely to need when we talk with you. At this point, another crop of questions usually appears, such as . . .

HOW FAR CAN I PUT THE MODULES FROM THE PUMP?

Often the best water source will be deep in a heavily wooded ravine. It's important that your PV modules have clear, shadow-free access to the Sun for as many hours as practical. Even a fist-size shadow will effectively turn off most PV modules. The hours of 10 a.m. to 2 p.m. are usually the minimum your modules want clear solar access, and if you can capture full Sun from 9 a.m. to 3 p.m., that's more power for you. If the pump is small, running off one or two modules, then distances up to 200 feet can be handled economically. Longer distances are always possible, but consult with our tech staff or check out the wire-sizing formula on page 585 first, because longer distances require large (expensive!) wire. Many pumps routinely come as 24- or 48-volt units now, as the higher voltage makes long-distance transmission much easier. The Grundfos sub pump accepts DC input up to 300 volts. If you need to run more than 300 feet for sunshine, this may be just the ticket.

WHAT SIZE WIRE DO I NEED?

This depends on the distance and the amount of power you are trying to move. See the Wire Sizing Chart on page 265 and formula that can properly size wire for any distance, at any voltage, AC or DC. Or just give our tech staff a call—we do this kind of consultation all the time. Going down a well, 10-gauge copper submersible pump wire is the usual, although some of the larger sub pumps we're using now require 8-gauge, or even 6-gauge, for very deep wells. If you need anything other than common 10-gauge wire, we'll let you know.

WHAT SIZE PIPE DO I NEED?

Many of the pumps we offer have modest flow rates of 4 gpm or less. At this rate, it's okay to use smaller pipe sizes such as ¾ inch or 1 inch for pumping delivery without increasing friction loss. However, that's for pumping delivery only. There's no reason you can't use the same pipe to take the water up the hill to the storage tank and also bring it back down to the house or garden. Pipes don't care which way the water is flowing through them. But if you do this, you'll want a larger pipe to avoid friction loss and pressure drop when the higher flow rate of the household or garden fixtures comes into play. We usually recommend at least 1¼-inch pipe for household and garden use, and 2-inch for fire hose lines. See the pipe friction loss charts on page 380 to be sure.

WHAT SIZE PV MODULES DO I NEED?

A number of the pumps listed in the product section have accompanying performance and wattage tables. For instance, the rotary-vane Solar Slowpump model 2503 lifting 40 feet will deliver 2.5 gpm and requires 60 watts. Hey, what's to figure here? All PV modules are rated by how many watts they put out, right? So you just need a 60-watt module. Wrong. PV modules are rated under ideal laboratory conditions, not real life. If you want your pump to work on hot or humid

Water Pumping Truths and Tips

1. Pumps prefer to *push* water, not *pull* it. In fact, most pumps are limited to 10 or 15 feet of lift on the suction side. Mother Nature has a theoretical suction lift of 33.9 feet at sea level but only if the pump could produce a perfect vacuum. Suction lift drops 1 foot with every 1,000 feet rise in elevation. To put it simply, **PUMPS DON'T SUCK, THEY PUSH.**

2. Water is heavy, 8.33 pounds per gallon. It can require tremendous amounts of energy to lift and move.

3. DC electric motors are generally more efficient than AC motors. If you have a choice, use a DC motor to pump your water. Not only can it be powered directly by solar modules, but your precious wattage will go further.

4. Positive-displacement pumps are far more efficient than centrifugal pumps. Most of our pumps are positive-displacement types. AC-powered submersibles, jet pumps, and booster pumps are centrifugal types.

5. As much as possible, we try to avoid batteries in pumping systems. When energy is run into and out of a battery, 25% is lost. It's more efficient to take energy directly from your PV modules and feed it right into the pump. At the end of the day, you'll end up with 25% more water in the tank.

6. One pound per square inch (psi) of water pressure equals 2.31 feet of lift, a handy equation.

days, then you'd better add 30% to the pump wattage. Heat reduces PV output; water vapor cuts available sunlight. So, in our example, we actually need 78 watts. Looking at available PV modules, we don't find exactly 78-watt modules, which means we buy an 80-watt module. The usual rule for sizing PV-direct arrays is to add about 20% to the pump wattage in a mild climate, or add 30% in a hot climate. And, of course, an LCB (linear current booster) of sufficient amperage capacity is practically standard equipment with any PV-direct system.

CAN I AUTOMATE THE SYSTEM?

Absolutely! Life's little drudgeries should be automated at every opportunity. The LCBs that we so strongly recommend with all PV-direct systems help us in this task. These units are all supplied with wiring for a remote-control option. This allows you to install a float switch at the holding tank, and a pair of tiny 18-gauge wires can be run as far as 5,000 feet back to the pump/ controller/PV modules area. Float switches can be used either to pump up or pump down a holding tank and will automatically turn the system off when satisfied. The LCBs, float switches, and other pumping accessories are presented in the product section. LCBs can also be used on battery-powered systems when remote sensing and control would be handy.

Where Do We Go from Here?

If you haven't found all the answers to your remote water pumping needs yet, please give our experienced technical staff a call. They'll be happy to work with you selecting the most appropriate pump, power source, and accessories for your needs. We run into situations occasionally where renewable energy sources and pumps simply may not be the best choice, and we'll let you know if that's the case. For 95% of the remote pumping scenarios, there is a simple, cost-effective, long-lived renewable energy-powered solution, and we can help you develop it.

Solar Water Supply Questionnaire

To thoroughly and accurately recommend a water supply system, we need to know the following information about your system. Please fill out the form as completely as possible. Please give us a daytime phone number, too.

Name: _____

Address: _____

City: _____ State: _____ Zip: _____

Email: _____ Phone: _____

DESCRIBE YOUR WATER SOURCE

Depth of well: _____

Depth to standing water surface: _____

If level varies, how much? _____

Estimated yield of well (gallons/minute): _____

Well casing size (inside diameter): _____

Any problems? (silt, sand, corrosives, etc.) _____

WATER REQUIREMENTS

Is this a year-round home? _____ How many people full time? _____

Is house already plumbed? _____ Conventional flush toilets? _____

Residential gallons/day estimated: _____

Is gravity pressurization acceptable? _____

Hard-freezing climate? _____

Irrigation gallons/day estimated: _____ Which months? _____

If you have a general budget in mind, how much? _____

Do you have a deadline for completion? _____

DESCRIBE YOUR SITE

Elevation: _____

Distance from well to point of use: _____

Vertical rise or drop from top of well to point of use: _____

Can you install a storage tank higher than point of use? _____

     How much higher? _____ How far away? _____

Complex terrain or multiple usage? _____ (Please enclose map)

Do you have utility power available? _____ How far away? _____

Can well pump be connected to nearby home power system? _____

     How far? _____ Home power system power battery voltage: _____

Mail your completed questionnaire to:

Tech Staff/Water Supply, Real Goods, 360 Interlocken Blvd. Ste. 300, Broomfield, CO 80021. Or fax to: 800-482-7602. Questions? Call us at 800-919-2400.

Real Goods Homestead Plumbing Recommendations

TYPES OF PIPE

Metals

Black iron and galvanized iron—used for gas (propane or natural gas) plumbing but little else. It is slow and difficult to cut and thread, and requires special, expensive tools. (Consider renting tools if needed.) A special plastic-coated iron pipe is used for buried gas lines. The insides of galvanized pipe corrode when used for water supply and eventually restrict water flow. Older houses often suffer from this affliction, which causes ultrasensitivity of shower temperature and other exciting problems.

Cast iron—used to be the material of choice for waste plumbing, but ABS plastic thankfully has replaced it. Simple conversion adapters to go from cast iron to ABS are available if you find yourself having to repair an existing older system.

Copper—the most common material for water supply plumbing in new home construction. Most plumbing codes require copper for indoor work. Type M, the most common grade for interior work, comes in 20-foot rigid lengths. Houses typically use ¾-inch lines for feeders and ½-inch lines for fixtures. If gravity-fed water pressure will be lower than the 20- to 40-psi city standard, consider increasing your supply pipes one size. This cures many low-pressure woes by eliminating pressure loss within the house (and provides a shower that's nearly immune to temperature fluctuations). The thicker-walled flexible tubing L and K grades can be used legally in some localities, but they generally cost considerably more. Copper does not tolerate freezing at all! Copper is relatively easy to work with, and you need only simple, inexpensive tools. Joints must be "sweat soldered," which takes about 10 minutes to learn and is kind of fun afterward. Be sure to use the newer lead-free solders for water supply piping! Some states allow the use of flexible copper tubing for gas supply lines, in which case special compression-type fittings are used.

Bronze/Brass—beautiful stuff, but too expensive for common use. Precut, prethreaded nipples are used occasionally for dielectric water tank protection (5 inches of brass qualifies as protection under code). Use brass only where plumbing will be exposed and you want it to look nice.

Plastics

Poly (polyethylene)—this black, flexible pipe is used widely for drip and irrigation systems. Available in utility or domestic grades. Never use utility grade for drinking water! It's made with recycled plastics, and you don't know what will leach out. Poly is almost totally freezeproof. It's easy to work with (when warm) and requires common hand tools. It's sold in 100-foot or occasionally longer rolls. Barbed fittings and hose clamp fittings *will* leak (why do you suppose it's used in *drip* systems?). Do not use this cheap plumbing for any permanently pressurized installations, unless you can afford to throw water away. Poly also degrades fairly rapidly in sunlight. Two to three years is the usual unprotected life span. One good use for poly is with submersible well pumps where the flexibility makes for easy installations. Poly pipe is the usual choice for sub pumps unless they're pumping hundreds of feet of lift at very high pressure. Special 100- and 160-psi high-pressure versions of poly pipe are used for sub pumps.

PVC (polyvinyl chloride)—one of the wonders of chemistry that makes homesteading easy. This is the white plastic pipe that should be used for almost everything outside the house itself. It's easy to work with and requires only common hand tools (although if you're doing a lot of it, a PVC cutter is a real timesaver over the hacksaw). PVC usually can survive mild freezing. Hard freezes will break joints, however, so use protection. It comes in 20-foot lengths. Sizes 1 inch and up generally are available with bell ends, saving you the expense and extra gluing of couplings. Use primer (the purple stuff), then glue (the clear stuff) when assembling. PVC must be buried or protected. If exposed to sunlight, it will degrade slowly from the UV, but it will also grow algae inside the pipe!

PVC grades: Schedule 40 is the standard PVC that is recommended for most uses. Some lighter-duty agricultural types are available. Class 125, class 160, and class 200 are graded by the psi rating of the pipe. Don't use these cheaper thin-wall classes if you're burying the pipe! Leaks from stones pushing through thinner walls are too hard to find and fix.

Try to arrange your house plan so that bathrooms, laundry room, and kitchen all fit back to back and can share a common plumbing wall. This saves thousands of dollars in materials costs, lots of time during construction, hours of time waiting for the hot water, and many gallons of expensive hot water.

CPVC (chlorinated polyvinyl chloride)—a PVC formulation specifically made for hot water. The fittings and the pipe are a light tan color so you can tell the difference from ordinary PVC. Some counties and states allow the use of CPVC; most don't. It works just like PVC.

ABS (acrylonitrile butadiene styrene)—a black rigid pipe universally used for unpressurized waste and vent plumbing. It's easy to work with, requires common hand tools, and glues up quickly just like PVC. Make sure that the glue you use is rated for ABS. Some glues are universal and will work on both PVC and ABS; others are specific. Buy a big can of ABS glue—these larger pipes use a lot more glue. Also, the bigger cans come with bigger swabs, which you'll need on larger pipe.

PB (polybutylene)—a flexible plastic pipe that is used occasionally for home plumbing, but mostly for radiant-floor heating systems. It requires special compression rings and tools for fittings. PB will survive considerable abuse during construction without springing leaks. It makes no reaction with concrete floors. PB is the best choice for radiant floors and has nearly infinite life expectancy when buried in concrete.

OTHER TIPS

Don't go small on supply piping. Long pipe runs from a supply tank up on the hill can produce substantial pressure drop. The typical garden needs 1¼-inch supply piping for watering and sprinklers. We recommend *at least* 1½-inch supply for most houses, and a 2-inch or larger line for a fire hose is often a very good idea.

Keep plumbing runs as short as possible. Try to arrange your house plan so that bathrooms, laundry room, and kitchen all fit back to back and can share a common plumbing wall. This saves thousands of dollars in materials costs, lots of time during construction, hours of time waiting for the hot water, and many gallons of expensive hot water.

Insulate *all* **hot water pipes,** even those inside interior walls before the walls are covered up. The savings and comfort over the life of the house quickly make up for the initial costs.

Store your glue cans upside down. This sounds wacky, but it works. Tighten the cap first, obviously. The glue seals any tiny air leaks and keeps all the glue inside and on the cap threads from drying solid. Don't store your primer upside down, though; it can't seal the tiny leaks and will disappear.

WATER PUMPING PRODUCTS

Fountain and Pond Pumps and Products

Pump and Panel for Solar Fountains

Use this nice little kit to create your own solar fountain or to replace the failed pump on existing fountains. Our kit includes an unbreakable PV panel that measures 4.7"x4.4" with a locking weatherproof electrical connector, 15 feet of flexible submersible cable, and a submersible DC pump that measures 2.5"x1.5"x2" (LxWxH) with a 5/16-inch tubing fitting and rubber suction-cup feet for quiet, stable operation. Under full Sun, it delivers ¾ gpm at zero lift, or a trickle at 24 inches of lift. The pump can be disassembled easily to clean the impeller. Motor brushes are not accessible or replaceable (expect about a two-year life). China.

40-0078 Small Solar Pump and Panel   **$59**

Solar Pumps for Fountains and Small Ponds

Our submersible 12-volt pumps are designed as continuous-duty marine bilge pumps. We've found they make very impressive solar-powered fountain pumps. They won't lift very high, but they move an incredible volume of water and use only a tiny amount of power. The pump motor twist-locks onto the pump base to allow easy cleaning and removal of debris. Can withstand dry running for a limited time. 3-yr. mfr. warranty. In full-time PV-direct use, these pumps can be expected to last about one summer. Attwood is great about honoring the warranty, so we recommend buying two pumps and being prepared to rotate your stock. Ballast resistor recommended. USA.

Model V625: A good match with any 20- to 30-watt module.
- Amps: 1.3 @ 3-ft. head
- Max. gpm: 7.5 @ 3-ft. head
- Max. head: 5 ft.

41157   V625 Solar Fountain Pump   **$30**

Model V1250: A good match with any 30- to 50-watt module.
- Amps: 2.6 @ 3-ft. head
- Max. gpm: 15.5 @ 3-ft. head
- Max. head: 8 ft.

41158   V1250 Solar Fountain Pump   **$44**

A Bigger Solar Fountain Pump

This Gold-Series solar-powered pump will make your waterfall or fountain even more impressive. Our largest submersible fountain pump can deliver up to 18 gpm at 1 meter lift, 11.5 gpm at 6½ ft., or 5 gpm at 9½ ft. Engineered to commercial standards with a long-life heavy-duty motor, this pump features an easy-to-clean snap-lock strainer, an extra-strong nylon/fiberglass impeller, a 11/8-inch outlet fitting, and the ability to run dry without burnout. It will run directly off a 60- to 130-watt PV module or a 12-volt power source and draws 4.8 amps at 12 volts, or 7 amps at 13.6 volts. To reach maximum listed flow, use a 120-watt or larger module. Life expectancy as a fountain pump is two to three years or longer with pump controller. 5-yr. mfr. warranty. 6"x4¼" (HxW). USA.

40-0161 Rule 04 Pump   **$114**
41-0117 Sharp 123W PV Module   **$729**
41-0177 Sharp 80W PV Module   **$495**
25-002   7A, 12/24V Pump Controller   **$119**

Laing Solar Surface Pump

Ideal for springs, ponds, and other surface water sources, this powerful, solar surface pump boasts high 8¾-foot maximum lift and 6½-gpm maximum flow rates that send water where you need it. A unique, shaft-free design provides an impressive 40,000-hour service life, and Maximum Power Point Tracking down-converts excess voltage to amperage so you get more power with less solar input. Its 8- to 24-volt DC compatibility prevents burnout from voltage variations and makes it a great match with a 12-volt nominal 9- to 30-watt PV module. Corrosion-free Noryl housing comes with ½-inch hose barbs. 4"x3½"x3½" (HxWxD). China.

40-0196 Laing Solar Surface Pump   **$249**

Ballast Resistor

Made of automotive-grade porcelain, this ceramic resistor lowers potentially damaging high voltage produced by PV modules to the 13.6 volts maximum comfortably handled by most pumps. Protects sensitive electronics and fan and pump motors from over-voltage in solar-direct applications. 3"x½"x5⁄8" (LxWxD). ⅓ lb. Mexico.

43-0235 Ballast Resistor $10

Alpine Blue and GreenClean Algae Control Additives

Enjoy a crystal-clear, healthy pond or fountain, free of unsightly pond scum. Alpine Blue and GreenClean additives safely control algae overgrowth and nuisance aquatic plants in ponds and fountains. Alpine Blue is a nontoxic aquatic dye that reduces the amount of light that penetrates the water, inhibiting plant growth. As an added benefit, it also enhances the water's color to a brilliant, natural blue. Fish and humans can safely swim in water treated with Alpine Blue, and 1 quart treats 4-8 acre-feet of water. GreenClean, one of the only non-copper-based algicides available, is classified as an organic algicide that can eliminate a broad spectrum of algae on contact. Reapplication of both products may be necessary throughout the season. USA.

40-0179 Alpine Blue $35
40-0180 GreenClean $185

Solar Island Fountain

Creating a healthy pond you can enjoy for years to come requires thorough oxygenation of the water to support fish and aquatic plant life. Relax to the sounds of falling water while bringing the advantages of aeration to your pond with our Solar Island Fountain. Using free solar energy, this floating fountain spouts water up to 17 inches high when in direct sunlight. For improved aesthetics, the new lily pad design sits lower in the water. No wiring needed; simply place on water's surface. Includes three fountain heads, pump, and filter. 3¾"x9½" (H x diameter). 1.1 lb. Thailand.

14-0336 Solar Island Fountain $55

Solaer Complete Solar Pond Aeration Kits

Solaer solar-powered aeration kits feature long-life, brushless DC air pumps on adjustable automatic daily

timers, with maintenance-free batteries. Solar arrays are sized to deliver 16-24 hours of operation in most locales. Pole-top mounts are provided. Components mounted in a weathertight, sound-deadening fiberglass box with legs. USA.

SB1A covers small ponds up to ½ acre. Includes an 80-watt PV module, one air pump, 50 feet of poly tubing, and one diffuser. Requires a 2½-inch steel pipe for mount.

40-0054 SB1A $3,150

SB1 covers ponds up to 1 acre. Includes a 120-watt PV module, one air pump, and one diffuser. Requires a 2½-inch steel pipe for mount.

40-0055 SB1 $5,165

SB2 covers ponds up to 2 acres. Includes two 80-watt PV modules, one high-output air pump, and one diffuser. Requires a 2½-inch steel pipe for mount.

40-0056 SB2 $5,565

SB3 covers ponds up to 3 acres. Includes two 120-watt PV modules, two air pumps, manifold, and three diffusers. Requires a 3-inch steel pipe for mount.

40-0057 SB3 $8,275

SB4 covers ponds and lakes up to 5 acres. Includes three 120-watt PV modules, two high-output air pumps, manifold, and four diffusers. Requires a 4-inch steel pipe for mount.

40-0058 SB4 $11,450
40-0194 ½-in. tubing (100-ft. roll) $125

Solar-Powered Pond Aeration Kit

Aeration improves the aesthetics and overall health of ponds and lakes, reducing the chance of having unsightly algae blooms, the accumulation of sludge, and fish kills. To bring these benefits to your pond, you can

The weighted air diffuser is mounted in 12" PVC with a float. Includes ½" and ⅜" air fittings.

either toss the water into the air or toss the air into the water. Since air weighs a lot less, your energy expenditure will be less and go further with an air pump—such as this 12-volt model that runs directly off a PV module or two. Our do-it-yourself kit supplies all the major pieces for ponds up to about 500 square feet and 12 feet deep. The bottom-mounted diffuser delivers air as millions of tiny bubbles for better oxygenation and stirring of the pond to discourage stratification.

The Thomas ⅒-hp air pump delivers 1.4 cubic feet per minute (cfm) at free discharge, or 0.8 cfm at 10 psi. It draws a maximum of 8.5 amps and can be driven to full output by a pair of 75- to 80-watt PV modules (sold separately below) with a linear current booster to help start-up. The diffuser is properly sized to work with this pump. On site, you'll need to supply a 2.5-inch pipe for the pole-top mount, weather (and possibly noise) protection for the pump, wiring, small parts, and weighed air line as needed by your site particulars. All products made in USA, except LCB made in Canada.

40-0072	Thomas 12V 1/10-hp Air Pump	$275
40-0073	Air Diffuser	$240
41-0177	Sharp 80W PV Module	$495
42-0185	Uni-Rac 500040 Pole-Top Rack	$134
25-003	Solar Converters 10A LCB	$135

Easy-Shape PondGard Liner

PondGard membrane is friendly to do-it-yourselfers. This EPDM synthetic rubber is a thick 45-mil waterproofing membrane that's highly flexible and stable. It's far easier to work with in small spaces than our standard pond liner material (next page), which tends to be quite stiff. PondGard is specially formulated to be safe for exposure to fish and plant life in decorative ponds. It can be shaped easily to fit the unique contours of any pond. Use it for pond lining, streams and fall features, erosion control, storage tanks, or wildlife ponds. PondGard has outstanding resistance to ultraviolet radiation (UV), ozone, and other environmental conditions. It requires little or no regular maintenance once installed. Available precut in sizes listed. Ships via standard UPS. USA.

40-0074	PondGard Liners 10'x10'	$105
	10'x15'	$149
	15'x15'	$225
	15'x20'	$280
	15'x25'	$375
	20'x20'	$390

Leak-Free Pond Liners

A pond is a major investment, requiring lots of earthwork, soil blending, and compaction. It's a gamble if it will hold water or not, or for how long. With a liner, earthwork can be kept to a minimum, and a leak-free pond can be established on any kind of soil. Our pond liners feature a very tough, UV-stabilized polyethylene that is highly resistant to punctures, tears, roots, and rodents. It's also nontoxic, nonleaching, and FDA/USDA approved for potable water supplies. Whether your plans call for a stocked aquaculture pond, an agricultural reservoir, or a garden feature, you need a liner that is absolutely leak-free to protect your investment of time and money. Our pond liners are custom-sized in any square or rectangular configuration. Heat-welded factory seams are stronger than the surrounding material. Liners are "accordion-folded" for easy spreading on large sites. Four workers can handle up to 10,000 square feet.

Available in tough 20-mil thickness, in super heavy-duty 30-mil, or for really extreme sites, 40-mil. Weight per 1,000 sq. ft.: 20 mil = 100 lb., 30 mil = 150 lb., 40 mil = 200 lb.

100 sq. ft. minimum order. $15 handling charge on orders under 500 sq. ft. Free continental U.S. freight on orders over 2,000 sq. ft. USA.

To figure size: Double the depth of your pond, add this to the width and length. Add an additional 2-5 feet for the buried apron. For instance, a 100'x50' pond that's 5 feet deep needs a 112'x62' liner at minimum (6,944 sq. ft.).

| 47-212 | Pond Liner Fab Tape 2"x30' | $36 |
| 47-227 | Handling Charge for under 500 sq. ft. | $15 |

Pond Liner Price and Comparison Chart

PRICE PER SQUARE FOOT

Item #	Weight	100-200	201-600	601-2,000	2,001-5,000	5,001-8,000	8,001-12,000	12,001-20,000
47-203	20-mil	.56/sq. ft.	.50/sq. ft.	.44/sq. ft.	.42/sq. ft.	.31 sq. ft.	.28 sq. ft.	.26 sq. ft.
47-204	30-mil	.66/sq. ft.	.59/sq. ft.	.54/sq. ft.	.53/sq. ft.	.37 sq. ft.	.34 sq. ft.	.32 sq. ft.
47-211	40-mil	.72/sq. ft.	.69/sq. ft.	.66/sq. ft.	.65/sq. ft.	.46 sq. ft.	.43 sq. ft.	.41 sq. ft.

Pond liners are made to order, so please have dimensions ready when ordering. 100 sq. ft. minimum order. Please call for pricing on larger liners.

Centrifugal Pumps

EL-SID Hot-Water Circulation Pumps

El-SID 5 and 10 PV models

El-SID 20 PV model

SID stands for Static Impeller Driver, the "EL" is just for fun. These completely solid-state "motors" have no moving parts! There are no brushes, bearings, or seals to wear out. The rustproof bronze pump section is magnetically driven and is an existing off-the-shelf pump head that can be replaced easily if ever necessary. Life expectancy is many times longer than any other DC circulation pump.

PV-driven models are offered at 2.0, 3.3, and 6.0 gpm maximum flow rates. The battery-driven model delivers 3.3 gpm max at 12 or 24 volts. The 2.0-gpm PV pump requires a 5-watt PV module, the 3.3-gpm pump requires a 10- to 20-watt module, and the 6.0-gpm pump requires a 30-watt module. These are circulation pumps for closed-loop systems and will not do any significant lift. See the specifications chart. Must be installed in a protected location (no rain exposure). Rated for temperatures up to 240°F. 1-yr. mfr. warranty. USA.

41134	EL-SID 5PV 2.0-gpm Pump	$215
41133	EL-SID 10PV 3.3-gpm Pump	$248
40-0082	EL-SID 20PV 6.0-gpm Pump	$305
40-0083	EL-SID 10B 12V 3.3-gpm Pump	$225
40-0084	EL-SID 10B 24V 3.3-gpm Pump	$275

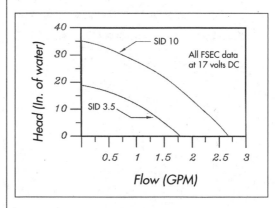

SunCentric Solar Surface Pumps

High-temperature option available

A reliable cast-iron DC pump with one moving part. Good for swimming pools, agriculture, hydronic and solar heating, pond aeration, and more. Can do lifts up to 80 feet or volumes up to 60 gpm, depending on

model; see performance charts. Maximum suction lift is 10 feet with a foot valve; centrifugal pumps will not self-prime. Mount the pump as close to the water source as possible; flooded inlet is best. Easy-starting on PV-direct systems; no electronic boost controls are needed. Both PV-direct and battery-powered models are available. Standard pumps use a glass-filled polycarbonate impeller with a temperature limit of 140°F. The $50 high-temperature option upgrades to a brass impeller with a 212°F limit. HT option reduces flow volume by 15%; power input is the same. Brush life is typically 3-10 years. 2-yr. mfr. warranty. USA.

41010	**SunCentric 7212 Pump**	**$705**
41011	**SunCentric 7213 Pump**	**$705**
41012	**SunCentric 7214 Pump**	**$705**
41013	**SunCentric 7321 Pump**	**$745**
41014	**SunCentric 7322 Pump**	**$745**
41015	**SunCentric 7323 Pump**	**$745**
41016	**SunCentric 7324 Pump**	**$705**
41017	**SunCentric 7325 Pump**	**$860**
41018	**SunCentric 7415 Pump**	**$860**
41019	**SunCentric 7423 Pump**	**$765**
41020	**SunCentric 7424 Pump**	**$765**
41021	**SunCentric 7425 Pump**	**$860**
41022	**SunCentric 7426 Pump**	**$855**
41023	**SunCentric 7442 Pump**	**$765**
41024	**SunCentric 7443 Pump**	**$765**
41025	**SunCentric 7444 Pump**	**$765**
41026	**SunCentric 7445 Pump**	**$860**
41027	**SunCentric 7446 Pump**	**$860**
41028	**SunCentric 7511 Pump**	**$810**
41029	**SunCentric 7521 Pump**	**$810**
41030	**SunCentric 7526 Pump**	**$905**
41031	**SunCentric 7622 Pump**	**$905**
41032	**SunCentric 7623 Pump**	**$905**
40-0103	**High Temp Option**	**$50**

Replacement Parts for SunCentric Pumps

41033	**SunCentric Brush Set**	**$45**
41034	**SunCentric 7xx1 to 4 Seal & Gasket Set**	**$18**
41035	**SunCentric 7xx5 or 6 Seal & Gasket Set**	**$18**
40-0104	**High Temp Gasket Set (specify model)**	**$38**

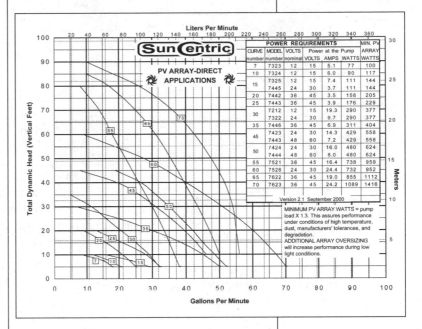

PV ARRAY-DIRECT APPLICATIONS

POWER REQUIREMENTS

CURVE number	MODEL number	VOLTS nominal	Power at the Pump VOLTS	AMPS	WATTS	MIN. PV ARRAY WATTS
7	7323	12	15	5.1	77	100
10	7324	12	15	6.0	90	117
15	7325	12	15	7.4	111	144
	7445	24	30	3.7	111	144
20	7442	36	45	3.5	158	205
25	7443	36	45	3.9	176	229
30	7212	12	15	19.3	290	377
	7322	24	30	9.7	290	377
35	7446	36	45	6.9	311	404
45	7423	24	30	14.3	429	558
	7443	48	60	7.2	429	558
50	7424	24	30	16.0	480	624
	7444	48	60	8.0	480	624
55	7521	36	45	16.4	738	959
60	7526	24	30	24.4	732	952
65	7622	36	45	19.0	855	1112
70	7623	36	45	24.2	1089	1416

Version 2.1 September 2000

MINIMUM PV ARRAY WATTS = pump load X 1.3. This assures performance under conditions of high temperature, dust, manufacturers' tolerances, and degradation.
ADDITIONAL ARRAY OVERSIZING will increase performance during low light conditions.

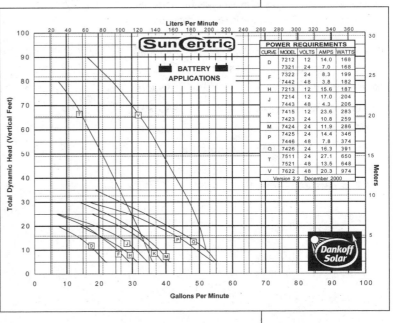

BATTERY APPLICATIONS

POWER REQUIREMENTS

CURVE	MODEL	VOLTS	AMPS	WATTS
D	7212	12	14.0	168
	7321	24	7.0	168
F	7322	24	8.3	199
	7442	48	3.8	182
H	7213	12	15.6	187
J	7214	12	17.0	204
	7443	48	4.3	206
K	7415	12	23.6	283
	7423	24	10.8	259
M	7424	24	11.9	286
P	7425	24	14.4	346
	7446	48	7.8	374
Q	7426	24	16.3	391
T	7511	24	27.1	650
	7521	48	13.5	648
V	7622	48	20.3	974

Version 2.2 December 2000

Dankoff Solar

Solar Pool-Pumping Systems

Pool pumping often can be a suburban home's largest single power use. Solar pool-pumping systems are moderately expensive initially but run for nothing and have life expectancies measured in decades. Our solar-powered pool-pumping systems consist of a high-effi-

ciency brushless motor coupled to a standard glass-filled polycarbonate pool pump head with a basket filter, clear cover, and 2-inch inlet/outlet fittings. The pump is driven directly from a solar-electric array for simplicity and next to no maintenance. When the Sun comes up, the pump runs.

An optional AC power converter provides back-up power from utility or generator power. The pump can draw power from solar and the converter simultaneously, with solar power as priority. Mounts sold separately. Manufacturers' warranties and countries of origin: 4 yr. on pump motor/control, China. 10 yr. on racks/mounts, USA. 20 yr. on PV modules, USA. Pump head, Germany.

Each system includes a pump assembly with controller plus BP Solar brand PV modules in 48-volt configuration. Choose a system wattage based on pool size and filter back pressure (see lower chart). Then choose your preferred mounting style for that size system.

40-0059	340W Sstem	$5,075
40-0060	480W System	$5,890
40-0061	600W System	$6,980
48-0015	GFI Option (for roof-mounted arrays)	$190
48-0016	AC Power Converter Option	$755

Solar Array Mounting Options for Pool Systems

Mounting Type	340-Watt System		480-Watt System		600-Watt System	
Flush Rooftop	42-0050	$169	42-0051	$214	42-0052	$214
Fixed Pole-Top	42-0053	$230	42-0054	$345	42-0055	$360
Tracking Pole-Top	42-0056	$475	42-0057	$995	42-0058	$995

Pool Pumping Systems Performance

Solar Array Watts	Back Pressure PSI	FIXED ARRAY		TRACKING ARRAY	
		Gallons per Day	Max. Pool Size**	Gallons per Day	Max. Pool Size**
340	3	17,600	25,100	26,600	38,000
480	3	23,400	33,500	35,400	50,600
600	3	27,500	39,300	39,600	56,600
340	6	11,100	15,900	16,900	24,200
480	6	17,200	24,500	26,000	37,200
600	6	21,100	30,200	31,700	45,300
340	8.5	6,400	9,200	9,800	14,000
480	8.5*	11,600	16,500	17,500	25,100
600	8.5	16,100	22,900	24,300	34,700
480	11	6,400	9,100	9,700	13,800
600	11	10,300	14,800	15,700	22,400

Daily performance based on solar irradiation of 6.0 peak Sun hours per day, and 17% degradation of array output due to heat, dirt, and manufacturer's tolerances.

*8 psi is typical back pressure for a system sized according to this selection table. Low friction (2") piping and a large cartridge-type filter can reduce it further.

**Max. pool size is based on a turnover of 70% per day. For faster turnover, choose a larger system, if possible.

30th ANNIVERSARY

Rotary-Vane Pumps

SOLAR SLOWPUMPS

Slowpumps are the original solar-powered pumps developed by Windy Dankoff in 1983. They set the standard for efficiency and reliability in solar water pumping where water demand is in the range of 50-3,000 gallons per day. The surface-mounted Slowpump is designed to draw water from shallow wells, springs, cisterns, tanks, ponds, rivers, or streams and to push it as high as 440 vertical feet for storage, pressurization, or irrigation.

These positive-displacement rotary-vane-type pumps feature forged brass bodies with carbon-graphite vanes held in a stainless steel rotor. No plastic! Motor brushes are externally replaceable, and last 5-10 years. Pump life expectancy is 15-20 years. Rebuilt/exchange pump heads are available for under $150. Slowpumps are NSF approved for drinking water. A wide variety of pump and motor combinations are available for a variety of lifts and delivery volumes. Consult the performance charts.

Rotary-vane pumps must have absolutely clear water. They will not tolerate any abrasives. A 10-micron cartridge prefilter is highly recommended for all installations. If your water is very dirty, improve the source or consider a diaphragm pump.

Fittings on 1300 and 1400 series pumps are ½-inch female; 2500 and 2600 series pumps have ¾-inch male fittings. Sizes and weights vary slightly with model but are approximately 6"x16" and 16 lb. 1-yr. mfr. warranty. USA.

¼-Horsepower PV or Battery-Powered Pumps, 12 or 24 Volts

See chart. Specify voltage when ordering.

41102	Slowpump 1322 pump	$475
41104	Slowpump 1308 pump	$475
41106	Slowpump 1303 pump	$475
41108	Slowpump 2507 pump	$475

½-Horsepower Slowpumps

See chart. State voltage when ordering.

41931	Slowpump 1404 Pump	$695
41932	Slowpump 1403 Pump	$695
41933	Slowpump 2605 Pump	$695
41934	Slowpump 2607 Pump	$695

AC Version of any Slowpump Model

41110	Slowpump AC Option	Add $145

¼-Horsepower 1300 & 2500 Models PV or Battery Power, 12V, 24V, or 115VAC

Total Lift in Feet	1322 GPM	1322 Watts	1308 GPM	1308 Watts	1303 GPM	1303 Watts	2507 GPM	2507 Watts
20	0.51	27	1.25	30	2.50	48	4.00	57
40	0.51	32	1.25	48	2.50	60	3.95	78
60	0.51	36	1.20	54	2.40	78	3.90	102
80	0.49	40	1.20	60	2.30	93	3.90	120
100	0.49	45	1.20	66	2.30	105	3.85	144
120	0.48	50	1.20	70	2.25	121	3.80	165
140	0.47	56	1.20	75	2.20	138	3.65	195
160	0.47	62	1.20	84	2.20	153		
180	0.47	66	1.18	93	2.15	165		
200	0.45	74	1.16	101	2.15	180		
240	0.44	90	1.14	117	2.15	204		
280	0.41	102	1.12	135				
320	0.41	120	1.10	153				
360	0.41	134	1.05	171				
400	0.40	150	1.00	198				
440	0.39	168						

Actual performance may vary ±10% from specifications. Performance listed at 15V or 30V (PV direct). Deduct 20% from flow and watts for battery. Watts listed are power use at pump.

½-Horsepower 1400 & 2600 Models, 24V or 48V Battery or 36V PV direct

Total Lift in Feet	1404 GPM	1404 Watts	1403 GPM	1403 Watts	2605 GPM	2605 Watts	2607 GPM	2607 Watts
160							4.30	283
180			3.35	280			4.25	305
200			3.33	296			4.20	338
240			2.55	266	3.30	331	4.05	396
280			2.50	302	3.25	373	4.00	444
320	1.66	255	2.50	338	3.20	410		
360	1.64	280	2.50	374	3.16	450		
400	1.62	312	2.50	406				
440	1.60	341	2.50	451				

Actual performance may vary ±10% from specifications. Watts listed are power use at pump.

Flowlight Booster Pumps

Sharing all the robust design features of the smaller Slowpumps, these larger rotary-vane pumps are specifically designed for household water pressurization. They use one-half to one-third as much energy as a similar-capacity AC pressure pump, and they eliminate the starting surges that are so hard on inverters. These

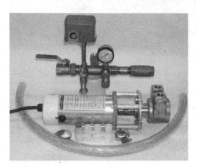

pumps are quieter, smoother-running, and far more durable than diaphragm-type pressure pumps. For full-time off-the-grid living, this is the pressure pump you want. Life expectancy is 15-20 years, and then they just need a simple pump head replacement.

Two models are available. Standard models feature higher flow rates but are noisier, require 1 inch or larger intake plumbing, and have less than 10 feet of suction capacity. Low-speed models are quieter, can accept ¾-inch intake plumbing and suction lifts greater than 10 feet, but have lower flow rates. Both models are available in 12 or 24 volts. Standard models also available in 48 volts DC or 115 volts AC.

A pressure tank must be used with this, or any, pressure pump. Forty-gallon capacity is the minimum size recommended; larger is better. The Booster Pump Installation Kit and the Inline Sediment Filter, as shown with the pump picture, are strongly recommended. Flexible hose input, and output hose assemblies and pressure relief valve are included with the pump. 1-yr. mfr. warranty. USA.

41141	Standard, 12V	$625
41142	Standard, 24V	$608
41000	Standard, 48V	$795
41147	Standard, 115VAC	$630
41145	Low-Speed, 12V	$555
41146	Low-Speed, 24V	$555

Shipped from manufacturer.

Booster Pump Performance								
	Standard Model				Low-Speed Model			
PSI	30	40	50	65	30	40	50	65
GPM	4.5	4.3	4.3	4.1	3.4	3.3	3.1	2.7
Amps @ 24V	6.5	7.5	8.0	11.0	5.0	5.5	6.0	7.5
W/H per Gal.	0.6	0.67	0.75	1.1	0.6	0.67	0.75	1.1

Booster Pump Installation Kit

This kit contains all the plumbing bits and pieces you'll need for installation of the Booster Pump. It includes a one-way check valve, a pressure switch, a pressure gauge, a drain valve, a shutoff valve, and the pressure tank tee all manifolded together. Shown with Booster Pump above. USA.

41-143	Booster Pump Install Kit	$99
41-137	Inline Sediment Filter	$49

Shipped from manufacturer.

Accessories for Slowpump Installation

Inline Sediment Filter

This is the recommended minimum filter for every Slowpump installation. This filter is designed for cold water only and meets NSF standards for drinking water supplies. It features a sump head of fatigue-resistant Celcon plastic, an opaque body to inhibit algae growth if installed outdoors, and a manually operated pressure-release button to simplify cartridge replacement. Rated for up to 125 psi and 100°F, it is equipped with ¾-inch female pipe thread inlet and outlet fittings and is rated for up to 6 gpm flow rates (with clean cartridge). It accepts a standard 10-inch cartridge with 1-inch center holes. Supplied with one 10-micron fiber cartridge installed. USA.

41137	Inline Sediment Filter	$49
42632	10-Micron Sediment Cartridge, 2 pk.	$13

30-Inch Intake Filter with Foot Valve

A great choice for any Slowpump with a surface water source. This large, triple-size 10-micron filter with no external cover can be lowered into wells or floated on a pond. Large surface area means less frequent replacements. Equipped with a foot valve so you won't lose your pump prime. Included bushing allows ½-inch or ¾-inch pipe thread outlet. Comes with assembled filter and first replacement cartridge as pictured. USA.

40-0086	30-in. Intake Filter/Foot Valve	$70
40-0087	30-in. Filter Replacement, 3-Pk.	$44

Dry-Run Shutdown Switch

Rotary-vane pumps will be damaged if allowed to run dry. This thermal switch clamps to the pump head of any new or older Slowpump. It will sense the temperature rise of dry running and shut off the pump before damage can occur. A manual reset is required to restart the pump. This switch will pay for itself many times over, even if used only once during the lifetime of your Slowpump. USA.

41135	Dry-Run Switch for 1300 Series Pumps	$85
41144	Dry-Run Switch for All Larger Series Pumps	$99

Replacement Motor Brush Sets

The Slowpump brushes are easy to inspect and replace. Simply unscrew and withdraw the old brushes, then screw in the new ones, if needed. Normal life expectancy is 5-10 years. Sold in sets of two. USA.

41111	12V Slowpump Motor Brush Set	$20
41112	24V Slowpump Motor Brush Set	$20

Replacement Pump Heads

Slowpump pump heads typically enjoy a 15- to 20-year life expectancy if the water is clean and they don't run dry. But hey, stuff happens, and some of the early Slowpumps have been out there for over 20 years now. Here's your replacement pump head, and it isn't going to hurt too bad. Specify your model when ordering. USA.

40-0910	Replacement Pump Head for 1300 Series	$138
41938	Replacement Pump Head for 2500 Series	$225

Diaphragm Pumps

Guzzler Pumps

Human-powered pumps

These high-volume, low-lift pumps are driven by simple, dependable muscle power. Both hand pump and foot pump models are available. These are simple self-priming, diaphragm-type pumps with reinforced flapper valves for durable, dependable pumping. All models can deliver about 12 feet of suction lift below the pump and 12 feet of delivery head above the pump. Guzzlers make excellent emergency sump pumps, bilge pumps, or water delivery systems. The Model 400 series delivers up to 10 gpm and uses 1-inch hose. Model 500 series delivers up to 15 gpm and uses 1½-inch hose. The foot pump models have a spring return stroke. USA.

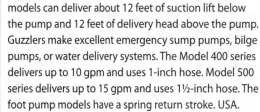

41151	400 Hand Pump	$55
41152	500 Hand Pump	$65
41936	400 Foot Pump	$133
41937	500 Foot Pump	$145

SHURflo Surface Pump Kit

Our surface pump kit features one of our most popular pumps, the SHURflo Low Flow, along with enough PV power to lift up to 130 feet. This kit includes a Sharp 123-watt PV module, a pole-top mount for ease of installation, LCB for the best delivery under marginal Sun conditions, and a screened foot valve for reliable operation. You supply a 10-foot length of 2¼-inch steel pipe for the PV mount, plumbing, and wiring as needed, and a surface water supply. Mount your pump as close to the water supply as practical with a maximum of 10-feet suction lift. Pumps would much rather push. Kit pricing saves $47. Delivers approximately 1½ gpm, depending on lift and sunlight availability. In a typical six-hour summer day, it puts more than 500 gallons into your garden or storage tank. All components made in USA, except LCB, which is made in Canada.

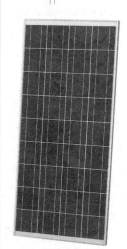

40-0197	SHURflo Surface Pump Kit	$999

SHURflo Pumps

SHURflo produces a high-quality line of positive-displacement diaphragm pumps for RV and remote household pressurization, and for lifting surface water to holding tanks. These pumps will self-prime with 10 feet of suction and a free discharge (no pressure on outlet). The three-chamber design runs quieter than other diaphragm pumps and will tolerate silty or dirty water or running dry with no damage. However, if your water is known to have sand or debris, it's best to use input filtering. Either SHURflo's small inline filter or a standard 10-inch cartridge will save you trouble by preventing sand or other crud from lodging in the check valves. No damage, but you'll have to disassemble the pump to flush it out. Who needs the hassle? Put a filter on it!

Motors are slow-speed DC permanent-magnet types (except AC model) for long life and the most efficient performance. All pumps are rated at 100% duty cycle and may be run continuously. There is no shaft seal to fail; the ball bearing pump head is separate from the motor. Pumps are easily field repaired and rebuild kits are available for all models. All pumps have a built-in pressure switch to prevent overpressurization. Some are adjustable through a limited range. Low-flow model has ⅜-inch MIPT inlets and outlets. Medium- and high-flow models are ½-inch MIPT. All pumps carry a full 1-yr. warranty. USA.

SHURflo Low-Flow Pump

For modest water requirements or PV-direct applications. Available in 12 volts DC only. Delivers a maximum of 60 psi or 135 feet of head. This is an excellent surface water delivery pump used with the 7-amp PV-Direct Pump Control (25002) and PV modules as necessary for your lift. (Add 30% to pump wattage to figure necessary PV wattage.) ⅜-inch MIPT inlet and outlet. (Model# 8000-443-136.)

41450   SHURflo Low-Flow Pump $109

SHURflo Low-Flow 12 Volts

PSI	GPM	Amps
10	1.66	3.9
20	1.57	4.2
30	1.48	4.9
40	1.38	5.6
50	1.30	6.9
60	1.23	7.2

SHURflo Medium-Flow Pump

The Medium Flow is available in 12- or 24-volt DC; performance specs are similar for both (the 24-volt pump does slightly less volume). Maximum pressure is 40 psi or 90 feet of lift for both models. This model is a good choice for low-cost household pressurization; it will keep up with any single fixture. You must use a pressure tank with any pressurization system; obtain locally. It has a built-in 20- to 40-psi pressure switch, perfect for most households. Buzzy output can be reduced by looping 24-inch flexible connectors on both inlet and outlet. ½-inch MIPT inlet and outlet. (12V Model# 2088-443-144; 24V Model# 2088-474-144.)

41451   SHURflo Medium-Flow Pump, 12V $149

41452   SHURflo Medium-Flow Pump, 24V $165

Medium-Flow 12V			Medium-Flow 24V		
PSI	GPM	Amps	PSI	GPM	Amps
10	2.83	5.8	10	2.80	2.41
20	2.56	7.0	20	2.25	2.63
30	2.31	8.0	30	1.75	2.73
40	2.02	9.1	40	1.25	2.71

SHURflo High-Pressure Pump

This model is higher pressure, 30- to 45-psi, for household pressurization systems. For those who need slightly higher pressure than the standard Medium-Flow pumps. Buzzy output can be reduced by looping 24-inch flexible connectors on both inlet and outlet. Available in 12 volts only. Maximum pressure is 45 psi or 104 feet of lift. ½-inch MIPT inlet and outlet. (Model # 2088-514-145.)

41453   SHURflo High-Pressure Pump $189

SHURflo High-Pressure 12V

PSI	GPM	Amps
10	2.90	5.6
20	2.60	7.1
30	2.30	8.4
40	2.07	9.0

SHURflo AC Pump, 115 Volts

For those who need to send power more than 150 feet away, this pump can save you big $$ on wire costs! The motor is thermally protected and will shut off automatically under heavy continuous use to prevent overheating. This should happen only at maximum psi after approximately 90 minutes of running. Pressure switch is adjustable from 25 to 45 psi. Maximum pressure is 45 psi. ½-inch MIPT inlet and outlet. Measures 8.6"x5"x4.4". (Model # 2088-594-154.)

115VAC Pump

PSI	GPM	Amps
10	2.60	0.58
20	2.34	0.67
30	2.08	0.76
40	1.85	0.85

41454   SHURflo AC Pump, 115V $179

SHURflo Extreme High-Pressure Pump

This is a special pump for very high pressures up to 100 psi. It can pump up to 230 feet of vertical lift. The built-in pressure switch is factory adjusted to turn on at 85 psi and off at 100 psi. Adjustment range is 80-100 psi. Not recommended for continuous operation. Has 3/8-inch female NPT inlet and outlet fittings. 12-volt only. (Model # 8002-793-238.) USA.

Extreme Pressure 12V

PSI	GPM	Amps
20	1.61	4.4
40	1.45	5.7
60	1.35	7.0
100	1.15	9.3

41357   SHURflo Extreme High-Pressure Pump, 12V $109

ULTRAflo Pump and Accumulator Combo

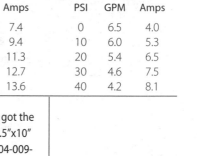

Delivering over 5 gallons per minute, SHURflo offers this pair of high-output pumps preassembled with a stainless steel 2-gallon accumulator tank. This is a complete water pressure delivery system for 12- or 24-volt DC operation. The diaphragm pumps deliver up to 5.6 gpm at 12 volts DC (15 amps), or 6.5 gpm at 24 volts DC (9 amps) for multifixture operation. The built-in pressure switch is on at 35 psi, off at 45 psi, with ±5 psi adjustment. Designed for marine and RV applications with limited space, these will perform better and live longer with the addition of a second larger pressure tank if you've got the space. ½-inch male pipe-thread outlet. 18.5"x14.5"x10" (LxHxD). 9 lb. 1-yr. mfr. warranty. (12V Model# 804-009-01; 24V Model# 804-009-02.) USA.

12V Pump/Accumulator			24V Pump/Accumulator		
PSI	GPM	Amps	PSI	GPM	Amps
0	5.6	7.4	0	6.5	4.0
10	5.1	9.4	10	6.0	5.3
20	4.4	11.3	20	5.4	6.5
30	3.7	12.7	30	4.6	7.5
40	2.9	13.6	40	4.2	8.1

40-0080   12V Pump/Accumulator $339

40-0081   24V Pump/Accumulator $345

SHURflo Inline Filter

For light-duty filtering, this stainless steel screen is all the highly tolerant SHURflo pumps need. It'll keep out anything big enough to cause trouble for the pump. Screws onto ½-inch male pump inlet; has ½-inch barbed fitting to water source. Disassemble for cleaning.

40-0088   SHURflo Inline Filter $9

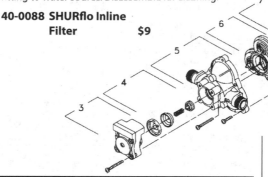

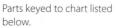

Parts keyed to chart listed below.

SHURFLO REPAIR AND REPLACEMENT PARTS

Pump Model	Pressure Switch 3	Check Valve 4	Upper Housing 3, 4, 5	Valve Kit 6	D&D Kit 7	Motor 8
Low Flow 12V 41450	41325 $20	41327 $10	— N/A	41331 $20	41333 $37	41335 $100
Medium Flow 12V 41451	41326 $20	41328 $10	41329 $25	41332 $15	41334 $30	41335 $100
Medium Flow 24V 41452	41326 $20	41328 $10	41329 $25	41332 $15	41334 $30	41337 $119
High-Pressure 12V 41453	N/A —	41328 $10	41330 $50	41332 $15	41334 $30	41339 $179
115VAC 41454	41326 $20	41328 $10	41329 $25	41332 $15	41334 $30	41338 $169

Solaram Pump

This is the highest yield, highest lift solar pump available. Using a large multipiston industrial diaphragm pump and a permanent-magnet DC motor with positive gear belt drive, the Solaram series of pumps features models capable of lifts up to 960 feet, or over 9 gallons per minute. The performance chart shows delivery and watts required. If running PV direct, the rated PV wattage must exceed the pump requirement by 25% or more. A pressure relief valve and flexible intake/outlet hoses are included.

In full-time daily use, yearly diaphragm replacement is recommended. Motor brushes typically last 10 years. The Diaphragm and Oil Kit provides a set of new diaphragms and the special nontoxic oil. The Long-Term Parts Kit contains three diaphragm and oil kits, plus a gear belt and motor brush kit.

Inlet fittings are 1.5 inches; outlet is 1 inch. Dimensions are 28"x16"x16.5" (LxDxH). Approximately 150 lb. (varies slightly with model) and is shipped in two parcels. 1-yr. mfr. warranty. USA.

Specify full 4-digit model number when ordering.

41480	Solaram 8100 Series Pump	$2,475
41481	Solaram 8200 Series Pump	$2,475
41482	Solaram 8300 Series Pump	$3,825
41483	Solaram 8400 Series Pump	$3,825
41484	Diaphragm and Oil Kit	$199
41485	Long-Term Parts Kit	$345

Solaram™ Surface Pump Performance Chart

Model Numbers: First 2 digits ———
Second 2 digits ———

TOTAL LIFT		_ _ 21			_ _ 22			_ _ 23			_ _ 41			_ _ 42			_ _ 43			Model #
Feet	Meters	GPM	LPM	Watts	GPM	LPM	Watts	GPM	LPM	Watts	GPM	LPM	Watts	GPM	LPM	Watts	GPM	LPM	Watts	Volts
0-80	24	3.0	11.4	170	3.7	14.0	207	4.6	17.4	285	6.2	23.5	258	7.5	28.4	339	9.4	35.6	465	81_ _ 24V
120	37	2.9	11.0	197	3.7	14.0	238	4.5	17.1	319	6.0	22.7	305	7.3	27.7	396	9.1	34.5	539	
160	49	2.9	11.0	225	3.6	13.6	268	4.5	17.1	352	5.8	22.0	354	7.2	27.3	453	8.9	33.7	619	
200	61	2.9	11.0	247	3.6	13.6	296	4.5	17.1	388	5.7	21.6	400	7.1	26.9	513	8.9	33.7	693	
240	73	2.8	10.6	265	3.6	13.6	327	4.5	17.1	427	5.6	21.2	453	7.0	26.5	572	8.6	32.6	724	82_ _ 24V
280	85	2.8	10.6	286	3.6	13.6	356	4.4	16.7	466	5.5	20.8	499	6.9	26.2	628	8.4	31.8	801	
320	98	2.8	10.6	315	3.5	13.3	388	4.4	16.7	496	5.4	20.5	548	6.8	25.8	686	8.3	31.5	869	
360	110	2.8	10.6	342	3.5	13.3	416	4.4	16.7	536	5.4	20.5	592	6.6	25.0	733	8.2	31.1	927	
400	122	2.7	10.2	363	3.4	12.9	450	4.4	16.7	572	5.3	20.1	649	6.5	24.6	782	8.7	33.0	1122	83_ _ 180V
480	146	2.7	10.2	416	3.4	12.9	505	4.3	16.3	649	5.3	20.1	717	6.5	24.6	900	8.5	32.2	1265	
560	171	2.7	10.2	456	3.3	12.5	570	4.3	16.3	693	5.2	19.7	800	6.5	24.6	1045	8.4	31.8	1397	
640	195	2.7	10.2	502	3.3	12.5	623	4.2	15.9	774	5.1	19.3	893	6.5	24.6	1116	8.2	31.1	1540	84_ _ 180V
720	220	2.6	9.9	551	3.2	12.1	690	4.1	15.5	856	5.1	19.3	1031	6.4	24.3	1287	8.1	30.7	1683	
800	244	2.6	9.9	589	3.2	12.1	715	4.1	15.5	931	5.1	19.3	1114	6.4	24.3	1408	8.0	30.3	1815	
880	268	2.6	9.9	647	3.2	12.1	774	4.0	15.2	1082	5.1	19.3	1206	6.3	23.9	1529	8.0	30.3	1958	85_ _ 180V
960	293	2.6	9.9	705	3.1	11.7	838	4.0	15.2	1190	5.0	18.9	1289	6.1	23.1	1650	8.0	30.3	2145	

Performance may vary ± 10 %

Piston Pumps

Solar Force Piston Pump

Built like a tank, the cast-iron Solar Force is an excellent choice for high-volume delivery of 4-9 gpm at lifts up to 220 feet. It can be used for water delivery to a higher storage tank, irrigation, fire protection, or pressurization. It has a 25-foot suction lift at sea level. The durable cast-iron and brass pump is designed to last for decades, and routine maintenance is required only every two to six years. A Linear Current Booster is needed for PV-direct drive. LCB sizing depends on model and voltage. Our tech staff will advise. Solar Force uses a nonslip gear belt drive on PV-direct models, and standard V-belt drive on battery models. Intake fitting is 1.25 inches; output is 1 inch. A pressure relief valve is included. 22"x16"x13" (WxHxD). Approximately 80 lb. maximum, shipped in two or three parcels depending on model. 2-yr. mfr. warranty. USA.

The Solar Force Pump is available in three basic models:
• Model 3010: available in 12 or 24 volts battery-driven only.
• Model 3020: available in 12, 24, or 48 volts, battery or PV direct, also 115VAC.
• Model 3040: available in 12, 24, or 48 volts, battery or PV direct, also 115VAC.
Please specify voltage (12, 24, 48, or 115) when ordering.

41254	**Model 3010 Solar Force, 12 & 24V Battery Only**	**$1,085**
41255	**Model 3020, Battery Powered**	**$1,350**
41256	**Model 3020, PV Powered**	**$1,525**
41257	**Model 3040, Battery Powered**	**$1,415**
41258	**Model 3040, PV Powered**	**$1,595**
41259	**Model 3020 or 3040, 115VAC Powered**	**$1,655**

SOLAR FORCE ACCESSORIES AND PARTS

Heavy-Duty Pressure Switch

A DC-rated pressure switch for using the Solar Force as a pressure pump. Will handle the heavy starting surge. Good for 12-, 24-, or 48-volt systems.

41264   Solar Force Pressure Switch $75

Surge Tank

This tank is included with the PV models. It helps absorb pressure pulsations when long piping runs are required between the pump and the tank. It keeps plumbing systems from being shaken apart.

41265   Solar Force Pulsation Tank $90

Seal and Belt Kit

Contains all the repair parts needed for routine service required every two to six years. Spare gaskets, rod and valve seals, new drive belt, and two sets of piston seals are included. PV kits are higher priced due to gear belt drive instead of V-belt.
Specify pump model number when ordering.

41267   Seal & Belt Kit for Battery Pumps $45

41268   Seal & Belt Kit for PV Pumps $75

Long-Term Parts Kit

Everything included in the Seal and Belt Kit above, plus a second drive belt, a replacement cylinder sleeve, motor brushes, and two oil changes.
Specify pump model number when ordering.

41269	**Long-Term Parts Kit for Battery Pumps**	**$160**
41270	**Long-Term Parts Kit for PV Pumps**	**$210**

Solar Force Piston Pump Performance

Vertical Lift Feet	Model 3110 GPM	Watts	Model 3020 GPM	Watts	Model 3040 GPM	Watts
20	5.9	77	5.2	110	9.3	168
40	5.6	104	5.2	132	9.3	207
60	5.3	123	5.1	154	9.2	252
80	5.1	152	5.1	182	9.2	286
100	5.1	171	5.0	202	9.1	322
120	4.9	200	5.0	224	9.1	364
140	4.9	226	5.0	252	9.1	403
160			4.9	269		
180			4.9	280		
200			4.8	308		
220			4.7	314		

Submersible Pumps

Amazon Submersible

Delivers a maximum of 4 gpm at zero lift, or a trickle at 30 feet of lift, this 1.1-pound pump has 13 feet of cable with battery clips, ½-inch barbed connections, and is 1.5" diameter. Draws 7 amps @ 12 volt maximum. Do not run dry or expect a long life. Intermittent use only. Great Britain.

41950	Amazon 12V Pump	$87
41963	Amazon 24V Pump	$87

Congo Submersible

Delivers a maximum of 7 gpm at zero lift, or a trickle at 30 feet of lift. Has 13 feet of cable with battery clips, ½-inch barbed connections, and a 1.6-inch diameter. Draws 8 amps at 12 volt maximum. Do not run dry or expect a long life. Intermittent use only. Great Britain.

41956	Congo 12V Pump	$120

SHURflo Solar Submersible Pump

This small, efficient, submersible diaphragm pump has been the standard for residential solar water pumping for over 10 years. For homesteads without large garden or orchard water needs, this is probably all the pump you need. The SHURflo Sub will deliver up to 230 feet of vertical lift and is ideal for deep well or freeze-prone applications. Remember, you're lifting only from the water surface, not from the pump. In a PV-direct application, it will yield 300-1,000 gallons per day, depending on lift requirements and power supplied. Solar submersible pumps are best used for slow, steady water production into a holding tank but may be used for direct pressurization applications as well. (See suggestions on

SHURflo Submersible Pump Performance Chart at 24 Volts

Vertical Lift Feet	Flow Rate Gal./Hr.	GPM	Minimum PV Watts	Amps
20	111	1.85	58	1.7
60	105	1.75	78	2.3
100	100	1.67	99	2.9
140	93	1.55	115	3.5
180	87	1.45	135	4.0
230	79	1.32	155	4.6

12-volt performance will be approximately 40% of figures listed above.

this subject beginning on page 349.) Flow rates range from .5 to 1.9 gpm depending on lift and power supplied. The pump can be powered from either 12- or 24-volt sources. Output is best with 24-volt panel-direct. Because it's a diaphragm-type, this pump is untroubled by running dry, as long as it doesn't last for more than two or three months. Yes, months! So this is a great choice for low-yielding wells. The best feature of this pump, other than proven reliability, is easy field serviceability. No special tools are needed. Experience has shown that the diaphragm needs replacement every four to five years, and the motor has about a 10- to 11-year life expectancy. Repair parts sold below.

The pump is 3.75 inches in diameter (will fit standard 4-inch well casings), weighs 6 pounds, and uses ½-inch poly delivery pipe, making easy one-person installation possible. Maximum pump submersion is 100 feet. Requires flat-jacketed style of submersible cable for the watertight splice kit. Not always available locally, we offer it below. 1-yr. warranty. USA.

In addition to the pump, you will also need:
- PV modules and mounting appropriate for your lift (see chart)
- A controller for PV-direct systems (PV-direct controller #25002 on page 000)
- ½-inch poly drop pipe or equivalent
- 10-gauge submersible pump cable (3-wire cable okay if flat style)
- Poly safety rope
- Sanitary well seal

Drop pipe, submersible cable, rope, and seal are all commonly available at most plumbing supply stores. Watertight splice kit for flat cable is included.

(12-volt performance will be approximately 40% of figures listed in performance chart.)

41455	SHURflo Solar Submersible	$695
26540	10-2 Jacketed Sub Pump Cable	$1.50/ft. (specify length)

Complete Solar Sub Pump Kit

Our Submersible Pump Kit features the popular and durable SHURflo submersible pump. We've put it together with a pair of high-powered 80-watt modules for maximum output, a pole-top mount for ease of installation and a current booster controller for good performance with marginal Sun. This package delivers full power performance down to 230 feet. USA.

41128	SHURflo Sub Pump Kit	$1,825
41638	Optional Float Switch	$35

SHURflo Solar Sub Service Parts

The drive and diaphragm on this hard-working little pump should be replaced at four- to five-year intervals. Do it *before* the diaphragm starts leaking, and you'll save yourself a $225 motor. If the lower canister (and motor) is full of water on disassembly, replace the motor. The carbon brushes soak up the water and soften. They'll last only a few days after you put it back into the well. Frustrating doing repairs twice, isn't it? O-rings should be replaced anytime you open the pump. We try very hard to keep all the common bits in stock for rush delivery. USA.

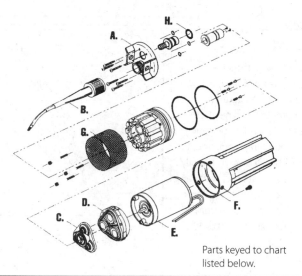

Parts keyed to chart listed below.

Replacement Parts Kit List for SHURflo Solar Sub Pump

Tool Kit	Lift Plate Kit A	Cable Plug Kit B	Valve Kit C	Drive & Diaphragm Kit D	Motor E	Canister F	Filter Screen G	Complete O-Ring Kit H
41-340	41-341	41-342	31342	41-344	41-345	Not available	41-437	41-348
$20	$80	$125	$40	$100	$225	—	$30	$25

Lorentz Submersible Pump Systems

The best solar pump just got better. An improved pump controller, usable either PV direct or from a battery bank, and a simplified pump selection handles virtually

any pumping need. This is a "helical rotor" pump, using a hardened stainless steel rotor in a durable flexible rubber sleeve. Lorentz pumps have all the electronic control parts aboveground, a water-filled brushless motor, and no diaphragms or brushes to replace. Lorentz simplifies their pump line by introducing three submersible pump systems: PS200 Mini, PS600, and PS1200. Select the pump and controller based on vertical lift and gallons per day for solar-direct systems or vertical lift and gallons per minute for battery systems. Three-phase variable-speed controllers feature Maximum Power

Point Tracking. Also, the controller is the same for PV-direct or battery systems, allowing flexibility for future applications. The optional low-water probe will shut the pump off to prevent dry-run damage. An optional PV Disconnect and Junction box is also available. The pump system includes a great installation manual to help walk you through every step.

Call or email our staff to size and price a system. Lorentz pump systems will lift up to 750 feet and can deliver as much as 25,000 gallons per day. All pump systems come with a 2-yr. mfr. warranty against defects in materials or workmanship.

Lorentz PS200 Mini Pump Systems
$2,400-$4,200

Lorentz PS600 Standard Pump Systems
$4600-$9,000

Lorentz PS1200 Standard Pump Systems
$5,800-$11,600

Please call for details.

Lorentz Pump Systems

	PS200	PS600	PS1200
Maximum Lift	165 feet	600 feet	760 feet
Maximum Volume	3,785 gpd	20,000 gpd	25,000 gpd
Well Diameter	4 in. or larger	4 in. or larger	4 in. or larger
Voltage Range	24-48VDC	48-72VDC	72-96VDC

Grundfos SQ Flex Submersible Pumps

Featuring incredible power adaptability, the submersible SQ Flex series comes from one of the most experienced, highest-quality pump manufacturers in the world. It features a single motor that can be fitted with seven different pump heads, depending on your

pumping requirements. The SQ Flex will happily accept any DC voltage from 30 to 300 volts, or any AC voltage from 90 to 260 volts (50 or 60 Hz). Just flip a switch on the optional AC/DC control box to go back and forth. This makes the SQ Flex pump extremely easy to back up with a generator or utility power. When running on DC, the SQ Flex features electronic Maximum Power Point Tracking to wring the best possible performance from your solar array. A low-water sensor is included and prewired to the pump input cable. Low-water shutoff and restart are automatic.

With a maximum of 900 watts input, the SQ Flex motor offers performance equivalent to a conventional ¾-1 horsepower AC pump. Three of the available pump heads are new positive-displacement helical rotor types featuring efficiencies that conventional AC pumps can't even dream about. Helical rotors are the newest pump type in the solar industry. They feature much greater efficiency than any other pump type and don't wear out or require any routine maintenance. Water delivery ranges from over 24,000 gallons per day at a minimal 20-foot lift, to 1,000 gallons per day at the maximum lift of 525 feet. The four higher-volume pump heads are conventional centrifugal types.

PV arrays can be either ground or pole mounted. An optional control box is required for switchable AC/DC operation or for remote float level sensing. The float control also features an LCD display of wattage consumption, operation indicator lights, on/off switch, and fault codes display.

Complete systems include pump, controller box(es), solar array, array wiring, and mounting structure. Complete system prices range from $3,100 to $10,900. Custom systems using 48-volt batteries, generators, or other power sources can be configured. 1-yr. mfr. warranty. Denmark/USA.

41980	Grundfos 3 SQF-2 Helical Rotor Pump	$1,661

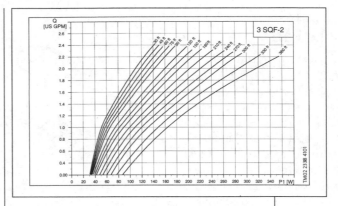

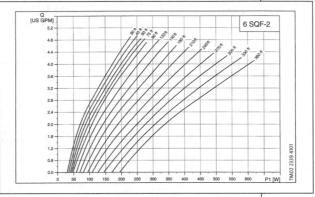

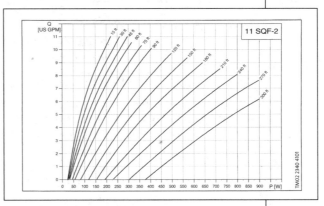

41981	Grundfos 6 SQF-2 Helical Rotor Pump	$1,661
41982	Grundfos 11 SQF-2 Helical Rotor Pump	$1,849
25815	Grundfos IO100 std. on/off Control	$140
25816	Grundfos IO101 AC/DC sw. Control	$369
25817	Grundfos CU200 Float/Monitor Control	$315

Complete Grundfos Pump Systems

$3,300-$8,000

Because of the wide variety of lifts, volumes, and power sources, you need to call or email our tech staff to size and price a Grundfos SQ Flex pumping system.

Submersible Pump Options

The Universal Solar Pump Controllers

Run ordinary AC pumps directly from PV power
From the same incredibly innovative company that produced GM's EV1 electric vehicle and the human-powered plane that crossed the English Channel comes an inverter/controller that will run any standard single- or three-phase AC submersible pump directly from PV power. The advanced USPC maintains a constant volts/Hz ratio that allows standard AC pumps to run at lower speeds without distress. This allows easy soft starts and the efficient use of lower-than-peak PV outputs with off-the-shelf pumps. Minimum speed is owner-selectable to avoid no-flow conditions. The USPC's power electronics are 97% efficient and will provide automatic shutdown in case of dry running, motor lock, or wiring problems. The USPC can drive any standard 50- or 60-Hz three-wire single-phase pump or three-phase induction motor that has a rated input voltage of 120, 208, or 230 volts AC. The model 2000 will run any 0.5-2.0 horsepower pump. The model 5000 is for pumps up to 5.0 horsepower, and a 10-horsepower model is available by special order. The enclosure is rugged outdoor-rated NEMA 3 steel. USPC complete systems, with PV array, controller, and pump run $5,000-$50,000. 2-yr. mfr. warranty. UL certified. USA.

Please call our tech staff for help with PV sizing, which will be unique for your installation.

41913	USPC-2000 Pump Controller	$2,320
41923	USPC-5000 Pump Controller	$4,300

PV-Direct Pump Controllers

These current boosters will start your pump earlier in the morning, keep it going longer in the afternoon, and give you pumping under lower light conditions when the pump would otherwise stall. Features include Maximum Power Point Tracking to pull the maximum wattage from your modules, a switchable 12/24-volt design in a single package, a float or remote switch input, a user-replaceable ATC-type fuse, and a weatherproof box.

Four amperage sizes are available in switchable 12/24 volts. Input and output voltages are selected by simply connecting or not connecting a pair of wires. Amperage ratings are surge power. The 7-amp model is the right choice for most pumping systems—it's the controller included in our submersible and surface pumping kits. 7- and 10-amp models are in a 4½"x2¼"x2" (LxWxD) plastic box. 15-amp

model is in a 5.5"x2.9"x2.9" (LxWxD) metal box. 30-amp model is supplied in a NEMA 3R raintight 10"x8"x4" (LxWxD) metal box. 1-yr. mfr. warranty. Canada.

25-002	7A, 12/24V Pump Controller	$119
25-003	10A, 12/24V Pump Controller	$159
25-004	15A, 12/24V Pump Controller	$239
25-005	30A, 12/24V Pump Controller	$395

Float Switches

These fully encapsulated mechanical float switches make life easier by providing automatic control of fluid level in the tank. You regulate the water level by lengthening or shortening the power cable and securing it with a band clamp or zip tie. Easily installed, the

switches are rated for 15 amps at 12 volts, or 13 amps at 120 volts. The "U" model will fill a tank (closed circuit when float is down, open circuit when up), and the "D" model will drain it (closed circuit when up, open circuit when down). When used with a PV-Direct Pump Controller, the "U" and "D" functions are reversed: "D" will fill and "U" will empty. The Goof-Proof Float Switch is your salvation. This three-wire float switch can be connected either way. There's a common wire, an "up" wire, and a "down" wire. 5-amp maximum current. Safe for domestic water. 2-yr. mfr. warranty. USA.

41637	Float Switch "U"	$35
41638	Float Switch "D"	$35
41-036	Goof-Proof Float Switch	$59

Flat Two-Wire Sub Pump Cable

This flat 2-conductor, 10-gauge submersible cable is necessary for the SHURflo Solar Sub splice kit and is sometimes difficult to find locally. Cut to length; specify how many feet you need. USA.

26540	Sub Pump Cable	$1.50/ft.

Please specify length.

Wind-Powered Pumps

Airlift Wind-Powered Water Pumps

Water pumping for windy sites

Airlift pumps use compressed air to lift water into a storage tank. Compressed air is delivered by a two- or four-cylinder air compressor driven by a wind turbine with tempered aluminum blades of 9 or 10 feet in diameter, depending on the model. The variable-pitch blades provide high-torque start-up with winds of 4-6 mph and reach optimum performance at 10-15 mph. In high wind conditions, the turbine will automatically turn out of the wind, then return as wind speed decreases. Yearly maintenance is a simple oil change and intake filter cleaning. The turbine mounts on a 2.5-inch steel pipe tower. The wind turbine can be hundreds of feet from the well. Air is routed to an air injection chamber submerged in the well. Air bubbles lift the water. Lifts up to 300 feet and 1-30 gpm are possible from well casings as small as 2 inches. An air injection chamber and 200 feet of air line are included with each pump. There are no moving or

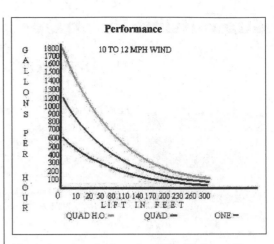

wearing parts in the well. Pumped water can be moved up to 500 feet horizontally, but the delivery line must slope upward at least slightly.

Airlift 1 will lift up to 300 feet maximum and can deliver up to 10 gpm maximum. The Quad model is limited to 315 feet of lift and can deliver up to 20 gpm. The Quad H.O. also lifts to 315 feet but can deliver up to 30 gpm.

Submergence—how far the injection chamber is below water level—is critical for these simple pumps. Too little submergence and the air bubbles will separate from the water; too much submergence and the pump won't lift water. Lift is the vertical distance from the static water level to the storage tank. For lifts up to 150 feet, 70% submergence is recommended. So a 100-foot

lift needs 70 feet of submergence. For lifts from 150 to 300 feet, 50% submergence is recommended.

For those lacking sufficient submergence depth or needing to move long horizontal distances, a positive pump option is offered that will run with only 5 feet of submergence. The positive pump option will deliver approximately 20% less water than air injection. These air-powered pumps require a 10- to 20-gallon air pressure tank teed into the air line so the pump can always complete its cycle and stop in the "start-up" position. Positive pumps can lift up to 300 feet.

A 22-foot folding pipe tower kit is available and highly recommended for ease of routine maintenance, or you can build your own. Tower is not included with pump. USA.

41713	Airlift 1	$3,895
41714	Airlift Quad	$4,595
43-0175	Airlift Quad H.O.	$5,095
42-0245	Airlift 22-ft. Hinged Tower Kit	$1,395
41-715	Airlift Positive Pump 3.5"x67"	$780
41-716	Airlift Positive Pump 4.5"x48"	$760
41-717	Airlift Positive Pump 6.0"x24"	$795

Water-Powered Pumps

Ram Pumps

The ram pump works on a hydraulic principle using the liquid itself as a power source. The only moving parts are two valves. In operation, the water flows from a source down a "drive" pipe to the ram. Once each cycle, a valve slaps shut, causing the water in the drive pipe to stop suddenly. This causes a water-hammer effect and high pressure in a one-way valve leading to the "delivery" pipe of the ram, thus forcing a small amount of water up the pipe and into a holding tank. Rams will pump only into unpressurized storage. In essence, the ram uses the energy of a large amount of water falling a short distance to lift a small amount of water a large distance. The ram itself is a highly efficient device; however, only 2%-10% of the liquid is recoverable. Ram pumps will work on as little as 2 gpm supply flow. The maximum head or vertical lift of a ram is about 500 feet.

SELECTING A RAM

Estimate amount of water available to operate the ram. This can be determined by the rate the source will fill a container. Make sure you've got more than enough water to satisfy the pump. If a ram runs short of water, it will stop pumping and simply dump all incoming water.

Estimate amount of fall available. The fall is the vertical distance between the surface of the water source and the selected ram site. Be sure the ram site has suitable drainage for the tailing water. Rams splash big-time when operating! Often a small stream can be dammed to provide the 1½ feet or more of head required to operate the ram.

Estimate amount of lift required. This is the vertical distance between the ram and the water storage tank or use point. The storage tank can be located on a hill or stand above the use point to provide pressurized water. Forty or 50 feet of water head will provide sufficient pressure for household or garden use.

Estimate amount of water required at the storage tank. This is the water needed for your use in gallons per day. As examples, a normal two- to three-person household uses 100-300 gallons per day, or much less with conservation. A 20- by 100-foot garden uses about 50 gallons per day. When supplying potable water, purity of the source must be considered.

Using these estimates, the ram can be selected from the following performance charts. The ram installation will also require pouring a small concrete pad, a drive pipe 5-10 times as long as the vertical fall, an inlet strainer, and a delivery pipe to the storage tank or use point. These can be obtained from your local hardware or plumbing supply store. Further questions regarding suitability and selection of a ram for your application will be promptly answered by our technical staff.

Aqua Environment Rams

We've sold these fine rams by Aqua Environment for over 15 years now with virtually no problems. Careful attention to design has resulted in extremely reliable rams with the best efficiencies and lift-to-fall ratio available. Working component construction is of all bronze with O-ring seal valves. The air chamber is PVC pipe. The outlet gauge and valve permit easy start-up. Each unit comes with complete installation and operating instructions.

41811	Ram Pump, ¾ in.	$339
41812	Ram Pump, 1 in.	$339
41813	Ram Pump, 1¼ in.	$415
41814	Ram Pump, 1½ in.	$415

Typical Performance and Specifications

Vertical Fall (feet)	Vertical Lift (feet)	PUMP RATE (gallons/day)			
		¾-in. Ram	1-in. Ram	1¼-in. Ram	1½-in. Ram
20	50	650	1,350	2,250	3,200
20	100	325	670	1,120	1,600
20	200	150	320	530	750
10	50	300	650	1,100	1,600
10	100	150	320	530	750
10	150	100	220	340	460
5	30	200	430	690	960
5	50	100	220	340	460
5	100	40	90	150	210
1.5	30	40	80	130	190
1.5	50	20	40	70	100
1.5	100	6	12	18	25

Water Required to Operate Ram

¾-in. Ram, 2 gpm 1¼-in. Ram, 6 gpm Maximum Fall, 25 ft. Maximum Lift, 250 ft.
1-in. Ram, 4 gpm 1½-in. Ram, 8 gpm Minimum Fall, 1.5 ft.

Folk Ram Pumps

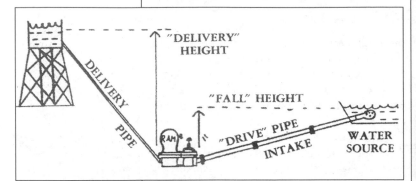

Folk Ram Pump Capacity

Folk Model Drive Pipe Size	Capacity in GPM Min.-Max.	Delivery Pipe
1"	2-4	1"
1¼"	2-7	1"
1½"	3-15	1"
2"	6-30	1¼"
2½"	8-45	1¼"
3"	15-75	1¼"

The Folk is the most durable and most efficient ram pump available. It is solidly built of rustproof cast aluminum alloy and uses all stainless steel hardware. With minimal care, it probably will outlast your grandchildren. It is more efficient because it uses a diaphragm in the air chamber, preventing loss of air charge and eliminating the need for a snifter valve in the intake that cuts efficiency.

Folk rams require a minimum of 2 gpm and 3 feet of drive pipe fall to operate. The largest units can accept up to 75 gpm and a maximum of 50 feet of drive pipe fall. Delivery height can be up to 15 times the drive fall, or a maximum of 500 feet lift. The drive pipe should be galvanized Schedule 40 steel and the same size the entire length. Recommended drive pipe length is 24 feet for 3 feet of fall, with an additional 3 feet length for each additional 1 foot of fall. Be absolutely sure you have enough water to meet the minimum gpm requirements of the ram you choose. Rams will stop pumping and simply dump water if you can't supply the minimum gpm.

To estimate delivery volume, multiply fall in feet times supply in gpm, divide by delivery height in feet, then multiply by 0.61. For example, a ram with 5 gpm supply, a 20-foot drive pipe fall, and delivering to a tank 50 feet higher: 20 ft. x 5 gpm ÷ 50 ft. x 0.61 = 2.0 gpm x 0.61 = 1.22 gpm, or 1,757 gallons per day delivery.

1-yr. mfr. warranty. USA. All Folk rams are drop-shipped via UPS from the manufacturer in Georgia.

41815	**Folk 1-in. Ram Pump**	**$1,095**
41816	**Folk 1¼-in. Ram Pump**	**$1,095**
41817	**Folk 1½-in. Ram Pump**	**$1,095**
41818	**Folk 2-in. Ram Pump**	**$1,295**
41819	**Folk 2½-in. Ram Pump**	**$1,295**
41820	**Folk 3-in. Ram Pump**	**$1,295**

High Lifter Pressure Intensifier Pump

The High Lifter Water Pump offers unique advantages for the rural user. Developed expressly for mountainous terrain and low summertime water flows, this water-powered pump delivers a much greater percentage of the input water than ram pumps can. This pump is available in either 4.5:1 or 9:1 ratios of lift to fall. The High

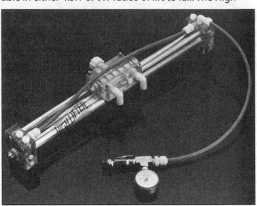

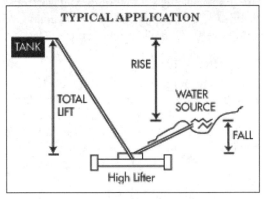

Lifter is self-starting and self-regulating. If inlet water flow slows or stops, the pump will slow or stop but will self-start when flow starts again.

The High Lifter pump has many advantages over a ram (the only other water-powered pump). Instead of using a "water hammer" effect to lift water as a ram does, the High Lifter is a positive-displacement pump that uses pistons to create a hydraulic lever that converts a larger volume of low-pressure water into a

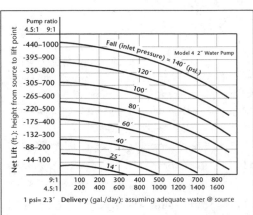

smaller volume of high-pressure water. This means that the pump can operate over a broad range of flows and pressures with great mechanical efficiency. This efficiency means more recovered water. While water recovery with a ram is normally about 5% or less, the High Lifter recovers 1 part in 4.5 or 1 part in 9, depending on ratio. In addition, unlike the ram pump, no "start-up tuning" or special drive lines are necessary. This pump is quiet and will happily run unattended.

The High Lifter pressure intensifier pump is economical compared with gas and electric pumps, because no fuel is used and no extensive water source development is necessary. A kit to change the working ratio of either pump after purchase is available, as are maintenance kits. Maintenance consists of simply replacing a handful of O-rings. The cleaner your input water, the longer the O-rings last. Choose your model High Lifter pump from the specifications and High Lifter performance curves. 1-yr. mfr. warranty on parts and labor. USA.

41801	**High Lifter Pump, 4.5:1 Ratio**	**$895**
41802	**High Lifter Pump, 9:1 Ratio**	**$895**
41803	**High Lifter Ratio Conversion Kit**	**$97**
41804	**High Lifter Rebuild Kit, 4.5:1 ratio**	**$65**
41805	**High Lifter Rebuild Kit, 9:1 ratio**	**$65**

Pump Accessories

Floating Suction Filters

Get water from the cleanest point in your storage system—just below the water surface. Improves the quality of your rainwater and protects pumps from drawing up sediment without a need for prefiltering. The floating ball lets the suction point rise and fall as well as preventing the filter from sinking to the bottom. Includes a float, fine filter housing with a nonreturn valve, and 1-inch barbed connection. Germany.

42908	**Float Suction Filter**	**$149**

Arkal 1-Inch Short Filter

This 200-mesh Arkal disc filter protects your expensive surface pump from abrasive grit and debris with filtration down to 55 microns. Recommended for agricultural applications from surface water sources such as springs and ponds, this filter is not NSF approved for potable water. Enlarged 49 square-inch cylindrical column of grooved discs prevents filter clogging. Filter can be field-cleaned, eliminating the need to order and stock replacement filter cartridges. Features sunlight- and frost-resistant reinforced polypropylene housing and all high-quality noncorrosive components. Inlet and outlet are 1-inch NPT male; maximum pressure is 145 psi; max flow is 26 gpm; maximum temperature is 158°F. Israel.

40-0199 Arkal Short Filter	**$69**

Inline Sediment Filter

This filter is designed for cold water only and meets NSF standards for drinking water supplies. It features a sump head of fatigue-resistant Celcon plastic, an opaque body to inhibit algae growth if installed outdoors, and a manually operated pressure-release button to simplify cartridge replacement. Rated for up to 125 psi and 100°F, it is equipped with ¾-inch female pipe-thread inlet and outlet fittings and is rated for up to 6 gpm flow rates (with clean cartridge). It accepts a standard 10-inch cartridge with 1-inch center holes. Supplied with one 10-micron fiber cartridge installed. USA.

41137	**Inline Sediment Filter**	**$49**
42632	**10-Micron Sediment Cartridge, 2 pk.**	**$13**

30-Inch Intake Filter with Foot Valve

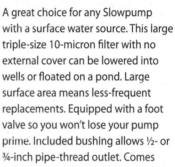

A great choice for any Slowpump with a surface water source. This large triple-size 10-micron filter with no external cover can be lowered into wells or floated on a pond. Large surface area means less-frequent replacements. Equipped with a foot valve so you won't lose your pump prime. Included bushing allows ½- or ¾-inch pipe-thread outlet. Comes with assembled filter and first replacement cartridge as pictured. USA.

40-0086 30-in. Intake Filter/Foot Valve	**$70**
40-0087 30-in. Filter Replacement, 3-pk.	**$44**

Nominal Pipe Size vs. Actual Outside Diameter for Steel and Plastic Pipe			
Nominal Size	Actual Size	Nominal Size	Actual Size
½"	0.840"	2½"	2.875"
¾"	1.050"	3"	3.500"
1"	1.315"	3½"	4.000"
1¼"	1.660"	4"	4.500"
1½"	1.900"	5"	5.563"
2"	2.375"	6"	6.625"

Friction Loss Charts for Water Pumping

HOW TO USE PLUMBING FRICTION CHARTS

If you try to push too much water through too small a pipe, you're going to get pipe friction. Don't worry, your pipes won't catch fire. But it will make your pump work harder than it needs to, and it will reduce your available pressure at the outlets, so sprinklers and showers won't work very well. These charts can tell you if friction is going to be a problem. Here's how to use them:

PVC or black poly pipe? The rates vary, so first be sure you're looking at the chart for your type of supply pipe. Next, figure out how many gallons per minute you might need to move. For a normal house, 10-15 gpm is probably plenty. But gardens and hoses really add up. Give yourself about 5 gpm for each sprinkler or hose that might be running. Find your total (or something close to it) in the "Flow GPM" column. Read across to the column for your pipe diameter. This is how much pressure loss you'll suffer for every 100 feet of pipe. Smaller numbers are better.

Example: You need to pump or move 20 gpm through 500 feet of PVC between your storage tank and your house. Reading across, 1-inch pipe is obviously a problem. How about 1¼ inches? 9.7 psi times 5 (for your 500 feet) = 48.5 psi loss. Well, that won't work! With 1½-inch pipe, you'd lose 20 psi . . . still pretty bad. But with 2-inch pipe, you'd lose ony 4 psi . . . ah! Happy garden sprinklers! Generally, you want to keep pressure losses under about 10 psi.

Friction Loss in PSI per 100 Feet of Scheduled 40 PVC Pipe

Flow GPM	½"	¾"	1"	1¼"	1½"	2"	3"	4"
1	3.3	0.5	0.1					
2	11.9	1.7	0.4	0.1				
3	25.3	3.5	0.9	0.3	0.1			
4	43.0	6.0	1.5	0.5	0.2	0.1		
5	65.0	9.0	2.2	0.7	0.3	0.1		
10		32.5	8.0	2.7	1.1	0.3		
15		68.9	17.0	5.7	2.4	0.6	0.1	
20			28.9	9.7	4.0	1.0	0.1	
30			61.2	20.6	8.5	2.1	0.3	0.1
40				35.1	14.5	3.6	0.5	0.1
50				53.1	21.8	5.4	0.7	0.2
60				74.4	30.6	7.5	1.0	.03
70					40.7	10.0	1.4	0.3
80					52.1	12.8	1.8	0.4
90					64.8	16.0	2.2	0.5
100					78.7	19.4	2.7	0.7
150						41.1	5.7	1.4
200						69.9	9.7	2.4
250							14.7	3.6
300							20.6	5.1
400							35.0	8.6

Friction Loss in PSI per 100 Feet of Polyethylene (PE) SDR-Pressure Rated Pipe

Flow GPM	0.5	0.75	1	1.25	1.5	2	2.5	3
1	0.49	0.12	0.04	0.01				
2	1.76	0.45	0.14	0.04	0.02			
3	3.73	0.95	0.29	0.08	0.04	0.01		
4	**6.35**	1.62	0.50	0.13	0.06	0.02		
5	9.60	2.44	0.76	0.20	0.09	0.03		
6	13.46	3.43	1.06	0.28	0.13	0.04	0.02	
7	17.91	4.56	1.41	0.37	0.18	0.05	0.02	
8	22.93	**5.84**	1.80	0.47	0.22	0.07	0.03	
9		7.26	2.24	0.59	0.28	0.08	0.03	
10		8.82	2.73	0.72	0.34	0.10	0.04	0.01
12		12.37	**3.82**	1.01	0.48	0.14	0.06	0.02
14		16.46	5.08	1.34	0.63	0.19	0.08	0.03
16			6.51	1.71	0.81	0.24	0.10	0.04
18			8.10	2.13	1.01	0.30	0.13	0.04
20			9.84	2.59	1.22	0.36	0.15	0.05
22			11.74	**3.09**	1.46	0.43	0.18	0.06
24			13.79	3.63	1.72	0.51	0.21	0.07
26			16.00	4.21	1.99	0.59	0.25	0.09
28				4.83	2.28	0.68	0.29	0.10
30				5.49	**2.59**	0.77	0.32	0.11
35				7.31	3.45	1.02	0.43	0.15
40				9.36	4.42	1.31	0.55	0.19
45				11.64	5.50	1.63	0.69	0.24
50				14.14	6.68	**1.98**	0.83	0.29
55					7.97	2.36	0.85	0.35
60					9.36	2.78	1.17	0.41
65					10.36	3.22	1.36	0.47
70					12.46	3.69	**1.56**	0.54
75					14.16	4.20	1.77	0.61
80						4.73	1.99	0.69
85						5.29	2.23	0.77
90						5.88	2.48	0.86
95						6.50	2.74	0.95
100						7.15	3.01	**1.05**
150						15.15	6.38	2.22
200							10.87	3.78
300								8.01

Storage

Rain Barrel and Diverter

Capture irrigation water at no cost with our virtually indestructible, 0.1875-inch-thick, recycled food-grade polyethylene Rain Barrel. Multiple barrels can be linked with an ordinary garden hose. Features overflow fitting, drain plug, screw-on cover, and threaded spigot. 39"x24" (H x diameter). 60-gallon capacity. 20 lb. Optional galvanized steel Rain Diverter (shown) fits any downspout (metal or plastic). When full, flip the diverter to a closed position to let downspout function as usual. USA.

14-9201 Rain Barrel and Diverter   **$149 ($50)**

14-0238 Rain Barrel only   **$134 ($50)**

46209   Diverter only   **$22**

Allow 4-6 weeks for delivery. Can be shipped to customers in the contiguous U.S. only. Cannot be shipped to P.O. boxes. Sorry, express delivery not available.

Kolaps-a-Tank

These handy and durable nylon tanks fold into a small package or expand into a very large storage tank. They are approved for drinking water, withstand temperatures to 140°F, and fit into the beds of several truck sizes. They will hold up under very rugged conditions, are self-supported, and can be tied down with D-rings. All tanks have a 1.5-inch plastic valve with standard plumbing threads for input/output, and a 2'x6" (L x diameter) filler sleeve at the top for filling. Our most popular size is the 525-gallon model, which fits into a full-size long-bed (5'x8') pickup truck. USA.

47-401	75 gal. (40"x50"x12")	$369
47-402	275 gal. (80"x73"x16")	$539
47403	525 gal. (65"x98"x18")	$649
47404	800 gal. (6'x10'x2')	$749
47405	1,140 gal. (7'x12'x2')	$1,029
47406	1,340 gal. (7'x14'x2')	$1,159

Water Conservation Products

Water-Saving Showerheads

Showers typically account for 32% of home water use

A standard old-style showerhead uses about 3-5 gallons of water per minute, so even a five-minute shower can consume 25 gallons. Since the 1990s, all showerheads sold in the U.S. must use 2.5 gpm or less. The average for the low-flows that we sell, however, is even less, a highly efficient 1.4 gpm! According to the U.S. Department of Energy, heating water is the second largest residential energy user. With a low-flow showerhead, water use, energy use, and costs for heating water for showers may drop as much as 50%. A recent study showed that changing to a low-flow showerhead saved 27¢ worth of water and 51¢ of electricity per day for a family of four. So, besides being good for Earth, a low-flow showerhead will pay for itself in about two months! Add one of our instantaneous water heaters for even greater savings.

Classic Low-Flow Showerhead

This is our longtime best-seller. Made in the USA of solid brass with a chrome-plate finish. Delivers a vigorous, well-controlled water spray pattern for a truly satisfying

shower. Our favorite feature of this showerhead is that it's O-ringed and threaded together for easy cleaning of the stainless steel diffuser when the spray pattern gets erratic. This is a no-tools fix you can perform in the middle of a shower. Built-in soap-up valve. Maximum flow is 2.25 gpm at 80 psi (1.2-1.4 gpm is about average for most folks). This head cuts water use by 50%-70% and can save a family of four up to $250 a year. Standard ½-inch pipe thread. 10-yr. mfr. warranty.

46104   Best Low-Flow Showerhead $12

2 or more $9.50 ea.

Oxygenics X-Stream

We've touted low-flow showerheads since our earliest catalogs because they're so invigorating that we never miss the water they save. This new model goes a step further by performing the same feat at the ultralow and variable water pressures found in RVs, cabins and camps, off-the-grid homes, and other places. Patented fluidic technology increases water velocity to produce a steady and powerful skin-tingling spray at pressures as low as 3 psi. Its 1.4-gallon-per-minute flow rate compares very impressively with the 2.5 gpm output of its competitors! China.

02-0332 Oxygenics X-Stream   **$19**

Water-Saving Toilet-Lid Sink—Back by Popular Demand!

One major water-wasting culprit in your house is the toilet. Precious clean water goes down the pipes every time you flush! Save water now with the Toilet-Lid Sink—BACK in the Real Goods catalog by popular demand. This unique sink is very popular in Japan and was inspired by inventor Carl Brown's travels to that country. The Toilet-Lid Sink conserves water and offers a convenient way to wash your hands.

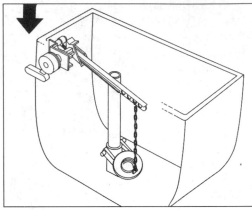

The lid sink re-routes incoming water from your home's fresh-water pipes directly up through a hand-washing spigot each time the toilet is flushed. Water does not go through any part of the toilet itself until it has run through the clean water spigot, allowing you to wash your hands with complete confidence. It then flows into the bowl where it is used appropriately for the next flush. Simultaneous with the flush, fresh water flows through the spigot again for your hands. Kids particularly delight in the novelty of washing their hands at the Toilet-Lid Sink, making it a handy "mother's helper" as well.

Turn Your Commode Top into a Water-Saving Sink

With each flush of your commode, clean water that would otherwise go straight down the toilet is first routed up through a chrome gooseneck spigot to dispense pure water for hand washing. The Toilet-Lid Sink installs easily without tools, is attractive for any bathroom, and is a great space saver. Shuts off automatically. Porcelain-like white plastic replaces your existing tank top and adjusts to fit standard toilets up to 8½" wide and 18-22" long. Built-in soap dish. Overhang varies up to 1½ inches. USA/China.

02-0334 Toilet-Lid Sink   **$89**

Controllable Flusher

Convert your standard toilet into a low-flow toilet—without tools or a plumber—and save up to 35,000 gallons of water a year. Because not all flushes need be the same, the dual-action Controllable Flusher controls the amount of water you use to flush waste matter. Push handle down for a conservative 1.5-gallon flush for liquid; lift for a powerful full flush for solids. Easily retrofits to standard front-flush toilets. 12.5"x5.5"x2" (HxWxD). 5 oz. USA.

02-0205 Controllable Flusher   **$36**

D'mand Water-Saving System

No more waiting for hot water!

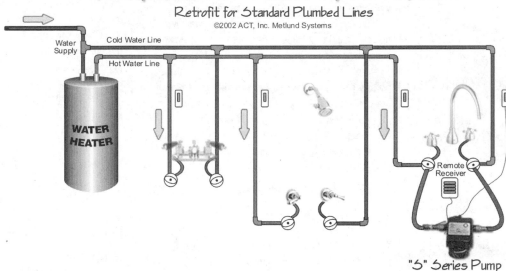

The Metlund® Hot Water D'MAND® System
Retrofit for Standard Plumbed Lines
©2002 ACT, Inc. Metlund Systems

Water Supply · Cold Water Line · Hot Water Line · WATER HEATER · Remote Receiver · "S" Series Pump

D'mand S-02-PF-R

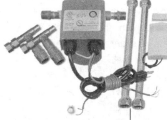

D'mand S-50-PF-R

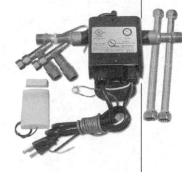

D'mand S-69-PF-R

Turn the hot-water tap on to wash your face, take a shower, or create a warm gravy for your dog's kibble, and you often have to wait a few minutes for the hot water to arrive at the faucet. Not only is it inconvenient, but it also wastes 3-7 gallons of water—about the same amount of water used to flush a toilet—every time you play this waiting game.

An on-demand system such as Real Goods' Hot Water D'mand System gets hot water to your fixtures four to five times faster with none wasted down the drain, and it doesn't require any plumbing modifications. While tankless water heaters are often referred to as "instantaneous," the D'mand System works with tank-type or tankless water heaters to send hot water right to the tap.

It's easy to install—simply place a small pump at the faucet furthest from your hot-water heater. This innovative pump redirects lukewarm water, which would have been wasted, back to the hot-water heater through the cold-water line while pumping hot water up the hot-water line. The pump is activated when a doorbell-type button is pushed at the tap.

With the push of a button, the D'mand System whisks hot water to the most remote faucet in seconds. The displaced cold water is pumped into the cold-water plumbing, returning to the heater, so there's no waste. It shuts off automatically when hot arrives

and costs less than $1 per year to operate. UL- and UPC-listed, this system is recognized by the Department of Energy as both a water- and power-saving device and has a life expectancy exceeding 15 years. It works with any type of water heater.

Our "no sweat" installation kit provides all the necessary plumbing bits and pieces, installs with common hand tools, and plugs into a standard outlet. Installed with a push button and pump at your furthest faucet, intervening faucets can be equipped with additional wireless push buttons to activate the system from up to 100 feet. Longer plumbing runs need a larger pump to keep wait time under 20-30 seconds. Use S-50 model for runs up to 50 feet, S-70 for runs up to 100 feet. S-02 model is for longer or commercial use. S-50 model has 3-yr. mfr. warranty; other models are 5-yr. USA.

45-0100 S-50-PF-R Kit	**$349**
45-0101 S-70-PF-R Kit	**$455**
45-0102 S-02-PF-R Kit	**$760**
45-0103 Additional Transmitter	**$20**

Books

Cottage Water Systems

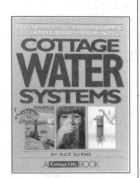

By Max Burns. An out-of-the-city guide to water sources, pumps, plumbing, water purification, and wastewater disposal, this lavishly illustrated book covers just about everything concerning water in the country. Each of the 12 chapters tackles a specific subject such as sources, pumps, plumbing how-to, water quality, treatment devices, septic systems, outhouses, alternative toilets, greywater, freeze protection, and a good bibliography for more info. This is the best illustrated, easiest to read, most complete guide to waterworks we've seen yet. 150 pages, softcover.

80098   Cottage Water Systems $24.95

Water Storage

This definitive work on water storage includes all you need to know to design, build, and maintain water tanks and ponds and to sustainably manage groundwater storage. It also delves into fire protection and disaster preparedness and includes the best building instructions we've ever seen for making your own ferrocement storage tanks. Like Art Ludwig's other books, the value-density of this book is exceptional. As the National Drinking Water Clearinghouse put it, "On the average water system, this book will pay for itself a hundred times over in errors avoided and maintenance savings." 125 pages, softcover.

21-0374 Water Storage $20

Rainwater Collection for the Mechanically Challenged

Revised and expanded in 2003. Laugh your way to a home-built rainwater collection and storage system.

This delightful paperback is not only the best book on the subject of rainwater collection we've ever found, it's funny enough for recreational reading and comprehensive enough to lead a rank amateur painlessly through the process. Technical information is presented in layman's terms and accompanied with plenty of illustrations and witty cartoons. Topics include types of storage tanks, siting, how to collect and filter water, water purification, plumbing, sizing system components, freezeproofing, and wiring. Includes a resources list and a small catalog. 108 pages, softcover.

80704   Rainwater Challenged Book $20

Rainwater Collection for the Mechanically Challenged Video

You say you're *really* mechanically challenged and want more than a few pictures? Here's your salvation. From the same irreverent, fun-loving crew that wrote the book above. See how all the pieces actually go together as they assemble a typical rainwater collection system and discuss your options. This is as close

as you can get to having someone else put your system together. 37 minutes and lots of laughs.

90-0168 Rainwater Challenged DVD $19.95

All About Hydraulic Ram Pumps

Utilizing the simple physical laws of inertia, the hydraulic ram can pump water to a higher point using just the energy of falling water. The operation sequence is detailed with easy-to-understand drawings. Drive pipe calculations, use of a supply cistern, multiple supply pipes, and much more are explained in a clear, concise manner. The

second half of the booklet is devoted to detailed plans and drawings for building your own 1- or 2-inch ram pump. Constructed out of commonly available cast-iron and brass plumbing fittings, the finished ram pump will provide years of low-maintenance water pumping for a total cost of $50-$75. No tapping, drilling, welding, special tools, or materials are needed. This pump design requires a minimum flow of 3-4 gallons per minute and 3-5 feet of fall. It is capable of lifting as much as 200 feet with sufficient volume and fall into the pump. The final section of the booklet contains a setup and operation manual for the ram pump. 25 pages, paperback. USA.

**80501   All About Hydraulic
             Ram Pumps $10.95**

CHAPTER 8

Water Heating:

The Most Cost-Effective Solar Alternative

MOST OF US TAKE HOT WATER at the turn of a tap for granted. It makes civilized life possible, and we get seriously annoyed by cold showers. Yet most people probably do not realize how much this convenience costs them. The average household spends an astonishing 20%-40% of its energy budget on water heating. And for most folks, all those energy dollars are given to an appliance that has a life expectancy of only 10-15 years and that throws away a steady 20% or more of the energy you feed it. Any efficiency improvements you provide for your water heater reduces your environmental impact and leaves a better world for your children.

There are better, cheaper, and more durable ways to heat your home water supply than using an electric water heater. This chapter describes all the common water heater types, discusses the good and bad points of each, and offers suggestions for efficiency improvements. A new feature of this 30th anniversary *Sourcebook* is a detailed examination of solar hot-water heating, which for most people is the simplest and most cost-effective renewable energy application around—indeed possibly the best investment available in America today! We're indebted to our old friend Bob Ramlow, former Real Goods store owner in Amherst, Wisconsin, longtime solar water system designer and installer, and author of *Solar Water Heating: A Comprehensive Guide to Solar Water and Space Heating Systems* (New Society, 2006), available on page 413. We've developed some good life-cycle cost comparison charts so you can examine the real operation costs over the lifetime of the appliance. Points to consider for each heater type are initial cost, cost of operation, recovery time (how fast does it heat water?), ease of installation or ability to retrofit, life expectancy, and finally, life-cycle cost.

> The average household spends an astonishing 20%-40% of its energy budget on water heating.

Common Water Heater Types

Storage or Tank-Type Water Heaters

Storage or tank-type water heaters are by far the most common kind of residential water heater used in North America. They typically range in size from 20 to 80 gallons and can be fueled by electricity, natural gas, propane, or oil.

Storage heaters work by heating up water inside an insulated tank. Their good points are the modest initial cost, the ability to provide large amounts of hot water for a limited time, and the fact that they're well understood by plumbers and do-it-yourself homeowners everywhere. Their bad points include constant standby losses, slow recovery times, and low life expectancy. Because heat always escapes through the walls of the tank (standby heat loss), energy is consumed even when no hot water is being used. This wastes energy and raises operation costs. Most manufacturers offer "high-efficiency" storage heaters now for a premium price. These just use more insulation to reduce standby losses slightly. Customer-installed water heater blankets do the same thing (although maybe not quite so well). Standby losses for gas and oil heaters are higher because the air in the internal flue passages is constantly being warmed, rising, and pulling in fresh cold air from the bottom. Although storage tanks

An "add-on" heat pump water heater.

An "intergral" heat pump water heater.

can deliver large volumes of hot water rapidly, once they're exhausted, the limited energy input means a long recovery time. Who hasn't had to wait to get a hot shower, or suffered through a quick cold one, after someone else used up all the hot water?

Life expectancy for tank-type water heaters averages 10-15 years. Usually the tank rusts through at this age, and the entire appliance has to be replaced—not an efficient way to design relatively expensive appliances. Life expectancy can be prolonged significantly if the sacrificial anode rod—installed by the tank manufacturer to keep the tank from rusting—is replaced at about five-year intervals.

Storage water heaters are fairly cost effective if you use natural gas, which is still something of an energy bargain. If you're heating with electricity or propane, then a storage water heater, though cheaper initially, is your highest-cost choice in the long run. See the Life-Cycle Comparison Chart on page 390.

Heat Pump Water Heaters

If you use electricity to heat water, heat pumps are three to five times more efficient than conventional tank-type resistive heaters. Heat pump water heaters use a compressor and refrigerant fluid to transfer heat from one place to another, like a refrigerator in reverse. Electricity is the only fuel option. The heat source is air in the heat pump vicinity, although some better models can duct in warm air from the attic or outdoors. The warmer the air, the better: Heat pumps work best in warm climates where they don't have to work as hard to extract heat from the air. In these applications, the upper element of the conventional electric water heater usually remains active for back-up duty.

The advantages of heat pump water heaters are that they use only 33%-50% as much electricity as a conventional electric tank-type water heater, they will provide a small cooling benefit to the immediate area, and life expectancy is a good 20 years. The disadvantages are that heat pumps are quite expensive initially (average installed cost is approximately $1,200), and recovery rates are modest, variable, and can be fairly low if the pump doesn't have a warm environment from which to pull heat. Like all storage tank systems, heat pumps have standby losses, and if installed in a heated room will rob heat from Peter (the furnace) to pay Paul (the water heater).

Heat pumps make better use of electricity because it's much more efficient to use electricity to *move* heat than to *create* it. Heat pump water heaters are available with built-in water tanks, called integral units, or as add-on units to existing water heaters. Add-on units may be a smarter investment, as the heat pump will probably outlive the storage tank. If you live in a warmer climate and use electricity to heat water, a heat pump is your best choice. In fact, heat pump units stack up quite favorably against everything but natural gas and solar if you can afford the initial purchase cost.

Heat pump water heaters are a bit complex to install, particularly the better units that take their heat from a remote site or use a separate tank. You'll probably require a contractor to install one. Call your local heating and air-conditioning contractor for more details on availability in your area, cost, and installation estimates.

Tankless Coil Water Heaters

This type of heater is probably found only in older oil- or gas-fired boilers. A tankless coil heater operates directly off the house boiler; it does not have a storage tank, so every time there is a demand for hot water, the boiler must run. This may be fine in the winter, when the boiler is usually hot from household heating chores anyway, but during the rest of the year, it results in a lot of start-and-stop boiler operation. Tankless coil boilers may consume 3 BTUs of fuel for every 1 BTU of hot water they deliver. This type of water-heating system is not recommended.

Indirect Water Heaters

Indirect water heaters use the home heating system's boiler. Hot water is stored in a separate insulated tank. Heat is transferred from the

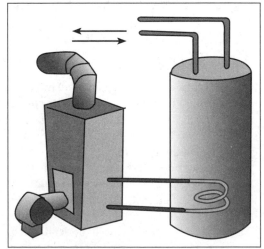

Indirect water heater.

boiler using a small circulation pump and a heat exchanger. The separate insulated storage tank adds reserve capacity. This means the boiler doesn't have to turn on and off as frequently, which greatly improves fuel economy. When used with the new high-efficiency, gas-fired boilers and furnaces, indirect water heaters are usually the least expensive long-term solution to providing hot water. They will add about $700 to the heating system cost. The life expectancy of the heat exchanger and circulation pump is an exceptional 30 years, although the storage tank may need replacement at 15-20 years.

The big disadvantage of an indirect water-heating system is that it is an integral part of the household boiler/heating system that is best installed during new construction by your heating and cooling contractor. In addition, they suffer the usual standby losses of storage tanks.

Demand, or Tankless, Water Heaters

The tankless strategy is so obvious that it's a no-brainer. Why keep 30-80 gallons of water at 120°F all the time? Do you leave your car idling 24 hours a day just in case you need to run an errand in the middle of the night? Of course not. You start it when you need it. Our water heaters should work the same way, starting up only when we need hot water. Otherwise, we heat up the water, it sits there in a big tank and

loses heat, then we heat it up again, it sits there . . . and so on, ad infinitum. In many cases, this waste heat must even be removed from the home with expensive air conditioning! Because of a long honeymoon with cheap power, America is one of the few countries where people still use tank-type water heaters. Nearly everybody else in the world figured out the virtues of demand water heating a long time ago. Due to heat loss, tank-type heaters use a minimum of 20% more energy than demand systems. If yours is a small one- or two-person household, your heat losses are even greater, because the hot water spends more time sitting around waiting to be used.

The advantages of demand water heaters are very low standby losses, lowest operation costs, unlimited amounts of hot-water delivery, a very long life expectancy, and ease of installation for do-it-yourselfers and retrofitting. One disadvantage is that they cost a bit more initially. Another is that the unlimited amounts of hot water can't be taken too rapidly, because flow rates are limited to the heater's abilities, so hot-water use may need to be coordinated at times. (People tend to be *very* sensitive and excitable about shower temperatures.)

HOW DEMAND WATER HEATERS WORK

Demand, or tankless, water heaters go to work only when someone turns on the hot water. When water flow reaches a minimum flow rate, the gas flame or heater elements come on, heat-

Bosch Aquastar 1600 —a demand, or tankless, water heater.

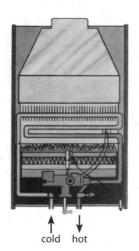

cold   hot

Basic demand water heater internal construction.

Go Tankless: Hot Water on Demand Feels Great and Saves Money

Nobody likes to take cold showers or waste money. Here's a great way to solve both of these problems.

HOT WATER ON DEMAND.

Tankless water heaters serve unlimited hot water on demand, which means that each member of your family who takes a shower in the morning gets a hot one. They'll thank you for it, and if you have teenagers, you'll thank yourself.

TANKLESS HEATERS SAVE YOU MONEY.

They're 15-30% more efficient than the best tank-type heaters because you're not continually heating (and paying for) water that sits in a storage tank 24 hours a day. If your gas water heater is more than 10 years old, you're probably wasting even more money, because its efficiency has deteriorated over time.

TAX CREDITS SAVE YOU EVEN MORE.

Right now you can receive a federal tax credit of up to $300 when you purchase a qualifying tankless hot-water heater (with an energy factor of 0.80 or greater) by December 31, 2007.

PLUMBERS GO TANKLESS.

Thanks to improvements in technology and growing customer demand, more plumbers are installing tankless water heaters than ever before. (If you call a plumber who says you don't want to go tankless, tell him people in most other countries use tankless hot-water heaters and you want to, too.) When retrofitting in an existing home, tell your plumber to make sure your vent pipe is 5 inches, compared with the standard 3 inches and 4 inches, and your gas line is ¾ inch. If you have questions about installing a tankless water heater, call our experienced technicians at 800-919-2400.

The great advantage of tankless heaters is that you run out of hot water only when either the gas or the water runs out.

Although solar water heaters are expensive initially, their long life expectancy and nearly zero cost of operation gives them the lowest life-cycle cost and one of the quickest paybacks of any water heating system.

ing the water as it passes through a radiator-like heat exchanger. Tankless heaters do not store any hot water for later use but heat water only as demanded at the faucet. The minimum flow rates required for turn-on prevent any possibility of overheating at very low flow and ensure that the unit turns off when the faucet is turned off. Minimum flow rates vary from model to model but are generally about 0.5-0.75 gpm for household units. Other safety devices include the standard pressure/temperature relief valve that all water heaters in North America are required to carry; tankless heaters use an additional overheat sensor or two on the heat exchanger.

The great advantage of tankless heaters is that you run out of hot water only when either the gas or the water runs out. On the other hand, tankless heaters will meter out the hot water at just so many gallons per minute. Excess water flow will result in lower temperature output. Some tankless heaters are limited to running just one fixture at a time. Larger household-sized heaters, such as the Bosch AquaStar 1600 or the Takagi T-K3, can run multiple fixtures simultaneously. Showers are a touchy issue, so what complaints we hear usually revolve around showers and multiple-fixture uses. Tankless heaters are probably the best choice for smaller homesteads of three people or fewer, where hot-water use can be coordinated easily. Larger homes with an intermittent use that's a long distance from the rest of the household hot-water plumbing, such as a master bedroom at the end of a long wing, are also good candidates for a tankless heater just to supply that isolated area.

See the More Details section below for a complete discussion of demand water heaters if this seems like a good choice for you.

CAN I GET AN ELECTRIC DEMAND WATER HEATER?

Yes, you can, but first a warning or two. Electricity is easily the most expensive power source for heating applications, unless you live in the Northwest where there are huge hydro projects. At average North American 2007 energy prices, natural gas is the least expensive choice for water heating, propane is about 60% more expensive, and electricity is about another 15% above that. It takes a shocking amount of electricity to heat water on the fly. Most households will need a 200-amp electric service at a minimum to support an instant electric water heater. That said, we've got a line of high-quality German electric water heaters with precise digital con-

trol and multiple temperature sensors that will do a great job with no standby losses.

Solar Water Heaters

THE BEST CHOICE FOR THE LONG RUN

Although solar water heaters are expensive initially, their long life expectancy and nearly zero cost of operation gives them the lowest life-cycle cost and one of the quickest paybacks of any water heating system. The solar option is environmentally friendly, and whether your system provides the bulk of your domestic hot-water needs or serves primarily as a preheater, it will be a smart investment that will increase the value of your house and pay for itself in relatively short order.

It's interesting to note that solar water heating has a long history that our society has generally ignored. The first solar water heating system was patented in 1891: a simple black water tank mounted in a box that had glass on one side, what we call today an ICS (integral collector storage) or batch water heater. In 1909, the first flat-plate collector was patented, and a few years later, the first closed-loop antifreeze system was introduced. This all happened before the development of electric and gas water heaters, meaning that the very first type of automatic water heating system was solar. Between 1935 and 1941, more than half the population of Miami, Florida, used solar water heaters.

Solar water heaters enjoyed a huge surge of interest following the Arab oil embargo during the mid-1970s and early 1980s, when they were supported by 40% federal tax credits. Unfortunately, those same tax credits encouraged a rash of sleazy door-to-door salesmen pushing less-than-the-best technology at astronomical prices. Many systems were installed in climates

Photo: Bob Ramlow

they were not designed for. A lot of people got burned in the process, and the solar hot-water industry still hasn't recovered from the bad image this circus left behind. However, the industry has matured, and reputable, experienced manufacturers and contractors stand ready to accelerate the adoption of solar water heating once again.

All the solar heaters carried by Real Goods were designed after the Carter solar tax-credit days and have benefited from lessons learned during this unfortunate period. The models available now cost far less and are much more reliable. So the first lesson was quality. Don't buy junk. All these modern water heaters are high-quality units warranted for 5-10 years by solid companies that are not going to disappear the day after the tax credits expire. In short, they stand on their own merits. These water heaters can reasonably be expected to last as long as your house. The second lesson was to make sure you choose a system that's a proven technology for your climate. A poor match will likely lead to system underperformance or failure. The third lesson was simplicity. The systems that don't use controllers, temperature sensors, or drain-down valves—all the complex hardware that experience showed us was most prone to breakdown—are preferred. These heaters use simple passive designs with sunlight and heat as the controls. However, more complex systems can function just fine if well cared for, and they are described in the sections that follow.

Solar water heating is a technology that nearly everyone can use toward the goal of reducing fossil fuel consumption. Heating domestic hot water uses 20%-40% of a home's annual energy budget. A solar water heating system can supply 60%-90% of that energy load. The return on your investment is higher than the return for any other renewable energy investment you can make. Depending on your particular situation, the savings in conventional fuel can pay

for the cost of the solar water heating system in as little as three years. Most often, the payback is around 5-10 years, but this is still a great investment, equating to a 10%-20% return on investment (ROI), which is better than any safe investment available in America today. From a climate change perspective, a typical two-panel system will offset between 1,500 and 2,000 pounds of CO_2 per year. That is the equivalent of driving between 1,300 and 1,750 miles in a car getting 22 miles per gallon. So how do you get started?

GETTING STARTED WITH SOLAR HOT WATER

The first step in the direction of solar hot water is the same first step in going solar for space heating or electricity: Conserve energy! As always, the three general principles are reduce losses, increase efficiency, and reduce consumption. To start, examine your heating system from top to bottom, and look for places where heat might leak out. When it comes to water, heat loss is nothing but waste. A little bit of cheap insulation can go a long way and make a noticeable difference. Thorough insulation of hot-water pipes is easiest in new construction, but it's a good idea, and usually not too hard, to insulate as much of the piping as you have access to. You should also insulate your existing water heater with a jacket made for that purpose. (Newer models are often well insulated to begin with.) Heat loss can also come from leaks. A faucet that leaks 30 drops of water a minute will waste almost 100 gallons a month. Fix leaky faucets promptly.

Next, try to increase the efficiency of everything in your home that uses hot water. Gener-

A solar water heating system can supply 60%-90% of that energy load. The return on your investment is higher than the return for any other renewable energy investment you can make.

Collector mounting locations.

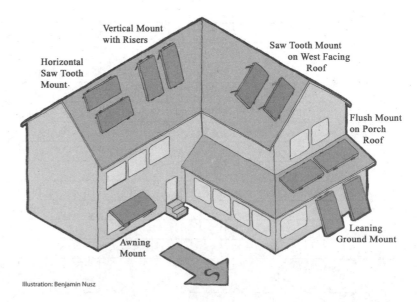

Vertical Mount with Risers

Horizontal Saw Tooth Mount

Saw Tooth Mount on West Facing Roof

Flush Mount on Porch Roof

Leaning Ground Mount

Awning Mount

Illustration: Benjamin Nusz

ally, you'll be limited to the washing machine and the dishwasher. If you upgrade these appliances to more energy-efficient models, you will see significant reductions in energy consumption. For instance, a front-loading washing machine uses half as much hot water as a standard top-loading model. This increase in efficiency saves 10-20 gallons of hot water per load, which can amount to thousands of gallons of hot water a year.

Finally, you can conserve energy by simply using less. A lot can be accomplished without significantly changing your daily habits. For example, when washing dishes in the sink, don't let the water run while rinsing; instead fill one sink with wash water and the other with rinse water and/or turn off the hot water between rinses. Soak pots and pans instead of letting the water run while you scrape them clean. If you use a dishwasher, run only full loads. Use cold water with the garbage disposal—it solidifies grease, allowing the disposal to get rid of it more effectively. Showers use less water than baths, but make sure you've installed low-flow showerheads. Standard showerheads use 3-4 gallons per minute; even a brief five-minute shower can consume 20 gallons of hot water. Low-flow showerheads will use half that amount, so that a family of four can save well over 1,000 gallons a month.

SOLAR WATER HEATING ECONOMICS

It's time to let you in on a little secret: Solar water heaters don't cost anything. They're FREE! That may sound absurd, but it's true. Now we're not recommending that you run over to the nearest solar distributor and just steal a system. Don't do that. We're just asking you to take a step back and think about solar in a different way. It takes a little work and a little change in perspective, but you'll see that in the end, solar water heaters have a net cost of zero dollars.

The first point to underscore is that when you install a solar water heater, you make an investment that will increase the value of your home. You gain in equity what you spent on the cost of installation. Solar water heaters typically have a life span of at least 30-40 years. In most cases, the collectors will outlast your roof. So if you decide to sell your house, you should get back most of what you paid to buy and install the system.

The second critical point is life-cycle costing. People often ask, "Why would I consider purchasing a solar water heater that costs several thousand dollars when I can purchase a gas or electric water heater for several hundred dollars?" The answer lies in life-cycle costing, which adds the original cost of a piece of equipment to its operating cost over its lifetime. Life-cycle cost analysis gives an accurate picture of the total cost of a purchase over time and takes into account long-term trends in energy prices. The following chart uses a realistic but conservative estimate that the rate of inflation for fossil fuels will average 10% over the next 30 years. (The historical average for 1970-1999 was around 7.5%.) Solar hot water compares favorably with its competitors.

Note that in this comparison of solar with natural gas and electric water heating, the solar heater has no operating cost, including zero inflation in the cost of the free solar fuel. Like any other piece of mechanical equipment, it does require some maintenance, but this amounts to around only $2 per month. As you can see, viewing the systems over the long term makes for a fair comparison. The cost of the solar hot water is equal to the operating cost of the electric water heater after only 4.5 years and the natural gas water heater after only 10 years. This number is commonly referred to as the pay-off date, since

LIFE-CYCLE COSTING			
	Electric Water Heater	Natural Gas Water Heater	Solar Water Heater
Energy Produced	4,300kWh (200 therms)	200 therms	200 therms
Cost per Unit	$0.15/kWh	$1.25/therm	$0
Energy Inflation Rate	10%	10%	—
Installed Cost Minus Federal Tax Credit	$1,500	$1,500	$5,000 ($7,000 – 30%)
Maintenance	$0	$0	$2/month
Cost to Operate per Year			
1	$645	$250	$0
2	$710	$275	$0
3	$780	$303	$0
4	$859	$333	$0
5	$9449	$366	$0
6	$1,039	$403	$0
7	$1,143	$443	$0
8	$1,257	$487	$0
9	$1,383	$536	$0
10	$1,521	$589	$0
10-Year Life-Cycle Cost	$11,780	$5,484	$5,240

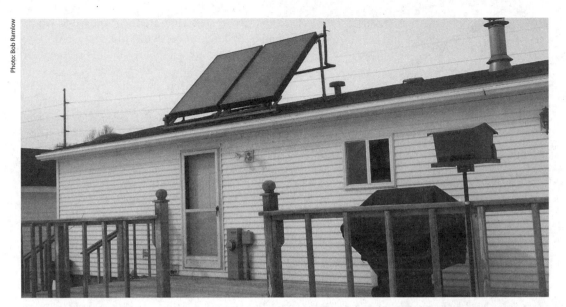

Photo: Bob Ramlow

after that amount of time, you would have paid for the system with the money saved from not purchasing energy from the utility. However, it should be stressed that this notion is something of a misstatement. As noted before, a solar water heater is paid off the second you install it because of what you gain in home equity.

Like other renewable energy investments, solar water heating is different from a conventional investment, because the free energy it harvests each month reduces—and eventually eliminates—a monthly bill that you pay to someone else. The savings gained from the solar water heater pays for the solar investment. The following chart reflects a cash-flow analysis that assumes you borrow the original system installed cost of $7,000, less the federal tax credit at 7% interest over 10 years. It shows your

monthly cash-flow impact. You can see that for the first five years, you have a slightly negative cash flow—but from year six onward, you have a positive cash flow. And, of course, this analysis does not take into account all the environmental advantages of having a solar water heating system.

TYPES OF SOLAR WATER COLLECTORS

ICS stands for Integral Collector Storage, also known as a batch heater. In a batch heater, the hot-water storage tank *is* the solar absorber. The tank or tanks are painted black or coated with a selective surface and mounted in an insulated box that has glazing on one side. The Sun shines through the glazing and hits the black tank, warming the water inside it. Some models feature a single large tank (30-50 gallons) while others feature a number of metal tubes plumbed in series (30- to 50-gallon total capacity). The single tanks are typically made of steel, while the tubes are typically made of copper. These collectors weigh 275-450 pounds when full, so

A tank-type, or batch, ICS system.

	Electric Water Heater	Solar Water Heater *Installed Cost:$7,000*	Cash Flow Impact per Month
Monthly Bill	Monthly Savings	Monthly Payment	
1st year	$35.83	$55.51	-$19.68
2nd year	$39.42	$55.51	-$16.09
3rd year	$43.36	$55.51	-$12.15
4th year	$47.69	$55.51	-$7.82
5th year	$52.46	$55.51	-$3.05
6th year	$57.71	$55.51	$2.20
7th year	$63.48	$55.51	$7.97
8th year	$69.83	$55.51	$14.32
9th year	$76.81	$55.51	$21.30
10th year	$84.49	$55.51	$28.98
11th year	$92.94	$0.00	$92.94

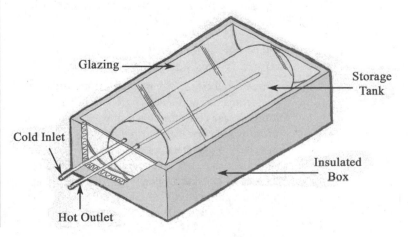

Glazing

Cold Inlet

Hot Outlet

Storage Tank

Insulated Box

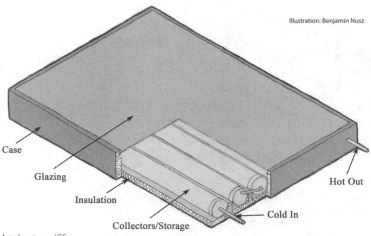

Illustration: Benjamin Nusz

Case

Glazing

Insulation

Collectors/Storage

Hot Out

Cold In

A tube-type ICS collector.

> Flat-plate collectors are the most widely used type of collector in the world for domestic solar water heating and solar space heating applications.

wherever they are mounted, the structure has to be strong enough to carry this significant weight; you may have to reinforce an existing roof. You should always mount the collectors tilted so that the system will drain properly.

Batch heaters are widely used around the world where the climate never experiences freezing conditions. They work great, given the climatic restrictions. They are a type of direct system, as the water heated in the collector is the water you actually use in your home. (Indirect systems are described below.) Batch heaters are also perfect in seasonal applications such as campgrounds and summer homes where they are used only during the warm months of the year and are drained before freezing conditions occur.

The tube type of batch heater will outperform the tank type because more surface area is exposed to the Sun. Another advantage of the tube type is that their profile is much smaller, thereby minimizing their aesthetic impact on a building. On the other hand, tube collectors cool off more quickly at night because of the larger surface area, so they lose efficiency, especially when the nights are cool. You can maximize the efficiency of this kind of system by using as much hot water as possible during the day and early evening hours.

Flat-Plate Collectors

Flat-plate collectors are the most widely used type of collector in the world for domestic solar water heating and solar space heating applications. These collectors have an operating range from well below 0°F to around 200°F, which is precisely the operating range required for these applications. They are durable and effective. Flat-plate collectors also have a distinct advantage over other types because they shed snow very well, which makes them a good choice in

colder climates. They are the standard to which all other kinds of collectors are compared.

Flat-plate collectors are shallow rectangular boxes that typically are 4 feet wide by 8 feet long and 4-6 inches deep, though they also come in 4x10-foot and 3x8-foot sizes. They consist of a strong frame, glazing fastened to the front of the collector, and a solid back. Just beneath the glazing lies an absorber plate. This absorber plate has manifolds that run across the top and bottom of the collector, just inside the frame. These manifolds are usually 1-inch-diameter copper pipe and extend out both sides of the collector through large rubber grommets. These internally manifolded collectors can be easily ganged together to make large arrays. Smaller riser tubes, typically ½-inch copper pipes, run vertically and are welded to the manifolds above and below, spaced 3-6 inches apart (the closer the better). Attaching a flat copper fin to each riser completes the absorber plate. The fin must make intimate contact with the riser tube to facilitate effective heat transfer from the fin to the tube; soldering or welding makes the best connection. The fins are usually made of copper and either plated with a selective surface or painted to maximize solar absorptivity. They also are usually dimpled or corrugated to increase absorbtivity. The absorber plates are not attached to the frame; they rest inside of it and can expand or contract as they are heated or cooled and not be restricted by the frame. The fins and tubes must be made of the same metal to reduce the chance of corrosion. (One absorber on the market has copper waterways with aluminum fins. They get away with this by laminating the metals together in such a way that galvanic reactions do not take place.)

The strength of a flat-plate collector is all in the frame, and strength is very important because the collectors must be able to withstand high wind conditions without breaking apart.

A flat-plate collector.

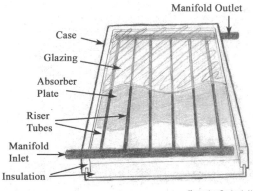

Manifold Outlet

Case

Glazing

Absorber Plate

Riser Tubes

Manifold Inlet

Insulation

Illustration: Benjamin Nusz

These frames are almost exclusively made of extruded aluminum, although some are made of rolled aluminum; we recommend the extruded aluminum. Mounting hardware fastens to channels or flanges built into the frame. Because these flanges go completely around the collector, great flexibility in mounting options is available.

Another important component to consider are the fasteners used to assemble the collector. All fasteners should be made of stainless steel. It is critical to always use compatible metals when attaching different kinds together. Aluminum and stainless steel are compatible, while aluminum and plain or galvanized steel are not. This principle must follow through to the mounting hardware as well as the collector construction. Because each manufacturer makes its own mounting hardware, and because that collector is tested with its specific hardware, you should always purchase the mounting hardware to match the collector.

While many materials have been used for the glazing of flat-plate collectors, only low-iron tempered glass has stood the test of time. This glass is usually patterned on the outside to reduce glare and reflection and to increase absorbtivity. An EPDM rubber gasket is fitted to the edges of the glass plate both to protect the edge and to create a good seal where it sets against the collector frame. Note that if you ever have to take the glazing off of a collector, the edge of tempered glass is very fragile. If you even tap the edge/side of a tempered glass pane, it can literally explode apart, so be very careful and always wear safety glasses and gloves when handling glass. All kinds of plastics have been used as glazing material, but they have all failed under direct and constant exposure to the Sun.

Evacuated-Tube Collectors

While flat-plate collectors are all made essentially the same way and perform similarly from one brand to another, evacuated-tube collectors vary widely in their construction and operation. Evacuated-tube collectors consist of a number of annealed glass tubes that each contain an absorber plate. During the manufacturing process, a vacuum is created inside the glass tube. The absence of air creates excellent insulation, allowing higher temperatures to be achieved at the absorber plate.

The similarity between different types ends there. Some evacuated-tube collectors have a riser tube fastened to the absorber plate that sticks out of each end of the tube and is attached

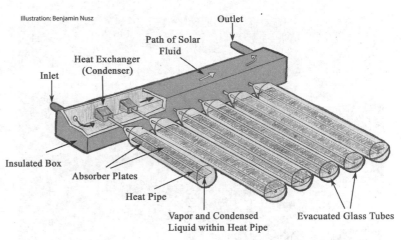

Illustration: Benjamin Nusz

An evacuated-tube collector.

to a manifold, much like that of a flat-plate collector absorber. The solar fluid circulates through each tube and is heated. Others use a hollow pipe attached to the absorber plate that is closed at one end and exits the glass tube on the top. A special liquid inside the pipe evaporates when heated, and the hot vapor rises to a heat exchanger manifold located along the top of the tubes. Solar fluid is heated in the exchanger and circulated throughout the system. As the solar fluid cools the vapor, it condenses and drops back down into the pipe. These liquid/vapor tubes often require a valve on each pipe. Still other models use a solid metal rod attached to the absorber that sticks out the top end of the glass tube and is inserted into a manifold. Sun shining on the absorber heats the rod, and when solar fluid is circulated through the manifold, it picks up heat from the rod ends.

Evacuated-tube collectors have several drawbacks that you need to be aware of. First, these collectors get hotter than flat-plate collectors and can generate temperatures above the boiling point of water. This can cause significant problems in a solar water heating or space heating system. It is therefore critically important to always make sure there is an adequate load on the system to keep the temperatures below 210°F within the system. One way you can keep the temperatures under control is by oversizing the solar storage tank or undersizing the collector, which is what is typically done in Asia and Europe where these collectors are most popular. It is also prudent to avoid using this type of collector if the system will sit idle for a period of time—for example, if you usually take long vacations. To avoid potential overheating with evacuated-tube collectors, it is best to use them in drainback systems, which are described below.

Second, the tubes are fragile. They are made of annealed glass, which is much more delicate

While flat-plate collectors are all made essentially the same way and perform similarly from one brand to another, evacuated-tube collectors vary widely in their construction and operation.

Photo: Bob Ramlow

The single largest application of active solar water systems is to heat outdoor swimming pools.

than tempered glass. Care must be taken when transporting and handling the glass tubes. Another issue is that collector performance depends upon the vacuum inside the tubes. Because a rod or pipe exits the tube on one end or both, a seal must be maintained at this junction. Most manufacturers of evacuated-tube collectors guarantee this seal for only 10 or 15 years, if at all. If the seal is broken, the collector performance is no better than that of a flat-plate collector, and probably worse. Another problem with evacuated-tube collectors is that they do not shed snow or melt frost. Because the evacuated tube is such a good insulator, little heat escapes from it, and the snow or frost that accumulates on the tubes can stick for a long time. Their surface is also irregular, so snow packs between the tubes as well. It is not uncommon for roof-mounted evacuated-tube collector arrays to become packed with snow in the early winter and stay that way until spring, which renders them completely useless for a good portion of the year. The problem is compounded because the fragility of the glass makes it difficult to scrape accumulated snow off the tubes.

Concentrating Collectors

A concentrating collector utilizes a reflective parabolic surface to reflect and concentrate the Sun's energy to a focal point where the absorber is located. To work effectively, the reflectors must track the Sun. These collectors can achieve very high temperatures because the diffuse solar resource is concentrated on a small area. In fact, the hottest temperatures ever measured on Earth's surface have been at the focal point of a massive concentrating solar collector. Concentrating collectors have been used to make steam that spins an electric generator in a solar power station. This is sort of like starting a fire with a magnifying glass on a sunny day.

Attempts have been made to use concentrating collectors in domestic water heating systems, but to date no successful or durable products have been developed. Problems have been encountered with the tracking mechanisms, the precision needed in the mechanisms, the whole mechanism freezing up, and the durability of the reflectors and linkages. They also tend to make the water too hot on a regular basis. It's a promising technology, however, so stay tuned.

Pool Collectors

The single largest application of active solar water systems is to heat outdoor swimming pools. Special collectors have been developed for this purpose, made of a special copolymer plastic. The collectors don't have to be glazed because they are used only when it is warm outside. Like any other solar water collector, they cannot withstand freezing conditions.

Solar pool heating collectors are direct systems because the pool water itself circulates through the collectors. This is the most efficient configuration, as there are no heat exchanger losses. Most pool water contains additives like chlorine, which is highly corrosive to copper, and this is one of the reasons the collectors and piping are made of plastic or polyvinyl chloride (PVC).

Plastic pool heating collectors are typically mounted flat on a roof. The collectors are held in place with a set of straps that go over the collectors but are not actually attached to them. The straps are often plastic-coated stainless steel and are threaded through special clips that are bolted to the roof. This method of holding down the collectors allows them to expand and contract on the roof without binding.

TYPES OF SOLAR WATER HEATING SYSTEMS

As we've been discussing, solar water systems can be either direct or indirect. In a **direct** solar water system, domestic water—the actual water you use in your home—enters the solar collector, where it is heated before being circulated into the house. In an **indirect** solar water heating system, a heat transfer fluid (also called solar fluid) is heated in the collectors and then circulated to a liquid-to-liquid heat exchanger where the solar heat is transferred from the solar fluid to your domestic water supply.

All systems except batch systems require some type of tank to store the water that was heated by the Sun. Most systems use a storage tank for the solar heated water, and also a back-up heater that can be either a tank-type heater or an on-demand (tankless) water heater. These are called **two-tank systems**. Some system designers attempt to use the solar storage tank as the back-up heating system as well by including an electric heating element or a gas burner to heat the water when insufficient solar energy is available to bring the water up to the desired temperature. These are called one-tank systems. We recommend two-tank systems because you usually need a second tank to have suitable storage capacity for the collectors. All the illustrations in this section are of two-tank systems.

Over the years, a number of different domestic solar water heating systems have been devised, but there are only three we would ever recommend: ICS or batch, drainback, and closed-loop antifreeze. These three are the only system types that have proven their reliability over the past 20 years.

ICS or Batch

Batch systems are considered passive because they require no pumps of any kind to operate. They are direct systems because the domestic water actually enters the collector. The batch unit is typically plumbed in series between the cold-water supply and the conventional water heater. Whenever a hot-water tap in the dwelling is opened, cold water from the supply enters the batch collector and forces the solar heated water stored in the batch collector into the conventional or back-up water heater. If the water from the batch collector is hotter than the setting on the back-up heater, that heater will not activate. If the water from the collector is warm but below the temperature setting of the back-up heater, the back-up will have to add only enough heat to raise the temperature up to the preset level. If no solar heating has taken place and all the water in the batch collector is cold, the back-up heater will have to deliver the whole load.

Because water is plumbed through the batch collector, this type of system is suitable only for climates or seasons where freezing conditions will not occur. Batch collectors are very popular along the extreme southern parts of the U.S. and in tropical climates. They have also been successfully installed in conjunction with vacation homes and recreational facilities like parks, where they are used only during the summer months and are drained the rest of the year.

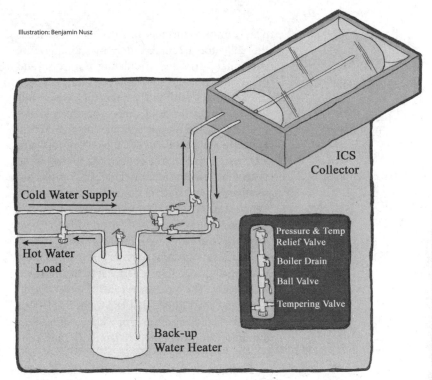

An ICS system.

Because of their simplicity, batch collectors are among the least expensive solar water heating systems available. A complete system consists of the batch collector and piping: No pumps, extra storage tanks, controllers, or other components are needed. Unfortunately, because it is simple and inexpensive, people who live in climates where freezing conditions do occur are sometimes tempted to use a batch system. They often think that if freezing conditions are forecast, they will simply drain the water out of the collector and piping and everything will be all right. In theory, this assumption is correct. In reality, Murphy's Law often comes into play (Murphy's Law states that if anything can go wrong, it will). Actually, all batch collectors can withstand mild freezing conditions. The mass of the water in the collector will keep it from freezing for quite some time (depending on the temperature). It is the piping going to and from the collector that is vulnerable. A well-insulated ¾-inch copper pipe will freeze in less than five hours at 29°F. Unless you plan to use it only seasonally, you're better off not installing a batch system if you live in a cold climate.

Drainback Systems

Drainback systems use flat-plate or evacuated-tube collectors to heat a solar fluid, usually distilled water or a weak antifreeze/water solution, that is circulated through them. Two insulated pipes connect the collectors to a specialized tank called a drainback tank. A high-

Because of their simplicity, batch collectors are among the least expensive solar water heating systems available.

The major limitation of drainback systems is their inability to prevent freeze-ups in climates that experience extended cold or snowy conditions.

head pump is installed on the pipe that feeds the collector array. A differential temperature controller turns the circulating pump on and off. When on, the solar fluid is pumped from the drainback tank to the collectors, where it is heated. The solar fluid then drops back down to the drainback tank, completing the circuit. When the system turns off, the pump stops and all the solar fluid in the collectors and the piping drains back into the tank. Drainback tanks are relatively small, so they do not store much heat. When the system is operating, various methods (detailed below) are used to transfer the heat from the drainback tank to a solar storage tank.

Drainback systems are one of the three most popular types of solar energy systems installed worldwide. They are an excellent choice for all climates except those that experience severe or extended cold conditions, or where a significant amount of snow is expected annually. In hot climates, they are the best choice.

Drainback systems work very well in warm climates, because when the water in the storage tank gets heated to its maximum desired temperature (its high limit), the system turns off and all the fluid drains out of the collector, which prevents the solar fluid from degrading due to overheating. This is especially important when systems experience idle periods like vacations during the summer months in warm or hot climates.

Drainback systems are classified as indirect solar water heaters because the domestic water is heated by a solar fluid and heat exchanger and not heated in the collector. They are classified as closed-loop because the solar fluid remains within a single circuit at all times. They are classified as active systems because they always use a pump or pumps to circulate fluids through the system.

The major limitation of drainback systems is their inability to prevent freeze-ups in climates that experience extended cold or snowy conditions. Some installers like to add antifreeze to the solar fluid in drainback systems to extend their range northward. This is problematic. Every time the system drains back, a thin film of antifreeze is left inside the collector, which dries and leaves a residue. Eventually, the film will build up enough to degrade the collector and diminish its efficiency. The only reason to add glycol to the solar fluid is to prevent freezing caused by the unlikely event that the pump would stay running because the controller is malfunctioning. In this scenario, the solar fluid would always be moving, and moving water is hard to freeze unless it is very cold outside. If it gets that cold where you live, you should be using a closed-loop antifreeze system.

Another limitation is that the collectors have to be located above the drainback tank. This eliminates the option of ground-mounted ar-

A drainback system.

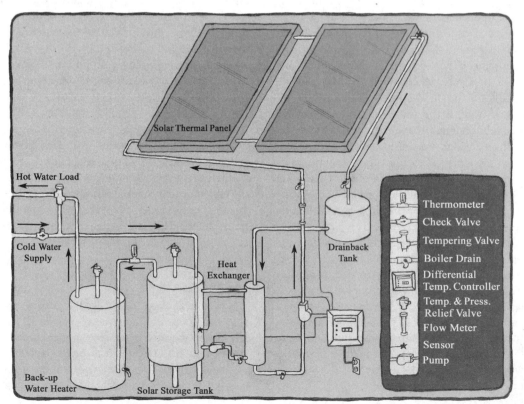

Solar Thermal Panel

Hot Water Load

Cold Water Supply

Heat Exchanger

Drainback Tank

Thermometer
Check Valve
Tempering Valve
Boiler Drain
Differential Temp. Controller
Temp. & Press. Relief Valve
Flow Meter
Sensor
Pump

Back-up Water Heater

Solar Storage Tank

Illustration: Benjamin Nusz

rays. It should also be noted that the high-head pumps that are required to operate these systems use considerably more electricity than the small pumps used on closed-loop antifreeze systems. Pump selection is more critical in drain-down systems, and there are height restrictions on how high a high-head pump can lift water. Unless the circulating pump has a motor over ½ hp, the distance between the water level in the drainback tank and the top of the collectors must be less than 28 feet.

Most manufacturers offer drainback kits that work well for the majority of installations. It is certainly true that most solar water heating systems are the same size and are installed the same way, so kits are fine for most situations. These kits are pre-engineered to take the guesswork out of designing a system. We highly recommend such kits if purchased from a reputable manufacturer. Sizing a drainback system collector array is exactly the same as sizing a closed-loop antifreeze system except you can deduct 5% from the array size to compensate for the higher efficiency of water as the heat exchange fluid versus glycol. To size a drainback tank, calculate all the liquid that would fill the collectors and all the piping above the tank and add 4 gallons.

The operation of the solar loop is straightforward and was described above. To make sure that a system functions properly, you need to follow a few simple rules. The system must be installed to facilitate complete and fast drainage when the system turns off. The collectors must be mounted so that they drain toward the inlet of the array. This would be the bottom manifold where the solar fluid enters the collectors. They should be mounted at a 15-degree angle sloping toward the feed inlet. Collectors should never be mounted so the riser tubes of the absorber plate are horizontal, as they will sag over time and prevent proper drainage. If sagging occurs, water can be trapped in the pipe and burst during freezing conditions. All piping that is not in conditioned space must be sloped at a 15-degree angle toward the drainback tank, and all horizontal pipe runs must be supported at least every 4 feet to prevent any sagging, which would inhibit proper drainage. Horizontal pipe runs that are within the conditioned space can be run at a minimum 10-degree slope toward the drainback tank. Use a pair of 45-degree elbows instead of a 90-degree elbow whenever practical to facilitate faster drainage. The minimum pipe size for drainback systems is ¾-inch hard copper. Never install any valves of any kind on the solar loop.

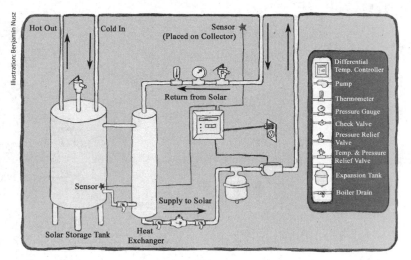

A drainback system with antifreeze, AC powered.

Drainback tanks should always be located inside the conditioned space and can be located at the highest point possible to reduce head pressure. This is an important consideration when the system is installed on a two-story house and the storage tank is in the basement. Drainback tanks should be unvented (except on large space heating systems or large commercial systems). They should be fitted with a sight glass to monitor fluid levels within the tank. These tanks should be well insulated to prevent heat loss.

The solar loop pump must be a high-head pump of sufficient size to raise the water from the drainback tank to the top of the collector array. When calculating the head, measure from the bottom of the drainback tank to the highest point of the collector array and add 4 feet. Your pump must be able to exceed that head. Once all the piping is full of solar fluid, the pump does not have to work very hard, because gravity pulling the fluid back down the return line helps pull fluid up the feed line. It will take a 120-volt AC pump to do this job, so the system must be powered by 120 volts AC and use a differential temperature controller to turn the pump on and off. These systems cannot be powered by photovoltaics, because when the system turns on, the pump must start with full force to overcome the head pressure, and PV-powered pumps don't work that way. The pump should be located at least 3 feet below the bottom of the drainback tank and set in a vertical pipe, pumping up to the collectors.

Drainback systems include a tank to store the solar heated water for later use. There are various methods of getting the heat from the drainback tank to the storage tank, but all use a liquid-to-liquid heat exchanger. One method circulates the hot solar fluid through an in-tank heat exchanger (including the wrap-around

Most solar water heating systems are the same size and are installed the same way, so kits are fine for most situations.

A double-pumped drainback system.

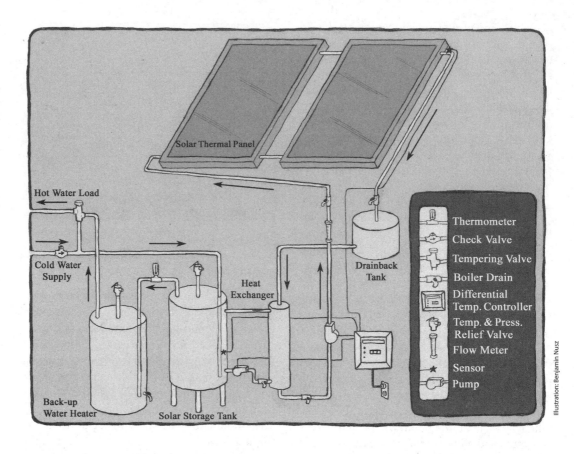

Solar Thermal Panel

Hot Water Load

Cold Water Supply

Drainback Tank

Heat Exchanger

Thermometer
Check Valve
Tempering Valve
Boiler Drain
Differential Temp. Controller
Temp. & Press. Relief Valve
Flow Meter
Sensor
Pump

Back-up Water Heater

Solar Storage Tank

Illustration: Benjamin Nusz

Closed-loop antifreeze systems are the most versatile of all solar water heating systems. They can be installed in virtually all climates and will not freeze.

type). In this design, one pump circulates the hot solar fluid throughout the system; hence, it's called a single-pumped system. Another method uses an external heat exchanger mounted to the storage tank. In this case, a second pump may be needed to circulate the water from the storage tank, through the heat exchanger, and back to the storage tank—this is called a double-pumped system. Both pumps are controlled by the differential temperature controller and turn on and off at the same time. The heat exchanger pump should be very small, because the flow through that circuit should be slow, and because there is very little total head to overcome. The third method of getting heat from the drainback tank to the storage tank is to have a heat exchanger below the minimum fluid level inside the drainback tank. A second pump is required with this system to circulate water from the storage tank, through the heat exchanger, and back to the tank. All these systems work very well when properly installed.

Closed-Loop Antifreeze Systems

Closed-loop antifreeze systems are the most versatile of all solar water heating systems. They can be installed in virtually all climates and will not freeze; collector arrays can be mounted in almost any imaginable way and can be located

at considerable horizontal distances from the exchanger.

Closed-loop antifreeze systems are similar to drainback systems in many ways, but they have a few significant differences. These antifreeze systems use flat-plate or evacuated-tube collectors to heat a solar fluid, usually a high-temperature propylene glycol/water mixture that is circulated through them. Two insulated copper pipes connect the collectors to a heat exchanger. A relatively small circulating pump is installed on the pipe that feeds the collector array, along with an expansion tank, some drain valves, a safety valve, and a gauge. A loop of piping starts at the heat exchanger, travels through the collector array, and then completes the loop traveling back to the heat exchanger. This piping loop is completely sealed and kept completely full of the solar fluid at all times.

These systems can be controlled and powered by a PV panel and a DC pump, or they can use a differential temperature controller operating a 120-volt AC pump. When the pump starts, the solar fluid circulates through the collector array, through the hot supply pipe to the heat exchanger, and then back to the collectors through the return line. When the pump stops, the solar fluid simply stops moving within the closed loop. Because the solar fluid stays in the entire closed loop, from the collector array to the piping that travels through unconditioned space, it must be able to act as a heat transfer fluid and also protect the system from freezing. Please study the previous section that describes the characteristics of antifreeze. Heat transfer from the solar fluid to the domestic water can be accomplished a couple of ways, which are detailed below.

Closed-loop antifreeze systems are the most widely distributed type of solar thermal system worldwide. They are an excellent choice for all climates except where it is hot and the application does not present a reliable and consistent load every day. They are not necessary in climates that do not experience freezing, but they are the only failsafe system for climates where freezing occurs and should be the only choice for climates that experience prolonged or severe cold weather and/or heavy snowfalls. As with drainback systems, most manufacturers make complete system kits that work just fine when properly installed.

The major limitation of closed-loop antifreeze systems is the limitation of the solar fluid. The best antifreeze solution available today is high-temperature formulated propylene glycol. This fluid can eventually break down and form a corrosive solution that can harm system components. High temperatures degrade the fluid, and the rate of degradation is directly proportional to the intensity of the overheating, so the hotter it gets, the quicker it will degrade. All overheating scenarios can be avoided, so this should not be considered a fatal flaw. The best way to reduce overheating is to make sure the system circulates fluid whenever it is sunny. If you have this type of system, it is critical that you check the condition of the solar fluid regularly. Propylene glycol solar fluid solutions need to be changed periodically. Think of it like changing the oil in your car or truck, except that you have to do it only once every 10-15+ years under normal conditions.

Taking note of a few important points will ensure that your closed-loop antifreeze system will perform effectively. The hot supply pipe between the collectors and the heat exchanger must always be made of copper and covered with high-temperature pipe insulation, because under rare circumstances it can get very hot for short periods of time. Copper is the only practical material that can withstand this heat. On residential solar water systems, the return line from the exchanger to the collectors should also be copper, but standard high-quality pipe

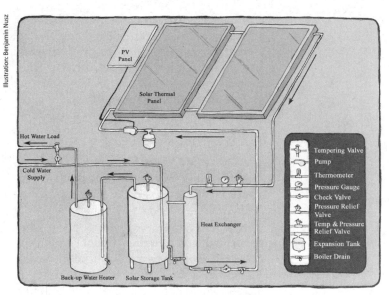

Illustration: Benjamin Nusz

A closed-loop antifreeze system.

insulation can be used on that line. (In large solar space heating systems, the return line can be made of PEX tubing, except for the final 10 feet where it attaches to the collectors.) Collector arrays should be mounted with a slight slope toward the return inlet where the solar fluid enters the array, and all piping should slope slightly toward the heat exchanger. For arrays that are remote, a drain should be placed at the inlet where the solar fluid enters the array to facilitate drainage. All horizontal copper pipe runs should be supported every 5-6 feet, and all vertical runs should be supported every 10 feet. Large space heating systems or any systems that are seasonal or may experience periodic idle times where no load is present require a heat diversion load, sometimes called the shunt load. This shunt diversion load is necessary to prevent the solar fluid from overheating. A shunt loop is typically a buried length of uninsulated pipe or a radiator located outside, preferably in a cool or windy spot. Outdoor hot tubs are also a common shunt load—not a bad idea!

Because fluid remains in the solar loop at all times, a method to prevent thermosiphoning must be included in the solar loop. Thermosiphoning can happen when collectors that are located above the storage tank are colder than the tank, as is the case with roof-mounted arrays. Heat from the storage tank and heat exchanger could rise up the supply pipe to the top of the array, and cold-heavy solar fluid could drop down the return pipe to the heat exchanger from the bottom of the array. Under this undesirable scenario, the fluid is circulating in the opposite direction from when it is heating. A check valve will stop this from happening. If a

The major limitation of closed-loop antifreeze systems is the limitation of the solar fluid. The best antifreeze solution available today is high-temperature formulated propylene glycol.

check valve is used in the system, it should be located between the two boiler drains that are used to charge the system. The check valve will also ensure that the fluid flows correctly when charging the system. If your collectors are not located above the storage tank, and no check valve is required, you may need to install a full-port ball valve between the drains to facilitate charging.

All that said, there are occasions when reverse thermosiphoning is a good thing. For instance, if you are going on an extended vacation (over two weeks), you may want to keep your tank cool to prevent overheating and speeding the degradation of the solar fluid. You can install a vacation bypass around the check valve to allow reverse thermosiphoning to occur. (This method would not be effective if your collectors are located below the storage tank.)

The solar fluid within the solar loop is put under pressure when the system is charged. The pressure is typically set at 32 pounds at 60°F fluid temperature. When the fluid is heated, it expands, raising the pressure within the closed loop. Eventually, the pressure would increase enough to burst the piping. To prevent a catastrophic event, an expansion tank is installed on the loop. The expansion tank will also provide some compensation when pressure in the loop drops as the fluid cools. It is not unusual to see the pressure in the solar loop fluctuate between 10 and 45 pounds seasonally, and sometimes daily. Set the pressure of the expansion tank 3-5 pounds below the pressure in the system at 60°F.

The best location for the expansion tank within the closed loop is directly before the circulating pump in relation to fluid flow. This maximizes flow by compensating for the negative pressure that is created on the suction side of the circulating pump. The expansion tank is the point of no pressure change within the closed loop, so it eliminates the negative pressure or vacuum caused by the pump. It is not absolutely necessary to follow this suggestion, but it will ensure optimum circulation in your system. The expansion tank should always hang below the pipe it is attached to. Heat will accelerate the deterioration of the bladder inside the tank, and by hanging it below the pipe, the tank will stay cooler.

A pressure gauge located in the loop should hang down from the pipe and be visible from the charging drain valves. By hanging below the pipe, no air will get trapped in the fitting or gauge. Unfortunately, most pressure gauges are set up to be above the pipe, so when you install the gauge hanging below it, the scale on the gauge will be upside down. Don't panic, you'll get used to it.

A pressure relief valve is also fitted to the closed loop. Notice that this is a pressure-only relief valve. The relief valve should be set to open at 80-90 pounds. Be sure to install a drain-pipe on the relief valve and terminate the drain near the floor.

There is no static head pressure the solar loop circulating pump must overcome, because all the pipes are filled with fluid at all times. So the only head pressure the pump has to overcome is friction head. Therefore, low-head circulating pumps can be used to circulate the solar fluid through the system. PV-powered pumps can be used here, because the pump can start slow and circulation will start immediately. In fact, a PV-powered system is optimal for several reasons. Most important, PV pumps run whenever the Sun is shining, so there is always circulation in the solar loop when the collectors are hot. This extends the life of the solar fluid by eliminating stagnation when the Sun is shining. PV-powered pumps are also naturally variable speed, so when the Sun's resource is low, the pump runs slower, but the thermal collector's output is also lower under this condition, so the pump speed is perfectly matched to the collector output. On the other side of the coin, under full Sun conditions, the pump runs faster and matches the temperature of the thermal collector by increasing the flow.

A differential temperature controller can also control closed-loop systems. With such a controller, a low-head 120-volt AC pump is used in the solar loop. If a differential temperature controller is used, it's a good idea to disable the high-limit function in the controller to assure

Inside an expansion tank.

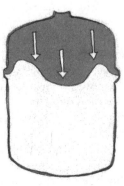

Illustration: Benjamin Nusz

that the pump will circulate at all times that the collector is hot. If the power goes out during a sunny period and a controller is used to regulate the pump, the pump will stop and stagnation can occur, which will promote deterioration of the solar fluid. This is another reason to periodically check the condition of the solar fluid.

Extra care needs to be taken when filling the system with solar fluid. It is essential to get all the air out of the system at that time. As already noted, the solar fluid, once installed into the closed loop, will be in there for at least 10 years under normal circumstances, and air can cause circulation problems, so a good job done at the start will result in many years of trouble-free service. Many installers over the years have specified placing an automatic air vent at the highest point in the closed loop. This practice is not recommended. Automatic air vents fail on a regular basis and should not be used on closed-loop systems. If you still feel that you require a port to bleed air from the system, install a short riser at the high point of your system on the collectors. Install a ball valve on that riser and a Schrader valve on the end of the riser. Air will accumulate in the riser, because it is the high point in the loop. You can open the ball valve, open the Schrader valve to expel any accumulated air, and then close the ball valve.

Some contractors that are used to installing traditional hydronic heating systems will want to put an automatic fill valve on a solar closed-loop system. They do this out of habit because most hydronic heating systems require this valve to keep the system pressurized. Never install an automatic fill valve on a closed-loop system! It will destroy a system by diluting the solar fluid, which can lead to decreased freeze protection.

Closed-loop systems can use a tank-integrated heat exchanger or an external heat exchanger. Tank-integrated heat exchanger systems are the simplest to install and are suited for situations where space is limited. These are always single-pumped systems, as the only pump required is the solar-loop pump. External heat exchanger systems are more efficient and can also be single pumped if the heat exchanger is of the thermo-siphon type. A plate-type heat exchanger or a coiled tube-in-shell heat exchanger can also be used. In these cases, the system must be double pumped. Double-pumped systems can either be PV powered or use a controller and two 120-volt AC pumps. Remember to calculate the correct wire size when running long low-voltage wires between the PV panel and the DC pumps.

Large space-heating systems and large commercial systems often use a closed-loop design. Because of their size, this may be the only option where the collectors cannot be located above the storage tank and heat exchanger. A shunt load must be provided on large systems to reduce overheating during nonload conditions. The world's largest flat-plate solar water heating system in Green Bay, Wisconsin, utilized more than 5,250 4x8-foot collectors (157,689 sq. ft. of collector area). This closed-loop system performed extremely well, producing over 37,500 million BTUs annually. Unfortunately, when natural gas prices fell to record lows during the 1990s, the system was shut down and dismantled. Aside from the issue of substituting a fossil fuel source for a renewable one, this was a shortsighted decision now that natural gas prices are near record-high levels.

Solar water heating makes tremendous ecological and financial sense, now more than ever. Check out the collectors and systems available (see pages 411-413), and for more information, especially if you're thinking about installing your own system, read Bob Ramlow's book *Solar Water Heating: A Comprehensive Guide to Solar Water and Space Heating Systems* (New Society, 2006).

Solar water heating makes tremendous ecological and financial sense, now more than ever.

More Details for Your Best Water Heating Choices

Demand or Tankless Water Heaters

Tankless heaters are of moderate cost initially, feature long (20 years or more) life expectancies, and next to solar, have the lowest operation costs, making them the best choice for most households. We covered operational basics above; what follows are more specifics of operation and some installation tips.

THERMOSTATIC CONTROL

All the demand heater units we sell are thermostatically controlled. At lower flow rates, the gas flame or electric element is reduced automatically to maintain a stable output temperature. The temperature is adjustable at the front panel from roughly 90°F to 140°F (110°-120°F is optimal for most uses). If the water flow rate increases, the heater will respond by increasing the heat input up to its BTU maximum. This makes it perform more like a storage tank heater. The system isn't perfect, because the sudden opening of a second tap still may cause the shower temperature to bobble. But this is also true of the plumbing in most homes with storage tank heaters. We recommend running only a single fixture with the tankless heaters we sell, if that fixture is a shower. We're all touchy about our shower temperature, and this is the best way to ensure a steady temperature. The larger Takagi and Bosch AquaStar units have built-in flow restrictors, so it's difficult to run more water through the unit than it can heat. If too many taps are opened simultaneously, the

water pressure will fall and the water will simply run lukewarm, rather than ice-cold.

LIFE EXPECTANCY OF A TANKLESS WATER HEATER

Buying a tankless heater is like buying a furnace. This appliance is going to give you decades of service. Demand heaters are designed so that all parts are repairable or replaceable. Bosch Aqua-Star warrants their heat exchanger for a full 15 years. No model has any corrosible parts that touch water. The manufacturers have toll-free 800 service numbers backed up with full-time technicians and fully stocked parts warehouses. These heaters should last the rest of your home's life. Compare this with tank-type heaters, which have a 10- to 15-year life expectancy at best.

WHERE SHOULD YOU INSTALL A TANKLESS HEATER?

As with tank-type heaters, the shorter the hot-water delivery pipe, the less energy is lost to warming up the plumbing. However, if you are willing to throw away the heat energy and wait for the hot water to reach the fixture, pipe length doesn't matter. A tankless heater does not need to be installed right at the fixture any more than a storage-type heater does. If you are replacing an existing tank heater, it is probably most convenient to install the tankless unit in the same space. Your water and gas piping are already in the vicinity and will require only minor plumbing changes. If this is new construction, just pick the most central location. As with all gas appliances, the heater requires venting to the outside. In a retrofit installation, the existing flue pipe will probably need to be upsized for the tankless heater. Do not reduce the flue size of the tankless unit to fit a smaller flue already in place. Do not install a tankless heater outside or in an unheated space, unless it never freezes in your climate. If the unit has a pilot light, it can be blown out easily if the unit is exposed to the wind.

SOLAR SYSTEM BACKUP

As we mentioned above, the standard tankless units sense the outgoing water temperature and adjust the heat input accordingly to maintain a steady output temperature. As flow rate or incoming water temperature varies, the heat input will be modulated up or down to compensate.

Energy Conversions

BTU: British Thermal Unit, an archaic measurement, the energy required to raise 1 lb. of water 1°F

1 gal. liquid propane	=	91,500 BTUs
1 gal. liquid propane	=	4 lb. (if you buy propane by the pound)
1 gal. liquid propane	=	36.3 cu. ft. propane gas @ sea level
1 therm natural gas	=	100,000 BTUs
1 cu. ft. natural gas	=	1,000 BTUs
1kW electricity	=	3,414.4 BTUs/hr.
1 hp	=	2,547 BTUs/hr.

The standard Bosch AquaStar units can modulate heat input down to only about 20,000 BTUs and are not particularly recommended for solar backup. The gas-fired Bosch AquaStar "S" units, and all Stiebel-Eltron electric units, can modulate all the way down to zero BTU and are highly recommended if there's any possibility you may use preheated water in the future. The Takagi and Paloma units will also shut off at their preset temperatures if the solar preheated water comes in at that temperature. These units do have high temperature limits, so it is best to limit (with a tempering valve) the incoming temperature to just under the tankless set-points. If the incoming water is already preheated to your selected output temperature, then the tankless heater will come on only briefly. If your preheated water is at 90°F and you've got a 120°F output set, then the water heater will come on just enough to give 120°F output. We think that's pretty slick, and it gracefully slides us into the next good choice, solar water heaters.

Tankless Water Heater Installation Tips

GAS PIPING

Most tankless heaters require a larger gas supply line than the tank-type heater you may be replacing. Just adapting an existing ½-inch supply line to ¾ inch at the heater will choke the fuel supply and limit the water heater output. If the installation manual says "¾-inch supply line," that means all the way back to the gas pressure regulator.

Most of these heaters have a small safety override regulator on the gas inlet. This is a backup for, and in addition to, the standard regulator at the tank or gas entrance. This secondary regulator is also the means for adjusting gas flow at higher altitudes (fully explained in the installation manual). The gas inlet is bottom center for all tankless heaters.

ELECTRIC SUPPLY

Even the smallest electric demand heater will require a dedicated heavy wiring circuit. This isn't something you can plug into an existing outlet, or tap off the existing water-heater circuit. Household-size units will require two or three 50-amp/240-volt circuits. So in a retrofit, you'll be pulling new 6-gauge wire. Make sure you've got the space for breakers and the electrical capacity to run this much power.

PRESSURE/TEMPERATURE RELIEF VALVE

Unlike conventional tank-type heaters, there may not be a P/T valve port built into the heater. This important safety valve must be plumbed into your hot-water outlet during installation. Simply tee into the hot-water outlet.

VENTING

Tankless heaters must be vented to the outside. All models come with the draft diverter installed. Bosch AquaStar tankless heaters use conventional double-wall Type B vent pipe. This is the same type of vent used with conventional water and space heaters. The Bosch AquaStar FX Model l, and all Takagi heaters, use inexpensive 4-inch single-wall vent pipe, due to their power-vented exhaust. Vent pipe is not included but is readily available at plumbing supply, building supply, or hardware stores. The vent piping used with most of these heaters is larger than tank-type vent piping. Do not adapt down to an existing vent size! Replace the vent and cut larger clearance holes as necessary if doing a retrofit. Venting may be run horizontally as much as 10 feet. Maintain 1 inch of rise per foot of run. Bosch AquaStar recommends 10 feet of vertical vent at some point in the system to promote good flow.

INSTALLATION LOCATION

Don't install your tankless heater outside unless it never freezes where you live. Freeze damage is the most common repair on these heaters. Don't tempt fate by trying to save a few bucks on vent pipe. If the heater is installed in a vacation cabin, put tee fittings and drain valves on both the hot and cold plumbing directly below the heater to ensure full drainage. The heat exchanger that works so well to get heat into the water works just as well to remove heat.

WARNING

It is illegal and dangerous to install a tankless water heater in an RV or trailer unless the heater is certified for RV service. The following chart will help you plan your tankless heater installation before you buy. Bosch AquaStar heaters have zero clearance on the back for wall mounting. The Takagi heaters need 1-inch clearance on back, provided by the included wall-mount brackets.

Tankless Water Heater Clearance to Combustibles Chart					
Heater	BTUs/hr.	Top	Bottom	Sides	Front
Bosch AquaStar 1000	40,000	12"	12"	4"	4"
Bosch AquaStar 1600	117,000	12"	12"	4"	4"
Takagi T-K Jr.	140,000	12"	6"	2"	24"
Takagi T-K3	199,000	12"	12"	2"	4"

Ask Real Goods

Is it possible to use one of the tankless heaters on my hot tub?

With some ingenuity and creativity, this has been done successfully. You need to bear in mind that these heaters are designed for installation into pressurized water systems. For safety reasons, they won't turn on until a certain minimum flow rate is achieved. The flow rate is sensed by a pressure differential between the cold inlet and the hot outlet. Tankless heater flow rates are 3 gallons per minute and less, which is a much lower flow rate than hot tubs. A hot tub system is very low pressure but high volume. What usually works is to tee the heater inlet into the hot tub pump outlet, and the heater outlet into the pump inlet. This diverts a portion of the pump output through the heater, and the pressure differential across the pump usually is sufficient to keep the heater happy. Bosch AquaStar has a very inexpensive "recirc kit" for its heaters that lowers the pressure differential required. This kit is mandatory! In addition, you'll need an aquastat to regulate the temperature by turning the pump on and off (your tub may already have one if a heating system already was installed). You'll also need a willingness to experiment and a good dose of ingenuity.

Fuel Choices for Water Heating: Electric vs. Gas

Propane costs about 1½ times as much as natural gas, but it's still less expensive than electricity in some areas.

Electricity is wonderfully useful for many tasks, but using it for larger-scale heating chores, such as household water and space heating, isn't a smart choice. While inexpensive and easy to install initially, electric tank heaters cost 25%-50% more to operate than natural gas-fired heaters. Over an average 13-year lifetime, this amounts to a $1,500-$2,500 difference. See our Cost Comparison chart below for details of electric, gas, tankless, and solar water heating options. If you have natural gas available, by all means use it. Over the lifetime of the heater, you can save double the initial purchase and installation cost. In many rural homes where natural gas is not available, the higher lifetime cost when using electricity may be an argument for bringing propane fuel on site or for investing in an efficient heat pump water heater. Propane costs about 1½ times as much as natural gas, but it's still less expensive than electricity in some areas. Since energy prices are skyrocketing and vary quite a bit depending on location and climate, the chart below should be used as a guide only. It is best to call a Real Goods technician to choose the most economical application for your situation. Note that the best way to mitigate the uncertainty of rising energy costs is to go solar, as the chart suggests.

COST COMPARISON FOR WATER HEATERS AT 20 YEARS

Type	Cost[1]	Cost per Year[2]	Life (Years)	20-Year Cost[3]	CO_2 (tons)[4]
Standard Electric Tank (90% Energy Factor)	$1,600	@10¢/kWh = $555 @12¢/kWh = $665 @14¢/kWh = $775	13	$14,300 $16,500 $18,700	85 — —
High-Efficiency Electric Tank (95% Energy Factor)	$1,400	@10¢/kWh = $526 @12¢/kWh = $631 @14¢/kWh = $736	13	$13,600 $15,700 $17,800	81 — —
Standard NG Tank (57% Energy Factor)	$1,750	@$1.40/therm = $419	13	$11,900	36
High-Efficiency NG Tank (65% Energy Factor	$1,600	@$1.40/therm = $367	13	$10,800	32
Standard LP Tank (57% Energy Factor)	$1,750	@$2/gal. = $654	13	$16,600	36
High-Efficiency LP Tank (65% Energy Factor)	$1,600	@$2/gal. = $573	13	$15,000	32
Electric Tankless (100% Energy Factor)	$1,850	@10¢/kWh = $499 @12¢/kWh = $599 @14¢/kWh = $699	20	$11,800 $13,800 $15,800	77 — —
NG Tankless (80% Energy Factor)	$2,550	@$1.40/therm = $298	20	$8,500	26
LP Tankless (80% Energy Factor)	$2,550	@$2/gal. = $466	20	$11,900	26
[5]Solar w/ Electric Tank Backup	$5,050	@10¢/kWh = $158 @12¢/kWh = $189 @14¢/kWh = $221	20	$8,200 $8,800 $9,500	24 — —
Solar w/ NG Tank Backup	$5,100	@$1.40/therm = $110	20	$7,300	10
Solar w/ LP Tank Backup	$5,100	@$2/gal. = $172	20	$8,500	10
Solar w/ Electric Tankless Backup	$4,900	@10¢/kWh = $150 @12¢/kWh = $180 @14¢/kWh = $210	20	$7,900 $8,500 $9,100	23 — —
Solar w/ NG Tankless Backup	$4,900	@$1.40/therm = $89	20	$6,700	8
Solar w/ LP Tankless Backup	$4,900	@$2/gal. = $140	20	$7,700	8

[1]Approximate installed cost (net of federal income tax credit where applicable).

[2]Energy cost based on family of four each using 20 gal./day of 60°F water heated to 130°F. The utility rates are 2006 averages.

[3]Future operating costs neither discounted nor adjusted for inflation. Includes replacement cost for systems with less than 20-year life expectancy.

[4]Estimated amount of CO_2 produced by the system over 20 years.

[5]Solar covers 70% of total energy required.

Source: Adapted from American Council for an Energy-Efficient Economy life-cycle charts.

WATER HEATING PRODUCTS

Tankless Water Heaters

BOSCH AQUASTAR TANKLESS WATER HEATING

Bosch, a leading manufacturer of heat exchangers for more than 60 years, sells more than 1.5 million tankless water heaters each year. These are our most popular units, providing hot water for average-sized homes with natural gas and propane water heaters. They are capable of delivering a constant supply of hot water at the temperature you select and have excellent energy-efficiency ratings that range from 80% on the 1000 and 1600 models to 87% on the 2400 models. Our smallest 1000P is appropriate for hand-washing applications, the 1600 series is great for one major hot-water application at a time (such as a shower or a load of laundry), and the 2400 units are the best for larger families requiring two major hot-water applications at a time. 12-yr. mfr. warranty on heat exchanger; 2 yr. on everything else.

Bosch AquaStar 1600

Model 1000P

The Bosch 1000P is the perfect water heater for a small cabin with a low-flow shower or anywhere the demand for hot water is low. Will run a shower so long as incoming water temperature is 55°F or above. The model 38 is equipped with a piezo igniter. Features an 80% efficiency rating. Specify propane (LP) or natural gas (NG).

45-101	**1000P**	**$339**

Model 125 and 1600 Whole-House Gas Tankless Water Heaters

The Bosch Model 1600 units can support a single shower or multiple smaller fixtures simultaneously. The first of its kind, the 1600H incorporates a revolutionary hydro-generated ignition system, allowing the heater to operate without a standing pilot, electricity, or a battery—using energy from the flowing water to ignite the spark. No electric plug necessary. 25¾"x16¾"x8½" (HxWxD). 33 lb. The 1600P has the same flow/output specifications but uses a standard pilot light. 24¾"x16¾"x8½" (HxWxD). 39 lb. The 125BS is the "solar" version of the 1600P, and it accepts solar or wood stove preheated water. 29¾"x18¼"x8¾" (HxWxD). 42 lb. The 125BO is an outdoor-rated unit. 36¾"x18¼"x10½" (HxWxD). 57 lb. The 125FX features a built-in power vent and electronic ignition. It is ideal for residential use when the installation calls for horizontal venting. Comes with FXHOOD in separate box. 29¾"x18¼"x8¾" (HxWxD). 44 lb.

45-114	**1600H**	**$669**
45-105	**1600P**	**$549**
45-113	**1600PS**	**$659**
45-0137	**125BO**	**$669**
45-003	**125FX**	**$839**

Model 2400E Whole-House Gas Tankless Water Heaters

The Bosch Model 2400E is for families that need to operate two major hot-water applications at a time, and your family will thank you for years to come for the endless supply of hot water. This heater has an average life expectancy of more than 20 years, offers you energy savings of up to 50%, is easy to install, and has multiple venting options. Also available in outdoor-rated unit 2400EO (no vent necessary on outdoor units). 23½"[26" on 2400EO]x15¾"x8½" (HxWxD). 49 lb.

45-0123	**2400E**	**$1,149**
45-0132	**2400EO**	**$1,149**

AquaStar Specifications

Model	1000P	125BO	1600H/ 1600P	1600PS	125FX	2400E/EO
BTU Input	40,000	117,000	117,000	117,000	130,000	175,000
GPM@55°F Temp. Rise	1.2	3.4	3.4	3.4	3.8	5.2
GPM@77°F Temp. Rise	0.8	2.4	2.4	2.4	2.7	3.7
GPM@90°F Temp. Rise	0.5	2.1	2.1	2.1	2.3	3.2
Min. Flow (gpm)	0.5	0.6	0.6	0.5	0.5	0.8
Vent Size	4"	N/A	5"	5"	4"	3"/N/A
Water Connections	½"	½"	¾"	½"	½"	¾"
Min. Gas Supply	½"	¾"	¾"	½"	¾"	¾"
Min. Water Pressure (psi)	13	30	30	30	30	30
Order #	45-101	45-0137	45-114/ 45-105	45-113	45-003	45-0123/ 45-0132
Price	$339	$669	$669/ $549	$659	$839	$1,149

TAKAGI TANKLESS WATER HEATERS

The most powerful, efficient, and adaptable water heaters available. The gas-fired Takagi water heaters feature electronic ignition, powered ventilation, computer-modulated gas flame for steady output temperatures, and the industry's best energy factors of up to 86. All waterways are noncorrosive copper or brass, burners are stainless steel, and the cabinet is powder-coated for long life. Automatic freeze protection will protect down to 5°F with no wind (don't tempt fate). The new T-K3 protects down to -22°F. The optional remote control allows greater temperature control and displays any self-diagnosing error codes if a problem develops. The stainless wall vent terminator is recommended for horizontal vents. The back-flow prevention kit is required for hard-freezing climates. AGA-approved. 10-yr. mfr. warranty on heat exchanger; 5 yr. on all else (requires licensed installer). Japan.

Model T-K Jr.

At just 20 inches high and weighing only 30 pounds, the T-K Jr. is the most compact unit in the Takagi line. Designed to produce endless hot water and radiant heating for smaller homes, the T-K Jr. uses the same innovative technology as the original Takagi units—only on an even smaller scale. Vent can be run up to 35 feet vertically or horizontally, with a 5-foot reduction for every elbow. Maximum of 3 elbows.

Model T-K3

Takagi's most versatile and powerful residential model. Connect up to four units to meet the demands of even the largest homes! The compact T-K3 measures under 14 inches wide, so it will fit snuggly between two wall studs to save space. But the T-K3 is also powerful. With Takagi's new easy-link technology, the T-K3 will meet the hot-water needs of most high-volume residential applications. Vent can be run up to 50 feet vertically or horizontally, with a 5-foot reduction for every elbow. Maximum of 5 elbows.

45-0108 LP	**Takagi T-K Jr. Propane**	**$849**
45-0108 NG	**Takagi T-K Jr. Natural Gas**	**$849**
45-0500 LP	**Takagi T-K3 Propane**	**$1,249**
45-0500 NG	**Takagi T-K3 Natural Gas**	**$1,249**
45-0501	**T-K Jr. Remote Control Option (TK-RE02)**	**$205**
45-0502	**T-K3 Remote Control Option (TK-RE10)**	**$205**
45-0503	**Takagi Wall Ventilator (TK-TV01)**	**$99**
45-0404	**Takagi Exhaust Backflow Prevention Kit (TK-TV03)**	**$69**
45-0405	**Takagi Outdoor Ventilator Cap (TK-TV04)**	**$69**
45-0406	**Takagi Wall Ventilation Terminator for DV (TK-TV05)**	**$159**

Model	T-K Jr.	T-K3
BTU Input Range	NG 19,500-140,000 LP 17,000-140,000	NG 11,000-199,000 LP 11,000-199,000
Flow Rating	0.75-5.8 gpm	0.5-7.0 gpm
First-Hour Rating	181 gal./hr. 3.0 gpm 77°F rise	258 gal./hr. 4.3 gpm 77°F rise
Vent Size	4" Category III s/s	4" Category III s/s
Water Connections	¾" NPT	¾" NPT
Minimum Gas Supply	¾" NPT	¾" NPT
Water Pressure	Min. 15 psi, Max. 150 psi	Min. 15 psi, Max. 150 psi
NG Pressure Inlet	Min. 5"WC, Max. 10.5"WC	Min. 5"WC, Max. 10.5"WC
LP Pressure Inlet	Min. 11"WC, Max. 14"WC	Min. 8"WC, Max. 13.5"WC
Manifold Pressure	NG 2.8"WC, LP 3.2"WC	NG 2.5"WC, LP 4.4"WC
Electrical Supply	120V, 60 Hz, 61W	120V, 60 Hz, 92W
Weight	30 lb.	40 lb.
Dimensions (HxWxD)	20"x14"x6"	20.5"x14"x8.5"
Energy Factor	NG 0.81, LP 0.81	NG 0.84, LP 0.86

Paloma Indoor and Outdoor Water Heaters

These tankless water heaters by Paloma, an industry leader in its fourth generation of ownership, are the ideal choice when larger demand or higher flow rates are a priority. The WaiWela tankless heaters feature two unique safety features. The first is an oxygen-depletion sensing system that provides continuous monitoring for carbon monoxide, emission control through automatic fan-speed adjustment, and system shutoff, if necessary. The second safety feature is a film-type heat-sensing system that will detect excessively high temperatures at any location around the combustion chamber. These power-vented heaters feature electronic ignition (no standing pilot light), an 84% energy rating, scale-buildup warning system, and freeze protection down to -30°F. Both units come with wall-mount hardware, draft-back-flow preventer, relief valve, and a digital, water-resistant remote control. The indoor unit also comes with a 10-foot 120-volt AC power cord and a washable air filter. Optional additional remote controllers are available (main and additional bath). 10-yr. mfr. warranty on heat exchanger; 3 yr. on parts. 25⅜"x14⅞"x11⅛" (HxWxD). 55 lb. Japan.

INDOOR UNITS

45-0127	PH-28RIFSN (Natural Gas)	$1,249
45-0127	PH-28RIFSP (LP Gas)	$1,249

OUTDOOR UNITS

45-0128	PH-28ROFN (Natural Gas)	$1,249
45-0128	PH-28ROFSP (LP Gas)	$1,249

Paloma Indoor and Outdoor Tankless Water Heaters

BTU Input Range	19,000-199,900
Max. Flow @ 45°F Temp. Rise	7.4 gpm
Max. Flow @ 50°F Temp. Rise	6.7 gpm
Max. Flow @ 60°F Temp. Rise	5.6 gpm
Max. Flow @ 70°F Temp. Rise	4.8 gpm
Max. Flow @ 80°F Temp. Rise	4.2 gpm
Max. Flow @ 90°F Temp. Rise	3.7 gpm
Minimum Flow	0.66 gpm
Hot-Water Adjustment Range	
Main	100°F~120°F (max. 140°F)
Shower	100°F~120°F
Vent Size	4" (N/A for outdoor)
Water Connections	¾" NPT Male
Gas Connection	¾" NPT Female (shutoff valve incl.)
Water Pressure	14 psi (min.); 150 psi (max.)
Inlet Gas Pressure	
Natural Gas	4.0"WC (min.); 10½"WC (max.)
LP Gas	8.0"WC (min.); 14.0"WC (max.)
Electrical Supply	120VAC/60Hz, less than 2A
Max. Vent Lengths	
With one 90° elbow	47.5 feet
With two 90° elbows	42.5 feet
With three 90° elbows	37.5 feet

For indoor unit only: Vent piping and wall vent terminator (for horizontal venting) or roof flashing kit (for vertical venting) not included. Ask for details.

Stiebel-Eltron Electric Water Heaters

When electricity is your choice for water heating, these high-quality tankless heaters will save you 15%-30% in energy by eliminating standing losses. We're offering four models, ranging from a small hand-washing heater to a large full-scale house heater. All feature automatic antiscald electronic controls that won't let the output temperature exceed 130°F. All but the smallest model have front-panel digital temperature control. Simply dial the output temperature you need from 86°-125°F. Advanced microprocessor control with multiple temperature sensors, flow sensors, and safety thermal cut-offs assures the output temperature never deviates. All electrical and water connections are clearly labeled and easy to access. These units require an adequately sized electrical service. Your electrician can determine this for you. 3-yr. mfr. warranty with North American distributor and parts warehouse. Germany.

Tempra 29

Stiebel-Eltron Specifications

Model	DHC 3-1	DHC-E 10	Tempra 20	Tempra 29
Recommended Use	Hand washing	Summer cabin	Modest house	Full-size house
Voltage	120V	240V	240V	240V
Amperage	25A	40A	80A	120A
Wattage	3kW	9.6kW	19.2kW	28.8kW
Min. Breaker Size	30A	50A	2x50A	3x50A
Wire Size AWG Copper	10	8	2x8	3x8
Min. Flow to Activate (gpm)	0.32	0.45	0.58	0.87
Max. Temp Increase @ gpm/°F	0.32/65	0.75/87	2.25/58	2.25/87
	0.5/41	1.0/65	3.0/44	3.0/65
	0.75/27	1.5/44	4.5/29	4.5/44
Water Connections	½" NPT male	½" NPT male	¾" NPT male	¾" NPT male
Size HxWxD (inches)	14.2x7.8x4.1	14.2x7.8x4.1	14.2x14.6x4.6	14.2x21.75x4.6
Weight (pounds)	4.6	5.9	21.0	24.25
Order #	45-0109	45-0110	45-0111	45-0112
Price	$190	$275	$495	$675

Other Water Heating Products

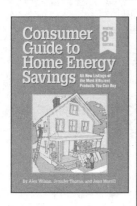

Consumer Guide to Home Energy Savings, 8th Ed.

Energy efficiency pays. Written by the nonprofit American Council for an Energy-Efficient Economy, the *Consumer Guide* has helped millions of folks do simple things around their homes that save hundreds of dollars every year.The new 8th Edition has up-to-date lists of all the most energy-efficient appliances by brand name and model number, plus an abundance of tips, suggestions, and proven energy-saving advice for home owners. 264 pages, softcover.

21-0344 Consumer Guide to Home Energy Savings, 8th Ed. $9.95

D'mand Water-Saving System

No more waiting for hot water!
Turn the hot-water tap on to wash your face, take a shower, or create a warm gravy for your dog's kibble, and you often have to wait a few minutes for the hot water to arrive at the faucet. Not only is it inconvenient, but it also wastes 3-7 gallons of water—about the same amount of water used to flush a toilet—every time you play this waiting game.

An on-demand system such as Real Goods' Hot Water D'mand System gets hot water to your fixtures four to five times faster with none wasted down the drain, and it doesn't require any plumbing modifications. While tankless water heaters are often referred to as "instantaneous," the D'mand System works with tank-type or tankless water heaters to send hot water right to the tap.

It's easy to install—simply place a small pump at the faucet furthest from your hot-water heater. This innovative pump redirects lukewarm water, which would have been wasted, back to the hot-water heater through the cold-water line while pumping hot water up the hot-water line. The pump is activated when a doorbell-type button is pushed at the tap.

With the push of a button, the D'mand System whisks hot water to the most remote faucet in seconds. The displaced cold water is pumped into the cold-water plumbing, returning to the heater, so there's no waste. It shuts off automatically when hot arrives and costs less than $1 per year to operate. UL- and UPC-listed, this system is recognized by the Department of Energy as both a water- and power-saving device and has a life expectancy exceeding 15 years. It works with any type of water heater.

Our "no sweat" installation kit provides all the necessary plumbing bits and pieces, installs with common hand tools, and plugs into a standard outlet. Installed with a push button and pump at your furthest faucet, intervening faucets can be equipped with additional wireless push buttons to activate the system from up to 100 feet. Longer plumbing runs need a larger pump to keep wait time under 20-30 seconds. Use S-50 model for runs up to 50 feet, S-70 for runs up to 100 feet. S-02 model is for longer or commercial use. S-50 model has 3-yr. mfr. warranty; other models are 5-yr. USA.

45-0100 S-50-PF-R Kit		**$299**
45-0101 S-70-PF-R Kit		**$399**
45-0102 S-02-PF-R Kit		**$549**
45-0103 Additional Transmitter		**$20**

D'mand S-02-PF-R

D'mand S-50-PF-R

D'mand S-69-PF-R

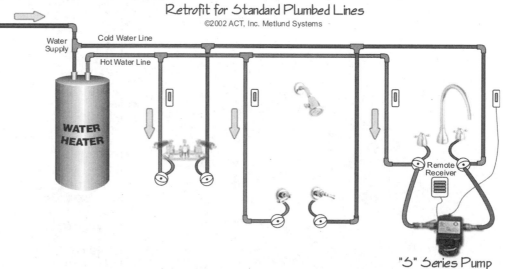

The Metlund® Hot Water D'MAND® System
Retrofit for Standard Plumbed Lines
©2002 ACT, Inc. Metlund Systems

"S" Series Pump

Snorkel Hot Tub Heaters

Snorkel stoves are wood-burning hot tub heaters that bring the soothing, therapeutic benefits of the hot tub experience into the price range of the average person. Snorkel stoves are simple to install in wood tubs, are easy to use, are extremely efficient, and heat water quite rapidly. They can be used with or without conventional pumps, filters, and chemicals. The average tub with a 450-gallon capacity heats up at the rate of 30 degrees Fahrenheit or more per hour. Once the tub reaches the 100°F range, a small fire will maintain a steaming, luxurious hot bath.

The stoves are made of heavy-duty, marine-grade plate aluminum that is powder coated for the maximum in corrosion resistance. This material is very lightweight, highly resistant to corrosion, and strong. Aluminum is also a great conductor of heat—three times faster than steel. The stoves are supplied with secure mounting brackets and a sturdy protective fence.

Two stove models are offered: the full-size 120,000 BTUs/hr. Snorkel and the smaller 60,000 BTUs/hr. Scuba. The Snorkel is recommended for 6- and 7-foot tubs. The Scuba is for smaller 5-foot tubs. These stoves are for wooden hot tubs only.

We also offer Snorkel's precision-milled, long-lasting western red cedar hot tub kits that assemble in about four hours. A 5-foot tub holds two to four adults; a 6-foot tub holds four to five adults. The 7-foot tub is for large families or folks who entertain frequently. USA.

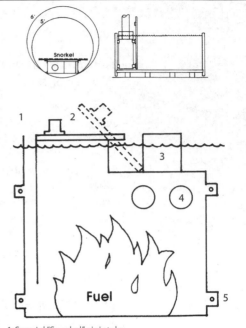

1. Special "Snorkel" air intake.
2. Sliding, cast aluminum, tilt-up door that services the fire box as well as regulates the air intake.
3. Sunken stack to increase efficiency.
4. Heat exchange tubes.
5. Mounting brackets for securing fence and stove.

45400	Scuba Stove	$763.89
45401	Snorkel Stove	$875
45002	Cedar Hot Tub 5'x3'	$1,500
47413	Cedar Hot Tub 6'x3'	$1,833.33
45208	Cedar Hot Tub 7'x3'	$2,595
45504	Drain Kit, PVC 1.5-in.	$48
45501	6-ft. Tub 3-Bench Kit	$316.67
45502	Snorkel Bench 5-ft. Set of 3	$284.72
45503	Snorkel Bench 7-ft. Set of 3	$333.33

Shipped freight collect from Washington State.

SOLAR WATER HEATING

4-Gallon Solar Shower

No more cold showers when you're camping. This incredible low-tech invention uses free solar energy to heat water for all your washing needs. Large 4-gallon capacity provides ample hot water for several hot showers. On a 70°F day, the Solar Shower will heat 60°F water to 108°F in only three hours for a tingling-hot shower. Great for camping, car trips, or emergency use. Taiwan.

17-0169 Solar Shower   **$19**

ProgressivTube Solar Water Heater

The best choice for mild climates

This passive batch-type heater is your most cost-effective water heater in climates that don't regularly freeze.* This is the simplest, yet most elegant and durable water heater we've seen. Plumbed inline ahead of your conventional water heater's cold inlet, it will preheat all incoming water, reducing your conventional heater's workload to near zero. Rated at approximately 30,000 BTUs/day, the 4x8-foot collector is housed in a top-quality, bronze-finished aluminum box with stainless and aluminum hardware. There are no rusting components. The dual glazing has tempered low-iron solar glass on the outside, and non-yellowing and nondegrading 96%-transmittance Teflon film on the inside. High-temperature, nondegrading phenolic foam board insulation is used on the sides (R-12.5), between internal tubes (R-12.5), and the bottom (R-16.7).The collector is composed of large 4-inch-

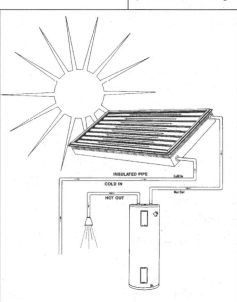

diameter selective-coated copper pipes aligned horizontally. They are connected in series, with cold water introduced at the bottom, and hot water taken off at the top. As the water warms, it stratifies and gradually works its way toward the top. The outgoing hot water is never diluted by incoming cold water. There are no pumps, sensors, controls, or other moving parts that can fail. 30-, 40-, and 50-gallon models are produced by varying the number of collector pipes inside. Exterior size is 3x8 feet for the 30-gallon model, and 4x8 feet for the 40 and 50 gallon models. BTUs collected per day under average North American Sun conditions are 22,100 for the PT-30, 28,400 for the PT-40, and 28,700 for the PT-50, according to FSEC.

The collector and mounting system have been officially tested and approved for 180-mph winds, and 250 units installed in St. Croix were tested unofficially by the 200+-mph winds of Hurricane Hugo in 1989. Only six units suffered minor damage, and that was due to flying debris.

Use caution if roof-mounting a ProgressivTube collector. Filled, they weigh close to 600 pounds. Rafters may need to be doubled up. Ground mounting is also a choice.

Both fixed (roof angle) and tilting mounts with 36 inches of cut-to-fit back leg tubes are offered. The two-way valve kit includes drain valves, a very important tempering valve, 150 psi pressure relief valve, and quality ball valves, allowing you to choose between solar preheating or solar bypass. ProgressivTube has a great 10-yr. warranty. USA. *An overnight freeze won't threaten this heater, but three to four days of temps in the 20s would. Supply and return plumbing is much more vulnerable. Plumbing runs need to be short and well-insulated, and/or you need to be prepared and willing to drain the system seasonally.*

45085	**PT-30**	**$1,550†**
45094	**PT-40**	**$1,799†**
45089	**PT-50**	**$1,989†**
45-525	**Basic Flush Mount Kit**	**$15**
45095	**Fixed Mounting**	**$75**
45096	**Tilt Mounting**	**$99**
45097	**Valve Kit**	**$250**

†Shipped freight collect from Sarasota, Florida. Collector pricing includes $50 crating charge.

Heliodyne Solar Hot Water Systems

Since 1976, Heliodyne has offered the highest-quality, closed-loop solar hot water systems and employs a food-grade glycol/water mix to prevent freezing in all climates. The PV-powered, grid-independent system operates directly from the Sun, and the brighter it shines, the faster the circulation pump runs. The AC system features thermostatically controlled pumps and a double-wall heat exchanger that meets even the strictest codes. Both systems' heat exchangers work with any off-the-shelf electric water heater tank of appropriate size (use 1.5 gal. of storage per sq. ft. of Gobi collector), or with our self-cleaning Energy Smart heater tanks.

Gobi collectors feature a black chrome surface and all-copper absorbers that bond to the risers with a non-delaminating bond, creating amazing heat transfer and high output in winter and summer. Dyn-O-Seal unions with no-solder connections make for quick and easy installation. Glazing is double-strength tempered glass with an antiglare finish held in place with a weather-tight, compression-fitted seal. Frames are exceptionally sturdy bronze-anodized aluminum, and the collectors are only 4 inches in profile.

Both systems include Gobi collectors, flush mount kits, Dyn-O-Seal unions, tempering valves and air vents (for scald prevention and ease of system filling/bleeding), and Dyn-O-Flo (inhibited, propylene glycol with an operating range of -50°F to 325°F). The PV system includes an HP Helix PV pump station (20W PV with DC pump and stainless steel, thermosiphon heat exchanger assembly). The AC system includes a preplumbed and wired HP 16 AC heat transfer appliance (double-wall heat exchanger assembly with differential control and sensors). Also included in both kits are an expansion tank, pump(s), pressure relief valve and gauge, check valve, thermometers, and installation instructions. Optional 50- and 80-gallon Energy Smart heaters feature a pair of 4.5kW elements, 3 inches of insulation, and a sophisticated controller that heats to usage patterns. The 120-gallon solar storage tank has a single 4.5kW element and 2-inch insulated walls. All systems include everything needed except ¾-inch copper supply and return piping and pipe insulation.

Systems are SRCC OG300 Certified to qualify for Federal and State tax credits.

STORAGE TANKS

45-0129	**50-gal. Solar Storage Tank**	**$586**
45-0130	**80-gal. Solar Storage Tank**	**$1,089**
45-0153	**120-gal. Solar Storage Tank**	**$1,399**

SOLAR COLLECTORS

45512	**3366 (3½'x7½') Gobi Collector**	**$975**
45498	**408 (4'x8') Gobi Collector (min. 2 per order)**	**$1,089**
45500	**410 (4'x10') Gobi Collector**	**$1,295**

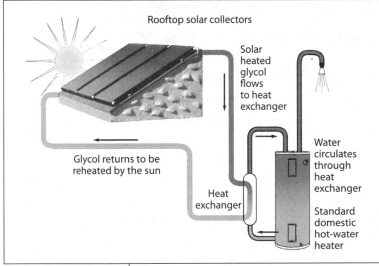

Rooftop solar collectors

Solar heated glycol flows to heat exchanger

Glycol returns to be reheated by the sun

Heat exchanger

Water circulates through heat exchanger

Standard domestic hot-water heater

Solar Water Heating Kits

	Helio-Pak Helix PV System	Helio-Pak 16 AC System	Includes	Recommended
1-2 People Light Use	$3,449 45-0147	$2,549 45-0150	1x3366 Collector	50-gal. Solar Storage Tank
2-4 People Average Use	$4,449 45-0148	$3,489 45-0151	2x3366 Collectors	80-gal. Solar Storage Tank
4-6 People Heavy Use	$4,649 45-0149	$3,689 45-0152	2x408 Collectors	120-gal. Solar Storage Tank

Heliodyne Solar Hot Tub Heaters

Our solar hot tub kit comes complete with a top-quality solar collector, flush-mount hardware, 120-volt

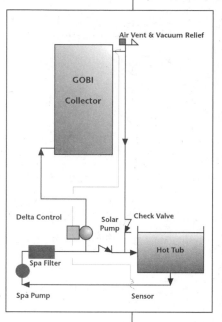

AC controller, pump, temperature sensors, and air vent. You add the tub and plumbing to suit your site. These kits are designed to mount the collector on a nearby roof and connect to existing hot tub plumbing in such a way that all water can drain from the collector and pipes when not operating. Automatic operation is provided by differential temperature sensors and the controller. The system will run only if the tub needs heat, and there's heat to be had at the collector. Tub temperature is user adjustable, with an upper limit of 105°F. The pump is not self-priming and must be installed below tub water level. It will lift water up to two stories. The smaller 3366 collector kit can deliver 18,000-30,000 BTUs/day.

With reasonable tub insulation, it will maintain daily tub temperature in North American climates and will bring smaller to medium-size tubs up to temperature within a few days after a water change. The larger 4'x10' kit can deliver 29,000-46,000 BTUs/day. Use it for larger tubs or faster unassisted recovery times. USA.

45000   **Solar Hot Tub Kit 3366**   **$1,516**

45209   **Solar Hot Tub Kit 410**   **$1,742**

Shipped by truck from San Francisco Bay area. System pricing includes $75 crating charge.

HELIODYNE OPTIONS

Solar Collection Fluid

A propylene-glycol solution with extra-strength corrosion inhibitors for high-temperature use. Nontoxic, nonflammable, FDA-approved, safe for human contact. Every system will use 1-2 gallons, diluted 40%-60%. Operating range is -50°F to 325°F. Should be changed about every five years. USA.

45100   **Dyn-O-Flo, four 1-gal. bottles**   **$114**

45515   **Dyn-O-Flo, single 1-gal. bottle**   **$40**

Blind Unions

Every system needs a pair of these to block off the two unused unions after the plumbing is assembled. (The complete PV-powered systems are supplied with these already.)

45495   **Heliodyne Blind Unions, 1 Pr.**   **$20**

Flush Mount Hardware

This hardware kit will mount one collector parallel to the roof, with an attractive stand-off clearance for code compliance and longest roof life. (The complete PV-powered systems are already supplied with these.) See the tilting rack options on pages 127-128 if your collector needs to be tilted in relation to the roof surface.

45496   **Heliodyne Flush Mount Kit**   **$57**

BOOKS

Solar Water Heating

Add solar water heating to your home and you'll reap huge economic and ecological benefits. This definitive guide shows how it's done. Covers system basics, including issues like system size and siting. This guide is also packed with detailed looks at components, installation, operation, and maintenance, not to mention topics like space and pool heating. It's the only book novices need and the only one professionals want. By longtime solar heating experts Bob Ramlow and Benjamin Nusz. 288 pages, softcover.

21-0395 Solar Water Heating   **$24.95**

Solar Hot Water Systems, Lessons Learned

Tom Lane probably has more hands-on practical experience from 1977 to 2002, and knowledge with all kinds of solar hot-water systems, than anyone on the planet. He is detailed, specific, and refreshingly honest about the good and bad points of commonly available solar hardware. Extensive use of pictures, drawings, and system schematics help make explanations clear. Beautifully organized with a detailed table of contents. Pick what you need. 120 pages, softcover.

21-0336 Solar Hot Water Systems,
           Lessons Learned   **$29**

30th
ANNIVERSARY

CHAPTER 9

Water and Air Purification

Clean Water and Air Are Human Rights!

IF YOU ARE CONCERNED about the degradation of our environment, you are probably also conscientious about the foods you eat. If you are conscientious about the foods you eat, then you may also want to pay attention to the quality of the air you breathe and the water you drink. Like food, air and water can carry into your body the contaminants that saturate the natural world. But you may not realize the extent to which your source of domestic water and the air inside your house may be polluted. This chapter examines the bad stuff that may be lurking in your air and water, the conditions that contribute to potentially unhealthful levels of these contaminants in your living space, and what you can do about these problems. At the end of the chapter, you'll find descriptions of our recommended air and water purification products and resources.

The Environmental Protection Agency estimates that one in five Americans, supplied by one-quarter of the nation's drinking water systems, consume tap water that violates safety standards under the Clean Water Act.

The Need for Water Purification

The Environmental Protection Agency (EPA) states that no matter where you live in the United States, some toxic substances are likely to be found in the groundwater. Indeed, the agency estimates that one in five Americans, supplied by one-quarter of the nation's drinking water systems, consume tap water that violates safety standards under the Clean Water Act. Even some of the substances that are added to our drinking water to protect us, such as chlorine, can form toxic compounds—for example, trihalomethanes, or THMs—and have been linked to certain cancers. The EPA has established enforceable standards for more than 100 contaminants. However, credible studies have identified more than 2,000 contaminants in the nation's water supplies.

The simplest way of dealing with water pollution is a point-of-use purification device. The tap is the end of the road for water consumed by our families, so this is the logical and most efficient place to focus water treatment. Different water purification technologies each have strengths and weaknesses and are particularly effective against specific kinds of impurities or toxins. So the system you need depends first and foremost on the nature of the problem you have, which in turn requires testing and diagnosis.

We will describe these systems in detail after discussing the contaminants you might encounter. Most treatment systems are point-of-use and deal with only drinking and cooking water, which is less than 5% of typical home use. Full treatment of all household water is a very expensive undertaking and is usually reserved for water sources with serious, health-threatening problems.

Contaminants in Water

Presumably your drinking water comes from a municipal system, a shallow dug well, or a deep drilled well. If you drink bottled water, you've already taken steps to control what you consume—but you should also read about the perversions of bottled water in Chapter 7. The fact is, most good purification systems provide tasty potable water for far less money than the cost of regularly drinking bottled water, and you won't be supporting the insidious movement perpetrated by Coca-Cola, Pepsi, and Nestlé to privatize water—a human right! Each type of water supply is more or less vulnerable to different kinds of pollutants, because surface water (rivers, lakes, reservoirs), groundwater (underground aquifers), and treated water each are ex-

The presence of toxic metals in drinking water is one of the greatest threats to human health. The major culprits include lead, arsenic, cadmium, mercury, and silver.

posed to environmental contaminants in different ways. The hydrogeological characteristics of the area you live in, and localized activities and sites that create pollution—factories, agricultural spraying, a landfill—potentially will impact your water quality. And some impurities, such as lead and other metals, can be introduced by the piping that delivers water to your tap.

Reliable and inexpensive tests are available to identify the biological and chemical contaminants that may be in your water. Here's what you may find.

Biological Impurities: Bacteria, Viruses, and Parasites

Microorganisms originating from human and animal feces, or other sources, can cause waterborne diseases. Approximately 4,000 cases of waterborne illness are reported each year in the U.S. Additionally, many of the minor illnesses and gastrointestinal disorders that go unreported can be traced to organisms found in water supplies.

Biological impurities largely have been eliminated in municipal water systems with chlorine treatment. However, such treated water can still become biologically contaminated. Residual chlorine throughout the system may not be adequate, and therefore microorganisms can grow in stagnant water sitting in storage facilities or at the ends of pipes.

Water from private wells and small public systems is more vulnerable to biological contamination. These systems generally use untreated groundwater supplies, which could be polluted due to septic tank leakage or poor construction.

Organic Impurities: Tastes and Odors

If water has a disagreeable taste or odor, the likely cause is one or more organic substances, ranging from decaying vegetation and algae to organic chemicals (organic chemicals are compounds containing carbon).

Inorganic Impurities: Dirt and Sediment, or Turbidity

Most water contains suspended particles of fine sand, clay, silt, and precipitated salts. This cloudiness or muddiness is called turbidity. Turbidity is unsightly, and it can serve as food and lodging for bacteria. Turbidity can also interfere with effective disinfection and purification of water.

Total Dissolved Solids (TDS)

Total dissolved solids consist of rock and numerous other compounds from the earth. The significance of TDS in water is a point of controversy among water purveyors, but here are some facts about the consequences of higher levels of TDS:

1. High TDS results in undesirable taste, which can be salty, bitter, or metallic.
2. Certain mineral salts may pose health hazards. The most problematic are nitrates, sodium, barium, copper sulfates, and fluoride.
3. High TDS interferes with the taste of foods and beverages.
4. High TDS makes ice cubes cloudy and soft, and they melt faster.
5. High TDS causes scaling on showers, tubs, and sinks, and inside pipes and water heaters.

Toxic Metals or Heavy Metals

The presence of toxic metals in drinking water is one of the greatest threats to human health. The major culprits include lead, arsenic, cadmium, mercury, and silver. Maximum limits for each of these metals are established by the EPA's Drinking Water Regulations.

Toxic Organic Chemicals

The most pressing and widespread water contamination problem results from the organic chemicals (those containing carbon) created by industry. The American Chemical Society lists more than 4 million distinct chemical compounds, most of which are synthetic (manmade) organic chemicals, and industry creates new ones every week. Production of these chemicals exceeds a billion pounds per year. Synthetic organic chemicals have been detected in many water supplies throughout the country.

They get into the groundwater from improper disposal of industrial waste (including discharge into waterways), poorly designed and sited industrial lagoons, wastewater discharge from sewage treatment plants, unlined landfills, and chemical spills.

Studies since the mid-1970s have linked organic chemicals in drinking water to specific adverse health effects. However, only a fraction of these compounds have been tested for such effects. Well over three-quarters of the substances identified by the EPA as priority pollutants are synthetic organic chemicals.

Volatile Organic Compounds (VOCs)

Volatile organic compounds are very lightweight organic chemicals that easily evaporate into the air. VOCs are the most prevalent chemicals found in drinking water, and they comprise a large proportion of the substances regulated as priority pollutants by the EPA.

Chlorine

Chlorine is part of the solution and part of the problem. In 1974, scientists discovered that VOCs known as trihalomethanes (THMs) are formed when the chlorine added to water to kill bacteria and viruses reacts with other organic substances in the water. Chlorinated water has been linked to cancer, high blood pressure, and anemia. Chlorine has been banned for years in Europe and most civilized countries due to its status as a known carcinogen.

The scientific research linking various synthetic organic chemicals to specific adverse health effects is not conclusive and remains the subject of considerable debate. However, given the ubiquity of these pollutants and their known presence in water supplies, and given some demonstrated toxicity associations, it would be reasonable to assume that chronic exposure to high levels of synthetic organic chemicals in water could be harmful. If they're in your water, you want to get them out.

Pesticides and Herbicides

The increased use of pesticides and herbicides in American agriculture since World War II has had a profound effect on water quality. Rain and irrigation carry these deadly chemicals into groundwater as well as into surface waters. These are poisonous, plain and simple.

Asbestos

Asbestos exists in water as microscopic suspended fibers. Its primary source is asbestos-cement pipe, which was commonly used after World War II for city water systems. It has been estimated that some 200,000 miles of this pipe are currently in use delivering drinking water. Because pipes wear as water courses through them, asbestos shows up with increasing frequency in municipal water supplies. It has been linked to gastrointestinal cancer.

Radionuclides

Earth contains naturally occurring radioactive substances. Certain areas of the U.S. exhibit relatively high background levels of radioactivity due to their geological characteristics. The three substances of concern to human health that show up in drinking water are uranium, radium, and radon (a gas). Various purification techniques can be effective at reducing the levels of radionuclides in water.

Testing Your Water

The first step in choosing a treatment method is to find out what contaminants are in your water. Different levels of testing are commercially available, including a comprehensive screening for nearly 100 substances. See the product section that follows.

Even before you test your water, a simple comparison can help you figure out what level of treatment your domestic water supply may need. Find some bottled water that you like. Note the normal levels of total dissolved solids (TDS), hardness, and pH, and also how high these levels range. If you don't find this information on the label, call the bottler and ask. Then get the same information about your tap water, which can be obtained by calling your municipal supplier. If your tap water has lower levels of these things than the bottled water, a good filter should satisfy your needs if you decide you want to treat it. If your tap water has higher levels than the bottled water you prefer, you may need more extensive purification.

If you're drinking water from a private well, you'll have to get it tested to know what's in it. Some common problems with well water, such as staining, sediment, hydrogen sulfide (rotten egg smell), and excess iron or manganese, should be corrected before you purchase a point-of-use treatment system, because they will interfere with the effective operation of the system. These problems can be identified easily with a low-priced test; many state water-quality agencies will perform such tests for a nominal fee, or we offer some tests in our product section.

Chlorinated water has been linked to cancer, high blood pressure, and anemia. Chlorine has been banned for years in Europe and most civilized countries due to its status as a known carcinogen.

If you have any reason to worry that your water may be polluted, we strongly advise you to conduct a more comprehensive test.

Water Choices: Bottled, Filtered, Purified

If you're not satisfied with your drinking water, you suspect it may be contaminated, or you've had it tested and you know it's polluted, you have two choices: You can buy bottled water, or you can install a point-of-use water treatment system in your home.

There are three kinds of bottled water: distilled, purified, and spring. Distillation evaporates the water, then recondenses it, thereby theoretically leaving all impurities behind (although some VOCs can pass right through this process with the water). Purified water is usually prepared by reverse osmosis, deionization, or a combination of both processes (see below for explanations). Spring water is usually acquired from a mountain spring or an artesian well, but it may be no more than processed tap water. Spring water generally will have higher total dissolved solids than purified water has. Distilled water and purified water are better for batteries and steam irons because of their lower content of TDS.

But why pay for bottled water forever when treatment will be cheaper? Bottled water typi-

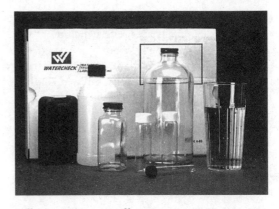

cally costs $2 per gallon or more. Many manufacturers claim that their purifiers produce clean water for just pennies a gallon; certainly these systems will pay for themselves in cost savings within a few years.

The two basic processes used to clean water are filtration and purification. The word "filter" usually refers to a mechanical filter, which strains the water, and/or a carbon filter system, which reduces certain impurities by chemically bonding them—especially chlorine, lead, and many organic molecules. Purification refers to a slower process, such as reverse osmosis, that greatly reduces dissolved solids, hardness, and certain other impurities, as well as many organics. Many systems combine both processes. Make sure that the claims of any filter or purifier you consider have been verified by independent testing.

> But why pay for bottled water forever when treatment will be cheaper? Bottled water typically costs $2 per gallon or more.

30th ANNIVERSARY

Methods of Water Filtration and Purification

The most efficient and cost-effective solution for water purity is to treat only the water you plan to consume. A point-of-use water treatment system eliminates the middleman costs associated with bottled water and can provide purified water for pennies per gallon. Devices for point-of-use water treatment are available in a variety of sizes, designs, and capabilities. Some systems improve only the water's taste and odor. Other systems reduce the various contaminants of health concern. Different systems work most effectively against certain contaminants. A system that utilizes more than one technology will protect you against a broader spectrum of biological pathogens and chemical impurities. Combining activated carbon filtration and reverse osmosis generally is considered the most complete and effective treatment.

When considering a treatment device, always pay careful attention to the independent documentation of the performance of the system for a broad range of contaminants. You should read the data sheets provided by the manufacturer carefully to verify its claims. Many companies are certified with the National Sanitation Foundation (NSF), whose circular logo appears on its data sheets.

The following is a brief analysis of the strengths and weaknesses of each option.

Mechanical Filtration

Mechanical filtration can be divided into two categories. Strainers are fine mesh screens that generally remove only the largest particles present in water. Sediment filters (or prefilters) remove smaller particles such as suspended dirt, sand, rust, and scale—in other words, turbidity. When enough of this particulate matter has accumulated, the filter is discarded. Sediment filters greatly improve the clarity and appeal of the water. They also reduce the load on any more expensive filters downstream, extending their useful life. Sediment filters will not remove the smallest particles or biological pathogens.

Activated Carbon Filtration

Carbon adsorption is the most widely sold method for home water treatment because of its ability to improve water by removing disagreeable tastes and odors, including objectionable chlorine. Activated carbon filters are

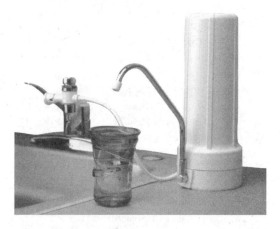

an important piece of the purification process, but they are only one piece. Activated carbon effectively removes many chemicals and gases, and in some cases, it can be effective against microorganisms. However, generally it will not affect total dissolved solids, hardness, or heavy metals. Only a few carbon filter systems have been certified for the removal of lead, asbestos, VOCs, cysts, and coliform.

Each type of carbon filter system—granular activated carbon and solid block carbon—has advantages and disadvantages (see below).

Activated carbon is created from a variety of carbon-based materials in a high-temperature process that creates a matrix of millions of microscopic pores and crevices. One pound of activated carbon provides anywhere from 60 to 150 acres of surface area. The pores trap microscopic particles and large organic molecules, while the activated surface areas cling to, or adsorb, small organic molecules.

The ability of an activated carbon filter to remove certain microorganisms and certain organic chemicals—especially pesticides, THMs (the chlorine byproduct), trichloroethylene (TCE), and PCBs—depends upon several factors:

1. The type of carbon and the amount used.
2. The design of the filter and the rate of water flow (contact time).
3. How long the filter has been in use.
4. The types of impurities the filter has removed previously.
5. Water conditions (turbidity, temperature, etc.).

GRANULAR ACTIVATED CARBON

Any granular activated carbon filter has three inherent problems. First, it can provide a base

> Carbon adsorption is the most widely sold method for home water treatment because of its ability to improve water by removing disagreeable tastes and odors, including objectionable chlorine.

for the growth of bacteria. When the carbon is fresh, practically all organic impurities (not organic chemicals) and even some bacteria are removed. Accumulated impurities, though, can become food for bacteria, enabling them to multiply within the filter. A high concentration of bacteria is considered by some people to be a health hazard.

Second, chemical recontamination of granular activated carbon filters can occur in a similar way. If the filter is used beyond the point at which it becomes saturated with the organic impurities it has adsorbed, the trapped organics can release from the surface and recontaminate the water, with even higher concentrations of impurities than in the untreated water. This saturation point is impossible to predict.

Third, granular carbon filters are susceptible to channeling. Because the carbon grains are held (relatively) loosely in a bed, open paths can result from the buildup of impurities in the filter and rapid water movement under pressure through the unit. In this situation, contact time between the carbon and the water is reduced, and filtration is less effective.

To maximize the effectiveness of a granular activated carbon filter and avoid the possibility of biological or chemical recontamination, it must be kept scrupulously clean. That generally means routine replacement of the filter element at 6- to 12-month intervals, depending on usage.

SOLID BLOCK CARBON

These filters are created by compressing very fine pulverized activated carbon with a binding medium and fusing the composite into a solid block. The intricate maze developed within the block ensures complete contact with organic impurities and, therefore, effective removal. Solid block carbon filters avoid the drawbacks of granular carbon filters.

Block filters can be fabricated with a porous structure fine enough to filter out coliform and other disease bacteria, pathogenic cysts such as giardia, and lighter-weight volatile organic compounds such as THMs. Block filters eliminate the problem of channeling. They are also dense enough to prevent the growth of bacteria within the filter.

Compressed carbon filters have two primary disadvantages compared with granular carbon filters. They have smaller capacity for a given size, because some of the adsorption surface is taken up by the inert binding agent, and they tend to plug up with particulate matter.

Thus, block filters may need to be replaced more frequently. In addition, block filters are substantially more expensive than granular carbon filters.

LIMITATIONS OF CARBON FILTERS

To summarize, a properly designed carbon filter is capable of removing many toxic organic contaminants, but it will fall short of providing protection against a wide spectrum of impurities.

1. Carbon filters are not capable of removing excess total dissolved solids (TDS). To gloss over this deficiency, many manufacturers and sellers of these systems assert that minerals in drinking water are essential for good health. Such claims are debatable. However, some scientific evidence suggests that minerals associated with water hardness may have some preventive effect against cardiovascular disease.
2. Only a few systems have been certified for the removal of cysts, coliform, and other bacteria.
3. Carbon filters have no effect on harmful nitrates or on high sodium or fluoride levels.
4. With both granular and block filters, the water must pass through the carbon slowly enough to ensure complete contact and filtration. An effective system therefore has to have an appropriate balance between a useful flow rate and adequate contact time.

Reverse Osmosis (Ultrafiltration)

Reverse osmosis (RO) is a water purification technology that utilizes normal household water pressure to force water through a selective semipermeable membrane that separates contaminants from the water. Treated water emerges from the other side of the membrane, and the accumulated impurities left behind are washed away. Sediment eventually builds up along the membrane and then it needs to be replaced.

Reverse osmosis is highly effective in removing several impurities from water: total dissolved solids (TDS), turbidity, asbestos, lead and other heavy metals, radium, and many dissolved organics. RO is less effective against other substances. The process will remove some pesticides (chlorinated ones and organophosphates, but not others) and most heavier VOCs. However, RO is not effective at removing lightweight VOCs such as THMs (the chlorine byproduct) and TCE (trichloroethylene), and certain pes-

> Reverse osmosis is highly effective in removing several impurities from water: total dissolved solids (TDS), turbidity, asbestos, lead and other heavy metals, radium, and many dissolved organics.

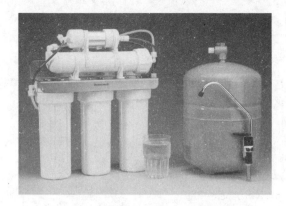

ticides. These compounds are either too small, too light, or of the wrong chemical structure to be screened out by an RO membrane.

Reverse osmosis and activated carbon filtration are complementary processes. Combining them results in the most effective treatment against the broadest range of water impurities and contaminants. Many RO systems incorporate both a prefilter of some sort and an activated carbon postfilter.

RO systems have two major drawbacks. First, they waste a large amount of water. They'll use anywhere from 3 to 9 gallons of water per gallon of purified water produced. This could be a problem in areas where conservation is a concern, and it may be slightly expensive if you're paying for municipal water. On the other hand, this wastewater can be recovered or redirected for purposes other than drinking, such as watering the garden or washing the car. Second, reverse osmosis treats water slowly: It takes about three to four hours for a residential RO unit to produce one gallon of purified water. Treated water can be removed and stored for later use.

Other Treatment Processes

Two other water treatment processes are worth knowing about. Distillation is a process that creates clean water by evaporation and condensation. Distillation is effective against microorganisms, sediment, particulate matter, and heavy metals; it will not treat organic chemicals. Good distillers will have a carbon filter to remove organic chemicals.

Ultraviolet (UV) systems use UV light to kill microorganisms. These systems can be highly effective against bacteria and other organisms; however, they may not be effective against giardia and other cysts, so any UV system you buy should also include a 0.5-micron filter. Other than some moderately expensive solar-powered distillers on the market, any distiller or UV unit will require a power source.

A Final Word on Water

Our best advice for those of you considering a drinking water treatment system consists of three simple points.

1. Test your water so you know precisely what impurities and contaminants you're dealing with.
2. Buy a system that is designed to treat the problems you have.
3. Carefully check the data sheets provided by the manufacturer to make sure that claims about what the system treats effectively have been verified by the National Sanitation Foundation (NSF) or another third party.

These actions will enable you to purchase a system that will most effectively meet your particular needs for safe drinking water.

WATER PURIFICATION PRODUCTS

NTL Water Check with Pesticide Option

This is the best analysis of your water available in this price range. Your water will be analyzed for 73 items, plus 20 pesticides. If you have any questions about your water's integrity, this is the test to give you peace of mind. The kit comes with five water sample bottles, a blue gel refrigerant pack (to keep bacterial samples cool for accurate test results), and easy-to-follow sampling instructions. You'll receive back a two-page report showing 93 contaminant levels, together with explanations of which contaminants, if any, are above allowed values. You'll also receive a follow-up letter with a personalized explanation of your test results, plus knowledgeable, unbiased advice on what action you should take if your drinking water contains contaminants above EPA-allowed levels. NTL, located outside of Detroit, must receive the package within 30 hours of sampling in order to provide accurate results. If you live in Alaska or Hawaii, you will need express shipping. Check with your shipper. This test is primarily for folks with private wells. Municipal water agencies are already doing this kind of testing; call them for results. USA.

01-0324 NTL Check & Pesticide Option   **$175**

Household Point-of-Use Water Filters

Handcrafted Terra-Cotta Crock

Now you can keep cool filtered water right on your countertop with our handcrafted Water Crock. Simply pour in a gallon of tap water, then let the natural force of gravity go to work. Water passes through a ceramic housing that contains a granular activated-charcoal filter made from coconut shells and a KDF media, then drips into the lower chamber where it's ready for drinking. Removes 95% of chlorine, pesticides, iron, aluminum, and lead, and 99% of cryptosporidium, giardia, and sediment. Filter lasts six months or 300 gallons. 17"x8" (H x diameter). 13 lb. Brazil.

01-0465 Water Crock	**$169**
01-0466 Replacement Filter	**$59**

Not available in California.

Wellness Water Filter

We rely on water filters for a safe supply, but often they strip out the good stuff, too, like the minerals that give water its taste and health benefits. This countertop filter solves that dilemma by putting some things back even as it takes others out. Entering water is first cleansed by special volcanic sand and carbon filter media that remove bacteria,

cysts, heavy metals, chlorine, many chemicals, sediment, and more. Then, natural magnetite and ceramic magnets recharge the water's solubility for better hydrating. Finally, multiple layers of unique minerals naturally reionize its antioxidant abilities, increase dissolved oxygen, and add trace levels of silica, potassium, and magnesium.

The result virtually turns your tap into a fresh mountain spring whose newly renourished water you have to taste to believe. One-gallon-per-minute flow. LCD readout monitors filter life. Installs in minutes with included 5-foot extension hose. Dual filters last for 1,000 gallons. 13½"x8"x5" (HxWxD). Japan.

01-0523 Wellness Water Filter	**$595**
01-0524 Replacement Filter Set	**$99**

Paragon Water Filtration Systems

Prevent a host of the most common water problems as well as eliminate bad taste and odor. Using a titanium silicate/carbon block filter, these filtration systems remove virtually all hazardous contaminants commonly found in tap water. The Basic Model (A) features extra-long tubing and a diverter system that sends filtered water out of your kitchen tap. The attractive Chrome Model (B) has a convenient spout. The Under-the-Counter Model (C) easily mounts under your sink with the included bracket and attaches to your cold-water line with quick-connect fittings. For cold-water use. USA.

01-0144 Basic (A) 12½"x4¼" (shown)
(H x diameter), 3 lb. **$109**

01-0145 Chrome (B) 12½"x5¾"
(H x diameter), 4 lb. **$159**

01-0150 Under the Counter (C) 14½"x7"
(H x diameter), 3½ lb. **$139**

01-0224 Replacement Filter
(replace every year) **$55**

Not available in California.

O-SO Pure 3-Stage UV System

Triple treatment to remove most contaminants, including bacteria and viruses. Sediment filters remove sand, silt, and other particulates, then the carbon second stage removes chlorine and other chemicals. Treated water is then exposed to high-intensity ultraviolet to destroy bacteria and viruses. Filter elements and UV are all replaceable. Operates on 120 volts AC or 12 volts DC—120-volt units include installation kit. 3-yr. warranty. These are residential units only.

42910 O-SO Pure 3-Stage UV POU
System **$450**

40-0067 O-SO Pure 3-Stage UV, 12V **$475**

40-0064 12V Installation Kit **$40**

REPLACEMENT FILTERS

42913 10-in. Sediment Filter **$9**

42914 10-in. Carbon Filter **$29**

42915 Replacement Bulb **$45**

Five-Stage Reverse-Osmosis Filter System

Five-stage reverse osmosis removes more contaminants than any other filtration method. Reverse osmosis has long been considered the most comprehensive and permanent solution for purer, healthier water. This powerful five-stage filtration system installs under your sink and reduces contaminants—parasites, chlorine, sediments, sand, debris, and more—to safe levels. The result is ultrafiltered water that's sent to the 3.2-gallon storage tank where it's ready for on-demand service. Features a flow rate of up to 50 gallons per day and auxiliary faucet. Professional installation recommended. 15"x17" (H x diameter). 25 lb. Taiwan.

01-0467 Five-Stage Reverse-Osmosis
Filter System **$299**

01-0469 Replacement Filter Kit (replace
every year) **$55**

01-0468 Replacement Membrane
(replace every 3 years) **$80**

Express delivery not available. Cannot be shipped to P.O. boxes. Can be shipped to customers in contiguous U.S. only.

Nonelectric Distiller

An economical, high-output stovetop water distiller that operates on a variety of heat sources for daily and emergency use. Distillate capacity, based on a 2,600-watt electric burner: 3.2 qt. in 1.2 hr.; up to 16 gal. per day. Stainless steel with no moving parts or fan; digital timer with alarm. 12"x12" (H x diameter). 9 lb. 3-yr. limited warranty. China/USA.

HOW IT WORKS

- The stainless steel boiler kills microbes in water by heating to 212°F (100°C).
- Steam rises, leaving behind dead microbes, dissolved solids, salts, heavy metals, and other substances.
- Low-boiling light gases are discharged through the gaseous vent.
- Steam is condensed in the three-level stainless steel tray.
- Purified/distilled water is captured in the collection cup assembly.
- The 100% steam-distilled water flows through the collection tube into the 3-liter stainless steel storage container.

01-0481 Nonelectric Distiller **$369**

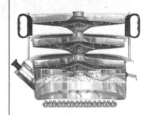

Waterwise® Distiller

The Waterwise Distiller turns tap water into purified steam, recondenses it, and then runs it through a carbon filter, leaving impurities behind and water that's 99% free of most tap water contaminants. It reduces 99% of inorganic/radiological contaminants such as lead, arsenic, and fluoride and removes cysts, bacteria, and viruses. Purifies a gallon of water in about four hours with an overall capacity of six gallons in a day. Filter status bar indicates when to change the carbon post filter. Includes a one-gallon nonleaching water bottle and programmable on/off settings. Regular cleaning recommended. 15"x15½" (HxW). 14.3 lb. USA.

01-0482 Waterwise Distiller   **$399**

01-0483 Carbon Post Filters (set of 6;
**   replace every 2-3 mo.)**   **$30**

01-0484 Distiller Cleaner/Descaler   **$20**

Inline Sediment Filter

This is the recommended minimum filter for every Slowpump installation. This filter is designed for cold water only and meets NSF standards for drinking water supplies. It features a sump head of fatigue-resistant Celcon plastic, an opaque body to inhibit algae growth if installed outdoors, and a manually operated pressure-release button to simplify cartridge replacement. Rated for up to 125 psi and 100ºF, it is equipped with a ¾-inch female pipe thread inlet and outlet fittings and is rated for up to 6 gpm flow rates (with clean cartridge). It accepts a standard 10-inch cartridge with 1-inch center holes. Supplied with one 10-micron fiber cartridge installed. USA.

41137   Inline Sediment Filter   **$49**

42632   10-Micron Sediment Cartridge,
**   2 pk.**   **$13**

Whole-House Filtration System

Here's one of the most affordable ways we've seen to deliver healthy filtered water to every fixture in your home. Attached to your water supply at its entry point, this Whole-House Filtration System uses two stages to provide pure water for drinking and bathing. An initial filtration preps your water by removing rust, dirt, debris, and other particles larger than 5 microns. Then a combination carbon and KDF media filter targets chlorine, VOCs, pesticides, and other chemicals and heavy metals while removing odors and improving your water's taste. 100,000-gallon filters are easy to change and last at least a year before requiring replacement. Flow rate up to 36 gpm. Max pressure 100 psi. Professional installation recommended. 18"x15"x8" (HxWxD). USA.

40-0077 Whole-House Filtration System $499

Portable Water Filters

Base Camp Water Filter

Ideal for emergencies and backcountry trips alike, this portable water filter serves whole families with ease. Just fill its 2.6-gallon water bag, hang it anywhere, and watch as a pumpless gravity-feed design effortlessly produces 16 ounces of safe drinking water per minute. A state-of-the-art glass fiber/carbon combination filter removes odors, tastes, and impurities down to 0.3 micron in size to make virtually any water source potable. Includes a 48-inch outlet hose with valve. Filters last for approximately 200 gallons. Switzerland.

01-0517 Water Filter   **$79**

01-0518 Replacement Filter   **$39**

Filtering Water Bottles

These polycarbonate, flip-top water bottles have an advanced fast-flow in-line filter so you can do more yet pack less. Our Filtering Bottle exceeds EPA standards to remove 99.9999% of all bacteria and cysts like giardia and cryptosporidium for 2,000 fill-ups. Our Purifying Bottle removes 99.9997% of all viruses as well, lasts for 12,000 fill-ups, and includes an adapter for use with a tap. Just fill and sip normally. It's that simple and that safe. In-line filters can also be used with CamelBak or gravity bags. 32 oz. capacity with filter. USA/China.

14-0407 Filtering Water Bottle   **$49**

14-0408 Purifying Water Bottle   **$129**

Stainless Steel Bottles

These containers won't impart unpleasant tastes and are nonleaching. Our stainless steel water bottle, with a white plastic cap, is 10"x3½" (H x diameter). China.

14-0436 40 oz. Stainless Steel Bottle   **$29**

14-0437 Stainless Steel Hoop Cap   **$7**

The Katadyn Drip Filter

With no moving parts to break down, superior filtration, and a phenomenal filter life, there is simply no safer choice for potentially pathogen-contaminated water. There are no better filters than Katadyn for removing bacteria, parasites, and cysts. Three 0.2-micron ceramic filters process 1 gallon per hour. Clean filters by brushing the surface. Ideal for remote homes, RV, campsite, and home emergency use. Food-grade plastic canisters stack to 25"x11" (H x diameter). 10 lb. 1-yr. mfr. warranty. Switzerland.

55-0083 Katadyn Drip Filter   **$289**

55-0082 Replacement Filter (needs 3) $75 ea.

Shower and Bath Filters

Shower Falls Showerhead

Enhanced by a soothing five-way massage spray, this handheld showerhead eliminates 99% of the chlorine in your water. Can be positioned at an extended height to create an invigorating waterfall-like effect. Showerhead and handle measure 17" long; hose is 72" long. 3 lb. USA.

01-0406 Shower Falls Showerhead   **$68**

01-0143 Replacement Filters (set of 2; replace every 3 mo.)   **$28**

High-Output Shower Filter

Hot shower water exposes you to chlorine that can harm your hair, skin, eyes, and nasal membranes, and it creates chloroform, a toxic gas that's inhaled during washing. The special noncarbon filter inside our High-Output Shower Filter removes 75% of chlorine and reduces lead, mold, mildew, iron deposits, sediments, and odors for healthier bathing. With an easy-open design for simple filter changes and filters that last up to a full year (depending on sediment levels), this is the most durable, highest capacity replaceable filter model made. Self-sealing threads prevent leaks. Use with your existing showerhead. Not suitable for homes with water pressure over 150 psi. Choose white or chrome-plated. 1-yr. warranty. USA.

01-0407 High-Output Shower Filter   **$54**

01-0408 Replacement Filter (replace every 3-6 mo.)   **$28**

Royale All-in-One Showerhead

With five adjustable spray options, the Royale All-in-One Showerhead is sheer indulgence. From fine mist to a full massage spray, this showerhead allows you to create just the shower pressure you desire. A triple-plated finish keeps the chrome band gleaming; antiscaling spray nozzle prevents clogging. 3½"x3½"x4½" (HxWxD). USA.

02-0208 Royale All-in-One Showerhead   **$59**

02-0212 Replacement Filter (replace every 6 mo.)   **$24**

Oxynator™ Shower Filter

We finally found a water-wise showerhead that delivers a powerful, wake-you-up spray. The April Shower® Oxynator Shower Filter oxygenates the water to deliver refreshing, invigorating showers, while its combination KDF media removes 98.9% of all chlorine and vapors, plus hydrogen sulfide, rust, heavy metal traces, sediments, and odors. 7"x3½" (HxW). USA.

01-0507 April Shower Oxynator Shower Filter $98

01-0506 Replacement Filter (replace every 6 mo.) $26

Sunflower Showerhead

Sulfates and phosphates in treated city water are on the rise, and these chemicals can greatly diminish both the effectiveness and life of regular filtration devices. Only the Sunflower Showerhead uses a proprietary patented filtering media that works to its full chlorine-reducing effectiveness. If your utility adds these chemicals to your water, this is the best filtering option for you. 12-inch-diameter head, 2.5 gpm. Adjustable arm included. USA.

01-0464 Sunflower Showerhead $78

01-0471 Replacement Filter (replace every 6-9 mo.) $25

Wellness Shower Filter

If you've ever wanted to shower in a pristine alpine waterfall, this one-of-a-kind shower filter is for you. It not only removes bacteria, cysts, heavy metals, chlorine, sediment, many chemicals, and other impurities from your shower water, it also increases its dissolved oxygen, reionizes it for an enhanced antioxidant effect, and puts back in healthy, trace amounts of silica, potassium, and magnesium. Your shower water is not only cleaner, it's healthier, too. (Your skin and hair will love it!) A special backwashing system recharges filter media with a simple turn of the cartridge. 2½-gpm flow rate. Universal fittings attach to any standard shower and accept your favorite showerhead or wand. Includes showerhead shown. Japan.

01-0521 Wellness Shower Filter $249

01-0522 Replacement Filter (replace every 2 yr.) $189

Bath Ball Faucet Filter

Now you can take a relaxing bath without chlorine (or without filling the tub with the showerhead). Our Bath Ball Faucet Filter is a plastic sphere filled with an ingenious technology that removes chlorine. Simply attach the sturdy vinyl strap under the tub faucet. Water flows into openings in the top and emerges 95% free of chlorine. You can notice the difference in the water quality with your very first bath. Filter lasts one year. Includes container to keep bath ball submerged when not in use. 4" diameter. USA.

01-1169 Bath Ball Faucet Filter $62

01-0132 Replacement Filter (replace once a year) $38

Pool and Spa Water Purification

Solar-Powered Pool Purifier

You can dramatically reduce dangerous chlorine use with the Floatron. It kills algae and bacteria and is a safer and economical alternative to the chemical marinade of most pools. It is solid-state, with no moving parts or batteries, and is portable and cost effective. The Floatron's solar panel generates a low-power electric current that energizes a specially alloyed electrode. The electrode delivers copper and silver ions into the water, which kill algae and bacteria cells on contact. The electrode element is consumed gradually and needs replacement every one to three seasons, depending on pool size. The element is available directly from the manufacturer for about $50 and takes a couple of minutes to replace. The solar panel lasts for the life of your pool. No more chlorine allergies, red eyes, discolored hair, or bleached bathing suits! Floatron typically will reduce chemical expenses by an average of 80%. Effective for pools up to approximately 40,000 gallons. 2-yr. mfr. warranty. USA.

14-0172 Floatron $269

Metal Foam Spa Purifier

Reduce chlorine or bromine usage by up to 90% with our safe, effective Metal Foam Spa Purifier. Our spa purifier destroys most of the same bacteria as standard chemical treatments, including E. coli and legionella pneumophilia. Once a month, remove the disc and backwash with a garden hose; replace disc twice annually. You'll still need to remove nonbacterial spa contaminants (sweat, skin products). Simply drop the disc into the skimmer or into the center hole of your spa's filter. Includes disc holder and two metal foam discs. USA.

42804   Metal Foam Spa Purifier $60

Books

The Water You Drink: Safe or Suspect?

By Julie Stauffer. This book is a clearly written, straightforward guide to North American tap water: what's in it, how it's treated, whether government regulations are strict enough, and whether consumers should be concerned about what they're drinking. It includes chapters on bottled water and home filter systems, as well as obtaining, treating, and storing private water supplies, and helpful resources. Julie Stauffer has authored several books and a video on water and pollution. 160 pages, paperback.

21-0531 The Water You Drink $14.95

The Drinking Water Book

How to Eliminate the Most Harmful Toxins from Your Water

By Colin Ingram. *The Drinking Water Book* takes a level-headed look at the serious issues surrounding America's drinking water supply. Unlike water-purifier manufacturers and public health officials, Ingram presents unbiased reporting on what's in your water and how to drink safely. Featuring all the latest scientific research, the book evaluates the different kinds of filters and bottled waters and rates specific products on the market. 185 pages, paperback.

21-0583 The Drinking Water Book $14.95

30th ANNIVERSARY

Air Purification

Most of us are familiar with the idea that the water we drink at home might be polluted or contaminated. But did you realize that the air inside your home (or place of work) potentially could be even more hazardous to your health than the water?

The Problem with Indoor Air

The problem of indoor air quality has attracted attention only recently. One reason is that the sources of indoor air pollution are more mundane and subtle than the sources of outdoor air and water pollution. In addition, before highly insulated, tightly constructed buildings became common in the 1970s, most American homes were drafty enough that the buildup of harmful gases, particles, and biological irritants indoors was unlikely. However, the combination of high-performance buildings that allow minimal infiltration of outside air and the continual increase of synthetic products that we bring into our homes has added up to a possible public health problem of significant proportions.

Little is known with relative certainty about the unhealthful consequences of constant exposure to polluted indoor air. Nonetheless, the EPA has concluded that many of us receive greater exposure to pollutants indoors than outdoors. EPA studies have found concentrations of a dozen common organic pollutants to be two to five times higher inside homes than outside, in both rural and industrialized areas. Furthermore, people who spend more of their time indoors and are thus exposed for longer periods to airborne contaminants are often the very people considered most susceptible to poor indoor air quality: very young children, the elderly, and the chronically ill.

SOURCES OF INDOOR AIR POLLUTION

You can probably identify some of the sources of indoor air pollution: new carpeting, adhesives, cigarette smoke, dust mites. But many things that contribute potentially irritating or harmful substances to indoor air may not be obvious.

Indoor air contaminants typically fall into one or more of the following categories:
- Combustion products
- Volatile chemicals and mixtures
- Respirable particulates
- Respiratory products
- Biological agents
- Radionuclides (radon and its byproducts)
- Odors

What kinds of substances are we talking about? Let's start with chemicals and synthetics. Building materials and interior furnishings emit a broad array of gaseous volatile organic compounds (VOCs), especially when they are new. These materials include adhesives, carpeting, fabrics, vinyl floor tiles, some ceiling tiles, upholstery, vinyl wallpaper, particleboard, drapery, caulking compounds, paints and stains, and solvents. In addition to producing emissions, many of these materials also act as "sponges," absorbing VOCs and other gases that can then be reintroduced into the air when the "sponge" is saturated. These kinds of building and furnishing materials also can be the source of respirable particulates, such as asbestos, fiberglass, and dusts.

Appliances, office equipment, and office supplies are another major source of VOCs and particulates. Among the culprits are leaky or unvented heating and cooking appliances; computers and visual display terminals; laser printers, copiers, and other devices that use chemical supplies; preprinted paper forms, rubber cement, and typewriter correction fluid; and routine cleaning and maintenance supplies, including carpet shampoos, detergents, floor waxes, furniture polishes, and room deodorizers.

Your own routine "cleaning and maintenance supplies" are not above suspicion, either. Many of the personal care and home cleaning products we use emit various chemicals (especially from the components that create fragrance), some of which may contribute to poor indoor

---

Appliances, office equipment, and office supplies are another major source of VOCs (Volatile Organic Compounds) and particulates.

Breathe Easier Without Indoor Pollutants

A five-year EPA study found the concentrations of 20 toxic compounds to be as much as 200 times higher in the air inside homes and offices than outdoors. Poor air circulation is partly to blame. Real Goods' air filters can remove hazardous pollutants, as well as dust, pollen, and pet dander, so the air you breathe is cleaner.

air quality. Clothes returning from the dry cleaner may retain solvent residues, and studies have shown that people do breathe low levels of these fumes when they wear dry-cleaned clothing. Even worse, the pesticides we use to control roaches, termites, ants, fleas, and other insects are by definition toxic. Sometimes these poisons are tracked indoors on shoes and clothing. Do you have any of this stuff stored underneath the sink? One EPA study has suggested that up to 80% of most people's exposure to airborne pesticides occurs indoors.

Other human activities can also have a dramatic impact on indoor air quality. Tobacco smoke from cigarettes, cigars, and pipes is obviously an unhealthy pollutant. Household heating and cooking activities may introduce carbon monoxide, carbon dioxide, nitrogen oxide and dioxide, and sulfur dioxide into the air. Candles and wood fires contribute CO and CO_2 as well as various particulates.

Many biological agents contribute to indoor pollution: We are surrounded by a sea of microorganisms. Bacteria, insects, molds, fungi, protozoans, viruses, plants, and pets generate a number of substances that contaminate the air. These range from whole organisms themselves to feces, dust, skin particles, pollens, spores, and even some toxins. Humans contribute too: We exhale microbes, and our sloughed skin is the primary source of food for the infamous dust mites. Furthermore, certain indoor environments, such as ducts and vents, or humid areas such as bathrooms and basements, provide ideal conditions for microbial growth.

Radon (and its decay products, known as "radon daughters") is one type of indoor pollutant that has received widespread notice. As noted above, indoor radon buildup can be a problem in certain parts of the U.S. where it is present in rocks and soil. A radioactive gas, radon has been implicated as a cause of lung cancer. Simple and inexpensive radon test kits are available at hardware stores. Contact your regional EPA office—you can look it up in the phone book—for more information about effective radon remediation strategies.

Odors are always present. Some are pleasant, others are unpleasant; some signal the presence of irritating or harmful substances, others do not. The perception of odors is highly subjective and varies from person to person. Most odors are harmless, more of a threat to state of mind than to health, but eliminating them from an indoor environment may be a high priority to keep people content and productive.

HEALTH EFFECTS OF INDOOR AIR POLLUTION

Before you become too alarmed about the health dangers that may be floating around in the air you breathe at home, remember the following proviso. As with most environmental pollutants, adverse health effects depend upon both the dose received and the duration of exposure. The combination of these factors, plus an individual's particular sensitivity to specific substances, will determine the potential health effect of the contaminant on that individual.

With that in mind, the health effects of indoor air pollution can be short term or long term and can range from mildly irritating to severe, including respiratory illness, heart disease, and cancer. Most often the effects of such pollutants are acute, meaning they occur only in the presence of the substance; but in some cases, and perhaps in many cases, the effects may be cumulative over time and may result in a chronic condition or illness.

The clinical effects of indoor air pollution can take many forms. The most common clinical signs include eye irritation, sneezing or coughing, asthma attacks, ear-nose-throat infections, allergies, and migraines. More seriously, prolonged exposure to radon, tobacco smoke, and other carcinogens in the air can cause cancer. Most volatile organic compounds can be respiratory irritants, and many are toxic, although little applicable toxicity data exists about low levels of indoor VOCs. One VOC, formaldehyde, is now considered a probable human carcinogen. Needless to say, the best way to protect your health is to avoid or reduce your exposure to indoor air contaminants.

Nearly all observers agree that more research is needed to better understand which health effects occur after exposure to which indoor air pollutants, at what levels, and for how long.

Improving Indoor Air Quality

There are three basic strategies for improving indoor air quality: source control, ventilation, and air cleaning.

Source control is the simplest, cheapest, and most obvious strategy for improving indoor air quality. Remove sources of air pollution from your home. Store paint, pesticides, and the like in a garage or other outbuilding. Read the labels on the products you bring into your home, and buy personal care, cleaning, and office products that contain fewer synthetic volatile compounds

The health effects of indoor air pollution can be short term or long term and can range from mildly irritating to severe, including respiratory illness, heart disease, and cancer.

or other suspicious and potentially irritating substances.

Other sources can be modified to reduce their emissions. Pipes, ducts, and surfaces can be enclosed or sealed off. Leaky or inefficient appliances, such as a gas stove, can be fixed.

Regular house cleaning is effective: Sweeping, vacuuming, and dusting can keep a broad spectrum of dust, insects, and allergens under control. You should be conscientious about cleaning out ducts and replacing furnace filters on a regular basis. Also, reduce excess moisture and humidity, which encourage the growth of microorganisms, by emptying the evaporation trays of your refrigerator and dehumidifier.

Ventilation is another obvious and effective way to reduce indoor concentrations of air pollutants. It can be as simple as opening windows and doors, operating a fan or air conditioner, and making sure your attic and crawl spaces are properly ventilated. Fans in the kitchen or bathroom that exhaust to the outdoors remove moisture and contaminants directly from those rooms. It is especially important to provide adequate ventilation during short-term activities that generate high levels of pollutants, such as painting, sanding, or cooking. Energy-efficient air-to-air heat exchangers that increase the flow of outdoor air into a home without undesirable infiltration have been available for several years. Such a system is especially valuable in a tight, weatherized house. See Chapter 6, page 288, for more information.

Unfortunately, indoor air cleaning or purification is more complicated than purifying water. Not many technologies are available for household use, most of them have limited effectiveness, and some of the more promising methods are highly controversial. Air filtration is problematic because it's only really effective if you draw all the indoor air through the filter. Even then, such filters reduce only particulate matter. Any furnace or whole-house air-conditioning system has a filter. Some of them, such as electrostatic filters or HEPA (High Efficiency Particulate Air) filters, can reduce levels of even relatively small particulates significantly. Portable devices incorporating these technologies have also been developed, which can be used to clean the air in one room. Some of them have EnergyStar ratings (including the PlasmaWave ionizer we sell, see below), which means they'll consume less power than other models. Follow the manufacturer's instructions about how frequently to change the filter on your furnace or standalone air cleaner. However, be aware that air filters are not designed to deal with volatile organic compounds or other gaseous pollutants.

Ionization is a second method of cleaning particulates out of the air. A point-source device generates a stream of charged ions (via radio frequencies or some other method), which disperse through the air and meet up with oppositely charged particles. As tiny particles clump together, they eventually become heavy enough to settle out of the air, and you won't breathe them in. But like filtration, ionization is not effective against VOCs.

What About Ozone?

Another method of air purification uses ozone, in combination with ionization. Ozone (O_3) is a highly reactive and corrosive form of oxygen. You know about the ozone layer in the upper atmosphere, which prevents a lot of ultraviolet light from reaching Earth's surface where it can damage living organisms. Ozone also exists naturally in small concentrations at ground level, where it has various effects. It is a component of smog, produced by the reaction of sunlight upon auto exhaust and industrial pollution. Ozone is also produced by lightning and is partially responsible for that fresh, clean smell you notice outside after a thunderstorm. Some studies suggest that slightly greater-than-natural outdoor levels of ozone can reduce respiratory, allergy, and headache problems, and possibly even enhance one's ability to concentrate. However, scientists agree that at slightly elevated levels in the air, ozone can be dangerous and unhealthful to people. In concentrations that exceed safe levels as determined by the EPA, ozone is considered an air pollutant.

Manufacturers of air cleaners that operate as ozone generators make strong claims about their products. They assert that these devices kill biological agents, such as molds and bacteria, that contribute to indoor air pollution; that the ozone penetrates into drapes and furniture and filters down to the floor into carpets and crevices to oxidize and destroy absorbed contaminants and those that fall out of the air; and that the ozone breaks down potentially harmful or irritating gases, including many VOCs, into carbon dioxide, oxygen, water vapor, and other harmless substances. They claim that their machines accomplish all this with levels of ozone that are safe, and that ozone production can be monitored, either automatically or manually, to ensure that ozone concentration does not become excessive.

It is especially important to provide adequate ventilation during short-term activities that generate high levels of pollutants, such as painting, sanding, or cooking.

The EPA strongly disputes many of these claims and considers them misleading. The agency's position is that "'available' scientific evidence shows that at concentrations that do not exceed public health standards, ozone has little potential to remove indoor air contaminants [and] does not effectively remove viruses, bacteria, mold, or other biological pollutants." While the EPA acknowledges that some evidence shows that ozone can combat VOCs, it points out that these reactions produce potentially irritating or harmful byproducts. In addition, the EPA is skeptical of manufacturer's claims that the amount of ozone produced by such devices can be controlled effectively so as not to exceed safe levels. The bottom line for the EPA is that "no agency of the federal government has approved ozone generation devices for use in occupied spaces. The same chemical properties that allow high concentrations of ozone to react with organic material outside the body give it the ability to react with similar organic material that makes up the body and potentially cause harmful health consequences. When inhaled, ozone will damage the lungs. Relatively low amounts can cause chest pain, coughing, shortness of breath, and throat irritation." (This information was obtained in March 2006 from a publication posted on the EPA website, at www.epa.gov/iaq/pubs/ozonegen.html.)

The ozone issue is still being debated, and at present, Gaiam Real Goods has limited the ozone purifiers in its product mix.

Resources

The best source of up-to-date information about indoor air pollution and residential air cleaning systems is the federal Environmental Protection Agency. The EPA maintains several useful telephone services and distributes, free of charge, a variety of helpful publications. You can view many of these publications online at www.epa.gov/iaq/pubs/. You can also request these publications from the regional EPA office in your area, which you can find in your local telephone book under U.S. Government, Environmental Protection Agency. The following resources may be particularly valuable (but note that while these remain the definitive EPA guides, several of them are more than 10 years old now):

The Inside Story: A Guide to Indoor Air Quality, publication #402K93007.

Indoor Air Facts, No. 7: Residential Air Cleaners, publication #20A-4001.

Residential Air Cleaning Devices: A Summary of Available Information, publication #400-1-90-002.

National Service Center for Environmental Publications and Information
800-490-9198
www.epa.gov/ncepihom/

Indoor Air Quality Information Clearinghouse
P.O. Box 37133, Washington, DC 20013-7133
800-438-4318
www.epa.gov/iaq/iaqxline.html

National Radon Hotline
800-SOS-RADON (767-7236); information recording operates 24 hours a day
www.epa.gov/radon/rnxlines.html

Manufacturers of air cleaners that operate as ozone generators claim that ozone kills biological agents that contribute to indoor air pollution; oxidizes and destroys contaminants absorbed by drapes and furniture; and breaks down potentially harmful or irritating gases, including many VOCs, all this with levels of ozone that are safe, and that ozone production can be monitored. The EPA strongly disputes many of these claims and considers them misleading.

AIR PURIFICATION PRODUCTS

Healthmate™ HEPA Air Filter

In independent tests of the top 21 air cleaners, our Healthmate HEPA Air Filter was rated best overall. Combining an ultralong-life HEPA filter that you change only twice a decade, a top-grade carbon filter, and a natural mineral zeolite filter, these units are most effective for serious contaminate removal, especially gaseous contaminants like sulfuric acid and ammonia. Both the standard and junior units feature ultraquiet motors that can operate 24 hours a day and offer three speeds and 360° intake. And the solid, welded-steel case and durable powder coating eliminate outgassing. Casters on standard model only. 14½"x14½"x23½" (LxWxH), 46 lb. Healthmate Junior: 11"x11"x16½" (LxWxH), 23 lb. Choose black, white, or sand. USA.

01-8005 Healthmate HEPA Air Filter   **$499**

01-0163 Healthmate HEPA Junior   **$299**

Allow 4-6 weeks for delivery. Can be shipped to customers in the contiguous U.S. only. Cannot be shipped to P.O. boxes. Sorry, express delivery not available.

NEW! Gaiam PlasmaWave™ Air Purifier

Relying on a plasma generator to split water molecules into positive hydrogen and negative oxygen ions, the hard-working PlasmaWave Air Purifier creates hydroxyl (OH) molecules to eliminate most airborne particles. A key byproduct of the positive and negative ions, OH is effective at removing viruses, bacteria, germs, and odors because it renders them completely inactive. When combined with the four-stage filtering process of a washable antibacterial prefilter and HEPA, nanosilver, and washable carbon filters, this advanced system purifies 400 square feet just like nature cleans the air outdoors. Graphic digital controls make it easy to identify when to change filters, adjust speed and time, sense when it's dark for sleep mode, and customize your operation settings. It's EnergyStar-qualified, too. For detailed information, see www.realgoods.com. 21.9"x16.5"x8.9" (HxWxD). 18.8 lb. South Korea.

01-0514 PlasmaWave Air Purifier   **$549 ($10)**

01-0515 Replacement Filters (replace every 12 mo.)   **$129**

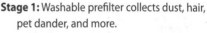

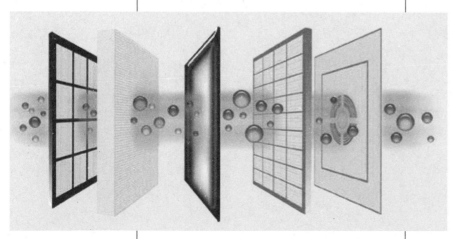

Stage 1: Washable prefilter collects dust, hair, pet dander, and more.

Stage 2: True HEPA filter is 99.99% effective in collecting microparticles up to 0.3 micron in size, including dust mites, pollen, and mold spores.

Stage 3: Nanosilver particles neutralize viruses and bacteria on contact.

Stage 4: Activated granular carbon collects and absorbs chemical vapors and odors.

Stage 5: PlasmaWave creates negative and positive ions that combine with water vapor to form hydroxyls, which change pollutants into harmless molecules.

Sun Pure UV Air Filter

This unit covers all the bases, from allergens to airborne viruses. The six-stage filtering process combines gas absorption, ultraviolet technology, and ionization with the standard HEPA and carbon filters, and throws in a macro prefilter for longer filter life. This is the system of choice for any environment where an active germ pool threatens family health, and it provides maximum relief for chemical sensitivities, asthma, and allergies in rooms as large as 50x40 feet. Infrared motion sensors control speed to reduce operating costs when rooms are unoccupied. It also features a "sleep" mode that provides superquiet operation while still purifying a room at 70% of normal capacity.

Step 1: A treated prefilter removes larger particles like dust (0.5 micron and above).

Step 2: A specially formulated gas absorption media substantially reduces exhaust fumes, organic hydrocarbons, pesticides, formaldehyde, and other noxious gases emitted from household cleaners, solvents, chlorine, and paint. Critical for those with chemical sensitivities!

Step 3: Strips air of odors and low-level gases with activated carbon.

Step 4: Sends air through a HEPA filter that removes allergens, smoke particles, and other pollutants as small as 0.3 micron (pollen, mold, fungal spores, dust mites, tobacco smoke, most bacteria).

3 Steps to Buying an Air Filter

1. Consider the range of pollutants each unit filters out, then buy according to budget.
2. Determine the room size (measure width and length, and see chart).
3. Consider what features are important (e.g., low noise, small profile, long filter life).

Step 5: High-intensity ultraviolet light kills viruses and bacteria on contact. This is what makes the system a "purifier" and not just a cleaner.

Step 6: An ionization chamber produces negative ions and activated oxygen, so indoor air tastes and smells like the real thing. Studies suggest that negative ions may even induce and enhance sleep.

A chemical sensor monitors air quality and alerts you to buildup of dangerous contaminants such as carbon monoxide. Indicator light signals filter change. Replacement kit includes all filters and UV replacement bulbs to last one year. 18"x21½"x8" (HxWxD). 23 lb. USA.

01-0137	**Sun Pure Air Purifier**	**$599**
01-0138	**Sun Pure Replacement Set (UV Lamp and Filters)**	**$135**

Item #	Description	Removes[1]	Type[2]	Cu. ft.	Room size/min. air exchange[3]	Special Feature
01-0163	Healthmate HEPA Jr.	B, G, P, D, M, F, S, VOC	HEPA/carbon/ zeolite	200	10'x25'/6x hr.	Best VOC removal
01-8005	Healthmate HEPA	B, G, P, D, M, F, S, VOC	HEPA/carbon/ zeolite	400	20'x25'/6x hr.	Best VOC removal
01-0137	UV Air Purifier	B, V, G, P, D, M, F, S, VOC	HEPA/UV/ion/ zeolite	467	24'x24'/6x hr.	Best overall
01-0514	Gaiam PlasmaWave	B, V, G, P, D, M, S, VOC	Plasma wave/ HEPA	200	10'x25'/7x hr.	Best virus/bacteria remover

Life expectancy of any filter media is dependent on the concentration of the contaminants to which the system is exposed. The chart is based on information provided by the manufacturer. Results may vary.

[1]**B** = bacteria; **V** = viruses; **G** = gases; **P** = pollen; **D** = dander; **M** = mold; **F** = fungi; **S** = smoke; **VOC** = Volatile Organic Compounds.

[2]**UV** = ultraviolet light—when exposed to UV, bacteria and viruses are killed on contact. **HEPA** = High Efficiency Particulate Air—filter traps tiny airborne particles. Rated 99.97% removal at 0.3 micron in size. **Ion** = ionization—ions attract and attach themselves to airborne irritants, including smoke, dander, and bacteria. The irritants are then drawn into the filter or dropped to the ground, out of the air. **Carbon** = removes common household odors from cleaners, paints, solvents, and other chemical substances. **Zeolite** = a volcanic material with absorptive properties that removes gases and odors. **Foam** = a low-level filter that removes large particulate matter and extends the life of internal media.

[3]The first set of numbers is the room size (width by length); the number after the slash indicates how many times filtered air is exchanged throughout the room.

Antibacterial Air Washer

There's nothing like a good rain to clear the air of dust, pollen, and odors. And that's exactly what this ingenious device will do—inside your home. Our Air Washer purifies up to 600 square feet of indoor air while also creating a healthy level of humidity. With 22 unique humidifying discs that constantly turn through water to remove particles as small as 0.3 micron, including dust, pollen, and odors, you'll never need replacement filters. Just place discs in the dishwasher to clean. A patented antibacterial protection system using soluble silver ions is proven to eliminate the need for chemical additives to prevent bacterial growth. Simply replace the Ionic Silver Stick once a year. Two whisper-quiet modes make it ideal for nighttime operation and draw only 20 watts. Portable water tank allows easy refills. 16½"x12½"x15" (LxWxH). 13 lb. Czech Republic.

01-0509 Air Washer $299

01-0510 Ionic Silver Stick Replacement (replace yearly) $49

Window Ventilator

Bring refreshing breezes into a room while keeping airborne pollutants out. The Window Ventilator has built-in baffles to filter out dust, pollen, dirt, smoke, and insects. Rustproof ventilator installs easily and adjusts to fit windows between 24 inches and 44 inches wide. USA.

08-0044 Window Ventilator 7½"H (1 lb. 2 oz.) $28

08-0044 Window Ventilator 11½"H (1 lb. 12 oz.) $32

Car Ionizer

Research shows that car interiors trap pollution and can often contain more contaminants than outside air. Our Car Ionizer plugs into your car's lighter socket to release healthy negative ions that bond to odors and pollutants in the air and remove them, creating an oasis of clean air. With an output of 30,000 ions per cubic centimeter, this is the most compact and quiet model available. 4½"L. 6 oz. China.

17-0331 Car Ionizer $29

Note: Operation may interfere with lower-frequency AM radio bands.

Composting Toilets and Greywater

Novel, Safe, and Clean Solutions to an Age-Old Human Problem

A Short History of Waste Disposal

The way humans dispose of their biological waste is a bellwether of civilized society. Does this society protect its members from the horrors of dysentery and even worse diseases? Over the 20th century, sanitation in North America evolved from almost every home having an outhouse in the backyard, and rivers in major cities serving as open sewers, to modern flush-and-forget-it systems where everything seems to simply disappear and rivers are running very much cleaner. Making distasteful items disappear and rivers run clean is certainly going to win public approval, and we won't be so foolish as to suggest that something was mystically good in the old days. Outhouses smell bad in the summer, are too close to the house, and allow flies to spread filth and disease. In the winter, they don't smell but are too far away.

However, in our rush to sanitize everything in sight, we end up throwing away a potentially valuable and money-saving resource, and we overdesign expensive and energy-intensive disposal systems that still may pollute groundwater. Our conventional plumbing systems mix a few gallons of heavily polluted blackwater from our toilet with hundreds of gallons of very lightly polluted greywater from our sink, shower, tub, and washer. By going for the simplest possible short-term solution, we've taken a small problem and made it much larger, more difficult, and more expensive.

In this chapter, we're going to talk about what happens if we separate the really nasty stuff that goes down the toilet from the just barely nasty stuff that goes down the shower, tub, or clothes-washer drain. Blackwater from toilets can and should be treated differently from greywater. Both are potential resources and should not

simply be thrown away. Greywater, with minimal or sometimes no treatment, can be used for landscape and garden watering. Human waste can be composted safely and odorlessly to kill pathogens, then used as a highly beneficial compost.

Composting toilets can close the nutrient cycle, turning a dangerous waste product into safe compost, without smell, hassle, or fly problems. They are usually less expensive than conventional septic systems, and they will reduce household water consumption by at least 25%. But like the venerable outhouse, composting toilets deal only with human excreta. Unlike a modern septic system, they won't provide greywater treatment.

> Actually, all pollution is simply an unused resource. Garbage is the only raw material that we're too stupid to use.
> —**Arthur C. Clarke**

A real two-story outhouse from the early 1900s.

What Is a Composting Toilet?

A composting toilet is a treatment system for toilet wastes that does not use a conventional septic system. Composting toilets were developed in Scandinavia, where thanks to recent (geologically speaking) glacier activity, almost no topsoils suitable for conventional septic systems exist. A composting toilet is basically a warm, well-ventilated container with a diverse community of aerobic microbes living inside that break down the waste materials. The process creates a dry, fluffy, odorless compost, similar to what you find in a well-maintained garden compost pile. Flowers and fruit trees love it, though we don't recommend using this compost on kitchen gardens, as some human pathogens possibly could survive the composting process.

The composting process in such a toilet does not smell. Rapid aerobic decomposition—active composting—which takes place in the presence of oxygen, is the opposite of the slow, smelly process that takes place in an outhouse, which works by anaerobic decomposition. Anaerobic microbes cannot survive in the presence of oxygen and the more energetic microbes that flourish in an oxygen-rich environment. If a composting toilet smells bad, it means something is wrong. Usually smells indicate pockets of anaerobic activity caused by inadequate mixing.

A modern Biolet composting toilet.

How Do Composting Toilets Work?

A composting toilet has three basic elements: a place to sit, a composting chamber, and an evaporation tray. Many models combine all three elements in a single enclosure, although some models have separate seating, with the composting chamber installed in the basement or under the house. In either case, the evaporation tray is positioned under the composting chamber to catch any liquids, and some sort of removable finishing drawer is supplied to carry off the finished compost material.

Ninety percent of what goes into a composting toilet is water. Compost piles need to be damp to work well, but many composting toilets suffer from too much water. Evaporation is the primary way a composting toilet gets rid of excess water. If evaporation can't keep up, then many units have an overflow that is plumbed to the household greywater or septic system. Heat and air flowing through the unit assist the evaporation process. Every composting toilet has a vertical vent pipe to carry off moisture.

> A composting toilet is basically a warm, well-ventilated container with a diverse community of aerobic microbes living inside that break down the waste materials.

Composting Toilet Basics

Composters use rapid aerobic decomposition, like a well-turned garden compost pile, to break down wastes. More than 95% of the material that goes into the composter disappears up the vent as water vapor or gases.

Why would I want one?

Composters can provide safe waste processing in locations where conventional septic systems are impossible or ill-advised, such as lakeside cabins or clay soils. They are the most environmentally friendly method of waste disposal.

Do they smell?

Forget your outhouse experiences. If a composting toilet smells, it's telling you something's wrong.

What's left, and what do I do with it at the end of composting?

There's a dry, fluffy, odorless material in the finishing drawer. Once every few months, you pull out the entire drawer and carry it out to your fruit trees or ornamental plants. It's compost! It's good for them!

When is a composter not a good idea?

Apartment dwellers, this isn't for you. Intermittent use in cold environments is a problem, too. The biological community in a composter likes 60°F or warmer. Below 50°F, the little critters go into hibernation and all composting activity stops. Composting toilets can freeze without harm, but obviously you can't use them then, except for very occasional use as a storage tank.

COMPOSTING TOILETS

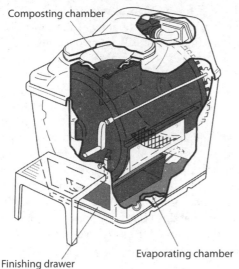

Composting chamber

Finishing drawer

Evaporating chamber

A typical Sun-Mar composting toilet.

Air flows across the drying trays, around and through the pile, then up the vent to the outside of the building. The low-grade heat produced by composting helps to provide sufficient updraft to carry vapor up the vent. However, like any passive vent with minimal heat, these are subject to downdrafts. Electric composters include vent fans and a small heating element as standard equipment. All manufacturers include, or offer, an optional vent fan with their nonelectric models that can be battery- or solar-driven. Every one to three days, the owner should add a cup of high-carbon-content bulking agent, such as peat moss, wood chips, or dry-popped popcorn. This helps soak up excess moisture, makes lots of little wicks to aid in evaporation, and creates air passages that prevent anaerobic pockets from forming.

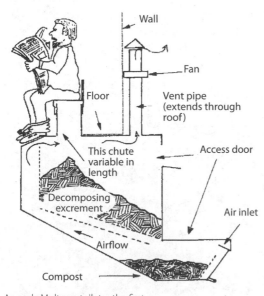

Wall

Fan

Floor

Vent pipe
(extends through
roof)

This chute
variable in
length

Access door

Decomposing
excrement

Air inlet

Airflow

Compost

An early Multrum toilet—the first generation of composting toilets.

The solids are treated with a diverse microbiological community—in other words, composted. Keeping this biological community happy and working hard requires warmth and plenty of oxygen, the same things needed for speedy evaporation. Smaller composters usually employ some kind of mixing or stirring mechanism to ensure adequate oxygen to all parts of the pile and faster composting action. Composting toilets work best at temperatures of 70°F or higher; at temperatures below 60°F, the biological process slows to a crawl, and at temperatures below 50°F, it comes to a stop. The composting action itself will provide some low-level heat, but not enough to keep the process going in a cold environment. It is okay to let a composting toilet freeze, like in a summer cottage over the winter, although it shouldn't be used when cold or frozen. Normal biological activity will resume when the temperature rises again.

The earliest composter designs, such as the Clivus Multrum system, use a lower-temperature decomposition process known as moldering, which takes place slowly over several years. These composters have air channels and fan-driven vents, but they lack supplemental heat or the capability of mixing and stirring. Therefore, these very large composters don't promote highly active composting. They have two other disadvantages. Liquids often have to be manually removed or pumped out of these units because the lower temperatures slow down the rate of evaporation. And, because temperatures don't get very high, there's a greater chance of pathogens or parasite eggs surviving the composting process. This type of large, slow composter is most effective in public access sites, and many are currently in successful use at state or national parks. Their large bulk lets them both absorb sudden surges of use, and weather long periods of disuse without upset.

Manufacturer Variations

Every composting toilet manufacturer will be delighted to tell you why their unit is the best. We'll attempt to take a more impartial attitude and give the pluses and minuses of each design.

First, some general comments. Smaller composters certainly cost less, but because the pile is smaller, they are more susceptible than larger models to all the problems that can plague any compost pile, such as liquid accumulation, insect infestations, low temperatures, and an unbalanced carbon/nitrogen ratio. Smaller composters require the user to take a more active

It is okay to let a composting toilet freeze, like in a summer cottage over the winter, although it shouldn't be used when cold or frozen.

role in the day-to-day maintenance of the unit. We have found that the smaller units with electric fans and thermostatically controlled heaters have far fewer problems than the totally non-electric units. So be aware that the less-expensive composting toilets have hidden and long-range costs.

Sun-Mar Composting Toilets

Sun-Mar is the largest and most experienced of the small residential-size composter manufacturers. Their composting toilets use a rotating drum design, like a clothes dryer, that allows the entire compost chamber to be turned and mixed easily—and remotely! This ensures that all parts of the pile get enough oxygen and that no anaerobic pockets form. Routine once- or twice-a-week mixing also tumbles the newer material in with older material, so the microorganisms can get to work on recent additions more quickly. Periodically, as the drum approaches one-half to two-thirds full, a lock is manually released and the drum is rotated backward. This action lets a portion of the composting material drop from the drum into a finishing drawer at the bottom of the unit. The next time the drum needs a portion moved along, the finishing drawer, which now contains a load of finished compost, is emptied onto your orchard, ornamental plants, or selected parts of your garden, reinstalled, and a fresh load dumped in.

Sun-Mar has introduced a large Centrex 3000 model that has a composting drum twice as long as anything we've seen previously, with stacked drying trays. This design doubles the solids capacity, triples the drying tray area, and doubles the time that material can spend fully composting, alleviating many of the more common complaints we've heard about Sun-Mar units.

The main disadvantage of the Sun-Mar drum system is that fresh material can be included with the material dumped into the finishing drawer, and as a result, some pathogens may survive the composting process. Also, the limited size of most Sun-Mar composters means they work best for intermittent-use cabins or small households. With the exception of the large Centrex 3000 model, these composters can on occasion be overwhelmed by full-time use in larger households.

Sun-Mar produces both fully self-contained composting toilets and centralized units, in which the compost chamber is located outside the bathroom and connected to either an air-flush toilet or an ultralow-flush toilet with standard 3-inch waste pipe. Electric models all feature thermostatically controlled heating elements to keep the microbe community happily active and assist with evaporation, and a small fan to ensure fresh airflow and negative air pressure inside the compost chamber. We strongly recommend the electric models for customers who have utility power available. For those with intermittent AC power, AC/DC models are available that take advantage of AC power when it's available but can operate adequately without it as well. Finally, Sun-Mar makes nonelectric models that don't need any power, although use should be limited to one or two people or a low-use cabin with these models.

BioLet Composting Toilets

BioLet composters use a composting chamber that sits upright like a bucket. Fresh material enters the top of the chamber, works its way down as composting progresses, and eventually is allowed to sift out the bottom of the chamber into the finishing drawer. Mixing and stirring rods rotate inside the drum to ensure oxygen supply and to sift finished material out the bottom screen. It is not possible for fresh material to pass through the composter to the finishing tray in this design, although the mixing and stirring may not result in as complete an aeration as with the tumbling-drum design.

BioLet produces two basic sizes, which, with various options, amount to four different models. The Standard is the smaller size and is available in a nonelectric model with manual mixing, an electric model with manual mixing, and an electric model with automatic mixing. The 40% larger XL size comes in one electric model with automatic mixing. The BioLet heater warms the

Some Sun-Mar models have the toilet separated from the composting chamber.

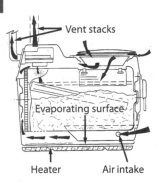

Cross-section of a typical Sun-Mar composter.

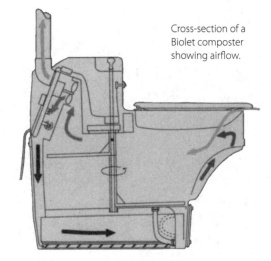

Cross-section of a Biolet composter showing airflow.

Carbon/Nitrogen Ratios

Proper composting requires a balance of carbon and nitrogen in the organic material being composted. Human excreta are not properly balanced, as they are too high in nitrogen. They require a carbon material to be added for the encouragement of rapid and thorough microbial decomposition. In the mid 1800s, the concept of balancing carbon and nitrogen was not known, and the high nitrogen content of humanure in dry toilets prevented the organic material from efficiently decomposing. The result was a foul, fly-attracting stench. It was thought that this problem could be alleviated by segregating urine from feces (which thereby reduced the nitrogen content of the fecal material), and dry toilets were devised to do just that. Today, the practice of segregating urine from feces is still widespread, even though the simple addition of a carbonaceous material to the feces/urine mix will balance the nitrogen of the material and render the segregation of urine unnecessary.

From *The Humanure Handbook*, 2nd ed.,
Joe Jenkins. Used with permission.

incoming air before it passes over the evaporation trays to speed evaporation. All models except the nonelectric require utility power or a plentiful renewable energy source such as a strong hydro system.

BioLet designs feature a number of innovative and user-friendly elements. A tasteful pair of clamshell doors immediately below the seat open only when weight is put on the seat. This means you don't have to look at your compost before sitting down. The heat level is adjustable, and the unit has a clear sight tube to see if liquid is accumulating (which means it's time to turn up the heat). All the fans, heaters, thermostat, and mechanical components are mounted on a single easy-access cassette in the top rear, so if repairs are ever needed, it's all easy to get to.

Disadvantages with these composters: They have a tendency to overly dry the compost, which leads it to clump up in big balls. This makes it really hard to stir. The shear pin or the nylon gears that drive the stirring rods are a common repair until folks learn to add water if clumping occurs. The tasteful clamshell doors

under the seat mean that guys may need to be resourceful, or take it outside.

Carousel Composting Toilets

The Carousel is a very large composter for full-time residential use. It features a round carousel design with four separate composting chambers. Like most composters, it has drying trays under the composting chamber and a fan-assisted venting system to ensure odorless operations. Each of the four chambers is filled in turn, which can take from two to six months; then a new chamber is rotated under the dry toilet.

This gives each of the four batches up to two years to finish composting without being disturbed. Batch composting has the advantage of not mixing new material with older, more advanced compost, allowing the natural ecological cascading of composting processes. This results in more complete composting and greatly reduces the risk of surviving disease organisms. The Carousel can handle full-time use and large families or groups. Because it does not have a heating element, it needs to be installed in a conditioned indoor space. But then, with only the small vent fan to run, the power use will be very low. Its disadvantages are obvious: It's big, and it's expensive initially.

Will Your Health Department Love It As Much As You Do?

Probably not, but this depends greatly on the enlightenment quotient of your local health official. Some health departments will welcome composting toilets with open arms and even encourage their use in some locales, while others will deny their very existence. We have often found that in lakeside summer cabin situations, health officials prefer to see composting toilets rather than pit privies or poorly working septic systems that leach into the lake. Several composters now carry the NSF (National Sanitation Foundation) seal of approval, which makes acceptance easier for local officials but does not mandate it. It is really up to your local sanitarian. He can play God within his own district, and there is nothing that you, Gaiam Real Goods, or the manufacturer can do about it. Our advice if you're trying to persuade your authorities? Be courteous. Health officials have to consider that a composting toilet might suit you to a tee and that you'll take good care of it, but what if you sell the house? Will the next homeowner be willing, or able, to take care of it, too?

The Carousel Composter for full-time and larger family use.

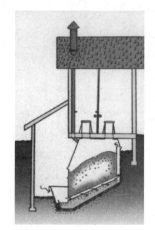

The CTS System for high-use public sites.

Compost piles like warmth because it makes all the little microbes work faster. Thus, composting toilets work best in warm environments.

Also, be aware that composting toilets deal only with toilet wastes, the blackwater. You still need a way to treat your greywater wastes—all the shower, tub, sink, and washer water. Unconventional greywater systems will face the same approval problems as composting toilets. Some localities happily will allow alternatives that use the greywater for landscape watering, but others may require a standard full-size septic system. In all cases, greywater systems must use subsurface disposal. Once the water goes down the drain, it can't see daylight, or the possibility of human contact, again. Greywater alternatives are covered in more detail below.

Composting Toilet Installation Tips

KEEP IT WARM

Compost piles like warmth because it makes all the little microbes work faster. Thus, composting toilets work best in warm environments. If your installation is in a summer-use cabin, then a composting toilet is ideal. A Montana outdoor installation in wintertime isn't going to work at all. If you plan to use the composter year-round, then install it in a heated space on an insulated floor. The small electric heater inside the compost chamber will not keep it warm in an outdoor winter environment, but it will run your electric bill up about $25 a month trying. Talk to one of Real Goods technicians if you have any doubts about proper model selection or installation.

PLUG IT IN

If you have utility power, by all means use one of the electrically assisted units; they have far fewer problems overall. If you have intermittent AC power from a generator or other source, use one of the AC/DC, or hybrid, units. Anything that adds warmth or increases ventilation will help these units do their work. We strongly recommend adding the optional vent fans to non-electric units.

LET IT BREATHE

The compost chamber draws in fresh air and exhausts it to the outside. If the composter is inside a house with a wood stove and/or gas water heater that are also competing for inside air, you may have lower air pressure inside the house than outside. This can pull cold air down the composter vent and into the house. Composters shouldn't smell bad, and something is wrong if they do, but you don't want to flavor your house with their gaseous byproducts either. Do the smart thing: Give your wood stove outside air for combustion.

DON'T CONDENSE

Insulate the composter's vent stack where it passes through any unheated spaces. The warm, humid air passing up the stack will condense on cold vent walls, and moisture will run back down into the compost chamber. Most manufacturers include vent pipe and some insulation with their kits. Add more insulation if needed for your particular installation.

KEEP EVERYTHING AFLOAT

With central units such as the Centrex models, where the composting chamber is separate from the ultralow-flush toilet, the slope of any horizontal waste-pipe run is critical. The standard 3-inch ABS pipe needs to have 1/8 to ¼ inch of drop per foot of horizontal run. More drop per foot than that allows the liquids to run off too quickly, leaving the solids high and dry.

Vertical runs are no problem, so if you want a toilet on the second floor, go ahead. Sun-Mar recommends horizontal runs of 18 feet maximum, but customers have successfully used runs of well over 20 feet. Play it safe and give yourself clean-out plugs at any elbow when assembling the ABS pipe. If you are using an air-flush unit, it must be installed directly over the composter inlet.

With thanks to Joseph Jenkins and *The Humanure Handbook.*

Greywater Systems

What Is Greywater?

Greywater is the mix of water, soap, and whatever we wash off at the shower, sink, dishwasher, or washing machine. Basically, it is any wastewater other than what comes from toilets. This is the majority of water used in most households. Most homes produce 20-40 gallons of greywater per person, per day. Greywater is also likely to contain bacteria, including small amounts of fecal coliform, protozoans, viruses, oil, grease, hair, food bits, and petroleum-based "whatevers."

WHY WOULD YOU WANT TO USE GREYWATER?

Utilizing greywater has many advantages. It greatly reduces the demands on your septic system, gives otherwise wasted nutrients to plants, reduces household water use, increases your awareness of natural cycles, provides very effective water purification, and probably reduces your use of energy and chemicals. How great is that? Greywater is used primarily for landscape irrigation, because soil is an excellent water purifier. By substituting for freshwater, greywater can make landscaping possible in dry areas. It may also keep your landscaping alive during drought periods. Sometimes a greywater system can make a house permit possible in sites that are unsuitable for a septic system.

WHEN WOULD GREYWATER BE A POOR CHOICE?

Greywater recycling isn't for everybody. High-rise apartment dwellers would be seriously challenged. But let's assume we're speaking to folks with at least a suburban plot around them. Even then, you may have insufficient yard and landscaping space, major bits of your drain plumbing may be encased in concrete, your climate may be too wet or too frozen, costs may outweigh benefits, or it might just be illegal where you live.

How Does a Greywater System Work?

Although the trace contaminants in greywater are potentially useful, even highly beneficial, for

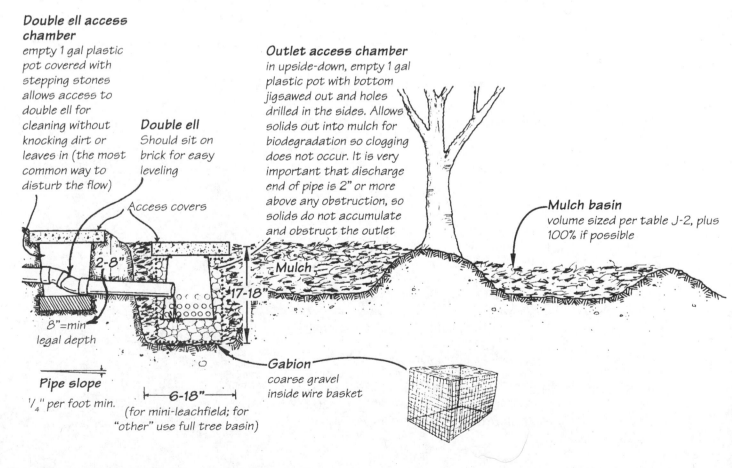

Double ell access chamber
empty 1 gal plastic pot covered with stepping stones allows access to double ell for cleaning without knocking dirt or leaves in (the most common way to disturb the flow)

Double ell
Should sit on brick for easy leveling

Access covers

Outlet access chamber
in upside-down, empty 1 gal plastic pot with bottom jigsawed out and holes drilled in the sides. Allows solids out into mulch for biodegradation so clogging does not occur. It is very important that discharge end of pipe is 2" or more above any obstruction, so solids do not accumulate and obstruct the outlet

Mulch basin
volume sized per table J-2, plus 100% if possible

2-8"

Mulch

17-18"

8"=min legal depth

Pipe slope
¼" per foot min.

6-18"
(for mini-leachfield; for "other" use full tree basin)

Gabion
coarse gravel inside wire basket

plants and landscaping, their presence demands some modest caution in handling and disposal. How do you recycle it safely? All responsible greywater designs stem from two basic principles:

- Natural purification occurs while greywater passes slowly through healthy topsoil.
- No human contact should occur before purification.

Greywater system designs display great diversity. At the simplest end of the spectrum, greywater can be carried over to plants. At the most complex, it can be finely filtered and automatically distributed as needed through subsurface drip irrigation tubing. Middle-of-the-road systems include connecting the washing machine to a hose that is moved from mulch basin to mulch basin, and branched drain systems, which take one flow and split it into several permanent outlets.

Like other aspects of ecological design, greywater systems are highly site-specific. Each system needs to suit the site, climate, and perhaps the commitment level of the owner. The one general rule is that there are no general rules for greywater systems. The optimum system depends primarily on how much water you've got, how much irrigation you need, the soil permeability, your budget, and legal constraints. At www.greywater.net, you can find a downloadable copy of the system selection chart from "Create an Oasis with Greywater," which lists the attributes of 20 different greywater system design options.

Will Your Health Department Love Greywater as Much as You Do?

As with composting toilets, acceptance of greywater depends entirely on the folks at your local health department. Some will give you a choice, some won't. Some may allow greywater systems so long as you also install a full-size, approved septic system. In most areas, the legality of greywater is ambiguous. Generally, official attitudes are moving toward a more accepting and realistic stance, and authorities will often turn a blind eye toward greywater use.

In the late 1970s, during a drought period, the state of California published a pamphlet that explained the illegality of greywater use, while showing detailed instructions of how to do it! Since then, California has become one of many places where you can install a greywater system legally but with rules so ridiculous that less than 1% of greywater systems go through the permitting process. Happily, there is a new trend toward reasonableness and realism in greywater regulation. Since 2001, the state of Arizona has allowed the installation of greywater systems of less than 300 gallons a day that meet a list of reasonable requirements without having to get a permit. For up-to-date information, and much more info on presenting a greywater system to your friendly local health department, consult the Greywater Policy Center at www.greywater.net.

Recycling and greywater use may suit you perfectly, but someday you're likely to sell your perfect house. Will a greywater system suit the next owners as well? It's the people who will buy the house from you that the health department is thinking about. The best advice is simply to call your health department and ask. But ask as a hypothetical example. Health officials recognize the "hypothetical" question readily. This allows you to ask specific pointed questions, and them to answer fully and honestly, without anyone admitting that any crime or bending of the rules has occurred or is likely to occur.

How to Make a Greywater System

Legal greywater systems tend toward engineering overkill, while many of the simple and economical methods that folks actually use are still

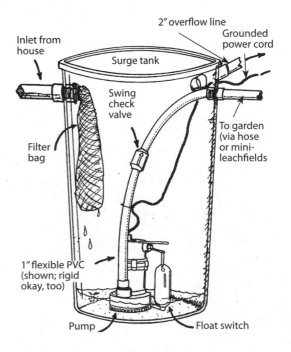

technically illegal. Start with the greywater guide series of booklets listed in the product section that follows. These provide the best overview of the subject and detail every type of greywater system, from the simple dishpan dump to fully automated designs. There are many possible ways to create a greywater system, and since the booklets cover them all with grace, style, a wealth of illustrations, and at minimal cost, we will not duplicate their work here. Aim for the best possible execution with the simplest pos-

sible system. During new home construction, it's fairly easy to separate the washer, tub, sink, and shower drains from the toilet drains. Retrofits are a bit more trouble, and in some cases impossible, like when the plumbing is encased in a concrete slab floor. Kitchen sinks are often excluded from greywater systems because of the high amount of oil, grease, and food particles. But this is a personal choice and depends on how much maintenance you're willing to do on a regular basis.

Legal greywater systems tend toward engineering overkill, while many of the simple and economical methods that folks actually use are still technically illegal. Aim for the best possible execution with the simplest possible system.

GREYWATER PRODUCTS

Greywater Systems

Earthstar Greywater Systems

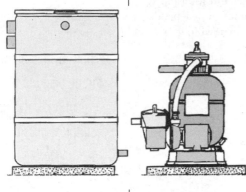

Automatic operation, sand filter with backwash cleaning. Simple, high quality, and easy maintenance. When greywater reaches a high level in the 55-gallon tank, a float switch starts a centrifugal pump that begins an irrigation cycle. A simple five-minute backwash cleaning process every two months keeps everything in top shape. And you don't even have to get your hands wet. Comes complete with everything you need to install. Color-coded components make assembly easy in less than an hour. The 12-gallon tank is good for tight spaces. Lead time 8-10 weeks. USA.

44831	Greywater System w/ 55-gal. tank	$1,199*
44832	Greywater System w/ 12-gal. tank	$1,099*

*Please allow 8-10 weeks for delivery; shipped directly from manufacturer.

Hard-to-Find Greywater Parts

Three-Way Valves. Installed at washers, kitchen sinks, or other drains that might not always be appropriate for greywater. Lets you choose to use your greywater system or your conventional septic system. Never needs lubrication. Guaranteed to never leak or break. Accepts two pipe sizes, one inside, one outside. CPVC material.

Double Ell with Cleanout. This specialized ell fitting with cleanout is required for the simple, low-maintenance branched-drain systems. See the book *Branched Drain Greywater Systems*. All other parts for these systems will be available locally. ABS material.

44833	3-Way Valve, 1.5 or 2 in.	$49
44834	3-Way Valve, 2 or 2.5 in.	$49
44835	Double Ell w/Cleanout 1.5 in.	$18
44836	Double Ell w/Cleanout 2 in.	$24

COMPOSTING PRODUCTS

Composting Toilets

Sun-Mar Composting Toilets: Saving the Planet One Toilet at a Time

With toilets on all seven continents—including sites in Antarctica and Mount Everest base camp—the Sun-Mar company has single-handedly conserved more than 2 billion gallons of water.

That's why, since 1985, Real Goods has been proud to feature products by Sun-Mar, the world leader in consumer composting. This remarkable company was founded in the mid-1960s in Sweden with a mission to create waterless toilets off the traditional septic grid, using Mother Nature instead of hazardous chemicals to compost human waste into a healthy and usable product.

Based in Canada, Sun-Mar has been in the composting toilet business longer than almost anyone. Sun-Mar designs are all based on a rotating drum, like a clothes dryer. Turning a crank on the outside of the composter rotates the drum, which ensures complete mixing and good aeration of the "pile." A well-turned compost pile works faster and more effectively. Mixing should happen once or twice a week, whenever the composter is in active use. When the drum is rotated in the normal mixing direction, the inlet flap swings shut and all material stays in the drum. Periodically, when the drum approaches one-half to two-thirds full, a lock is released that allows the drum to rotate backward. This lets some compost drop into the finishing tray at the bottom. It finishes composting here until the next time the drum needs to dump some material.

The fully self-contained Sun-Mar composters, the Excel and Compact models, are good choices for summer camps, sites with intermittent use, sites without electricity (Excel-NE only), or full-time residences with winter heating (composting drums must be in a 65°F or warmer space to function properly).

The Centrex series puts the composting drum outside or in the basement for those who want a more traditional-looking toilet or toilets in the bathroom. Centrex models are available for a full range of uses from intermittent summer camps to full-time residences. The entire line is ANSI/NSF listed now, which will make it easier for local health officials to accept their use.

If you have utility electricity available at your site, we strongly recommend using one of the standard AC models with fan and heater. They compost faster, are more tolerant of variable uses, and are much less likely to suffer any performance problems. This electrical advice applies to all composting toilet models.

SMELLING LIKE A ROSE

As you might imagine, Sun-Mar's customer service team gets some interesting queries and comments—most notably, "What about the smell?" Answer: It's a nonissue with Sun-Mar composting toilets. Aerobic decomposition produces no sewer gases, and a partial vacuum is maintained by the venting system to ensure odor-free operation.

SUN-MAR CENTRAL COMPOSTING SYSTEMS

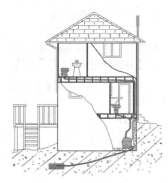

For those who want the benefits of a composting toilet but prefer a traditional type of toilet in the bathroom, the Sun-Mar Centrex low-flush systems are ideal. There are two types of central units available: low-flush and water-free systems. Centrex low-flush systems are designed for use with a porcelain 1-pint flush toilet. The low-flush toilet installed in the bathroom is connected by a standard 3-inch plumbing pipe to the composting unit, which is below the toilet and normally within 15-20 feet (horizontal distance). Canada.

CENTREX 1000 SERIES

Our smallest central composter
The Centrex 1000 series is designed for light to medium seasonal use. It is not recommended for continuous residential use and is available in a standard AC model or a nonelectric model. The Centrex 1000 has all controls, access, and venting easily accessible for service. The vent location makes cottage installation easier, even when using the NE (nonelectric) model, which requires a straight vent with no bends. The Centrex 1000 uses an RV toilet with an ultralow flush of approximately 1 pint. Toilets are sold separately. The 1-inch overflow drain should be connected to the household greywater system.
Dimensions: 27.5"x31.5"x26.25" (HxWxD)
(35.5" width needed to turn handle)
Weight: 50 lb.
(Shipping weight: 100 lb. approx.; varies slightly with model)
Mfr. warranty: 3 yr.
Ships via freight from Buffalo, NY.

Centrex 1000

This light- to medium-capacity AC electric unit is designed for use with a low-flush toilet (sold separately). 2-inch vent stack with a 35-watt fan and thermostatically controlled 260-watt heater.
Capacity: Residential use: *not recommended*. Seasonal use: 5-7 people.

44105   Centrex 1000 $1,595

Centrex 1000 NE

This nonelectric version of the Centrex 1000 incorporates a 4-inch vent for airflow with no heater or fan. The optional 1.4-watt DC vent fan is recommended. It is designed for use with a low-flush toilet (sold separately). Capacity: Residential use: *not recommended*. Seasonal use: 4-6 people.

44106   Centrex 1000 NE $1,395

Stack Fans

44803	**12V/1.4W 4-in. Stack Fan**	**$55**
44804	**24V/1.4W 4-in. Stack Fan**	**$55**

CENTREX 2000 SERIES

For medium to heavy seasonal use or light residential use

The Centrex 2000 series is probably the most popular of the central composting units. Good for medium to heavy seasonal use or light residential use, the 2000 series is the refined result of over 25 years of composting experience. Available in a standard AC version or a nonelectric version. In addition, each version can be ordered for either a low-flush, RV-type toilet(s) or for a dry, air-flush toilet. Multiple low-flush toilets can serve a single central composter, but only a single air-flush toilet can be used with the A/F models. Toilet options are sold separately.

A 1-inch overflow drain is included and should be connected to an approved facility. Suggested capacities vary slightly with model, but generally, the 2000 series is good for 3-6 adults in residential use or 5-8 adults in vacation use. Canada.
Dimensions: 27.5"x44.75"x26.5" (HxWxD)

(48.75" width needed to turn handle). Weight: 100 lb. (Shipping weight: 125 lb. approx.; varies slightly with model). Mfr. warranty: 3 yr.
Ships via freight from Buffalo, NY.

Centrex 2000

This standard high-capacity AC electric unit is designed for use with a low-flush toilet (sold separately) and comes with a 370-watt thermostatically controlled heater, a 2-inch vent, and a 35-watt fan. Capacity: Residential use: 4-6 people. Seasonal use: 7-9 people.

44817   Centrex 2000 $1,795

Centrex 2000 NE

This nonelectric version of the Centrex 2000 incorporates a 4-inch vent and has no heater. It is designed for use with a low-flush toilet (sold separately). 2.4-watt DC vent fans are included.
Capacity: Residential use: 3-5 people. Seasonal use: 6-8 people.

44818   Centrex 2000 NE $1,595

Sun-Mar water-free systems use the handsome Sun-Mar Dry Toilet, fitted with a removable bowl liner and connected to the central composting unit by means of a 10-inch-diameter pipe section. Odor-free operation is maintained by a fan that pulls air down through the toilet seat and up the vent stack. Sun-Mar water-free toilet systems can be used only where the toilet can be placed directly over the composting unit.

Centrex 2000 A/F

This high-capacity electric unit is designed for use with a Sun-Mar Dry Toilet (sold separately).
Capacity: Residential use: 4-6 people. Seasonal use: 6-8 people.

44820   Centrex 2000 A/F $1,895

Centrex 2000 NE A/F

This nonelectric version of the Centrex 2000 A/F incorporates a 4-inch vent and a 12-volt fan for airflow. It is designed for use with a Sun-Mar Dry Toilet (sold separately).
Capacity: Residential use: 3-5 people. Seasonal use: 5-7 people.

44821   Centrex 2000 NE A/F $1,695

Stack Fans

44803	**12V/2.4W 4-in. Stack Fan**	**$55**
44804	**24V/3.12W 4-in. Stack Fan**	**$55**

CENTREX 3000 SERIES

Our highest-capacity composting toilet

The Centrex 3000 series is designed for heavy cottage use or medium continuous residential use. The double-length composting drum in the 3000 series has 40% more capacity than any other Sun-Mar unit. This allows

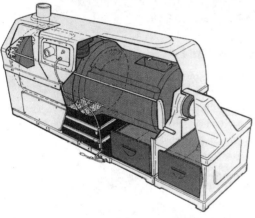

more time for composting and handles more volume or surge loads, and compost now cleverly drops into the finishing drawer automatically when the drum is turned. The evaporating chamber for excess liquids now has an added evaporation tray above the composter bottom, greatly increasing available evaporation area.

The Centrex 3000 is available in a standard AC version or a nonelectric version. All versions are further available for use with either 1-pint, ultralow-flush RV toilet(s), or an air-flush (dry) toilet. Toilets are purchased separately. Suggested capacities vary slightly with model, but generally, the 3000 series is good for 5-8 adults in residential use or 7-10 adults in vacation use. Canada.
Dimensions: 30.25"x71"x27.5" (HxWxD) all versions (45.5" depth needed to remove finishing drawer). Weight: 120 lb. approx. (Shipping weight: 200 lb.; some models slightly less). Ships via freight from Buffalo, NY. Mfr. warranty: 3 yr.

Centrex 3000

Electric models have a pair of 370-watt thermostatically controlled heaters to boost evaporation rates and keep the compost chamber warm and friendly. A 1-inch overflow security drain hose is included with all models, which should be connected to the household greywater system. Use a standard version if you have utility power. It comes with a 2-inch vent and 35-watt fan.
Capacity: Residential use: 6-8 people. Seasonal use: 9-11 people.

44-118	Centrex 3000	$1,995

Centrex 3000 NE

This nonelectric version uses a 4-inch vent with DC fan.
Capacity: Residential use: 5-7 people. Seasonal use: 8-10 people.

44-119	Centrex 3000 NE	$1,795

Centrex 3000 A/F

Capacity: Residential use: 5-7 people. Seasonal use: 8-10 people.

44-121	Centrex 3000 A/F	$2,095

Centrex 3000 NE A/F

Capacity: Residential use: 5-7 people. Seasonal use: 7-9 people.

44-122	Centrex 3000 NE A/F	$1,895

Stack Fans

44815	12V/2.4W 4-in. Stack Fan	$55
44816	24V/3.12W 4-in. Stack Fan	$55

Toilets for Use with Centrex Composters

We offer the fine china Sealand 1-pint flush toilet for use with the Centrex Series composters. The Dry Toilet includes a bowl liner that can be removed for cleaning as required. The Dry Toilet is used only with the Centrex Air Flush models. Canada.

44823	Sealand 510 China Toilet, White or Bone	$399*
44-116	Sun-Mar Dry Toilet, White	$295*
44-117	Extra-Dry Toilet 45-in. Pipe Extension	$65

*All toilets shipped freight when ordered with a Sun-Mar composter; shipped UPS when ordered alone.

Sealand 510

Sun-Mar Dry Toilet

SUN-MAR EXCEL SELF-CONTAINED COMPOSTING TOILETS

Excel self-contained composters offer a toilet and horizontal composting drum all in one assembly and are ANSI/NSF listed. Available in AC, AC/DC, or NE (nonelectric) models, with identical exterior dimensions. The black bowl liner can be easily removed for cleaning. If you have utility power, use the AC model—the vent fan and thermostatically controlled heater normally eliminate all liquids. It requires no water connections. Installed in a protected, moderate-temperature indoor location, this high-capacity unit can handle 3-5 people full-time, or even more intermittently, making it ideal for year-round or seasonal use. The drum crank handle is recessed. White or bone. Canada.
Dimensions: 32"x22½"x33" (HxWxD)
(48" depth needed to remove finishing drawer)
Dimensions for Compact: 27½"x22½"x33" (HxWxD)
(45" depth needed to remove finishing drawer)
Weight: 60 lb. (50 lb. for Compact)
Mfg. warranty: 3 yr.
Ships anywhere in the contiguous U.S. for $158.

Sun-Mar Excel AC

This popular, NSF-approved model is suitable for 3-5 people full-time and for more people intermittently. Has a 35-watt fan and a 260-watt thermo-statically controlled heater. The Excel AC is our most popular composter for installations with 120-volt AC electricity. Plugs into a standard outlet with included 6-foot cord. Normally, the Excel will evaporate all liquids. A ½-inch emergency drain is fitted and should be connected if heavy use or prolonged power outages are expected. Black bowl liner can be easily removed for cleaning. The 2-inch vent pipe exits at the top rear and can be installed invisibly through the wall. The AC fan speed control is an option to install into the fan door cover and is recommended for intermittent winter use or where slight fan noise is a concern.
Capacity: Residential use: 3-5 people. Seasonal use: 6-8 people.

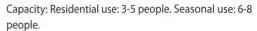

| 44-102 | Excel AC | $1,545 |
| 44824 | AC Fan Speed Control | $40 |

Sun-Mar Excel AC/DC

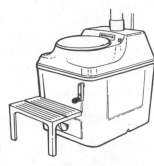

The Excel AC/DC hybrid was co-developed by Real Goods and Sun-Mar specifically for Real Goods' off-the-grid customers who derive part of their power from generators. The AC/DC is identical to the Excel model described above, except that it is fitted with an NE drain and with an additional 4-inch NE vent installed next to the Excel's 2-inch vent stack. For off-grid homes where power is intermittent or where 120-volt power will soon become available. Works with or without AC power. Supplied with a 1.4-watt DC stack fan. Please specify 12- or 24-volt.
Capacity: Residential use: 2-4 people. Seasonal use: 5-8 people.

| 44-103 | Excel AC/DC | $1,645 |

Sun-Mar Excel NE

The Non-Electric is perfect for many of our customers living off-the-grid and not wanting to be dependent on their inverters. This nonelectric model has a single 4-inch vent exiting from the rear center. In full-time residential use, an optional vent stack fan is recommended to improve capacity, aeration, and evaporation. Fans can be run on a 12-volt battery and a 10-watt PV panel combination. A 1-inch drain is supplied and must be connected to an approved facility. Must be installed in a warm location for winter use. Recommended for 2-3 people full-time.
Capacity: Residential use: 2-3 people. Seasonal use: 5-7 people.

| 44-101 | Excel NE | $1,345 |

Sun-Mar Compact

This is a scaled-down version of the Excel model. For those who don't need the larger capacity of the Excel but do have access to 120-volt power, this is the ideal unit. The working components are the same as the Excel: a bio-drum for

mixing and aeration, a thermostatically controlled base heater, and a small fan for positive air movement. What makes the Compact model different is a smaller, variable-diameter bio-drum, which allows a smaller overall size; an attractive rounded design; and most important, no more footrest! Also, the handle for rotating the bio-drum is now hinged and folds into the body when not being used. Same venting and electrical requirements and benefits as Excel AC.

Capacity: Residential use: 1-2 people. Seasonal use: 2-4 people. Dimensions: 28½"x21½"x33" (HxWxD) (45" depth needed to remove finishing drawer). Seat height: 21.5". Weight: 50 lb. (Shipping weight: 90 lb.) Ships anywhere in the contiguous U.S. for $158.

| 44-104 | Compact | $1,445 |

Stack Fans

44-824	AC Fan Speed Control	$40
44-803	12V/1.4W 4-in. Stack Fan	$55
44-804	24V/3.12W 4-in. Stack Fan	$55

Composting Accessories

Compost Sure™ Starter Mulch

A mix of coarse peat moss and chopped hemp stalk. Designed to maintain porosity and retain oxygen while keeping compost moist, mixed, and well supplied with available carbon. 30-liter/8-gallon bag. Canada.

| 44253 | Starter Mulch | $25 |

Compost Quick™

Natural enzyme solution facilitates the action of aerobic bacteria. Also effective as a cleaner for bowl liners (self-contained toilets) or 1-pint toilets (central composting systems). 16 oz. with spray top. Canada.

| 44838 | Compost Quick | $20 |
| | | Or 2 for $36 |

Microbe Mix™

Special dried microbes and enzymes initiate and accelerate composting in all Sun-Mar composting toilets and systems. Packaged on a corn grit carrier; contains citronella to discourage insects. 500g (17.6 oz.) jar with scoop. Canada.

| 44839 | Microbe Mix | $18 |
| | | Or 2 for $32 |

BioLet Composting Toilets

Made in Sweden, BioLet composting toilets are well-designed and some of the best-selling toilets in the world. They offer excellent performance and a compact design for homes that have utility power. Features include:

- Clamshell doors below the seat that hide the compost chamber until weight is put on the seat.
- A composting process that works with gravity. Fresh material enters the top; only the oldest, most completely composted material drops out the bottom into the finishing tray.
- An adjustable thermostat to control liquid buildup. Won't use electricity unless actually needed.
- A mechanical cassette with all fans, heaters, and thermostats in the top rear makes any service quick, easy, and clean.

BioLet toilets are supplied with an installation kit consisting of 14 feet of 2-inch vent pipe; 39 inches of insulation; 39 inches of outer pipe jacket to cover insulation; vent cap; and a flexible roof flashing. All Biolet models are made in Sweden.

BioLet has done an exceptional job of packaging their products, so they will ship via FedEx ground at very reasonable rates. BioLets usually ship for half or less what other composter brands will cost to deliver.

BioLet produces two basic sizes, which, with various options, produces four different models. The Standard is the smaller size and is available in a nonelectric manual-mix model, an electric manual-mix model, and an electric automatic-mixing model. The 40% larger XL size comes in one electric model with automatic mixing. All models except the nonelectric require utility power or a plentiful RE source, such as a strong hydro system.

BioLet XL

BioLet XL Model

This is BioLet's top-of-the-line model with about 40% more composting capacity than the Deluxe or Standard models. The thermostatically controlled heating element is 305 watts. The compost-stirring mixer motor runs automatically for 30 seconds every time the lid is lifted and replaced. The quiet 25-watt fan runs continuously; the heating elements run as required by thermostat setting and room temperature. This model is for homes with utility power. The XL can handle 4 people full-time or 6 people for intermittent vacation use. Sweden.
Dimensions: 26"x26"x32" (HxWxD) (26"x54" floorspace needed for finishing tray to slide out the front). Weight: 50 lb. approx. (Shipping weight: 75 lb. approx.) Ships via FedEx ground. Ships from Massachusetts; $90 to continental U.S. Mfr. warranty: 3 yr.

44109   BioLet XL Model $1,559

BioLet Deluxe Model

The Deluxe model, using BioLet's 40% smaller ABS plastic body, features the top-of-the-line automatic-stirring system, fans, and thermostatically controlled heaters like the XL model above. The heater for this smaller model is 250 watts and will run as required by thermostat setting and room temperature. The quiet 25-watt fan runs continuously. The Deluxe can handle 3 people full-time or 5 people for intermittent vacation use. This model is for homes with utility power. Sweden.
Dimensions: 26"x22"x29" (HxWxD) (22"x40" footprint needed for finishing tray to slide out). Weight: 50 lb. approx. (Shipping weight: 75 lb. approx.) Ships via FedEx ground. Ships from Massachusetts; $50 to continental U.S. Mfr. warranty: 3 yr.

44124   BioLet Deluxe Model $1,495

BioLet Standard Model

This model, BioLet's bestseller, has the 40% smaller ABS plastic body with a thermostatically controlled 250-watt heating element. Compost stirring is done manually with the handle on top; instead of flushing, you simply twirl the handle a few times. The mechanism is geared down 10-to-1, so little effort is needed. The quiet 25-watt fan runs continuously; the heater element runs as required by thermostat setting and room temperature. This model is for homes with utility power. The Standard model can handle 3 people full-time or 5 people for intermittent vacation use. Sweden.
Dimensions: 27"x22"x29" (HxWxD) (22"x40" floor space needed for finishing tray to slide out the front). Weight: 50 lb. approx. (Shipping weight: 75 lb. approx.) Ships via FedEx ground. Ships from Massachusetts; $50 to continental U.S. Mfr. warranty: 3 yr.

44110   BioLet Standard Model $1,390

BioLet Basic Non-Electric

The Basic Non-Electric model uses BioLet's smaller ABS body and the manual mixing system just like the Standard model. Unlike the Standard model, there is no heater, thermostat, or fan. An overflow fitting is supplied with this model, which needs to be routed to the home's greywater or septic system. It needs a nice warm environment to live in, as there's no heater to help out the microbes. Temperatures below 70°F will slow it way down. In a friendly environment, this composter can handle up to 3 people full-time. An optional 12-volt vent fan is available and recommended. Sweden.
Dimensions: 26"x22"x29" (HxWxD) (22"x40" footprint needed to allow finishing tray to slide out). Weight: 50 lb. approx. (Shipping weight: 75 lb. approx.) Price includes only FedEx ground shipping anyplace in the continental U.S. Ships via freight prepaid within continental U.S. Ships from Massachusetts. Mfr. warranty: 3 yr.

44125   BioLet Basic Non-Electric Model $995

44126   Optional BioLet 12V Vent Fan $60

BioLet Composting Toilet Mulch

BioLet strongly recommends using its own mulch for starting and ongoing maintenance, rather than commercial peat moss. We're increasingly finding that commercial peat moss has been ground almost to dust and has very little structure or ability to open air channels. This leads to poor composting and material clumping. BioLet's mulch is formulated specially for composting toilets. It has more tiny air channels, speeds composting by introducing oxygen, and resists clumping. Your composting toilet needs only a cup a day, or less, so it lasts a long time. 8-gallon bag, approximately 12 lb. USA.

44827   BioLet Composting Toilet Mulch $39

Incinerating Toilets

We love the composting toilet's ability to manage waste (and even recycle nutrients!) very efficiently. But, let's face it, they just don't work everywhere you might need sanitary facilities, for instance in an unheated cabin that sees only a few weekends use every winter. ECOJOHN™ provides efficient incinerating toilets as an alternative to composting toilets. They are designed to handle most types of climates and can run off the grid. By use of propane, these units incinerate waste into a sterile ash that needs to be emptied only a few times per year. The ECOJOHN™ products provide environmental, logistical, and economical benefits.

ECOJOHN™ Sr

A waterless incinerating toilet (propane)

ECOJOHN™ Sr is a self-contained toilet that is ideal in any remote application. By installing a chimney system, a propane tank, and either 12-volt DC or 110-volt AC, it is now possible to have a very efficient waterless toilet that requires minimal maintenance.

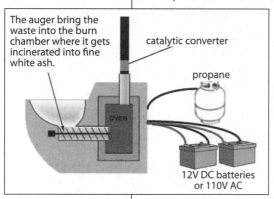

The auger bring the waste into the burn chamber where it gets incinerated into fine white ash.

catalytic converter

propane

OVEN

12V DC batteries or 110V AC

After usage, an auger moves the waste into a burn chamber where all the paper and waste get incinerated into a sterile ash. After six months of usage, the only remaining ashes will fit into a soda can. The toilet is equipped with a small container of water and a rinse button. For cleaning purposes, by pressing the rinse button, one can rinse a little bit of water in the bowl if needed. Its unique patented burn chamber makes it possible to use this model in a colder climate; there is no risk for freezing and malfunctioning parts. The catalytic converter, which is located inside the chimney system, cleans the air from pollution and eliminates any bad septic odor. A regular 20-pound propane tank will last for approximately 200 flushes.

Parts included: toilet unit, 8-foot (6-inch-diameter) double-wall chimney system, catalytic converter, chimney fan.
25.2"x23.2"x33.5" (HxWxD). 18.9" sitting depth. 132 lb.
Capacity: Residential use: 6 people. Seasonal use: 12 people.

44-0500 ECOJOHN™ Sr   **$3,995**

ECOJOHN™ Jr WC5

A waste incinerator (propane)
with low-flush toilet

ECOJOHN™ Jr WC5 system provides a new way of dealing with waste handling. A propane incinerator (Jr) connects with a waste tank and incinerates all the black-water. Therefore, with this system, it is now possible to have a water toilet in a remote application and not have to deal with any expensive pump-out processes.

A complete ECOJOHN™ Jr WC5 system includes a water tank, low-flush toilet with macerator pump, waste tank, waste pump, the incinerator (Jr), and a chimney system. After the toilet has been used, the blackwater goes through a macerator pump and is then delivered to the waste tank. There are sensors in the waste tank, which run on either 12-volt DC or 110-volt AC, that signal the Jr that there is blackwater in the tank that needs to get incinerated. Automatically, the waste pump will portion blackwater into the Jr, and the incineration process begins. The Jr continues to incinerate the blackwater until the waste tank is empty. It will then shut down and wait until it gets a new signal that new blackwater is in the waste tank. The Jr WC5 models incinerates 1.25 gallon of waste per hour. After six months of usage, the only remaining ashes will fit into a soda can. A regular 20-pound propane tank will last for approximately 200 flushes.
Dimensions:
Incinerator 24"x14"x27" (HxWxD). 70 lb.
15-gallon Waste tank 24"x15.5"x21" (HxWxD). 35 lb.

44-0501 ECOJOHN™ Jr WC5   **$5,295**

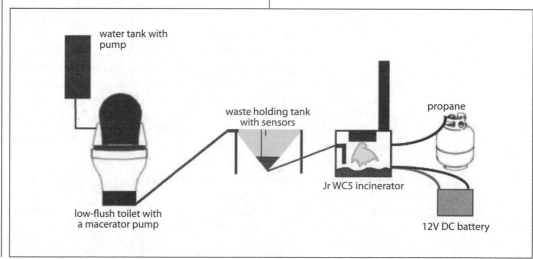

water tank with pump

waste holding tank with sensors

propane

Jr WC5 incinerator

low-flush toilet with a macerator pump

12V DC battery

Books

THE GREYWATER GUIDE SERIES

Create an Oasis with Greywater, Revised 4th Ed.

Choosing, building, and using greywater systems

Describes how you can save water, save money, help the environment, and relieve strain on your septic tank or sewer by irrigating with reused wash water. Describes 20 kinds of greywater systems. The concise, readable format is long enough to cover everything you need to consider, and short enough to stay interesting. Plenty of charts and drawings cover health considerations, greywater sources, designs, what works and what doesn't, bio-compatible cleaners, maintenance, preserving soil quality, and the list goes on. 51 pages, softcover. USA.

82440   Create an Oasis with Greywater $14.95

Builder's Greywater Guide

A companion to Create an Oasis with Greywater

Will help you work within or around codes to successfully include greywater systems in new construction or remodeling. Includes reasons to install or not to install a greywater system, flowcharts for choosing an appropriate system, dealing with inspectors, legal requirements checklist, design and maintenance tips, how Earth purifies water, and the complete text of the main U.S. greywater codes with English translations. 46 pages, softcover.

80097   Builder's Greywater Guide $14.95

Branched Drain Greywater Systems

A companion to Create an Oasis with Greywater

Detailed plans and specific construction details for making a branched drain greywater system in any context. This ultrasimple, robust system is the best choice for over half of residential installations. It provides reliable, sanitary, and low-maintenance distribution of household greywater to downhill plants without filtration, pumping, or a surge tank. It is one of the few systems that is both practical and legal, and has been described as "the best greywater system since indoor plumbing." Note: The area to be irrigated has to be downhill from the house, and laying out the lines must be done with fanatical attention to proper slope. 52 pages, softcover.

82494   Branched Drain Greywater $14.95

Greywater Guide Booklet Set

Save over $5 by picking up all three Greywater booklets together.

21-0360 Greywater Guide Booklet Set $39

The Humanure Handbook, 2nd Ed.

A Guide to Composting Human Manure

Deemed "Most Likely to Save the Planet" by the Independent Publishers in 2000. For those who wish to close the nutrient cycle and don't mind getting a little more personal about it than our manufactured composting toilets require, here's the book for you. Provides basic and detailed information  about the ways and means of recycling human excrement, without chemicals, technology, or environmental pollution. Includes detailed analysis of the potential dangers involved and how to overcome them. The author has been safely composting his family's humanure for the past 20 years, and with humor and intelligence has passed his education on to us. 302 pages, softcover.

82308   The Humanure Handbook $25

Off-Grid Living, Farming, and Homesteading Tools

Growing Your Own and Living the Good Life

IN CELEBRATION OF REAL GOODS' 30th anniversary, and in keeping with this *Sourcebook*'s focus on relocalization of energy and food production, we have expanded this chapter on homesteading tools to look closely at two relevant approaches to gardening and farming. With substantial help from two practitioners—Benjamin Farher and Jim Fullmer—we are pleased to offer this introduction to the concepts and techniques of Permaculture and Biodynamic Agriculture. Both are holistic systems for designing growing spaces and integrating food production into larger ecological, social, and spiritual contexts. Both are fundamentally grounded in principles and practices drawn from close observation and deep understanding of the natural world. Permaculture and Biodynamics offer us profound guidelines for the right way to practice agriculture on any scale in a world increasingly defined by peak oil and global climate change.

Permaculture and Biodynamics are conceptual tools that are particularly appropriate for off-grid living and any kind of homesteading. Real Goods can provide you with many of the physical tools that will enable you to do the "hard work of simple living" efficiently and effectively. The better the quality of your tools, the better your craftsmanship is likely to be. Jobs will also be easier and more enjoyable. Humans are a tool-using, tool-loving species, and good tools improve the quality of our lives and contribute to making them more fulfilling.

Tools for better lives and more efficient living are what this entire *Sourcebook* is about. We've managed to compartmentalize most of them neatly into chapters about solar power, water pumping, composting toilets, or the like, but we've always carried a selection of useful gizmos, such as our solar-powered hat fan, a motion-detecting deer chaser, and our push lawn mower, that defy nice, neat categorization. So, after you've read about Permaculture and Biodynamic Agriculture, check out our wild, diverse, and fun collection of things that defy being compartmentalized, in the product section following the essays. Enjoy browsing!

> Permaculture and Biodynamics are conceptual tools that are particularly appropriate for off-grid living and any kind of homesteading.

Permaculture: A Holistic Design System for Your Site

You may have heard of Permaculture, and chances are what comes to mind is a style of gardening. That's true—but Permaculture is much more than a style or even a philosophy of gardening. Fundamentally, Permaculture is a holistic system of design that embraces the totality of a place. Grounded in the ethical intentions of Earth Care, People Care, and Fair Share, Permaculture design is a system of assembling conceptual, material, and strategic components

> Fundamentally, Permaculture is a holistic system of design that embraces the totality of a place.

in patterns that provide mutually beneficial, sustainable, and secure places for all forms of life on this Earth.[1] Permaculture pulls from the depths of indigenous wisdom as well the patterns found in Nature. It is a way to garden and a way to build our homes, but it is also a design system that can be applied to larger-scale economic and social institutions.

History and Background

The word *Permaculture* comes from marrying the two words *permanent* and *agriculture*, or just *culture*. But while agriculture is a foundation, limiting Permaculture to simply a way of gardening would be like limiting the concept of energy production to the burning of fossil fuels. The concept is much broader, invoking all aspects of a healthy food system and a holistic way of thinking and living. Permaculture is truly about design, connectivity, and relationships.

The term *Permaculture* was coined in the late 1970s by two Australians, Bill Mollison and David Holmgren, who were studying the unstable and unsustainable characteristics of Western industrialized culture. Asking questions about how the human species had existed for so long in harmony and balance with nature yet was now contributing so much to the destruction of Earth, they were drawn to indigenous worldviews and the self-sustaining and regenerative qualities of natural processes and systems. They began to think about how human processes and systems could be patterned after the designs found in nature. Other people were doing similar work at the same time, and Permaculture drew from and contributed to a larger conversation among pioneers of ecological thinking:

Permaculture begins with an ethic of personal responsibility for taking care of the Earth by our actions and the choices we make.

P. A. Yeomans' Keyline Plan, J. Russell Smith's work on tree crops, Masonobu Fukuoka's *One Straw Revolution*, James Lovelock's Gaia Hypothesis, and Lawrence Halprin's work on ecological design. Holmgren brought all these currents together in a thesis he wrote under Mollison's tutelage, which was published in 1978 as *Permaculture One*. Mollison published *Permaculture Two* (1979), and then the book that became the definitive text of Permaculture design, *Permaculture: A Designer's Manual* (1988), which covers the ecology of all living systems and how to design our human place in them. Holmgren focused on applying and testing the theories and concepts that he and Mollison had developed. He established a successful design firm that has created projects all over the world during the past 25 years, the majority of them in Australia and Europe. Most recently, he published *Permaculture: Principles and Pathways Beyond Sustainability* (2003), which elaborates a comprehensive, integrated set of Permaculture principles.

Permaculture Ethics, Principles, and Practices

Permaculture begins with an ethic of personal responsibility for taking care of the Earth by our actions and the choices we make. As Mollison says, "The only ethical decision is to take responsibility for our own existence and that of our children. **Make it now.**" This foundational ethic is elaborated in three simple concepts:

Earth Care. Allowing provisions and resources for all life systems to continue and multiply.

People Care. Allowing provisions for people to access those resources necessary to their existence.

Fair Share. Return of unneeded surplus and respect for the idea that there are limits to population and consumption. By governing our own individual needs, we can set resources aside for Earth Care and People Care.

In *Permaculture: A Designer's Manual*, Mollison articulated five primary organizing principles for the practice of Permaculture. Other practitioners have since articulated related sets of ideas. The core design ideas of Permaculture can be distilled into the following 12 principles:

1. Work with nature, rather than against natural elements, forces, processes, dynamics, and evolutions, so that we can assist rather than impede natural developments. On a practical level, this translates into using gravity, native species, and the Sun and wind, for example, in designing homescapes, landscapes, gardens, and other environments.

2. The problem is the solution; everything works both ways. How we see things determines whether situations or elements are advantageous or not. Everything is a valuable resource.

3. Make the least change for the greatest possible effect. Allow work to be a source of your energy, not a sink.

4. The yield of the system is theoretically unlimited. The limits on the number of uses for a given resource within a system are the information available and the imagination of the designer.

5. Everything gardens, or has an effect on its environment; everything is connected.

6. Relinquishing power. The role of beneficial authority is to return function and responsibility to life and people.

7. Unknown good benefit. If we start with good intentions, other good things follow naturally.

8. Succession of evolution. Natural design inherently evolves toward stability and resiliency in a system.

9. Cyclical opportunity. Every cyclical event increases the opportunity for increased yield; therefore, increased cycling of resources increases yield.

10. Functional design. All functions are supported by many elements, while each element performs many functions.

11. Stability is created by a number of beneficial connections between diverse beings.

12. Information as a resource. Information is the critical potential resource. Bad information can result in a poor design, whereas good information increases the opportunity for a good design.

Observation, Cycles, Relationships, and Patterning

While these ethics and principles supply the framework for the Permaculture design process and govern the choices designers make, careful observation is the crucial tool that informs those choices.

Observation is by definition the attentive watching of somebody or something. Permaculture goes deeper in calling for Protracted And Thoughtful Observation (PATO). It is thoughtful because, as a designer, you are observing the relationships and cycles that are happening before you, and it is protracted because you are observing them over a period of time. One

Photo: Benjamin Fahrer

An herb spiral embodies the spirit of Permaculture.

can only assume and project patterns and flows upon first impressions of a situation or site. Observations over time are necessary in order to glean information that can reveal probable patterns of ecological succession and evolution. Many Permaculturalists will tell you that that the hammock is one of their best design tools; a raft floating on a pond is another. By melting into the surroundings and spending time in focused observation, nature begins to reveal itself to you, and this unveiling is a true blessing and an invaluable resource.

Considering cycles, relationships, and patterns brings us to the idea that Permaculture as a whole design science has both visible and invisible structures. A simple home garden, for example, encompasses many cycles and patterns that are visible: the nutrient cycle, the life cycle, the hydrological cycle, plant cycles, seasonal cycles, and so on. A huge variety of patterns are present within the elements of these cycles. Whether it is the web of a garden spider, the spiral of a sunflower, or the branching of a leaf or tree, recognizing those patterns is essential to creating integral concepts that work with

Permaculture is whole-system design, in which the social or invisible aspects are just as integral as the physical and visible ones.

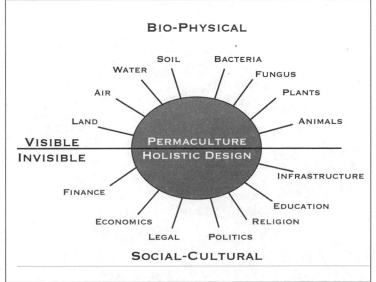

Illustration source: Permaculture Drylands Institute, rendered by Brock Dolman.

nature. Form is the envelope within which life pulsates. And in Nature's design, form follows function. Permaculture relies on biomimicry, or emulating nature's patterns and processes, to create efficient and elegant designs for human use that also enable other living things to flourish.

In the realm of human relations and social institutions, the principles of Permaculture can help to orient us to the power of invisible structures. Take self-governance within a community, for example. The cycling of roles and responsibilities, the flow of money within larger financial cycles, yearly cycles of meetings and fundraisers: The character of these kinds of interactions is strongly influenced by people's habitual patterns of behavior, emotional and physical, that surround them in a working environment—and they are largely invisible to the people involved. The practice of Protracted And Thoughtful Observation enables conscious participants to sustain ethical intention and help keep everyone positively connected in pursuit of the tasks and goals at hand. Permaculture is about mutually beneficial relationships, in human affairs as well as in the ways humans relate to nature.

Zone and Sector Analysis

One of the most unique aspects of Permaculture design is the concept of zone and sector analysis, which centers on the conservation of energy. Analyzing zones and sectors within an environment literally energizes the third major principle noted above, gaining the greatest effect from the least amount of work.

Zones refer to the energy resources that are on-site and are somewhat controllable through

Zones reflect frequency of use, or how often an element within the system is visited and needs attention. Zone 0 is the point of origin, such as a house. Things placed in Zone 1 are visited intensively, and frequency decreases through Zone 5, a wild place that provides inspiration. The concept is cyclical, so that energy from Zone 5 is brought back into the other zones.

Permaculture is a kind of relational intelligence, wherein all the parts of a system relate to each other organically.

human action. Sectors refer to the energy resources that come in from off-site and are more or less uncontrollable because they are produced at the command of nature.

Permaculture looks at zones as a series of edgy concentric circles moving outward from a point of origin that represents the focal point of the system—you, the subject. These zones are determined and defined by frequency of use. The point of origin, called Zone 0, is typically the inside of your house. It can also be thought of as the interior, subjective place where your being and spirit reside.

Zone 1 begins when you step outside of your house or personal space. Items or areas within Zone 1 are visited multiple times throughout the day and usually require continual observation and work, as they are areas of intensely cultivated space, such as a kitchen garden. Here nature is arranged in such a way as to fit specific human needs.

Zone 2 is not as intensely cultivated and is visited one or two times a day. The components of Zone 2 do not need as much attention as those within Zone 1, yet they still requires daily interaction. Animals such as chickens or goats are housed here along with the more robust garden and small fruit trees. Zone 2 is designed to express a greater balance between human needs and nature.

Zone 3 is considered the "farm zone," and objects or elements in this area are meant to supply the site with an abundance of surplus to either sell or reinvest. Larger orchards, row crops, and grazing pasture located in Zone 3 require less than daily attention, perhaps visitation only one or two times a month.

Zone 4 is a managed area of land that borders on wilderness. Including tree crops grown for fuel and food, large-scale water collection in ponds and lakes, and wild crafting and foraging

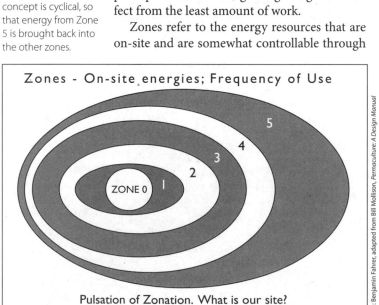

Zones - On-site energies; Frequency of Use

ZONE 0 1 2 3 4 5

Pulsation of Zonation. What is our site?
What level are we working at? Sense of scale.

Illustration: Benjamin Fahrer, adapted from Bill Mollison, *Permaculture: A Design Manual*

Photo: Benjamin Fahrer

of native plants, Zone 4 is managed about once a year.

Zone 5 is the area considered wilderness and has no management strategies except those imposed by nature. In this zone, we are visitors, not managers. Zone 5 is where we look for examples to emulate and the true teaching of natural design. Although you might think this zone is the least used area of land on the site, and thus frequency of use would be very low, that is not necessarily so. In fact, Zone 5, as with all the zones, can actually cross over to or border one's own home zone. Observations within Zone 5 can happen every morning over tea as one looks out over pristine forest and beholds nature's masterpiece. Zone 5 actually feeds back into Zone 0 and provides constant inspiration and rejuvenation. It is a feedback loop adding a pattern of connectivity and wholeness between the core of the system and the rest of the universe.

Sectors are external forces like the Sun, wind, water, fire, and sound, to name a few, that flow through the site. The incoming energy from various sectors can be blocked, deflected, or channeled, depending on the characteristics of the site and the goals of the design. Wind blocks are very common, for example trees or fences built to help shelter a field or home from a prevailing wind, or to hide a road from view. Trees and windscreens can also be planted to channel wind certain ways and use it as a resource, for example, to provide evaporative cooling to a house or to create a Venturi effect for a wind turbine. Many approaches to landscape or site design utilize the blocking or channeling of sector energies. What's unique about Permaculture is its holistic and integrated view that seeks to optimize the use of energy flows throughout the environment of a site. In many cases, the energy of a sector is used to do work that would normally be performed with an on-site energy resource. Examples would be siting a house to utilize passive solar heating, thereby reducing the demand for firewood from Zone 4, and planting deciduous trees on the south side of the house to provide summer shade and passive cooling.

Once zones and sectors are identified and outlined, they are mapped out on the site. Such mapping pays particular attention to how various forces and elements overlap and interact. Determining an appropriate Permaculture design usually requires a lot of jostling around of buildings, plantings, and elements. Ultimately, an effective energy conservation placement pat-

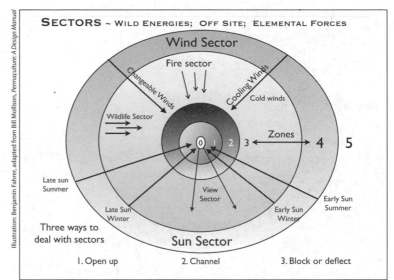

tern takes form that efficiently utilizes the resources at hand.

Permaculture Design Process

The general core model is a graphic aid that demonstrates how all the components of Permaculture fit and work together in the design process. As just noted, designing a site requires the participants to be open to change, trial and error, and revision, as their understanding of energy flows and resource interactions evolves and matures through sustained observation. Representation as a mandala suggests that the process is a never-ending loop of assessment, visioning, conceptual planning, and master planning. This mandala forms a lens of natural-systems thinking that is used to view all dimensions of the site in question. What works on paper does not necessarily work in the field, and small changes in

Sector analysis considers the wild, or external, energies that flow through a site. These energies may be harnessed, channeled, deflected, or blocked.

The seed is the idea or site that is being designed. Permaculture uses the method of Protracted and Thoughtful Observation, grounded in ethical intentions, to generate a process of assessment, visioning, conceptual planning, implementation, and further assessment.

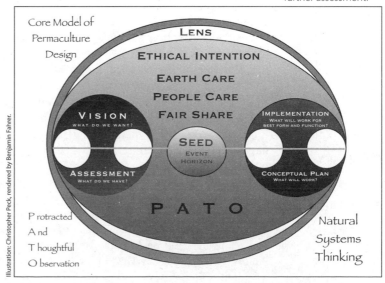

Swales are commonly used by Permaculture practitioners to direct water flows through a site.

Photo: Benjamin Fahrer

one aspect can ripple into dramatic changes in overall structure and design.

This process would probably drive traditional designers and conventional clients crazy. But in embracing a way of whole-systems thinking, Permaculture practitioners accept the ever-evolving design as a living, creative entity. Designer Benjamin Fahrer tells the story of a three-day winter rain that revealed previously unrecognized aspects of a client's site and prompted a complete rethinking of the original

homestead and garden design. What seemed like a problem at first soon became a wonderful opportunity to fulfill the client's deepest desires—but only because the people involved were willing to observe natural dynamics, reassess initial assumptions, and reimagine their plans. Flooding and drainage problems became a pond, gardens were relocated to a more advantageous spot, and those changes provided more sun for the house.

The essential purpose of design is to create the possibility for events to happen. A perfect design is fixed and stagnant, and requires no further input, human or natural. An imperfect or incomplete design accepts change and is enhanced by chance occurrences—enriched by weeds and edge, by the changing patterns of sunlight and shade, even by a branch falling on the paths or terrace. A habitation is a natural system, and it should be designed and treated as such.

In its most profound sense, the Permaculture design process is not limited to land-based projects: It can be applied to any aspect of life. It is inclusive, it is participatory, it is a navigational compass for designing one's entire life—and it is Permaculture when it lives up to the ethical intentions of Earth Care, People Care, and Fair Share.

Biodynamic® Agriculture: Food and the Cosmos

History and Current Relevance

Created by Rudolf Steiner (also the founder of the Waldorf education movement), Biodynamic Agriculture is one of the original foundations of what we today call organic agriculture.

Biodynamic Agriculture is a holistic, some would say spiritual, approach to growing food, which understands a farm or garden as a living entity that is situated within a web of other living entities, both large-scale and small-scale. Created in 1924 by the German philosopher and scientist Rudolf Steiner (also the founder of the Waldorf education movement), Biodynamic Agriculture is one of the original foundations of what we today call organic agriculture. Steiner laid out the principles of Biodynamic Agriculture in a series of lectures, in response to solicitations from farmers who were alarmed both at the growing use of synthetic chemical fertilizers and pesticides, and at what they perceived as a noticeable and relatively rapid decline in crop and animal vitality. As explained by Dr. Ehrenfried Pfeiffer, a student of Steiner's who introduced the Biody-

namic method to the United States in 1938, the name *Biodynamic* refers to "working with the energies which create and maintain life." (Biodynamic® is a registered trademark of Demeter Association Inc., which does educational work and is the certifying agency for Biodynamic Agriculture.)

Biodynamic farm management relies on close attention to the interrelation of the constituent parts of an agricultural process—soil fertility, water, pests, animals—rather than concentrating on the individual parts in isolation from each other. A critical aspect of this practice is to provide necessary agricultural inputs from within the living dynamics of the farm itself, rather than importing them from the outside. In theory, this concept is fundamental to "organic" agriculture as well, but in practice, the focus on this fundamental has diminished greatly. Increasingly, as the scale of organic agriculture expands, the organic *industry* seems

Plant growth forms, spiraling and enveloping.

likely to forget its true and integral foundations. Biodynamic Agriculture offers a way to restore this focus in modern times, with deep roots in history, and the modern-day organic *movement* would do well to take heed. Humanity cannot afford to lose sight of these fundamentals.

The original idea of "organic" agriculture requires co-operation with living systems inherent to this planet. Fertility, for example, is based on recycling organic material that is generated on the farm. In one way or another, raw organic materials are fed to an army of soil life that uses it as food. The organisms die and, in return, provide nutrition to growing crops via the dynamics of soil humus formation and the related soil biology and chemistry.

These systems operate within a biological concept of time that is rhythmic and is based on seasons, weather patterns, sunrise, and sunset. This biological timeframe dictates how quickly a farm can reach a level of fine-tuned efficiency and maximum productivity. The biological networks that support agriculture can be intensified and accelerated to a degree, but the system cannot be pushed beyond its means without bringing in help in the form of imported materials.

At some point in time, an "organic" agriculture arose that, out of necessity, was guided by the pressures of supply and demand economics. At that point, the foundational principle of internally provided inputs began to break down and become abandoned. Bringing in external materials to an organic system reintroduces some of the same problems that conventional agriculture presents, namely dependence on Earth's natural resources to mine, refine, and transport a myriad of products that are shipped all over the world. By its nature, this practice puts pressure on natural resources and the natural systems where these materials are mined or harvested. When the demand for organic food exploded, the only way to meet it was to fortify the existing farms with imported materials. This

reality has created a new form of organic farming that has diverged from its roots.

The Philosophy of Biodynamic Agriculture: A Farm as a Living Organism

In principle, a Biodynamic farm is formed in the image of a living organism. A good example of a self-contained living organism as seen in its natural state is an undisturbed forest. The fertility, pest control, and water dynamics a healthy forest needs for thriving all come from within the forest ecosystem itself. Most farming systems are a few steps removed from a forest ecosystem, but their vitality and production results from the farm ecosystem that is created.

Biodynamic farming is based on *holistic* observations of living nature itself. When one looks at nature holistically, rather than trying to understand it as separate parts, it becomes evident that *forces* are at work that are not physically manifest. In the plant world, for instance, unseen forces are literally molding matter into an infinite myriad of often delicate and pristine forms, colors, and aromas that defy the laws of gravity.

Consider the relationship between the plant world and the Sun. The Sun is a celestial body very distant from our Earth, and Earth is only one element of the Sun's reality, which includes other solar systems. If you watch closely the growing tips and resulting morphology of most green plants, you will see an expression of grasping at sunlight, spiraling, enveloping—a pulling of a plant species into form, not a pushing out into space.

Rudolf Steiner's concept of Biodynamic Agriculture was based on a worldview he described as "spiritual science." In his view, the physical world around us cannot be understood without also understanding the life forces that form and decay it. The modern concern that Biodynamic

These aerial photos (right, at the bottom of this page, and on the facing page) show how a single farm is embedded in and part of a series of ecosystems, watersheds, and bioregions.

Animals, manure, crop rotation, and green manures are integral to Biodynamic agriculture.

Farming seeks to address is that Earth, a living being herself, is not in a state of healthy balance; Biodynamic Farming is a method for deep healing of Earth.

To paint in your imagination the scope of a Biodynamic farm, consider the images above, below, and on the facing page.

The first three photos above are aerial views (increasing in altitude) of a Biodynamic farm in western Oregon. Within the boundaries of this farm (circled) is a dynamic system complete with biodiverse habitat (as forest, wetland, river riparian, hedgerow, diverse cropping, and developed insectaries) and internal humus-based fertility dynamics (livestock/ composting manure, careful crop rotation, green manures). The development of this system also provides the farm with its crop-nutrition, pest-control, and water-conservation inputs without having to import them on a truck or in a bag. Instead of seeking this or that remedy for specific management problems, single inputs that address numerous management concerns are implemented to more or less kick-start the biological wisdom that is already inherent to the farm ecology. For instance, by implementing a fertility program to develop soil humus, water management (by increasing the soil's ability to hold water) and pest control (through balanced crop growth that receives nutrition from the dynamic relationship between roots and soil life) are also addressed.

As one ascends skyward, it is clear that this organism within the fence line is itself part of a wider ecosystem. The farm is part of the Douglas fir forest ecosystem that surrounds it.

A myriad of birds, amphibians, mammals, and insects move out of the surrounding forest and flow through the farm; some move in and contribute.

The next three photos carry the theme further. In the first, it can be seen that the farm is situated on the flood plain of a river valley. On a clear night, the cold air from the highlands pours down onto the valley floor where the farm resides and creates a chilly microclimate that is quite cooler than areas a few miles downstream. Next, that valley can be seen in relation to the other drainages that shed water down the east slope of the Willamette Valley in western Oregon. The farm is an element of an even broader system that drains copious amounts of water.

As the next images show, the farm and its valley are situated in the Pacific Northwest of North America, an environment that is conditioned by the ocean to the west and the winds it brews. A great river of air ebbs and flows across the face of the North American continent. The farm, always a microcosm within a greater macrocosm, is an element of the identity of Earth, itself a living organism with a cold top and bottom and a warm middle. It spins and thus the farm (and the farmer) contends with the forces of gravity.

Earth is an element of a wider solar system. Astronomy tells us that Earth and its moon are in a rhythmic dance with each other and also with other celestial bodies, circling the Sun and each other. This dance goes on with the utmost mathematic precision and predictability. The star that Earth and its comrades dance around

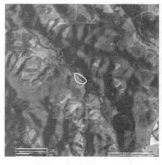

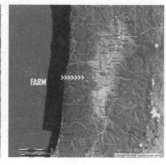

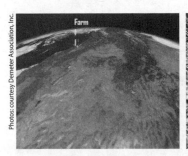

Farm

is one of countless others in a larger system, all a microcosm of yet a greater macrocosm, the Milky Way galaxy. And galaxies abound in the universe.

The critical point for Biodynamic Agriculture is that even the tiniest aspect of a farming system, a seed or a bud for instance, contains the archetypal imprint of the widest celestial sphere. The premise is that Earth (including its inhabitants) has lost elements of its ability to be "awake" and in concert with the activity of the wider spheres. The result is basically a state of illness, which can be healed only by reharmonizing Earth with the celestial.

Biodynamic Preparations

(courtesy of Hugh Courtney)

The good agronomy practices that are at the base of Biodynamic Farming are not sufficient to achieve the process of healing. Steiner developed a set of unique "Biodynamic preparations," which are a critical element of Biodynamic Farming. In the U.S., the Biodynamic preparations have been most successfully created and distributed by Hugh Courtney at the Josephine Porter Institute, for many years now. Some regional groups and individual farmers have also been making the preparations.

In total, there are nine BD "preps." Horn manure (BD #500) and horn silica (BD #501) are field sprays, diluted and stirred vigorously for one hour in water, which relate to earth and light forces, respectively. The yarrow (BD #502), chamomile (BD #503), stinging nettle (BD #504), oak bark (BD #505), dandelion (BD #506), and valerian (BD #507) preparations are known as the compost preps and are applied to compost and manure piles. The horsetail herb, *Equisetum arvense* (BD #508), is sprayed in liquid form and has a regulatory function with respect to fungus and other manifestations of the watery element.

A different way of thinking is required to approach agriculture from the Biodynamic perspective. In our "conventional chemical" or even in the "organic" approach, we are condi-

tioned to think in terms of substances—or, rather, in terms of chemical requirements that can be met by particular substances. In chemical agriculture, we bring nitrogen (N) to the soil via ammonia or urea, and in organic agriculture, we do so via manure. For phosphorus (P), the choice is superphosphate or rock phosphate. Regardless, we are still thinking in terms of chemical substances or, as many understand it, in NPK terms. Biodynamic Agriculture and its preparations ask us to think in terms of *forces* rather than substances. One need not throw out all knowledge of soil chemistry to think in Biodynamic terms—but we must go beyond the chemical point of view. Just as the effects of the forces of gravity or magnetism can be observed without actually being able to *see* these forces, so too we can recognize through their effects the forces that are released through use of the soil-enlivening Biodynamic preparations. For instance, Steiner describes the plant as a being that exists between Sun and Earth, or between the polarities of silica and limestone with clay as the mediating factor. In present-day agriculture, silica is totally ignored, and the variables associated with limestone are barely recognized. When Steiner refers to silica, limestone, or clay, he doesn't mean the substances but rather the cosmic forces that permeate the substances. The beneficial effects of cosmic forces are noticeable in greater seed viability, improved food quality, and healthier livestock and crops.

Practicing Biodynamic Agriculture

To understand the forces behind the minute amounts of substances used in Biodynamic preparations, here is Hugh Courtney's analysis of the preps as they relate to silica/limestone/clay:

Limestone	Clay (or clay-humus)	Silica
Polarity	Mediating factor	Polarity
BD #500	BD #502-507	BD #501 & #508

Earth itself is also embedded in and is part of larger celestial systems.

> The critical point for Biodynamic Agriculture is that even the tiniest aspect of a farming system, a seed or a bud for instance, contains the archetypal imprint of the widest celestial sphere.

For the enlivening cosmic forces standing behind the silica and limestone to work with full effect, the first step is to introduce the mediating forces of the BD Compost Preparations to the soil. This is usually done via properly made BD compost using BD #502-507. Proper composting using the preps is a key feature of Biodynamic Agriculture. Another means for applying the compost preps (BD# 502-507), particularly when beginning a conversion, is by use of the BC (Biodynamic Compound Preparation), a product containing BD #502-507 that can be applied as a spray. BD-made compost itself, however, has the virtue of imparting a longer-lasting effect to the soil than does the BC. Once the forces of BD #502-507 are working in the soil, the BD #500 (horn manure) is applied, followed by BD #501 (horn silica) and BD #508 (horsetail herb/*Equisetum arvense*) according to the need.

The Biodynamic Principles

1. To restore to the soil the organic matter it needs to hold its fertility in the form of the very best humus.

2. To restore to the soil a balanced system of functions. This requires looking at the soil not only as a mixture or aggregation of chemicals, mineral or organic, but as a living system. Biodynamics concerns itself with the conditions under which soil microlife can be fully established, maintained, and increased.

3. While the Biodynamic method does not deny the role and importance of the mineral constituents of the soil, especially the so-called fertilizer elements and compounds—including nitrogen, phosphate, potash, lime, magnesium, and the trace minerals—it sponsors the most skillful use of organic matter as the basic factor for soil life. (Interestingly, the importance of trace minerals for normal and healthy plant growth was actually pointed out by Rudolf Steiner as early as 1924.) Advocates of organic farming were among the first champions of the value of manure and compost.

4. In the Biodynamic method, life and health depend on the interaction of matter and energies as well as chemical substances. A plant grows under the influence of light and warmth—that is, energies—and transforms them into chemically active energies by way of photosynthesis. A plant consists not only of mineral elements (inorganic matter that makes up only 2%-5% of its substance) but also of organic matter such as protein, carbohydrates, cellulose, and starch, all of which derive from the air (carbon dioxide, nitrogen, oxygen) and make up the major part of the plant mass aside from water (15%-20%). The greater part of plant mass, some 70% or more, consists of water.

5. The interaction of the substances and the energy factors forms a balanced system. Only when a soil is balanced can a healthy plant grow and transmit both substance and energy as food. The Biodynamic method aims to establish a system that brings into balance all factors that maintain life.

6. Were we to concentrate only on nitrogen, phosphate, and potash, we would neglect the important role of biocatalysts (trace minerals), enzymes, growth hormones, and other transmitters of energy. The Biodynamic way of treating manure and composts includes the knowledge of enzymatic, hormonal, and other energy factors.

7. Restoring and maintaining balance in a soil requires proper crop rotation. Soil-exhausting crops with heavy demands on fertilizing elements should alternate with neutral or even fertility-restoring crops—on the farm as well as in the garden and even in the forest. A soil that has been put to maximum effort—producing corn, potatoes, tomatoes, peppers, and cabbage (all greedy crops), for example—should have a rest period with restorative soil-building crops such as legumes. Temporary cover with grass and clover helps to improve humus and nitrogen levels. Greedy crops and arable cultivation consume humus.

8. The entire environment of a farm or garden is important. One must pay attention to the air, the functioning biological system, the quality of the hillsides, and the water balance.

9. The soil also has a physical structure. The maintenance of a crumbly, friable, deep, well-aerated structure is an absolute must if one wants to have a fertile soil. All factors that lead to structural disintegration of the soil (like plowing soil that is too wet) must be considered.

The preparations must be made with exactitude, following Steiner's directions, so they are able to convey forces to enliven the soil and heal the Earth. Equally careful attention must be paid to the processes of stirring, spraying, and storing the preps by the person intending to put them to use. Various of the preparations require stirring in a particular way for specific lengths of time. *The stirring is the most important step* and cannot be circumvented if the forces of the preparations are to be transferred to the water and brought to the soil as a spray. Proper equipment to do these tasks should be on hand before obtaining large quantities of preps.

Ultimately, the BD farmer should plan to make his or her own preparations so as to regain control over the economic destiny of the farm. Further help in making preps is available from the Biodynamic Farming and Gardening Association (see below) and through workshops offered at the Josephine Porter Institute for Applied Bio-Dynamics in Woolwine, Virginia (see below). JPI also sells preparations to those new to Biodynamic Agriculture, or otherwise unable to make their own.

If you're serious about practicing the art of Biodynamic Agriculture, it takes a serious commitment. The Earth, a living being, can be healed, and our foods can be endowed with spiritual and cosmic forces, which are otherwise lacking in foods grown using NPK thinking. The nine BD preparations need to be applied as a totality and should be orchestrated by the farmer or gardener out of an attempt to understand the dynamic forces and influences working into nature out of the cosmos.

End Note

1. Bill Mollison, *Permaculture: A Designer's Manual* (Tagari, 1988).

Resources

Demeter Association
PO Box 1390
Philomath, OR 97370
541-929-7148
www.demeter-usa.org
(Education and certification of Biodynamic Agriculture)

The Josephine Porter Institute for Applied Biodynamics
PO Box 133
Woolwine, VA 24185
276-930-2463
info@jpibiodynamics.org
www.jpibiodynamics.org
(Product catalog available for biodynamic preps)

Bio-Dynamic Farming and Gardening Association
25844 Butler Road
Junction City, OR 97448
541-998-0105
biodynamic@aol.com
www.biodynamics.com

Bio-Dynamic Farming and Gardening Association of Northern California
PO Box 453
Fair Oaks, CA 95628
916-965-0389
biodynamic@aol.com
www.biodynamics.com

HOME AND GARDEN PRODUCTS

Most jobs need tools. The better the quality your tools are, the better your craftsmanship is likely to be. The job will also be easier and more enjoyable. Humans are a tool-using, tool-loving species—that's one of the primary traits that sets *Homo sapiens* apart from the rest of the animal kingdom. Good tools improve the quality of our lives and contribute to making them more fulfilling.

Tools for better lives and more efficient living are what this entire *Sourcebook* is about. We've managed to compartmentalize most of them neatly into chapters about solar power, water pumping, composting toilets, and the like, but we've always carried a selection of useful gizmos, such as our solar-powered hat fan, a motion-detecting deer chaser, and our push lawn mower, that defy nice, neat categorization. So here, in this section on off-the-grid living and homesteading tools, you'll find our wild, diverse, and fun collection of things that defy being compartmentalized. Enjoy browsing!

GARDEN, BIODYNAMICS, COMPOST, AND PERMACULTURE TOOLS

Our Gaiam Real Goods color catalogs offer our widest selection of gardening tools and gizmos, but we have a few perennial favorites that deserve year-round space here in the *Sourcebook*. To order one of our periodic Gaiam Real Goods catalogs, call 800-919-2400 or check out our website: www.realgoods.com.

The Best Hemp Work Gloves

Knit from 100% sustainable hemp, these work gloves are the best we've used. Highly durable, naturally rot-resistant hemp fibers shrug off moisture and tough work conditions. They insulate and protect better than cotton or synthetics, and last far longer. Even when they get wet and muddy, dirt just dries and falls off, leaving these gloves as pliant as ever. And because their natural fibers absorb skin oils, they only get softer with use. S, M, L. USA.

14-0288 Hemp Work Gloves   **$10**

Topsy Turvy Planter

Forget the hassle you think is required to grow a home garden. With the Topsy Turvy Planter, there's no more digging, weeding, bending, stakes, or cages needed. It instantly turns any space with ample sunlight into a garden—even patios and balconies. As plants grow down, all you do is water conveniently from the top with the included funnel. Works wonders for tomatoes, cucumbers, peppers, and flowers. 35½"x9½" (H x diameter). 4 lb. China.

14-0345 Topsy Turvy Planter   **$17**

Extra-Wide-Brim Garden Hat

Garden in comfort with 360° Sun protection. Our lightweight hat uses an extra-wide brim, a long 3-foot rear veil, and fabric with an SPF rating of 45 to keep harmful rays away from your head and neck. An oversized crown with mesh panels provides ample ventilation to keep you cool and dry, and an internal sweatband absorbs perspiration. For all-day comfort, a rear sizing band, locking chin strap, and ponytail hole work together to create a custom fit. It's even water resistant! M, L. USA.

14-0287 Garden Hat   **$34**

Biodegradable and Compostable Bags

For years, we've searched for an alternative to polyethelene-based plastics for the kitchen, dog waste, and the yard. We're proud to finally say that we've found

an answer in Bio-Bags™ and Garden Bio-Film. Made from cornstarch, vegetable oil, and other renewable resources, all are 100% biodegradable and compostable. The bags are tough enough to handle your household and yard needs—and break down in about a month. For your garden, use Bio-Film between the soil and your plantings to protect from weeds and frost and to stimulate growth. Within 90 days, 95% of the film will be decomposed, adding its nutrients to the soil. Each product meets strict U.S. and international certification standards. USA.

14-9331 Lawn & Leaf Bio-Bags, 33 gal.
(3 boxes of 5); 39"x32½" (LxW)   **$20**

14-9333 Doggy Waste Bio-Bags, full size
(3 boxes of 50); 11½"x74/5" (LxW) **$18**

14-9332 Kitchen Compost Bin Bio-Bags,
15 qt. (3 boxes of 25)   **$18**

14-0334 Garden Bio-Film (1 sheet);
30'x5' (LxW)   **$12**

O'Connor's Strawberry Clover

This hardy bright green clover has made up the lawn of our Solar Living Center for more than 12 years, and even after thousands of people stampede it at the yearly SolFest, it still looks great. Popular in southern California as an erosion-preventing ground cover, this clover can be seeded by itself or with flower seeds. Perfect for xeriscaping, the nitrogen-rich legume tolerates a variety of soil conditions and thrives in Sun to partial shade. Works as a lawn when mowed to a ½-inch height. Aromatic red and white strawberry-shaped flowers. One ounce covers 125 sq. ft. Australia.

14-0350 8 oz. $10

14-0350 1 lb. $18

Compost

The Fastest and Most Effective Composter We've Found

About 10 years ago, Mike Wilkinson, president of North America's top composting toilet maker, Sun Mar Corporation, started thinking about making a better kitchen- and garden-waste composter. From Wilkinson's frustration with traditional composting methods came Sun Mar's new Auto Flow 200 series composter. Wilkinson drew on his company's experience designing environmentally friendly composting toilets, using the same mechanics to develop the Auto Flow.

How It Works

- The composter uses a patented, rotating Bio-Drum. As the drum is rotated, the compost material moves along an outer drum and then back toward a collection chamber through an inner drum. By the time it reaches the output port, it has become a light, fluffy end product.
- You can load the composter with kitchen scraps continuously, making the Auto Flow unlike typical composters that require you to stop adding material at some point.
- Minimal assembly is required.

An easy-to-assemble, continuously loading composter that provides a superior finished product. Green polyethylene. 50-gal. capacity. 32"x24"x33½" (HxWxL). 38 lb. 90-day limited warranty. Canada.

44-0108 Auto Flow Composter $299 ($20)

Allow 4-6 weeks for delivery. Can be shipped to customers in the contiguous U.S. only. Cannot be shipped to PO boxes. Sorry, express delivery not available.

Can-O-Worms

With our user-friendly composting system, a team of red worms digests kitchen wastes into worm castings—an organic, nutrient-rich garden amendment. The perforated stacking

trays separate the worms from their castings automatically, making it very easy to gather the worm castings left behind. The handy spigot allows you to capture the "worm tea"—a rich liquid amendment your plants will love. Sturdy, odorless, and pest-resistant, this composting system is made from 100% postconsumer recycled plastic. With complete instructions. 29"x20" (H x diameter). Can: Australia; worms: USA.

14-0176 Can-O-Worms Stacking Tray $149
Requires FedEx ground delivery.

14-0156 Worms, 1 lb. $37

Spinning Composter

Built-in rollers in our urban-smart composter let you effortlessly mix in oxygen and speed decomposition—and it's compact enough to store on a balcony or patio. Made of 50% postconsumer recycled plastic, it produces up to 85 pounds of rich, fertile compost in as little as 30 days without odors. Removable base collects compost drips, a nutritious byproduct for gardens and plants. Secure latching door is hinged for filling; detaches for emptying. Arrives assembled. 31"x21"x26" (HxWxL). 7 cu. ft. 14 lb. USA.

14-0020 Spinning Composter $179 ($15)
Requires FedEx ground delivery.

Garden Composter and Turner

Today's enlightened gardener recognizes composting as a simple, efficient way to reduce waste and create nutrient-rich soil for plants. Made of 100% recycled plastic, our unobtrusive bin traps solar heat to accelerate the production of compost. Includes adjustable air vents, easy-access hinged lid for adding materials, and sliding bottom door for compost removal. Snaps together in minutes without using tools. Includes a guide to successful backyard composting and has a 5-yr. mfr. warranty. 10.2-cu. ft. capacity. 40"x23" (HxW). Canada. Compost turner: 36½" long; handle: 7¾" across. Mexico.

Plunge in

Twist and pull out

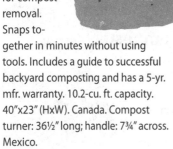

14-0023 Garden Composter $70 ($15)
Requires FedEx ground delivery.

14-0174 Compost Turner $20

Compost Activator

Start compost fast, reduce odors, and deactivate weed seeds and pathogens. Instructions included. 2 lb. USA.

14-0147 Compost Activator $18

Paulownia Trees

In appropriate growing zones, Paulownia trees can be the foundation for ecologically sustainable communities. Fast-growing deciduous trees, they provide shade, supply valuable hardwood, and create ideal microclimates for growing a wide variety of organic food crops. By growing and thriving in areas that have severe dust, smoke, and sulfur, they reduce air pollution. To avoid overgrowth, these trees are not available in most states east of the Mississippi River.

ORNAMENTAL BENEFITS
- Shade for home starting in just 16-18 months
- Grows to 30 feet in three years

FOOD SOURCE BENEFITS
- Leaves are 18%-20% protein, with high nitrogen; excellent fodder or fertilizer
- Flowers are tasty and beautiful in salads
- Bees love the flowers—excellent source of organic honey

FORESTRY BENEFITS
- Drought-resistant, disease-resistant
- Hardwood does not warp, twist, or crack; mahogany-like grain
- Lightweight, with excellent weight-to-strength ratio
- Great for furniture, doors, moldings, poles, paper, and more
- Fire-resistant and high-water tolerant
- Regenerates from the stump after harvesting

AIR AND WASTE WATER BENEFITS
- Significantly reduces waterway contamination and odors
- Increases nitrogen uptake
- Absorbs dust, smoke, sulfur. and other chemicals

Kawakamii variety grows 30 feet high in 3 years; Elongata grows 75 feet high in 15-18 years. Both thrive in growing zones 6-10. Six- to 12-inch seedling trees are priority-mailed with complete planting instructions. Place in 1-gallon pots for a few months before permanent planting. USA.

81441	**Kawakamii, 30-ft.**	**$29 ea.**
81442	**Elongata, 75-ft.**	**$29 ea.**

Trees are guaranteed to thrive or will be replaced at no charge.

We urge caution if planting these trees in the southeastern United States, where they may propagate without human help, and have been identified as an invasive species.

Garden, Biodynamic, and Permaculture Books

Designing and Maintaining Your Edible Landscape Naturally

First published in 1986, this classic is the authoritative text on edible landscaping, featuring a step-by-step guide to designing a productive environment using vegetables, fruits, flowers, and herbs for a combination of ornamental and culinary purposes. It includes descriptions of plants for all temperate habitats, methods for improving soil, tree pruning styles, and gourmet recipes using low-maintenance plants. 382 pages, paperback.

21-0524 Designing and Maintaining Your Edible Landscape **$49.95**

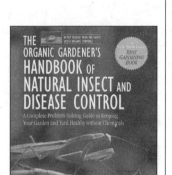

The Organic Gardener's Handbook of Natural Insect and Disease Control

A Complete Problem-Solving Guide to Keeping Your Garden and Yard Healthy Without Chemicals

End your worries about garden problems with safe, effective solutions from this handbook! Easy-to-use, problem-solving encyclopedia covers more than 200 vegetables, fruits, herbs, flowers, trees, and shrubs. Complete directions on how, when, and where to use preventive methods, insect traps and barriers, biocontrols, homemade remedies, botanical insecticides, and more. Newly revised with the latest, safest organic controls. 544 pages, paperback.

21-0536 Handbook of Natural Insect and Disease Control **$19.95**

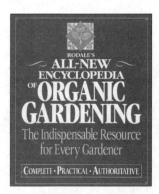

Rodale's All-New Encyclopedia of Organic Gardening

The Indispensable Resource for Every Gardener
Over 400 entries of the most practical, up-to-date gardening information ever, collected from garden experts and writers nationwide! "Gardens are places to renew yourself in mind and body, to reawaken to the truth and beauty of the natural world, and to feel the life force inside and around you. And the organic way to garden is safer, cheaper, and more satisfying. Organic gardeners have shown that it's possible to have pleasant and productive gardens in every part of this country without using toxic chemicals. They make their home grounds an island of purity." —Robert Rodale. 704 pages, paperback.

21-0538 Rodale's All-New Encyclopedia of Organic Gardening **$19.95**

The Vegetable Gardener's Bible

Discover Ed's High-Yield W-O-R-D System for All North American Gardening Regions

Ed Smith, an experienced vegetable gardener from Vermont, has put together this amazingly comprehensive and commonsensical manual, *The Vegetable Gardener's Bible*. This book, filled with step-by-step info and color photos, breaks it all down for you. Ed's system is based on W-O-R-D: Wide rows, Organic methods, Raised beds, Deep soil. With deep, raised beds, vegetable roots have more room to grow and expand. 320 pages, paperback.

21-0539 The Vegetable Gardener's Bible **$24.95**

Gardening for Life— The Biodynamic Way

Biodynamic techniques are based on the recognition that plants are intimately connected with and depend on live, vital soil. These methods also attempt to harmonize with the influences of the Moon phases and cosmos as a whole. You will learn the best times to plant various vegetables, how to deal with pests and diseases, ways to build the soil, and the effects of planets and stars on plant growth. 128 pages, paperback.

21-0540 Gardening for Life— The Biodynamic Way **$30**

A Biodynamic Farm

By Herbert H. Koepf. The inventory of knowledge that is generally warehoused under the classification of *biodynamic* is rich and timeless, and yet very few farmers have even a nodding acquaintance with the subject. This book performs a rescue operation. A practical, how-to guide to making all the biodynamic preparations, this book will provide what you need for putting these proven techniques to work in your fields. Perhaps the best of the many biodynamic titles currently available. Further explains how to achieve success through CSA-style market garden marketing. 215 pages, paperback.

21-0594 A Biodynamic Farm **$15**

Grasp the Nettle

Making Biodynamic Farming and Gardening Work

Author Peter Proctor tells the reader how to apply biodynamic methods of farming and gardening to a wide range of conditions in New Zealand and in other countries. Peter gives tips on how to recognize healthy soil and pasture and how to make your own biodynamic preparations, and also gives examples of farms that are successfully using biodynamic methods. This book aims to assist biodynamic farmers and gardeners to observe the processes of life and growth and to understand how these processes are governed by cosmic forces, so they can use this knowledge in applying their practical skills. 176 pages, paperback.

21-0595 Grasp the Nettle   **$20**

The Bio-Gardener's Bible

Bio-gardening combines the resources of science and nature to produce yield after yield of safe, abundant, and nutritious crops. Noted agriculturist Lee Fryer tells the home gardener how to build perpetual soil fertility by: providing a comfortable home for the organisms that extract vital nutrients from the soil and supply them to the plants; growing fertilizer right in the garden by encouraging the mechanisms that capture nitrogen from the air; analyzing specific soil needs, then mixing and applying a well-balanced diet; borrowing the best, most productive organic farming and agribusiness methods; controlling pests with healthy plants, not with pesticides. The book includes an A-to-Z vegetable-growing guide and special tips for getting the most out of patio and mini-gardens. 240 pages, paperback.

21-0541 The Bio-Gardener's Bible   **$9.95**

Moon Rhythms in Nature

How Lunar Cycles Affect Living Organisms

This brilliant book distills wide-ranging observations of lunar influences on the Earth. Following an introduction to the astronomy of the Moon rhythms is a study of how the tides and other intricate ocean movements are connected with the life processes of numerous organisms. Richly detailed and clearly written for the general reader, chapters lead up to the spectrum of human rhythms and a description of the whole concept of time. 17 color plates, 37 illustrations. 308 pages, paperback

21-0542 Moon Rhythms in Nature   **$36**

The Basics of Permaculture Design

First published in Australia in 1996, this is an excellent introduction to the principles, processes, and tools of permaculture. Packed with useful tips, clear illustrations, and a wealth of experience, it guides you through designs for gardens, urban and rural properties, water harvesting systems, animal systems, permaculture in small spaces like balconies and patios, farms, schools, and ecovillages. This is both a do-it-yourself guide for the enthusiast and a useful reference for permaculture designers. 170 pages, paperback.

21-0543 The Basics of Permaculture Design   **$25**

Gaia's Garden

A Guide to Home-Scale Permaculture

Picture your backyard as one incredibly lush garden, filled with edible flowers, bursting with fruit and berries, and carpeted with scented herbs and tangy salad

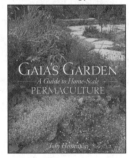

greens. The flowers nurture endangered pollinators and songbirds gather twigs for their nests. The plants themselves are grouped in natural communities, where each species plays a role in building soil, deterring pests, storing nutrients, and luring beneficial insects.

There is nothing technical, intrusive, secretive, or expensive here. All you need is some botanical knowledge (find it here) and a mindset that defines a backyard paradise as something other than a carpet of grass fed by MiracleGro. 240 pages, paperback.

21-0544 Gaia's Garden: A Guide to Home-Scale Permaculture   **$24.95**

Permaculture: A Designers' Manual

By Bill Mollison. This is the definitive permaculture design manual in print since 1988. It covers design methodologies and strategies for both urban and rural applications, describing property design and natural farming techniques. 286 pages, paperback.

21-0382 Permaculture: A Designers' Manual   **$83.95**

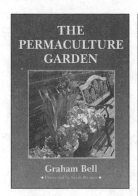

The Permaculture Garden

Working entirely in harmony with nature, *The Permaculture Garden* shows you how to turn a bare plot into a beautiful and productive garden. Learn how to plan your garden for easy access and minimum labor; save time and effort digging and weeding; recycle materials to save money; plan crop successions for year-round harvests; save energy and harvest water; and garden without chemicals by building up your soil and planting in beneficial communities. Full of practical ideas, this perennial classic, first published in 1995, is guaranteed to inspire, inform, and entertain. 170 pages, paperback.

21-0545 The Permaculture Garden $25

Permaculture: Principles and Pathways Beyond Sustainability

David Holmgren brings into sharper focus the powerful and still evolving Permaculture concept he pioneered with Bill Mollison in the 1970s. It draws together and integrates 25 years of thinking and teaching to reveal a whole new way of understanding and action behind a simple set of design principles. The 12 design principles are each represented by a positive action statement, an icon, and a traditional proverb or two that captures the essence of each principle. 286 pages, paperback.

**21-0546 Permaculture: Principles and
          Pathways Beyond Sustainability $30**

Food Not Lawns

By H. C. Flores. Forget sterile lawns! Grow a vibrant garden filled with food and beauty instead. That's the advice of one of our favorite new books. It's a virtual encyclopedia of wisdom and how-to advice about ecological design, organic gardening, food and eating, health, self-reliance, community-building, habitat, and biodiversity. Learn how city dwellers and rural denizens alike can use a nine-step permaculture strategy to create a piece of paradise that feeds both body and soul. 344 pages, softcover.

21-0392 Food Not Lawns $25

OUTDOOR AND YARD TOOLS

Marie's Poison Oak Soap

This truly amazing product prevents poison oak or poison ivy outbreaks even after exposure, and helps to prevent and heal many types of common rashes when used on a daily basis. Even after an irritation has started, healing is greatly accelerated. The soap's natural blend of oils, herbs, extracts, clay, oatmeal, and glycerin pulls poison oak and poison ivy oils off the skin, promotes healing, and helps stop itching and redness due to the soothing anti-histamine properties of the herbs. The Oregon Highway Department buys more than 2,000 bars every year for its employees. Contains no dyes, artificial ingredients, or scents. USA.

08-0346 3 bars $15
**   6 bars $28**

Recycled-Tire Buckets

Having the right tool on hand always makes your job easier. These long-lasting buckets are made of recycled tires and waste plastic, using a manufacturing process that is gentle to the Earth. Unlike metal buckets, they won't corrode or buckle, and the handsome galvanized steel handles hold up to the hardest knocks. To suit every need, we offer them in a great value set of three: 8-quart, 12-quart, and 18-quart. A product line that delivers where the rubber meets the road. USA.

06-9341 Recycled-Tire Buckets, set of 3 $29

Recycled-Tire Doormats

Made from sliced-up used tires, these mats will last a lifetime. Years of scuffing boots will only soften the surface and make them better mats.

They clean easily, as accumulated material simply falls through them when you shake them out. They work great as doormats, or as antifatigue mats in front of your tool bench or laundry folding table. Held together with heavy-duty galvanized steel wire. Three sizes available. USA.

18-0064 **Recycled-Tire Doormat**
23"x13.5" (LxW) **$20**

18-0065 **Recycled-Tire Doormat**
28"x18.5" (LxW) **$28**

18-0066 **Recycled-Tire Doormat**
36"x24" (LxW) **$56**

Push Reel Lawn Mower

Made by the oldest lawn mower manufacturer in the U.S., our mower is safe, light-weight, and easy to push. It's perfect for small lawns and hard-to-cut landscaping. The reel mower provides a better cut than power mowers, and it doesn't create harmful fumes or noise. Grass clippings from the mower can be left on the lawn as natural fertilizer, or purchase the optional grass catcher and compost the clippings. Cutting width is 16 inches. It has 10-inch adjustable wheels, five blades, and a ball bearing reel. USA.

14-0187 Push Reel Lawn Mower **$149 ($10)**

14-0188 Grass Catcher **$34**

Mighty Mule Gate Openers

These amazing gate openers are designed for home, ranch, or farm gates up to 16 feet in length or weighing up to 350 pounds per leaf. Designed for do-it-yourself installation with a video. Powered by a battery that's immune to power outages, it can be charged by the included AC transformer or by the optional solar panel with mounting arm. Closes automatically with an adjustable 0- to120-second delay. Kit includes opener arm, AC transformer, 12-volt battery, control box, receiver with 10-foot cable, mounting hardware, power cables, warning signs, and one entry transmitter. The dual gate kit includes a second opener arm and control cable.

Options include: Push-to-Open kit converts opener to swing out from property, outdoor digital keypad will accept up to 15 entry codes, an automatic gate lock (a MUST to actually lock the gate), and a buried gate opening sensor to automatically open the gate for outbound vehicles. 1-yr. mfr. warranty. UL-listed. USA.

63-126 **Single Gate Opener** **$595**

14-0245 Dual Gate Opener **$925**

14-0246 Solar Panel Kit **$175**

63-144 **Extra Transmitter** **$33**

14-0247 Key Ring Transmitter **$33**

63-146 **Digital Keypad** **$60**

63-147 **Automatic Gate Lock** **$199**

64-500 **Push-to-Open Kit** **$18**

14-0248 Buried Gate Opening Sensor **$209**

Parmak Solar Fence Charger

The 6-volt Parmak Fence Charger will operate for 21 days in total darkness and will charge up to 25 miles of fence. It includes a solar panel and a 6-volt sealed, leakproof, low-internal-resistance gel battery. It's made of 100% solid-state construction with no moving parts. Totally weatherproof, it has a full 2-yr. warranty.

63128 **Parmak Solar Fence Charger** **$210**

63138 **6V Replacement Battery**
for Parmak **$50**

Pest Control

Bug-Off Instant Screen

Hands-free pass-through, automatic closure, and ease of installation make our Bug-Off Instant Screen an excellent choice wherever an insect barrier is required. The durable fiberglass mesh hangs from a spring tension rod at the top of the door jamb, the sides are secured to the jamb with Velcro® strips, and vinyl-reinforced magnets automatically snap the screen shut after you pass through. It installs in minutes without tools and

will not damage the door frame. Custom fitted with channels for doors 80 inches high. Choose 30-inch, 32-inch, 36-inch, or 48-inch width; each fits a door up to 2 inches smaller or 1 inch larger than stated size. They also can be used on one half of a double or French door, fitting between the jamb and one closed door (this configuration requires that the screen be removed before shutting both doors). USA.

14-0205 Instant Screen, 30-in. W		**$45**
Instant Screen, 32-in. W		**$48**
Instant Screen, 36-in. W		**$48**
Instant Screen, 48-in. W		**$54**

The World's Most Effective Deer Deterrent

Hook the scarecrow up to your garden hose and stake it. The smart motion sensor detects an intruder (up to 35 feet away) and sends a full-pressure blast of water right at it. It switches off immediately, using a conservative 2 cups of water per discharge. You preset

the detection area (protects a 1,000-square-foot area) and sensitivity to prevent triggering by household pets. Sensor sensitivity is dampened automatically in windy conditions to avoid false triggers. Up to 1,000 discharges on a single 9-volt battery (not included). Reinforced nylon stake with sturdy step and hose flow-through. 24 inches high overall. Canada.

16-0023 Scarecrow Sprinkler **$89**

Bat Conservatory

Because they devour up to 600 mosquitoes in an hour and rarely bother humans, it makes sense to welcome bats to take up nearby residence. Made of western red cedar sawmill trim, our slatted bat conservatory provides shelter for approximately 40 bats. Attaches easily to your barn or nearby tree. Put out the welcome mat for our native North American bats, and these fascinating creatures will reward your hospitality year after year. No more nasty mosquito repellents! 24½"x16"x5¼" (Hx-WxD). Learn more about bats with Merlin Tuttle's book *America's Neighborhood Bats*. 96-page, softcover. USA.

16-0060 Bat Conservatory **$50**

21-0306 America's Neighborhood Bats $10.95

Solar Ant Charmer

The red imported fire ant was introduced accidentally in Mobile, Alabama, from South America in the 1930s. The species spread through the lower United States and continues to expand northward and westward into areas with mild climates and plentiful food. Fire ants attack animals, people, and even electrical devices. They cause failures in air-conditioning units and traffic control boxes. Fire ants kill livestock and also have been known to kill humans. The Solar Ant Charmer controls fire ants without poisons by eliminating a major part of the ant colony's defenses. It collects thousands of fire ants in just minutes. Fire ant poisons pollute the Earth, and fire ant baits take up to six weeks to become effective. Use the Charmer over and over again for years without replacing any batteries—the solar cell does it all! USA.

63853   Solar Ant Charmer **$119**

Mosquito Dunks

Mosquitoes lay eggs in standing water around your home. Doughnut-shaped dunks float on water and slowly release bacillus thuringiensis—a natural pesticide lethal to mosquito larvae. Place all-natural dunks anywhere water is left standing for more than a week. Safe for birdbaths, rain barrels, ponds, ditches, tree holes, roof gutters, unused swimming pools, or anywhere else water collects. A single dunk destroys mosquito larvae in up to 100 square feet of water for a month. Package of 6. USA.

16-0013 Mosquito Dunk **$12**

Anti-Bug CF Light Bulb

The warm yellow glow of our 15-watt compact fluorescent Anti-Bug Light Bulb welcomes guests instead of bugs to your door. Mexico.

16-0095   **Anti-Bug CF Light Bulb**   **$14**

Solar Mosquito Guard

We've received countless raves from satisfied Solar Mosquito Guard owners, all with the same theme: "These amazing things really work. We're mosquito-free, while our friends are swatting away." Our pocket-sized devices put out an audible, high-frequency wave that repels some mosquito species. No scientific evidence proves that mosquitoes can hear, yet this really seems to work! There is an on/off switch so you don't have to activate until the mosquitoes arrive. Battery fully recharges in three hours of sunlight. China.

17-0329   **Solar Mosquito Guard**   **$10**

      3-5   **$9 ea.**

      6 or more   **$8 ea.**

Recycled Glass Wasp Catcher

Relax in your yard without wasps—or exposure to chemical repellents—with our beehive-style trap. Based on a design used since the Middle Ages, the "beehive" catcher traps insects with the lure of sugar water. Hangs from a hook. 100% recycled glass. Green, aqua, or orange. 8½"x5½" (HxW). 1 lb. Imported.

16-0120 **Wasp Catcher**   **$12**

Skeeter Skatter

Skeeter Skatter is the natural, effective way to chase away pesky bugs during your summertime outdoor activities. Commercial bug repellents often contain environmentally destructive and synthetic additives, such as DEET, that are harmful to humans and animals. Skeeter Skatter uses citronella, eucalyptus, lavender, catnip, and pennyroyal oils to safely keep bugs at bay. Simply spritz the pine-scented spray on clothing or directly on skin for a strong layer of potent protection. Safe for children. Handy 2.66-oz. spray pump. USA.

16-0015 **Skeeter Skatter**   **$14**

Orange Guard

Water-based Orange Guard relies on natural orange peel extract to kill insects on contact and repel others for weeks, including cockroaches, ants, and fleas. Certified for use in organic production. Great for both the garden and kitchen. Not hazardous to insect-eating birds or lizards. USA.

16-0041 **Trigger Spray Bottle, 32 oz.**   **$12**

      Heavy-Duty Bottle with Sprayer, 1 gal.   **$30**

Ultrasonic Pest Control Deluxe

Get rid of rats, mice and roaches without exposing your children and pets to dangerous poisons. This plug-in unit has two electromagnetic/ultrasonic settings (high for roaches and small rodents; low for rats) that penetrate walls and send pests scurrying. Covers 5,000 sq. ft. Harmless to humans, dogs, cats, birds and non-rodent pests. 6"x4"Wx1½" (HxWxD). Imported.

16-0129 **Ultrasonic Pest Control Deluxe**   **$79**

Pets such as hamsters, gerbils or guinea pigs may experience discomfort with this device..

4-Pack Pest Chaser*

Mice and other rodents aren't welcome in any home, and neither are poisons. Protect your home without dangerous chemicals with our 4-pack Pest Chaser. Improved sonic pest-repelling technology is effective, is eco-safe, and costs pennies to operate. Each 5-watt plug-in unit protects up to 800 square feet and can be placed throughout the home for total coverage of up to 3,200 square feet. Completely harmless to humans, dogs, cats, birds, and other nonrodent pets. Approved by the National Pest Management Association. 1½"x2" (H x diameter). 10 oz. USA.

16-0096 **4-Pack Pest Chaser**   **$49**

*If you have pets such as hamsters, gerbils, or guinea pigs, this device may cause them discomfort.

Cedar/Lavender Set

While natural cedar wood has been used for generations to discourage moths in clothes and linens, it also absorbs moisture and staves off mustiness. Convenient

storage accessories in the Cedar/Lavender Set take this time-tested practice one step further. Imbued with natural lavender oil, they add a sweet scent to clothing and linens. Includes three hang-ups to fit over any standard closet bar, eight blocks to hide in drawers, and 48 balls to nestle among smaller items. USA.

06-9448 Cedar/Lavender Set   **$22**

06-9510 Cedar/Lavender Drawer Liners
12"x3½"x3⁄8" (LxWxD); set of 10   **$30**

Cedar Shoe Rack

Bring a breath of fresh air and order to your closet with the Cedar Shoe Rack. Naturally aromatic cedar repels moths and pests. Each rack holds about five pairs of women's shoes or four of men's. Add more shelf space with the predrilled Cedar Shoe Rack Topper, made from cedar/medium-density fiberboard, which easily attaches to the top of the shoe rack. The natural hues of cedar vary from blonde to red and even purple, making the color of each rack different. Refresh the scent of cedar

blocks, chests, and more with 100% natural cedarwood oil (below). USA.

06-0511 Cedar Shoe Rack Topper
     2"x37½"x12" (HxWxD); 12 lb.   **$35**

06-0334 Cedar Shoe Rack 5½"x37½"x12"
     (HxWxD); 7 lb.   **$39**

06-9334 Cedar Shoe Rack, set of 2   **$69**

Cedarwood Oil

With its powerful antimoth and antiflea formula, Cedarwood Oil can refresh the scent of cedar blocks and chests, treat pet beds, and remove musty smells. 100% natural cedar oil in a mineral oil base. Two 8-oz. pump bottles. USA.

06-0048 Cedarwood Oil   **$18**

Pro-Ketch Mouse Trap

Protect your home and give yourself peace of mind with our humane trap that professionals use. Made from tough galvanized steel, Pro-Ketch captures multiple mice without ever resetting. Place bait inside to entice mice to enter. The teeter-totter-style ramp lowers to allow mice in but snaps back once they step off so they can't exit. Clear-view lid allows easy monitoring. Simply take trap far from your home, open lid, and let mice scurry out. Check and empty trap frequently. USA.

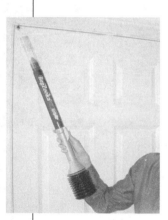

16-0083 Pro-Ketch Mouse Trap   **$15**

BugZooka™

Our BugZooka battery-free wand generates 10 times the suction power of motorized models to humanely capture insect invaders for outdoor release. Just push the bellows to compress some air, extend the telescoping arm up to 24 inches, and push the trigger. Wasps, spiders, flies, and more are whisked inside a removable catch tube for transport outside. No unpleasant handling or squashing. No poisonous sprays. Kids can even get an educational close-up look at whatever you've caught before it's set free. Includes two clear catch tubes and wall mount. 38 inches long. 2 lb. Taiwan/China.

16-0267 BugZooka™   **$29**

"The BugZooka looks, sounds, and is priced like a kids' toy but is highly efficient and evolved. Just cock the bellows, point the end to within an inch of the offending insect, and thumb the trigger. FWOOP, the insect is gone."

—Stewart Brand
Founder, *Whole Earth Catalog*

Solar Cooking

In addition to the big job of making worldwide weather, or the small local job of heating your domestic hot water or pool, the thermal energy from sunlight can be used to heat buildings (which we cover fully in Chapter 2, "Land and Shelter") and to cook food. Solar-powered cooking has enjoyed a huge surge in interest over the past two decades. In many developing countries, gathering fuel wood for cooking can occupy half the family members full-time and has resulted in barren moonscapes where every available scrap of fuel gets snatched up—including any reforestation efforts. Solar cookers are quickly taking the place of the family fire pit. A simple reflective box with a clear cover can easily attain 300°-400°F temperatures on an average sunny day, making an excellent oven with no fuel cost and no time spent collecting fuel. The nonprofit Solar Cookers International group provides cookers and simple cardboard and aluminum foil construction plans to developing countries. Some of our solar cooker product sales support their good efforts.

Solar cookers can be as simple as a cardboard box with aluminum foil reflectors and a dark pot with a clear turkey roasting bag wrapped over it, or they can be as fancy as our Sun Oven with insulated sides, self-leveling tray, and folding reflectors. Both will get the job done. Obviously, the Sun Oven will last longer and has more user-friendly features, but it doesn't necessarily cook lunch any faster.

Sun Oven

A great portable solar cooker weighing only 21 pounds, our Sun Oven is ruggedly built with a strong, insulated fiberglass case and tempered-glass door. The reflector folds up and secures for easy portability on picnics, etc. It's completely adjustable and comes with a built-in thermometer. The interior oven dimensions are 9"x14"x14" (HxWxD) and temperatures range from 350° to 400°F. This is a very easy oven to use, and it will cook most anything! Includes a handy 16-page recipe booklet. USA.

63421   Sun Oven $249

HotPot Simple Solar Cooker

This is the most convenient and portable solar cooker we sell. It heats up fast and can produce cooking temperatures within an hour and be used for up to six hours, perfect for baked goods or slow-cooking stews. The HotPot's features include a 5.3-quart enameled steel pot. Respected institutions such as Florida Solar Energy Center and Energy Laboratories, Inc., tested and refined the product's design over a six-year period. Sales of the HotPot help subsidize the distribution of this device in developing countries to reduce deforestation and respiratory diseases caused by traditional cooking methods. 12½"x12½"x15" (HxWxD). USA/Mexico.

01-0478 HotPot Simple Solar Cooker $99

Sport Solar Oven

We love the look on friends' faces when we present fresh bread and other home-baked treats while camping. So will you when you take our lightweight solar oven on your next trip. It concentrates the renewable heat of the

Sun to effortlessly (and deliciously!) roast meats, steam vegetables, bake breads and cookies, and prepare rice, soups, and stews. Ideal for backyard use, camping, boating, and picnicking. Requires only minimal Sun aiming to cook most foods in two to four hours. Our complete kit includes two pots, oven thermometer, water pasteurization indicator, and recipe book. 12"x27"x16" (HxWxD). 11 lb.

01-0454 Sport Solar Oven $150

Cooking with the Sun
How to Build and Use Solar Cookers

By Beth and Dan Halacy. Solar cookers don't pollute, they use no wood or any other fuel resource, they operate for free, and they run when the power is out. And a simple and highly effective solar oven can be built for less than $20.

The first half of this book, generously filled with pictures and drawings, presents detailed instructions and plans for building a solar oven that will reach 400°F or a solar hot plate that will reach 600°F. The second half has 100 tested solar recipes. These simple-to-prepare dishes range from everyday Solar Stew and Texas Biscuits to exotica like Enchilada Casserole. 116 pages, softcover.

21-0273 Cooking with the Sun $9.95

KITCHEN AND HOME TOOLS

This is a truly eclectic mix of products. Now we're really getting down to things that defy compartmentalization! The only common thread here is that these are all things you would use inside your home (probably).

Convection Oven

Convection ovens are one of the most eco-efficient cooking methods around—cutting cooking time up to 30%. Our Convection Oven allows you to broil, warm, defrost, rotisserie, or bake a pizza in a small amount of counter space. The patented Heat Circ™ technology heats uniformly, and the enameled Durastone™ interior distributes it evenly. Provides an extra-large 1.1-cubic-foot capacity to fit a roasting chicken. Two-tier cooking capability allows you to cook two dishes at once. Includes two cooking racks, two bake pans, and a rotisserie kit with pan. Exclusive pizza function ensures a perfect pizza (even from frozen), while the drop-down top heating element allows for simple cleanups. 13"x21"x16½" (HxWxD). 33 lb. Italy.

09-0246 Convection Oven $399 ($25)

French Butter Keeper

In the centuries-tested French tradition, our artisan-made butter bell preserves butter at a spreadable temperature without refrigeration or rancidity. Pack a single stick of softened butter into its cone-shaped lid, fill the base 2/3 full of pure water, and then invert the lid into the base. A seal keeps oxygen out and freshness in. If you want, you can melt and serve the butter right from the lid. The French Butter Keeper is microwavable and made from dishwasher-safe stoneware with 100% lead-free glazes. 4½"x5" (W x diameter). Choose sage or terra. USA.

09-0170 French Butter Keeper $44

Glass Butter Crock

Our Glass Butter Crock uses the same strategy as its handcrafted French cousin (below left) to keep butter fresh at room temperature for weeks, so it's always soft and ready to serve. Just fill the serving dish with a stick of butter, and the base with water. Invert the dish into the base, and the water will seal out the oxygen that causes rancidity. Glass design lets you monitor butter and water levels. Dishwasher- and microwave-safe. 3½"x3½"x4" (LxWxH). China.

09-0247 Glass Butter Crock $14

Stainless Steel Grain Flaker

Bring food processing home—and eat healthy. Most of us know that whole foods are tastier and more nutri-

tious than processed ones. Our Grain Flaker moves your diet in this better, more delicious direction. Clamped to any counter, its stainless steel rollers crush and roll fresh oats, spelt, wheat, flax, sunflower seeds, and other whole grains and seeds to release healthy oils so you can enjoy their nutrients and flavors before time and air destroy them. Detachable hopper for easy cleaning. 4"x10" (WxH). Crank handle is 8½ inches. Germany.

09-0267 Stainless Steel Grain Flaker $259

Hand-Cranked Blender

The human-powered Vortex blender is perfect for picnics, tailgate parties, and camping. It features a 48-ounce (1.5-liter) graduated Lexan pitcher with an O-ring sealing top, a removable pour spout that's a 1-ounce shot glass, a brushed stainless steel finish, soft stable rubber feet, an ergonomic handle, two-speed operation, and a C-clamp for stable operation. The base fits inside the pitcher for easy packing. High-speed operation is noisy so your camping neighbors will know you're making margaritas! Weighs 4.5 lb. China.

63867   Hand-Cranked Blender $88

Stovetop Waffle Maker

Creates delicate, crisp waffles with extra-deep pockets to capture your favorite toppings. And faster than an electric waffle iron! Long-lasting, heavy cast aluminum with a nonstick coating and cool Bakelite handles. 16½"x8½"x1½", baking surface 7¼"x7¼".

51647   Stovetop Waffle Maker $59

Countertop Compost Bin

Put kitchen scraps conveniently out of sight—and significantly reduce odors—using our easy-to-clean compost bin before transporting to your garden compost. Its trim design fits handily under the sink or at the back of the counter. Carbon-activated filters effectively prevent odors for up to six months. Made of durable plastic; lid snaps snugly to eliminate spills. 5.5 qt., 7¾"x8"x7½" (HxWxD). 9.6 qt., 12"x9"x8" (HxWxD). Canada.

09-0069 Compost Bin, 5.5 qt. $17

09-0068 Compost Bin, 9.6 qt. $19

09-0079 Replacement Filters (3) $8

Kitchen Compost Crock

Our fully glazed ceramic crock looks beautiful and prevents stains and odor absorption. The 1-gallon interior holds a week's worth of scraps, and a lid with an activated carbon filter traps odors and lets air circulate to prevent rotting. The removable stainless steel handle makes carrying easy. Dishwasher safe. Replacement filters available in a set of 6. 7"x10½" (WxH) with lid. 6 lb. China.

09-1051 Compost Crock $42

09-9023 Replacement Filters (6) $10

Bag Dryer

Crafted from sustainably harvested birch and ash woods, our Countertop Plastic Bag Dryer makes it easy to dry and reuse plastic bags. Includes a hanging hook. Folds for storage. 14"x7¼" (HxW). 6.3 oz. USA.

06-0007 Bag Dryer $19

Flour Sack Towels

Stock your kitchen with our all-purpose towels that work better and last longer than almost any other. Just like those your grandmother used in the days before paper towels, oversized Flour Sack Towels are made without dyes or chlorine bleach from a hardy, lint-free, 100% cotton material that's ideal for washing, streak-free dish drying, and many other kitchen tasks. 36"x24" (LxW). India.

09-0195 Flour Sack Towels, set of 3 $10

09-9195 2 sets of 3 (total of 6) $18

30th ANNIVERSARY

Cleaning and Paper Products

SEVENTH GENERATION

Seventh Generation cleaners and household products, the best household products for the job … and the planet. As the largest cataloger of Seventh Generation's eco-friendly cleaners and household products, we know they make a difference. In fact, in the past two years alone, our customers have saved:

• 27,000 living trees, 70,000 cubic feet of landfill space and 9.8 million gallons of water by purchasing Seventh Generation recycled paper products.

• 525 barrels of petroleum and 3,900 pounds of chlorine from entering the environment by purchasing Seventh Generation laundry products.

Bathroom Tissue

Tissue feels cushiony-soft and smooth against your skin. Each roll is made of 100% recycled paper with a minimum 80% postconsumer content and is safely bleached with sodium hydrosulfite. Septic system safe. 500 2-ply sheets. 58.6 sq. ft. USA.

Safe for your septic system, 1-Ply Bathroom Tissue is made from 100% recycled fibers with a minimum 80% postconsumer content and bleached with environmentally safe sodium hydrosulfite and hydrogen peroxide. With 1,000 sheets per roll, our 1-ply bathroom tissue lasts an extra-long time. 116 sq. ft. Canada.

2-PLY BATHROOM TISSUE

28-1016	12 Rolls	$13.95
28-1014	Case, 48 Rolls	$52.95
28-1012	2 Cases, 96 Rolls	$96.95

1-PLY BATHROOM TISSUE

28-1006	12 Rolls	$15.95
28-1004	Case, 48 Rolls	$56.95

Paper Towels

Our brown towels are made from unbleached, undyed, 100% postconsumer "low-grade" recycled paper. Each roll packs more than twice the sheets of most brands, cutting down on packaging and fuel waste. 120 two-ply sheets per roll. 11"x9". 123.75 sq. ft. USA.

28-1036	4 Rolls	$7.95
28-1034	8 Rolls	$15.50
28-1032	15 Rolls	$29.50
28-1038	Case, 30 Rolls	$54.95

Facial Tissue

Facial tissue made from 100% recycled paper with a minimum 20% postconsumer content and bleached with environmentally safe sodium hydrosulfite and hydrogen peroxide. 175-count floral print box. USA.

28-1047	9 Boxes	$20.95
28-1048	Half-Case, 18 Boxes	$40.95
28-1049	Case, 36 Boxes	$74.95

Natural Napkins

The convenience of paper napkins with the lowest possible impact on the environment. Natural napkins are made from an unbleached, dioxin-free 100% postconsumer mix of newspaper, corrugated boxes, and phone books. 1 ply. 6"x6½" folded, 500/pack. USA.

28-1056	2 Packs	$11.95
28-1054	6 Packs	$34.95
28-1052	Case, 12 Packs	$68.95

Recycled Plastic Trash Bags

Sturdy and dependable trash bags are made from 100% recycled plastic with a minimum 30% postconsumer content and require 25% less energy and 90% less water to produce than virgin plastic bags. The 13-gallon white bags are ideal for kitchen waste. Use the 33-gallon bags for 32-gallon containers. USA.

13-GAL. TALL KITCHEN BAGS (30/BOX)

28-6006	3 Boxes	$11.95
28-6004	6 Boxes	$22.95
28-6002	Case, 12 Boxes	$42.95

33-GAL. LARGE TRASH BAGS (15/BOX)

28-6026	3 Boxes	$11.95
28-6024	6 Boxes	$22.95
28-6022	Case, 12 Boxes	$42.95

Baby Wipes

Lightly scented, extra-thick baby wipes are moistened with aloe vera and water—not irritating alcohol. Just one does the job of two to three ordinary wipes. Bleached with hydrogen peroxide and chlorine dioxide, which is much less harmful than chlorine. 80 wipes per travel/refill pack. USA.

28-7005	2 Packs	$9.50
28-7007	4 Packs	$18.95
28-7008	Case, 12 Packs	$49.95

Chlorine-Free Diapers

Made with a unique blend of unbleached materials that are gentler on baby's tender skin and the environment. With resealable tapes, stretchy leg gathers, and a clothlike covering. 4 packs per case. Germany.

28-9037	Stage 1 (8-14 lb.), 56/pk.	$59.95
28-9038	Stage 2 (12-18 lb.), 48/pk.	$59.95
28-9039	Stage 3 (16-28 lb.), 40/pk.	$59.95
28-9040	Stage 4 (22-37 lb.), 34/pk.	$59.95
28-9041	Stage 5 (27+ lb.), 30/pk.	$59.95
28-9042	Stage 6 (35+ lb.), 26/pk.	$59.95

Dish Liquid

Mild on your hands and gentle on the environment, this dish liquid cuts grease and loosens caked-on food with the power of coconut oil, not petroleum. No artificial colors, and lightly scented with natural fragrances. Not for use in automatic dishwashing machines. USA/Canada.

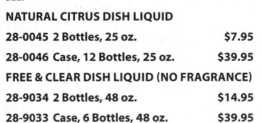

NATURAL CITRUS DISH LIQUID

28-0045	2 Bottles, 25 oz.	$7.95
28-0046	Case, 12 Bottles, 25 oz.	$39.95

FREE & CLEAR DISH LIQUID (NO FRAGRANCE)

28-9034	2 Bottles, 48 oz.	$14.95
28-9033	Case, 6 Bottles, 48 oz.	$39.95

Laundry Products

Laundry Liquid

All-temperature laundry liquid uses powerful coconut-oil-based detergents to remove dirt, grime, and stains effectively in cold or warm water. Fresh, natural citrus scent or fragrance-free. Includes convenient measuring cap. 33 loads per 100 oz. bottle; 16 loads per 50 oz. bottle. USA.

28-2034	2 Bottles, 50 oz.	$12.95

Lavender Fabric Softener

Ultraconcentrated fabric softener uses renewable canola oil (rapeseed) to make laundry fluffy, downy soft, and static-free. And with 40 loads per 40 oz. bottle, it's also economical. Biodegradable. Canada.

28-2068	2 Bottles	$13.95
28-2069	Case, 6 Bottles	$39.95

Chlorine-Free Bleach

Get whites and colors their brightest without chlorine. Safe for most colors; testing recommended for naturally dyed fabrics. 21 loads per 64 oz. bottle. Canada.

28-2048	2 Bottles, 64 oz.	$10.50
28-2049	Case, 6 Bottles, 64 oz.	$30.50

James Washer

The James hand-washing machine is made of high-grade stainless steel with a galvanized lid. A pendulum agitator swings on nylon bearings, sweeps in an arc around the bottom of the tub, and prevents clothes from lodging in the corner or floating on the surface. This ensures that hot

suds are mixed thoroughly with the clothes. The James is sturdily built. The corners are electrically spot-welded. All moving parts slide on nylon surfaces, reducing wear. The faucet at the bottom permits easy drainage. Capacity is about 17 gallons. 30 lb. USA.

63411   James Washer $449
Hand-constructed to order. Please allow up to 12 weeks for delivery.

Hand Wringer

The hand wringer will remove 90% of the water, while automatic washers remove only 45%. It has a rustproof, all-steel frame and a very strong handle. Hard maple bearings never need oil. Pressure is balanced over the entire length of the roller by a single adjustable screw. We've sold these wringers without a problem for more than 17 years. USA.

63412   Deluxe Hand Wringer $169

Static Eliminator

The first time you use our incredibly effective reusable, chemical-free dryer sheets, you'll notice a difference in your

laundry. Unlike disposable dryer sheets you throw away after every load, Static Eliminator can be used up to 500 times—saving money and reducing waste. What's more, these hypoallergenic sheets are ideal for allergy sufferers because they're free of the harmful chemicals— three of which are on EPA's Hazardous Waste List—used in conventional brands. Works wonders at removing pet fur from laundry. Two per box. Guaranteed never to spot or stain. 12"x7½" (LxW). Canada.

06-0531 Static Eliminator $20

Nellie's Dryer Balls

Designed to reduce drying time and soften fabrics without the use of chemical fabric softeners, Dryer Balls are an eco-friendly solution to landfill-clogging dryer sheets and chemical-laden liquid softeners. As your wet laundry tumbles in the dryer, these bumpy balls lift and separate fabric to make it soft and fluffy and allow air to flow more efficiently— proven to reduce drying time up to 25% in independent testing. Set of 2. 4½"x2" (L x diameter). China.

06-0598 Dryer Balls $18

Dryer Vent Cleaner

Save money on utility bills, spend less time doing laundry, and most important, avoid dryer fires with the Rotary Dryer Vent Cleaner. Damp clothes at the end of a typical drying cycle, longer drying times than normal, or excessive lint in the laundry room are all signs of lint blocking your clothes dryer's exhaust system. With the assistance of your shop vacuum/blower and a cordless drill, this brush cleaner navigates the twists and turns in your venting systems with a powerful spinning action. Ultrasoft bristles reach up to 15 feet to clean behind, underneath, and all around your dryer. Effective on metal, foil, or white vinyl vents. Includes five 36-inch-long flexible rods, 4-inch auger brush, 2¼-inch lint trap brush, vacuum adapter, dryer adapter, blockage removal tool, and instructions. 4 lb. USA.

Before and after

06-9605 Rotary Dryer Vent Cleaner $64

50 Feet of Drying Space

This is the most versatile clothes dryer we've seen yet. Don Reese and his family are still making them by hand, from 100% sustainably harvested New England pine and birch. Don's sturdy dryer can be positioned to create either a peaked top or a flat top ideal for sweaters. A generous 50 feet of drying space—and you can fold it up to carry back in the house (or to the next sunny spot) still loaded with laundry. Measures 53"x30"x33" (HxWxD) in peaked position, 46"x30"x40" (HxWxD) in flat-top position, and holds an average washer load. Folds to 58"x4½" (L x diameter). USA. *Requires FedEx ground delivery.*

10-8002 Wooden Dryer $89

Wall Shelf Drying Rack

Our wooden Wall Shelf Drying Rack is easy to install on any wall surface. Position unit on porch for solar drying, near back door for easy access, or in laundry room for indoor drying. You can also use it as a drying rack for flowers, herbs, or pasta. Constructed of sturdy Eastern white pine with durable birch hardwood dowels and four Shaker-style pegs. Comes fully assembled. Includes mounting hardware. 12"x27"x6" (HxWxD). Extends to 22 inches when in use. USA.

10-0023 Wall Shelf Drying Rack $76

Umbrella Clothes Dryer

Having beautifully fragrant clothes with the scent of the fresh outdoors is easier than ever! Our easy-to-assemble folding clothes dryer is a classic energy-efficient method for drying. It rotates easily, and has 164-feet of PVC-coated line with 4" spacing to allow easy hanging. The line holes in the arms are protected to prevent line chafing. It can accommodate four to five loads of laundry, yet folds to only 4.5"W x 78.5"H. The diagonal width is an impressive 9'6" when open. The height conveniently adjusts from 4' to 6'. Polished aluminum 1.75" post, galvanized steel arms. Easy opening and closing for stowing between uses. Cover for folded dryer prevents bird spots and UV degradation of line. Netherlands.

06-0229 Umbrella Cothes Dryer $159 ($5)

10-0034 Cover $16

Retractable Clothesline

Our Retractable Clothesline makes air drying laundry easy in any location. Extendable up to 40 feet, the tangle-free line hooks to a bracket you secure at any desired distance and retracts into the impact-resistant cover for out-of-the-way storage. Supports a load's worth of sheets and towels while remaining taut. Easily installs to wood, metal, or drywall for indoor or outdoor use. Stainless steel springs won't rust in outdoor applications. Includes mounting hardware. Wipes clean with a damp cloth. 6" diameter. USA.

10-0037 Retractable Clothesline $20

Cleaning Products

Rotary Sweeper

Effectively cleaning floors and low-pile rugs without power, our Rotary Sweeper is a cleaning solution that exceeds sustainability standards. With a low profile, it easily reaches beneath furniture. Includes simple bottom-mount dustpan and exceptional design and craftsmanship. Lightweight and compact, it weighs only 5 lb. 11"x11"x37" (LxWxH). Germany.

06-0482 Rotary Sweeper $70

Floor Cleaning Kits

Keep polyurethane-finished and factory prefinished wood floors shining radiantly with the nontoxic cleaner used and recommended by professional floor refinishers. Our Premium Hard Surface Floor Cleaning Kits for sealed floors include a mop head with an improved telescoping handle, 32 ounces of floor cleaner, two washable microfiber cleaning pads, 32 ounces of refresher to restore gloss to 500 square feet of dull floors, a refresher application pad, and two microfiber cleaning cloths with a potent electrostatic charge to capture dirt like a magnet. USA.

06-0575 Hardwood Floor Cleaning Kit $72

06-0650 Hard Surface Floor Cleaning Kit $68

06-9481 Floor Mop Replacement Pads (2) $22

06-0577 Hardwood Floor Cleaner (1 gal.) $24

**06-0651 Hard Surface Floor Cleaner
               (1 gal.) $24**

Super Pine Cleaner

This is no ordinary pine cleaner. No trees were cut to extract its all-natural pine oil; and to save packaging and money, this Super Pine Cleaner is highly concentrated to make 20 times more cleaner than most brands. With an aromatic pine scent, it deodorizes and sanitizes laundry, floors, wood, walls, bath surfaces, and even garbage cans. Two 16-oz. bottles. USA.

06-9016 Super Pine Cleaner $14

White Wizard

Completely safe for any surface or fabric, from leather to upholstery, White Wizard is an odorless, biodegradable, nontoxic formula that virtually erases old or new stains like grease, blood, pet stains, and more, even on dry-clean-only items. Don't let a stain ruin anything. Just a dab of the wizard and it's gone! 10-oz. tub. USA.

06-0028 White Wizard $8.50

Odor Absorbers

Our nontoxic air refreshers contain volcanic minerals with a natural ionic charge to completely erase problem odors. Simply hang the packs up and walk away. A single pack deodorizes up to 600 square feet of space. Placed in the air return of your central A/C or heating system, Odor Absorber House (not shown) keeps 1,000 square feet naturally odor-free. Recharge in sunlight for a lifetime of use in any climate. USA.

06-0622 Closet Odor Absorbers (four 6 oz. and two 32 oz. packs) $35

06-0623 House Odor Absorber (light blue, 32 oz.) $12

Air Freshener

Mist Air Freshener has a wonderfully fresh fragrance that actually removes odors instead of just masking them with synthetics. Nontoxic and biodegradable formula. Choose lime, orange or grapefruit. 7-oz. pump-spray bottle. USA.

06-0072 Try 1 Scent $8
06-9072 Set of 3 (1 of ea.) $23

Citra-Solv

Just a splash of highly concentrated Citra-Solv in water makes a stellar all-purpose cleaner for kitchen or bath. Mix a stronger solution to clean and deodorize floors, or pour it on undiluted when you need a stain remover for fabrics, an oven cleaner that cuts grime, or a biodegradable cleaner for your bike chain. Safe for porcelain, metal, tile, and fabric. USA.

06-0063 32 oz. $17
      1 Gal. $49

Lime-Eater

Lime-Eater eats away hard water stains, calcium deposits, soap scum, and rust from sinks, tubs, tile, and glass—even fiberglass and cultured marble—without hazardous solvents. Nonabrasive and Green Cross certified. Great for stainless steel and cooking gear, too! 21 oz. trigger-spray bottle. USA.

06-9015 Lime Eater, 2 bottles $14

ChemFree Toilet Cleaner

Keep your toilet clean and clear without toxins. First scrub your tank, then drop in ChemFree. Special "mineral magnets" kill bacteria on contact and prevent mineral stains, mildew, and fungus buildup. Lasts 5 years or 50,000 flushes. USA.

06-9029 ChemFree Toilet Cleaner, set of 2 $16

Mildew Stain Away

The biodegradable formula used in Mildew Stain Away washes away mildew and its stains in the bathroom without chlorine or other toxic chemicals. Safe for any surface, even book covers. Just spray and rinse. Ideal for pretreating shower curtains. Convenient 32-oz. trigger-spray bottle. USA.

06-0046 Mildew Stain Away $16

Smells Begone

Smells Begone is a heavy-duty odor-eater that neutralizes rather than masks odors of all kinds. It's nontoxic, nonstaining, and unscented. Spray it in the air or mix it with water and sponge on areas. Lavender or Rain. USA.

06-0061 Smells Begone Trigger Sprayer, 24 oz.   **$23**

    Smells Begone Pump, 12 oz.   **$13**

Septic Treatment

Imagine a septic treatment so effective you could significantly reduce costly, time-consuming tank pump-outs. Organica Cesspool/Septic Treatment uses an exclusive blend of naturally occurring microbes and enzymes that thrive in all septic conditions. With regular use, the treatment biodegrades septic wastes, preventing clogs in the system. Follow directions and simply pour down drain or toilet every couple of weeks to naturally break down organic wastes like grease, fat, paper, starch, and protein. Biodegradable, noncorrosive, and safe for industrial and home use. Tub lasts 7-8 months. USA.

06-0224 Septic Treatment   **$16**

Organic Bed and Bath

Organic Cotton Percales

Per your request, our 100% organic cotton percales now feature 250 thread count. Available in five new spring colors, created from low eco-impact dyes. Set includes one solid flat sheet, one 16-inch-deep solid fitted sheet, two pillowcases, and a matching bag with button closure. Butter, French Blue, Natural, Rose Petal, or Celery. India.

03-0734 Solid Sheet Set

Twin (1 standard pillowcase)	$89
Full (2 standard pillowcases)	$99
Queen (2 standard pillowcases)	$109
King (2 king pillowcases)	$119

03-0736 Solid Pillowcases (2)

Standard, 30"x20" (LxW)	$20
King, 40"x20" (LxW)	$22

03-0735 Solid Duvet Cover

Twin, 88"x72" (LxW)	$69
F/Q, 90"x90" (LxW)	$89
King, 90"x108" (LxW)	$99

Organic Wool Mattress Pad

Enjoy a more comfortable mattress naturally—and experience a deeper, more restful sleep with our Organic Wool Mattress Pad. The natural insulating properties of the Natura Grow Wool™ filling help regulate your body's temperature in warm and cool weather so you spend more time in REM sleep and less time waking to adjust your covers. Naturally nonallergenic. Covered in 100% natural cotton. Spot or dry clean. Canada.

03-0497

Twin, 80 oz.; 75"x39" (LxW)	$268
Full, 120 oz.; 75"x54" (LxW)	$288
Queen, 134 oz.; 80"x60" (LxW)	$308
King, 160 oz.; 80"x76" (LxW)	$358

Natural Cotton Mattress Pad

Designed to rest on top of your mattress like a feather bed, our untreated Natural Cotton Mattress Pad is free of dyes and chemical finishes throughout. Tufted natural cotton cover is filled with natural cotton batting to comfortably cushion your body. Spot or dry clean. Canada.

03-0565

Twin, 50 oz.; 75"x38" (LxW)	$100
Full, 70 oz.; 75"x54" (LxW)	$120
Queen, 83 oz.; 80"x60" (LxW)	$130
King, 105 oz.; 80"x76" (LxW)	$140

Organic Cotton & Wool Pillow

Our Organic Cotton & Wool Pillow is filled with nonitchy, pure wool and covered in soft organic cotton twill. It's breathable, nonallergenic, and naturally resistant to mold, mildew, dust mites, and bacteria. Soft wool fibers won't clump or shift. Standard size only. 30 oz. Canada.

03-0376 Organic Cotton & Wool Pillow $76

Organic Cotton Towels

Make bath time more enjoyable with these soft 500-gram-weight towels. Made from long-staple 100% organic cotton, they support the ever-growing organic cotton movement. Rich colors are created with low eco-impact dyes. Chlorine- and softener-free finishing preserves the purity and softness of the cotton. Set includes bath towel, hand towel, and two washcloths. 10 colors. Imported.

02-9091	Towel Set	$54 (a $64 value!)
02-0331	**Organic Cotton Towels**	
	Washcloth, 12"x12" square	$8
	Hand Towel, 30"x20" (LxW)	$16
	Bath Towel, 55"x28" (LxW)	$32
	Bath Sheet, 71"x39" (LxW)	$58

Organic Buckwheat Pillow

Our uniquely shaped Organic Buckwheat Pillow helps ensure ideal spinal alignment for a rejuvenating rest. Filled with conforming, supportive organic buckwheat hulls and covered with wool for comfort and breathability. Outer shell is made from 100% natural cotton. 21"x15"x4" (LxWxH). USA.

03-0370 Organic Buckwheat Pillow $70

Kapok Pillow

Considered sacred by many indigenous cultures, the Kapok tree is sustainably harvested for the hypoallergenic fibers of our Kapok Pillow. 180 thread count organic cotton cover. Spot clean. 16"x12½" (LxW). Brazil.

03-0686

	Regular, 30 oz.	$46
	Extra Fill, 36 oz.	$50

Organic Cotton Pillow

Rest easy as your head sinks into the supportive softness of our Organic Cotton Pillow. 100% organic cotton covered with 180-thread-count organic cotton. 26"x20" (LxW). USA.

03-0096

	Regular, 30 oz.	$48
	Soft, 23 oz.	$46

30th
ANNIVERSARY

CHAPTER 12

Sustainable Transportation

Toward a Future of Carbon-Free Mobility

IT'S A SHAME THAT THIS CHAPTER on sustainable transportation is so far back in the *Sourcebook*, because personal use of motor vehicles is as much or more of a global warming culprit as operating a household. It's hard to generalize, because so many factors come into play, but it seems safe to say that for most Americans, at least a good quarter of our personal CO_2 emissions are attributable to how we get around, and for many of us, that figure easily is closer to one-half. Of course, our cultural addiction to fossil fuels creates problems beyond global warming, most notably air pollution, sprawl, a large military establishment predicated on protecting access to Middle East oil, and vulnerability to both supply interruptions and terrorism.

However, we take up transportation near the end of the book for a very practical reason: Gaiam Real Goods is not in the business of selling vehicles. We offer some excellent resources but only a few relevant products. Nevertheless, because the energy and climate change implications of transportation for long-term sustainable living on Earth are so critical, we think it is important to offer you our take on the subject. Thanks to Steve Heckeroth and Dave Blume for assistance with this chapter.

The concept of sustainable transportation should be understood holistically, just like the concepts of sustainable shelter, energy, and agriculture. As defined by the Global Development Research Center, "Sustainable transportation concerns systems, policies, and technologies. It aims for the efficient transit of goods and services and for sustainable freight and delivery systems. The design of vehicle-free city planning, along with pedestrian- and bicycle-friendly design of neighborhoods, is a critical aspect for grassroots activities, as are telework and teleconferencing" (www.gdrc.org/uem/sustran/sustran.html). For most of us who live in automobile-centered societies, where personal mobility is considered a civil and almost moral right, the key conceptual shift is to understand that transportation should be about access to people and goods, rather than freedom of mobility. Urban design and transportation policy should favor people and healthy communities over motorized vehicles.

In the here and now, though, sustainable transportation basically comes down to vehicles. The problems associated with fossil fuel consumption for transportation can be addressed in a number of ways. Probably the most painless (for everyone except Detroit's auto manufacturers, and even they won't suffer too much), and certainly the most critical first step, is to increase the fuel efficiency of our vehicle fleet. The United States managed to do this once before, in the wake of the oil crisis of the 1970s. But due to corporate lobbying and government neglect, most of the fuel-efficiency gains of the past 30 years have been wiped out, as the popularity of energy-hog SUVs has soared. According to the EPA, the average fuel economy of all model year 2005 cars, passenger vans, SUVs, and pickups was 21.0 miles per gallon, lower even than 1982 and well below the 1987-88 peak of 22.1 mpg (which is also shameful). 40 mpg is feasible and realistic in all passenger cars, and such an increase would have significant benefits. According to Paul Roberts, author of *The End of Oil: On the Edge of a Perilous New World* (Houghton

> At least a good quarter of our personal CO_2 emissions are attributable to how we get around, and for many of us, that figure easily is closer to one-half.

> Urban design and transportation policy should favor people and healthy communities over motorized vehicles.

By doubling the average fuel economy of cars and trucks to 40 miles per gallon (which existing hybrid technology can easily do), we could save 5 million barrels of oil a day by 2015—or more than twice our current imports from the Middle East.

Mifflin, 2004), "By doubling the average fuel economy of cars and trucks to 40 miles per gallon (which existing hybrid technology can easily do), we could save 5 million barrels of oil a day by 2015—or more than twice our current imports from the Middle East" ("Energy Policy Should Look Beyond Oil," *Baltimore Sun*, July 13, 2004). That would represent an enormous savings in CO_2 emissions, too. (5 million barrels x 55 gal./barrel x 365 days/yr. x 10 years x 23.8 lb. CO_2/gal. = nearly 12 billion tons of CO_2!)

The past two or three years have proven beyond a doubt that alternative-fuel vehicles are here to stay. Automakers are scrambling to keep up with growing consumer demand for hybrid electric cars. Constantly evolving battery technologies and emerging plug-in hybrids (PHEVs) promise to make superlow-emission driving for commuting and short trips—by far the majority of driving that Americans do—a reality in the foreseeable future. And biofuels are starting to catch on in a big way, with the infrastructure for cost-effective and energy-efficient production of biodiesel and ethanol beginning to be put in place.

Switching to zero-emission vehicles that don't burn any fossil fuels will eventually make a big difference, and someday our economy might be powered primarily by hydrogen, which is itself produced without greenhouse gas emissions. But those days are a way off, and many of us feel considerable urgency to act now to begin withdrawing from our destructive fossil fuel habit. What can a mobility-loving, energy-conscientious individual do?

- Drive a vehicle that gets good gas mileage.
- Drive a vehicle that can burn biodiesel or ethanol.

Fueling Up with Biomass

The biofuels industry in the U.S. is growing at a rate of 25%-50% each year. The U.S. Department of Energy's Energy Information Administration projects that:

- At least 7.5 billion gallons of ethanol and biodiesel will be used annually by 2012.

- U.S. ethanol consumption will be above 10 billion gallons annually by around 2013.

- In 2030 U.S. demand for ethanol in transportation fuels should range—depending on world oil prices—somewhere between 10 and 15 billion gallons annually.

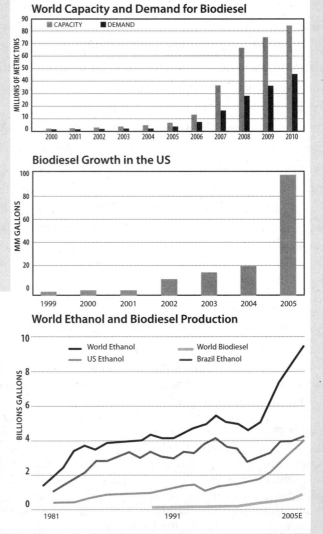

World Capacity and Demand for Biodiesel

Biodiesel Growth in the US

World Ethanol and Biodiesel Production

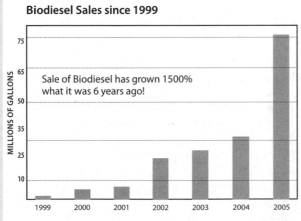

Biodiesel Sales since 1999

Sale of Biodiesel has grown 1500% what it was 6 years ago!

Table 1. Estimated Number of Alternative-Fueled Vehicles in Use in the United States, by Fuel, 1995-2004

Fuel	1995	1996	1997	1998	1999	2000	2001	2002	2003	2004 (Projected)	Av. Ann. Growth Rate (Percent)
Liquefied Petroleum Gases (LPG)	172,806	175,585	175,679	177,183	178,610	181,994	185,053	187,680	190,438	194,389	1.3
Compressed Natural Gas (CNG)	50,218	60,144	68,571	78,782	91,267	100,750	111,851	120,839	132,988	143,742	12.4
Liquefied Natural Gas (LNG)	603	663	813	1,172	1,681	2,090	2,576	2,708	3,030	3,134	20.1
Methanol, 85 Percent (M85) [A]	18,319	20,265	21,040	19,648	18,964	10,426	7,827	5,873	4,917	4,592	-14.3
Methanol, Neat (M100)	386	172	172	200	198	0	0	0	0	0	0.0
Ethanol, 85 Percent (E85) [A,B]	1,527	4,536	9,130	12,788	24,604	87,570	100,303	120,951	133,776	146,195	78.8
Ethanol, 95 Percent (E95) [A]	136	361	347	14	14	4	0	0	0	0	0.0
Electricity [C]	2,860	3,280	4,453	5,243	6,964	11,830	17,847	33,047	45,656	55,852	39.1
Non-LPG Subtotal	74,049	89,421	104,526	117,847	143,692	212,670	240,404	283,418	320,367	353,515	19.0
Total	246,855	265,006	280,205	295,030	322,302	394,664	425,457	471,098	510,805	547,904	9.3

A. The remaining portion of 85 percent methanol and both ethanol fuels is gasoline.

B. In 1997, some vehicle manufacturers began including E85-fueling capability in certain model lines of vehicles. For 2002, the EIA estimated that the number of E-85 vehicles that are capable of operating on E85, gasoline, or both, is about 4.1 million. Many of these alternative-fueled vehicles (AFVs) are sold and used as traditional gasoline-powered vehicles. In this table, AFVs in use include only those E85 vehicles believed to be intended for use as AFVs. These are primarily fleet-operated vehicles.

In 1997, some vehicle manufacturers began including E85-fueling capability in certain model lines of vehicles. For 2002, the EIA estimated that the number of E-85 vehicles that are capable of operating on E85, gasoline, or both, is about 4.1 million. Many of these alternative-fueled vehicles (AFVs) are sold and used as traditional gasoline-powered vehicles. In this table, AFVs in use include only those E85 vehicles believed to be intended for use as AFVs. These are primarily fleet-operated vehicles.

C. Excludes gasoline-electric hybrids.

Notes: Estimates for 2003, in italics, are based on plans or projections. Estimates for historical years may be revised in future reports if new information becomes available.

Sources: 1995: Science Applications International Corporation, "Alternative Transportation Fuels and Vehicles Data Development," unpublished final report prepared for the Energy Information Administration (McLean, VA, July 1996) and U.S. Department of Energy, Office of Energy Efficiency and Renewable Energy. 1996-2004: Energy Information Administration, Office of Coal, Nuclear, Electric, and Alternate Fuels. Beginning in 2000, Federal data were derived from the DOE/GSA Federal Automotive Statistical Tool (FAST).

- Buy a hybrid gas-electric vehicle.
- Consider getting an electric vehicle.
- Use mass transit as much as you can.
- Take advantage of carpooling or car sharing in communities where that system exists.
- Walk or ride your bike for a commuting change.
- Plant trees to offset your personal CO_2 emissions. Organizations such as American Forests (www.americanforests.org) and Trees for the Future (www.treesftf.org) have excellent reforestation programs in the United States and around the world; all you need to do is plunk down a little cash. The American Forests Web site even has an easy-to-use Personal Climate Change Calculator that tells you how many trees have to be planted to offset your annual contribution to global warming.

Let's take a look at the near and far horizons of vehicle development and see what choices are now on the market and likely to be available in the future.

Hybrids and Other Low-Emission Vehicles

The fact that hybrid electric vehicles are becoming wildly popular with consumers and that the auto manufacturers are having trouble keeping up with demand is good news all around, because gas-electric hybrids represent the simplest, most effective, most affordable short-term way to reduce fossil fuel consumption in passenger cars and the environmental impact of driving motor vehicles.

The Toyota Prius hybrid is taking the country by storm.

In response to more-stringent state and federal emissions standards, the manufacturers are also beginning to market Ultra-Low Emission and Super Ultra-Low Emission gasoline vehicles (ULEVs and SULEVs) that feature redesigned, more-efficient engines and more-effective tailpipe emission controls.

A hybrid combines the best features of internal combustion and electric drive, while eliminating the worst parts of each. Hybrid vehicles have both an internal combustion engine and an electric motor, with a highly sophisticated automatic control system that chooses which one is running under what conditions. The electric motor is used at low speeds and for acceleration boost; the internal combustion motor delivers cruising speed and long range but automatically shuts off at stoplights. Deceleration and braking actually capture energy by generating electricity that recharges the batteries.

Hybrid vehicles are available right now and getting better all the time. Several Real Goods employees at our Hopland Solar Living Center drive one to work every day. The 2006 Toyota Prius gets 60 mpg in the city and 51 mpg on the highway—a 15% improvement over the 2003 model. The 2006 Honda Civic hybrid gets 49 and 51 mpg, respectively. Even better, Honda's two-seater hybrid car, the Insight, rates at 57/56 mpg. These three models, all qualifying as SU-LEVS, have the highest "Green Scores" issued by the American Council for an Energy-Efficient Economy (except for a Civic model powered by compressed natural gas; these vehicles are available, but the fuel is hard to come by and currently more appropriate for fleets). These early-generation hybrids continue to lean heavily on the internal combustion engine, because it's the technology we understand best. Future hybrids will rely more on electric storage and drive and will increase fuel mileage greatly.

The other good news is that hybrids are affordable: The Prius, Civic, and Insight each cost between $20,000 and $22,000. Furthermore, the automakers are finding hybrids to be profitable, indicating that this technology is poised for longevity in the marketplace. Honda now makes an Accord hybrid with a V6 engine that lists for around $31,000.

Hybrids are the technology that is most immediately accessible to manufacturers right now. Every major manufacturer has hybrid models in the works. You've probably noticed that they're even advertising hybrid SUVs! The number of models that meet the toughest emissions standards now stands at nearly 90, in all classes of cars, trucks, vans, and SUVs.

Biodiesel, Ethanol, and Other Biofuels

Believe it or not, the original engines invented by Rudolph Diesel in 1895 were designed to run on vegetable oil.

Like gas-electric hybrids, biodiesel is a readily available transportation choice that is more environmentally friendly than driving a conventional gasoline-powered car. The increasing public awareness and use of biodiesel actually represents a "back to the future" development. Believe it or not, the original engines invented by Rudolph Diesel in 1895 were designed to run on vegetable oil. That is the promise of biodiesel and other biofuels, such as ethanol, in the 21st century and beyond. These are clean-burning, relatively nonpolluting fuels made from renewable resources.

Biodiesel is a biodegradable, nontoxic fuel derived either from vegetable oils such as soybean oil or canola oil, or from recycled waste cooking oil. Biodiesel is made in a refinery process that removes glycerin from the oil. The process is not highly energy intensive, and the glycerin byproduct can be sold for use in the manufacturing of soap and cosmetics. Biodiesel can be burned in compression-ignition (diesel) engines, as a pure fuel or blended with petroleum diesel (aka dino-diesel), with little or no modification to the vehicle.

Any use of biodiesel instead of petroleum diesel reduces emissions of pollutants and greenhouse gases. As demonstrated in Environmental Protection Agency tests, burning pure biodiesel reduces particulate emissions by 47%, polycyclic aromatic hydrocarbons by 75%-85%, unburned hydrocarbons by 68%, carbon monoxide by 48%, and sulfur oxides and sulfates by nearly 100% compared with petroleum diesel. The only exception is nitrogen oxide emissions, which may increase slightly with biodiesel. The positive impact on climate change is even better. Not only does burning biodiesel reduce carbon dioxide emissions by 78%, but biodiesel (and other biofuels) is carbon-neutral: While combustion of any biofuel releases carbon dioxide into the atmosphere, growing the agricultural crop used in that fuel captures a similar amount of CO_2 through the process of photosynthesis. More and more people today run B20, a blend of 20% biodiesel with 80% petroleum diesel. Burning B20 reduces tailpipe emissions by one-fifth of the percentages just noted. All the Volkswagen TDI (turbo diesel injection) models of the last several years run just fine on B100 or 100% biodiesel.

Biodiesel also has a positive energy balance. For every unit of energy consumed in the process of growing and manufacturing it, the resulting fuel contains anywhere from 2.5 to 3.2 units of energy. Fossil fuels have a negative energy balance, because extracting and refining them is highly energy intensive. While a given unit of biodiesel contains slightly less energy than the same unit of petroleum diesel, it has a slightly higher combustion efficiency, so using biodiesel in place of petroleum does not noticeably affect vehicle performance. You can even have a hybrid electric vehicle that uses a diesel motor instead of a conventional gasoline motor, and several major automakers are currently developing diesel hybrids.

While it's true that you can use biodiesel without making any engine modifications, it does have a solvent effect that may release accumulated deposits on tank and pipe walls, which may, in some vehicle models, initially clog filters and necessitate their replacement. Some experts also recommend replacing rubber hoses with ones made of synthetic materials in certain makes of vehicles. However, B20 does not appear to present a problem to those rubber components. Biodiesel also clouds and gels at higher temperatures than petroleum diesel, which could present a problem in colder climates. As with petroleum, however, winterizing agents added to biodiesel enable it to be used trouble-free to at least -10°F. Here at the Solar Living Center, we dispense B100—100% biodiesel—to numerous cars traveling the northern California Highway 101 corridor. Six Real Goods employees currently drive biodiesel vehicles using B100.

A number of town vehicle and school bus fleets in the Northeast have successfully piloted using B20 year-round. The momentum of the fuel switch is growing, especially since the price of B20 continues to be competitive with petroleum diesel. As of this writing, the retail price is nearly the same. Many school districts and municipalities are able to achieve even greater savings because they purchase their diesel in bulk. We all know that the price of oil is destined to rise drastically in our lifetime, and every price increase makes biodiesel that much more competitive in the marketplace. The list of cities, towns, and schools that are turning to biodiesel, out of concern for the health of schoolchildren and municipal employees as well as the global environment, continues to grow. Keene, New Hampshire, uses it in their truck fleet, as do several towns in Vermont; Warwick, Rhode Island, uses it in 60 school buses (and to heat school buildings); Deer Valley School District in Phoenix, Arizona, uses it in 100 buses; and the states of Michigan and Minnesota have created incentive programs to encourage the switch. Companies such as L.L.Bean and Alcoa are experimenting with biodiesel, and even the federal government is getting into the act, testing biodiesel in its vehicle fleet in Yellowstone National Park.

Other biofuels may have even more promise than biodiesel produced from oil seeds or recycled cooking oil. Ethanol is familiar to many people, a fuel primarily made from corn today. It's controversial in some quarters because it is expensive to make from corn; the process most commonly used in the U.S. yields a small positive energy balance, and it is commercially viable right now only because of huge government subsidies. (Of course, the oil companies have been the beneficiaries of huge government subsidies that make their product commercially viable, too.) But ethanol can be produced more efficiently and sustainably from corn and other biomass feedstocks, including a variety of non-traditional crops like buffalo gourd and nopales cactus, paper mill sludge, perennial grasses such as switch grass, and rice or wheat straw. Just converting the lawn clippings currently going into landfills would yield 11 billion gallons of fuel per year. In Brazil and India, the energy input is not fossil fuel but sugar cane bagasse or self-produced methane. As alcohol plants in the U.S. retrofit to burn biomass or self-produced methane—this is beginning to happen—the en-

As demonstrated in Environmental Protection Agency tests, burning pure biodiesel reduces particulate emissions by 47%, polycyclic aromatic hydrocarbons by 75%-85%, unburned hydrocarbons by 68%, carbon monoxide by 48%, and sulfur oxides and sulfates by nearly 100%.

Not only does burning biodiesel reduce carbon dioxide emissions by 78%, but biodiesel (and other biofuels) is carbon-neutral.

Biomass Energy Content

The highest yield feedstock for biodiesel is algae, which can produce 250 times the amount of oil per acre as soybeans.

1 acre of algae yields 10,000 gallons of biodiesel
1 bushel of soybeans (60 lb.) yields 11 pounds of soybean oil, which makes 1.5 gallons of biodiesel
1 acre of soybeans yields 40 gallons of biodiesel

1 bushel of corn (56 lb.) yields about 2.5 gallons of ethanol
1 ton of corn stover yields 80-90 gallons of ethanol
1 ton of switch grass yields 75-100 gallons of ethanol

Biogas from animal waste (methane) The number of animals it takes to produce methane with an energy content of 1 gallon of gas per day: 46 pigs or 5.5 cows or 810 chickens

ergy return on energy input will rise significantly. Scientists working to improve the production of ethanol from cellulose predict that foreseeable technological advances implemented on a large scale could lower the cost of ethanol to 50¢ a gallon. Venture capital is flowing into cellulosic ethanol projects in North America, and factories are moving from the drawing board to pilot plants. The potential environmental benefits are enormous. Burning ethanol instead of gasoline reduces carbon emissions by more than 80% and eliminates sulfur dioxide. Ethanol also reduces emissions of carbon monoxide, nitrous oxides, and hydrocarbons by over 90%. And unlike biodiesel, ethanol can be used in both gasoline and diesel vehicles.

Brazil offers an intriguing case study of the potential positive impact of biofuel use on a massive scale. In the 1970s, Brazil's military dictatorship spurred the creation of an ethanol industry, based on sugarcane, to fuel the country's transportation needs. Today all motor fuel is at least a blend of 25% ethanol with 75% gasoline and 45% of all cars run exclusively on alcohol.

Flex-fuel vehicles, which can switch back and forth between different types of fuel carried onboard, are one key. Roughly one-third of all vehicle models now built in Brazil, and three-quarters of new cars sold, are flex-fuel. Ethanol accounts for nearly half of the fuel that Brazilians consume in their passenger cars. As a result, Brazil no longer imports oil, and its indigenous ethanol industry is a significant part of the country's economic landscape. Brazil uses

> Roughly one-third of all vehicle models now built in Brazil, and three-quarters of new cars sold, are flex-fuel.

less than 1% of its land for alcohol production and is rapidly expanding production for export to China and other countries. The U.S. could supply all its transportation fuel from ethanol on less than 2% of its agricultural land if our municipal sewage was diverted to cattail-based energy farms (cattail is a temperate crop similar to sugar cane), producing ethanol without petroleum-based fertilizers.

In the U.S., ethanol production currently stands at 5 billion gallons annually, projected to jump to 8 billion in the next two years. Over a thousand filling stations (out of a total of 175,000) now sell E85. And the number of flex-fuel vehicles on the road has gone from zero in 1993 to 5 million today, though consumers don't always know they have them. Any gasoline car can be converted to alcohol for as little as $50-$300 in parts. More important, all fuel-injected cars can run on at least 50% alcohol without any modifications! Most hybrids can run on E85 without changes even though they are not advertised as being E85 capable.

With such tremendous potential upside, the domestic biofuels industry is clearly poised for takeoff. Current legislation proposes that the U.S. mandate an increase in the Renewable Fuels Standard that would require the production of 60 billion gallons of ethanol per year by 2030, and that almost all new vehicles be flexible fuel by 2012.

As a matter of fact, scientists—including some at the National Renewable Energy Laboratory—are even investigating the possibility of biofuel made from algae. Algae are the most efficient organisms on the planet at consuming carbon dioxide and producing oil. According to NREL estimates, a 1,000-square-meter pond of algae can produce more than 2,000 gallons of oil per year. In comparison, the same crop area of high-yield canola can produce 50 gallons of oil. Even if only 10% of the algal oil can be extracted for use as fuel, it still outperforms the canola by a factor of 4. A half-million acres of algae ponds (that's equivalent to less than 1% of the fallow cropland in the U.S.) could produce something on the order of 10 billion gallons of oil annually, which could then be processed into biodiesel. In the grandest scenario, marine algae farms off the California coast and in the Dead Zone in the Gulf of Mexico could supply all the transportation fuel for the entire country, without planting a single seed in the ground. Talk about renewable energy! Stay tuned.

Electric Vehicles

Wither the electric car? After a serious upsurge of interest among the major automakers during the 1990s, those same companies have recently acted—some would say conspired—to drive a stake through the heart of the all-electric vehicle. General Motors developed the prototype EV-1 in 1990, and as a result, the California Air Resources Board issued its groundbreaking Zero Emission Vehicle Mandate. But fairly quickly, the auto manufacturers and the oil companies mobilized to attack and undermine the mandate, and lo and behold, by 2003, trucks, vans, and SUVs made up more than half of all passenger vehicles sold in the U.S., and average fuel efficiency of the nation's cars had declined by nearly 50%. Several hundred happy consumers who had bucked the trend and leased EV-1s from General Motors were shocked when the company called in the leases, demanded the return of every last vehicle, and literally crushed them, thus ending its electric vehicle development program. Meanwhile, GM filed a lawsuit against the state of California to destroy the Zero Emission Vehicle Mandate. (For the full story, we recommend the movie *Who Killed the Electric Car?*)

In spite of these depressing developments, improvements in battery technology, the continuing growth of electric power generated with renewable resources, and increasing public concern about climate change and instability in the fossil fuel markets add up to real hope that electric vehicles will become commercially available in the foreseeable future. When you consider that according to the EPA, around 80% of all auto trips in North America involve a round

A Clean and Quiet Revolution

The technologies exist to clean the air, stabilize the climate, and maintain our standard of living all at the same time. By relying on clean renewable technologies, we can eliminate much of the U.S. trade deficit and the reason for war while achieving energy independence.

A quick study of the chart below shows the overwhelming advantages of plug-in hybrid (PHEV) and battery electric vehicles (EV). EVs are zero emission and can be charged from zero emission renewable energy sources like the Sun and wind. By adding more batteries to hybrid electric vehicles (HEV), plug-in hybrids can be built which offer the range of gas vehicles (400 miles) with the zero emission and cost-saving benefits of battery electric vehicles for short trips.

The main assumptions used to produce the values in the chart are:

1. The average cost of gasoline over the next year will be approximately $3.50/gallon.

2. The Time of Use (TOU) rate for nighttime charging is approximately $0.05/kWh.

3. There are about 40kWh of energy in a gallon of gasoline.

4. Burning 1 gallon of gasoline produces approximately 23 lb. of CO_2.

—*Steve Heckeroth*

Steve Heckeroth (www.renewables.com) has been building and driving EVs for 15 years.

Vehicle Type	$ Gas 25 mi./ day	kWh 25 mi./ day	$/year 25 mi./ day	Gal./year 25 mi./ day	Tons of CO_2/year Tailpipe	*+Tons of Upstream CO_2/year
10 mpg Gas	8.75	100	$3,200	915	10.5	13.7
20 mpg Gas	4.37	50	$1,600	460	5.3	6.8
30 mpg Gas	2.93	34	$1,050	305	3.5	4.5
40 mpg HEV	2.20	25	$800	230	2.6	3.4
50 mpg HEV	1.75	20	$640	180	2.1	2.8
Plug-in HEV 25-mi. Range	0	10	$200	0	0	0.7
Battery EV	0	6	$100	0	0	0.4
Solar/Pedal/Electric	0	1	0	0	0	ZERO

*This column includes upstream CO_2 emissions for exploration, extraction, transport, refining, and distribution of gasoline, as well as CO_2 emissions from the California mix of power plants that produce electricity to charge electric vehicles.

trip of less than 20 miles—well within the range of zero-emission battery-electric vehicles—it's crystal clear that EVs have to become the wave of the future. Imagine the impact on greenhouse gas emissions if most Americans commuted and ran short errands in battery-powered cars.

An electric vehicle runs on an electric motor and contains a battery pack to store electrical energy. The great advantages of EVs are emissions and operating cost. EVs have no point-of-use emissions, and research has shown that even taking into account the power plants that generate the electricity, an EV releases 90% less emissions than a comparable gasoline-powered car. An EV can also be powered by solar energy, which of course would be the ideal scenario. EVs also are cheaper to operate than internal combustion vehicles. They don't need tune-ups, fuel and oil filters, oil changes, or mufflers, and because they have fewer moving parts, they require less maintenance. (This is probably one reason why Big Auto ganged up on the EV.) Research has demonstrated that the overall life-cycle cost of operating an EV can be approximately 60% less than that of a comparable conventional vehicle (see sidebar).

The liabilities of electric vehicles are what present the problem. The high cost of conversion ($5,000 or more for a typical passenger vehicle) and the limitations in range imposed by current battery technology (still only 50-60 miles before a recharge is needed) continue to conspire against EVs. In 2001 we wrote, "We

haven't reached the performance level of jumping into your EV at a moment's notice, driving 300 miles, and recharging it in 10 minutes. In other words, the kind of vehicle performance we're used to." While that's still largely true, range and performance are improving every year. If you're interested in knowing more about the nuts and bolts of electric vehicle technology, see pages 440-48 of the 11th edition *Solar Living Sourcebook*.

The current phase of EV development is called the Plug-in Hybrid Electric Vehicle, or PHEV. It's just what it sounds like, a hybrid car that you can plug in and recharge. PHEVs offer zero-emission performance for everyday short-distance driving as well as the convenience of traveling 400 miles on one tank of fuel for trips over the battery's capacity. Because long trips are relatively infrequent, PHEVs can increase the feasibility of large-scale use of fuels like biodiesel, ethanol, and hydrogen produced from renewable sources. The current buzz focuses on nanotechnology to increase battery range, and the concept of "vehicle-to-grid" use, in which parked PHEVs could "rent" their networked batteries for distributed energy storage, thereby benefiting utilities, fleet owners, and consumers alike. Even the automakers are again cautiously, grudgingly acknowledging the potentially immense advantages of plug-in hybrids.

For more information on the cutting edge in electric vehicles, check out www.hybridcars.com and www.pluginpartners.com.

> PHEVs offer zero-emission performance for everyday short-distance driving as well as the convenience of traveling 400 miles on one tank of fuel for trips over the battery's capacity.

Fuel Cell and Hydrogen Vehicles

> We are persuaded by the analysts who argue that fuel cell and hydrogen vehicles have a long way to go before they're commercially viable, and therefore represent a long-term solution when more-realistic short-term fixes are needed urgently now.

Fuel cells! The Hydrogen Economy! Now there's the magic bullet. Not. Since we first wrote about hydrogen fuel cells in the 12th edition of this *Sourcebook*, the hype about hydrogen has continued to mount. And over that same period of time, our skepticism about that hype has deepened. Hydrogen—the most common element in the universe—probably is the wave of the future. It may well become the primary storage medium of the 21st century, and that would be a good thing for people and the planet. However, even more today than in 2004, we are persuaded by the analysts who argue that fuel cell and hydrogen vehicles have a long way to go before they're commercially viable, and therefore represent a long-term solution when more-realistic short-term fixes are needed urgently now. A strategically targeted, well-funded crash hydrogen development program is advisable, but other steps can and should be taken immedi-

ately to reduce the greenhouse gases and other pollutants generated by the transportation sector of the economy.

Functionally, a fuel cell is close kin to a gas-powered generator. A fuel cell takes a hydrogen-rich fuel source and delivers electricity, heat, and some kind of exhaust through chemical conversion. The electrical output from the fuel cell is delivered to an electric motor, and you've got your basic fuel cell-powered vehicle. A modest battery pack sometimes is used to boost acceleration and help even out the electrical load on the cell. Fuel cells are quiet, are two to three times more efficient than an internal combustion engine, and produce less than half the emissions. In the ideal scenario, a fuel cell running on pure hydrogen will deliver only pure water vapor as exhaust. But on a practical level, hydrogen is found bound up with other elements to make water or hydrocarbons, so pure hydrogen is expensive. Most fuel cell vehicles under development use some kind of hydrocarbon fuel, such as natural gas, alcohol, or even gasoline, which runs through an onboard "fuel processor" that boosts the hydrogen content. If you run something other than perfectly pure hydrogen into your fuel cell, you'll get something other than perfectly pure water vapor for exhaust. It still beats the pants off any kind of combustion process for cleanliness.

Hydrogen is often considered a nonpetroleum fuel, but it should be noted that it is not a source of energy, only a way of storing and transporting energy. Hydrogen fuel must be "manufactured" by extracting it from fossil fuel or water. Electricity can be used to split water into hydrogen and oxygen, but this process, known as electrolysis, is expensive. Cheaper processes produce toxic byproducts or significant greenhouse gas emissions. Currently, the bulk of federal fuel cell research is dedicated to making hydrogen from fossil fuels rather than renewable sources. This means that renewables will not be a part of the hydrogen economy, and even if the rest of the world switches to hydrogen made from water, Americans will end up yet more dependent on fossil fuels for generations.

Even if produced with renewable resources, hydrogen is still far from an ideal automobile fuel. In its densest form (liquid), hydrogen has only one-third as much energy per liter as gasoline. If stored as compressed gas at 300 atmospheres (a more practical option), it delivers less than one-fifth the energy per volume of gasoline. Such low energy density means that fuel storage would take up lots of room in a hydrogen-powered car; alternatively, a modest-sized fuel tank would severely restrict the vehicle's range between fill-ups.

The crucial difference between pure hydrogen and hydrocarbons points to the underlying problem with fuel cell technology: It's still in an early research-and-development stage. Joseph J. Romm, executive director of the Center for Energy and Climate Solutions, assistant secretary of energy for Energy Efficiency and Renewable Energy in the Clinton Administration, and author of *The Hype About Hydrogen: Fact and Fiction in the Race to Save the Climate* (Island Press, 2004), is a leading hydrogen skeptic. He points out that the current cost of producing hydrogen from natural gas is three times more than the price of gasoline, and producing it from water is six to nine times more expensive than gasoline. A February 2004 report from the National Academies' National Academy of En-

An electric vehicle, outstanding in its field.

gineering and National Research Council concluded that in the best-case scenario, the transition to a hydrogen economy will take "many decades" and achieve insignificant reductions in oil imports and CO_2 emissions over the next 25 years. Another government report estimated that it will cost at least $500 billion to create and implement the infrastructure necessary to provide hydrogen fuel for just 40% of the cars on the road.

Romm and others agree that hydrogen fuel cell vehicles cannot make a significant contribution to slowing climate change until after 2030. The problem is, global CO_2 emissions are projected to rise by more than 50% by 2030. It just doesn't make ecological sense to burn fossil fuels to produce hydrogen to run fuel cells. Even in 2030, Romm notes, using natural gas to displace coal in the generation of electricity rather than using it to displace gasoline for transportation will be a better strategy for lowering greenhouse gas emissions. And using renewables to produce hydrogen for vehicular transportation is not especially advantageous, either. Using renewable energy to generate electricity directly, rather than using it to produce hydrogen for fuel cells, has a more positive impact on climate change.

The bottom line for Romm: Most gas-electric hybrids available today are nearly as efficient as what is projected for fuel cell vehicles. His argument is that running a fuel cell car on fossil fuel-produced hydrogen in 2020 has no life-cycle greenhouse gas advantage over today's Prius hybrid running on gasoline. If you're concerned about the environmental impact of your personal transportation choices, don't hold your breath waiting for fuel cell vehicles to save the day. Get a gas-electric or diesel-electric hybrid today, make the switch to biodiesel or ethanol, or place your bet on the rapid commercialization of plug-in hybrids.

Using renewable energy to generate electricity directly, rather than using it to produce hydrogen for fuel cells, has a more positive impact on climate change.

The Hypercar

One other cutting-edge transportation concept is worth keeping an eye on. Amory Lovins of the Rocky Mountain Institute has been developing the Hypercar for the past decade. As explained on the RMI Web site: "A Hypercar vehicle is designed to capture the synergies of ultralight construction; low-drag design; hybrid-electric drive; and efficient accessories to achieve 3- to 5-fold improvement in fuel economy, equal or better performance, safety, amenity, and affordability, compared to today's vehicles. Rocky Mountain Institute's research has shown that the best (possibly, the only) way to achieve this is by building an aerodynamic vehicle body using advanced composite materials and powering it with an efficient hybrid-electric drivetrain." Lovins's idea is ultimately to power the Hypercar with a hydrogen fuel cell. A for-profit collaboration has been established between RMI, automobile companies, and others to bring the Hypercar to market in a commercially feasible way.

If anyone is capable of packaging together the most-advanced strategies for vehicle design, materials, and fueling, and devising a product that can capture the attention of consumers at a reasonable price, these guys can. Keep your eyes and ears peeled for more news about the Hypercar. In the meantime, ride your bike, take the bus, and buy a hybrid or biodiesel car!

Sources of Further Information

Electric Auto Association (EAA)
PO Box 6661
Concord, CA 94514
www.eaaev.org
The national electric vehicle group since 1967. Has local chapters in many states. Yearly membership is $39 and includes an informative monthly newsletter with tech articles, industry news, and want ads. The Web site has links to practically everything and everybody EV related.

Electrifying Times
63600 Deschutes Market Road
Bend, OR 97701
www.electrifyingtimes.com
A commercial EV magazine that comes out three times a year. Runs 60 pages or more. Much up-to-date information, and ads from suppliers and builders. Edited by Bruce Meland. $12/year.

Department of Energy Office of Transportation Technologies Hybrid Electric Vehicle Program
www.nrel.gov/vehiclesandfuels/hev/
The DOE's site contains a lot of good information: What hybrids are, how they work, where you can buy one.

Marshall Brain's How Stuff Works Web site, "How Hybrid Cars Work"
www.howstuffworks.com/hybrid-car.htm
The How Stuff Works Web site is great fun, is highly informative, and has good interactive graphics, but you'll need to put up with some commercial content. Their hybrid site is very complete.

National Biodiesel Board
www.biodiesel.org

The Hypercar
www.hypercar.com

International Institute for Ecological Agriculture
www.permaculture.com/alcohol/index.shtml

SUSTAINABLE TRANSPORTATION PRODUCTS

eGo Cycle: The Electric Scooter with Moxie

Our employees road-tested this remarkable electric cycle on everything from commutes to grocery runs—and quickly declared it the little bike that could. The eGo has what it takes to get you where you need to go, at a running cost of just a penny a mile. This ingenious mode of transport is built on a quasi-bicycle frame—making it affordable as well as light enough to efficiently transport you and a surprisingly sizable payload. A simple turn of a key and a quick throttle twist engage either of two electronically controlled operating modes ("Go Far" or "Go Fast"), letting you tailor your ride for greater distance or maximum speed. The eGo's regenerative braking system captures your momentum as you ride, recharging the battery and optimizing the DC motor's range. The built-in 5-amp charger plugs into any standard AC outlet for an 80% recharge in just three hours—and a full recharge from empty in only four to six hours. The included 24-volt/34-amp-hour sealed battery pack lasts four to five years. The eGo's cargo features are both accommodating and versatile: Built-in rear rack holds briefcases, picnic baskets, and extra gear; optional folding rear baskets each hold a full bag of groceries; optional front basket converts to a shopping basket; and optional cargo trailer attaches in a snap, folds flat for storage, can haul an additional 100 pounds, and keeps the brake light visible. New LX is fully equipped for registration in any state and includes front and rear turn signals; left grip lighting, signal and horn controls; and backlit speedometer/odometer. Silver with four accent color options. 63"x40" (LxH). 120 lb. USA.

65-0037 **eGo Classic Cycle**		$1,399
65-0037 **eGo LX Cycle**		$1,749

Ships anywhere in the contiguous U.S. for $120. Please allow 4-6 weeks.

65-0038 **Front Basket**		$29
65-0039 **Rear Folding Baskets**		$49
65-0041 **Rain Cover**		$49
65-0042 **Extended-Range Battery Pack**		$189

Two speed/distance options:
"Go Fast"—15-20 miles at 24 mph
"Go Far"—20-25 miles at 17 mph

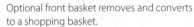

Optional front basket removes and converts to a shopping basket.

Optional rear baskets fold flat.

Optional cargo trailer.

Optional rain cover.

Extended battery pack.

Automated and Manual Biodiesel Processors

The BioPro 190 is a fully automated, stainless steel processor designed for people who want to economically produce their own clean-burning 100% biodiesel fuel. It's compact, portable, and self-contained, with individual compartments for each ingredient required to produce biodiesel. Fill the compartments with the chemical ingredients (new or used vegetable oil, methanol, and small amounts of sulfuric acid and lye) and press

Start. The BioPro 190 does not require transfer of the finished fuel into storage or distribution tanks. Simply pump the biodiesel directly into your vehicle or equipment fuel tank using the integrated, 12-gpm filler pump.

The BioPro 190 does not require titration or measuring. Simply fill everything up to indicated levels and turn on your processor. The two-stage esterification and transesterification reaction removes uncertainty and eliminates delicate, time-consuming chemical tests. 1-yr. mfr. warranty. USA.

63-0025 BioPro 190   **$7,995**

Specifications

Capacity	50 gallons
Total Process Time	48 hours
Size	67"x21"x21" (HxWxD)
Weight	350 lb. (empty)
Electrical Requirements	120VAC, 60 Hz, 15-amp circuit
Construction	Polyethylene reaction vessel, 304 stainless steel body, painted carbon-steel structure, powder-coated, carbon-steel covers and component covers

FuelMeister™ Biodiesel Processor

With a FuelMeister™, you don't have to be a plumber or mechanic to make your own biodiesel. This is not a hobby kit but a complete system engineered with quality materials. You get all the specialized equipment and supplies you need to start making biodiesel that same day. All you need to add is vegetable oil, methanol, electricity, and tap water. The whole setup fits comfortably in the corner of your garage. This is a "closed" system, so it won't smell, and you won't have to pour or stir any liquids. It will turn even heavily used cooking oil into clean burning, biodegradable, and smooth-running biodiesel. Whatever you own that runs on diesel fuel, you can power with the biodiesel you produce with your own FuelMeister biodiesel production system.

The FuelMeister produces approximately 40 gallons of biodiesel

Optional oil preheat kit fits any 55-gallon barrel.

every 48 hours, with about one hour of operator time required per batch. Supplies typically cost 70¢ per gallon or less. Methanol is widely available as racing fuel; the lye catalyst is available at any hardware store. A safe, manual methanol pump and a precise digital scale for the lye is included, along with personal safety items (rubber gloves, safety glasses, etc.) and a titration kit for testing your raw oil to see precisely how much catalyst is required. This well-engineered package is designed for safety, quality, and ease of use with ball valves and see-through tanks and tubing.

The optional Oil Pre-Heater with thermostat accelerates production and improves yields in cold climates. It includes wrap-on insulation, heater, and tape to fit any standard 55-gallon barrel you provide. 120VAC. All components are UPS shippable. USA.

63-0015 FuelMeister Biodiesel Kit,
**   62"x25" (H x diameter) $2,999**

63-0022 FuelMeister Plus Biodiesel Kit $3,995

63-0017 Oil Pre-Heater Kit, 40"x20"
**   (H x diameter) $399**

BOOKS

How to Live Well Without Owning a Car: Save Money, Breathe Easier, and Get More Mileage out of Life

Perhaps the first practical, accessible, and sensible guidebook to living in North America without an automobile, this book shows it's possible for nearly everyone to discard their cars and live a better, richer, car-free life. Chris Balish writes from personal experience, having lived (accidentally at first) for four years with hardly a moment behind the wheel (amounting to a savings of nearly $40,000!!), and has the persuasive authority to convince even the most dedicated drivers that cutting back on time behind the wheel is at least a possibility. 216 pages, softcover.

21-0397 How to Live Well Without Owning a Car **$12.95**

Plug-In Hybrids: The Cars That Will Recharge America

Written by Sherry Boschert, cofounder and president of the San Francisco Electric Vehicle Association, *Plug-In Hybrids* is the book over which our politically polarized America can come

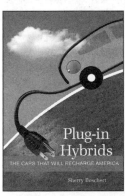

together. It is widely recognized that something's got to change if we're to succeed in our attempts to avoid the devastation that comes with dependence on oil. Plug-in hybrids are a part of the solution, and the technology to put them into mass production is already existent. Demand for plug-ins is nearly at a fever pitch already, and this book examines how you can get one, along with just what they are and why you want one. 231 pages, softcover.

21-0394 Plug-In Hybrids **$16.95**

Fuel from Water: Energy Independence with Hydrogen

By Michael Peavey. An in-depth and practical book on hydrogen fuel technology that details specific ways to generate, store, and safely use hydrogen. Hundreds of diagrams and illustrations make it easy to understand, very informative, and practical. If you're interested in using hydrogen in any way, you will find this book indispensable. 244 pages, softcover.

80210 Fuel from Water **$25**

Alcohol Can Be a Gas

Alcohol, the first automotive fuel, is poised to replace gasoline and diesel in the rapidly approaching peak oil world. It's 98% pollution free and reverses global warming, and you can make it for less than a buck a gallon. Originally written in 1982 to accompany the 10-part PBS series *Alcohol as Fuel*, this book has been completely revised and expanded for 2007. Explores the history of alcohol as fuel, conversion of gasoline and diesel vehicles, and more. Shows how to organize driver-owned alcohol stations and harvest federal tax credits. 576 pages; 500 photos, illustrations, and charts; softcover.

21-0402 Alcohol Can Be a Gas **$47**

Biodiesel Basics and Beyond

By William Kemp. Here's everything you need to know about making and using biodiesel for home heating, transportation, farming, and other uses. Separating truth from myth and misinformation, it's a one-stop volume packed with history, background, and detailed instructions on how to build and operate a 40-gallon-per-day, high-quality biodiesel "refinery" safely and sustainably. From detailed lists of gear and resources to recycling and waste treatment advice, this is the one to read. 588 pages, softcover.

21-0398 Biodiesel Basics and Beyond **$29.95**

Who Killed the Electric Car?

A must-see documentary for anyone mindful of the skyrocketing cost of gasoline, impending fossil fuel shortages and the spectacular instability in oil producing regions. Produced to educate and enlighten audiences with the story of the newly resurrected electric car, this DVD entertains as well. Beginning with a funeral for a car and flashing back over the 100-year development and eventual demise of the electric car, director Chris Paine crafts a story that makes complex science and politics utterly understandable. DVD, 91 min.

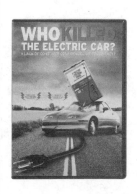

90-0186 Who Killed the Electric Car? **$26.96**

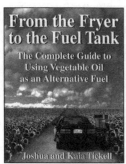

From the Fryer to the Fuel Tank

The Complete Guide to Using Vegetable Oil as an Alternative Fuel

New updated and expanded version by Joshua and Kaia Tickell, who, with sponsorship from Real Goods, drove their unmodified Veggie Van more than 10,000 miles around the United States during the summer of 1997, while stopping at fast-food restaurants for fuel fill-ups of used deep-fat fryer oil. Vegetable oil provided 50% of their fuel. They towed a small portable fuel-processing lab in a trailer behind their Winnebago. This inexpensive and environmentally friendly fuel source can be modified easily to burn in diesel engines by following the simple instructions. A complete introduction to diesel engines and potential fuel sources is included. Over 130 photographs, diagrams, charts, and tables. 162 pages, softcover.

80961　**From the Fryer to the
　　　　　 Fuel Tank**　　　　　 **$24.95**

Biodiesel America

How to Achieve Energy Security, Free America from Middle-East Oil Dependence, and Make Money Growing Fuel

Our tale begins more than 100 years ago with Rudolf Diesel's invention of an engine running on vegetable oil, and carries us through to modern times where energy security is a major concern and many of us are trying to decrease our dependence on foreign oil. Biodiesel is not only the way of the future, it is well grounded in the past. Joshua Tickell crafts a book that weaves the history of America's relationship with oil together with a plan for the future, and steps we can take to change that relationship. Tickell has created a book few can afford to not read. 356 pages, hardcover.

21-0383 Biodiesel America　　　　　 **$29.95**

Biodiesel Power

The Passion, the People, and the Politics of the Next Renewable Fuel

Biodiesel Power is the history of biodiesel in the making. It will appeal to a wide audience concerned about the end of oil. It lightly touches on the technical aspects of the fuel, its qualities, and its specifications, and focuses on the people and stories of the biodiesel movement. It explores the tensions between grass-roots activists and their altruistic co-ops, the profit-minded commercial producers and the voices of agribusiness, and the current administration—or "the coalition of the drilling." 272 pages, paperback.

21-0533 Biodiesel Power　　　　　 **$16.95**

Biodiesel: Growing a New Energy Economy

Greg Pahl's essential new book explores the history and technology of biodiesel, its current use around the world, and its exciting potential in the United States and beyond. A crop-derived liquid fuel, biodiesel can be made from a wide range of renewable, locally grown plant sources—even from recycled cooking oils or animal fats. While it is not the answer to all our energy problems, it is an important step in the long overdue process of weaning ourselves from fossil fuels. 282 pages, paperback.

**21-0372 Biodiesel: Growing a
　　　　　 New Energy Economy**　　　　　 **$18**

Natural Burial

The Ultimate Back-to-the-Land Movement

HERE'S SOMETHING NEW for our special 30th-anniversary edition of the *Solar Living Sourcebook*. You're probably "dying to do the right thing" during your time here on Earth, whether it's taking good care of your family, seeking the right livelihood, reducing your carbon footprint, working for social justice, or doing what you can to live by the precepts of sustainability. But you probably haven't realized that you can also act on these ideas when you die, for the greater good of your friends and family and the planet. We invited Cynthia Beal, who has been working on natural burial for years, to help us out with our new chapter. Cynthia is working on a book profiling the burgeoning natural burial movement, *Be a Tree: The Manual for Turning Yourself into a Forest*. In addition to introducing our readers to this relatively new concept, Real Goods is also offering a small selection of natural burial products for the first time (at the end of this chapter). Far from macabre or depressing, we think you will find this information about home funerals and natural burial to be inspiring and uplifting. It's not too soon to start planning ahead, and we invite you to join the "ultimate back-to-the-land movement"!

> Natural burial is an exciting solution, with win/win scenarios for individuals, communities, and our wildlife friends at every turn.

Dying to Do the Right Thing

In the United Kingdom, a compelling new consumer movement is underway. "Woodland" burial grounds (we prefer the term natural)—where people are buried in biodegradable containers, without embalming fluid or synthetics, and returned to the earth to compost into soil nutrients with a forest of trees marking the spot—are springing up across this island nation.

Driving the Change: Land Stewardship, Home Funeral Services, and Green Grave Goods

Concerns about pollution, appropriate land and energy use, and the depersonalization of the dying process, as well as a baby-boom demographic that puts 80 million Americans over the edge in the next couple of decades, are driving the natural burial trend. Two distinct groups stand at the forefront of this sea change: the natural burial grounds proponents themselves and their vocal citizen counterpart, the home funeral movement. The latter is spurred on by educational and nonprofit consumer organizations who focus on presenting a full spectrum of end-of-life options to the public, like the Natural Death Centre in London, the USA's Funeral Consumer Alliance, the Memorial Society of British Columbia, and alternative funeral service providers like ARKA Original Funerals in the U.K., Thresholds in San Diego, and the home-funeral guide-training classes of Jerri Lyons' Final Passages in Sebastopol, California.

Home funeral advocates focus on returning control over the death and dying process to individuals and families, encouraging and teaching them to take charge of their own end-of-life affairs in a proactive manner that engages family and friends, returning dignity and meaning to what has become, for many, a sterile and uncomfortable commercial process. For increasing numbers of people, in addition to hands-on participation, that means a natural burial, too.

> Home funeral advocates focus on returning control over the death and dying process to individuals and families, encouraging and teaching them to take charge of their own end-of-life affairs in a proactive manner that engages family and friends, returning dignity and meaning to what has become, for many, a sterile and uncomfortable commercial process

Natural Burial Companies Fill Real Needs with Style

The ARKA Acorn and Ecopod (made of recycled fiber) represent the leading edge in green burial products.

Side by side with this public front are an equally dynamic group of green business entrepreneurs and artisan-manufacturers. Some are producing attractive biodegradable burial vessels made from natural and recycled materials, while other forward-thinking land stewards are pioneering green burials by starting modern eco-cemeteries around the world. These companies and individuals are doing for the industrialized funeral sector what organic farmers and food producers have done for the agricultural sector: serving an unmet but very real consumer demand, and working to change the conventional practices of a huge industry that has a detrimental environmental impact.

The new green grave goods are stimulating a renaissance in the once-thriving burial arts: Recycled paper and alternative fibers are made into caskets and coffins. Handcrafted woven items are making a comeback in the form of willow, bamboo, and sea-grass burial boxes, while fabric artists fashion imaginative shrouds of organic cotton and hemp.

Unique new burial containers like the Ecopod recycled paper coffin, traditionally woven bamboo and willow caskets, and the ARKA Acorn ash-burial urn appeal to environmentally minded folks who want to depart from life as naturally as they've lived it. The Natural Burial Company and Gaiam Real Goods are pleased to be at the forefront of this education and distribution network, bringing our customers the information and products you need to make even your final act a positive and self-reliant one.

When It's Time to Leave No Trace

For decades, the end of a human life in American society has been managed by a cadre of corporate professionals who package our experience of death just as rigidly as others have packaged our living. Prior to the modern era, death was the exclusive province of the family. Burials were done according to custom and tradition. Respect was a matter of course, for strangers were not in charge, and dignity was conferred in the sincere, if sometimes clumsy, acts of caring for and carrying our dead.

Today, however, life moves rather mechanically—and for a hefty fee—out of the raft of boxes above ground and into more boxes below. Once you have died, your body is expertly managed by a professional funeral services team, plumped and preserved from immediate decay with environmentally harmful embalming fluid, and interred in a metal, chipboard, or plastic decorative box (with an optional airtight seal, at extra cost, to prevent contact with the elements). The casket is placed into a rigid liner (deemed unnecessary in most non-American cemeteries but considered mandatory by U.S. cemetery managers for lawn maintenance) and is usually buried on government-regulated grounds on high-priced real estate that easily commands upwards of 1 million dollars or more an acre for its owners. (1,000-2,000 bodies per acre at a minimum of $1,000 each for the plot is standard, sometimes stacked two or more high.) For lots of people, that double-box process looks like litter, and upon closer examination, it's not the dignified and simple close to a grateful life that most of us wish to have.

Minimize Impacts, Plan, and Respect

A popular outdoor ethics campaign, the Leave No Trace program (www.lnt.org), took backcountry garbage to heart in the 1970s and '80s, thoughtfully outlining objectives for individual

A traditional cemetery.

Photo: © C.A. Beal, 2007

"waste management" and behavior when visiting natural and wilderness areas: plan ahead, dispose of waste properly, minimize impact, respect wildlife. Many folks think that we're "just visiting" here on Earth, and that when it's your time to go, "Leave No Trace" doesn't seem like such a bad idea.

Today, people are beginning to question the wisdom of leaving toxic burial chemicals and synthetic substances in the ground and atmosphere for future generations to clean up. If pressed, many of these same folks would prefer their bodies to "Leave No Trace," as well.

Should We Be Burying This Stuff?

We think it's appropriate to ask these questions about toxic burial chemicals and promote alternatives. Under most state and federal regulations, any company would be hard-pressed to get permits to bury almost 2 million gallons of embalming fluid in the soil annually, and yet that's exactly what happens with most of the 1.5 million bodies that are embalmed and buried in U.S. cemeteries every year. The European Union is considering banning embalming fluid altogether, as reported in the October 6, 2006, issue of the *Wall Street Journal* (see www.enviroblog.org). Conventional embalming fluid contains a number of chemicals, including methanol, ethanol, and formaldehyde, the latter a suspected carcinogen linked to nasal and lung cancer in mortuary workers and other occupational groups. Used by a large number of funeral homes to slow the body's decomposition, formaldehyde arrests natural breakdown processes by "fixing" cellular proteins. It stiffens the body's tissues and, with the help of added colorants, is used to make a corpse more "attractive."

Guess what else is buried along with these embalmed bodies every year? More than 100 thousand tons of steel; 10 tons of copper and brass; 30 million board-feet of hardwood timber; uncounted tons of plastic, vinyl, and fiberglass; and 1.5 million tons of reinforced concrete accompany Americans to the afterlife—an unattainable disposal permit indeed, unless you're burying caskets one at a time!

The chemical-intensive double-box casket-and-liner system used in this unnatural burial is an industrial solution to the uncomfortable and very human experience that usually accompanies our Big End. In an often compassionate but increasingly commercial and profitable effort to ease our emotional pain, offending sights

What's Buried Along with Our Loved Ones in U.S. Cemeteries Every Year

- 827,060 gallons of embalming fluid
- 90,272 tons of steel (caskets)
- 2,700 tons of copper and bronze (caskets)
- 1,636,000 tons of reinforced concrete (vaults)
- 14,000 tons of steel (vaults)
- 30+ million board-feet of hardwoods, much tropical (caskets)

Source: Mary Woodsen, Greensprings Natural Cemetery FAQ, March, 2007; http://naturalburial.org/index.php?option=com_content&task=view&id=25&Itemid=46

and smells are whisked away, and the vacuum is filled with services that temporarily mask the reality of death. The box is pretty, the lawns are neat, and nature can't get a word in edgewise.

Natural Burial: The Traditional Alternative

Until recently, the environmentalist's response to this product-intensive and cumbersome process has been to opt for cremation. However, as baby boomers age and the actual experience of managing our deaths (and those of our parents and friends) comes upon us, cremation—now complicated by legitimate questions around energy use, mercury and carbon emissions, the lack of meaningful and enforceable environmental standards, and aging industrial crematoria—is not always the modern funeral panacea most wish it to be.

The urgent creativity of the 21st century is rising to these challenges, and other solutions for our bodies' final end that support the values of green space preservation, carbon sequestration, habitat creation, nutrient cycling, and brownfields reclamation are becoming apparent. Our understanding of the importance of forests and the usefulness of trees, along with the power of the soil to transform natural elements and return them to utility for the web of life itself, grows every year. The demand to "Leave No Trace" is increasing.

On the heels of these developments, and the emergence of a consumer who is looking for a clean, "green" death to accompany a conscious, low-environmental-impact life, natural burial is an exciting solution, with win/win scenarios for individuals, communities, and our wildlife friends at every turn.

Under most state and federal regulations, any company would be hard-pressed to get permits to bury almost 2 million gallons of embalming fluid in the soil annually, and yet that's exactly what happens with most of the 1.5 million bodies that are embalmed and buried in U.S. cemeteries every year.

Preserve, Disappear, or Reintegrate

Once you're done with your body, only one thing happens to it next: It goes away. Well, it never "goes away"; in the words of anthropologist and garbage guru William Rathje, "there is no away." So you do go somewhere. Fortunately, where and how you go is still up to you, and if you want to take control of the situation, you'll need to decide to manage your physical remains for preservation, disappearance, or return—in other words, reintegration into the interdependent biological organism that cycles elements on planet Earth.

According to modern biologists, living systems depend upon a complex network that coordinates independent cells so that they function together as skin, organs, blood, and nerves working as a team—i.e., You—to repel the invasion of external bacteria and fungi that would

Tarn Moor is a pioneering meadow burial ground in the United Kingdom.

Photo: © C.A. Beal, 2007

otherwise colonize and consume the weaker individuals. "Life" is a constant struggle to resist turning into something else's dinner, and as long as you're around, your side is winning.

As soon as you check out, that system collapses, and your cells, no longer your very clever and fun-loving collective that is turning sunlight into ATP (biochemical energy) and ATP into gardens, solar arrays, and microbrew-festivals, bid each other a fond farewell and take their turn as food. In the natural world, a whole host of creatures—animals, insects, fungi, and microbes—then get their spot in the Sun, so to speak, and take on the very necessary work of breaking you down into smaller component parts, putting you back into the system that you built yourself from in the first place. It's an amazing cycle—or it can be, if we'd just leave it alone.

But no, WE have IDEAS.

Preservation: It's Not All It's Cracked Up to Be

The ancient Egyptians were masters of the Slow, learning to pickle and preserve human remains for reasons that are still somewhat obscure today. Modern embalming came into fashion in the mid-1800s, spurred by the sale of early forms of embalming fluid through the emerging military supply industry to Union and Confederate armies, preserving the bodies of dead soldiers for positive identification and burial during and after the Civil War.

Today embalming is a common practice in the U.S., generally required at the larger funeral home chains, while independent funeral directors often have more discretion. Contrary to industry opinion, embalming is not necessary to prevent decomposition in the first few days after death; refrigeration does the job efficiently. Embalming does not prevent the spread of disease. Its use as a sanitizer is overrated, and according to the federal Centers for Disease Control, embalming serves little appreciable sanitizing or public health purpose that couldn't be handled with more natural techniques.[1] Embalming fluid is a formaldehyde-based chemical that slows decomposition by killing some of the microbes that begin the first stages of body breakdown, resulting in the hardening of tissues. It's not used to kill human pathogens, and one of its biggest "dangers," outside of the toxicity of its primary ingredients, is the myth that it does. It is rarely required by law or regulation—no state in the U.S. requires embalming, although surveys conducted by the Funeral Consumers' Alliance and others show that a number of funeral tradespeople regularly imply to their customers (and to legislators) that it's "necessary" for public health and safety.[2]

Exposure to formaldehyde may pose significant health risks to people who work in the funeral industry. Nasal and lung cancers have been indicated in scientific studies, though some industrial research disputes these claims. (For the CDC evidence of carcinogenicity in formaldehyde, see www.cdc.gov/niosh/81111_34.html.) A related issue is that because embalming fluid is used to replace blood in the body, that blood has to go somewhere. You guessed it. According to the Federal Trade Commission, more than 2 million embalming procedures are performed in the U.S. each year, producing 2½-3 gallons of blood and excess embalming fluid per body. That fluid, along with the organs and internal parts suctioned out of the corpse during the process, goes down the drain and into the water supply. Not a pretty picture.

In the cause of preservation, embalming fluid, with preservative effects that begin to fade in a matter of days (although still leaving the natural decomposition process compromised because the life cycles of the decomposers have been destabilized), takes a back seat to the double-box casket system in common practice in the U.S. since the last half of the 20th century.

Steel, fiberglass, and chipboard-with-veneer caskets are designed to resist decomposition, and that resistance is a selling point, with the more durable boxes bringing in the highest prices. Caskets touted for their resistance to degradation are often fitted with an optional rubber or plastic seal installed between the bottom and the lid, designed to prevent mold and rot (our friendly decomposing fungi and bacteria at work). In reality, this seal fosters an anaerobic environment and causes the body to putrefy rather than break down. This practice has even led to exploding crypts in cemeteries that inter bodies above ground—embarrassing for the gasket manufacturers, and messy for the groundskeepers.

After the packaging is complete, the casket isn't lowered directly into the ground but is instead placed inside a concrete, steel, or fiberglass liner. This keeps the casket itself from deteriorating, forestalling what's known in cemetery landscape-maintenance parlance as "subsidence," a slight sinking of the casket that creates bumpy ground that's inconvenient for heavy tractor lawn-mowing and causes headstones to topple. In addition, the caskets and headstones are often placed in cemeteries or churchyards "in perpetuity" and require ongoing maintenance, sometimes "forever." By contrast, in a natural burial ground, meadow grass is not mowed and the headstone is a tree—a lazy landscaper's dream if there ever was one.

Modern body preservation is expensive, environmentally destructive, and perhaps more important, begs the real question: What, or whom, are you being preserved for?

Disappear: To Burn or Not to Burn

It's not certain which tradition is older, burning or burial. Evidence exists for both scenarios, and each has a longer history than preservation. The oldest arts we know of are the burial arts, and the practice of cremation is thousands of years old. In times of disease and mass death, cremation has often been the method of choice, especially in landscapes where the soil was not suitable for rapid breakdown. But cremation takes fuel—wood, gas, or electricity today—and fuel is often scarce in times of disaster or, as in our time, an unwise choice given the devastating effects of global climate change.

Some religions teach about the impermanence of life and back that up with a ritualized display of public burning, proving to the community that the person cannot come back—once they're burned to ash, they're truly "gone." Other groups consider burning the harshest of punishments, depriving the soul of a body to either return to or use in an afterlife. In either case, the goal of cremation is to disappear, and for those who desire to unburden the world, this may seem the logical choice. The appeal of "disappearance" has led many people who are disenchanted with modern industrial burial to opt for cremation.

Cremation, however, has its own array of problems. Chief among them are the energy used for complete combustion; the emissions that result from burning synthetic materials and body implants, and mercury fillings (vaporized dental amalgam accounts for 16% of the airborne mercury pollution in the U.K.)[3]; the lack of crematoria standards worldwide (since

When the churchyard is full.

Photo: © C.A. Beal, 2007

Cremation has its own array of problems. Chief among them are the energy used for complete combustion; the emissions that result from burning synthetic materials, body implants, and mercury fillings; the lack of crematoria standards worldwide; and the carbon footprint that cremation entails. Cremation is a viable alternative. If you want it to be clean, however, you should shop around.

Cremation Trends

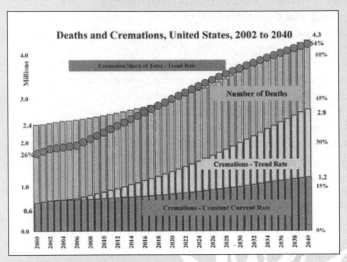

Deaths and Cremations, United States, 2002 to 2040

This chart shows for USA:

1. The percentage of cremations from 2000 projected to 2040 (circles).

2. Total number of deaths each year, 2000 projected to 2040 (3 colors).

3. The increase in cremation trends in both percentages and numbers from 2002 to 2040 (middle section).

4. Cremation rate if it were to remain constant. 2000-2040 in number of cremations and percentages.

Source: www.cremationassociation.org/html/photo-deaths-cremation.html

A natural burial that makes one's body available as a full-spectrum nutrient source for the soil web does more for the planet's biological system than cremation.

air travels and we have a shared atmosphere); and the carbon footprint that cremation entails. Crematorium makers are quick to tout their progress: Increasingly efficient filtration systems capture more emissions than they once did. However, the filters are very expensive and the trapped pollutants must still be disposed of. Agreements on emissions standards are difficult to achieve, especially given that over 80% of operating crematoria in the U.S. are the older, polluting models. Cremation *is* a viable alternative. If you want it to be clean, however, you should shop around.

When a green crematorium can be located and cremation is still your method of choice, the cremated remains offer a chance for survivors to honor a loved one after they're gone. With the rise in cremation rates, scattering ashes has become popular—so popular in some national parks, for example, that visitors have to be reminded not to spread the highly alkaline ash because it interferes with plant growth in high concentrations.

Many people are surprised at the amount of ash, and even bone, that remains after a crema-

tion, and scattering it around can feel awkward. Creative ways of memorialization—like casting a pinch of ash in blown glass, compressing it into an artificial diamond, or placing the ashes in a beautiful handmade urn or jar—allow one to continue to express and maintain a meaningful connection. However, some estimates suggest that perhaps as much as a quarter of all cremated ashes are still "on the shelf" somewhere, perhaps even in the original crematorium box or bag. Biodegradable burial urns like the ARKA Acorn Urn make personalized forest burials of the bulky ash—especially in a conservation forest setting—a thoughtful option.

But the least talked about, and perhaps most compelling, argument against cremation (or disappearance at all, for that matter), even if the energy use was negligible and harmful emissions nonexistent—may be that, in disappearing, we actually lose our chance to continue to participate in planetary life in a meaningful way. As ash, we can be scattered to the winds or on the waters, or remain cherished and elemental, a comforting presence in our descendants' lives. But planted in a forest and becoming food for the regenerative Gaian system, we can still do one last thing with our bodies that may be much more significant than a disappearing act: We can remain fully present, albeit transformed, nourish the soil, and rekindle life as a forest or a tree.

Reintegration: Making the Case for a Biological Return

The trend worldwide is toward cremation, and there is little ground for argument if inputs and emissions are managed properly and the only other available method is the resource-intensive conventional industrial model. For many people—especially those whose deaths involve complex organ donation, serious infectious disease, limited funds that preclude supporting a forestland, or dying far away from your chosen place of burial—cremation is the best option. Green cremation "wins" over an embalmed body and nondegradable casket system any day.

On the continuum of processes, a natural burial that makes one's body available as a full-spectrum nutrient source for the soil web does more for the planet's biological system than cremation. According to Dorian Sagan, author of *Into the Cool* and student/teacher of the thermodynamics of living systems, the longer that our biological web can keep life forms "in play," transferring energy from one creature to another in the great chain of being, the more resilient our planetary system can remain. The complex and self-organizing, self-regulating biological and geophysical systems that help to balance temperature, moisture, and atmospheric gases and support life as we know it on Earth are created and maintained by the continuous recycling of the organic and inorganic matter that are the elemental building blocks of all animate beings. Sterilization (from the formaldehyde) and the combustion of cremation destroys the integrity of fundamental molecules, enzymes, and microbes present in your body and the soil it's buried in. In contrast to the chemical-intensive practice of preservation or the energy-intensive process of combustion, returning to Earth's natural system arguably makes the best use of our parts—us—for the greatest number of beings, over the longest period of time.

With the human contributions to climate change looming large on the horizon, our continually increasing understanding of the science of Gaia may generate additional compelling reasons that we might want to return to the web in a literal, as well as a figurative, sense. Becoming a tree, if for no other reason than to offset our own personal CO_2 emissions during our lifetime, might be just the ticket we need.

All We Leave Is Energy

A number of methods are available to us for reintegration with Earth's biological systems in natural ways. Some of them, such as the Tibetan "Sky Burial," or the more familiar "burial at sea" (as long as the body is in a weighted shroud and not a nondegradable casket), are older than our recorded histories.

One of the newer methods is a creative option called "Promession," which was developed by the Swedish soil scientist (and former organic gardener) Susanne Wiigh-Mäsak, now independently marketed by Promessa Organic AB company (www.promessa.se/index_en.asp). This method uses a cryogenic technique to freeze-dry the body immediately after death. Frozen solid, the body is then vibrated by sound at an amplitude that reduces it to powder. The moisture—70% of a body's mass—is evaporated off, and the various metals and nondegradables are sifted out (very helpful in situations where there are numerous implants that can't be removed prior to burial). What remains afterward is a dry, silt-like and nutrient-dense substance highly suitable for burial and fertilization.

This new technique has a few disadvantages when compared with a truly natural burial. The

process uses liquid nitrogen, which is expensive to purchase and store, and is costly in energy to produce since it's generated with electricity. Liquid nitrogen can't be handled without the use of specialized equipment and knowledge of the potential hazards, and does pose some danger to its handlers. At this time, it's hard to say where this process stacks up with respect to cremation in "sustainability" terms, especially if solar crematorium efforts like those now underway in India bear fruit. However, given the fact that a "freeze-dried" body remains in a form that can still nourish the landscape, it is superior to ash as a fertilizing nutrient and likely takes second place only to natural decomposition.

Once such facilities are built, freeze-drying and sifting may end up being the solution for communities that do not wish to have new crematoria but don't yet have natural burial options. And as with cremation, there are certain circumstances—medically complicated deaths, implants, some diseases—that make natural burial less than ideal. In these situations, freeze-drying may be the sensible choice—and you still remain useful as something else's dinner!

Becoming a tree, if for no other reason than to offset our own personal CO_2 emissions during our lifetime, might be just the ticket we need.

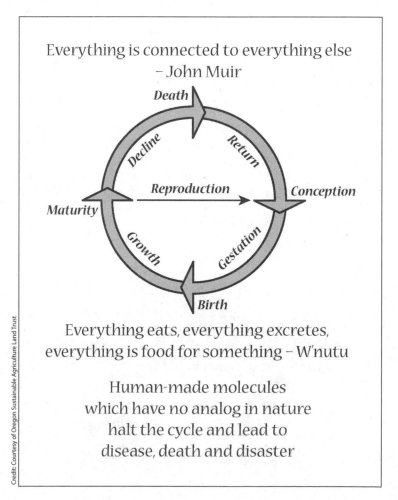

Everything is connected to everything else
– John Muir

Death · Return · Conception · Gestation · Birth · Growth · Maturity · Decline

Reproduction

Everything eats, everything excretes, everything is food for something – W'nutu

Human-made molecules which have no analog in nature halt the cycle and lead to disease, death and disaster

Credit: Courtesy of Oregon Sustainable Agriculture Land Trust

Is it really such a big leap to imagine that your own death can be a doorway back into that natural and elemental world?

Ideally, natural death and burial should support and sustain the cycle of life, not compromise it. Modern biology is only now beginning to deeply connect with other scientific disciplines—geology, climatology, physiology, and thermodynamics—to quantify the energy transfer that interdependent living systems generate and manage in the complex soup of life. Is it really such a big leap to imagine that your own death can be a doorway back into that natural and elemental world? For those of us who've been frustrated by the difficulty of living an in-

tegral life in this forest of synthetic industrial marvels, a natural death may be the easiest lifestyle choice we'll ever make.

Ideally, all we leave is energy. Good, useful energy still available in the form of complex molecules—fat, bone, and blood—there to be wrestled apart and turned into good little worms and beetles (who eventually also take their turn in feeding the small). Or as one organic gardener insisted he wanted on his headstone—"WORM PARTY!"

What You Need for a Natural Burial

Because natural burial is innovative and as yet unfamiliar in our society, it is advisable to make advance arrangements to ensure that your end is a more natural one.

The key elements of a natural burial are:
• A clean/green body-management process
• A fully biodegradable container
• A place to go—air, water, or sky
• People to put you there in the manner you determine
• Laws to support your right to be there
• A community to help you stay there

Whether you're buried in a container or wrapped in a fiber shroud, the first thing to insist on is the use of biodegradable materials in everything that accompanies you "out the door" or into the earth, no matter where you end up. Just by using a natural container, you'll minimize your impact on the environment because of all the conventional casket materials you *won't* be buying or burying, along with any polluting or energy-inefficient processes used to make them. Your box or "vessel" is a great place to start. Even if you (or your parents) are buried in a conventional cemetery, you'll less-

The hand-woven SAWD bamboo coffin is free-trade certified.

en the ecological footprint of burial boxes just by choosing the natural ones, and it just gets greener from there.

What's in the Box (Besides You)

While the debate evolves as to what will and won't biodegrade in the presence of healthy soil microbes and hungry trees, a few guidelines readily appeal to our common sense. The list below is progressive. We suggest that you start with synthetic-free items and, as new products come into being, go "up the ladder," choosing enhanced biodegradability, recycled and non-virgin materials, and sustainable production characterized by local handicrafting, family businesses, fair trade, and economic justice.

• Avoid synthetic and nonnatural materials in your container and clothing
• Choose products designed to enhance breakdown in the soil web
• Favor items from recycled and waste material instead of virgin resources
• Support sustainably produced burial goods with organic, fair trade, and eco-certifications as they begin to appear in the marketplace if you're not making your own

Any additional requirements can be spelled out in your final instructions and should include asking the family to leave your favorite gadget at home (or better, give it away!) and not burying you in synthetic clothing.[4]

Modern bodies tend to go out with more than they came in with. Teeth are often filled with mercury amalgam—stable when cool and in the ground, but still buried nonetheless. Silicone and artificial joint implants are increas-

Give your pet a green burial.

ingly common, and bodies may have pacemakers (they explode in crematoriums, and silicone pools in the kiln). Unless you leave instructions, it's unlikely that these items will be removed prior to your death, and dealing with them just afterward might be a bit awkward. This is generally one of the least pleasant tasks left to be managed and should, if at all possible, be arranged for completely in advance, by you.

A number of U.K.-based makers and artisans, in conjunction with the Natural Burial Company, are teaming up with Real Goods to bring the first comprehensive selection of market-tested biodegradable caskets, urns, shrouds, and other useful products to the U.S. With over a decade of experience serving nature-minded folks with green grave goods in the U.K., their experience means that we can make a quick transition to natural practices—we've got the goods, and all you need is a place to go!

We especially like the woven wicker—whether bamboo, willow, sea grass, jute, or wood splits. Unlike solid wood, the woven caskets biodegrade rapidly, and the materials are often renewable. Their production keeps an important suite of artisan skills alive. Most of these coffins are suitable for cremation as well as burial; some have wooden runners on the bottom, and others include cotton liners or headrests. A wide range of makers, styles, and sizes are available, including ones for pets.

Indeed, managing the disposition of deceased pets can be challenging today, especially for those who live in cities. Given the propensity of vets, shelters, and animal hospitals to sell the unclaimed remains of pets to pet-food factories for use as protein in animal feeds[5], it's advisable to consider taking control of your pets' ashes or bodies and returning them to the earth yourself (though consult regulations where you live, and work to change them if they haven't caught up with the times).

To avoid introducing "anonymous proteins" into the food chain at all levels, we recommend either burying your pets yourself or having them cremated to guard against this unsavory practice. Pet cemeteries are on the rise, and there's no reason that "green" ones can't spring up alongside their human counterparts. Our biodegradable burial urns and smaller-sized caskets and coffins are suitable for this purpose. More-detailed information on home pet burial can be found in the forthcoming book, *Be a Tree*.

Since the establishment of the Funeral Rule by the U.S. Federal Trade Commission (www.ftc.gov/bcp/conline/pubs/services/funeral.htm), consumers have been able to acquire caskets and coffins from any independent source and use them without extra charge in the funeral home or cemetery of their choice. Consequently, the natural products marketplace can become a ready source of biodegradable caskets, as it's more likely to be responsive to requirements for environmental rites. Don't be surprised, now that Real Goods is carrying biodegradable caskets, if your local natural foods co-op or garden center or favorite online eco-retailer begins to offer a selection of "final furnishings" in your own not-too-distant future.

The Healer's Maxim: "First, Do No Harm"

You don't have to be buried in a natural burial ground to make your last moments greener ones. By planning ahead (so others don't have to guess or make the tough decisions for you), by choosing your process and your container in advance, and by making it clear that you want no synthetic materials on your last ride, you'll go a long way toward improving what might otherwise be quite the opposite of what you'd wish, if someone could ask you after the fact.

Not all of us will have the luxury of being conveniently located near our ideal forest

Since the establishment of the Funeral Rule by the U.S. Federal Trade Commission, consumers have been able to acquire caskets and coffins from any independent source and use them without extra charge in the funeral home or cemetery of their choice.

sanctuary. Natural burial grounds are just now forming, and some of us may go down before then. But you do have the option to seek out a conventional cemetery that has sustainable landscape management practices, as well as one that allows liner-free burials, a practice that is usually up to the cemetery's discretion.

Whether you're required by law to be buried in a container at all varies from state to state in the U.S., although generally this decision is left up to the cemetery. Because sustainable landscape management has yet to catch on, most cemeteries have long-standing rules favoring the double-box casket-and-liner package. If in doubt, consult the Funeral Consumers' Alliance website, www.funerals.org/; they have chapters in every state and can point you to sources of local information. Legislatures are constantly updating the regulations, however, so your best bet may be to find a natural burial ground or cemetery that will do what you want, make advance arrangements by reserving the plot, stipulate a biodegradable casket, put your final affairs in order, and get on with your life.

Shroud burial, where the body is wrapped in fabric of some kind, is a perfectly acceptable form of natural burial that is still common in much of the world. But there are practical considerations, especially when it comes to handling and moving a body, so plan ahead. Rigor mortis fades after 24 hours, the body softens again, and lowering a shrouded person gracefully into even a shallow grave takes some skill and forethought. Just be aware that you'll have some awkward logistics to manage that can be done respectfully if you think it through in advance.

Although embalming is rarely the law, refrigeration is almost universally required within 24 hours of death. Strategically packing a body with dry ice serves this cooling function, but that, too, can be difficult with only a shroud unless you've made other arrangements. Consequently, many people find a solid container with a sturdy bottom easier to manage, and the majority of natural caskets now available adequately convey the body with dignity, respecting Earth at the same time.

A Place to Go

Once you've decided upon your vessel of choice, finding the right place to plant you is next on the list. The number of burial grounds that operate naturally is likely to increase yearly, but it may take awhile for one to convert or open up in your area unless you help to start it.

Seeking out or working to create a local green cemetery option may make more sense, however, especially if you live far away from an existing natural burial ground. Regulations currently require the embalming of bodies when

crossing some state lines, and common carriers like airlines will not currently transport an unembalmed body. These rules will hopefully change over time as more people wake up to different options, and the mythologies about embalming and public safety are dispelled.

Nonexistent less than 15 years ago, the number of natural burial sites in the U.K. has risen to more than 250 in 2007, and over half of the woodland burial grounds are owned and run by city councils with public funds. The first modern woodland burial site was established by the city of Carlisle in 1993 at the urging of cemetery manager Ken West as an environmentally sound alternative to conventional burial that was less expensive to maintain and sustainable.[6]

The original Carlisle scheme planted an oak whip on each grave, with the aim of creating a biodiverse oak forest on the site. Graves are deliberately not marked with an inscribed memorial. Embalming is discouraged, but nonbiodegradable containers are allowed. Anonymity is seen as a benefit, and there are no paths to graves, deterring people from visiting specific gravesites and damaging the ground flora.

Established in 2001, Memorial Woodlands in Bristol is a popular site with a comprehensive set of offerings that include Community plots (for consecration and religious burial), Genera-

> Nonexistent less than 15 years ago, the number of natural burial sites in the U.K. has risen to more than 250 in 2007, and over half of the woodland burial grounds are owned and run by city councils with public funds.

The first woodland burial ground in the U.K. was established by the city of Carlisle.

Photo: © C.A. Beal, 2007

tional plots, individual and shared plots, and ashes burial. The site offers either "leasehold" (you lease the land) or "freehold" (you own the land) options. The land is held in perpetuity and will be placed under the care of a charitable trust when full. A unique feature of Memorial Woodlands' arrangement is that they hold title to the first three feet of soil and all that grows on top of it. This allows the management to tend the landscape in a natural and uniform manner while allowing plot users the security of ownership if desired. They advocate choice and have no restrictions regarding embalming or biodegradable caskets.

Memorial Woodlands encourages families to be actively involved in the organization of the funeral. They permit only one funeral per day on the grounds to make sure the family gets ample time, space, and privacy for the closure and celebration they need. Nineteenth-century buildings on the site have been converted into reception rooms, and a private chapel, not consecrated to any particular faith, can be used for services of any sort. Other woodland burial grounds like Tarn Moor, Mayfield's Remembrance Park, and the City of Brighton's Woodvale vary widely in their structures, some with detailed conditions of use and allowed materials, and others accommodating family desires without restriction.

Natural Burial Grounds Come in All Shapes and Sizes

As of 2007, less than a dozen dedicated natural burial grounds have been created in North America, but a number are in the planning stages. The Natural Burial Company maintains a link to online directories of cemeteries, including U.K. woodland burial grounds, that state they accept biodegradable caskets without liners.

We can learn a lot from the U.K. pioneers. To get an idea of what you should negotiate for and expect when choosing your site, consulting the Web sites and terms of various U.K. burial grounds can provide you with a good overview of what's already been successfully done. Your desires should be spelled out in writing with the people you pay to manage your interment for the long haul. The U.S. trend is toward greater environmental protection than the U.K., however, so don't hesitate to look for and insist upon that.

What happens to you and your tree, if you choose to plant one, should be governed by a contract signed between you and the burial ground proprietors when you purchase your

Memorial Woodlands is in Bristol (top) and Remembrance Park is in Mayfield, U.K. (bottom).

plot—meaning that you need to think this one through. For example, many, but not all, contracts allow for a tree to be planted at the grave to serve as a marker. In smaller grounds with tighter budgets, this isn't always practical, but any diverse woodland consists of shrubs and meadow areas as well as trees, and some people really would enjoy just pushing up daisies!

Because everyone is fairly new at natural burial in the U.S., the details of these eco-cemetery contracts will probably vary widely, and many questions will arise. Some contract terms will be governed by federal, state, and local regulations. In certain cases, though—if the burial grounds are run by a recognized religious organization—the grounds may be exempt from such rules. Jewish and Muslim burial customs are "green" by tradition, and cemeteries that serve these populations will be familiar with green burial concepts.

In addition, historic pioneer cemeteries often have more-lenient regulations than conventional ones. Many of these older cemeteries are managed by fraternal orders like the Oddfellows or small local boards of directors. However, while these may provide hopeful opportunities, most cemetery officials of any stripe are probably not familiar with the natural burial option. Visiting them with a copy of this article might be just the impetus they need for considering a change.

Some burial grounds may have provisions for cutting the timber after a certain number of years, harvesting any produce, or rotating the plots as a means of paying for the land and services and maintaining the site. Others put the land into permanent trust and let your tree or sod grow undisturbed, in perpetuity. Small sites with no room for expansion, and conventional cemeteries with high-density plot schemes find it problematic to plant a tree for each individual; larger conservation grounds with the goals of reforestation and habitat creation are happy with a low-density plan and are more likely to support plans to be a tree.

For some, being part of an orchard or a garden and turning into dinner makes perfect sense. For others, the thought of being harvested is an abomination. Only you can know what will work for you and your friends and relations, and it should all be spelled out in the contract when you purchase your plot. Alternatively, you can let those terms remain vague and be released to the needs of future generations.

The Ultimate Back-to-the-Land Movement

Probably the most compelling model for greenspace advocates is the "conservation cemetery," a piece of land dedicated to natural burial and protected by a land trust (see Trust for Natural Legacies at www.naturallegacies. org) with conservation easements. Favored by the mission-oriented ideals of staunch natural burial advocates like Memorial Ecosystems' founder Dr. Billy Campbell and Greensprings Natural Cemetery trustee Mary Woodsen, this new kind of "conservation cemetery" may reasonably include the partnership of a private (or public) landowner who holds title and puts the land into trust, perhaps even contracting with a

conservation group for the ecological oversight. Such a trust would be run by a board of directors with a land management plan that includes written guidelines about what can be buried, when, where, and how.

Additionally, the funeral operations concession could be managed by an experienced, local funeral-services provider known to the community who supports the environmental mission. Such a collaboration could enable a conservation cemetery to form quickly, with the expertise and knowledge necessary to create a functional burial ground with a minimum of delay and complication. Within this type of structure, larger opportunities for research, soil and carbon banking, greenspace preservation, multiple land use, habitat creation, and the storage of public asset dollars in arable soils rather than in buildings with capital maintenance costs can exist and even flourish.

Soil: The Living Web

Burial in a biodegradable container presumes and encourages decomposition. Decomposition requires a living Earth, and according to soil scientists, the same conditions that are necessary for proper decomposition—nutrients cycling at the right rates for complete breakdown to occur—are required for healthy plant systems, too. Organic carbon is the key to these processes, as it is constantly recycled from organism to organism, including trees and other plants that absorb it out of the air. But it's not enough to just plant the tree. The soil web has to be healthy enough to grow the tree well.

The "ideal" natural burial site of the future will probably have a healthy mix of treed areas, hedges, and open grasslands. Perhaps we'll see the return of "family plots" after all, with parcels of land dedicated to a family's or group's use but at a fraction of the expense in a conventional cemetery. Maybe you'll elect to go into a vineyard or an orchard, contributing to the *terroir* of a truly individual jam or wine. Wherever you do choose to settle, one thing is relatively certain: As long as the rain falls, the Sun shines, and the worms go in and out, and as long as the rules of the conservation trust you've employed to manage your site remain unbroken, you'll continue to play a role in the web of life that you've chosen.

> Probably the most compelling model for greenspace advocates is the "conservation cemetery," a piece of land dedicated to natural burial and protected by a land trust

The Home Funeral Movement: Genesis of Natural Burial

For those who really want to do it yourself, a home funeral may be the ideal "way to go." Until recently, most funeral directors were reluctant to let the family get involved. However, as alternative funeral service providers, ritualists, and celebrants have begun to make themselves available to serve deeply personal and nontraditional needs, the "dismal trade" is beginning to get on board.

In the early 1990s, more than 90% of people in the U.K. died in a hospital as opposed to the home, providing some of the original impetus behind the founding of the Natural Death Centre in London. It began as the project of three psychotherapists, spearheaded by Nicholas Albery, the Centre's founder, with the mission of enabling a person to die a more natural death in personal surroundings, tended by loved ones, receiving treatments that they—not the hospital system—desired.

The Natural Death Centre quickly became a main source of inspiration and information for self-reliance in death, primarily through its guide, the *Natural Death Handbook*. The Internet has helped spread the Centre's work around the world, picked up and expanded over the last decade by home funeral advocates such as Jerri Lyons, who began the Natural Death Care Project (www.finalpassages.org) and now trains home funeral guides in the U.S.

This home funeral renaissance, with its desire to return the funeral back to the purview of the family and reinstate affordable simplicity, has led to public calls for natural burial: no embalming, the "plain pine box," the shroud, memorialization with a tree, or even anonymity. That call, in turn, has engendered the modern green burial movement.

Crafting the Fond Farewell

When a loved one dies, a couple of key arenas need to be managed properly, and the services of an experienced helper are useful. In fact, the emerging profession that sits at this cusp between hospice and gardening is increasingly becoming known as "death midwifery," for the required emotional and physical management skills are similar to those needed at the beginning of life, and having the assistance of a competent person—a home funeral guide—can ease the passage for all concerned.

YES, MORE PAPERWORK!

Death has paperwork. There's no getting around it. But since other people have to do it, make it easy on them. While the specifics may vary slightly from state to state (and some states still have onerous laws prohibiting personal involvement with a loved one's body, so check this out with your state first), some general principles apply.[7]

First, the doctor or physician certifies the cause of death and signs a death certificate. Funeral directors often have blank copies, as do county registrars. Obtaining a blank one shortly before it's needed is not a bad idea. Only in suspicious circumstances is legal supervision required; otherwise, if the death was natural and expected, the management of the body is generally up to you—or rather, it's up to the person you've designated as your "Personal Funeral Director," the "person in charge of interment," who manages the "disposition of the body," officially, in advance, on a notarized piece of paper. Really.

If you haven't named a personal funeral director, still allowed in most states (and this is different from the executor of your will or the person with a general power of attorney), the task falls to your "official next of kin." Absent the next of kin, the only others who are legally able to transport your body around are certified licensed professionals. Anyone else caught dead with you—sorry, you dead with them—could have a problem.

This person handles the multiple copies of your death certificate, one of which acts as the permit to transport you in your not-living state, one of which is left with the cemetery or crematorium handling your body, and one of which is filed with the registrar when the process is completed. Knowing what has happened to your body, and why, *is* a matter of public record, and should be. Ensuring you've not met with foul play is high on their list, but once that's ascertained, the professionals don't really need you anymore, and they should let you go home without too much fuss.

Finally, don't forget the other aspects of bureaucratic closure: a living will, a personal will, and an advance directive, at minimum, along with a comprehensive listing of all the bits someone needs to know if they're going to have to dig through your files and piece together what you

This home funeral renaissance, with its desire to return the funeral back to the purview of the family and reinstate affordable simplicity, has led to public calls for natural burial: no embalming, the "plain pine box," the shroud, memorialization with a tree, or even anonymity. That call, in turn, has engendered the modern green burial movement.

were supposed to be paying for next week but couldn't. Yes, it's a big job, but someone's gotta do it, and it ought to be you.

CREATE AND WORK WITH YOUR PERSONAL GROUP

Aside from the paperwork, the most demanding part of your death (provided you've arranged for everything else in advance) is the actual handling of your body, since you're no longer very good at it.

If you can, put together a group of committed friends and loved ones who are willing to handle you properly when the time comes and support your wishes, and make this known to your main family. This is where the help of an experienced consultant can come in—someone trained as a home-funeral guide, or a sympathetic funeral director if you need more extensive service—since your personal group will need to understand how to bathe and dress you, how to carry you and when to move you, where and how to place you, and, in general, to be there to help others feel okay about being there with you when the time comes. It's not rocket science, but it helps to have the guidance of those who've been through it before.

Neutral groups like Hospice can be helpful, but they tend to shy away from advocacy for specific needs like those of green burial, especially when those needs deviate from traditional death-management practices and utilize alternative service providers like celebrants and home-funeral guides. Generally, the hospice role is to assist until just before you die, and then turn the final step over to professionals who complete the work in the conventional manner. Funeral homes traditionally manage this part of the work for a range of fees, but as alternatives emerge, a new range of end-of-life consultants is filling the gap.

> If you can, put together a group of committed friends and loved ones who are willing to handle you properly when the time comes and support your wishes, and make this known to your main family.

The Natural Burial Company's Caskit is a simple, nontoxic, and affordable burial option.

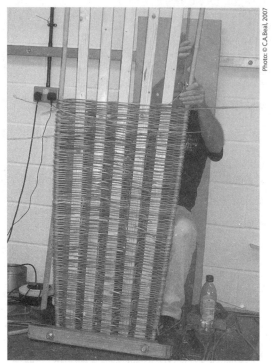

A Somerset willow casket being woven.

And, like a growing number of people today, if you're facing a terminal illness and are choosing to be cognizant of (or even determining the time of) your end, discussing this process openly with your group will be a relief for all concerned, especially those who may not be able to cope easily with your passing. Your wishes will be known, your group will become as comfortable as possible under the circumstances, and in the process they will become guides for the rest of your friends and family, turning a typically disengaged experience into a fully empowered one. Here's one revealing testimonial: "I know the discussions of funerals may sound a bit morbid to many out there. However, you cannot believe the change in my father's attitude once my mom, dad, and I sat down and discussed some of this stuff. Suddenly, he was able to discuss everything regarding his cancer more easily, which eventually led him to realizing that his chances for survival are very good." *CatieJayBee, 2002, Online Organic Gardening Forum.*[8]

SAVING MONEY

In 1959, according to *Time* magazine (1960 issue), $1.5 billion was spent on burial annually, at an average of about $900 per death. As of 2006, the average funeral with burial in America is running between $5,500 and $6,500, funeral and cremation cost around $3,500. Additional costs include opening and closing the grave, placing the liner and lowering the casket, and a host of other service options.

Natural caskets tend to cost less than the conventional ones, primarily because a lot of effort goes into making sure that the conventional ones never degrade, and nondegradable materials are expensive. Making them shiny, dent- and blemish-free, and available in a variety of designer finishes with curly handles and sateen lining also comes at a price. Physical imperfections in a material are the hallmark of natural products, especially when they have handmade components. The most expensive biodegradable casket offered by The Natural Burial Company, still well below the average retail price of a conventional casket, is the recycled-paper Ecopod, beautifully colored in hand-papered mulberry bark and silk fiber.

Shrouds are very affordable. The Jewish community traditionally buries their dead ritually wrapped in an unfinished (unhemmed) shroud, in the classic "plain pine box" that is, per orthodox rule, "unadorned." Hemmed shrouds in creative designs, made of organic hemp, cotton, and other natural fibers are also available.

Home funerals can be very inexpensive, especially when negotiated in advance. Burial in a cemetery involves basic fees, but maintenance of a natural burial site is much less intensive than a conventional plot, so expect direct costs to be lower. Some cemeteries (and even some states) require the graveside presence of a certified professional during burial, but if the requirement is a law, expect it to be challenged in the future by the growing funeral consumer movement as an unjustifiable cost, and if the requirement is a business practice, expect the free market and competition to change it.

Funerals don't have to be expensive, but they don't have to be cheap, either. They should, however, be as affordable as possible, with services or products required only in ways that preserve public health, safety, and the environment. With freedom to choose, you can put your money where your values are, and with the money you save, someone might be able to throw a darned good party in your honor!

> Funerals should be as affordable as possible, with services or products required only in ways that preserve public health, safety, and the environment. With freedom to choose, you can put your money where your values are, and with the money you save, someone might be able to throw a darned good party in your honor!

Last Acts That Make a Difference:

How Your Support of Environmentally Friendly Products Can Turn an Industry Around

Typically, about 2 million people die annually in the U.S. The post-World War II baby-boom generation turned 60 in 2006, creating a bulge in the upcoming death demographic that will put over 20% of Americans over age 65 by 2030. That means that more of us will be hitting the end of the line for the next 20-30 years, which will cause our society to focus more intently on how we die, and what we do as we do it, than ever before.

Over $20 billion dollars a year is spent on death-management in the U.S., much of it for industrial burial packaging products (caskets and liners), cemetery land purchase, and maintenance.[9] Industry estimates place funeral sales at $11 billion annually, but this does not include cemetery fees and burial plot sales. The true environmental costs of aging cemeteries have not yet been factored into many equations, and city planners, corporate cemetery stockholders, and their insurers are only now beginning to appreciate the expense accruing as they run out of space and are faced with tighter regulatory controls on the burial and discharge of pollutants and nondegradables into the environment.

When we change our purchasing behavior, we send a signal to the industries we want to change. Requesting that our caskets be free of toxins and pollutants, that our cemeteries get creative and end the use of liners and nonsustainable land management practices, or that our communities follow the lead of the U.K. and provide low-cost burial options as the public utility that the service rightfully could be, are not unreasonable requests.

Tarn Moor: Be a Tree.

Photo: © C.A. Beal, 2007

Taking the Natural Step

Natural burial grounds are possible today. Organic agriculturalists have provided the rationale and technology for chemical-free landscape management. Conservation groups can provide long-term land planning, oversight, and land-trust mechanisms. Cities and counties can buy land and hold it for future generations, creating greenspace and habitat (and soil-bank insurance) close to the urban core. Brownfields and industrial sites can be reclaimed, and valuable class I and II soils can be preserved from unnecessary development.

When public demand makes itself known, alternative-minded funeral directors around the world, already chafing at the corporate bridle that has forced so many of them into a business nothing like what they once knew, will respond. Just like farmers did for organics, and machinists did for renewable technology, there is a sector of funeral service professionals who are ready to serve the environmental interest. Most conventional casket manufacturers are out of touch with what their customers really want, and attentive funeral businesses know this.

With urgent calls to reduce our carbon footprint and end polluting activities that put (and store) toxins in the earth and atmosphere, a new breed of eco-aware consumers are changing behaviors on every front, including their last. Rather than be artificially preserved or disappear, folks choosing to reintegrate into a forest, bind some carbon, and kick some oxygen back into the system can take steps to do so well in advance.

The community of people who look ahead is increasingly putting its money where its mouth is, and now we're putting our bodies there, too. It's hard to do the daily things right, every day, all the time. No one can. But dying is a once-in-a-lifetime experience, and each of us can take the time to plan it out and do it right. The tasks are fairly straightforward: 1) Green your process; 2) Use a clean container; 3) Face the music and call your own tune; 4) Get your family and friends onboard; 5) Pick your plot and tend to its well-being and future until you get there; and 6) Don't leave a mess.

Planting forests is a lot of work, and every community needs a natural burial sanctuary, in our opinion. Check out your local arboretum; are they strapped for cash? Does your city have an urban growth boundary or brownfield areas that could use some healthy greenspace? Is there a pioneer cemetery nearby with room for a Bioneer or two? Consider adopting Arbor Day as one of your "memorial holidays" and join the Go Zero campaign, gifting trees in honor of your friends and loved ones whenever you can afford to.

Our wiser folk talk about the need for a shift in consciousness that brings us all into greater awareness about the material consequences that loom for future beings in our too-often thoughtless wakes. Is it possible that taking responsibility for our own deaths may make us more aware of the unintended deaths we bring to others throughout the webs of life? And could we, in managing our own ends properly and in advance, plan an exit that reduces, or even reverses, the toll our lives have taken on natural resource systems up to this point?

It's an exciting thought. The emerging natural burial movement offers unexpected and overlooked opportunities to make choices that just might nudge our culture in another direction, if we make the time to do what no one else can do for us—plan ahead to exit stage left, and do it right.

All that's left to make the leap is you.

End Notes

1. Funeral Consumers Alliance, www.funerals.org/faq/myths.htm.
2. Ibid.
3. Memorial Society of British Columbia, *Public Health Impact of Crematoria*: www.memorialsocietybc.org/c/g/cremation-report.html.
4. Natural Burial FAQ: www.naturalburialfaq.com/final_instructions/.
5. Animal Protection Institute, *Rendering, the Invisible Industry*, www.api4animals.org/articles?p=378&more=1.
6. Natural Death Centre UK, www.parliament.the-stationery-office.co.uk/pa/cm200001/cmselect/cmenvtra/91/91m06.htm.
7. The Funeral Consumers' Alliance, Directory of State Organizations, www.funerals.org/directry.htm.
8. http://forums.organicgardening.com/eve/forums. See "Caring for the Dead, Your Final Act of Love," by Lisa Carlson, an excellent reference, sold through the FCA site www.funerals.org.
9. www.bloomberg.com/apps/news?pid=2060188&sid=aQS1ODb4kQSg&refer=home

Just like farmers did for organics, and machinists did for renewable technology, there is a sector of funeral service professionals who are ready to serve the environmental interest.

NATURAL BURIAL PRODUCTS

Get on board with the "ultimate back-to-the-land movement" and go out as simply as you came in—in something handmade, from natural materials, that turns you back into good clean dirt.

One look at the recycled paper Ecopod, and they'll know this isn't your ordinary funeral—and for some, it's the last thing they'll want to be seen in. With Fair Trade Bamboo caskets, you cast your vote for social justice. Our woven artisan willow coffins let you remind us all one last time about the importance of tradition and handicraft. And, just in case you've still got a loved one's ashes in a box on the closet shelf, there are ash-burial urns ready to plant in their favorite forest.

No matter what method of "moving on" you choose, there's a clean way to go. The Natural Burial Company's biodegradable coffins and burial urns are suitable for both clean cremation and eco-burial, ensuring that the container you go out in makes the lightest environmental footprint possible, and that all YOU leave is energy.

Biodegradable Burial and Cremation Caskets

ARKA Ecopod Recycled Paper Coffins

The recycled fiber ARKA Ecopod, used in woodland burial grounds, conventional cemeteries, and crematoria throughout the U.K. since 1996, is handmade from 100% recycled postconsumer newsprint and office paper, pulped and molded into shape. Natural hardeners make the Ecopod extremely robust, able to carry up to 250 pounds with the aid of three canvas straps that pass through simple buckles and under the Ecopod, secure the lid, and attach to six bamboo carrying handles. Each Ecopod

is hand-finished with naturally dyed, screen-printed recycled silk and mulberry leaf paper. The interior is painted with solvent-free natural paint and lined with a simple quilted cotton pad. Available in two colors and sizes. Call for availability and shipping information. Made in the U.K.

Please note: Products ship freight-collect. Rapid delivery direct to residence, funeral home, or cemetery may be available. Due to the artisan nature of these items, lead times can vary. Please plan accordingly.

61-0500 ARKA Ecopod Recycled Paper Coffin, Small (to 5'8"), red or blue $3,200

61-0501 ARKA Ecopod Recycled Paper Coffin, Large (to 6'1"), red or blue $3,200

Somerset English Willow Coffin

Traditional willow basket-making techniques are carefully translated into modern natural burial coffin-making by the English crafters at Somerset Willow, who have supplied the U.K.'s woodland burial and cremation customers for over a decade. Renewable English willow can be sustainably harvested annually from crowns that, when well tended, will last for up to 40-50 years. Woven

willow retains its natural strong but flexible character and, when buried, decomposes much more quickly than hardwood.

Somerset's artisan willow coffins are fully functional and are a common choice in England today due to their conventional appearance and sturdy top-of-the-line construction. Able to carry up to 350 pounds, these coffins are weather-beaten gold with woven willow handles. The sides are inlaid with colored, water-based dye bands; choose green or blue. United Kingdom.

Please note: Products ship freight-collect. Rapid delivery direct to residence, funeral home, or cemetery may be available. Due to the artisan nature of these items, lead times can vary. Please plan accordingly.

61-0502 Willow Casket, gold, willow handle, 5'8"x20"x14", green or blue band $3,600

61-0503 Willow Casket, gold, willow handle, 6'4"x22"x14", green or blue band $3,600

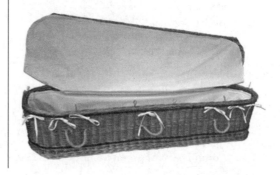

Bamboo Coffin

Our Fair Trade-certified bamboo coffin, handwoven by the SAWD Partnership, a small, family-run factory based in the Hunan Province of China, which shares both

Chinese and U.K. heritage, is probably the most popular biodegradable coffin sold in the U.K. at this time. For over a decade, SAWD has supplied thousands of affordable bamboo, willow, and fiber coffins and woven urns to the general public and to funeral directors in the U.K. for both burial and cremation. The bamboo used in these coffins is a variety that pandas do not eat, grown and cut under license from the Chinese government. Wooden runners on the bottom facilitate ease of use in crematoria. All bamboo coffins include cotton liners and bamboo headrests.

Please note: Products ship freight-collect. Rapid delivery direct to residence, funeral home, or cemetery may be available. Due to the artisan nature of these items, lead times can vary. Please plan accordingly.

**61-0504 Bamboo Fine-Weave Coffin,
5′9″x20″x11″ (int. dim.) $1,300**

**61-0505 Bamboo Fine-Weave Coffin,
6′2″x23″x13″ (int. dim.) $1,300**

**61-0506 Bamboo Fine-Weave Coffin,
6′8″x26″x13″ (int. dim.) $1,300**

Bamboo Pet Caskets

These bamboo pet caskets are made in the same fashion as the larger ones for people. Each comes with a leakproof cotton liner, and these are suitable for either cremation or burial.

**61-0507 Pet Casket, Small (18″x13″ int.
dim.), natural green bamboo $350**

**61-0508 Pet Casket, Medium (24″x18″ int.
dim.), natural green bamboo $350**

**61-0509 Pet Casket, Large (36″x20″ int.
dim.), natural green bamboo $350**

Biodegradable Ash-Burial Urns

With cremation trends on the rise, and a number of our relatives already cremated (and perhaps even still on the shelf), biodegradable ash-burial urns make a lot of sense, especially if your loved ones would have enjoyed the idea of being part of a forest.

Ashes do possess plant nutrients, mainly minerals and potassium. Scattered ashes can burn foliage and new plant roots, so strew with care. Burying ash is preferred to scattering, placed at least a few inches away from new plant roots. Whether sizing an urn for a pet or a person, you can calculate your urn size based on original body mass. As a rough measure, when a body is cremated, approximately 1 pound of body weight results in about 1 cubic inch of ash. Both of our urns are large enough for most people.

ARKA Acorn Ash-Burial Urn

The ARKA Acorn ash-burial urn is manufactured from recycled paper and cardboard fibers. The urn is intended to contain ashes after cremation and is suitable for burial. Produced to U.K. regulation size for licensed crematoria, with interior dimensions of 8″x10″ (WxH) (suitable for most individuals up to 380 lb.). The Acorn urn is overlaid with a moss green handmade paper, with a chestnut brown top and a fiber-twist. Fully biodegradable. United Kingdom.

**61-0510 Handfinished Acorn Ash-Burial
Urn, green $170**

Bamboo Ash-Burial Urn— Fair Trade

These woven bamboo ash-burial urns are made by a family-owned and run business based in the U.K. and China. The urns come complete with impenetrable natural cotton liners and have been tested and sized for U.K. crematoria use, with an interior capacity of 8″x8″ (WxH) (suitable for most individuals up to 400 lb.). Fully biodegradable. Suitable for forest burial. Fair-trade certified by IFAT. China.

61-0511 Bamboo Ash-Burial Urn, SAWD $180

The Real Goods Sustainable Living Library

Reading for a Renewable Future

"KNOWLEDGE IS POWER" the old saying goes, but at Real Goods, knowledge is even more: It's our most important product.

The *Solar Living Sourcebook* provides essential products and technical information about renewable energy, but if that's all that was contained in these pages, this would be little more than a glorified catalog. The real value of this book is that it is a hub that connects you to the entire world of solar living, including renewable energy, green building, ponds, permaculture gardens, sustainable transportation, and all points in between. Because our basic mission is to move the world toward the paradigm shift of low-impact, small-footprint, carbon-neutral sustainable living, ideas are just as important as equipment.

Even in the midst of an ongoing information revolution, the traditional book still reigns as the heavyweight champ. The book is amazingly efficient, portable, durable, and even sensuous in its ability to catalog, organize, and present data. Data accumulates into information, information transforms into knowledge, and knowledge matures into wisdom. No offense intended to all those crazy new technologies—they all shine in one way or another—but none of them has come close to challenging the book's grip on the championship belt. That is why we have put significant energy into making this *Sourcebook* a "book of books." The library we have assembled represents the end result of countless hours of research and selection. Each title, in its turn, leads you to a new journey into the world of knowledge.

The previous edition of the *Sourcebook* was the first time we collected our Sustainable Living Library in one place. This time around, as we celebrate 30 years of work and optimism, we have greatly expanded our offerings, in service to you and the goal we share of building a sustainable future. You won't find such a comprehensive library anywhere, not on Amazon, not in all the Borders and Barnes & Noble combined. If you come to our Solar Living Center in Hopland, however, please drop by and browse through the shelves, where our library comes alive. But if you can't, then do your browsing here. You will be giving yourself the gift of power through the product of knowledge, which will ultimately result in wisdom.

LAND AND SHELTER

Sensible Design

Design Like You Give a Damn
Architectural Responses to Humanitarian Crises

Edited by Architecture for Humanity. Truly a rallying cry for rethinking humanitarian assistance, as well as a blueprint for pursuing innovative solutions to contemporary housing crises, *Design Like You Give a Damn* demonstrates the power that design has to genuinely improve lives. Showcasing more than 80 solutions to situations from around the world such as basic shelter, healthcare and education, and access to clean water, energy, and sanitation, this book is bursting with intriguing and beautiful examples of how designers and architects have created innovative housing for those in need. Proceeds from the sale of this book go to support the work of Architecture for Humanity. 333 pages, softcover.

21-0404 Design Like You Give a Damn $35

The Natural House
A Complete Guide to Healthy, Energy-Efficient Environmental Homes

By Daniel D. Chiras. Aimed at builders as well as dreamers, this extraordinary book explores environmentally sound, natural, and sustainable building methods,

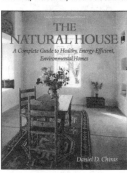

including rammed earth, straw bale, Earthships, adobe, cob, and stone. Besides sanctuary to shelter our families, today we want something more—we want to live free of toxic materials. Chiras is an unabashed advocate of these new technologies, yet he openly addresses the pitfalls of each, while exposing the principles that make them really work for natural, humane living. This book is devoted to analysis of the sustainable systems that make these new homes habitable. This is the short course in passive solar technology, green building material, and site selection, addressing more-practical matters—such as scaling your dreams to fit your resources—as well. 469 pages, softcover.

82-474   The Natural House $35

The Solar House:
Passive Heating and Cooling

By Daniel D. Chiras. Passive solar design is basically simple—the devil is in the details. This book admits, even highlights, ideas from the '70s that didn't work, but mostly highlights the good design ideas that DO work. Dan Chiras explains in detail how homebuilders can design excellent solar buildings attuned to regional climates—buildings that are warm in the winter and cool in the summer, and require a minimum amount of energy to stay that way. This is simply the best passive solar design book that's ever been available. Includes experienced reviews of current software programs for solar design, and extensive resource lists and appendixes. 284 pages, softcover.

21-0335 The Solar House $30

The New Ecological Home

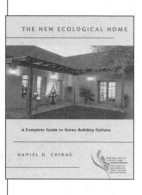

By Daniel D. Chiras. Shelter, while crucial to our survival, has come at a great cost to the land. But, as author Dan Chiras shows in his follow-up book to *The Natural House*, this doesn't have to be the case. With chapters ranging from green building materials and green power to landscaping, he covers every facet of today's homes. A good starting point for decreasing the environmental impact of new construction. 352 pages, softcover.

21-0346 The New Ecological Home $35

The Homeowner's Guide to
Renewable Energy

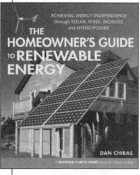

By Dan Chiras. *The Homeowner's Guide to Renewable Energy* outlines the likely impacts of fossil fuel shortages and some basic facts about energy, and discusses energy conservation as a way to slash energy bills and prepare for renewable energy options. Focusing on strategies needed to replace fuels, the book examines each practical energy option available to homeowners: solar hot water; cooking and water purification; space heating (solar and wood); passive cooling; solar, wind, and microhydro electricity; and hydrogen, fuel cells, methane digesters, and biodiesel. The book gives readers sufficient knowledge to hire and communicate effectively with contractors, and for those wanting to do installations themselves, it recommends more-detailed manuals. With a complete resource listing, this well-illustrated and accessible guide is a perfect companion for illuminating the coming dark age. 352 pages, softcover.

**21-0385 The Homeowner's Guide to
           Renewable Energy $27.95**

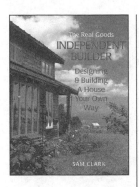

Independent Builder

Designing & Building a House Your Own Way

REAL GOODS INDEPENDENT LIVING BOOKS

By Sam Clark. This is *the* book for anyone thinking about building their own home. It is comprehensive and detailed, and covers subjects I have never before seen covered in homebuilding books, like how to make a small house seem bigger, incorporating ergonomics and accessibility, doing your own drawings and scale models, making contracts that work, and working effectively with professional designers and builders. 522 pages, softcover.

21-0508 Independent Builder   **$40**

The Sun-Inspired House

House Designs Warmed and Brightened by the Sun

By Debra Rucker Coleman. *Sun-Inspired House* is a unique combination of passive solar information and house plans designed to use the Sun to maximum efficiency. The information is educational while the designs are inspirational. The elegant yet functional homes capture the nonpolluting heat and wonderful light of the Sun in a wide range of warm and welcoming styles. Blueprints are available for most of the designs. 248 pages, softcover.

21-0550 Sun-Inspired House   **$29.95**

Green by Design

By Angela Dean. *Green by Design* provides a thorough analysis of what it means to build green and offers advice on what to consider when designing a sustainable home. The book features full-color photographs, and line drawings of floor plans show different examples of successful sustainable homes. It also includes in-depth case studies of more than a dozen homes, so readers planning a green home can see what worked for others. By providing people with knowledge, inspiration, and the ability to ask the right questions (and understand the answers), *Green by Design* puts home builders and owners on a path to creating beautiful, environmentally responsible homes that they can be proud to live in. Angela Dean, AIA, is principal architect of AMD Architecture in Salt Lake City. She specializes in environmentally responsible designs to create healthy, comfortable buildings that are in harmony with the environment. 160 pages, softcover.

21-0552 Green by Design   **$24.95**

Ecological Design and Building Schools

Green Guide to Educational Opportunities in the United States and Canada

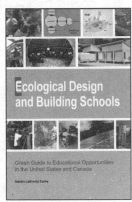

By Sandra Leibowitz Earley. The only directory of its kind in North America, this comprehensive guide features an annotated listing of schools and educational centers offering programs in ecological architecture and construction. Included also are a 10-year overview of sustainable design education, tables comparing school programs, and listings of instructors, green building organizations, selected textbooks, and publicly available curricula. Sandra Leibowitz Earley is a registered accredited architect. She founded Sustainable Design Consulting and provides green design specifications for commercial, institutional, and multifamily residential projects throughout the country. Earley has coauthored several guidance documents, including the *HOK Guidebook to Sustainable Design* and the *U.S. Green Building Council Toolkit*. 167 pages, softcover.

21-0551 Ecological Design and Building Schools   **$19.95**

More Small Houses

By *Fine Homebuilding* editors. In *More Small Houses*, you'll find that smaller is beautiful—and more intimate and affordable. Twenty homes, each less that 2,000 square feet, each a study in craft and efficiency, are used to explore space-saving design ideas. 159 pages, hardcover.

21-0217 More Small Houses   **$24.95**

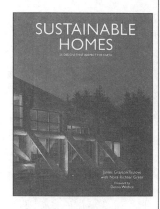

Sustainable Homes: 26 Designs that Respect the Earth

By James Grayson Trulove. *Sustainable Homes* puts to rest the stigma that "green architecture" is unattractive and unsuitable for residential building, for the houses shown in this volume represent design equal to or superior to most conventional houses. Each of these "eco" houses is unique. Some tread lightly on the site by nestling into the land. Others are designed to be extremely well suited to the climatic zones in which they are built. Many use materials that do no harm to the environment—materials that are recycled, salvaged, or harvested. 208 pages, softcover.

21-0553 Sustainable Homes $29.95

Building with Vision

Optimizing and Finding Alternatives to Wood (Wood Reduction Trilogy)

By Dan Imhoff. Part green-building primer, part architectural photo essay, this is an essential resource for professionals and homeowners interested in the leading edge of environmental building. Imhoff traveled extensively to document and photograph beautiful and novel alternatives to wood-intensive building. 136 pages, softcover.

21-0511 Building with Vision $22

Patterns of Home

By Max Jacobson, Murray Silverstein, and Barbara Winslow. Looking for inspiration and direction in the design or remodel of your home? Clearly written and full of photographs, *Patterns of Home* brings you the timeless lessons of residential design. The 10 patterns described in the book—among them, "capturing light" and "the flow through rooms"—are presented with clarity by the authors, renowned architects who have designed homes together for more than 30 years. This book will jump-start the design process and make all the difference between a house that satisfies material requirements and one that is a "home." *Patterns of Home* promises to become the "design bible" for homeowners and architects alike. 288 pages, hardcover.

21-0363 Patterns of Home $35

Green Remodeling

Changing the World One Room at a Time

By David Johnston and Kim Master. Millions of North Americans are renovating their homes every year. How do you remodel in a healthy, environmentally friendly way? *Green Remodeling* is a comprehensive guide. Buildings are responsible for 40% of worldwide energy flow and material use, so how you remodel can make a difference. Green remodeling is more energy-efficient, more resource-conserving, healthier for

occupants, and more affordable to create, operate, and maintain. The book discusses simple green renovation solutions for homeowners, focusing on key aspects of the building, including foundations, framing, plumbing, windows, heating, and finishes. Room by room, it outlines the intricate connections that make the house work as a system. Then, in an easy-to-read format complete with checklists, personal stories, expert insights, and an extensive resource list, it covers easy ways to save energy, conserve natural resources, and protect the health of loved ones. Addressing all climates, this is a perfect resource for conventional homeowners as well as architects and remodeling contractors. 368 pages, softcover.

21-0365 Green Remodeling $29.95

The Passive Solar House

By James Kachadorian. In the expanded 2nd edition of the best-selling guide to building homes that heat and cool themselves, James Kachadorian delivers a book that does not fail to impress. Kachadorian utilizes techniques that translate the essentials of solar design into practical, on-the-ground wisdom for builders of all types. Includes CSOL passive

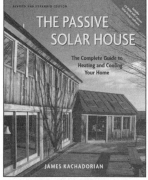

solar design software to help you analyze the efficiency of your passive solar house design and the passive solar potential of your current home. 240 pages, hardcover, includes CD-ROM.

80372 The Passive Solar House $40

The Art of Natural Building

Design, Construction, Resources

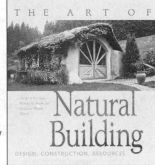

Edited by Joseph F. Kennedy, Michael G. Smith, and Catherine Wanek. For architects, designers, and the rest of us, this introduction to natural building covers how to build economically and environmentally sound houses that are as beautiful as they are comfortable. Organized as an anthology of articles from leaders in various building techniques—from straw bale and cob to recycled concrete and salvaged materials. Five sections take the reader from an overview of natural building through planning and design, specific techniques, and example structures, and then into complementary systems to complete your natural home. 288 pages, softcover.

82669 The Art of Natural Building **$27**

Blueprint Affordable

How to Build a Beautiful House Without Breaking the Bank

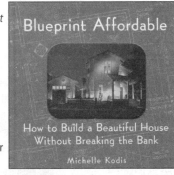

By Michelle Kodis. Dreaming of an architecturally distinctive house, filled with graceful, beautiful materials, that has been customized to fit your lifestyle—but think you can't afford it? *Blueprint Affordable* will dispel those fears and put builders on the path to a home that is personal, beautiful, and affordable. Discover the secrets of cost-conscious design and construction through the examples of 10 cutting-edge residential designs, and see how building a unique home doesn't have to be prohibitively expensive. Author Michelle Kodis includes an amazing list of helpful information, guidelines, and tips to follow throughout the planning and building process. 192 pages, hardcover.

21-0554 Blueprint Affordable **$24.95**

The Big Book of Small House Designs

75 Award-Winning Plans for Your Dream House, All 1,250 Square Feet or Less

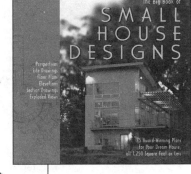

By Don Metz, Catherine Tredway, Lawrence Von Banford, and Kenneth R. Tremblay. *The Big Book of Small House Designs* is a collection of more than 500 drawings illustrating all aspects of 75 small homes of various styles, including a New England farmhouse, a sophisticated steel frame, and a Santa Fe ranch. Each design includes detailed floor plans, section drawings, elevations, and perspectives, as well as a description of the materials used and landscaping around the home. Keeping in mind that a chief priority for a small home is often energy efficiency, most of the plans incorporate some energy-efficient element. There are dozens of plans suitable for every environment and climate in the country. The designs are all a direct result of several international competitions that solicited from architects the best homes of 1,250 square feet or less. Contact information for the architects is provided in the back of the book. 368 pages, hardcover.

21-0555 Big Book of Small House Designs **$14.95**

Building with Awareness

The Construction of a Hybrid Home

HOME DVD AND GUIDEBOOK

By Ted Owens. Constructing a straw bale solar home requires the merging of solar design with alternative building techniques. Books are useful for accessing information on these topics quickly and easily, while videos are a great way to show actual building methods and techniques firsthand. The *Building with Awareness* DVD/guidebook combination brings you the best of both worlds. The award-winning DVD is inspiring and informative, with over five hours of material on every aspect of building a green home. The handy reference guidebook complements the DVD, with color photographs, diagrams, suggestions, and step-by-step methods, all condensed into a nuts-and-bolts format. Used together, these two valuable resources provide a visually dynamic and easy-to-understand library of information.

21-0378 Building with Awareness DVD **$35**

Designing Your Natural Home

By David Pearson. From studs to refrigerators, this book offers a practical, how-to guide on creating an eco-home for the do-it-yourselfer. Handy for remodeling, expanding, starting from scratch, or even just coffee table fodder. Photo essays of 10 homes exemplify eco-living for every style and budget. Architect and author David Pearson has been at the forefront of integrating home design with the green living movement. 160 pages, softcover.

21-0381 Designing Your Natural Home $29.95

New Organic Architecture: The Breaking Wave

By David Pearson. *New Organic Architecture* is a manifesto for building in a way that is both aesthetically pleasing and kinder to the environment. It illuminates key themes of organic architects, their sources of inspiration, the roots and concepts behind the style, and the environmental challenges to be met. The organic approach to architecture has an illustrious history, including Celtic design, Art Nouveau, Arts and Crafts, and the work of Antoni Gaudi and Frank Lloyd Wright. 224 pages, hardcover.

21-0556 New Organic Architecture $60

XS: Small Structures, Green Architecture

By Phyllis Richardson. A follow-up to the highly successful *XS: Big Ideas, Small Buildings*, this book features contemporary solutions to two of today's most challenging problems—how to conserve space and help save the environment. The design goals of the 40 houses included here are to build as small as possible, to harmonize with the site, to use natural heating and cooling techniques, and, above all, to combine aesthetic beauty with ecological sensitivity. The houses are striking in appearance, inexpensive to build, and totally functional, and will serve as inspiration for architects and potential owners. 224 pages, hardcover.

**21-0509 XS: Small Structures,
            Green Architecture $29.95**

Off the Grid

By Lori Ryker. *Off the Grid* confronts the ecological and cultural problems associated with the way we get and use energy, and explains how it is possible to live in a beautifully designed home using much less—no matter where your home is located. 160 pages, hardcover.

21-0557 Off the Grid $29.95

Good Green Homes

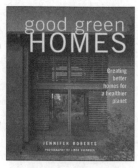

By Jennifer Roberts. *Good Green Homes* is a guide for people who want to live in comfortable, healthy, environmentally conscious homes. With some simple steps outlined in this book, you can save money and do your part to help save the environment. Perfect for homeowners, remodelers, renters, architects, builders, and interior designers, this book lays out seven fundamental principles of green building, illustrated with more than 150 color and 20 black-and-white photographs of more than 25 homes. 160 pages, hardcover.

21-0507 Good Green Homes $39.95

Redux

By Jennifer Roberts. *Redux* offers a host of solutions for creating a green home with recycled, reused, and environmentally healthy materials, whether remodeling, redecorating, or building from the ground up. This book combines extensive salvage use with the larger goal of efficiency and environmentalism—and the results are simply stunning. 160 pages, hardcover.

21-0558 Redux $29.95

The Japanese Bath

By Bruce Smith and Yoshiko Yamamoto. In the West, a bath is a place one goes to cleanse the body. In Japan, one goes there to cleanse the soul. Bathing in Japan is about much more than cleanliness: It is about family and community. It is about being alone and contemplative, time to watch the Moon rise above the garden. Along with 60 full-color illustrations of the light and airy baths themselves, *The Japanese Bath* delves into the aesthetic of bathing Japanese-style and the innate beauty of the steps surrounding the process. The authors explain how to create a Japanese bath in your own home. A Zen meditation, the Japanese bath indeed cleanses the soul, and one emerges refreshed, renewed, and serene. 96 pages, hardcover.

21-0559 The Japanese Bath $21.95

The Green House

New Directions in Sustainable Architecture

By Alanna Stang and Christopher Hawthorne. From the arid deserts of Tucson, Arizona, to the icy forests of Poori, Finland; from the tropical beaches of New South Wales, Australia, to the urban jungle of downtown Manhattan; critics Alanna Stang and Christopher Hawthorne have traveled to the farthest reaches of the globe to find all that is new in the design of sustainable, or "green," homes. The result: more than 35 residences in 15 countries—and nearly every conceivable natural environment. Each chapter features a series of homes that show the diversity and possibility of sustainable design. Projects are presented with large color images, plans, drawings, and an accompanying text that describes their green features and explains how they work with and in the environment. 196 pages, hardcover.

21-0560 The Green House **$45**

Rustic Retreats

A BUILD-IT-YOURSELF GUIDE

By David and Jeanie Stiles. This inspiring book presents 20 step-by-step plans for buildings including a log cabin, floating water gazebo, yurt, grape arbor, sauna hut, wigwam, river raft, and fold-up tree house. Many of these low-cost outdoor buildings can be built in a few hours or in less than a day and do not require high levels of building skills. Individual creativity can make any of the projects unique personal statements. Includes detailed illustrations. By David and Jeanie Stiles. 160 pages, softcover.

21-0362 Rustic Retreats **$20**

The Not So Big House

A Blueprint for the Way We Really Live

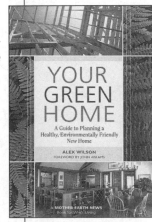

By Sarah Susanka, with Kira Obolensky. This is one of the best, most inspiring home design books that's ever been written. Sarah's sense of what delights us, what makes us feel secure, and what simply works well for all humans is flawless. Is it a surprise that people gather in the kitchen? It shouldn't be; human beings are drawn to intimacy, and this book proposes housing ideas that serve our spiritual needs while protecting our pocketbook. Commonsense building ideas are combined with gorgeous home design and dozens of small simple touches that delight the senses. This is an inspiring book on building houses that make us feel safe, protected, and comfortably at home. 199 pages, softcover.

21-0132 The Not So Big House **$23**

Design for Life: The Architecture of Sim Van der Ryn

By Sim Van der Ryn. *Design for Life* surveys the work and principles of the author, a world leader in the field of sustainable architecture. Sharing his years of experience as a teacher and using his building designs as examples, Van der Ryn shows us that buildings are not objects but organisms, and cities are not machines but complex ecosystems. 192 pages, hardcover.

21-0561 Design for Life **$39.95**

Your Green Home

A Guide to Planning a Healthy, Environmentally Friendly New Home

MOTHER EARTH NEWS WISER LIVING SERIES

By Alex Wilson and John Abrams. More and more homeowners today want houses that are healthy to live in and cause minimal damage to the environment. That's what green building is all about. *Your Green Home* is written for homeowners planning a new home—whether you are working with an architect or builder, or serving as your own general contractor. Intended to improve the overall environmental performance of new houses being built, the book sets out to answer some of the big-picture questions relating to having a home designed and built—and getting what you want. 237 pages, softcover.

21-0562 Your Green Home **$17.95**

BUILDING TYPES AND MATERIALS

Straw Bale Construction

Building a Straw Bale House: The Red Feather Construction Handbook

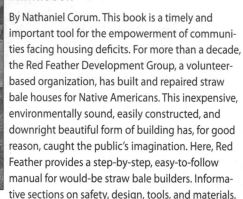

By Nathaniel Corum. This book is a timely and important tool for the empowerment of communities facing housing deficits. For more than a decade, the Red Feather Development Group, a volunteer-based organization, has built and repaired straw bale houses for Native Americans. This inexpensive, environmentally sound, easily constructed, and downright beautiful form of building has, for good reason, caught the public's imagination. Here, Red Feather provides a step-by-step, easy-to-follow manual for would-be straw bale builders. Informative sections on safety, design, tools, and materials, and case studies picked from over 35 Red Feather projects give a comprehensive overview to straw bale building. But this book is much more than a construction manual. It is also the inspiring story of Red Feather itself, a tale of community action and cooperation that suggests a can-do solution to the growing housing crisis on America's Native American reservations. 192 pages, paperback.

21-0515 Building a Straw Bale House $24.95

Design of Straw Bale Buildings: The State of the Art

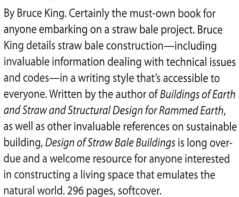

By Bruce King. Certainly the must-own book for anyone embarking on a straw bale project. Bruce King details straw bale construction—including invaluable information dealing with technical issues and codes—in a writing style that's accessible to everyone. Written by the author of *Buildings of Earth and Straw and Structural Design for Rammed Earth*, as well as other invaluable references on sustainable building, *Design of Straw Bale Buildings* is long overdue and a welcome resource for anyone interested in constructing a living space that emulates the natural world. 296 pages, softcover.

21-0391 Design of Straw Bale Buildings $40

Buildings of Earth and Straw
Structural Design for Rammed Earth and Straw Bale

By Bruce King. Straw bale and rammed earth construction are enjoying a fantastic growth spurt in the United States and abroad. When interest turns to action, however, builders can encounter resistance from mainstream construction and lending communities unfamiliar with these materials. *Buildings of Earth and Straw* is written by structural engineer Bruce King and provides technical data from an engineer's perspective. Information includes special construction requirements of earth and straw, design capabilities and limitations of these materials, and most important, the documentation of testing that building officials often require. This book will be an invaluable design aid for structural engineers and a source of insight and understanding for builders of straw bale or rammed earth. 190 pages, softcover with photos, illustrations, and appendixes.

80-041   Buildings of Earth and Straw $25

Serious Straw Bale
A Home Construction Guide for All Climates

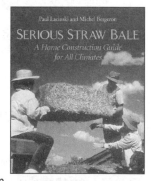

By Paul Lacinski. Until the publication of *Serious Straw Bale*, straw bale was assumed appropriate only for dry, arid climates. This book does not advocate universal use of straw bale but provides practical solutions to the challenges brought on by harsh climates. This is the most comprehensive, up-to-date guide to building with bales ever published. Builders, designers, regulators, architects, and owner/builders will benefit from learning bale-building techniques, even for climates and locales once thought impossible. 371 pages, softcover.

21-0180 Serious Straw Bale $29.95

Straw Bale Details

By Chris Magwood and Chris Walker. Nothing conveys a wealth of information like a good drawing by a professional architect, which is what this book is all about. It's the nitty-gritty of how to assemble a well-built, long-lasting straw bale structure. All practical foundation, wall, door and window, and roofing

options are illustrated, along with notes and possible modifications. Has a clever ring binding that lays flat for easy worksite reference. 68 pages, softcover.

21-0341 Straw Bale Details **$32.95**

More Straw Bale Building

A Complete Guide to Designing and Building with Straw

By Chris Magwood, Peter Mack, Tina Therrien, and Dale Brownson. Straw bale houses are easy to build, affordable, super energy efficient, environmentally friendly, attractive, and can be designed to match the builder's personal space needs, esthetics, and budget.

Despite mushrooming interest in the technique, however, most straw bale books focus on "selling" the dream of straw bale building but don't adequately address the most critical issues faced by bale house builders. Moreover, since many developments in this field are recent, few books are completely up to date with the latest techniques. *More Straw Bale Building* is designed to fill this gap. A completely rewritten edition of the 20,000-copy best-selling original, it leads the potential builder through the entire process of building a bale structure, tackling all the practical issues: finding and choosing bales; developing sound building plans; roofing; electrical, plumbing, and heating systems; building code compliance; and special concerns for builders in northern climates. 276 pages, paperback.

21-0516 More Straw Bale Building **$32.95**

The Beauty of Straw Bale Homes

By Athena and Bill Steen. This book is a visual treat, a gorgeous pictorial celebration of the tactile, timeless charm of straw bale dwellings. A diverse selection of building types is showcased—from personable and inviting smaller homes to elegant large homes and contemporary institutional buildings. The lavish photographs are accompanied by a brief description of each building's unique features. Interspersed throughout the book are insightful essays on key lessons the authors have learned in their many years as straw bale pioneers. 128 pages, softcover.

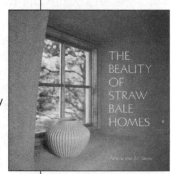

21-0158 The Beauty of Straw Bale Homes **$22.95**

Small Strawbale

Natural Homes, Projects & Designs

By Bill Steen, Athena Steen, and Wayne Bingham. This practical guide is filled with rich photos of homes, greenhouses, studios, sheds, open-air structures, and more, each pulsating with unique yet subtle creativity. Both a pragmatic construction manual and a philosophical, artistic guidebook, *Small Strawbale* is an inspirational starting point for a straw bale dreamer and a great source of information for those who are ready to get bailing. 240 pages, paperback.

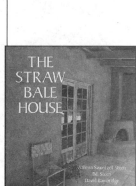

21-0518 Small Strawbale **$29.95**

The Straw Bale House

A REAL GOODS INDEPENDENT LIVING BOOK
Authors Athena Swentzell Steen and Bill Steen founded the Canelo Project, which promotes innovative building; David Bainbridge is a California restoration ecologist; and David Eisenberg is an alternative-materials builder who pioneered straw bale wall testing. Between them, they have encyclopedic knowledge of their subject. The book is comprehensive, broadly covering why and how to build with straw and then focusing on the details, which are both intellectually and aesthetically delightful. Besides being cheap, clean, and lightweight, straw also provides advantages such as energy efficiency and resistance to seismic stresses. 336 pages, paperback.

21-0514 The Straw Bale House **$30**

The New Strawbale Home

By Catherine Wanek. *The New Strawbale Home* compiles floor plans and images from 30 cutting-edge homes across North America, from California to Quebec, New Mexico to New England, showcasing a spectrum of regional styles and personal aesthetic choices. This practical guide discusses varying climate considerations and essential design details for problem-free construction and low maintenance, and also points out the ecologically friendly, energy-saving aspects of straw bale construction. 176 pages, hardcover.

21-0517 The New Strawbale Home $39.95

Earth, Stone, and Concrete

The Cob Builders Handbook
You Can Hand-Sculpt Your Own Home

By Becky Bee. This how-to book includes information on all aspects of cob building. Topics range from choosing a building site to drainage, floors, plaster, and finishing touches. An informal style and numerous hand-drawn illustrations make this book easy to read. 173 pages, paperback.

21-0522 The Cob Builders Handbook $23.95

Concrete at Home

By Fu-tung Cheng. In his follow-up book to *Concrete Countertops*, Fu-Tung Cheng expands from countertops to floors, walls, fireplaces, and more, showing us that truly anything is possible with concrete. His beauti-

fully designed book includes step-by-step guidance in forming, pouring, and polishing concrete throughout your home. You'll find yourself amazed to discover how well concrete can emulate natural materials and create a feeling of warmth and beauty. Packed with hundreds of color photos and drawings, this book is part art, part instruction manual. 224 pages, softcover.

21-0376 Concrete at Home $32

Concrete Countertops

By Fu-Tung Cheng and Eric Olsen. Here at last is a start-to-finish book on creating concrete countertops for your kitchen and bathroom. This sustainable material works equally well in traditional and modern homes. Fu-Tung Cheng takes you step by step through the entire process. You'll be inspired by the 350 color photos that showcase this versatile medium and the various creative options for customizing with color, polish, stain, stamp, and embedded objects. Throughout this well-organized book, Cheng offers valuable troubleshooting advice and useful tips on maintaining your countertop. 208 pages, softcover.

21-0361 Concrete Countertops $30

Planting Green Roofs and Living Walls

By Nigel Dunnett and Noel Kingsbury. This book introduces a revolutionary new concept to gardeners. Planting on roofs and walls began in Europe, but it is now becoming popular all over the world. Green roofs and walls reduce pollution and run-off, and also help insulate and reduce the maintenance needs of buildings. *Planting Green Roofs and Living Walls* discusses the practical techniques required to make planting on roofs and walls a reality. It describes how roofs can be modified to bear the weight of vegetation, considers the different options for drainage layers and growing media, and lists the plants suitable for different climates and environments. This informative book will encourage gardeners everywhere to consider the enormous benefits to be gained from planting on their roofs and walls. 256 pages, hardcover.

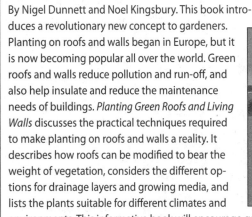

**21-0563 Planting Green Roofs and
          Living Walls $34.95**

Ceramic Houses & Earth Architecture
How to Build Your Own

By Nader Khalili. How to build, step by step, an affordable adobe and ceramic architecture. How to build arches, vaults, and domes and utilize the natural energy of wind, sun, and shade to help save forests and create a sustainable architecture. How to fire and glaze an entire building after it is constructed from clay-earth on site. A new update chapter introduces the Superadobe technology, building with almost any on-site soil using sandbags and barbed wire. 233 pages, paperback.

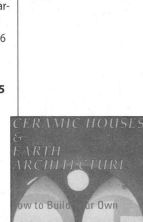

**21-0564 Ceramic Houses and
          Earth Architecture $24.95**

The Natural Plaster Book

By Cedar Rose Guelberth and Dan Chiras. This comprehensive step-by-step guide to choosing, mixing, and applying natural plasters takes the mystery out of using earth, lime, and gypsum plasters. Includes plaster recipes, tips on pigmentation, common mistakes, and thorough insight on the tools and techniques for professional results. 252 pages, hardcover.

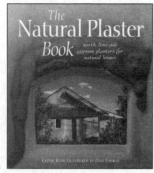

21-0350 The Natural Plaster Book $29.95

Earthbag Building
The Tools, Tricks, and Techniques

By Kaki Hunter and Donald Kiffmeyer. *Earthbag Building* is the first comprehensive guide for building with bags filled with earth—or earthbags. The authors developed this "Flexible Form Rammed Earth Technique" over the last decade. A reliable method for constructing homes, outbuildings, garden walls, and much more, this enduring, tree-free architecture can also be used to create arched and domed structures of great beauty in any climate. This profusely illustrated guide first discusses the many merits of earthbag construction and then leads the reader through the key elements of an earthbag building, including special design considerations; foundations, walls, and floors; electrical, plumbing, and shelving; lintels, windows, and door installations; roofs, arches, and domes; and exterior and interior plasters. With dedicated sections on costs, making your own specialized tools, and building code considerations, as well as a complete resources guide, *Earthbag Building* is the long-awaited, definitive guide to this uniquely pleasing construction style. 288 pages, softcover.

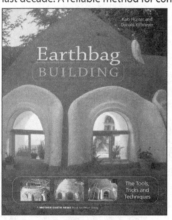

21-0367 Earthbag Building $29

Earth-Sheltered Houses
How to Build an Affordable Underground Home

By Rob Roy. An earth-sheltered, earth-roofed home has the least impact upon the land of all housing styles, leaving almost zero footprint on the planet. *Earth-Sheltered Houses* is a practical guide for those who want to build their own underground home at moderate cost. It describes the benefits of sheltering a home with earth, including the added comfort and energy efficiency from the moderating influence of the earth on the home's temperature (keeping it warm in the winter and cool in the summer), along with the benefits of low maintenance and the protection against fire, sound, earthquakes, and storms afforded by the earth. The book covers all the various construction techniques involved, including details on planning, excavation, footings, floor, walls, framing, roofing, waterproofing, insulation, and drainage. 253 pages, paperback.

21-0565 Earth-Sheltered Houses $26.95

Green Roof Plants
A Resource and Planting Guide

By Edmund C. Snodgrass and Lucie L. Snodgrass. In just a few years, green roofs have gone from a horticultural curiosity to a booming growth industry, primarily because the environmental benefits of extensively planted roofs are now beyond dispute, whether for industrial or governmental complexes or for private homes in urban or suburban settings. Despite the high level of interest in green roofs, until now there has been no reliable reference devoted exclusively to the various species of drought-tolerant plants that are suitable for use on extensive green roofs. *Green Roof Plants* fills that void. The book contains photographs and cultural information for more than 200 species and cultivars of plants, including valuable data on moisture needs, heat tolerance, hardiness, bloom color, foliage characteristics, and height. Concise, accurate, and easy to use, this book is destined to become an indispensable practical reference guide, not just for architects, landscape designers, engineers, and environmentalists, but also for environmentally conscious home gardeners. 220 pages, hardcover.

21-0566 Green Roof Plants $29.95

Grow Your Own House

Simone Velez and Bamboo Architecture

By Simon Velez, Jean Dethier, and Klaus Steffens. Bamboo, which has been used as building material for centuries and is widely available (35 million acres worldwide are covered in bamboo), is being rediscovered today. Its cost-effectiveness and ability to endure adverse environmental forces make it one of the preeminent construction materials on the planet. Bamboo's unique aesthetic appearance has been exploited in design and furniture building. The highly visual and engaging *Grow Your Own House* will open your eyes to the beauty and lightness of bamboo structures and designs. Bamboo—a widely available and renewable resource almost as strong as steel, yet very light—lends itself to architectural experiments. This lavishly and colorfully illustrated volume is published in dual languages (German and English). Contributors to this volume include Jean Dethier, Walter Liese, Eda Schaur, Frei Otto, and Mateo Kries. 265 pages, paperback.

21-0317 Grow Your Own House   **$29.95**

Building with Cob

A Step-by-Step Guide

By Adam Weismann and Katy Bryce. Learn how to create your dream home with clay subsoil, aggregate, straw, and water in *Building with Cob*, one of the most practical and well-illustrated books on earth building ever published. From designing, planning, and siting your home to adding the roof, insulation, and floors, authors Adam Weismann and Katy Bryce impart wisdom gained from having built and restored many cob buildings and run their own building company for years. More than 400 photographs and illustrations make

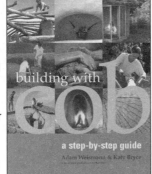

it easy for even a novice to understand and construct a cob home to call your own. 256 pages, paperback.

21-0387 Building with Cob   **$45**

Building with Earth

A Guide to "Dirt Cheap" and "Earth Beautiful"

By Paulina Wojciechowska. *Building with Earth* is the first comprehensive guide to the re-emergence of earthen architecture in North America. Even inexperienced builders can construct an essentially tree-free building, from foundation to curved roof, using recycled tubing or textile grain sacks. Featuring beautifully textured earth- and lime-based finish plasters for weather protection, "earthbag" buildings are being used for retreats, studios, and full-time homes in a wide variety of climates and conditions. This book tells (and shows in breathtaking detail) how to plan and build beautiful, energy-efficient earthen structures. Author Paulina Wojciechowska is founder of Earth, Hands, and Houses, a nonprofit that supports building projects that empower indigenous people to build shelters from locally available materials. 200 pages, softcover.

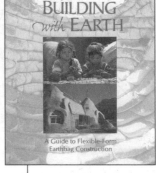

21-0323 Building with Earth   **$24.95**

Wood

Logs, Wind, and Sun

Handcraft Your Own Log Home . . . Then Power It with Nature

By Rex A. Ewing and Lavonne Ewing. Many dream of getting back to nature and living self-sufficiently in a house built with their own hands. The Ewings show readers how in this account of how they built a log house and then powered it using Sun and wind. This monumental undertaking is pragmatically described by the Ewings in an easygoing prose style. They take a clear, step-by-step approach to log building—a project one should not undertake without considerable building experience. Fully a third of the text is devoted to explaining how to run one's home completely off the power grid. Readers are offered a wealth of information about solar modules and wind generators, charge controllers, batteries, and inverters. This section makes their work stand out from other log-building books. 304 pages, paperback.

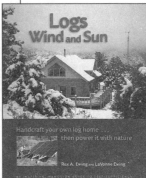

21-0520 Logs, Wind, and Sun   **$28.95**

The Cabin: Inspiration for the Classic American Getaway

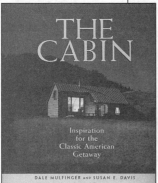

By Dale Mulfinger. Cabins are simple, sometimes primitive structures, but the heart of each cabin—a treasury of feelings, sensations, and memories of family and friends—makes them special. *The Cabin* presents 37 inspiring examples, showing how people are building, reclaiming, transforming, or buying this basic form of American residential architecture for a chance at the good life. The book includes 248 color photos and 50 color illustrations, plus site plans and floor plans, and covers the four basic styles: rustic, traditional, transformed, and modern. In the process, it celebrates the possibilities and pleasures of cabins as both shelter and a way of life. 256 pages, paperback.

82617   The Cabin $24.95

Graphic Guide to Frame Construction

Details for Builders and Designers (For Pros by Pros)

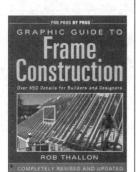

By Rob Thallon. Teaching through hundreds of meticulous drawings, Rob Thallon's guide covers foundations, roofs, building systems and materials, durability, energy efficiency, and more. 240 pages, paperback.

**21-0567 Graphic Guide to Frame
            Construction $34.95**

Cordwood Building

The State of the Art

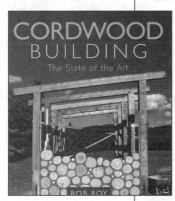

By Rob Roy. Cordwood masonry is an ancient building technique whereby walls are constructed from "log ends" laid transversely in the wall. It is easy, economical, aesthetically striking, energy-efficient, and environmentally sound. *Cordwood Building* collects the wisdom of more than 25 of the world's best practitioners, detailing the long history of the method and demonstrating how to build a cordwood home using the latest and most up-to-date techniques, with a special focus on building code issues. 240 pages, paperback.

21-0519 Cordwood Building $26.95

Yurts

The Complete Yurt Handbook

By Paul King. The yurt is a low-impact structure that

causes no permanent damage to the land on which it is pitched. It is easy to erect and can be taken down in an hour. This book gives fully illustrated and detailed instructions on how to make several of the most popular types of yurts, including the "weekend yurt." Featuring a thorough history of the yurt and the principles behind its construction, it explores modern life in a Mongolian ger (yurt) and the culture and etiquette of ger living. For centuries, people throughout central Asia have made robust and versatile yurts their homes, and it is the ultimate portable dwelling perfect for offices, summer houses, meditation spaces, spare rooms, or just beautifully satisfying spaces to be in! With a few common woodworking tools, even an absolute beginner could build the frame for this simple, elegant structure. This handbook shows you how. 121 pages, paperback.

21-0512 The Complete Yurt Handbook $21.95

Mongolian Cloud Houses

How to Make a Yurt and Live Comfortably

By Dan Frank Kuehn. Written for those interested in alternative lifestyles, outdoor living, camping, and

do-it-yourself projects, this lively book recounts the author's experiences building his first yurt. Dan Frank Kuehn carefully guides readers through every step of the creation of a 13-foot-diameter, 10-foot-tall model. He covers topics that include the poles and lattice that form the basic structure, the pluses and minuses of various materials, and the "luxuries" like solar windows and lofts. The book highlights new building techniques and contains detailed lists of commercial yurt manufacturers, tools, and materials. 152 pages, paperback.

21-0513 Mongolian Cloud Houses $16.95

General Building, Other

The Hand-Sculpted House

By Ianto Evans, Michael G. Smith, and Linda Smiley. Learn how to build your own home from a simple earthen mixture. Derived from the Old English word meaning "lump," cob has been a traditional building method for millennia. Authors Ianto Evans, Michael G. Smith, and Linda Smiley teach you how to sculpt a cozy, energy-efficient home by hand without forms, cement, or machinery. Includes illustrations and photography. 384 pages, softcover.

21-0331 The Hand-Sculpted House   **$35**

Rocket Mass Heaters

By Ianto Evans and Leslie Jackson. Newly updated (a companion volume to the very popular *Rocket Stoves*), *Rocket Mass Heaters* teaches readers how to construct a rocket stove in a single weekend for less than $100. As heating costs spiral upward, learn how to use locally procured materials to build and warm your home with this extra-efficient heating system. Expanded case studies and color photographs, extended troubleshooting, and a Q&A section help explain the smallest details. 100 pages, paperback.

21-0373 Rocket Mass Heaters   **$18**

Creative Country Construction
Building & Living in Harmony with Nature

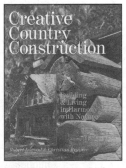

By Robert Inwood and Christian Bruyere. Recommended for academic collections, public library collections, and individuals interested in homesteading in a nature-friendly manner. A copious compendium of old-time skills and ambitious projects. Beautifully illustrated. 152 pages, paperback.

21-0568 Creative Country Construction $19.95

The Book of Masonry Stoves
Rediscovering an Old Way of Warming

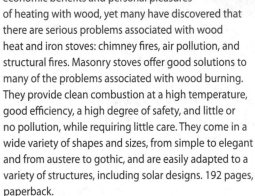

By David Lyle. *The Book of Masonry Stoves* represents the first comprehensive survey ever published of all the major types of masonry heating systems, ancient and modern. Detailed plans and building information are included in the book. As a complete introduction to masonry stoves, it will help many people, as the title states, rediscover an old way of warming. Within the past decades, millions of Americans have discovered the economic benefits and personal pleasures of heating with wood, yet many have discovered that there are serious problems associated with wood heat and iron stoves: chimney fires, air pollution, and structural fires. Masonry stoves offer good solutions to many of the problems associated with wood burning. They provide clean combustion at a high temperature, good efficiency, a high degree of safety, and little or no pollution, while requiring little care. They come in a wide variety of shapes and sizes, from simple to elegant and from austere to gothic, and are easily adapted to a variety of structures, including solar designs. 192 pages, paperback.

21-0524 The Book of Masonry Stoves   **$30**

Building with Structural Insulated Panels

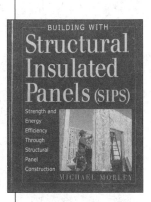

By Michael Morley. Structural Insulated Panels (SIPs) are replacing the stick-framed, fiberglass-insulated houses and light commercial buildings that were the postwar norm. SIPs produce a structurally superior, better insulated, faster to erect, and more environmentally friendly house than ever before possible. Experienced SIP builder Michael Morley covers choosing the right panels for the job, tools and equipment, wall and roof systems, and mechanical systems. This is an indispensable reference for anyone who wants to work with this innovative, field-tested construction system. 186 pages, hardcover.

82621 Building with Structural Insulated Panels   **$35**

The Good House Book

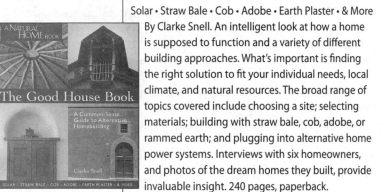

A Common-Sense Guide to Alternative Homebuilding

Solar • Straw Bale • Cob • Adobe • Earth Plaster • & More
By Clarke Snell. An intelligent look at how a home is supposed to function and a variety of different building approaches. What's important is finding the right solution to fit your individual needs, local climate, and natural resources. The broad range of topics covered include choosing a site; selecting materials; building with straw bale, cob, adobe, or rammed earth; and plugging into alternative home power systems. Interviews with six homeowners, and photos of the dream homes they built, provide invaluable insight. 240 pages, paperback.

21-0521 The Good House Book $19.95

References/Resources

Code Check Electrical, 4th Ed.

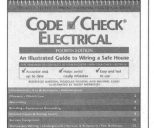

An Illustrated Guide to Wiring a Safe House

By Michael Casey, Redwood Kardon, and Douglas Hansen. *Code Check Electrical, 4th Ed.*, outlines the principles behind the various accepted electrical codes and provides information on the code requirements for all types of residential electrical systems. 332 pages, spiral-bound.

21-0280 Code Check Electrical $17.95

Pocket Ref

By Thomas J. Glover. Almost every conversion factor and bit of technical data you could ever imagine wanting to know packed into a pocket-sized book you can easily take with you anywhere. The BTUs in a gallon of propane or a kilowatt-hour, wire and lumber sizing, area and zip codes, mixing ratios, wind and water formulas, exchange rates, maps, charts; it goes on and on. 3rd Ed. 544 pages, softcover.

21-0312 Pocket Ref $12.95

Habitat for Humanity: How to Build a House

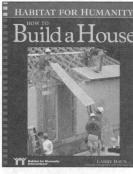

By Larry Haun. The world-famous volunteer organization Habitat for Humanity now offers its best nuts-and-bolts expertise in an easy-to-follow home construction handbook. Whether the reader contemplates building his or her own house or plans on volunteering to build for others, *Habitat for Humanity: How to Build a House* delivers on the promise of its title. Step-by-step instructions by author and veteran crew supervisor Larry Haun are provided in plain, simple English, with lots of encouragement and no condescension to beginners. Practically every page contains an extra "helping hand" tip on materials, tools, building codes, or safety precautions. In hundreds of color photographs and black-and-white line drawings, the book follows the construction of a single house—from choosing its location to the final step of installing its exterior door locks. 288 pages, paperback.

21-0569 Habitat for Humanity $24.95

Architectural Resource Guide

From the Northern California Architects/ Designers/Planners for Social Responsibility

Ever wondered where to find all those nontoxic, recycled, sustainable, or just plain better building materials? Well, here it is. This absolute gold mine of information has a huge wealth of products listed logically and intelligently following the standard architect's Uniform Specification System (i.e., concrete, thermal and moisture protection, windows and doors, finishes, etc.). Each section is preceded by a few pages explaining the pros and cons of particular materials or appliances. Each listing says what it is and gives the brand name, a brief explanation of the product or service, and access info, both national and NorCal local, for the manufacturer. Although written for the professional architect or builder, there's plenty of info for everyone here. Updated continuously, we order small quantities to ensure shipping the latest edition. 152 pages (approx.), softcover.

80-395   Architectural Resource Guide $35

Green Building Products

The GreenSpec Guide to Residential Building Materials

By Alex Wilson and Mark Piepkorn. *Green Building Products* includes more than 1,600 products from the GreenSpec database, which the company has maintained since 1998. Includes everything from precast concrete foundation systems to recycled-plastic roofing shingles and top-efficiency heating equipment. Products are organized according to building component and indexed under manufacturer or product name. Photos are included for about 300 products. A must for anyone—builder, designer, architect, or homeowner—who wants to know the unbiased and unembellished truth about what's really green. Products are selected on criteria that *Environmental Building News* editors have developed over more than 10 years, including recycled content, FSC-certified wood, avoidance of toxic constituents, reduction of construction impacts, energy or water savings, and contribution to a safe, healthy indoor environment. Not a paid directory, "we base selections on careful in-house review by our editorial staff." Details on the product selection process and the full list of criteria are included. 338 pages, paperback.

21-0523 Green Building Products   **$34.95**

RENEWABLE ENERGY IN THEORY AND PRACTICE

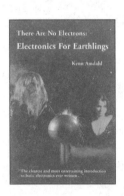

There Are No Electrons

By Kenn Amdahl. Do you know how electricity works? Want to? Read this book and learn about volts, amps, watts, and resistance. So much fun to read, you won't even notice you're learning. And if you already "know" about all this, you'll enjoy it even more—just be prepared to rethink everything. 322 pages, hardcover.

82646   There Are No Electrons   **$16**

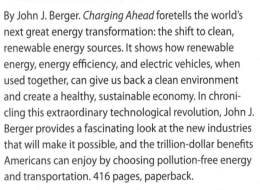

Charging Ahead

The Business of Renewable Energy and What It Means for America

By John J. Berger. *Charging Ahead* foretells the world's next great energy transformation: the shift to clean, renewable energy sources. It shows how renewable energy, energy efficiency, and electric vehicles, when used together, can give us back a clean environment and create a healthy, sustainable economy. In chronicling this extraordinary technological revolution, John J. Berger provides a fascinating look at the new industries that will make it possible, and the trillion-dollar benefits Americans can enjoy by choosing pollution-free energy and transportation. 416 pages, paperback.

21-0573 Charging Ahead   **$24.95**

Who Owns the Sun?

People, Politics, and the Struggle for a Solar Economy

By Daniel M. Berman and John T. O'Connor. Environmental activists Berman and O'Connor offer a scathing explanation of why solar technology has played such an insignificant role in meeting America's energy needs. Politicians, utility companies, and even many mainstream environmental groups come under attack for either their lack of leadership on this issue or for their downright hostility to solar possibilities. The most interesting solutions include public ownership of utilities, enlightened building codes favorable to solar technology, utility company buy-backs of excess electricity generated by homeowners, tax breaks for the installation of nonpolluting sources of power, removal of massive governmental subsidies of fossil fuels, and equalization of governmental research dollars for renewable and nonrenewable sources of energy. 352 pages, paperback.

21-0574 Who Owns the Sun?   **$17.95**

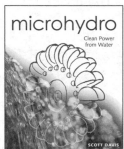

Microhydro

By Scott Davis. Hydroelectricity is the world's largest source of renewable energy. But until now, there were no books that covered residential-scale hydropower, or micro-hydropower. This excellent guide covers the principles of micro-hydro, design and site considerations, and equipment options, plus legal, environmental, and economic factors. Author Scott Davis covers all you need to know to assess your site, develop a workable system, and get the power transmitted to where you can use it. 157 pages, softcover.

21-0340 Microhydro   **$19.95**

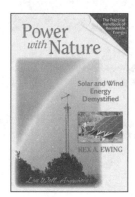

Power with Nature

Solar and Wind Energy Demystified

Author Rex Ewing is right up to date, explaining all the latest offerings like utility intertie systems and Maximum Power Point Tracking charge controls—all with extensive drawings, pictures, sidebars, and witticisms. You'll be enlightened by the wealth of practical, hands-on information for solar, wind, and micro-hydropower; plus common-sense explanations of the necessary components such as batteries, inverters, charge controllers, and more. 255 pages, softcover.

21-0339 Power with Nature   **$24.95**

The Solar Electric Independent Home Book

By Paul Jeffrey Fowler. A good, basic primer for getting started with PV, written for the layperson. Lots of good charts, clear graphics, a solid glossary, and appendix. Includes 75 detailed diagrams, and the text includes recent changes in PV technology. This is one of the best all-round books on wiring your PV system. 174 pages, paperback.

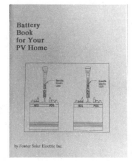

80102 The Solar Electric Independent Home Book   **$20**

Battery Book for Your PV Home

This booklet by Fowler Electric Inc. gives concise information on lead-acid batteries. Topics covered include battery theory, maintenance, specific gravity, voltage, wiring, and equalizing. Very easy to read and provides the essential information to understand and get the most from your batteries. Highly recommended by Dr. Doug for every battery-powered household. 22 pages, softcover.

80104 Battery Book for Your PV Home   **$8**

Battery Care Kit (The Battery Book)

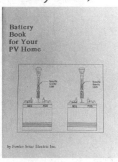

By New England Solar Electric. Our kit includes the informative *Battery Book for Your PV Home* and a full-size graduated battery hydrometer. The *Battery Book* explains how batteries work and shows how to care for your batteries and get the best performance. Every battery-powered household needs a copy. The hydrometer is your most important tool for determining battery state of health. Detect and fix small problems before

they require complete battery bank replacement. Use tip: The color-coded red, yellow, and green zones on the hydrometer are calibrated for automotive starting batteries: Ignore them. Just look for differences among cells. 22 pages, softcover.

15754 Hydrometer/Book Kit   **$14**

Wind Energy Basics

A Guide to Small and Micro Wind Systems

By Paul Gipe. A step-by-step manual for reaping the advantages of a windy ridgetop as painlessly as possible. Chapters include wind energy fundamentals, estimating performance, understanding turbine technology, off-the-grid applications, utility intertie, siting, buying, and installing, plus a thorough resource list

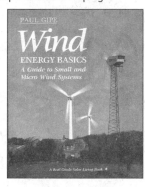

with manufacturer contacts and international wind energy associations. A Real Goods Solar Living Book. 222 pages, softcover.

80928 Wind Energy Basics   **$19.95**

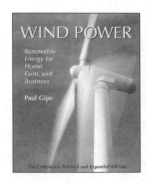

Wind Power: Renewable Energy for Home, Farm, and Business

By Paul Gipe. Wind energy technology has truly come of age—with better, more reliable machinery and a greater understanding of how and where wind power makes sense. In addition to expanded sections on gauging resources and turbine siting, this thoroughly updated edition contains a wealth of examples, case studies, and human stories. Covers everything from small off-grid to large grid-tie, plus water pumping. This is the wind energy book for the industry. Everything you ever wanted to know about wind turbines. Has an abundance of pictures, charts, drawings, graphs, and other eye candy, with extensive appendixes. 504 pages, softcover.

21-0343 Wind Power: Renewable Energy for Home, Farm, and Business $45

Energy Power Shift:
Benefiting from Today's New Technologies

By Barry J. Hanson. *Energy Power Shift* describes a host of exciting emerging technologies enabling communities to produce their own transportation fuels, electricity, and heat. The book argues that the U.S. has five times the renewable energy needed for all energy services using local resources and waste streams, and that money currently spent on energy could create instead thousands of new jobs. This analysis of the science, engineering, and economics of renewable energy provides a blueprint for an elegant energy solution spurring a new wave of economic development. Barry J. Hanson, a chemist and mechanical design engineer, is renewable energy consultant in Wisconsin. 208 pages, paperback.

21-0575 Energy Power Shift $16.95

Residential Microhydro Power Video

By Don Harris. A program we highly recommend to anyone considering residential hydro. Covers basic technology and hands-on applications, with easy-to-follow explanations of micro-hydro power. Also covers performance and safety issues. With micro-hydro pioneer Don Harris of Harris Hydroelectric. VHS, 42 min.

80362   Microhydro Power Video $39

The Renewable Energy Handbook
A Guide to Rural Energy Independence, Off-Grid and Sustainable Living

By William Kemp. *The Renewable Energy Handbook* focuses completely on off-grid, sustainable living and energy independence in a rural setting. Author William Kemp and his wife designed their own high-efficiency off-grid home in 1991. They worked methodically to produce a home that has all the standard "middle-class" creature comforts while using six times less heating, cooling, and electrical energy than the average Ontario home. 567 pages, paperback.

21-0571 The Renewable Energy Handbook $29.95

Energy Switch
Proven Solutions for a Renewable Future

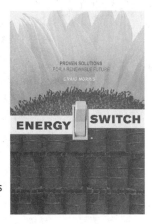

By Craig Morris. Declining oil supplies and the environmental impact of coal dictate a switch to renewable energy sources for a sustainable future. Written for a popular audience, *Energy Switch* details this momentous transition and proposes that the remaining nonrenewable resources be used to develop a long-term supply of renewable energy. 209 pages, paperback.

21-0576 Energy Switch $16.95

National Electrical Code® 2005 Handbook

Produced by the NFPA, the popular *National Electrical Code® 2005 Handbook* contains the complete text of  the 2005 edition of the NEC® supplemented by helpful facts and figures, full-color illustrations, real-world examples, and expert commentary. An essential reference for students and professionals, this handbook is the equivalent of an annotated edition of the 2005 NEC® that offers insights into new and more-difficult articles to guide users to success in interpreting and applying current code requirements to all types of electrical installations. A valuable information resource for anyone involved in electrical design, installation, and inspection, the *NEC® 2005 Handbook* is updated every three years and provides 100% of the information needed to "meet code" and avoid costly errors. 1,333 pages, hardcover.

21-0525 National Electrical Code® 2005 Handbook $125

For the full NEC, please visit www.realgoods.com/nec.pdf.

Energy for Keeps

Electricity from Renewable Energy: An Illustrated Guide for Everyone Who Uses Electricty

By Marilyn Nemzer, Deborah Page, and Anna Carter. *Energy for Keeps* offers an introduction to renewable energy for everyone who uses electricity—from students to energy policy makers. This book explains both renewable and nonrenewable energy resources, with an emphasis on renewables. However, it does not promote any one particular technology. *Energy for Keeps* aims at furthering 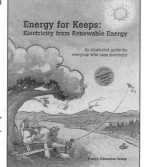 energy literacy for the general public and also serves as a great supplementary text for middle and high school students. 176 pages, hardcover.

21-0577 Energy for Keeps $24.95

The Citizen-Powered Energy Handbook

By Greg Pahl. Solar roof panels, backyard wind turbines, and biofuel stills: In this how-to vision of a future without hydrocarbon fuels, small really is beautiful. Faced with the paired (and frightfully imminent) dangers of global warming and the point at which half the total recoverable oil on Earth has been extracted and production begins to decline, Pahl champions a spectrum of alternative energy sources. Separate chapters on water, geothermal, and biomass (firewood and plant matter) energies in addition to solar, wind, and biofuel (the distillate of corn, soy, and other crops) sources are both practical and inspirational. First comes technical information; then Pahl reports on community and cooperative alternative-energy successes. 336 pages, paperback.

21-0578 The Citizen Powered Energy Handbook $21.95

Natural Home Heating

The Complete Guide to Renewable Energy Options

By Greg Pahl. *Natural Home Heating* is the first comprehensive guide to heating your home with renewable energy sources. Greg Pahl offers a well-organized, easy-to-understand tour of all available renewable home-heating options, including wood, pellet, corn, and grain-fired stoves; fireplaces, furnaces, and boilers; and masonry heaters, active and passive solar systems, and heat pumps. Learn how to burn environmentally friendly biodiesel fuels, not just in your car but in your furnace or boiler. Included is everything you need to know about the fuels, systems, technologies, costs, advantages, and disadvantages of each option. 281 pages, paperback.

92-0266 Natural Home Heating $30

Windpower Workshop

By Hugh Piggott. *Windpower Workshop* provides all the essential information for people wanting to build and maintain a wind power system for their own energy. 160 pages, paperback.

21-0572 Windpower Workshop $15.95

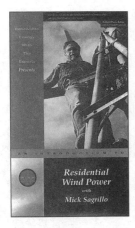

Residential Wind Power with Mick Sagrillo

Mick Sagrillo, the former owner of Lake Michigan Wind and Sun, has been in the wind industry for more than 16 years, with over 1,000 turbine installations/repairs under his belt. This professionally produced video includes basic technology introductions, hands-on applications, and performance and safety issues. The action moves into the field to review actual working systems, including assembling a real system. VHS, 63 min.

80363   Renewable Experts Tape—Wind $39

Wiring 12 Volts for Ample Power

By David Smead and Ruth Ishihara. The most comprehensive book on DC wiring to date, out of print until recently updated and self-published by the authors. This book presents system schematics, wiring details, and troubleshooting information not found in other publications. Leans slightly toward marine applications. Chapters cover the history of electricity, fundamentals of DC and AC, battery loads, electric sources, wiring practices, system components, tools, and troubleshooting. 100 pages, softcover.

80111   Wiring 12 Volts $19.95

Photovoltaics

By Solar Energy International staff. Used as a textbook in Solar Energy International workshops, *Photovoltaics* delivers the critical information to successfully design, install, and maintain PV systems. It covers solar electricity basics; PV applications and system components; site analysis and mounting; how to size standalone and PV/generator hybrid systems; utility-interactive PV systems; system costs; safety issues; and much more. 317 pages, softcover.

21-0366 Photovoltaics $59

The Solar Electric House

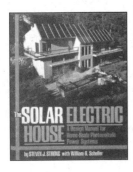

By Steven J. Strong. The author has designed more than 75 PV systems. This fine book covers all aspects of PV, from the history and economics of solar power to the nuts and bolts of panels, balance of systems equipment, system sizing, installation, utility intertie, standalone PV systems, and wiring instruction. A great book for the beginner. 276 pages, paperback.

80800   The Solar Electric House $21.95

Renewable Energy with the Experts

A professionally produced video series that covers basic technology introductions as well as hands-on applications. Each tape is presented by the leading expert in that field and tackles a single subject in detail. Includes classroom explanations of how the technology works, with performance and safety issues to be aware of. Then action moves into the field to review actual working systems and finally assemble a real system on camera as much

as possible. Other tapes in this series cover storage batteries with system sizing and water pumping. You'll find these tapes in appropriate chapters. This is the closest to hands-on you can get without getting dirt under your fingernails. All tapes are standard VHS format.

80360   Set of 3 Renewable Experts Tapes $100

Don Harris, the owner of Harris Turbines, practically invented the remote microturbine industry and has manufactured over 1,000 in the past 15 years. 42 min.

80362   Renewable Experts Tape—Hydro $39

Mick Sagrillo, the former owner of Lake Michigan Wind and Sun, has been in the wind industry for over 16 years with over 1,000 turbine installation/repairs under his belt. 63 min.

80363   Renewable Experts Tape—Wind $39

Johnny Weiss has been designing and installing PV systems for over 15 years. He teaches hands-on classes at Solar Energy International in Colorado. 48 min.

80361   Renewable Experts Tape—PV $39

GARDEN, FARM, FOOD, AND COMMUNITY

Garden and Farm

Ecological Aquaculture
A Sustainable Solution

By Barry A. Costa-Pierce. *Ecological Aquaculture* offers a design framework for successful ecological aquaculture in all but the most extreme climates and regions. The systems described are self-sustaining. While primarily aimed at people with a freshwater resource who want to make use of it in a

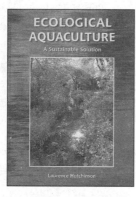

sustainable way, the book is also for those interested in environmental management and aquatic ecosystem enhancement and repair. It serves as a reference work for academic research and a guide for planning authorities and conservation programs. 160 pages, hardcover.

21-0587 Ecological Aquaculture   **$45**

Moon Rhythms in Nature
How Lunar Cycles Affect Living Organisms

By Klaus Peter Endreys, Wolfgang Schad, and Christian von Arnim. This brilliant book distills wide-ranging

observations of lunar influences on Earth. Following an introduction to the astronomy of the Moon rhythms is a study of how the tides and other intricate ocean movements are connected with the life processes of numerous organisms. Richly detailed and clearly written for the general reader, chapters lead up to the spectrum of human rhythms and a description of the whole concept of time. 17 color plates, 37 illustrations. 308 pages, paperback.

21-0542 Moon Rhythms in Nature   **$36**

The Organic Gardener's Handbook of Natural Insect and Disease Control
A Complete Problem-Solving Guide to Keeping Your Garden and Yard Healthy Without Chemicals

By Barbara W. Ellis and Fern Marshall Bradley. End your worries about garden problems with safe, effective solutions from *The Organic Gardener's Handbook of Natural Insect and Disease Control*! Easy-to-use problem-solving encyclopedia covers more than 200 vegetables, fruits, herbs, flowers, trees, and shrubs. Complete directions on how, when, and where to use preventive methods, insect traps and barriers, biocontrols, homemade remedies, botanical insecticides, and more. Newly revised with the latest, safest organic controls. 544 pages, paperback.

21-0588 The Organic Gardener's Handbook   **$19.95**

Food Not Lawns

By H. C. Flores. Forget sterile lawns! Grow a vibrant garden filled with food and beauty instead. That's the advice of one of our favorite new books. It's a virtual encyclopedia of wisdom and how-to advice about ecological design, organic gardening, food and eating, health, self-reliance, community-building, habitat, and biodiversity. Learn how city dwellers and rural denizens alike can use a nine-step permaculture strategy to create a piece of paradise that feeds both body and soul. 344 pages, softcover.

21-0392 Food Not Lawns   **$25**

The Bio-Gardener's Bible

By Lee Fryer. Bio-gardening combines the resources of science and nature to produce yield after yield of safe, abundant, and nutritious crops. Noted agriculturist Lee Fryer tells the home gardener how to build perpetual soil fertility by providing a comfortable home for the organisms that extract vital nutrients from the soil and supply them to the plants; growing fertilizer right in the garden by encouraging the mechanisms that capture nitrogen from the air; analyzing specific soil needs, then mixing and applying a well-balanced diet; borrowing the best, most productive organic farming and agribusiness methods; controlling pests with healthy plants, not with pesticides. The book includes an A-to-Z vegetable growing guide and special tips for getting the most out of patio and mini-gardens. 240 pages, paperback.

21-0541 The Bio-Gardener's Bible   **$9.95**

How to Grow More Vegetables

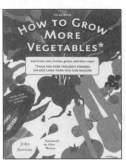

By John Jevons. First published in 1974, this book has become the go-to guide for anyone trying to create or sustain a home garden. Author John Jevons shows how to use Alan Chadwick's biointensive techniques to reduce daily watering and general maintenance—from the beginning stages of setting up the garden beds to creating insect harmony. Includes charts and illustrations. 192 pages, softcover.

21-0345 How to Grow More Vegetables $17.95

Designing and Maintaining Your Edible Landscape Naturally

By Robert Kourik and Rosalind Creasy. First published in 1986, this classic is the authoritative text on edible landscaping, featuring a step-by-step guide to designing a productive environment using vegetables, fruits, flowers, and herbs for a combination of ornamental and culinary purposes. It includes descriptions of plants for all temperate habitats, methods for improving soil, tree pruning styles, and gourmet recipes using low-maintenance plants. 382 pages, paperback.

**21-0534 Designing and Maintaining
Your Edible Landscape $49.95**

Landscaping Earth Ponds

By Tim Matson. The guru of earth ponds explains how to site, design, shape, and plant these beloved fixtures of rural landscapes and make them fit your property and your life. In *Landscaping Earth Ponds*, he shares what he

has learned to make these captivating ponds truly fit into their landscapes and into the lives and lifestyles of their owners. With dozens of color photographs, Matson shows you how to site a pond in right relation to your house, offering surprisingly simple ways to visually link the two. 304 pages, paperback.

21-0589 Landscaping Earth Ponds $30

Rodale's All-New Encyclopedia of Organic Gardening

The Indispensable Resource for Every Gardener

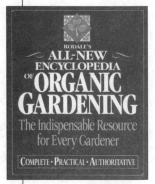

Over 400 entries of the most practical, up-to-date gardening information ever, collected from garden experts and writers nationwide! "Gardens are places to renew yourself in mind and body, to reawaken to the truth and beauty of the natural world, and to feel the life force inside and around you. And the organic way to garden is safer, cheaper, and more satisfying. Organic gardeners have shown that it's possible to have pleasant and productive gardens in every part of this country without using toxic chemicals. They make their home grounds an island of purity." —Robert Rodale. 704 pages, paperback.

**21-0538 Rodale's All-New Encyclopedia
of Organic Gardening $19.95**

Successful Small-Scale Farming: An Organic Approach

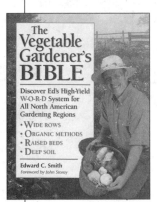

By Karl Schwenke. This inspiring handbook contains everything small-farm owners need to know, from buying land to organic growing methods and selling cash crops. 144 pages, paperback.

**21-0591 Successful Small-Scale
Farming $14.95**

The Vegetable Gardener's Bible

Discover Ed's High-Yield W-O-R-D System for All North American Gardening Regions

By Ed Smith. An experienced vegetable gardener from Vermont, author Ed Smith has put together this amazingly comprehensive and commonsensical manual. This book, filled with step-by-step info and color photos, breaks it all down for you. Ed's system is based on W-O-R-D: Wide rows, Organic methods, Raised beds, Deep soil. With deep, raised beds, vegetable roots have more room to grow and expand. 320 pages, paperback.

**21-0539 The Vegetable Gardener's
Bible $24.95**

Gardening When It Counts
Growing Food in Hard Times

MOTHER EARTH NEWS WISER LIVING SERIES

By Steve Solomon. *Gardening When It Counts* helps readers rediscover traditional low-input gardening methods to produce healthy food. Designed for readers with no experience and applicable to most areas in the English-speaking world except the tropics and hot deserts, this book shows that any family with access to 3-5,000 square feet of garden land can halve their food costs using a growing system requiring just the odd bucketful of household wastewater, perhaps $200 worth of hand tools, and about the same amount spent on supplies—working an average of two hours a day during the growing season. Steve Solomon is a well-known West Coast gardener and author of five previous books, including *Growing Vegetables West of the Cascades*, which has appeared in five editions. 340 pages, paperback.

21-0592 Gardening When It Counts   **$19.95**

Gardening for Life— The Biodynamic Way
A Practical Introduction to a New Art of Gardening, Sowing, Planting, Harvesting

By Maria Thun and Angelika Throll-Keller. A practical introduction to a revolutionary art of gardening, sowing, planting, and harvesting. Biodynamic techniques are based on the recognition that plants are intimately connected with and depend on live, vital soil. These methods also attempt to harmonize with the influences of the Moon phases and cosmos as a whole. You will learn the best times to plant various vegetables; how to deal with pests and diseases; ways to build the soil; and the effects of planets and stars on plant growth. 128 pages, paperback.

21-0540 Gardening for Life   **$30**

Weeds: Control Without Poisons

By Charles Walters. Specifics on a hundred weeds, why they grow, what soil conditions spur them on or stop them, what they say about your soil, and how to control them without the obscene presence of poisons. All cross-referenced by scientific and various common names, and a pictorial glossary. 352 pages, paperback.

21-0593 Weeds: Control Without Poisons   **$25**

Biodynamics

A Biodynamic Farm

By Herbert H. Koepf. The inventory of knowledge that is generally warehoused under the classification of *biodynamic* is rich and timeless, and yet very few farmers have even a nodding acquaintance with the subject. This book performs a rescue operation. A practical, how-to guide to making all the biodynamic preparations, this book will provide what you need for putting these proven techniques to work in your fields. Perhaps the best of the many biodynamic titles currently available. Further explains how to achieve success through CSA-style market garden marketing. 215 pages, paperback.

21-0594 A Biodynamic Farm   **$15**

Grasp the Nettle
Making Biodynamic Farming and Gardening Work

Author Peter Proctor tells the reader how to apply biodynamic methods of farming and gardening to a wide range of conditions in New Zealand and in other countries. Peter gives tips on how to recognize healthy soil and pasture and how to make your own biodynamic preparations, and also gives examples of farms that are successfully using biodynamic methods. This book aims to assist biodynamic farmers and gardeners to observe the processes of life and growth and to understand how these processes are governed by cosmic forces, so they can use this knowledge in applying their practical skills. 176 pages, paperback.

21-0595 Grasp the Nettle   **$20**

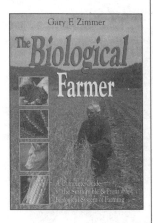

The Biological Farmer

A Complete Guide to the Sustainable & Profitable Biological System of Farming

By Gary F. Zimmer. Biological farming works with nature, feeding soil life, balancing soil minerals, and tilling soils with a purpose. It improves the environment, reduces erosion, reduces disease and insect problems, and alters weed pressure, all by working in harmony with nature. This is the farming consultant's bible. It schools the interested grower in methods of maintaining a balanced, healthy soil that promises greater productivity at lower costs, and it covers some of the pitfalls of conventional farming practices. 352 pages, paperback.

21-0596 Biological Farmer   **$25**

Permaculture

The Permaculture Garden

By Graham Bell and Sarah Bunker. Working entirely in harmony with nature, *The Permaculture Garden* shows you how to turn a bare plot into a beautiful and productive garden. Learn how to plan your garden for easy access and minimum labor; save time and effort digging and weeding; recycle materials to save money; plan crop successions for year-round harvests; save energy and harvest water; and garden without chemicals by building up your soil and planting in beneficial communities. Full of practical ideas, this perennial classic, first published in 1995, is guaranteed to inspire, inform, and entertain. 170 pages, paperback.

21-0545 The Permaculture Garden   **$25**

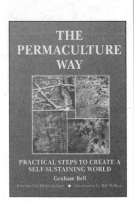

The Permaculture Way

Practical Steps to Create a Self-Sustaining World

By Graham Bell. *The Permaculture Way* shows us how to consciously design a lifestyle that is low in environmental impact and highly productive. It demonstrates how to meet our needs, make the most of resources by minimizing waste and maximizing potential, and still leave the Earth richer than we found it. 239 pages, paperback.

21-0597 The Permaculture Way   **$29.95**

Gaia's Garden

A Guide to Home-Scale Permaculture

By Toby Hemenway and John Tod. Picture your backyard as one incredibly lush garden, filled with edible flowers, bursting with fruit and berries, and carpeted with scented herbs and tangy salad greens. The flowers nurture endangered pollinators. Bright-

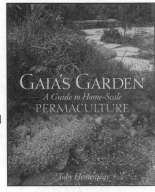

featured songbirds feed on abundant berries and gather twigs for their nests. The plants themselves are grouped in natural communities, where each species plays a role in building soil, deterring pests, storing nutrients, and luring beneficial insects. There is nothing technical, intrusive, secretive, or expensive about this form of gardening. All that is required is some botanical knowledge (which is in this book) and a mindset that defines a backyard paradise as something other than a carpet of grass fed by MiracleGro. 240 pages, paperback.

21-0544 Gaia's Garden   **$24.95**

Permaculture: Principles and Pathways Beyond Sustainability

David Holmgren brings into sharper focus the powerful and still evolving permaculture concept he pioneered with Bill Mollison in the 1970s. It draws together and integrates 25 years of thinking and teaching to reveal a whole new way of understanding and action behind a simple set of design

principles. The 12 design principles are each represented by a positive action statement, an icon, and a traditional proverb or two that captures the essence of each principle. 286 pages, paperback.

21-0546 Permaculture: Principles and Pathways   **$30**

Edible Forest Gardens

Ecological Design and Practice for Temperate-Climate Permaculture

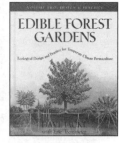

By Dave Jacke and Eric Toensmeier. *Edible Forest Gardens* focuses all its attention on effective design and practice. It offers a unique "pattern language" for forest garden design and provides detailed advice for how to design, prepare the site for, plant, and maintain your forest garden. 1,068 pages, hardcover.

21-0598 Edible Forest Gardens   **$75**

The Basics of Permaculture Design

By Ross Mars and Martin Ducker. *The Basics of Permaculture Design*, first published in Australia in 1996, is an excellent introduction to the principles of permaculture, design processes, and the tools needed for designing sustainable gardens, farms, and larger communities. Packed with useful tips, clear illustrations, and a wealth of experience, it guides you through designs for gardens, urban and rural properties, water harvesting systems, animal systems, permaculture in small spaces like balconies and patios, farms, schools, and ecovillages. This is both a do-it-yourself guide for the enthusiast and a useful reference for permaculture designers. 170 pages, paperback.

21-0543 The Basics of Permaculture Design   **$25**

Permaculture: A Designers' Manual

By Bill Mollison. This is the definitive permaculture design manual in print since 1988. It covers design methodologies and strategies for both urban and rural applications, describing property design and natural farming techniques. 576 pages, hardcover.

21-0382 Permaculture: A Designers' Manual   **$83.95**

Permaculture Plants: A Selection

By Jeff Nugent and Julia Boniface. This is an easy-to-use guide to selecting hundreds of perennial species. It is indispensable for growers and designers working in subtropical and warm temperate/arid climates and also includes some cool-climate-tolerant species. *Permaculture Plants: A Selection* details hundreds of common and unusual edible, medicinal, and useful plants. 160 pages, paperback.

21-0599 Permaculture Plants: A Selection   **$29.95**

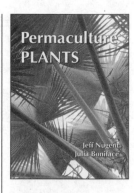

The Earth Care Manual

A Permaculture Handbook for Britain & Other Temperate Climates

By Patrick Whitefield. The long-awaited exploration of permaculture specifically for cooler Northern Hemisphere climates is finally here! Already regarded as the definitive book on the subject, *The Earth Care Manual* is accessible to the curious novice as much as it is essential for the knowledgeable practitioner. 482 pages, hardcover.

21-0600 The Earth Care Manual   **$75**

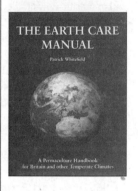

Solar Cuisine and Sustainable Diet

Cooking with Sunshine

The Complete Guide to Solar Cuisine with 150 Easy Sun-Cooked Recipes

By Lorraine Anderson and Rick Palkovic. Solar cooking, a safe, simple cooking method using the Sun's rays as the sole heat source, has been known for centuries and can be done at least during the summer in just about any place where there is Sun. In *Cooking with Sunshine*, Lorraine Anderson and Rick Palkovic provide everything you need to know to cook great Sun-fueled meals. They describe how to build your own inexpensive solar cooker, explain how solar cooking works and its benefits over traditional methods, offer more than 100 tasty recipes emphasizing healthy ingredients, and suggest a month's worth of menu ideas. 224 pages, paperback.

21-0601 Cooking with Sunshine   **$16.95**

The Post-Petroleum Survival Guide and Cookbook

Recipes for Changing Times

By Albert Bates. *The Post-Petroleum Survival Guide and Cookbook* provides useful, practical advice for preparing your family and community to make the transition. This book takes a positive, upbeat, and optimistic view of "the Great Change," promot-

ing the idea that it can be an opportunity to redeem our essential interconnectedness with nature and with each other. The many rifts that have grown up since oil became the world's prime commodity can be mended: between cities and their food sources; the design of the suburban-built environment and its car-oriented sprawl; runaway greenhouse warming; and the clearing of forests and toxification of rivers, oceans, and land. 237 pages, paperback.

21-0602 The Post-Petroleum Survival Guide and Cookbook $19.95

Foodwise

Understanding What We Eat and How It Affects Us: The Story of Human Nutrition

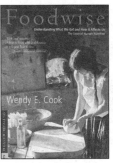

By Wendy E. Cook. In *Foodwise*, Wendy Cook presents a remarkable cornucopia of challenging ideas, advice, and commentary, informed by the seminal work of the scientist Rudolf Steiner. She begins the volume with biographical glimpses relating to her experience of food and how it has influenced her life. She then presents an extraordinary perspective on the journey of human evolution, relating it to changes in consciousness and the consumption of different foods. In the following section, she considers the importance of agricultural methods, the nature of the human being, the significance of grasses and grains, the mystery of human digestion, and the question of vegetarianism. In the next section, she analyzes the "building blocks" of nutrition, looking in some detail at the nutritional (or otherwise) qualities of many foodstuffs, including carbohydrates, minerals, fats and oils, milk and dairy products, herbs and spices, salt and sweeteners, stimulants, legumes, the nightshade family, bread, water, and dietary supplements. 352 pages, paperback.

21-0603 Foodwise $34

The Encyclopedia of Country Living

An Old-Fashioned Recipe Book

By Carla Emery. This definitive classic on food, gardening, and self-sufficient living is a complete resource for living off the land, with over 800 pages of collected wisdom from country maven Carla Emery—how to cultivate a garden, buy land, bake bread, raise farm animals, make sausage, milk a goat, grow herbs, churn butter, catch a pig, make soap, work with bees, and more. Encyclopedia of Country Living is so basic, so thorough, so reliable, it deserves a place in every home—whether in the country, the city, or somewhere in between. 864 pages, paperback.

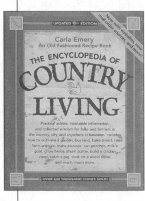

21-0604 Encyclopedia of Country Living$29.95

The Solar Food Dryer

How to Make and Use Your Own Low-Cost, High Performance, Sun-Powered Food Dehydrator

By Eben Fodor. *The Solar Food Dryer* describes how to use solar energy instead of costly electricity to dry your food. With your own solar-powered food dryer, you can quickly and efficiently dry all your extra garden veggies, fruits, and herbs to preserve their goodness all year long—with free sunshine! Applicable to a wide geography—wherever gardens grow—this well-illustrated book includes complete step-by-step plans for building a high-performance, low-cost solar food dryer from readily available materials. 121 pages, paperback.

21-0605 The Solar Food Dryer $14.95

Cooking with the Sun

How to Build and Use Solar Cookers

By Beth Halacy and Dan Halacy. Solar cookers don't pollute, use no wood or other fuel resources, operate for free, and still run when the power is out. A simple and highly effective solar oven can be built for less than $20 in materials. The first half of this book, generously filled with pictures and drawings, presents detailed instructions and plans for building a solar oven that will reach 400°F or a solar hot plate that will reach 600°F. The second half has 100 tested solar recipes. These simple-to-prepare dishes range from everyday Solar Stew and Texas Biscuits, to exotica like Enchilada Casserole. 116 pages, softcover.

21-0273 Cooking with the Sun $9.95

Eating Fossil Fuels

Oil, Food, and the Coming Crisis in Agriculture

By Dale Allen Pfeiffer. The miracle of the Green Revolution was made possible by cheap fossil fuels to supply crops with artificial fertilizer, pesticides, and irrigation. Estimates of the net energy balance of agriculture in the United States show that 10 calories of

hydrocarbon energy are required to produce 1 calorie of food. Such an imbalance cannot continue in a world of diminishing hydrocarbon resources. *Eating Fossil Fuels* examines the interlinked crises of energy and agriculture and highlights some startling findings: The worldwide expansion of agriculture has appropriated fully 40% of the photosynthetic capability of this planet. 127 pages, paperback.

21-0606 Eating Fossil Fuels   **$11.95**

Fed Up! DVD

Genetic Engineering, Industrial Agriculture and Sustainable Alternatives

By Angelor Sacerdote. About 70% of the food we eat contains genetically engineered ingredients, and the biotech industry is spending $50 million a year to convince us that this technology is our only hope. Using hilarious and disturbing archival footage and featuring interviews with farm-

ers, scientists, government officials, and activists, *Fed Up!* presents an entertaining and compelling overview of our current food production system, from the Green Revolution to the Biotech Revolution, and what we can do about it. The DVD includes English closed captions, a chapter menu, and three archival films in their entirety. The archival films are "Chicken of Tomorrow," "Death to Weeds," and "Man and His Culture." 57 mins. 40 sec.

21-0607 Fed Up! DVD   **$15**

Wild Fermentation

The Flavor, Nutrition, and Craft of Live-Culture Foods

By Sandor Ellix Katz and Sally Fallon. Sandor Ellix Katz has experimented with wild fermentation, and his book explains to others how to take advantage of natural fermentation processes to produce bread, yogurt, cheese, beer, wine, miso, sauerkraut, kimchi, and other fermented foods. A gold mine for science-fair projects, Katz's work presents properly supervised young people ample opportunity to explore both the science and the art of fermented foods (alcoholic beverages excepted). 200 pages, paperback.

21-0608 Wild Fermentation   **$25**

The Sustainable Kitchen

Passionate Cooking Inspired by Farms, Forests, and Oceans

By Stu Stein, Judith H. Dern, and Mary Hinds. *The Sustainable Kitchen* brings home the thrill of tasting fine cuisine made from the best seasonal ingredients grown locally. Designed for people who want to make food choices that promote the economic, environmental, and social health of their communities, this book gives seasonal cuisine new flair, using recipes adapted for exciting home cooking. 231 pages, paperback.

21-0630 The Sustainable Kitchen   **$22.95**

The Ethical Gourmet

By Jay Weinstein. If you are concerned about the environment but unsure how to make a difference, here is a handbook for finding and cooking environmentally friendly and ethically produced foods. Chef and environmentalist Jay Weinstein has written the bible for those who care about both the well-being of the world and flavorful food. He informs us about when organics really matter, where to source humanely raised meats and other ethically produced foods, and how to make choices with a clean conscience when dining out. He also explores subjects ranging from genetically modified foods to being savvy about farmed fish, and why to avoid disposable wooden chopsticks and bottled water. 368 pages, paperback.

21-0609 The Ethical Gourmet   **$18.95**

Food and Food Security

Recipes from America's Small Farms

Edited by Joanne Hayes and Lori Stein. Nothing beats a farmhouse meal, and with this incredible cookbook in your kitchen, you'll enjoy them every day. It's overflowing with the favorite recipes of farmers, CSA members, and renowned chefs. The chapters, organized by food type, discuss history, characteristics, and nutrition while providing cooking techniques, useful sidebars, farmer profiles, and (of course!) dozens of delicious original dishes you have to eat to believe, including vegan, meat, and dairy recipes. 304 pages, softcover.

**21-0399 Recipes from America's
Small Farms** $17.95

The Revolution Will Not Be Microwaved
Inside America's Underground Food Movements

By Sandor Ellix Katz. The struggle between constant convenience and rich, vibrant food is one that plagues many of us. In his latest book, Sandor Ellix Katz explores this topic far beyond the organic label and the lack of GM (genetically modified) food labeling and challenges the way we think about food. At the same time, he offers a rich history and analysis of profit-driven policy and a delightful recipe book of individual actions from seed-saving and sprouting to gardening and foraging. An inspired read that profiles the heroes of the growing renaissance of vibrant, healthy, living food. 378 pages, softcover.

**21-0393 The Revolution Will Not Be
Microwaved** $20

Animal, Vegetable, Miracle
A Year of Food Life

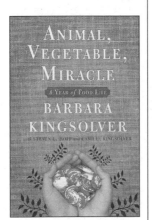

Overflowing with the same illuminating insight and graceful beauty that mark her award-winning fiction, acclaimed author Barbara Kingsolver's first foray into non-fiction details her family's year-long quest to eat only locally-produced foods. It's an endlessly fascinating, frequently funny, rumination on everything from the industrial food chain to turkey mating. Its fact- and recipe-filled pages are part memoir, part exposé, and part passionate call to restore the American kitchen and the farms that feed it to sustainable health. 384 pages, hardcover.

21-0415 Animal, Vegetable, Miracle $26.99

The Omnivore's Dilemma

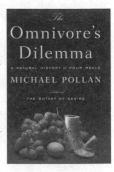

Follow award-winning writer Michael Pollan on this fascinating journey through the American food chain. From the pitfalls of industrial agriculture and the promise of organic farming to strange but true supermarket tales and his experiences foraging in the wild, Pollan looks at what we eat, where it comes from, and the good, the bad, and the unappetizing hiding in between. It's a thoughtful and lucid, yet hugely entertaining volume that leaves no plate unturned. 288 pages, hardcover.

21-0401 The Omnivore's Dilemma $26.95

The Way We Eat
Why Our Food Choices Matter

By Peter Singer and Jim Mason. Eating is a lot more complicated than it used to be. Is it better to buy fair trade? Go organic? Eat local? Avoid animal products? Meticulously researched, this landmark book resolves today's dietary dilemmas. Dissecting the eating habits of three typical American families, renowned ethicist Peter Singer and ag expert Jim Mason follow our food from farm to table and explore all the moral considerations involved—from animal welfare to environmental stewardship. 328 pages, hardcover.

21-0406 The Way We Eat $29.95

Plenty: One Man, One Woman, and a Raucous Year of Eating Locally

It began with a spontaneous dinner created completely from foraged edibles. It ended with a question: Is it possible live only on foods produced within 100 miles of home? With that, a year-long experiment to try began, one that asked vital questions about globalization, agribusiness, the oil economy, ecological collapse, and community. This is the story of that thought-provoking journey. From initial struggle to deep contentment, follow as the authors learn about themselves, our food, and our world, and start a movement that's changing how we eat and live. 272 pages, hardcover. USA.

21-0414 Plenty $24

PEAK OIL, POLITICS, AND RELOCALIZATION

Peak Oil and Climate Change

Preparedness Now!

By Aton Edwards, Executive Director of the NYC-based nonprofit organization International Preparedness Network (IPN). IPN has worked with the Red Cross, Centers for Disease Control, New York City Police Department, and other organizations to train thousands domestically and overseas to prevent and respond to emergencies and disasters. *Preparedness Now!* provides information and techniques that can help mitigate the destructive effects of disasters, whatever the cause. With illustrations, photographs, and step-by-step instructions, this manual delivers practical advice. 336 pages, paperback.

21-0526 Preparedness Now!   **$14.95**

The Weather Makers

How Man Is Changing the Climate and What It Means for Life on Earth

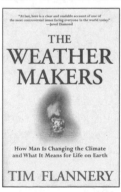

By Tim Flannery. Sometime this century, the day will arrive when the human influence on the climate will overwhelm all other natural factors. Over the past decade, the world has seen the most powerful El Niño ever recorded, the most devastating hurricane in 200 years, the hottest European summer on record, and one of the worst storm seasons ever experienced in Florida. With one out of every five living things on this planet committed to extinction by the levels of greenhouse gases that will accumulate in the next few decades, we are reaching a global climatic tipping point. *The Weather Makers* is both an urgent warning and a call to arms, outlining the history of climate change, how it will unfold over the next century, and what we can do to prevent a cataclysmic future. 384 pages, hardcover.

21-0579 The Weather Makers   **$24**

An Inconvenient Truth

A relatively easy and entertaining read, *An Inconvenient Truth* is a follow-up to Al Gore's best-selling *Earth in Balance: Ecology and the Human Spirit* and is inspired by a series of multimedia presentations on global warming that Gore created and delivers to groups around the world. Highly recommended for your personal collection and as a gift that gives to the environment by educating and entertaining your friends and family. 328 pages, with 330 photographs, illustrations, and charts; softcover. DVD run time, 100 min.

21-0389 An Inconvenient Truth (book)   **$21.95**

90-0169 An Inconvenient Truth (DVD)   **$29.95**

The Last Hours of Ancient Sunlight

Revised and Updated: The Fate of the World and What We Can Do Before It's Too Late

By Thom Hartmann, Joseph Chilton Pearce, and Neale Donald Walsch. *The Last Hours of Ancient Sunlight* details what is happening to our planet, the reasons for our culture's blind behavior, and how we can fix the problem. Thom Hartmann's comprehensive book, originally published in 1998, has become one of the fundamental handbooks of the environmental activist movement. Now, with fresh, updated material and a focus on political activism and its effect on corporate behavior, *The Last Hours of Ancient Sunlight* helps us understand—and heal—our relationship to the world, to each other, and to our natural resources. 400 pages, paperback.

21-0501 The Last Hours of Ancient Sunlight   **$14.95**

The Party's Over

Oil, War, and the Fate of Industrial Societies

By Richard Heinberg. Without oil, what would you do? How would you travel? How would you eat? The world is about to change dramatically and forever as the global production of oil reaches its peak. Thereafter, even with a switch to alternative energy sources, industrial societies will have less energy available to do all the things essential to their survival. Author Richard Heinberg deals head-on with this imminent decline of cheap oil in *The Party's Over*, a riveting wake-up call that does for oil depletion what Rachel Carson's *Silent Spring* did for the issue of chemical pollution—raise to consciousness a previously ignored global problem of immense proportions. 288 pages, softcover.

92-0267 The Party's Over   **$17.95**

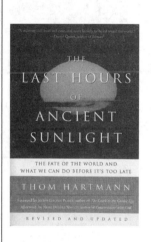

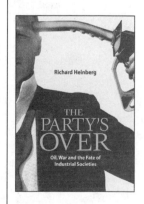

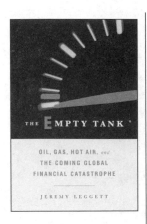

The Empty Tank

Oil, Gas, Hot Air, and the Coming Global Financial Catastrophe

Geologist and ex-Greenpeace official Jeremy Leggett, author of *The Carbon War*, argues incisively that oil production has peaked, with dwindling supplies and soaring prices in the offing. News? No. The more pertinent story told here is his suggestion that, worse than the possibility that the world cannot cope, is the threat that it will cope all too well by burning more coal and coal-derived synthetic fuels, thus exacerbating atmospheric carbon dioxide buildup and global warming. This will lead the planet down "the road to horror," illustrated in a sketchy montage of the usual environmental doomsday scenarios: rising sea levels, extreme weather, famine, and war. 256 pages, hardcover.

92-0269 The Empty Tank   **$24.95**

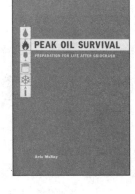

Peak Oil Survival

Preparation for Life After Gridcrash

By Aric McBay. Oil and energy are not limitless resources, and someday the supply will be depleted. *Peak Oil Survival* shows readers how to plan for the future: how to survive and thrive when the food, transport, and energy industries sputter out. Author Aric McBay gives an essential crash course complete with clear, simple instructions and easy-to-read diagrams. *Peak Oil Survival* will explain how people can protect their families and strengthen their communities in the event of a crisis—and live comfortably off the grid. 128 pages, softcover.

21-0580 Peak Oil Survival   **$12.95**

The Oil Depletion Protocol

A Plan to Avert Oil Wars, Terrorism, and Economic Collapse

By Richard Heinberg. *The Oil Depletion Protocol* describes a unique accord whereby nations would voluntarily reduce their oil production and oil imports according to a consistent, sensible formula. This would enable energy transition to be planned and supported over the long term, providing a context of stable energy prices and peaceful cooperation. The protocol will be presented at international gatherings, initiating the process of country-by-country negotiation and adoption, and mobilizing public support. 195 pages, paperback.

21-0581 The Oil Depletion Protocol   **$16.95**

Heat: How to Stop the Planet From Burning

We don't think it's exaggerating to call this the most important book of our time. With fierce intelligence and optimistic wit, activist/journalist George Monbiot looks at climate crisis science and comes to one conclusion: to avoid catastrophe we must cut global CO2 emissions 90% by 2030. Sound impossible? Guess again. Monbiot shows us how we can do it -- without sending humanity back to the Stone Age. Here's what will work, what won't, and how it all fits together to create a solution whose time has come. Read it and act. 304 pages, softcover.

21-0420 Heat: How to Stop the Planet From Burning   **$22**

Relocalization, Growth, and Technology

Ecovillages

A Practical Guide to Sustainable Communities

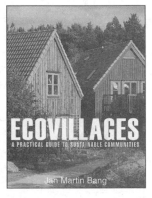

By Jan Martin Bang. *Ecovillages* explores the new departures in personal, social, and ecological living represented by this phenomenon. This book explores the background and history of the ecovillages movement and provides a comprehensive manual for planning, establishing, and maintaining a sustainable community using a permaculture approach. Ecovillages will appeal to a broad range of people interested in living in community—or planning to do so. Jan Martin Bang has spent a lifetime working in alternative communities, ranging from the kibbutz in Israel to his present home in the camphill community of Solborg in Norway. He leads training and development for new ecovillage projects around the world. 284 pages, paperback.

21-0610 Ecovillages   **$24.95**

Better Off: Flipping the Switch on Technology

A decade ago, author Eric Brende was pursuing a graduate degree at MIT, studying technology's influence on society, and he reached conclusions that disturbed both him and his faculty mentors. Brende and his new wife moved to a religious, "Mennonite-type" community (that in many respects makes the Amish seem worldly) where he hoped to pare his environment down to "a baseline of minimal machinery" that could sustain human comfort while allowing him to stay off the power grid. The pervasive back-to-basics sentiment will surprise few, but Brende's nostalgia for a simpler way of life is far from rabid. His gentle case for simple living will easily resonate with the converted and may inspire skeptics to grapple more intimately with the issue. 272 pages, paperback.

21-0505 Better Off: Flipping the
Switch on Technology $13.95

The End of Suburbia

Oil Depletion and the Collapse of The American Dream

Written by Dan Chiras and David Wann. As we enter the 21st century and the global demand for fossil fuels begins to outstrip supply, serious questions are beginning to emerge about the sustainability of the American way of life. With brutal honesty and a touch of irony, *The End of Suburbia* explores our prospects as the planet approaches a critical era—world oil peak and the inevitable decline of fossil fuels. A must-see DVD for anyone—suburbanite or not—interested in the issue of world oil peak. DVD, 78 min.

90-0047 The End of Suburbia $27.75

Collapse: How Societies Choose to Fail or Succeed

In his Pulitzer Prize-winning bestseller *Guns, Germs, and Steel*, geographer Jared Diamond laid out a grand view of the organic roots of human civilizations. That vision takes on apocalyptic overtones in this fascinating comparative study of societies that have undermined their own ecological foundations. Diamond examines storied examples of human economic and social collapse, and even extinction. He explores patterns of population growth, overfarming, overgrazing, and overhunting, often abetted by drought, cold, rigid social mores, and warfare, that lead inexorably to vicious circles of deforestation, erosion, and starvation. Extending his treatment to contemporary environmental trouble spots, from Montana to Australia, he finds today's global, technologically advanced civilization very far from solving the problems that plagued primitive, isolated communities in the remote past. Diamond is a brilliant expositor of everything from anthropology to zoology, providing a lucid background of scientific lore to support a stimulating, incisive historical account of many declines and falls. Readers will find his book an enthralling, and disturbing, reminder of the indissoluble links that bind humans to nature. 592 pages, paperback.

21-0504 Collapse: How Societies Choose
to Fail or Succeed $17

Ethical Markets: Growing the Green Economy

By Hazel Henderson and Simran Sethi. With insight, clarity, warmth, and enthusiasm, Hazel Henderson announces the mature presence of the green economy. Mainstream media and big-business interests have sidelined its emergence and evolution to preserve the status quo. Throughout *Ethical Markets*, Henderson weaves statistics and analysis with profiles of entrepreneurs, environmentalists, scientists, and professionals. Based on interviews conducted on her longstanding public television series, these profiles celebrate those who have led the highly successful growth of green businesses around the world. *Ethical Markets* is the ultimate sourcebook on today's thriving green economy. 300 pages, paperback.

21-0611 Ethical Markets: Growing the
Green Economy $30

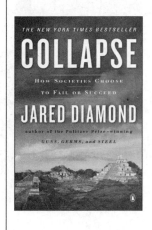

Rural Renaissance

Renewing the Quest for the Good Life

By John D. Ivanko, Caroline Bates, and Bill McKibben.

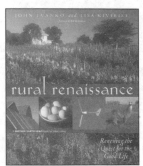

Rural Renaissance captures the American dream of country living for contemporary times. Journey with the authors and experience their lessons, laughter, and love for the land as they trade the urban concrete maze for a 5-acre organic farm and bed and breakfast in southwestern Wisconsin. Rural living today is a lot more than farming. It's about a creative, nature-based, and more self-sufficient lifestyle that combines a love of squash, solar energy, skinny-dipping, and serendipity. *Rural Renaissance* will appeal to a wide range of Cultural Creatives, free agents, conservation entrepreneurs, and both arm-chair and real-life homesteaders regardless of where they live. Lisa Kivirist and John Ivanko are innkeepers, organic growers, and copartners in a marketing consulting company, and have previously published books. 255 pages, paperback.

21-0612 Rural Renaissance **$22.95**

Fostering Sustainable Behavior

An Introduction to Community-Based Social Marketing

EDUCATION FOR SUSTAINABILITY SERIES

By Doug McKenzie-Mohr and William Smith. Our consumption patterns are threatening to outstrip Earth's ability to support humanity and other species. A sustainable future will require sweeping changes in public behavior. While conventional marketing can help create public awareness, social marketing identifies and overcomes barriers to long-lasting behavior change. This groundbreaking book is the primary resource for the emerging new field of community-based social

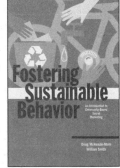

marketing and is an invaluable guide for anyone involved in designing public education programs with the goal of promoting sustainable behavior, from recycling and energy efficiency to alternative transportation. 176 pages, paperback.

21-0613 Fostering Sustainable Behavior$14.95

Deep Economy

By Bill McKibben. McKibben argues that the world doesn't have enough natural resources to sustain endless economic expansion. Drawing the phrase "deep economy" from the expression "deep ecology," a term environmentalists use to signify new ways of thinking about the environment, he suggests we need to explore new economic ideas. He gives examples of promising ventures of this type, such as a community-supported farm in Vermont and a community biosphere reserve, or large national park-like area, in Himalayan India. McKibben's proposals for new, less growth-centered ways of thinking about economics are intriguing and offer hope that change is possible. 272 pages, hardcover.

21-0413 Deep Economy **$25**

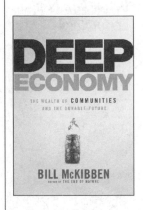

Limits to Growth: The 30-Year Update

By Donella Meadows, Jorgen Randers, and Dennis Meadows. Updated for the second time since 1992, this book, by a trio of professors and systems analysts, offers a dark view of the natural resources available for the world's population. Using extensive computer models based on population, food production, pollution, and other data, the authors demonstrate why the world is in a potentially dangerous "overshoot" situation. The consequences may be catastrophic: "We . . . believe that if a profound correction is not made soon, a crash of some sort is certain. And it will occur within the lifetimes of many who are alive today." The book discusses population and industrial growth, the limits on available resources, pollution, technology, and, importantly, ways to avoid overshoot. The authors do an excellent job of summarizing their extensive research with clear writing and helpful charts illustrating trends in food consumption, population increases, grain production, etc., in a serious tome likely to appeal to environmentalists, government employees, and public policy experts. 368 pages, paperback.

21-0503 Limits to Growth **$22.50**

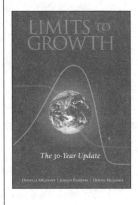

It's a Sprawl World After All

The Human Cost of Unplanned Growth—and Visions of a Better Future

By Douglas E. Morris. *It's a Sprawl World After All* is the first book to link America's increase in violence and the corresponding breakdown in society with the post-World War II development of suburban sprawl. Without small towns to bring people together, the unplanned growth of sprawl has left Americans isolated, alienated, and afraid of the strangers that surround them. Suburbia has substituted cars for conversation, malls for main streets, and the artificial community of television for authentic social interaction. 245 pages, paperback.

21-0614 It's a Sprawl World After All **$17.95**

Toward Sustainable Communities
Resources for Citizens and Their Governments

By Mark Roseland and Stacy Mitchell. Local governments are increasingly caught between rising expectations that development initiatives be sustainable and the fact that more and more services are being downloaded to the municipal level. The 3rd edition of this classic text offers practical suggestions and innovative solutions to a range of community problems—including energy efficiency, transportation, land use, housing, waste reduction, recycling, air quality, and governance. In clear language, with updated tools, initiatives and resources, and a new preface and foreword, this sustainable practices resource is for both citizens and governments. 239 pages, paperback.

**21-0615 Toward Sustainable
Communities $22.95**

Future Hype

By Bob Seidensticker. *Future Hype* debunks the popular myth of technology growth and teaches that technology does not have to control you, nor do you need to feel overwhelmed by the ubiquitous white wires streaming from the ears of nearly everybody. The key lies in understanding how technology works and allowing yourself to better evaluate, anticipate, and control it. New technology comes out every day, and yet, as Bob Seidensticker shows us, the popular view of technology is wrong and the future won't be so shocking. 240 pages, softcover.

21-0388 Future Hype $15.95

The Small-Mart Revolution
How Local Businesses Are Beating the Global Competition

Author Michael Shuman (*Going Local*) once again explores the issue of relocalization and the benefits—from the nostalgic return of small-town America's "main street," through the economic stability of our nation, to the very environmental stability of our planet—of this growing movement. In *Small-Mart*, his analysis turns to the realm of Mom & Pop in a world of big-box giants, as he presents a thorough analysis and how-to guide for consumers, investors, entrepreneurs, policymakers, and more. Refreshing in tone, this is an inspired guide to the revolution that is the resurgence of locally owned business and is not another big-box-bashing tome. A must-read for anyone interested in prioritizing real actions toward reclaiming a unique and vibrant local landscape. 200 pages, hardcover.

21-0400 The Small-Mart Revolution $24

EcoVillage at Ithaca
Pioneering a Sustainable Culture

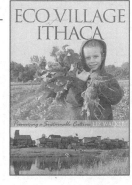

By Liz Walker. In a world filled with stories of environmental devastation and social dysfunction, *EcoVillage at Ithaca* is a refreshing and hopeful look at a modern-day village that is taking an integrated approach to addressing these problems. This book tells the story of life at EcoVillage at Ithaca, an internationally recognized example of sustainable development. It transports the reader into the midst of a vibrant community that includes co-housing neighborhoods, small-scale organic farming, land preservation, green building, energy alternatives, and hands-on education. 236 pages, paperback.

21-0616 EcoVillage at Ithaca $17.95

Gaviotas: A Village to Reinvent the World

By Alan Weisman. Twenty-five years ago, the village of Gaviotas was established in one of the most brutal environments on Earth, the eastern savannas of war-torn, drug-ravaged Colombia. Read the gripping, heroic story of how Gaviotas came to sustain itself agriculturally, economically, and artistically, eventually becoming the closest thing to a model human habitat since Eden. These incredibly resourceful villagers invented superefficient pumps to tap previously inaccessible sources of water, solar kettles to sterilize drinking water, wind turbines to harness the mildest breezes—they even regrew a rain forest! To call this book inspiring is akin to calling the Amazon a winter creek. You won't be able to put it down. 240 pages, softcover.

82306   Gaviotas $14.95

Edens Lost & Found

How Ordinary Citizens Are Restoring Our Great American Cities

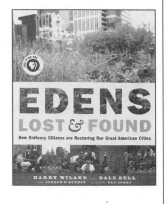

By Harry Wiland and Dale Bell. Part celebration and part inspiration, *Edens Lost & Found*, the companion book to the PBS series, chronicles the forward-looking transformation of America's urban landscapes and communities. With *Edens Lost & Found*, award-winning filmmakers Harry Wiland and Dale Bell herald an exciting sea change in the relationship between ordinary citizens, environmental groups, and government. From across America, they gather evidence of a new spirit of cooperation among neighbors, planners, architects and builders, city officials, and government agencies. 285 pages, hardcover.

21-0617 Edens Lost & Found $40

Sustainable Transportation

How to Live Well Without Owning a Car

Save Money, Breathe Easier, and Get More Mileage Out of Life

By Chris Balish. Perhaps the first practical, accessible, and sensible guidebook to living in North America without an automobile, this book shows it's possible for nearly everyone to discard their cars and live a better, richer, car-free life. Chris Balish writes from personal experience, having lived (accidentally at first) for four years with hardly a moment behind the wheel (amounting to a savings of nearly $40,000!!), and has the persuasive authority to convince even the most dedicated drivers that cutting back on time behind the wheel is at least a possibility. 216 pages, softcover.

21-0397 How to Live Well Without
**   Owning a Car $12.95**

Plug-in Hybrids

The Cars That Will Recharge America

Written by Sherry Boschert, co-founder and president of the San Francisco Electric Vehicle Association, *Plug-in Hybrids* is the book over which our politically polarized America can come together. It is widely recognized that something's got to change if we're to succeed in our attempts to avoid the devastation that comes with dependence on oil. Plug-in hybrids are a part of the solution, and the technology to put them into mass production is already existent. Demand for plug-ins is nearly at a fever pitch already, and this book examines how you can get one, along with just what they are and why you want one. 231 pages, softcover.

21-0394 Plug-in Hybrids $16.95

Alcohol Can Be a Gas

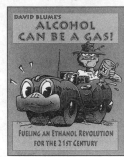

Original Foreword by R. Buckminster Fuller. Alcohol, the first automotive fuel, is poised to replace gasoline and diesel in the rapidly approaching peak oil world. It's 98% pollution free and reverses global warming, and you can make it for less than a buck a gallon. Originally written in 1982 to accompany the 10-part PBS series *Alcohol as Fuel*, this book has been completely revised and expanded for 2007. Explores the history of alcohol as fuel, conversion of gasoline and diesel vehicles, and more. Shows how to organize driver-owned alcohol stations and harvest federal tax credits. 500 photos, illustrations, and charts; 576 pages, softcover.

21-0402 Alcohol Can Be a Gas $47

21-0549 Alcohol Can Be a Gas DVD $47

Cutting Your Car Use

Save Money, Be Healthy, Be Green!

By Randall Howard Ghent, Anna Semlyen, and Axel Scheffler. Imagine living where every child can walk or cycle to school in safety, where local businesses thrive, and where a car is not essential for enjoying life. Picture how different your neighborhood would be with fewer moving cars. Think how much better your life might be if you were in a car less often. Cutting down on the time you spend in cars will save you money, benefit your health, and improve everyone's quality of life. *Cutting Your Car Use* is a practical guide for people who want to reduce their car dependency by making simple changes to their travel habits or by sharing or giving up their car. It helps readers to understand the real cost of car use, evaluate alternatives to personal car ownership, and plan trips using a mix of transportation modes. This book also contains a helpful directory of further resources. Randall Ghent is car-free by choice. A past editor of *Auto-Free Times*, he co-directs World Carfree Network's international coordination center in Prague, has edited two other books for Car Busters Press, and is a co-editor of *Car Busters* magazine. Anna Semlyen is a traffic reduction consultant in the U.K. who rides a folding bicycle with basket or trailer, ride shares, takes taxis, and sometimes hires a car. 118 pages, paperback.

21-0618 Cutting Your Car Use $9.95

Biodiesel Power
The Passion, the People, and the Politics of the Next Renewable Fuel

By Lyle Estill. *Biodiesel Power* is the history of biodiesel in the making. It will appeal to a wide audience concerned about the end of oil. It lightly touches on the technical aspects of the fuel, its qualities, and its specifications, and focuses on the people and stories of the biodiesel movement. It explores the tensions between grass-roots activists and their altruistic co-ops, the profit-minded commercial producers and the voices of agribusiness, and the current administration—or "the coalition of the drilling." 272 pages, paperback.

21-0533 Biodiesel Power **$16.95**

Biodiesel Basics and Beyond

By William Kemp. Here's everything you need to know about making and using biodiesel for home heating, transportation, farming, and other uses. Separating truth from myth and misinformation, it's a one-stop volume packed with history, background, and detailed instructions on how to build and operate a 40-gallon-per-day, high-quality biodiesel "refinery" safely and sustainably. From detailed lists of gear and resources to recycling and waste treatment advice, this is the one to read. 588 pages, softcover.

21-0398 Biodiesel Basics and Beyond **$29.95**

Biodiesel: Growing a New Energy Economy

Greg Pahl's essential new book explores the history and technology of biodiesel, its current use around the world, and its exciting potential in the United States and beyond. A crop-derived liquid fuel, biodiesel can be made from a wide range of renewable, locally grown plant sources—even from recycled cooking oils or animal fats. While it is not the answer to all our energy problems, it is an important step in the long-overdue process of weaning ourselves from fossil fuels. 282 pages, paperback.

21-0372 Biodiesel: Growing a New Energy Economy **$18**

Fuel from Water
Energy Independence with Hydrogen

By Michael Peavey. An in-depth and practical book on hydrogen fuel technology that details specific ways to generate, store, and safely use hydrogen. Hundreds of diagrams and illustrations make it easy to understand, very informative, and practical. If you're interested in using hydrogen in any way, you will find this book indispensable. 244 pages, softcover.

80210 Fuel from Water **$25**

From the Fryer to the Fuel Tank
The Complete Guide to Using Vegetable Oil as an Alternative Fuel

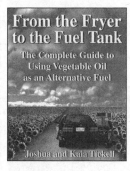

By Joshua Tickell, Kaia Tickell, and Kaia Roman. New updated and expanded version by Joshua and Kaia Tickell, who, with sponsorship from Real Goods, drove their unmodified Veggie Van more than 10,000 miles around the United States during the summer of 1997, while stopping at fast-food restaurants for fuel fill-ups of used deep-fat fryer oil. Vegetable oil provided 50% of their fuel. They towed a small portable fuel-processing lab in a trailer behind their Winnebago. This inexpensive and environmentally friendly fuel source can be modified easily to burn in diesel engines by following the simple instructions. A complete introduction to diesel engines and potential fuel sources is included. Over 130 photographs, diagrams, charts, and tables. 162 pages, softcover.

80961 From the Fryer to the Fuel Tank **$24.95**

Biodiesel America
How to Achieve Energy Security, Free America from Middle-East Oil Dependence, and Make Money Growing Fuel

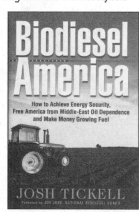

By Joshua Tickell. Our tale begins more than 100 years ago with Rudolf Diesel's invention of an engine running on vegetable oil, and carries us through to modern times where energy security is a major concern and many of us are trying to decrease our dependence on foreign oil. Biodiesel is not only the way of the future, it is well grounded in the past. Joshua Tickell crafts a book that weaves the history of America's relationship with oil together with a plan for the future, and steps we can take to change that relationship. Tickell has created a book few can afford to not read. 356 pages, hardcover.

21-0383 Biodiesel America **$29.95**

CONSERVATION AND ECOLOGY

Green Volunteers

The World Guide to Voluntary Work in Nature Conservation, 6th Ed.

By Fabio Ausenda. *Green Volunteers* lists over 200 projects worldwide for those who want to experience active conservation work as a volunteer. Projects are in a variety of habitats and countries, lasting from one week to one year or more. Projects involve volunteer work in wildlife rehabilitation centers, national parks, and protected areas, and general conservation work with a variety of animal species. The guide has three indexes set up by geographical area, animal species, and project cost, all to help prospective volunteers choose their favorite conservation project. Almost all the projects are open to anyone without previous conservation experience, and most projects do not have language requirements. 256 pages, paperback.

21-0619 Green Volunteers **$14.95**

The Consumer's Guide to Effective Environmental Choices

Practical Advice from the Union of Concerned Scientists

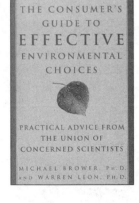

By Michael Brower and Warren Leon. Paper or plastic? Cloth or disposable? Regular or organic? Every day, environmentally conscious consumers are faced with the overwhelming catch-22 of a capitalist society—reconciling the harm we do by consuming, while still providing ourselves and our families with the goods and services we need. It's enough to make a city dweller crazy. Fret no more! The Union of Concerned Scientists has put together a well-researched and eminently practical guide to the decisions that matter. The authors hope that the book will help you set priorities, stop worrying about insignificant things, and understand the real environmental impacts of household decisions. For instance, you may be surprised to learn that buying and eating meat and poultry is much more harmful to the environment than the packaging the meat is wrapped in, even if it's Styrofoam. If you're confused and overwhelmed by all the environmental decision-making in the modern world, you'll find new inspiration in this book. 304 pages, paperback.

21-0620 The Consumer's Guide to Effective Environmental Choices $15

Heal the Ocean

Solutions for Saving Our Seas

By Rodney M. Fujita and Peter Benchley. *Heal the Ocean* provides a refreshing change in the literature by emphasizing success stories in the struggle to save the seas. Rodney M. Fujita—a marine ecologist dedicated to protecting and restoring ocean ecosystems—first describes the nature of ocean environments and then discusses current and emerging threats, including pollution, overfishing, poor land use, deep sea mining, and the search for new energy sources. Upbeat and inspiring, *Heal the Ocean* will appeal across the board. 240 pages, paperback.

21-0621 Heal the Ocean **$16.95**

50 Ways to Save the Ocean

(INNER OCEAN ACTION GUIDE)

By David Helvarg, Philippe Cousteau, and Jim Toomey. The oceans, and the challenges they face, are so vast that it's easy to feel powerless to protect them. *50 Ways to Save the Ocean*, written by veteran environmental journalist David Helvarg, focuses on practical, easily implemented actions everyone can take to protect and conserve this vital resource. Well-researched, personal, and sometimes whimsical, the book addresses daily choices that affect the ocean's health: what fish should and should not be eaten; how and where to vacation; storm drains and driveway run-off; protecting local water tables; proper diving, surfing, and tide pool etiquette; and supporting local marine education. Helvarg also looks at what can be done to stir the waters of seemingly daunting issues such as minimizing toxic pollutant runoff, protecting wetlands and sanctuaries, keeping oil rigs off shore, saving reef environments, and replenishing fish reserves. 208 pages, paperback.

21-0622 50 Ways to Save the Ocean **$12.95**

The Difference a Day Makes

365 Ways to Change Your World in Just 24 Hours

By Karen M. Jones. This timely compilation features 365 simple actions people can take to change the world, one day—or even five minutes—at a time. Each suggested action, in 16 "helping" categories, can be started and finished in a day or less, and none requires a cash donation. Readers may choose to accomplish a different altruistic step each day of the year, activate the same tool every day, or take actions that address a personally favored issue, such as animal welfare or the pursuit of peace. 320 pages, paperback.

21-0623 The Difference a Day Makes $12.95

The New Atlas of Planet Management

By Jennifer Kent and Norman Myers. *The New Atlas of Planet Management* was regarded as the most groundbreaking survey of the state of our planet when it was first published in 1984. After over 20 years in print, it has become the bible of the environmental movement and the definitive guide to a planet in critical transition. Regularly featured among the top 10 books on the environment, the *Atlas* has been read by millions of people and translated into more than a dozen languages. This enlarged edition brings the classic reference up to date. Thoroughly revised with the latest figures and analysis, fresh full-color and easy-to-read graphics, an expanded format, and a wealth of current environmental and political topics that have arisen during the previous two decades, *The New Atlas of Planet Management* will equip a further generation of readers with information to face the challenges of the new millennium. 304 pages, paperback.

**21-0624 The New Atlas of Planet
        Management $39.95**

Frozen Earth

The Once and Future Story of Ice Ages

By Douglas Macdougall. In this engrossing and accessible book, Doug Macdougall explores the causes and effects of ice ages that have gripped our planet throughout its history, from the earliest known glaciation—nearly 3 billion years ago—to the present. Following the development of scientific ideas about these dramatic events, Macdougall traces the lives of many of the brilliant and intriguing characters who have contributed to the evolving understanding of how ice ages come about. As it explains how the great Pleistocene Ice Age has shaped Earth's landscape and influenced the course of human evolution, *Frozen Earth* also provides a fascinating look at how science is done, how the excitement of discovery drives scientists to explore and investigate, and how timing and chance play a part in the acceptance of new scientific ideas.
267 pages, paperback.

21-0625 Frozen Earth $15.95

Genetic Roulette

The Documented Health Risks of Genetically Engineered Foods

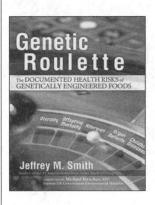

By Jeffrey M. Smith. This book, prepared in collaboration with a team of international scientists, is for anyone wanting to understand GM technology, learn how to protect themselves, or share their concerns with others. As the world's most complete reference on the health risks of GM foods, *Genetic Roulette* is also ideal for schools and libraries. 312 pages, hardcover.

21-0626 Genetic Roulette $23.95

Worldchanging

User's Guide for the 21st Century

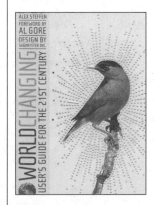

By Alex Steffen, Al Gore, and Bruce Sterling. *Worldchanging* is poised to be the "whole earth catalog" for this millennium. Written by leading new thinkers who believe that the means for building a better future lie all around us, *Worldchanging* is packed with the information, resources, reviews, and ideas that give readers the tools they need to make a difference. Brought together by Alex Steffen, co-founder of the popular and award-winning Web site Worldchanging.com, this team of top-notch writers includes Architecture for Humanity founder Cameron Sinclair, Geekcorps founder Ethan Zuckerman, sustainable food expert Anna Lapp, and many others. 608 pages, hardcover.

21-0627 Worldchanging $37.50

It's Easy Being Green

A Handbook for Earth-Friendly Living

By Crissy Trask. Surveys find that over 80% of Americans agree with the goals of the environmental movement. Sadly, most Americans admit to doing little more than basic recycling when it comes to acting on that disposition. What is the reason for this great divide between environmental sentiment in this country and individual actions? Author and environmental consultant Crissy Trask seeks to answer this question—and solve the disparity—with a new book that makes it easy to be an environmentalist, no matter how busy or hectic your lifestyle. This is a day-to-day guide with simple, practical suggestions that anyone can put into action. 168 pages, paperback.

21-0628 It's Easy Being Green   **$12.95**

Stormy Weather

By Paula L. Woods. In the 60s, *Silent Spring* forced us to pay attention to the problem with pesticides. The 70s galvanized the nation to conserve. In the 90s, our communities came together to "reduce, reuse, and recycle." And now, the first decade of the new millennium will focus our attention on our most urgent environmental challenge yet: global warming and the potential for irreversible climate change. In a clear and understandable style, *Stormy Weather* explains why we and the planet have reached this overheated situation and how scientists predict that "runaway" climate change will affect the Earth and our lives. The solutions to global warming revolve around 12 core methods of reducing our use of fossil fuels and filling our energy needs with solar, wind, tidal, and bio fuels. Each user-friendly solution is organized on two facing pages with a description, illustrations, quotations, resources, and a detailed "how-to" section. 352 pages, hardcover.

21-0629 Stormy Weather   **$19.95**

Energy Conservation

Consumer Guide to Home Energy Savings, 8th Ed.

Energy efficiency pays. Written by the nonprofit American Council for an Energy-Efficient Economy, the *Consumer Guide* has helped millions of folks do simple things around their homes that save hundreds of dollars every year. The new 8th edition has up-to-date lists of all the most energy-efficient appliances by brand name and model number, plus an abundance of tips, suggestions, and proven energy-saving advice for homeowners. 264 pages, softcover.

21-0344 Consumer Guide to Home
   Energy, 8th Ed.   **$9.95**

The Fuel Savers

A Kit of Solar Ideas for Your Home

By Bruce Anderson. What can you do right now, with your funky existing home and a minimum of cash, to save energy and live more comfortably? This book has a wealth of ideas, from insulating curtains to solar hot-water heaters. Each idea is examined for materials costs, fuel reduction, advantages, disadvantages, and cost effectiveness. A great step-by-step guide for improving the comfort and efficiency of any building. 83 pages, softcover. Updated in 2002.

21-0357 The Fuel Savers   **$15**

The Home Energy Diet

How to Save Money by Making Your House Energy-Smart

(MOTHER EARTH NEWS WISER LIVING SERIES) By Paul Scheckel. With rising energy costs, homeowners are beginning to examine the energy efficiency of their own homes, asking questions about where energy comes from and how much it costs, how to choose new appliances, and what options exist for renewable energy. *The Home Energy Diet* answers all these questions and more while helping readers take control of their personal energy use and costs so they can save money, live more comfortably, and help the environment. Energy auditor Paul Scheckel first explores energy literacy, and then describes how your home uses—and loses—energy you pay for electricity, hot water, heating, air conditioning, windows, walls, and insulation. 308 pages, paperback.

21-0570 The Home Energy Diet   **$18.95**

The Carbon Buster's Home Energy Handbook

There's no need to wait for governmental leadership or the Kyoto Protocol. Start your own grassroots carbon reduction program and potentially save $15,000 in energy costs over five years. This empowering book helps you assess a detailed carbon accounting of your own emissions and prioritizes solutions based on your highest reductions and returns – often with surprising results! 171 pages, softcover.

21-0418 The Carbon Busters Home Energy Handbook $12.95

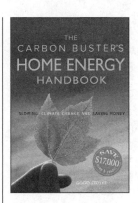

WATER DEVELOPMENT

Water Systems, Pumping, and Storage

Rainwater Collection for the Mechanically Challenged

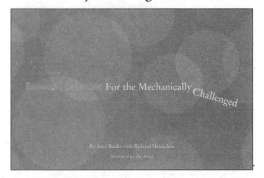

By Suzy Banks, Richard Heinichen, and Tre Arenz. Revised and expanded in 2003. Laugh your way to a home-built rainwater collection and storage system. This delightful paperback is not only the best book on the subject of rainwater collection we've ever found, it's funny enough for recreational reading and comprehensive enough to lead a rank amateur painlessly through the process. Technical information is presented in layman's terms and is accompanied with plenty of illustrations and witty cartoons. Topics include types of storage tanks, siting, how to collect and filter water, water purification, plumbing, sizing system components, freezeproofing, and wiring. Includes a resources list and a small catalog. 108 pages, softcover.

80-704 Rainwater Collection for the Mechanically Challenged $20

Rainwater Collection for the Mechanically Challenged DVD

By Suzy Banks, Tre Arenz, and Richard Heinchen. You say you're really mechanically challenged and want more than a few pictures? Here's your salvation. From the same irreverent, fun-loving crew that wrote the book above. See how all the pieces actually go together as they assemble a typical rainwater collection system and discuss your options. This is as close as you can get to having someone else put your system together. Lots of laughs. 37 min.

90-0168 Rainwater Challenged DVD $19.95

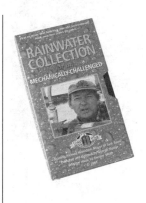

Cottage Water Systems

By Max Burns. An out-of-the-city guide to water sources, pumps, plumbing, water purification, and wastewater disposal, this lavishly illustrated book covers just about everything concerning water in the country. Each of the 12 chapters tackles a specific subject, such as sources, pumps, plumbing how-to, water quality, treatment devices, septic systems, outhouses, alternative toilets, greywater, freeze protection, and a good bibliography for more info. This is the best illustrated, easiest to read, most complete guide to waterworks we've seen yet. 150 pages, softcover.

80098 Cottage Water Systems $24.95

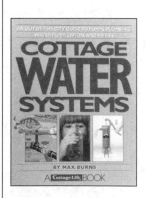

A Great Water Pumping Video

Part of the Renewable Energy with the Experts series, this video features solar-pumping pioneer Windy Dankoff, who has more than 15 years of experience in the field. Windy demonstrates practical answers to all the most common questions asked by folks facing the need for a solar-powered pump, and offers a number of tips to avoid common pitfalls. This is far and away the best video in this series and offers some of the best advice and knowledge available for off-the-grid water pumping. This grizzled old technician/copyeditor even learned a few new tricks about submersible pump installations. Highly recommended! 59 min.

80368   RE Experts Water Pumping Video $39

Water Storage

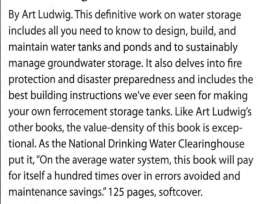

By Art Ludwig. This definitive work on water storage includes all you need to know to design, build, and maintain water tanks and ponds and to sustainably manage groundwater storage. It also delves into fire protection and disaster preparedness and includes the best building instructions we've ever seen for making your own ferrocement storage tanks. Like Art Ludwig's other books, the value-density of this book is exceptional. As the National Drinking Water Clearinghouse put it, "On the average water system, this book will pay for itself a hundred times over in errors avoided and maintenance savings." 125 pages, softcover.

21-0374 Water Storage $20

Flowforms: The Rhythmic Power of Water

Working with his remarkable invention, the Flowform, John Wilkes has uncovered the hidden secrets of water. Among those secrets, says Wilkes, is that water is the universal bearer of whatever character we put into it, and that for this reason, the way we treat water is of fundamental importance to our health and to the well-being of our planet. This lavishly illustrated book documents a lifetime of inquiry into the true nature of water. It includes a history of flowform research as well as the most important up-to-date developments in this research throughout the world. It also includes informative appendixes on metamorphosis, flowform designs and applications, and the scientific and technical aspects of flowform research. 208 pages, paperback.

21-0582 Flowforms $35

All About Hydraulic Ram Pumps

By Don Wilson. Utilizing the simple physical laws of inertia, the hydraulic ram can pump water to a higher point using just the energy of falling water. The operation sequence is detailed, with easy-to-understand drawings. Drive pipe calculations, use of a supply cistern, multiple supply pipes, and much more are explained in a clear, concise manner. The second half of the booklet is devoted to detailed plans and drawings for building your own 1- or 2-inch ram pump. Constructed out of com-

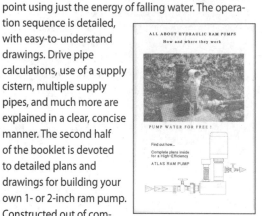

monly available cast iron and brass plumbing fittings, the finished ram pump will provide years of low-maintenance water pumping for a total cost of $50-$75. No tapping, drilling, welding, special tools, or materials are needed. This pump design requires a minimum flow of 3-4 gallons per minute, and 3-5 feet of fall. It is capable of lifting as much as 200 feet with sufficient volume and fall into the pump. The final section of the booklet contains a setup and operation manual for the ram pump. 25 pages, paperback.

80501   All About Hydraulic
**   Ram Pumps $10.95**

Water Heating

Solar Hot Water Systems, Lessons Learned

Tom Lane probably has more hands-on practical experience from 1977 to 2002, and knowledge about all kinds of solar hot-water systems, than anyone on the planet. He is detailed, specific, and refreshingly honest about the good and bad points of commonly available solar hardware. Extensive use of pictures, drawings, and system schematics help make explanations clear. Beautifully organized, with a detailed table of contents. Pick what you need. 120 pages, softcover.

21-0336 Solar Hot Water Systems,
**   Lessons Learned $29**

Solar Water Heating

By Bob Ramlow and Benjamin Nusz. Add solar water heating to your home, and you'll reap huge economic and ecological benefits. This definitive guide shows how it's done. Covers system basics, including issues like system size and siting. This guide is also packed with detailed looks at components, installation, operation, and maintenance, not to mention topics like space and pool heating. It's the only book novices need and the only one professionals want. By longtime solar heating experts Bob Ramlow and Benjamin Nusz. 288 pages, softcover.

21-0395 Solar Water Heating $24.95

Drinking Water

The Drinking Water Book

How to Eliminate the Most Harmful Toxins from Your Water

By Colin Ingram. *The Drinking Water Book* takes a level-headed look at the serious issues surrounding America's drinking water supply. Unlike water purifier manufacturers and public health officials, Ingram presents unbiased reporting on what's in your water and how to drink safely. Featuring all the latest scientific research, the book evaluates the different kinds of filters and bottled waters and rates specific products on the market. 185 pages, paperback.

21-0583 The Drinking Water Book $14.95

The Water You Drink: Safe, or Suspect?

By Julie Stauffer. This book is a clearly written, straightforward guide to North American tap water: what's in it, how it's treated, whether government regulations are strict enough, and whether consumers should be concerned about what they're drinking. It includes chapters on bottled water and home filter systems, as well as on obtaining, treating, and storing private water supplies, and helpful resources. Julie Stauffer has authored several books and a video on water and pollution. 160 pages, paperback.

21-0531 The Water You Drink $14.95

Composting Toilets and Greywater

THE GREYWATER GUIDE SERIES

Create an Oasis with Greywater, Revised 4th Ed.

Choosing, building, and using greywater systems

Describes how you can save water, save money, help the environment, and relieve strain on your septic tank or sewer by irrigating with reused wash water. Describes 20 kinds of greywater systems. The concise, readable format is long enough to cover everything you need to consider, and short enough to stay interesting. Plenty of charts and drawings cover health considerations, greywater sources, designs, what works and what doesn't, bio-compatible cleaners, maintenance, preserving soil quality, and the list goes on. 51 pages, softcover.

82440   Create an Oasis with Greywater $14.95

Builder's Greywater Guide

A companion to Create an Oasis with Greywater

Will help you work within or around codes to successfully include greywater systems in new construction or remodeling. Includes reasons to install or not install a greywater system, flowcharts for choosing an appropriate system, dealing with inspectors, legal requirements checklist, design and maintenance tips, how the earth purifies water, and the complete text of the main U.S. greywater codes with English translations. 46 pages, softcover.

80097   Builder's Greywater Guide $14.95

Branched Drain Greywater Systems

A companion to Create an Oasis with Greywater

Detailed plans and specific construction details for making a branched drain greywater system in any context. This ultrasimple, robust system is the best choice for over half of residential installations. It provides reliable, sanitary, and low-maintenance distribution of household greywater to downhill plants without filtration, pumping, or a surge tank. It is one of the few systems that is both practical and legal, and it has been described as "the best greywater system since indoor plumbing." Note: The area to be irrigated has to be downhill from the house, and laying out the lines must be done with fanatical attention to proper slope. 52 pages, softcover.

82494   Branched Greywater $14.95

Greywater Guide Booklet Set

Save over $5 by picking up all three Greywater booklets together.

21-0360 Greywater Guide Booklet Set $39

Lifting the Lid

By Peter Harper and Louise Halestrap. This book examines alternatives to the flush toilet and how to manage your household's output of sewage, conserve water, and turn a potential pollutant into a nutrient for use in improving the environment. Includes chapters on the biology of breakdown, how to choose a composting toilet, maximizing septic tanks, utilizing greywater in the garden, and designing twin-vault toilets. 160 pages, paperback.

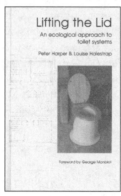

21-0532 Lifting the Lid $15.95

The Humanure Handbook, 2nd Ed.

A Guide to Composting Human Manure

By Joseph Jenkins. Deemed "Most Likely to Save the Planet" by the Independent Publishers in 2000. For those who wish to close the nutrient cycle and who don't mind getting a little more personal about it than our manufactured composting toilets require, here's the book for you. Provides basic and detailed information about the ways and means of recycling human excrement, without chemicals, technology, or environmental pollution. Includes detailed analysis of the potential dangers involved and how to overcome them. The author has been safely composting his family's humanure for the past 20 years, and with humor and intelligence has passed his education on to us. 302 pages, softcover.

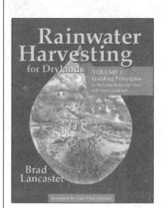

82308   The Humanure Handbook $25

Rainwater Harvesting for Drylands

Guiding Principles to Welcome Rain into Your Life and Landscape

By Brad Lancaster. *Rainwater Harvesting for Drylands: Guiding Principles to Welcome Rain into Your Life and Landscape* is about how to conceptualize, design, and implement sustainable water-harvesting systems for your home, landscape, and community. This book enables you to assess your on-site resources, gives you a diverse array of strategies to maximize their potential, and empowers you with guiding principles to create an integrated, multifunctional water-harvesting plan specific to your site and needs. 200 pages, paperback.

**21-0584 Rainwater Harvesting for
          Drylands: Guiding Principles $24.95**

Rainwater Harvesting for Drylands

Water-Harvesting Earthworks

By Brad Lancaster. Volume 2 of this book shows you how to select, place, size, construct, and plant your chosen water-harvesting earthworks. It presents detailed how-to information and variations on a diverse array of earthworks, including chapters on mulch, vegetation, and greywater recycling so you can customize the techniques to fit the unique requirements of your site. Real-life stories and examples permeate this volume. XXX pages, paperback.

**21-0585 Rainwater Harvesting for
          Drylands: Water-Harvesting $32.95**
[image needed]

How to Shit in the Woods

By Katheleen Meyer. A humorous name for a useful book. Chock-full of practical tips that are useful in a variety of situations—and the great sense of humor makes learning fun. Written by a woman, it includes lots of extra information for women in the wild. This 2nd edition covers health concerns, feedback from the 1st edition, and much more. 128 pages, softcover.

82640   How to Shit in the Woods $7

The Toilet Papers
Recycling Waste and Conserving Water

By Sim Van der Ryn. A classic is back in print! One of the favorite books of the back-to- the-land movement, *The Toilet Papers* provides an informative and irreverent look at how people have dealt with human wastes over the centuries, and what safe designs are available today that reduce water consumption and avert the necessity for expensive treatment systems. Van der Ryn provides homeowner plans for several types of dry toilets, compost privies, and greywater systems, and also discusses the history and philosophy of turning organic wastes into a rich humus, linking us to the fertility of the soil and ensuring our ultimate well-being. Van der Ryn is a former architect, and his designs for compost privies are downright elegant as well as environmentally sound. 124 pages, paperback.

21-0586 The Toilet Papers $14

Appendix

The Real Goods and Solar Living Institute staffs at their 2006 Christmas Party.

Real Goods' Mission Statement

Through our products, publications, and educational demonstrations, Real Goods promotes and inspires an environmentally healthy and sustainable future.

Part I: Who We Are and How We Got Here

Who Put the "Real" in Real Goods

As did many of his contemporaries in the 1960s and early 1970s, John Schaeffer, founder of Real Goods, experimented with an alternative lifestyle. After protracted exposure to nearly every strand of the lunatic fringe, he graduated from U.C. Berkeley in 1971 and moved to a commune called "Rainbow" outside of Boonville, California. There, in an isolated 290-acre mountain community, John pursued a picturesque life of enlightened self-sufficiency.

John Schaeffer,
founder of Real Goods

Despite the idyllic surroundings, John soon found that certain key elements of life were missing. After several years of reading bedtime stories to his children by the flickering light of a kerosene lamp, John began to squint. He grew tired of melted ice cream and lukewarm beer. He began to miss the creature comforts his family was lacking due to their "off-the-grid" lifestyle. He yearned for just a tiny amount of energy to strike a balance between the lifestyle he had grown up with and complete deprivation. In other words, John came to the realization that self-sufficiency was much more appealing as a concept than a reality.

Then he discovered 12-volt power. John hooked up an extra battery to his car that he charged while commuting to work, with just enough juice to power lights, a radio, and the occasional television broadcast. Despite his departure from a pure ascetic lifestyle, each and every time that *Saturday Night Live* aired, John's home became the most popular place on the commune. Eventually, when the 12-hour community work days began to take their toll, John took a job as a computer operator in Ukiah, some 35 twisty miles from Boonville.

Once the word got out that John would be making the trek over the mountain to the "big city" daily, he became a one-man pick-up and delivery service, procuring the wood stoves, fertilizer, chicken wire, bone meal, gardening seeds, tools, and supplies needed for the commune. As a conscientious and naturally frugal person, John spent hours scrutinizing the hardware stores and home centers of Ukiah, searching for the best deals on the real goods needed for the communards' close-to-the-earth lifestyle.

One day, while driving his VW bug back to the commune after a particularly vexing shopping trip, a thought occurred to John. "Wouldn't it be great," he mused, "if there was one store that sold all the products needed for independent, off-the-grid living, and sold them at fair prices?" The idea of Real Goods was born. The company thrived opening up retail stores and eventually morphed into a mail-order company.

Real Goods in the New Millennium

From its humble beginnings in 1978, Real Goods became a Real Business, with Real Employees serving Real Customers. In the 1990s, Real Goods pioneered the "direct public offering" process, whereby it raised investment capital from its customers without the need for investment bankers or other financial middlemen. Before the Internet had really caught on, Real Goods was selling stock electronically and allowing its customers to print virtual stock certificates in the privacy of their own homes. The company now can lay claim to the title of the Oldest and Largest catalog firm devoted to the sale and service of renewable energy products in the world. Real Goods, now Gaiam Real Goods since its January 2001 merger with Gaiam, Inc., of Colorado, is still devoted to the same principles that guided its founding—quality, innovative, well-made products for fair prices, and unsurpassed customer service with courtesy and dignity.

Early on, John managed to turn his personal commitment to right livelihood into company policy, pioneering the concept of a socially conscious and environmentally responsible business. The company consistently has been honored and awarded for its ethical and environmental business standards. Plaudits include Corporate Conscience Awards (from the Council on Economic Priorities); inclusion in *Inc.* magazine's 1993 list of America's 500 Fastest-Growing Companies; three consecutive Robert Rodale Awards for Environmental Education; Northern California Small Business of the Year Winner for 1994; finalist for Entrepreneur of the year two years running; news coverage in *Time*, *Fortune*, *The Wall Street Journal*, and *Mother Earth News*; numerous TV appearances, countless Japanese magazines; and many thick scrapbooks full of press clippings.

Five Principles to Live By

Gaiam Real Goods is considered newsworthy not because our methods reflect the latest trends in corporate or business-school thinking, but because we unwittingly have helped to birth an astonishingly healthy "baby"—an ethical corporate culture based on environmental and social responsibility. Led by a certain naiveté and affection for simplicity, Gaiam Real Goods has discovered some simple principles that, by comparison to the "straight" business world, are wildly innovative. This has not been the work of commercial gurus or public relations mavens, but rather the result of realizing that business need not be so complicated that the average person cannot understand its workings. Our business is built around five simple principles.

PRINCIPLE #1: THIS IS A BUSINESS

A business is, first and foremost, a financial institution. You can have the most noble social mission on the planet, but if you can't maintain financial viability, you cease to exist. And so does your mission. The survival instinct is very strong at Gaiam Real Goods, and that reality governs many decisions. To say it another way, you can't be truly sustainable if your business isn't economically sustainable (read profitable). To ensure the continued flourishing of our mission, we pursue profitability through our catalogs, retail store, residential solar sales and installation business, internet presence, and our design and consulting business. We learned long ago that nonprofits aren't driven by profit. That's why we spun off the Solar Living Institute in 1998. The SLI furthers the original educational mission of Real Goods without the constraints of the profit motive. Both organizations have a symbiotic relationship and help each other out immensely but there are no financial or legal ties between them.

PRINCIPLE #2: KNOW YOUR STUFF

Knowledge seldom turns a profit, yet our social and environmental missions cannot be achieved without it. The independent lifestyle we advocate relies largely on technologies that often require a high degree of understanding, and a level of interaction that has been largely forgotten during our nation's half-century binge on cheap power. We aren't interested in selling things to people that they aren't well-informed enough about to live with comfortably and happily with. We want people to understand not only what we are selling and how it is used, but how a particular piece of hardware contributes to the larger goal of a sustainable lifestyle.

It goes against the grain of mainstream business to give anything away. Even a "loss leader" is designed to suck you into the store to buy other, higher-profit items. At Gaiam Real Goods, knowledge is our most important product, yet we give it away daily, through our catalogs (there's a lot in there to learn, even if you never buy a thing); through our Solar Living Center (free self-guided and group tours), now run by the nonprofit Solar Living Institute; through free workshops at our Hopland store, and through workshops we support through the SLI's annual SolFest renewable energy celebration. Our website acts as a launching pad for renewable energy research, leading you to fascinating information on sustainability topics of all kinds. We believe that as our collective knowledge of sustainability principles and renewable energy technology increases, the chance of achieving the Gaiam Real Goods mission increases, too.

PRINCIPLE #3: GIVE FOLKS A WAY TO GET INVOLVED

Just about all of our employees are also our customers and more than half of them live with solar—many live completely off-the-grid. Our parking lot looks like an advertisement for biodiesel with all the VW diesel vehicles driven by our employees. Our Lifetime Membership program honors our community of customers with financial and educational benefits, and interns and volunteers are welcomed warmly by the SLI staff. Gaiam Real Goods has become a real community, acting in concert toward the common goal of a sustainable future. With the knowledge that the age of oil is likely soon coming to an end, it's comforting to know that we are all in this together.

PRINCIPLE #4: WALK THE WALK

At Gaiam Real Goods, we conduct our business in a way that is consistent with our social and environmental mission. We use the renewable energy systems we sell, and we sell what works. Our merchandising team makes absolutely sure the merchandise we sell performs as expected, is safe and nontoxic when used as directed, and is made from the highest quality sustainable materials. In 1990, we challenged our customers to help us rid the atmosphere of one billion pounds of CO_2 by the year 2000, and we achieved our goal three years ahead of schedule. Again in 2007 we set an ambitious goal with all of Gaiam to offset the production of another one billion

pounds of CO_2—this time even more quickly. We think we can achieve this new billion pound goal before 2010. Real Goods was recognized by being awarded the Rodale Award, as the business making the most positive contribution to the environment in America, for three years running. We don't just talk the talk at Gaiam Real Goods, we walk the walk. Come visit us at the Solar Living Center in Hopland, California, and see some of our innovative practices. Fill up your biodiesel vehicle on site, see water pumped from the sun, learn from hydrogen fuel cell demonstrations, and see 150kW of solar powering the site and much more.

PRINCIPLE #5: HAVE FUN!

We strongly support the best party of the year for 10,000 of our closest friends every summer in August on the grounds of our Solar Living Center in Hopland. We look forward to the Solar Living Institute's annual SolFest renewable energy celebration all year. If we've learned anything since 1978, it's that all work and no inspiration makes Jack and Jill a couple of burnt-out zombies. SolFest is our little reminder to take care of ourselves with some good, clean fun, so we'll be rejuvenated and reinvested in the hard work of creating a sustainable future.

Part II: Living the Dream:
The Real Goods Solar Living Center

Imagine a destination where ethical business is conducted daily amidst a diverse and bountiful landscape, where the gurgle of water flowing through its naturally revitalizing cycle heightens your perception of these ponds, these gardens, these living sculptures. You follow the sensuous curve of the hill and lazy meanders of the watercourse to a structure of sweep-

The Solar Living Center from the air during SolFest 2005. Note the shadow of the airplane on the solar array at the bottom!

APPENDIX

ing beauty, where floor-to-ceiling windows and soaring architecture clearly proclaim this building's purpose—to take every advantage of the power of the sun throughout its seasonal phases. A few more steps and the spidery legs of a water-pumping windmill come into view, and the top of a tree that looks as though it might be planted in the rusted shell of a vintage e Cadillac. An awesome sense of place begins to reveal itself to you. Inside the building, sunlight and rainbows play across the walls and floors of a 5,000-square-foot showroom, and you begin to understand that all of this, even the offices and cash registers, are powered by the energy of sun and wind. Welcome to the Solar Living Center in Hopland, California, the crowning achievement of the Gaiam Real Goods mission.

Our Solar Living Center began as the vision of Real Goods founder and president, John Schaeffer. His dream was to create an oasis of biodiversity, where the company could demonstrate the culture and technology of solar living, where the grounds and structures were designed to embody the sustainable living philosophy of Real Goods' catalogs and business. With the opening of the Solar Living Center in April 1996, John's vision is now a reality. As of mid-2007, close to two million people have visited the center, and have left this place with an overwhelming sense of inspiration and possibility.

In 1998, the Real Goods Solar Living Institute split off from its parent Real Goods Trading Corporation and became a legal 501(c)(3) nonprofit called the Solar Living Institute (SLI). Since then the SLI has nurtured and developed the 12-acre permaculture site that has flourished with fecundity.

Form and Function United: Designing for the Here and Now

If the "weird restrooms" sign doesn't grab them first, the 40,000 daily passersby on busy Highway 101 are bound to notice the striking appearance of the company showroom. This does not look like business as usual! The building design and the construction materials were selected with an eye toward merging efficiency of function, educational value, and stunning beauty.

The architect chosen to design the building was Sim Van der Ryn of the Ecological Design Institute of Sausalito, California. His associate, David Arkin, served as project architect, and Jeff Oldham of Real Goods managed the building of the project (both remain on the SLI's board of directors). Their creation is a tall and gracefully curving single-story building that is so adept in its capture of the varying hourly and seasonal angles of the sun that additional heat and light are virtually unnecessary. Wood-burning stoves provide back-up heating for the coldest winter mornings and solar-powered fluorescent lighting is available, but is rarely used. Through a combination of overhangs and manually controlled hemp awnings, excess insolation during the hot-weather months has been avoided. Solar-powered evaporative coolers provide a low-energy alternative to air conditioning, and are also used to flush the building with cool night air, storing "coolth" in the 600 tons of thermal mass of the building's walls, columns, and floor. Grape arbors and a central fountain with a "drip ring" for evaporative cooling are positioned along the southern exposure of the building to serve as a first line of defense against the many over-one-hundred-degree days that occur during the summer in this part of California.

Many of the materials used in the construction of the building were donated by companies and providers with a commitment similar to Real Goods'. As an example, the walls of the SLC were built with more than 600 rice straw bales donated by the California Rice Industries Association. Previously, rice straw has been disposed of by open burning, a practice that contributes to the production of carbon dioxide, the so-called "greenhouse gas" that is the leading cause of global warming. By using this agricultural by-product as a building material, everyone benefits. The farmers receive income for their straw bales, no carbon dioxide is produced, and the builder benefits from a low-cost, highly efficient building material that minimizes energy consumption.

At the SLC, visitors experience the practicality of applied solar power technology, including the generation of electricity and solar water pumping. The electrical system for the facility comprises nearly 150 kilowatts of photovoltaic power. Through an intertie with the Pacific Gas and Electric Company, the SLI sells the excess power it generates to the electric company, making the SLC 100% independent from the grid. Once again, like-minded companies have shared in the costs of developing the Solar Living Center as a demonstration site. Siemens Solar (now Solar World) donated more than 10 kilowatts of the latest state-of-the-art photovoltaic modules to the center, and periodically uses the SLC as a test site for new modules. Trace

The central oasis and "stonehenge of the future" at the Hopland Solar Living Center (picture taken with a camera on a kite!).

Engineering (now Xantrex) contributed four intertie inverters, which are on display behind the glass window of the SLC's "engine room" so that visitors can see the inner workings of the electrical system. More recently, Beacon Power donated a 5,000-watt intertie inverter to the electrical room.

In November 1999, a partnership between GPU, Astropower, Real Goods, and the Solar Living Institute installed a 132-kilowatt PV array on campus, one of the largest in power-hungry Northern California. This direct-intertie array delivers 163,000 kWh of power annually, enough to power fifty average California homes. On tours, either self-guided or with Solar Living Institute tour guides, visitors learn about the guiding principles of sustainable living, and are offered a chance to appreciate the beauty that lies in the details of the project. The site also provides a wonderful space for presentations by guest speakers and special events, and serves as the main campus and classroom for the workshop series staged by the Solar Living Institute, a nonprofit dedicated to education and inspiration toward sustainable living.

The Natural World Reclaimed: The Grounds and Gardens

Learning potential is intrinsic to the award-winning landscape, designed by Chris and Stephanie Tebbutt of Land and Place. For this project, the design of the grounds, gardens, and water-works was the first phase of construction and contributed much to establishing the character of the site. This is a radically different approach than most commercial building projects, where the landscaping appears to be a cosmetic afterthought. At the SLC, the gardens are a synthesis of the practical and the profound. Most of the plantings produce edible and/or useful crops, and the vegetation is utilized to maximize the site's energy efficiency while portraying the dramatic aspects of the solar year. Plantings and natural stone markers follow the lines of sunrise and sunset for each equinox and solstice, emanating from a sundial at the exact center of the oasis. More sundials and unique solar calendars scattered throughout the site encourage visitors to establish a feeling for the relationship between this specific location and the Sun. Those of us who work and play here daily have discovered an almost organic connection with the seasonal shifts of the solar year and the natural rhythms of the Earth.

The gardens themselves follow the sun's journey through the seasons, with zones planted to represent the ecosystems of different latitudes. Woodland, Wetland, Grassland and Dryland zones are manifested through plantings moving from north to south, with the availability of water the definitive element. Trees are planted to indicate the four cardinal directions. The fruit garden, perennial beds, herbs, and grasses reflect the abundance and fertility of a home-based garden economy. Visitors discover

Bicycle generators at the Solar Living Center demonstrate that the average human can generate 150 sustained watts per hour. Therefore, it would take six people peddling for 24 hours to produce 21.6kWh per day, which is ⅔ of the energy that the average American home uses. This is the most popular interactive display on site.

aesthetic statements in design and landscape tucked into nooks and crannies all over the grounds, and unexpected simple pleasures, too, like shallow water channels for cooling aching feet, and perfect hidden spots for picnics and conversation.

Unique to these gardens are the "Living Structures," which reveal their architectural nature according to the turn of the seasons. Through annual pruning, plants are coaxed into various dynamic forms, such as a willow dome, a hops tipi, and a pyramid of timber bamboo. These living structures grow, quite literally, out of the garden itself. Visitors unaccustomed to the heat of a Hopland summer find relief inside the "agave cooling tower," where the turn of the dial releases a gentle mist into the welcome shade of vines and agave plants. By the time visitors leave the center, they've begun to understand the subtle humor of the "memorial car grove," where the rusting hulks of '50s and '60s "gas hog muscle" cars have been turned into planter boxes for trees. These "grow-through" cars make a fascinating juxtaposition to the famous Northern California "drive-through" redwood trees!

A Place to Play

In case this all sounds awfully serious, it should be pointed out that the SLC is a wonderful place to play! Upon entering the showroom, one is greeted by a delightful rainbow spectrum created by a large prism mounted in the roof of the building. Visitors need not understand on a conscious level that this rainbow functions throughout the year as a "solar calendar," or that the prism's bright hues mark the daily "solar noon"—this is a deeper learning that ignites the place where inspiration happens, not a raw scientific dissertation. Outside, interactive games and play areas tempt the young at heart to forget about the theory and enjoy the pleasure of pure exploration. A six-station bicycle "generator" allows riders to see how much energy humans are capable of generating compared to the average American home's energy needs. It doesn't take long for riders to really feel how much energy is required to create the tiniest bit of energy; almost everyone takes the time to compare the aching results of muscle power to the ease with which the same amount of energy is harvested from the sun with a solar panel.

The hands-down favorite for kids is the sand and water area. A solar-powered pump provides a water source that can then be channeled, diverted, dammed, and flooded through whatever sandy topography emerges from the maker's imagination. A shadow across the solar panel stops the flow of water, and it doesn't take long for kids to become immersed in starting and stopping the flow at will. Without even realizing it, these little scientists are learning about engineering, hydrology, erosion, and renewable energy theory!

A Global Warming displays shows a map of the greater San Francisco Bay Area as it sits today and what the water level would look like with a 1-meter rise in the ocean's level. If current levels of carbon dioxide production continue, a 1-meter rise is expected by the mid- to late twenty-first century. The display shows the Sacramento valley and much of the oceanside and bayside towns inundated.

A hydrogen fuel cell display shows visitors how solar power electrolyzes water into hydrogen and oxygen and how those two elements are fed back into a fuel cell to power a small fan. In 2005, the SLI installed the world's first and only solar-powered carousel, where kids can choose to ride a variety of hand-carved indigenous Mendocino County critters, including a skunk, Coho salmon, wild boar, and deer. These are only a very few of the dozens of hands-on interactive displays available to the public at the Solar Living Center.

Where to Find the Solar Living Center

The Real Goods Solar Living Center is located 94 miles north of San Francisco on Highway 101 and is open every day except Thanksgiving and Christmas. There is no admission charge for regularly scheduled or self-guided tours, and picnicking is strongly encouraged.

Customized group tours for students, architects, gardeners, or others with special interests are available on a fee basis by advanced reservation, through the nonprofit Solar Living Institute. The Institute also offers a variety of structured learning opportunities, including intensive, hands-on, one-day to one-week seminars on a variety of renewable energy, sustainable living, and permaculture gardening topics. Please call the Institute at 707-744-2017 for more information, or visit its website at www. solarliving.org.

Eureka (200 mi.)

Portland

Highway 20

Ukiah

Hopland

Solar Living Center

Sacramento

Cloverdale

Interstate 5

US 101

SF/Oakland (90 mi.)

Los Angeles

Part III: Spreading the Word: The Solar Living Institute

In April 1998, the Solar Living Institute separated from Real Goods Trading Corporation to become a legal nonprofit 501(c)(3). By severing financial ties to a for-profit corporation, the Institute is now free to develop its educational mission without the constraint of profitability, and to focus its efforts solely on environmental education.

Since Real Goods' merger with Gaiam, Inc., in January 2001, the Solar Living Institute (SLI) has received several generous donations from Gaiam, which have enabled it to greatly expand its interactive displays and programs and heighten its impact on environmental education. This generosity is a part of the SLI's annual funding with the bulk coming from donations and partnerships from individuals like you.

The SLI's Four-Fold Mission

The Institute's mission is to teach people of all ages how to live more sustainably on the Earth, to teach interdependence between people and the environment, to replace fossil fuels with renewable energy, and to honor biodiversity in all its forms. To further its mission, the Institute is focusing its resources on three endeavors: Sustainable Living Workshops, the Solar Living Center's Interactive Displays, Exhibits, and Educational Tours, and the SolFest Energy Fair, Educational Celebrations, and Earth Day for Kids events. In short, the Institute's mission is to promote sustainable living through inspirational environmental education.

#1: Sustainable Living Workshops

The Solar Living Institute is a leader in educating our society to prepare for a fossil fuel-free, sustainable future. The Institute's workshops approach the themes of shelter, energy, transportation, and food from an ecologically sustainable perspective, teaching students from around the world how to rethink these basic necessities in a manner that is in harmony with our planet. Workshop topics include green and natural building; biofuels and alternative transportation; solar, wind and hydro energy; and permaculture and organic farming. Now in its

A Solar Living Institute workshop entitled "Women in Solar" teaches solar energy hands-on in a supportive environment by women and for women.

15th season with a few thousand students annually, Institute workshops provide an ideal opportunity to learn hands-on skills, meet people of similar interests, work beside them, and learn the most cutting-edge techniques in sustainability.

As the issues of peak oil and climate change become more accepted by the mainstream, entrepreneurial and career opportunities in various sectors of the green economy abound, creating a need for a workforce trained in regenerative and sustainable techniques. One of the main focuses of the Institute's workshop program is to prepare its students to be forward-thinking individuals who take the skills they've learned and utilize them to find sustainable solutions to the challenges our society faces. After graduating from the Solar Living Institute's classes, many of the Institute's workshop students go on to make career changes into a green jobs, or found their own environmentally friendly businesses. In 2006, the Institute hosted its first ever, hugely successful Green Career Conference, connecting students with opportunities in various sectors of the green economy.

Workshops are offered mostly at the main campus at the Solar Living Center (SLC), but also take place throughout California, and are expanding across the country to places like the San Francisco Bay Area, Los Angeles, New Jersey, and New York. The main campus is not only an immensely inspirational setting, but a living, breathing model for sustainable development,

Build a straw bale house from scratch, including mixing the natural plaster.

restorative permaculture landscaping, and renewable technologies. Many of the hands-on workshops at the SLC enable students to participate in actual projects like the building of a straw bale structure in the intern village, the erecting of a windmill to pump water, or the installation of a PV array to power the volunteer kitchen. Students often camp at the SLC, and are not only offered the opportunity to practice what they've learned, but also live temporarily in a sustainable setting. One graduate expressed it best, "I can't decide if I've just had the best short vacation of my life, or the best learning experience! Could it be both?"

Product demonstrations, cost-benefit analyses, and a tour of applied technologies at the SLC provide students with a practical grasp on the tools and techniques available for taking their energy lives in hand. Those who arrive with some technical skill leave qualified to put together an independent energy system, to begin the design of an energy-efficient home made of alternative and sustainable building materials, to know what to look for in a piece of property, or to create balance and harmony

in their own garden. The mission of each workshop is to share excitement, sense of purpose, and knowledge with others interested in living sensibly and lightly on the Earth.

The Solar Living Institute's expert faculty has decades of experience and a strong passion for passing along its expertise to information-hungry students. The practical skills that they teach inspire students to become informed consumers and practitioners, and able to ask the right questions and to understand the answers. They report that the experience inspires them to go further forward along the path toward independent living.

TYPICAL CLASSES OFFERED AT THE SOLAR LIVING INSTITUTE

Classes range from one-day introductions to week-long intensives, and cost between $95 and $135 per day. Workshops are scheduled year round. Here is a representative list of workshops that were scheduled for the 2007 workshop season. Because the Institute is strongly committed to expanding its workshop series, this list will continue to develop. Potential students are encouraged to visit the Solar Living Institute's beautifully redesigned website at www.solarliving.org and to let them know which additional topics you might be interested in pursuing.

Intro to Green Renovations
Intro to LEED and Commercial Green
     Building
Organizing Sustainable Communities
The Great Energy Transition
Payback: The Financial Case for Solar
Breakthrough Solar Sales and Marketing
Photovoltaics for Homeowners
Off-Grid Photovoltaic Design and Installation
Advanced Off-Grid Energy Systems
PV Design & Installation Boot Camp for
     Electricians

Green Career Conference

As the global issues of climate change and peak oil become increasingly urgent, we must rethink the way we occupy the planet and satisfy our basic needs, including food, shelter, transportation and energy. Redesigning these basic needs has created a plethora of new jobs, entrepreneurial ventures, and exciting career opportunities.

The Solar Living Institute and its cadre of sustainable living professionals offers an interactive day of practical information at its annual **Green Career Conference** that will help you find a rewarding career and enable you to make a living while making a difference!

PV Design & Installation Boot Camp for
    Beginners
Commercial PV
Find Your Dream Job in Solar
Find Your Dream Job in Biodiesel
Find Your Dream Job in Green Building
Power Your Home with Hydro
Wind Energy Basics
Solar Hot Water
Women's Survey of Renewable Energy
Grid-Tied PV for Women
Carpentry for Women
Advanced Carpentry Project for Women
Biodiesel—Fuel from Vegetables
Build Your Own Biodiesel Processor
Biodiesel Intensive: From the Processor
    to the Tank
Veggie Oil Basics
Alcohol Can Be a Gas
Electric Vehicle Hands-on Clinic
5-Day Natural Building Intensive
Ecological Design
Carpentry for Natural Building
Introduction to Cob Building
Natural and Earthen Plasters
Build a Straw Bale Home
Do-It-Yourself Home Repairs
Old Ways Survival Skills
Introduction to Permaculture
Permaculture First Responder
Biointensive Backyard Gardening
Ecological Urban Gardening
Design and Construct a Greywater System
Rural Water Development

**FEEDBACK FROM SOLAR LIVING
INSTITUTE GRADUATES**

■ "I am so excited to get home and get started
that I can't sit still. One night I was sizing my
system all night. Last night I dreamt I had five
different PV systems laid out around the top of
a green meadow. All night I was choosing the
best one—a combination—for my needs. Jeff
and Ross and Nancy were there talking to other
people then answering all my questions!"

■ "An honest possibility even for a conservative
Republican that voted twice for Reagan. It is not
as difficult as others want you to believe."

■ "Good spiritual company, great food, wonderful setting, excellent information."

■ "You guys are great people, and your knowledge and experience is invaluable to the rest of
us—that is why we came to you. Please concen-

A hands-on wiring
lesson during a solar
"boot camp" at the
Solar Living Center.
One-week boot
camps teach the
complete course in
solar energy for those
embarking on solar
careers.

trate on conveying what is unique to you and
your experience. The rest we can get from the
books you sell."

■ "Very informative introductory course…I
feel rejuvenated and motivated to take action!"

■ "Super course—lots of practical information. Instructor is excellent, his enthusiasm is
catching!"

To register for classes, or for more information or to request a workshop catalog, call
707-744-2017, write the Solar Living Institute at
P.O. Box 836, Hopland, CA 95449, or check out
our website at www.solarliving.org.

#2: Interactive Displays and Curriculum

The Solar Living Institute is continually expanding and enriching the Solar Living Center's
interactive and educational displays. The staff
won't rest until the SLC is recognized globally
as a major learning campus and educational
tour destination. As a nonprofit organization,
the SLI seeks sources of funding in keeping with
its educational goals. For example, the Institute
has applied for grant monies to fund additional
interactive exhibits on hydro power, solar water
heating, photovoltaics, wind power, and hydrogen fuel cells at the SLC. The SLI has engaged
professional designers to create more engaging,
effective exhibits and interactive displays on the
site. The SLI is also in the process of designing
renewables curriculum materials for schools,
and further promoting the SLC as a destination
for school groups. The SLI is also developing
partnerships with educators to bring sustainable living education into local schools, with
the vision to expand the program across the

country.. In addition to improving the on site educational demonstrations, the SLI is working towards bringing it's educational programs to the world, recognizing that the future limits of readily available cheap fossil fuels will make it more difficult for people to come to the SLC.

#3: SolFest—Annual Summer Energy Festival and Educational Fair

The Solar Living Institute's largest annual event is the SolFest energy fair, scheduled each year for the third weekend in August. The first ever SolFest took place in June 1996, celebrating the Grand Opening of the Real Goods Solar Living Center in Hopland, California. The gala three-day event was attended by 10,000 people, inaugurating the Solar Living Center as the premier destination for those interested in learning about renewable energy and other sustainable living technologies. The educational theme of the event, in concert with the uniquely beautiful setting of the Solar Living Center, has inspired thousands of visitors. World-class speakers, unique entertainment, educational workshops, exhibitor booths with a renewable energy orientation, along with a parade and display of electric vehicles, all helped to create a lively and very successful event.

Over the years, SolFest has featured incredible speakers, including Amory Lovins of the Rocky Mountain Institute, Paul Hawken, Wes Jackson of the Land Institute, Ralph Nader, Jim Hightower, Ben Cohen of Ben and Jerry's Homemade, Julia Butterfly-Hill, Alice Walker, Helen Caldicott, Amy Goodman (Democracy Now), Ed Begley Jr., Darryl Hannah, and many others. World-class entertainers who have graced the SolFest stage include Bruce Cockburn, Michelle Shocked, Spearhead, Michael Franti, David Grisman, Charlie Hunter, and more. SolFest also features over 75 ongoing educational workshops exploring topics like solar energy, straw bale construction, building with bamboo, biodiesel, electric vehicles, eco-design, socially responsible investing, unconventional financing, climate change, industrial hemp, creative reuse, preparing for peak oil, solar cooking, and much, much more. The family stage and educational area provides fun and educational games, performances, rides on the solar carousel, and

Bruce Cockburn has brought his unique blend of politics and amazing music to two SolFests—2005 and 2007.

interactive children's workshops. SolFest provides a unique opportunity for renewable energy and sustainable living aficionados to gather and swap stories, and to have a great time for two days in the sun while learning and playing. Call the SLI at 707-744-2017 or visit the website (www.solarliving.org) to find the exact date and bill of entertainment—the annual SolFest occurs on the third weekend in August.

The SLI Internship and Volunteer Program

The Solar Living Institute has a well-developed and well-established internship program for students wanting to continue their studies in organic and bio-dynamic gardening, renewable energy, green and natural building, alternative fuels, and other sustainability education topics. SLI interns are honored and appreciated for their contributions, and come away with a lasting sense of achievement. The SLI has an extensive internship program with students of all ages coming to the Solar Living Center from all over the world to live, work, learn, and experience the joys of sustainable living. Interns come for 12- to 24-week stints and consistently depart with the remark that their internships have been one of their life's peak experiences. For more information on the SLI's internship program check out the website at www.solarliving.org.

It takes a lot of energy to manifest the SLI's vision, and is not the kind that is measured in kilowatts. The Institute extends a warm welcome to any volunteers willing to give their time and energy to promoting sustainable living through inspirational environmental education. You do not need to live in California to help; volunteering could take the form of consultation, publicity, physical labor, day-to-day support, or wherever your talents lie and however you'd like to help.

The Institute's Partnership Program

The other kind of energy needed to sustain the SLI is the green kind. The SLI asks each of you seriously to consider partnering with it in building the best sustainable living education program on the planet. They have applied for and received grants and are seeking large-scale donations from like-minded businesses and individuals, and hope to find additional support. However, it has always been grassroots mom-

Solar Living Institute intern class of 2004.

and-pop generosity that got any nonprofit ball rolling. As little as $35 per year (for a Student/ Senior Partnership) will help expand the Sustainable Living Workshop Series, build more educational exhibits at the Solar Living Center, and help reach schoolchildren with environmental educational programs.

Partnership includes concrete benefits like: free magazine subscriptions, dvds, books, and Solfest tickets. Donors of larger amounts receive a free copy of *A Place in the Sun* (the story of the evolution of the Solar Living Center), and a personalized tour of the Solar Living Center. But the real benefits of Institute Partnership will never be counted in quantifiable units. Becoming a Partner in the Solar Living Institute means joining a growing community of individuals whose vision and dedication have the potential to effect truly far-reaching change for our planet. To become a Partner in the Institute, to make a donation, or to volunteer, please contact the SLI at 707-744-2017 or email sli@solarliving.org.

Solar Living Institute interns manage the organic farm and harvest and market the produce.

Flood of 2005: Disaster Creates New Opportunity

A historic flood on December 31, 2005 caused over $150,000 damage to our educational facilities, and caused a fire to erupt in the interns' dome dwelling forcing a middle-of-the-night evacuation and subsequent boat rescue. The response from the community to this difficult situation has been extraordinary. Individuals and compa-

High water from the Russian River leaves three trucks and the biodiesel tank submerged at the Solar Living Center.

nies have provided over $250,000 to help us restore and redevelopment our site projects and replace damaged equipment.

Damage included our brand new yurt floating 150 yards away and the top was crushed; two pickups, a van, an intern's and an employee's car were submerged and destroyed; much of our SolFest signage and event materials were destroyed including the SolFest stage; most of our natural buildings were damaged or destroyed; all of our storage areas and our maintenance shed were gone; many of our site tools were damaged or destroyed; and additional

Volunteers help move the dome frame after it was destroyed by fire.

damage was done to landscape, fences, the dock, and the road. The good news was no one was hurt, and nothing was damaged to prevent our workshop program or SolFest in 2006 from continuing uninterrupted.

Since one of the founding principles of permaculture is to view every problem as an opportunity, we looked at this disaster as an opportunity to build a better and more flood-proof Solar Living Center. Using the outpouring of human and financial energy, we held a charette on March 18, 2006, to develop plans for rebuilding and improving the site. Among the many changes and enhancements include:

- **Classroom Yurt Relocation** – Our deluxe, 30-foot yurt donated by Colorado Yurt Company has been moved to a beautiful new location adjacent to the ponds and the organic vegetable garden.
- **New Intern Village** – A new intern village, including an industrial-strength kitchen, will be located adjacent to the Central Oasis.
- **New Straw Bale Wall** – The largest straw bale wall in North America has been built connecting the new intern village to the new SolFest stage area.
- **New Storage Areas** – A mobile storage unit has been installed to allow for flood-season relocation of supplies and equipment.
- **New and Improved SolFest Facilities** – A new stage and storage unit will be built.

Part IV: Getting Down to Business

Remember our first principle? We never forget it. When the profitability principle is met, we are free to pursue our educational mission effectively. We achieve profitability through our catalogs, retail store, design and consulting group, and our website (www.realgoods.com). We support the principles of knowledge, involvement, credibility (and even fun!) through a variety of innovative programs, from educational opportunities to networking. When our business goals and principles work in tandem with our educational goals and principles, we sleep easier at night. These are the elements and programs which, taken together, make our business real.

Catalogs

The Gaiam Real Goods catalogs include products and information geared to folks who have been thinking a lot about bringing elements of sustainable living into their lives. Our focus is on high-impact, user-friendly merchandise like compact fluorescent lighting, air and water purification systems, energy-saving household implements, and mainstream products that introduce the concepts behind renewable energy without being dauntingly technical. We also include lots of information on sustainable living, environmental responsibility, and books, books, books. The full-color Gaiam Real Goods catalog is a lot like the Gideon's Bible in a hotel drawer. We don't expect it to produce any wholesale conversions, but we hope it steers a few people in the right direction. The Real Goods catalog also has evolved into our technical publication, for individuals who have made a serious commitment to reducing their impact on the planet, and need the tools to get on with it. Along with renewable energy system components (like PV modules, wind generators, hydroelectric pumps, cables, inverters, and controllers), our catalog offers off-the-grid and DC appliances, energy-saving climate control products, solar ovens, air and water purification systems, a smattering of green building materials, and even more books.

Since our merger with Gaiam in 2001, our customers now have access to the Gaiam Harmony catalog, which includes a huge eco-selection of products for the home and outdoors and the Gaiam Living Arts catalog, which includes great products for yoga and "mind-body-fitness." You can order either of these fine catalogs at 800-919-2400 or through www.gaiam.com.

The Gaiam Real Goods catalog in early spring 2007.

Retail Store

As a catalog and Internet business, Gaiam Real Goods can reach almost every nook and cranny in America. Even so, there will always be peo-

APPENDIX

The staff at the Real Goods store have over 30 years' combined experience, providing customers with knowledgeable support.

ple who want to "kick the tires" before making a purchase. Our Hopland Solar Living Center flagship retail store is second to none. Not only do customers have an opportunity to visit 12 gorgeous acres of permaculture gardens and enjoy dozens of interactive displays, but they get to see all the products in this *Sourcebook* and our other catalogs come to life. Our retail store offers an opportunity to snag curious passersby and show them how sustainable living can enhance their lives.

Real Goods Technical Staff

The Real Goods Technical staff are experts at designing, procuring, and installing residential renewable energy systems—both off-the-grid and line intertie systems to your utility. You don't want your project to be a test bed for unproved technology or experimental system design. You want information, products, and service that are of the highest quality. You want to leave the de-

tails to a company with the capability to do the job right the first time. We welcome the opportunity to design and plan entire systems. Here's how the Real Goods technical services work:

1. To assess your needs, capabilities, limitations, and working budget, we ask you to complete a specially created worksheet (see page xxx [SLS12, pp. 431-432]). The information required includes a list of your energy needs, an inventory of desired appliances, site information, and potential for hydroelectric and wind development. Note that we only need this detailed energy use information for off-the-grid systems.

2. A member of our technical staff will determine your wattage requirements, and design an appropriate system with you. Or, for intertie systems, we will determine either how much you want to spend or how much utility power you wish to offset, and then design an appropriate system.

3. At this point, we will begin tracking the time we spend in helping you plan the details of your installation. Your personal tech rep will work with you on an unlimited time basis until your system has been completely designed and refined. He/she will order parts and talk you through assembly, assuring you that you get precisely what you need. He will also consult and work with your licensed contractor, if need be. The first hour is free and part of our service. Beyond this initial consultation, time will be billed in 10-minute intervals at the rate of $75 per hour. If the recommended parts and equipment are purchased from Gaiam Real Goods, however, this time will be provided at no charge.

Real Goods has been providing clean, reliable, renewable energy to people all over the planet for 30 years and has provided solar energy systems for over 60,000 homes and businesses in that time. With the creation of our separate residential solar installation division, our catalog renewables technicians now have a tighter focus on off-the-grid residential and farm systems; this specialization means we are better prepared than ever to design either a utility intertie or independent (off-grid) home system that meets your unique needs.

You will leave your Real Goods technical consultation with the information you need to make informed decisions on which technologies are best suited for your particular application. You'll learn how to utilize proven technologies and techniques to reap the best environmental advantages for the lowest possible cost. And we won't leave you hanging once your system

Log onto www.
solarliving.org for
the latest on events
like the Green
Career Conference,
workshops and Earth
Day for Kids programs,
and lots more
happenings at the
Solar Living Institute.

is installed; our technicians will help you with troubleshooting, ongoing maintenance, and future upgrades. The Real Goods commitment to custom design, proven technology, and ongoing support has resulted in more than two decades of exceptionally high customer satisfaction. Real Goods specializes in residential and commercial systems of 100 watts to 100 kilowatts output, including:

- Utility Intertie with Renewable Energy
- Off-the-Grid Renewable Energy Systems Design and Installation (solar, wind, hydro)
- Large-Scale Uninterruptible Back-up Power
- High-Efficiency Appliances
- Biological Waste Treatment Systems
- Power Quality Enhancement
- Whole Systems Integration
- Water Quality and Management

Real Goods technical services and sales for residential and commercial renewable energy systems are available by phone 800-919-2400, from 7:30 a.m. to 6 p.m., Pacific Time, Monday through Friday, and 9 a.m. to 6 p.m. on Saturday. We endeavor to answer technical email within 48 hours, and snail mail (USPS) within one week.

Cyberspace Is a Tree-Free World: The Real Goods Website

The biggest conundrum we've ever faced as a company has been trying to live our environmental mission while being in the catalog business, which by definition survives by consuming trees. We have been, and will continue to be, as environmentally responsible as possible in a print media business. Our print catalogs use maximum postconsumer content recycled paper and soy-based inks, and we mail only to customers who have the highest potential for buying from us. Still you just can't beat a virtual catalog for environmental responsibility. In fact, our website uses only postconsumer recycled electrons of the finest quality. Since we first established an Internet presence in 1995, our online business has grown by leaps and bounds. One unexpected benefit of our "tree-free catalog" is the almost limitless amount of space available. Our website has evolved from a simple online catalog to a compendium of renewables resources. You'll find links to fascinating sustainability and renewable energy sites, and you can find out just about anything about our company. It's quite possible that virtual catalogs like ours will reduce (or even eliminate) the more wasteful aspects of the mail-order business in the not-too-distant future.

Real Goods Wedding Registry

Do your dreams of nuptial bliss lean toward the practical, rather than the extravagant? Does your vision of unwrapping wedding gifts reveal inverters and compact fluorescent light bulbs, rather than sterling silver tea service and a ten-slice toaster? Why not register with Real Goods, so that well-wishing friends can honor you with the gifts that won't end up in the closet. Is this a joke? No way. A number of young married couples have embarked upon their dream of living lightly on the planet with a running start, thanks to the Gaiam Real Goods Wedding Registry. Then there are those reluctant folks who need a little nudge in the direction of sustainability, and those who prefer simply to let their friends and family choose for themselves. We'll happily send a gift certificate in any amount to the person of your choice, along with our latest edition of the Gaiam Real Goods catalog.

SYSTEM SIZING WORKSHEET

AC device	Device watts	×	Hours of daily use	×	Days of use per week	÷	7	=	Average watt-hours per day
		×		×		÷	7	=	
		×		×		÷	7	=	
		×		×		÷	7	=	
		×		×		÷	7	=	
		×		×		÷	7	=	
		×		×		÷	7	=	
		×		×		÷	7	=	
		×		×		÷	7	=	
		×		×		÷	7	=	
		×		×		÷	7	=	
		×		×		÷	7	=	
		×		×		÷	7	=	
		×		×		÷	7	=	
		×		×		÷	7	=	
		×		×		÷	7	=	
		×		×		÷	7	=	
		×		×		÷	7	=	
		×		×		÷	7	=	
		×		×		÷	7	=	
		×		×		÷	7	=	

1   Total AC watt-hours/day

2   × 1.1 = Total corrected DC watt-hours/day

DC device	Device watts	×	Hours of daily use	×	Days of use per week	÷	7	=	Average watt-hours per day
		×		×		÷	7	=	
		×		×		÷	7	=	
		×		×		÷	7	=	
		×		×		÷	7	=	
		×		×		÷	7	=	
		×		×		÷	7	=	
		×		×		÷	7	=	
		×		×		÷	7	=	
		×		×		÷	7	=	

3   Total DC watt-hours/day

SYSTEM SIZING WORKSHEET

3 (from previous page)	Total DC watt-hours/day	
4	Total corrected DC watt-hours/day from Line 2 +	
5	Total household DC watt-hours/day =	
6	System nominal voltage (usually 12 or 24) ÷	
7	Total DC amp-hours/day =	
8	Battery losses, wiring losses, safety factor × 1.2	
9	Total daily amp-hour requirement =	
10	Estimated design insolation (hours per day of sun, see map on p. 434) ÷	
11	Total PV array current in amps =	
12	Select a photovoltaic module for your system	
13	Module rated power amps ÷	
14	Number of modules required in parallel =	
15	System nominal voltage (from line 6 above)	
16	Module nominal voltage (usually 12) ÷	
17	Number of modules required in series =	
18	Number of modules required in parallel (from Line 14 above) ×	
19	Total modules required =	

BATTERY SIZING

20	Total daily amp-hour requirement (from line 9)	
21	Reserve time in days ×	
22	Percent of useable battery capacity ÷	
23	Minimum battery capacity in amp-hours =	
24	Select a battery for your system, enter amp-hour capacity ÷	
25	Number of batteries in parallel =	
26	System nominal voltage (from line 6)	
27	Voltage of your chosen battery (usually 6 or 12) ÷	
28	Number of batteries in series =	
29	Number of batteries in parallel (from line 25 above) ×	
30	Total number of batteries required	

Power Consumption Table

Appliance	Watts	Appliance	Watts	Appliance	Watts
Coffeepot	200	Electric blanket	2,000	Compact fluorescent	
Coffee maker	800	Blow dryer	1,000–1,500	Incandescent equivalents	
Toaster	800–1,500	Shaver	15	40 watt equiv.	11
Popcorn popper	250	WaterPik	100	60 watt equiv.	16
Blender	300	Computer		75 watt equiv.	20
Microwave	600–1,700	Laptop	50–75	100 watt equiv.	30
Waffle iron	1,200	PC	200–600	Ceiling fan	10–50
Hot plate	1,200	Printer	100–500	Table fan	10–25
Frying pan	1,200	System (CPU, monitor, laser printer)	up to 1,500	Electric mower	1,500
Dishwasher	1,200–1,500	Fax	35	Hedge trimmer	450
Sink disposal	450	Typewriter	80–200	Weed eater	450
Washing machine		DVD Player	25	¼" drill	250
Automatic	500	TV 25" color	150+	½" drill	750
Manual	300	19" color	70	1" drill	1,000
Vacuum cleaner		12" b&w	20	9" disc sander	1,200
Upright	200–700	VCR	40–100	3" belt sander	1,000
Hand	100	CD player	35–100	12" chain saw	1,100
Sewing machine	100	Stereo	10–100	14" band saw	1,100
Iron	1,000	Clock radio	1	7¼" circular saw	900
Clothes dryer		AM/FM car tape	8	8¼" circular saw	1,400
Electric NA	4,000	Satellite dish/Internet	30–65	Refrigerator/freezer— Conventional Energy Star	
Gas heated	300–400				
		CB radio	5	23 cu. ft.	540 kWh/yr
Heater		Electric clock	3	20 cu. ft.	390 kWh/yr
Engine block NA	150–1,000	Radiotelephone		16 cu. ft.	370 kWh/yr
Portable NA	1,500	Receive	5	Sun Frost	
Waterbed NA	400	Transmit	40–150	16 cu. ft. DC (7)	112
Stock tank NA	100	Lights		12 cu. ft. DC (7)	70
Furnace blower	300–1,000	100W incandescent	100	Freezer—Conventional	
Air conditioner NA		25W compact fluorescent	28	14 cu. ft. (15)	440
Room	1,500	50W DC incandescent	50	14 cu. ft. (14)	350
Central	2,000–5,000	40W DC halogen	40	Sun Frost freezer	
Garage door opener	350	20W DC compact fluorescent	22	19 cu. ft. (10)	112

Solar Insolation Maps

The maps below show the sun-hours per day for the U.S. Charts courtesy of the D.O.E.

YEARLY AVERAGE

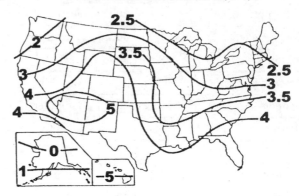

FOUR-WEEK AVERAGE 12/7-1/4

Solar Insolation by U.S. City

This chart shows solar insolation in kilowatt-hours per square meter per day in many U.S. locations. For simplicity, we call this figure "Sun Hours/Day." To find average Sun hours per day (last column) in your area, check local weather data, look at the maps above, or find a city in the table that has similar weather to your location. If you want year-round autonomy, use the figure in the Low column. If you want 100% autonomy only in summer, use the figure in the High column.

State	City	High	Low	Avg.	State	City	High	Low	Avg.	State	City	High	Low	Avg.
AK	Fairbanks	5.87	2.12	3.99	KS	Manhattan	5.08	3.62	4.57	NY	Schenectady	3.92	2.53	3.55
AK	Matanuska	5.24	1.74	3.55	KS	Dodge City	6.5	4.2	5.6	NY	Rochester	4.22	1.58	3.31
AL	Montgomery	4.69	3.37	4.23	KY	Lexington	5.97	3.6	4.94	NY	New York City	4.97	3.03	4.08
AR	Bethel	6.29	2.37	3.81	LA	Lake Charles	5.73	4.29	4.93	OH	Columbus	5.26	2.66	4.15
AR	Little Rock	5.29	3.88	4.69	LA	New Orleans	5.71	3.63	4.92	OH	Cleveland	4.79	2.69	3.94
AZ	Tucson	7.42	6.01	6.57	LA	Shreveport	4.99	3.87	4.63	OK	Stillwater	5.52	4.22	4.99
AZ	Page	7.3	5.65	6.36	MA	East Wareham	4.48	3.06	3.99	OK	Oklahoma City	6.26	4.98	5.59
AZ	Phoenix	7.13	5.78	6.58	MA	Boston	4.27	2.99	3.84	OR	Astoria	4.76	1.99	3.72
CA	Santa Maria	6.52	5.42	5.94	MA	Blue Hill	4.38	3.33	4.05	OR	Corvallis	5.71	1.9	4.03
CA	Riverside	6.35	5.35	5.87	MA	Natick	4.62	3.09	4.1	OR	Medford	5.84	2.02	4.51
CA	Davis	6.09	3.31	5.1	MA	Lynn	4.6	2.33	3.79	PA	Pittsburgh	4.19	1.45	3.28
CA	Fresno	6.19	3.42	5.38	MD	Silver Hill	4.71	3.84	4.47	PA	State College	4.44	2.79	3.91
CA	Los Angeles	6.14	5.03	5.62	ME	Caribou	5.62	2.57	4.19	RI	Newport	4.69	3.58	4.23
CA	Soda Springs	6.47	4.4	5.6	ME	Portland	5.23	3.56	4.51	SC	Charleston	5.72	4.23	5.06
CA	La Jolla	5.24	4.29	4.77	MI	Sault Ste. Marie	4.83	2.33	4.2	SD	Rapid City	5.91	4.56	5.23
CA	Inyokern	8.7	6.87	7.66	MI	East Lansing	4.71	2.7	4.0	TN	Nashville	5.2	3.14	4.45
CO	Granby	7.47	5.15	5.69	MN	St. Cloud	5.43	3.53	4.53	TN	Oak Ridge	5.06	3.22	4.37
CO	Grand Lake	5.86	3.56	5.08	MO	Columbia	5.5	3.97	4.73	TX	San Antonio	5.88	4.65	5.3
CO	Grand Junction	6.34	5.23	5.85	MO	St. Louis	4.87	3.24	4.38	TX	Brownsville	5.49	4.42	4.92
CO	Boulder	5.72	4.44	4.87	MS	Meridian	4.86	3.64	4.43	TX	El Paso	7.42	5.87	6.72
DC	Washington	4.69	3.37	4.23	MT	Glasgow	5.97	4.09	5.15	TX	Midland	6.33	5.23	5.83
FL	Apalachicola	5.98	4.92	5.49	MT	Great Falls	5.7	3.66	4.93	TX	Fort Worth	6	4.8	5.43
FL	Belle Isle	5.31	4.58	4.99	MT	Summit	5.17	2.36	3.99	UT	Salt Lake City	6.09	3.78	5.26
FL	Miami	6.26	5.05	5.62	NM	Albuquerque	7.16	6.21	6.77	UT	Flaming Gorge	6.63	5.48	5.83
FL	Gainesville	5.81	4.71	5.27	NB	Lincoln	5.4	4.38	4.79	VA	Richmond	4.5	3.37	4.13
FL	Tampa	6.16	5.26	5.67	NB	North Omaha	5.28	4.26	4.9	WA	Seattle	4.83	1.6	3.57
GA	Atlanta	5.16	4.09	4.74	NC	Cape Hatteras	5.81	4.69	5.31	WA	Richland	6.13	2.01	4.44
GA	Griffin	5.41	4.26	4.99	NC	Greensboro	5.05	4	4.71	WA	Pullman	6.07	2.9	4.73
HI	Honolulu	6.71	5.59	6.02	ND	Bismarck	5.48	3.97	5.01	WA	Spokane	5.53	1.16	4.48
IA	Ames	4.8	3.73	4.4	NJ	Seabrook	4.76	3.2	4.21	WA	Prosser	6.21	3.06	5.03
ID	Boise	5.83	3.33	4.92	NV	Las Vegas	7.13	5.84	6.41	WI	Madison	4.85	3.28	4.29
ID	Twin Falls	5.42	3.42	4.7	NV	Ely	6.48	5.49	5.98	WV	Charleston	4.12	2.47	3.65
IL	Chicago	4.08	1.47	3.14	NY	Binghamton	3.93	1.62	3.16	WY	Lander	6.81	5.5	6.06
IN	Indianapolis	5.02	2.55	4.21	NY	Ithaca	4.57	2.29	3.79					

Magnetic Declinations in the United States

Figure indicates correction of compass reading to find true north. For example, in Wasington state when you compans reads 22°E, it is pointing due notrth.

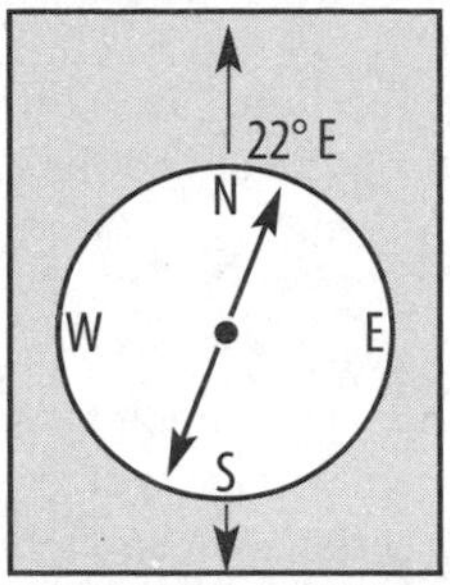

TRUE SOUTH IF STANDING IN NORTH WASHINGTON

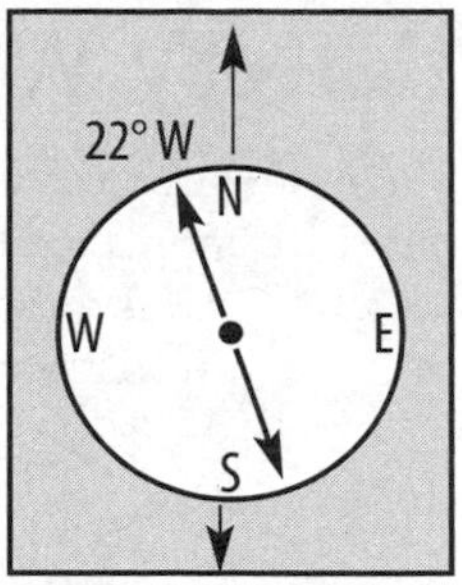

TRUE SOUTH IF STANDING IN NEW BRUNSWICK

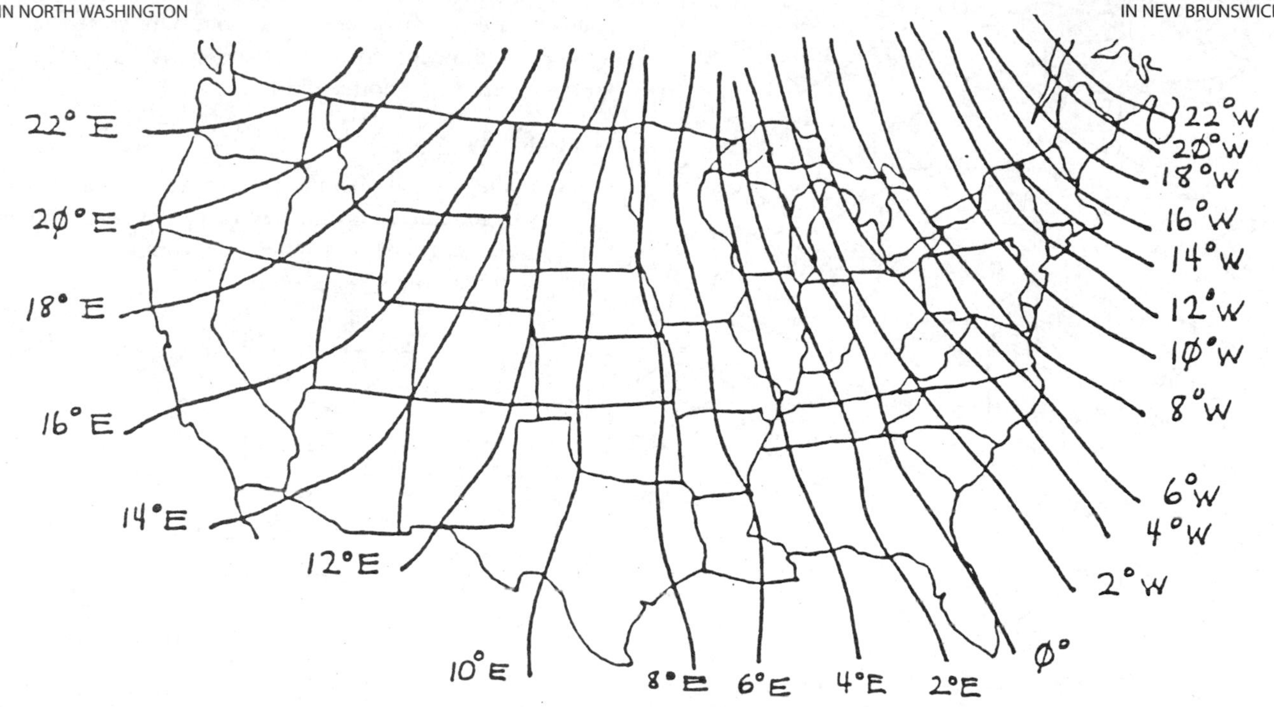

MAXIMUM NUMBER OF CONDUCTORS A FOR A GIVEN CONDUIT SIZE.

Conduit size		½"	¾"	1"	1¼"	1½"	2"
Conductor size	#12	10	18	29	51	70	114
	#10	6	11	18	32	44	73
	#8	3	5	9	16	22	36
	#6	1	4	6	11	15	26
	#4	1	2	4	7	9	16
	#2	1	1	3	5	7	11
	#1		1	1	3	5	8
	#1/0		1	1	3	4	7
	#2/0		1	1	2	3	6
	#3/0		1	1	1	3	5
	#4/0		1	1	1	2	4

Battery Wiring Diagrams

The following diagrams show how 2-, 6- and 12-volt batteries are connected for 12-, 24- and 48-volt operation.

Fuse symbol (always use appropriate fusing at your battery).

Wiring Basics

We answer a lot of basic wiring questions over the phone, which we're always happy to do, but there's nothing like a picture or two to make things apparent.

Battery Wiring

The batteries for your energy system may be supplied as 2-volt, 6-volt, or 12-volt cells. Your system voltage is probably 12, 24, or 48 volts. You'll need to series wire enough batteries to reach your system voltage, then parallel wire to another series group as needed to boost amperage capacity. See our drawings for correct series wiring. Paralleled groups are shown in dotted outline.

PV Module Wiring

PV modules are almost universally produced as nominal 12-volt modules. For smaller 12-volt systems this is fine. Most larger residential systems are configured for 24- or 48-volt input now.

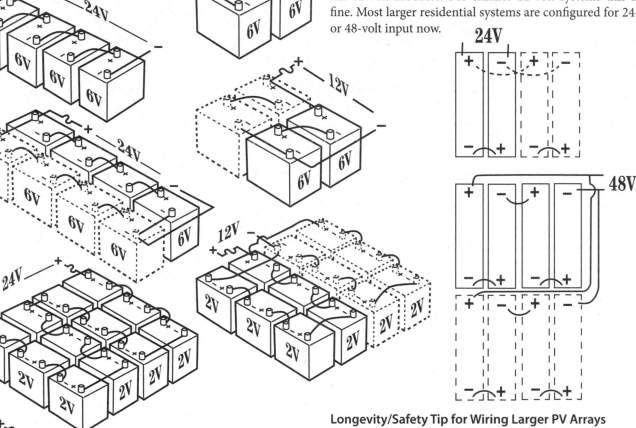

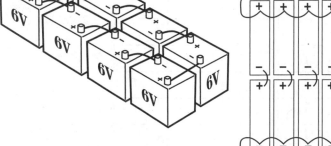

Longevity/Safety Tip for Wiring Larger PV Arrays

If you have a large PV array that produces close to, or over, 20 amps, multiple power take-off leads are a good idea. They may prevent toasted terminal boxes. Instead of taking only a single pair of positive and negative leads off some point on the array, take off one pair at one end, and another pair off the opposite end. Then join them back together at the array-mounted junction box where you're going to the larger wire needed for transmission. This divides up the routes that outgoing power can take, and eases the load on any single PV junction box.

Wire Sizing Chart/Formula

This chart is useful for finding the correct wire size for any voltage, length, or amperage flow in any AC or DC circuit. For most DC circuits, particularly between the PV modules and the batteries, we try to keep the voltage drop to 3% or less. There's no sense using your expensive PV wattage to heat wires. You want that power in your batteries!

Note that this formula doesn't directly yield a wire gauge size, but rather a "VDI" number, which is then compared to the nearest number in the VDI column, and then read across to the wire gauge size column.

1. Calculate the Voltage Drop Index (VDI) using the following formula:

   VDI = AMPS x FEET ÷ (% VOLT DROP x VOLTAGE)

   Amps = Watts divided by volts

   Feet = One-way wire distance

   % Volt Drop = Percentage of voltage drop acceptable for this circuit (typically 2% to 5%)

2. Determine the appropriate wire size from the chart below.

   A. Take the VDI number you just calculated and find the nearest number in the VDI column, then read to the left for AWG wire gauge size.

   B. Be sure that your circuit amperage does not exceed the figure in the Ampacity column for that wire size. (This is not usually a problem in low-voltage circuits.)

Example: Your PV array consisting of four Sharp 80-watt modules is 60 feet from your 12-volt battery. This is actual wiring distance, up pole mounts, around obstacles, etc. These modules are rated at 4.63 amps x 4 modules = 18.5 amps maximum. We'll shoot for a 3% voltage drop. So our formula looks like:

VDI = (18.5A x 60 ft) ÷ (3% x 12V) = 30.8

Looking at our chart, a VDI of 31 means we'd better use #2 wire in copper, or #0 wire in aluminum. Hmmm. Pretty big wire.

What if this system was 24-volt? The modules would be wired in series, so each pair of modules would produce 4.4 amps. Two pairs x 4.63 amps = 9.3 amps max.

VDI = (9.3A x 60 ft) ÷ (3% x 24V) = 7.8

Wow! What a difference! At 24-volt input you could wire your array with little ol' #8 copper wire.

Wire Size	Copper Wire		Aluminum Wire	
AWG	VDI	Ampacity	VDI	Ampacity
0000	99	260	62	205
000	78	225	49	175
00	62	195	39	150
0	49	170	31	135
2	31	130	20a	100
4	20	95	12	75
6	12	75	•	•
8	8	55	•	•
10	5	30	•	•
12	3	20	•	•
14	2	15	•	•
16	1	•	•	•

Chart developed by John Davey and Windy Dankoff. Used with permission.

Why There's No National Eleectric Code in This *Sourcebook*

Previous editions of this *Sourcebook* have printed in full the Suggested Practices of the National Electric Code with respect to Photovoltaic Power Systems. This time around, to save paper, trees, and unnecessary costs and because the NEC is so easily accessible (and free!) on the Internet, we recommend you download it at: http://www.nmsu.edu/~tdi/Photovoltaics/Codes-Stds/Codes-Stds.html. If you don't have access to the Internet, you can purchase the most recent handbook (see p. 535).

APPENDIX

Friction Loss Charts for Water Pumping

HOW TO USE PLUMBING FRICTION CHARTS

If you try to push too much water through too small a pipe, you're going to get pipe friction. Don't worry, your pipes won't catch fire. But it will make your pump work harder than it needs to, and it will reduce your available pressure at the outlets, so sprinklers and showers won't work very well. These charts can tell you if friction is going to be a problem. Here's how to use them:

PVC or black poly pipe? The rates vary, so first be sure you're looking at the chart for your type of supply pipe. Next, figure out how many gallons per minute you might need to move. For a normal house, 10-15 gpm is probably plenty. But gardens and hoses really add up. Give your-self about 5 gpm for each sprinkler or hose that might be running. Find your total (or something close to it) in the "Flow GPM" column. Read across to the column for your pipe diameter. This is how much pressure loss you'll suffer for every 100 feet of pipe. Smaller numbers are better.

Example: You need to pump or move 20 gpm through 500 feet of PVC between your storage tank and your house. Reading across, 1-inch pipe is obviously a problem. How about 1¼ inches? 9.7 psi times 5 (for your 500 feet) = 48.5 psi loss. Well, that won't work! With 1½-inch pipe, you'd lose 20 psi . . . still pretty bad. But with 2-inch pipe, you'd lose ony 4 psi . . . ah! Happy garden sprinklers! Generally, you want to keep pressure losses under about 10 psi.

Friction Loss in PSI per 100 Feet of Scheduled 40 PVC Pipe

Flow GPM	Nominal Pipe Diameter in Inches							
	½"	¾"	1"	1¼"	1½"	2"	3"	4"
1	3.3	0.5	0.1					
2	11.9	1.7	0.4	0.1				
3	25.3	3.5	0.9	0.3	0.1			
4	43.0	6.0	1.5	0.5	0.2	0.1		
5	65.0	9.0	2.2	0.7	0.3	0.1		
10		32.5	8.0	2.7	1.1	0.3		
15		68.9	17.0	5.7	2.4	0.6	0.1	
20			28.9	9.7	4.0	1.0	0.1	
30			61.2	20.6	8.5	2.1	0.3	0.1
40				35.1	14.5	3.6	0.5	0.1
50				53.1	21.8	5.4	0.7	0.2
60				74.4	30.6	7.5	1.0	.03
70					40.7	10.0	1.4	0.3
80					52.1	12.8	1.8	0.4
90					64.8	16.0	2.2	0.5
100					78.7	19.4	2.7	0.7
150						41.1	5.7	1.4
200						69.9	9.7	2.4
250							14.7	3.6
300							20.6	5.1
400							35.0	8.6

Friction Loss in PSI per 100 Feet of Polyethylene (PE) SDR-Pressure Rated Pipe

Flow GPM	Nominal Pipe Diameter in Inches							
	0.5	0.75	1	1.25	1.5	2	2.5	3
1	0.49	0.12	0.04	0.01				
2	1.76	0.45	0.14	0.04	0.02			
3	3.73	0.95	0.29	0.08	0.04	0.01		
4	**6.35**	1.62	0.50	0.13	0.06	0.02		
5	9.60	2.44	0.76	0.20	0.09	0.03		
6	13.46	3.43	1.06	0.28	0.13	0.04	0.02	
7	17.91	4.56	1.41	0.37	0.18	0.05	0.02	
8	22.93	**5.84**	1.80	0.47	0.22	0.07	0.03	
9		7.26	2.24	0.59	0.28	0.08	0.03	
10		8.82	2.73	0.72	0.34	0.10	0.04	0.01
12		12.37	**3.82**	1.01	0.48	0.14	0.06	0.02
14		16.46	5.08	1.34	0.63	0.19	0.08	0.03
16			6.51	1.71	0.81	0.24	0.10	0.04
18			8.10	2.13	1.01	0.30	0.13	0.04
20			9.84	2.59	1.22	0.36	0.15	0.05
22			11.74	**3.09**	1.46	0.43	0.18	0.06
24			13.79	3.63	1.72	0.51	0.21	0.07
26			16.00	4.21	1.99	0.59	0.25	0.09
28				4.83	2.28	0.68	0.29	0.10
30				5.49	**2.59**	0.77	0.32	0.11
35				7.31	3.45	1.02	0.43	0.15
40				9.36	4.42	1.31	0.55	0.19
45				11.64	5.50	1.63	0.69	0.24
50				14.14	6.68	**1.98**	0.83	0.29
55					7.97	2.36	0.85	0.35
60					9.36	2.78	1.17	0.41
65					10.36	3.22	1.36	0.47
70					12.46	3.69	**1.56**	0.54
75					14.16	4.20	1.77	0.61
80						4.73	1.99	0.69
85						5.29	2.23	0.77
90						5.88	2.48	0.86
95						6.50	2.74	0.95
100						7.15	3.01	**1.05**
150						15.15	6.38	2.22
200							10.87	3.78
300								8.01

Temperature Conversions

°C = Degrees Celsius. 1 degree is $\frac{1}{100}$ of the difference between the temperature of melting ice and boiling water.

°F = Degrees Fahrenheit. 1 degree is $\frac{1}{180}$ of difference between the temperature of melting ice and boiling water.

Temperature Conversion Chart							
°C	°F	°C	°F	°C	°F	°C	°F
200	392	140	284	80	176	15	59
195	383	135	275	75	167	10	50
190	374	130	266	70	158	5	41
185	365	125	257	65	149	0	32
180	356	120	248	60	140	−5	23
175	347	115	239	50	122	−10	14
170	338	110	230	45	113	−15	5
165	329	105	221	40	104	−20	−4
160	320	100	212	35	95	−25	−13
155	311	95	203	30	86	−30	−22
150	302	90	194	25	77	−35	−31
145	293	85	185	20	68	−40	−40

Nominal Pipe Size versus Actual Outside Diameter for Steel and Plastic Pipe

Nominal Pipe Size vs. Actual Outside Diameter for Steel and Plastic Pipe			
Nominal Size	Actual Size	Nominal Size	Actual Size
½"	0.840"	2½"	2.875"
¾"	1.050"	3"	3.500"
1"	1.315"	3½"	4.000"
1¼"	1.660"	4"	4.500"
1½"	1.900"	5"	5.563"
2"	2.375"	6"	6.625"

The Real Goods Resource List

Although this *Sourcebook* is a complete source of renewable energy and environmental products, we can't be everything to everyone. Here is our current list of other trusted resources for information and products. We've selected these organizations carefully, and since we have worked directly with many of them, we are giving you the benefit of our experience. Still, we'll offer the standard disclaimer that Gaiam Real Goods does not necessarily endorse all the actions of each group listed, nor are we responsible for what they say or do.

This list can never be complete. One good place to look for additional resources is www.wiserearth.org, an amazing compendium of tens of thousands of nonprofits compiled by Paul Hawken over many years (also see his book *Blessed Unrest*, p. 29). We apologize for resources we may have overlooked and for contact information that may have changed since this 30th-Anniversary Edition of the *Sourcebook* went to press. We welcome your suggestions. Please mail a brief description of the organization and access info to Gaiam Real Goods, Attention: Resource List.

AAP Automation
Website: www.aapautomation.com/
Frames/indexAAPAuto.html
If you're into high-level engineering, automation, motors, pneumatics, controls, shop talk, and the like, this site might be for you!

Advanced Buildings
Website: www.advancedbuildings.org/
Email: advancedbuildings@
enermodal.com
Tremendous cache of resources and case studies, geared toward building professionals interested in technologies and practices for environmentally appropriate and energy-efficient construction.

American Council for an Energy-Efficient Economy (ACEEE)
1001 Connecticut Avenue, Suite 801
Washington, DC 20036
Website: www.aceee.org
Email: info@aceee.org
Phone (Research and Conferences): 202-429-8873; Publications: 202-429-0063
Publishes books, papers, yearly guides, and comparisons of appliances and vehicles based on energy efficiency. Their website is an excellent source of efficient appliance info.

The American Hydrogen Association
2350 W. Shangri La
Phoenix, AZ 85028
Website: www.clean-air.org/
Email: contact@clean-air.org, question@clean-air.org
Phone: 602-328-4238
A nonprofit organization that promotes the use of hydrogen for fuel and energy storage. Publishes Hydrogen Today, *a bimonthly newsletter.*

American Society of Landscape Architects (ASLA)
636 Eye Street NW
Washington, DC 20001-3736
Website: www.asla.org
Phone: 202-898-2444
This professional organization advocates on public policy issues such as livable communities, surface transportation, the environment, historic preservation, and small business affairs.

American Solar Energy Society (ASES)
2400 Central Avenue, Suite A
Boulder, CO 80301
Website: www.ases.org/
Email: ases@ases.org
Phone: 303-443-3130
ASES is the United States section of the International Solar Energy Society, a national organization dedicated to advancing the use of solar energy for the benefit of U.S. citizens and the global environment. Publishers of the bimonthly Solar Today magazine for members, and sponsors of the yearly National Tour of Solar Homes.

The American Wind Energy Association
1101 14th Street NW, 12th Floor
Washington, DC 20005
Website: www.awea.org/
Email: windmail@awea.org
Phone: 202-383-2500
A national trade association that represents wind-power plant developers, wind turbine manufacturers, utilities, consultants, insurers, financiers, researchers, and others involved in the wind industry. These folks primarily work with utility-level wind systems. Not a good source for residential info.

Architects/Designers/Planners for Social Responsibility (ADPSR)
PO Box 9126
Berkeley, CA 94709-0126
Website: www.adpsr.org
Email: forum@adpsr.org
Phone: 510-845-1000
ADPSR works for peace, environmental protection, ecological building, social justice, and the development of healthy communities. It has chapters around the country and publishes books on community arts and development through New Village Press (www.newvillage.net).

Artha Sustainable Living Center
9784 County Road K
Amherst, WI 54406-9355
Website: www.arthaonline.com
Email: info@arthaonline.com
Phone: 715-824-3463
Offers solar thermal (water and/or space heating) site assessments for large applications worldwide. Offers hands-on workshops and training at their site or yours on solar thermal and sustainable living.

Battery Recycling (for NiCads)
Rechargeable Battery Recycling Corp.
1000 Parkwood Circle, Suite 450
Atlanta, GA 30339
Website: www.rbrc.org
Email: recycling@rbrc.com
Phone: 678-419-9990
A nonprofit public service organization to promote the recycling of NiCad batteries. Just type your zip code in

the website and get a list of local stores that will accept your old NiCads for recycling.

Bioneers
Website: www.bioneers.org
Bioneers, whose slogan is Revolution from the Heart of Nature, is "a forum for connecting the environment, health, social justice, and spirit within a broad progressive framework." Their annual conference, focused on practical and visionary solutions for restoring imperiled ecosystems and healing human communities, is extraordinary.

The Borrower's Guide to Financing Solar Energy Systems
Website: www.nrel.gov/docs/fy99osti/26242.pdf
A downloadable publication of the National Renewable Energy Lab.

BuildingGreen, Inc.
122 Birge Street, Suite 30
Brattleboro, VT 05301
Website: www.buildinggreen.com
Email: info@buildinggreen.com
Phone: 802-257-7300
A leading source of objective information on green building and renewable energy. Publishes the journal Environmental Building News *and the* GreenSpec Directory *of green building products.*

California Energy Commission
Media and Public Communications Office
1516 Ninth Street, MS-29
Sacramento, CA 95814-5512
Website: www.energy.ca.gov
Email: mediaoffice@energy.state.ca.us
Phone: 800-555-7794 (inside CA); 916-654-4058 (outside CA)
The state of California's primary energy policy and planning agency. Strongly supports energy efficiency and small-scale utility intertie projects. A good source of honest intertie information.

California Solar Center
Website: www.californiasolarcenter.org/downloads.html
Lots of great downloads of California-specific information, reports, etc.

California Straw Building Association (CASBA)
PO Box 1293
Angels Camp, CA 95222
Website: www.strawbuilding.org
Email: mbennjr@mac.com
Phone: 209-785-7077
CASBA is a nonprofit project of the Tides Center. Its primary objective is to further the practice of straw bale construction by exchanging current information and practical experience, conducting research and testing, and making that body of knowledge available to working professionals and the public at large.

CalStart
Northern California Office:
1160 Brickyard Cove, Suite 101
Richmond, California 94801
Phone: 510-307-8700
Southern California Office:
48 South Chester Avenue
Pasadena, CA 91106
Phone: 626-744-5600
Colorado Office:
1536 Wynkoop Street, Suite 600
Denver, CO 80202
Phone: 303-825-7550
Website: www.calstart.org
Email: calstart@calstart.org
The latest information on electric, natural gas, and hybrid electric vehicles, and other intelligent transportation technologies.

Carbohydrate Economy Clearinghouse
Institute for Local Self-Reliance
1313 5th Street SE
Minneapolis, MN 55414-1546
Website: www.carbohydrateeconomy.org/
Email: info@ilsr.org
Phone: 612-379-3815
Biofuels is one aspect of the emerging "carbohydrate economy," which is based on biomass rather than petroleum. Lots of great information here about biofuels and other biomass issues. The parent organization, the Institute for Local Self-Reliance, has been around for more than 30 years, leading the relocalization movement before it was a movement with a name.

Center for Energy and Climate Solutions/Cool Companies
Website: www.cool-companies.org/homepage.cfm
CECS promotes clean and efficient energy technologies as a money-saving tool for reducing greenhouse gas emissions and other pollutants. Its website contains good information, including more skeptical views about a hydrogen-based economy.

Center for Renewable Energy and Sustainable Technology (CREST)
See Renewable Energy Policy Project

Chelsea Green Publishing
85 N. Main Street, Suite 120
PO Box 428
White River Junction, VT 05001
Website: www.chelseagreen.com
Email: seaton@chelseagreen.com
Phone: 802-295-6300
One of the world's premier publishers of books on renewable energy, natural building, and sustainable living.

Clean Energy First!
706 North Ivy Street
Arlington, VA 22201
Website: www.CleanEnergyFirst.com
Email: CleanEnergyFirst@aol.com
This site hosts a variety of clean energy web links.

Electric Auto Association
PO Box 6661
Concord, CA 94514
Website: www.eaaev.org
Email: contact@eaaev.org
The national electric vehicle association, with local chapters in most states and Canada. Dues are $39/year with an excellent monthly newsletter. The website has links to practically everything in the EV biz.

Energy Efficiency and Renewable Energy Clearinghouse (EREC)
Mail Stop EE-1
Department of Energy
Washington, DC 20585
Website: www.eere.energy.gov/
Email: EEREMailbox@EE.DOE.Gov
Phone: 877-337-3463

The best source of information about renewable energy technologies and energy efficiency. These folks seem to be able to find good information about anything energy related. One of the best overall websites about energy.

The Energy & Environmental Building Association (EEBA)
6520 Edenvale Boulevard, Suite 112
Eden Prairie, MN 55346
Website: www.eeba.org
Email: inquiry@eeba.org
Phone: 952-881-1098
Provides education and resources to transform the residential design and construction industry to profitably deliver energy-efficient and environmentally responsible buildings and communities.

Energy Information Administration
1000 Independence Avenue SW
Washington, DC 20585
Website: www.eia.doe.gov
Email: infoctr@eia.doe.gov
Phone: 202-586-8800
Official energy statistics from the U.S. government. Gas, oil, electricity, you name it, there's more supply, pricing, and use info here than you can shake a stick at.

Energy Star®
Environmental Protection Agency (EPA)
ENERGY STAR Hotline (6202J)
1200 Pennsylvania Avenue NW
Washington, DC 20460
Website: www.energystar.gov/
Email: info@energystar.gov
Phone: 888-STAR-YES
(888-782-7937)
Good, up-to-date listings of the most-efficient appliances, lights, windows, home and office electronics, and more.

Environment California
Lobby office: 1107 9th Street, Ste, 601
Sacramento, CA 95814
Main office: 3435 Wilshire Boulevard,
Suite 365
Los Angeles, CA 90010
Website: www.environmentcalifornia.
org

Email: info@environmentcalifornia.
org
Phone: 213-251-3688
Maintains a website committed to providing California consumers with up-to-date information on solar electric and solar water heating systems for new and existing homes.

The Environmental and Energy Study Institute
122 C Street NW, Suite 630
Washington, DC 20001
Website: www.eesi.org
Email: eesi@eesi.org
Phone: 202-628-1400
This nonprofit dedicated to sustainability has programs on climate and energy, agriculture and energy, transportation and energy, and energy and smart growth. The website contains many publications, press releases, and updates and offers a great way to keep current with relevant public policy deliberations.

Environmental News Network
Website: www.enn.com
ENN rounds up the most important and compelling environmental news stories of the week.

Environmental Protection Agency (EPA)
Ariel Rios Building
1200 Pennsylvania Avenue NW
Washington, DC 20460
Website: www.epa.gov
Email: public-access@epamail.epa.
gov
Phone: 202-272-0167

Findsolar.com
Website: www.Findsolar.com
This is a free public service to help home and building owners estimate the cost and benefits of a solar energy system, and to help them select a qualified installer, designer, or other solar professional.

Florida Solar Energy Center (FSEC)
1679 Clearlake Road
Cocoa, FL 32922-5703
Website: www.fsec.ucf.edu/en
Email: info@fsec.ucf.edu

Phone: 321-638-1000
Provides highly regarded independent third-party testing and certification of solar hot-water systems and other solar or energy-efficiency goods.

Gas Appliance Manufacturers Association
2107 Wilson Boulevard, Suite 600
Arlington, VA 22201
Website: www.gamanet.org
Email: info@gamanet.org
Phone: 703-525-7060

Geothermal Resources Council
2001 Second Street, Suite 5
PO Box 1350
Davis, CA 95617-1350
Website: www.geothermal.org
Email: grc@geothermal.org
Phone: 530-758-2360
A nonprofit organization dedicated to geothermal research and development.

Go Solar California
California Public Utilities Commission
505 Van Ness Avenue
San Francisco, CA 94102
Website: www.gosolarcalifornia.
ca.gov
Phone: 415-703-2782
Joint website of the California Public Utilities Commission and the California Energy Commission, providing a portal into California's Million Solar Roofs Initiative, with resources for homeowners, businesses, schools, and public buildings.

Green Living Journal: A Practical Journal for Friends of the Environment
100 Gilead Brook Road
Randolph, VT 05060
Website: www.greenlivingjournal.com
Phone: 802-234-9101
Green Living is a quarterly, grass-roots publication that is published in local editions (currently Vermont, New Hampshire, Massachusetts, southern Oregon) to serve "friends of the environment," businesses, and individuals who value the protection of natural resources. Local publishers needed for new editions.

GreenMoneyJournal.com
www.greenmoneyjournal.com/
A leading periodical about socially responsible investing and green business, "From the stock market to the supermarket." They've been doing it for 15 years.

The Green Power Network
Net Metering and Green Power information
Website: www.eere.energy.gov/greenpower/
Information on the electric power industry's green power efforts. Includes up-to-dat-e info on green power availability and pricing. Also the best state-by-state net metering details on the Net.

Grist.org
Website: www.grist.org/
An excellent online periodical dedicated to environmental news and commentary.

The Heartwood School for the Homebuilding Crafts
Johnson Hill Road
Washington, MA 01223
Website: www.heartwoodschool.com
Email: request@heartwoodschool.com
Phone: 413-623-6677
Offers a variety of hands-on building workshops. Specializes primarily in timber frames.

Home Power Magazine
PO Box 520
Ashland, OR 97520
Website: www.homepower.com
Email: hp@homepower.com
Phone: 800-707-6585
The journal of the renewable energy industry. Published bimonthly. U.S. subscription is $24.95/yr.

International Institute for Ecological Agriculture (IIEA)
309 Cedar Street #127
Santa Cruz, CA 95060
Website: www.permaculture.com
Email: info@permaculture.com
Phone: 888-Permaculture (737-6228)

Provides courses, consulting, and books on permaculture design and alcohol fuel. Publishers of Alcohol Can Be a Gas, the only comprehensive manual on farm-scale alcohol production and its use. Also provides services for design and planting of high-value timber retirement or endowment forests that mature in 15-20 years.

The Last Straw Online
PO Box 22706
Lincoln, NE 68542-2706
Website: www.thelaststraw.org
Email: thelaststraw@thelaststraw.org
Phone: 402-483-5135
An excellent hub site with links to many other straw bale and sustainable building sites. This is also the online home of The Last Straw, an excellent straw bale quarterly journal ($32/yr. in U.S.).

Mendocino Organic Network (MON)
680 Blue Oak
Ukiah, CA 95482
Website: www.mendocinoorganicnetwork.com
Email: mendorenegade@gmail.com
Phone: 707-462-1776
A nonprofit organization that promotes organic farming and business; a project of the Cloud Forest Institute. Also includes Mendocino Renegade (www.mendocinorenegade.com), an affordable, no-hassle nonprofit, nongovernmental "beyond organic" certification program for Mendocino County.

Midwest Renewable Energy Association
7558 Deer Road
Custer, WI 54423
Website: www.the-mrea.org
Email: info@the-mrea.org
Phone: 715-592-6595
A regional organization that promotes renewable energy, energy efficiency, and sustainable living through education and demonstration.

Mother Jones
Website: www.motherjones.com

A leading progressive magazine of politics and analysis: "Smart, fearless journalism."

National Association of Home Builders (NAHB)
1201 15th Street NW
Washington, DC 20005
Website: www.nahb.org
Email: info@nahb.com
Phone: 800-368-5242
Now has an active green building section.

National Association of State Energy Officials (NASEO)
1414 Prince Street, Suite 200
Alexandria, VA 22314
Website: www.naseo.org
Email: mnew@naseo.org
Phone: 703-299-8800
An excellent portal site with access info for all state energy offices.

National Biodiesel Board
3337A Emerald Lane
PO Box 104898
Jefferson City, MO 65110-4898
Website: www@biodiesel.org
Email: info@biodiesel.org
Phone: 800-841-5849
This advocacy organization has an excellent website brimming with all things biodiesel.

National Center for Appropriate Technology
PO Box 3838
Butte, MT 59702
Website: www.ncat.org
Email: info@ncat.org
Phone: 800-275-6228
A nonprofit organization and information clearinghouse for sustainable energy, low-income energy, resource-efficient housing, and sustainable agriculture.

National Renewable Energies Laboratory (NREL)
1617 Cole Boulevard
Golden, CO 80401-3393
Website: www.nrel.gov
Email: webmaster@nrel.gov
Phone: 303-275-3000

The nation's leading center for renewable energy research. Tests products, conducts experiments, and provides information on renewable energy.

New Society Publishers
PO Box 189
Gabriola Island, BC
Canada V0R 1X0
Website: www.newsociety.com
Email: info@newsociety.com
Phone: 250-247-9737
Leading publisher of books for renewable energy, alternative fuels, and sustainable living, including the Solar Living Sourcebook. New Society has just gone carbon-neutral!

The North American Board of Certified Energy Practitioners (NABCEP)
10 Hermes Road, Suite 400
Malta, NY 12020
Website: www.nabcep.org
Phone: 518-899-8186
Offers certification for renewable energy installers.

Ode Magazine
Website: www.odemagazine.com
This inspirational magazine focuses on positive new people and ideas that are changing the world for the better.

Oikos Green Building Source
Website: www.oikos.com/index.php
Oikos is a website geared to professionals whose work promotes sustainable design and construction. Lots of good books, resources, and links here.

Plenty Magazine
Website: www.plentymagazine.com
This green living site is more than a magazine, including podcasts, information on "green gear," blogs, and more.

The Public Press
100 Gilead Brook Road
Randolph, VT 05060
Website: www.thepublicpress.com
Phone: 802-234-9101

The Public Press is a publisher of books that are too special, too controversial, or too experimental for conventional publishers. The company goal is to protect freedom of speech "word by word."

The Rahus Institute
Website: www.rahus.org
A nonprofit research and educational organization that focuses on resource efficiency. Website has news, products, and books, and offers consulting.

RenewableEnergyAccess.com
Website: www.renewableenergyaccess. com/rea/home
News, podcasts, products, job opportunities, events, and all things renewable energy.

Renewable Energy Policy Project and Center for Renewable Energy and Sustainable Technology (CREST)
1612 K Street NW, Suite 202
Washington DC 20006
Website: www.crest.org
Email: info2@repp.org
Phone: 202-293-2898
An excellent online resource for sustainable energy information. Provides Internet services, software, databases, resource lists, and discussion lists on a wide variety of topics.

Rocky Mountain Institute
2317 Snowmass Creek Road
Snowmass, CO 81654
Website: www.rmi.org
Email: outreach@rmi.org
Phone: 970-927-3851
A terrific source of books, papers, and research on renewable energy, energy-efficient building design and components, and sustainable transportation.

Sandia National Laboratory's Photovoltaic Systems Program
Website: www.sandia.gov/pv/
Check out the Design Assistance Center.

Small-Scale Sustainable Infrastructure Development Fund
#800, 14th Cross, J.P. Nagar 1st Phase, Bangalore
560078, Karnataka, India
Website: www.s3idf.org
Email: info@s3idf.org
Phone: 91-80-56902558
Known affectionately as S3IDF, this is a "social merchant bank" that helps small enterprises provide modern energy and other infrastructural services to poor people in developing countries in ways that are financially sustainable and environmentally responsible.

Solar Cookers International
1919 21st Street #101
Sacramento, CA 95814
Website: http://solarcookers.org/
Email: info@solarcookers.org
Phone: 916-455-4499
A nonprofit organization that promotes and distributes simple solar cookers in developing countries to relieve the strain of firewood collection. Check out an amazing online archive: http://solarcooking.org; newsletter and small products catalog available.

Solar Electric Light Fund
1612 K Street NW, Suite 402
Washington, DC 20006
Website: www.self.org/
Email: info@self.org
Phone: 202-234-7265
A nonprofit organization promoting PV rural electrification in developing countries.

Solar Energy Industries Association (SEIA)
805 15th Street NW, Suite 510
Washington, DC 20005
Website: www.seia.org
Email: info@seia.org
Phone: 202-628-0556

Solar Energy Info for Planet Earth
Website: http://eosweb.larc.nasa. gov/sse/
Complete solar energy data for anyplace on the planet! Thanks to NASA's Earth Science Enterprise program, there's more solar data here

than even the Gaiam Real Goods techies can find a use for. Just point to your location on a world map.

Solar Energy International
76 S. 2nd Street
PO Box 715
Carbondale, CO 81623
Website: www.solarenergy.org
Email: sei@solarenergy.org
Phone: 970-963-8855
Offers hands-on and online classes on renewable technology.

Solar Living Institute
13771 S. Highway 101
Hopland, CA 95449
Website: www.solarliving.org
Email: sli@solarliving.org
Phone: 707-744-2017
A nonprofit educational organization that offers world-class workshops on renewable energy, sustainable living, green building, and permaculture. Sponsors annual SolFest energy fair. Sign up for electronic newsletter: www. solarliving.org.

Solar Today
Website: www.solartoday.com
A leading magazine in the solar field: "Delivering today's news on solar energy technology."

SustainableABC.com, Sustainable Architecture, Building, and Culture
PO Box 30085
Santa Barbara, CA 93130
Website: www.SustainableABC.com
Email: royprince@sustainableabc.com
A unique compendium of links and content oriented to the global community of ecological and natural building proponents. Has a free newsletter.

Sustainable Buildings Industry Council
1112 16th Street NW, Suite 240
Washington, DC 20036
Website: www.sbicouncil.org
Email: sbic@sbicouncil.org
Phone: 202-628-7400
This trade organization has a nice website with lots of information about workshops, publications, and links to solar energy organizations. Especially good for professionals, but useful for home owners, too.

SustainableBusiness.com
Website: www.sustainablebusiness.com/
On online newsletter and news service dedicated to green business, including sections about progressive investing, green dream jobs, and a great resource directory.

Sustainable By Design
3631 Bagley Avenue N
Seattle, WA 98103
Website: http://susdesign.com/tools.php
Email: christopher@susdesign.com
Phone: 206-925-9290
Provides shareware tools for sustainable design.

The Union of Concerned Scientists
2 Brattle Square
Cambridge, MA 02238-9105
Website: www.ucsusa.org
Email: ucs@ucsusa.org
Phone: 617-547-5552
Organization of scientists and citizens concerned with the impact of advanced technology on society. Programs focus on energy policy, climate change, and national security.

Voice of the Environment (VOTE)
270 Beach Road
Belvedere, CA 95420
Phone: 707-467-0329
1330 Boonville Road
Ukiah, CA 95482
Phone: 415-250-4115
Website: www.voiceoftheenvironment.org
Email: vote@pacific.net
Voice of the Environment educates the public regarding environmental, political, and social justice issues.

Wind Energy Maps for U.S.
Website: http://rredc.nrel.gov/wind/pubs/atlas
The complete Wind Energy Resource Atlas for the United States in downloadable format. Offers averages, regions, states, seasonal; more slicing and dicing than you can imagine. Includes Alaska, Hawaii, Puerto Rico, and Virgin Islands.

WiserEarth
Website: www.wiserearth.org
Paul Hawken's new website is a community directory and networking forum for people working on the critical issues of the day and all things sustainable. The network is growing rapidly and there are many different ways to plug in.

World Business Council for Sustainable Development
1744 R Street NW
Washington, DC 20009
Website: www.wbcsd.org/templates/TemplateWBCSD5/layout.asp?MenuID=1
Email: info@wbcsd.org
Phone: 202-420-7745
A CEO-led, industry-oriented global organization, WBCSD was recently rated "By far the best website for information on sustainable development."

Worldwatch Institute
1776 Massachusetts Avenue NW
Washington, DC 20036-1904
Website: www.worldwatch.org
Email: worldwatch@worldwatch.org
Phone: 202-452-1999
A nonprofit public policy research organization dedicated to informing policymakers and the public about emerging global problems and trends. Offers a magazine and publishes a variety of books.

Yestermorrow Design/Build School
189 VT Route 100
Warren, VT 05674
Website: www.yestermorrow.com
Email: designbuild@yestermorrow.org
Phone: 888-496-5541
Yestermorrow teaches intensive hands-on courses in sustainable design, building, woodworking, and traditional crafts, offering over 100 hands-on courses per year.

State Energy Offices Access Information

Here is all the straight, up-to-date information on utility intertie and on any state programs that might help pay for a renewable energy system. All states have provided mail and phone access. Many also provide fax, website, and email access. We've listed everything available as of press time. The ✹ symbol denotes a state that allows net metering (as of mid 2007). In addition, the states of Colorado, Florida, Kentucky, Illinois, Idaho, and Arizona have some utilities that support net metering without a statewide law.

Also check www.dsireusa.org for the most up-to-date information on rebates and other state programs.

Alabama Energy Office
Department of Economic and Community Affairs
401 Adams Avenue
PO Box 5690
Montgomery, AL 36103-5690
Phone: 334-242-5290
Fax: 334-242-0552
Website: www.adeca.alabama.gov/EWT/default.aspx

Alaska Energy Authority
Alaska Industrial Development and Export Authority
813 W. Northern Lights Boulevard
Anchorage, AK 99503
Phone: 907-269-3000
Fax: 907-269-3044
Website: www.aidea.org/aea.htm

American Samoa Energy Office
Territorial Energy Office
American Samoa Government
Samoa Energy House, Tauna
Pago Pago, American Samoa 96799
Phone: 684-699-1101
Fax: 684-699-2835

✹Arizona Energy Office
Arizona Department of Commerce
1700 West Washington, Suite 220
Phoenix, AZ 85007
Phone: 602-771-1194
Fax: 602-771-1203
Email: energy@ep.state.az.us
Website: www.azcommerce.com/Energy/

✹Arkansas Energy Office
Arkansas Industrial Development Commission
One Capitol Mall
Little Rock, AR 72201
Phone: 501-682-1370
Fax: 501-682-2703
Email: energy@1800arkansas.com
Website: www.1800arkansas.com/Energy/

✹California Energy Office
California Energy Commission
1516 Ninth Street, MS #32
Sacramento, CA 95814-5512
Phone: 916-654-4287
Fax: 916-654-4420
Email: mediaoffice@energy.state.ca.us
Website: www.energy.ca.gov

✹Colorado Energy Office
Governor's Energy Office
225 E. 16th Avenue, Suite 650
Denver, CO 80203
Phone: 303-866-2100
Fax: 303-866-2930
Email: geo@state.co.us
Website: www.state.co.us/oemc

✹Connecticut Energy Office
Policy Development and Planning—Energy Management
Connecticut Office of Policy and Management
450 Capitol Avenue
Hartford, CT 06106-1379
Phone: 860-418-6374
Fax: 860-418-6495
Email: john.mengacci@po.state.ct.us
Website: www.opm.state.ct.us/pdpd2/energy/enserv.htm

✹Delaware Energy Office
1203 College Park Drive, Suite 101
Dover, DE 19904
Phone: 302-735-3480
Fax: 302-739-1840
Website: www.delaware-energy.com

✹District of Columbia Energy Office
District Department of the Environment Energy Office
2000 14th Street NW, Suite 300E
Washington, DC 20009
Phone: 202-673-6700
Fax: 202-673-6725
Email: dceo@dc.gov
Website: www.dceo.dc.gov

✹Florida Energy Office
Florida Department of Environmental Protection
2600 Blairstone Road, MS 19
Tallahassee, FL 32399
Phone: 850-245-8002
Fax: 850-245-2947
Email: alexander.mack@dep.state.fl.us
Website: www.dep.state.fl.us /energy/

✹Georgia Energy Office
Division of Energy Resources
Georgia Environmental Facilities Authority
233 Peachtree Street NE, Harris Tower, Suite 900
Atlanta, GA 30303
Phone: 404-584-1000
Fax: 404-584-1069
Website: www.gefa.org/index.aspx?page=32

Guam Energy Office
548 N. Marine Corps Drive
Tamuning, Guam 96913
Phone: 671-646-4361
Fax: 671-649-1215
Email: energy@ns.gov.gu
Website: www.guamenergy.com/

✹Hawaii Energy Office
Department of Business, Economic Development, and Tourism
PO Box 2359
Honolulu, HI 96804
Phone: 808-587-3807
Fax: 808-586-2536
Email: gmishina@dbedt.hawaii.gov
Website: www.hawaii.gov/dbedt/ert/energy.html

✳**Idaho Energy Office**
Energy Division
Idaho Department of Water
Resources
322 E. Front Street
PO Box 83720
Boise, ID 83720-0098
Phone: 208-287-4800
Fax: 208-287-6700
Email: energyguys@idwr.state.id.us
Website: www.idwr.state.id.us/energy/

✳**Illinois Energy Office**
Bureau of Energy & Recycling
Illinois Department of Commerce
and Economic Opportunity
620 East Adams
Springfield, IL 62701
Phone: 217-785-3416
Website: www.commerce.state.il.us/
dceo/Bureaus/Energy_Recycling/

✳**Indiana Energy Office**
Energy Division
Office of Energy & Defense
Development
101 W. Ohio Street, Suite 1250
Indianapolis, IN 46204
Phone: 317-232-8939
Fax: 317-232-8995
Email: rbrown@oed.in.gov
Website: www.in.gov/energy/
ENERGY/energydiv.html

✳**Iowa Energy Center**
2521 Elwood Drive, Suite 124
Ames, IA 50010-8229
Phone: 515-294-8819
Fax: 515-294-9912
Email: iec@energy.iastate.edu
Website: www.energy.iastate.edu/

Kansas Energy Office
[according to the dsire site, KS no
longer has a net metering program]
Kansas Corporation Commission
1500 S.W. Arrowhead Road
Topeka, KS 66604-4027
Phone: 785-271-3100
Email: public.affairs@kcc.ks.us
Website: www.kcc.state.ks.us/energy/
index.htm

✳**Kentucky Energy Office**
Governor's Office of Energy Policy

500 Mero Street, 12th Floor, Capital
Plaza Tower
Frankfort, KY 40601
Phone: 502-564-7192
Email: marie.anthony@ky.gov
Website: www.energy.ky.gov

✳**Louisiana Energy Office**
Technology Assessment Division
Department of Natural Resources
617 N. Third Street
PO Box 44156
Baton Rouge, LA 70804-4516
Phone: 225-342-1399
Email: techasmt@la.gov
Website: www.dnr.louisiana.gov/sec/
execdiv/techasmt

✳**Maine Energy Office**
Maine State Energy Program
Maine Public Utilities Commission
18 State House Station
Augusta, ME 04333-0018
Phone: 866-376-2463
Email: bundlemeupinfo@yahoo.com;
betsy.elder@maine.gov
Website: www.maine.gov/msep; see
also http://bundlemeup.org/; www.
maineenergyinfo.com

✳**Maryland Energy Office**
Maryland Energy Administration
1623 Forest Drive, Suite 300
Annapolis, MD 21403
Phone: 410-260-7655
Email: mea@energy.state.md.us
Website: www.energy.state.md.us

✳**Massachusetts Energy Office**
Division of Energy Resources
Executive Office of Energy and
Environmental Affairs
100 Cambridge Street, Suite 1020
Boston, MA 02114
Phone: 617-727-4732
Email: doer.energy@state.ma.us
Website: www.magnet.state.ma.us/
doer

✳**Michigan Energy Office**
Michigan Public Service Commission
Michigan Department of Labor and
Economic Growth
PO Box 30221
Lansing, MI 48909
Phone: 517-241-6180

Email: mpsc_commissioners@
michigan.gov
Website: www.michigan.gov/mpsc

✳**Minnesota Energy Office**
Energy Info Center
Minnesota Department of Commerce
85 7th Place E, Suite 500
St. Paul, MN 55101
Phone: 651-296-5175
Email: energy.info@state.mn.us
Website: www.state.mn.us/portal/
mn/jsp/content.do?subchannel=-
536881511&id=-536881350&agency=
Commerce

Mississippi Energy Office
Energy Division
Mississippi Development Authority
510 George Street, Suite 300
PO Box 849
Jackson, MS 39205
Phone: 601-359-6600
Email: energydiv@mississippi.org
Website: www.mississippi.org/
content.aspx?url=/page/78&

Missouri Energy Office
Energy Center
Department of Natural Resources
PO Box 176
Jefferson City, MO 65102
Phone: 573-751-3444
Email: energy@dnr.mo.gov
Website: www.dnr.mo.gov/energy/
deprograms.htm

✳**Montana Energy Office**
Energize Montana
Department of Environmental
Quality
1100 N. Last Chance Gulch
PO Box 200901
Helena, MT 59620-0901
Phone: 406-841-5240
Website: www.deq.state.mt.us/
energy/

Nebraska Energy Office
1111 "O" Street, Suite 223
Lincoln, NE 68508
Phone: 402-471-2867
Email: energy@mail.state.ne.us
Website: www.neo.ne.gov/

✳Nevada Energy Office
Nevada State Office of Energy
727 Fairview Drive, Suite F
Carson City, NV 89701
Phone: 775-687-9700
Email: dhoward@dbi.state.nv.us
Website: http://energy.state.nv.us

✳New Hampshire Energy Office
NH Office of Energy and Planning
57 Regional Drive, Suite 3
Concord, NH 03301-8519
Phone: 603-271-2155
Email: oepinfo@nh.gov
Website: www.nh.gov/oep/index.htm

✳New Jersey Energy Office
Office of Clean Energy
New Jersey Board of Public Utilities
Two Gateway Center
Newark, NJ 07102
Phone: 973-648-2026
Email: energy@bpu.state.nj.us
Website: www.bpu.state.nj.us

✳New Mexico Energy Office
Energy Conservation and
Management Division
New Mexico Energy, Minerals, and
Natural Resources Department
1220 S. St. Francis Drive
PO Box 6429
Santa Fe, NM 87505
Phone: 505-476-3310
Website: www.emnrd.state.nm.us/
ecmd/

✳New York Energy Office
New York State Energy Research and
Development Authority
17 Columbia Circle
Albany, NY 12203-6399
Phone: 518-862-1090
Website: www.nyserda.org

✳North Carolina Energy Office
State Energy Office
North Carolina Department of
Administration
1340 Mail Service Center
Raleigh, NC 27699-1340
Phone: 919-733-2230
Email: energyinfo@ncmail.net
Website: www.energync.net

✳North Dakota Energy Office
Division of Community Services
North Dakota Department of
Commerce
1600 E. Century Avenue, Suite 2
PO Box 2057
Bismarck, ND 58503
Phone: 701-328-5300
Email: kchristianson@nd.gov
Website: www.state.nd.us/dcs/energy

**North Mariana Islands Energy
Office**
Commonwealth of the Northern
Mariana Islands
2121 "R" Street NW
Washington, DC 20008
Phone: 670-673-5869

✳Ohio Energy Office
Office of Energy Efficiency
Ohio Department of Development
77 S. High Street
PO Box 1001
Columbus, OH 43216-1001
Phone: 614-466-6797
Website: www.odod.state.oh.us/cdd/
oee/

✳Oklahoma Energy Office
Division of Community Development
Oklahoma Department of Commerce
900 N. Stiles Avenue
Oklahoma City, OK 73104-3234
Phone: 405-815-5249
Website: www.okcommerce.gov/
index.php?option=com_content&tas
k=category&sectionid=4&id=164&It
emid=712

✳Oregon Energy Office
Oregon Department of Energy
625 Marion Street NE
Salem, OR 97301-3737
Phone: 503-378-4040
Email: energy.in.internet@state.or.us
Website: www.oregon.gov/energy

✳Pennsylvania Energy Office
Office of Energy and Technology
Deployment
Department of Environmental
Protection
400 Market Street, RCSOB
Harrisburg, PA 17101
Phone: 717-783-2300

Website: www.depweb.state.pa.us/
energy/cwp/view.asp?a=3&q=482723

Puerto Rico Energy Office
Energy Affairs Administration
Puerta de Tierra Stateion
PO Box 9066600
San Juan, Puerto Rico 00906-6600
Phone: 787-724-8777
Email: quintanaj@caribe.net

✳Rhode Island Energy Office
Office of Energy Resources
One Capital Hill
Providence, RI 02908
Phone: 401-574-9100
Website: www.riseo.state.ri.us/

South Carolina Energy Office
South Carolina Budget and Control
Board
1201 Main Street, Suite 430
Columbia, SC 29201
Phone: 803-737-8030
Website: www.state.sc.us/energy/

South Dakota Energy Office
Energy Management Office
523 E. Capitol Avenue
Pierre, SD 57501-3182
Phone: 605-773-3899
Email: boaeneralinformation@state.
sd.us
Website: www.sdreadytowork.com/

Tennessee Energy Office
Department of Economic &
Community Development
Energy Division
312 8th Avenue N, 10th Floor
Nashville, TN 37243-0405
Phone: 615-741-2373
Email: brian.hensley@state.tn.us
Website: www.state.tn.us/ecd/energy.
htm

✳Texas Energy Office
State Energy Conservation Office
Texas Comptroller of Public Accounts
111 E. 17th Street, #1114
Austin, TX 78701
Phone: 512-463-1931
Website: www.seco.cpa.state.tx.us/

✳Utah Energy Office
State Energy Program

1594 W. North Temple, Suite 3110
PO Box 146100
Salt Lake City, UT 84114-6100
Phone: 801-537-3365
Email: dbeaudoin@utah.gov
Website: www.energy.utah.gov

✳Vermont Energy Office
Energy Efficiency, Conservation, and
Renewable Energy
Vermont Department of Public
Service
112 State Street, Drawer 20
Montpelier, VT 05620-2601
Phone: 802-828-2811
Email: vtdps@state.vt.us
Website: http://publicservice.
vermont.gov/energy-efficiency/
energy-efficiency.html

✳Virginia Energy Office
Division of Energy
Department of Mines, Minerals &
Energy
202 N. Ninth Street, 8th Floor
Richmond, VA 23219-3402
Phone: 804-692-3200
Email: dmmeinfo@dmme.virginia.
gov
Website: www.dmme.virginia.gov/
divisionenergy.shtml

✳Virgin Islands Energy Office
Department of Planning and Natural
Resources
45 Estate Mars Hill
Frederiksted, VI 00840
Phone: 340-773-1082
Email: energydir@viaccess.net
Website: www.vienergy.org/

✳Washington Energy Office
Washington State University Energy
Program
925 Plum Street SE, Bldg. #4
PO Box 43165
Olympia, WA 98504-3165
Phone: 360-956-2000
Website: www.energy.wsu.edu

✳West Virginia Energy Office
Energy Efficiency Program
West Virginia Development Office
Capitol Complex, Bldg. 6, Room 533
1900 Kanawha Boulevard E
Charleston, WV 25305-0311

Phone: 304-558-2234
Website: www.wvdo.org/community/
eep.htm

✳Wisconsin Energy Office
Division of Energy Services
Department of Administration
101 E. Wilson Street
Madison, Wisconsin 53702
Phone: 608-266-8234
Email: heat@wisconsin.gov
Website: www.doa.state.wi.us/index.
asp?locid=5

✳Wyoming Energy Office
Mineral, Energy, and Transportation
Division
Wyoming Business Council
214 W. 15th Street
Cheyenne, WY 82002-0240
Phone: 307-777-2800
Email: info@wyomingbusiness.org
Website: www.wyomingbusiness.
org/business/energy.aspx

Glossary

A

AC: alternating current, electricity that changes voltage periodically, typically 60 times a second (or 50 in Europe). This kind of electricity is easier to move.

activated stand life: the period of time, at a specified temperature, that a battery can be left stored in the charged condition before its capacity fails.

active solar: any solar scheme employing pumps and controls that use power while harvesting solar energy.

A-frame: a building that looks like the capital letter A in cross-section.

air lock: two doors with space between, like a mud room, to keep the weather outside.

alternating current: AC electricity that changes voltage periodically, typically 60 times a second.

alternative energy: "voodoo" energy not purchased from a power company, usually coming from photovoltaic, micro-hydro, or wind.

ambient: the prevailing temperature, usually outdoors.

amorphous silicon: a type of PV cell manufactured without a crystalline structure. Compare with single-crystal and multi- (or poly-) crystalline silicon.

ampere: an instantaneous measure of the flow of electric current; abbreviated and more commonly spoken of as an "amp."

amp-hour: a one-ampere flow of electrical current for one hour; a measure of electrical quantity; two 60-watt 120-volt bulbs burning for one hour consume one amp-hour.

angle of incidence: the angle at which a ray of light (usually sunlight) strikes a planar surface (usually of a PV module). Angles of incidence close to perpendicularity (90°) are desirable.

anode: the positive electrode in an electrochemical cell (battery) toward which current flows; the earth ground in a cathodic protection system.

antifreeze: a chemical, usually liquid and often toxic, that keeps things from freezing.

array: an orderly collection, usually of photovoltaic modules connected electrically and mechanically secure; array current: the amperage produced by an array in full sun.

avoided cost: the amount utilities must pay for indepen dently produced power; in theory, this was to be the whole cost, including capital share to produce peak demand power, but over the years supply-side weaseling redefined it to be something more like the cost of the fuel the utility avoided burning.

azimuth: horizontal angle measured clockwise from true north; the equator is at 90°.

B

backup: a secondary source of energy to pick up the slack when the primary source is inadequate. In alternatively powered homes, fossil fuel generators are often used as "backups" when extra power is required to run power tools or when the primary sources—sun, wind, water—are not providing sufficient energy.

balance of system: (BOS) equipment that controls the flow of electricity during generation and storage baseline: a statistical term for a starting point; the "before" in a before-and-after energy conservation analysis.

baseload: the smallest amount of electricity required to keep utility customers operating at the time of lowest demand; a utility's minimum load.

battery: a collection of cells that store electrical energy; each cell converts chemical energy into electricity or vice versa, and is interconnected with other cells to form a battery for storing useful quantities of electricity.

battery capacity: the total number of ampere-hours that can be withdrawn from a fully charged battery, usually over a standard period.

battery cycle life: the number of cycles that a battery can sustain before failing.

berm: earth mounded in an artificial hill.

bioregion: an area, usually fairly large, with generally homogeneous flora and fauna.

biosphere: the thin layer of water, soil, and air that supports all known life on Earth.

blackwater: what gets flushed down the toilet.

blocking diode: a diode that prevents loss of energy to an inactive PV array (rarely used with modern charge controllers).

boneyard: a peculiar location at Gaiam Real Goods where "experi enced" products may be had for ridiculously low prices.

Btu: British thermal unit, the amount of heat required to raise the temperature of 1 pound of water 1 degree Fahrenheit. 3,411 Btus equals one kilowatt-hour.

bus bar: the point where all energy sources and loads connect to each other; often a metal bar with connections on it.

bus bar cost: the average cost of electricity delivered to the customer's distribution point.

buy-back agreement or contract: an agreement between the utility and a customer that any excess electricity generated by the customer will be bought back for an agreed-upon price.

C

cathode: the negative electrode in an electrochemical cell.

cell: a unit for storing or harvesting energy. In a battery, a cell is a single

GAIAM REAL GOODS

chemical storage unit consisting of electrodes and electrolyte, typically producing 1.5 volts; several cells are usually arranged inside a single container called a battery. Flashlight batteries are really flashlight cells. A photovoltaic cell is a single assembly of doped silicon and electrical contacts that allow it to take advantage of the photovoltaic effect, typically producing 0.5 volts; several PV cells are usually connected together and packaged as a module.

CF: compact fluorescent, a modern form of lightbulb using an integral ballast.

CFCs: chlorinated fluorocarbons, an industrial solvent and material widely used until implicated as a cause of ozone depletion in the atmosphere.

charge controller: device for managing the charging rate and state of charge of a battery bank.

controller terminology:

> **adjustable set point:** allows adjustment of voltage disconnect levels.
>
> **high-voltage disconnect:** the battery voltage at which the charge controller disconnects the batteries from the array to prevent overcharging.
>
> **low-voltage disconnect:** the voltage at which the controller disconnects the batteries to prevent overdischarging.
>
> **low-voltage warning:** a buzzer or light that indicates low battery voltage.
>
> **maximum power tracking:** a circuit that maintains array voltage for maximal current.
>
> **multistage controller:** a unit that allows multilevel control of battery charging or loading.
>
> **reverse current protection:** prevents current flow from battery to array.
>
> **single-stage controller:** a unit with only one level of control for charging or load control.
>
> **temperature compensation:** a circuit that adjusts setpoints to

ambient temperature in order to optimize charge.

charge rate: the rate at which a battery is recharged, expressed as a ratio of battery capacity to charging current flow.

clear-cutting: a forestry practice, cutting all trees in a relatively large plot.

cloud enhancement: the increase in sunlight due to direct rays plus refracted or reflected sunlight from partial cloud cover.

compact fluorescent: a modern energy-efficient form of light bulb using an integral ballast.

compost: the process by which organic materials break down, or the materials in the process of being broken down.

concentrator: mirror or lens-like additions to a PV array that focus sunlight on smaller cells; a very promising way to improve PV yield.

conductance: a material's ability to allow electricity to flow through it; gold has very high conductance.

conversion efficiency: the ratio of energy input to energy output across the conversion boundary. For example, batteries typically are able to store and provide 90% of the charging energy applied and are said to have a 90% energy efficiency.

cookie-cutter houses: houses all alike and all-in-a-row. Daly City, south of San Francisco, is a particularly depressing example. Runs in direct contradiction to our third guiding principle, "encourage diversity."

core/coil-ballasted: the materials-rich device required to drive some fluorescent lights; usually contains radioactiveAmericium (see electronic ballasts).

cost effectiveness: an economic measure of the worthiness of an investment; if an innovative solution costs less than a conventional alternative, it is more cost effective.

cross section: a "view" or drawing of a slice through a structure.

cross-ventilation: an arrangement of openings allowing air to pass through a structure.

crystalline silicon: the material from which most photovoltaic cells are made; in a single crystal cell, the entire cell is a slice of a single crystal of silicon, while a multicrystalline cell is cut from a block of smaller (centimeter-sized) crystals. The larger the crystal, the more exacting and expensive the manufacturing process.

cut-in: the condition at which a control connects its device.

cutoff voltage: in a charge controller, the voltage at which the array is disconnected from the battery to prevent overcharging.

cut-out: the condition at which a control interrupts the action.

cycle: in a battery, from a state of complete charge through discharge and recharge back to a fully charged state.

D

days of autonomy: the length of time (in days) that a system's storage can supply normal requirements without replenishment from a source; also called days of storage.

DC: direct current, the complement of AC, or alternating current, presents one unvarying voltage to a load.

deep cycle: a battery type manufactured to sustain many cycles of deep discharge that are in excess of 50% of total capacity.

degree-days: a term used to calculate heating and cooling loads; the sum, taken over an average year, of the lowest (for heating) or highest (for cooling) ambient daily temperatures. Example: if the target is 68° and the ambient on a given day is 58°, this would account for 10 degree-days.

depth of discharge: the percent of rated capacity that has been withdrawn from the battery; also called DOD.

design month: the month in which the combination of insolation and loading require the maximum array output.

diode: an electrical component that permits current to pass in only one direction.

direct current: the complement of AC, or alternating current, presents one unvarying voltage to a load.

discharge: electrical term for withdrawing energy from a storage system.

discharge rate: the rate at which current is withdrawn from a battery expressed as a ratio to the battery's capacity; also known as C rate.

disconnect: a switch or other control used to complete or interrupt an electrical flow between components.

doping, dopant: small, minutely controlled amounts of specific chemicals introduced into the semiconductor matrix to control the density and probabilistic movement of free electrons.

downhole: a piece of equipment, usually a pump, that is lowered down the hole (the well or shaft) to do its work.

drip irrigation: a technique that precisely delivers measured amounts of water through small tubes; an exceedingly efficient way to water plants.

dry cell: a cell with captive electrolyte.

duty cycle: the ratio between active time and total time; used to describe the operating regime of an appliance in an electrical system.

duty rating: the amount of time that an appliance can be run at its full rated output before failure can be expected.

E

earthship: a rammed-earth structure based on tires filled with tamped earth; the term was coined by Michael Reynolds.

Eco Desk: an ecology information resource maintained by many ecologically-minded companies.

edison base: a bulb base designed by (a) Enrico Fermi, (b) John Schaeffer, (c) Thomas Edison, (d) none of the above. The familiar standard residential light bulb base.

efficiency: a mathematical measure of actual as a percentage of the theoretical best. See conversion efficiency.

electrolyte: the chemical medium, usually liquid, in a battery that conveys charge between the positive and negative electrodes; in a lead-acid battery, the electrodes are lead and the electrolyte is acid.

electromagnetic radiation: EMR, the invisible field around an electric device. Not much is known about the effects of EMR, but it makes many of us nervous.

electronic ballasts: an improvement over core/coil ballasts, used to drive compact fluorescent lamps; contains no radioactivity.

embodied: of energy, meaning literally the amount of energy required to produce an object in its present form. Example: an inflated balloon's embodied energy includes the energy required to blow it up.

EMR: electromagnetic radiation, the invisible field around an electric device. Not much is known about the effects of EMR, but it makes many of us nervous.

energy density: the ratio of stored energy to storage volume or weight.

energy efficient: one of the best ways to use energy to accomplish a task; for example, heating with electricity is never energy efficient, while lighting with compact fluorescents is.

equalizing: periodic overcharging of batteries to make sure that all cells are reaching a good state of charge.

externalities: considerations, often subtle or remote, that should be accounted for when evaluating a process or product, but usually are not. For example, externalities for a power plant may include downwind particulate fallout and acid rain, damage to lifeforms in the cooling water intake and effluent streams, and many other factors.

F

fail-safe: a system designed in such a way that it will always fail into a safe condition.

feng shui: an Asian system of placement that pays special attention to wind, water, and the cardinal directions.

ferro-cement: a construction technique; an armature of iron contained in a cement body, often a wall, slab, or tank.

fill factor: of a photovoltaic module's I-V (current/voltage) curve, this number expresses the product of the open circuit voltage and the short-circuit current, and is therefore a measure of the "squareness" of the I-V curve's shape.

firebox: the structure within which combustion takes place.

fixed-tilt array: a PV array set in a fixed position.

flat-plate array: a PV array consisting of nonconcentrating modules.

float charge: a charge applied to a battery equal to or slightly larger than the battery's natural tendency to self-discharge.

FNC: fiber-nickel-cadmium, a new battery technology.

frequency: of a wave, the number of peaks in a period. For example, alternating current presents 60 peaks per second, so its frequency is 60 hertz. Hertz is the standard unit for frequency when the period in question is one second.

G

gassifier: a heating device which burns so hotly that the fuel sublimes directly from its solid to its gaseous state and burns very cleanly.

gassing: when a battery is charged, gasses are often given out; also called outgassing.

golf cart batteries: industrial batteries tolerant of deep cycling, often used in mobile vehicles.

gotcha!s: an unexpected outcome or effect, or the points at which, no matter how hard you wriggle, you can't escape.

gravity-fed: water storage far enough above the point of use (usually 50 feet) so that the weight of the water provides sufficient pressure.

greywater: all other household effluents besides blackwater (toilet water); greywater may be reused with much less processing than blackwater.

grid: a utility term for the network of transmission lines that distribute electricity from a variety of sources across a large area.

grid-connected system: a house, office, or other electrical system that can draw its energy from the grid; although usually grid-power-consumers, grid-connected systems can provide power to the grid.

groundwater: as distinct from water pumped up from the depths, groundwater is run off from precipitation, agriculture, or other sources.

H

heat exchanger: device that passes heat from one substance to another; in a solar hot water heater, for example, the heat exchanger takes heat harvested by a fluid circulating through the solar panel and transfers it to domestic hot water.

high-tech glass: window constructions made of two sheets of glass, sometimes treated with a metallic deposition, sealed together hermetically, with the cavity filled by an inert gas and, often, a further plastic membrane. High-tech glass can have an R-value as high as 10.

homeschooling: educating children at home instead of entrusting them to public or private schools; a growing trend quite often linked to off-the-grid-powered homes.

homestead: the house and surrounding lands.

homesteaders: people who consciously and intentionally develop their homestead.

hot tub: a quasi-religious object in California; a large bathtub for several people at once; an energy hog of serious proportions.

house current: in the United States, 117 volts root mean square of alternating current, plus or minus 7 volts; nominally 110-volt power; what comes out of most wall outlets.

HVAC: heating, ventilation, and air conditioning; space conditioning.

hydro turbine: a device that converts a stream of water into rotational energy.

hydrometer: tool used to measure the specific gravity of a liquid.

hydronic: contraction of hydro and electronic, usually applied to radiant in-floor heating systems and their associated sensors and pumps.

hysteresis: the lag between cause and effect, between stimulus and response.

I

incandescent bulb: a light source that produces light by heating a filament until it emits photons—quite an energy-intensive task.

incident solar radiation: or insolation, the amount of sunlight falling on a place.

indigenous plantings: gardening with plants native to the bioregion.

inductive transformer/rectifier: the little transformer device that powers many household appliances; an "energy criminal" that takes an unreasonably large amount of alternating electricity (house current) and converts it into a much smaller amount of current with different properties; for example, much lower-voltage direct current.

infiltration: air, at ambient temperature, blowing through cracks and holes in a house wall and spoiling the space conditioning.

infrared: light just outside the visible spectrum, usually associated with heat radiation.

infrastructure: a buzz word for the underpinnings of civilization; roads, water mains, power and phone lines, fire suppression, ambulance, education, and governmental services are all infrastructure. *Infra* is Latin for "beneath." In a more technical sense, the repair infrastructure is local existence of repair personnel and parts for a given technology.

insolation: a word coined from incident solar radiation, the amount of sunlight falling on a place.

insulation: a material that keeps energy from crossing from one place to another. On electrical wire, it is the plastic or rubber that covers the conductor. In a building, insulation makes the walls, floor, and roof more resistant to the outside (ambient) temperature.

Integrated Resource Planning: an effort by the utility industry to consider all resources and requirements in order to produce electricity as efficiently as possible.

interconnect: to connect two systems, often an independent power producer and the grid; see also intertie.

interface: the point where two different flows or energies interact; for example, a power system's interface with the human world is manifested as meters, which show system status, and controls, with which that status can be manipulated.

internal combustion engines: gasoline engines, typically in automobiles, small stand-alone devices like chainsaws, lawn mowers, and generators.

intertie: the electrical connection between an independent power producer—for example, a PV-powered household—and the utility's distribution lines, in such a way that each can supply or draw from the other.

inverter: the electrical device that changes direct current into alternating current.

irradiance: the instantaneous solar radiation incident on a surface; usually expressed in Langleys (a small amount) or in kilowatts per square meter. The definition of "one sun" of irradiance is one kilowatt per square meter.

irreverence: the measure, difficult to quantify, of the seriousness of Gaiam Real Goods techs when talking with utility suits.

IRP: See Integrated Resource Planning.

I-V curve: a plot of current against voltage to show the operating characteristics of a photovoltaic cell, module, or array.

K

kilowatt: 1,000 watts, a measure of instantaneous work. Ten 100-watt bulbs require a kilowatt of energy to light up.

kilowatt-hour: the standard measure of household electrical energy use. If the 10 bulbs left unfrugally burning in the preceding example are on for an hour, they consume 1 kilowatt-hour of electricity.

L

landfill: another word for dump.

Langley: the unit of solar irradiance; 1 gram-calorie per square centimeter.

lead-acid: the standard type of battery for use in home energy systems and automobiles.

LED: light-emitting diode. A very efficient source of electrical lighting, typically lasting 50,000 to 100,000 hours.

life: You expect an answer here? Well, when speaking of electrical systems, this term is used to quantify the time the system can be expected to function at or above a specified performance level.

life-cycle cost: the estimated cost of owning and operating a system over its useful life.

line extensions: what the power company does to bring their power lines to the consumer.

line-tied system: an electrical system connected to the power lines, usually having domestic power generating capacity and the ability to draw power from the grid or return power to the grid, depending on load and generator status.

load: an electrical device, or the amount of energy consumed by such a device.

load circuit: the wiring that provides the path for the current that powers the load.

load current: expressed in amps, the current required by the device to operate.

low-emissivity: applied to high-tech windows, meaning that infrared or heat energy will not pass back out through the glass.

low-flush: a toilet using a smaller amount (usually about 6 quarts) of water to accomplish its function.

low pressure: usually of water, meaning that the head, or pressurization, is relatively small.

low-voltage: usually another term for 12- or 24-volt direct current.

M

maintenance-free battery: a battery to which water cannot be added to maintain electrolyte volume. All batteries require routine inspection and maintenance.

maximum power point: the point at which a power conditioner continuously controls PV source voltage in order to hold it at its maximum output current.

ME: mechanical engineer; the engineers who usually work with heating and cooling, elevators, and the other mechanical devices in a large building.

meteorological: pertaining to weather; meteorology is the study of weather.

microclimate: the climate in a small area, sometimes as small as a garden or the interior of a house. Climate is distinct from weather in that it speaks for trends taken over a period of at least a year, while weather describes immediate conditions.

micro-hydro: small hydro (falling water) generation.

millennia: 1,000 years.

milliamp: one thousandth ($\frac{1}{1000}$) of an ampere.

module: a manufactured panel of photovoltaic cells. A module typically houses 36 cells in an aluminum frame covered with glass or acrylic, organizes their wiring, and provides a junction box for connection between itself, other modules in the array, and the system.

N

naturopathic: a form of medicine devoted to natural remedies and procedures.

net metering: a desirable form of buy-back agreement in which the line-tied house's electric meter turns in the utility's favor when grid power is being drawn, and in the system owner's favor when the house generation exceeds its needs and electricity is flowing into the grid. At the end of the payment period, when the meter is read, the system owner pays (or is paid by) the utility depending on the net metering.

NEC: the National Electrical Code, guidelines for all types of electrical installations including (since 1984) PV systems.

NiCad: slang for nickel-cadmium, a form of chemical storage often used in rechargeable batteries.

nominal voltage: the terminal voltage of a cell or battery discharging at a specified rate and at a specified temperature; in normal systems, this is usually a multiple of 12.

GAIAM REAL GOODS

nontoxic: having no known poisonous qualities.

normal operating cell temperature: defined as the standard operating temperature of a PV module at 800 W/m°, 20° C ambient, 1 meter per second wind speed; used to estimate the nominal operating temperature of a module in its working environment.

N-type silicon: silicon doped (containing impurities), which gives the lattice a net negative charge, repelling electrons.

O

off-peak energy: electricity during the baseload period, which is usually cheaper. Utilities often must keep generators turning, and are eager to find users during these periods, and so sell off-peak energy for less.

off-peak kilowatt: a kilowatt hour of off-peak energy.

off-the-grid: not connected to the power lines, energy self-sufficient.

ohm: the basic unit of electrical resistance; I=RV, or Current (amperes) equals Resistance (ohms) times Voltage (volts).

on-line: connected to the system, ready for work.

on-the-grid: where most of America lives and works, connected to a continent-spanning web of electrical distribution lines.

open-circuit voltage: the maximum voltage measurable at the terminals of a photovoltaic cell, module, or array with no load applied.

operating point: the current and voltage that a module or array produces under load, as determined by the I-V curve.

order of magnitude: multiplied or divided by 10. 100 is an order of magnitude smaller than 1,000, and an order of magnitude larger than 10.

orientation: placement with respect to the cardinal directions north, east, south, and west; azimuth is the measure of orientation.

outgas: of any material, the production of gasses; batteries outgas during charging; new synthetic rugs outgas when struck by sunlight, or when warm, or whenever they feel like it.

overcharge: forcing current into a fully charged battery; a bad idea except during equalization.

overcurrent: too much current for the wiring; overcurrent protection, in the form of fuses and circuit breakers, guards against this.

owner-builder: one of the few printable things building inspectors call people who build their own homes.

P

panel: any flat modular structure; solar panels may collect solar energy by many means; a number of photovoltaic modules may be assembled into a panel using a mechanical frame, but this should more properly be called an array or subarray.

parallel: connecting the like poles to like in an electrical circuit, so plus connects to plus and minus to minus; this arrangement increases current without affecting voltage.

particulates: particles that are so small that they persist in suspension in air or water.

passive solar: a shelter that maintains a comfortable inside temperature simply by accepting, storing, and preserving the heat from sunlight.

passively heated: a shelter that has its space heated by the Sun without using any other energy.

patch-cutting: clear-cutting (cutting all trees) on a small scale usually less than an acre.

pathetic fallacy: attributing human motivations to inanimate objects or lower animals.

payback: the time it takes to recoup the cost of improved technology as compared to the conventional solution. Payback on a compact fluorescent bulb (as compared to an incandescent bulb) may take a year or two, but over the whole life of the CF the savings probably will exceed the original cost of the bulb, and payback will take place several times over.

peak demand: the largest amount of electricity demanded by a utility's customers; typically, peak demand happens in early afternoon on the hottest weekday of the year.

peak kilowatt: a kilowatt-hour of electricity taken during peak demand, usually the most expensive electricity money can buy.

peak load: the same as peak demand but on a smaller scale, the maximum load demanded of a single system.

peak power current: the amperage produced by a photovoltaic module operating at the "knee" of its I-V curve.

peak sun hours: the equivalent number of hours per day when solar irradiance averages one sun (1 kW/m°). "Six peak sun hours" means that the energy received during total daylight hours equals the energy that would have been received if the sun had shone for six hours at a rate of 1,000 watts per square meter.

peak watt: the manufacturer's measure of the best possible output of a module under ideal laboratory conditions.

Pelton wheel: a special turbine, designed by someone named Pelton, for converting flowing water into rotational energy.

periodic table of elements: a chart showing the chemical elements organized by the number of protons in their nuclei and the number of electrons in their outer, or valence, band.

PG&E: Pacific Gas and Electric, the local and sometimes beloved utility for much of northern California.

phantom loads: "energy criminals" that are on even when you turn them off: instant-on TVs, microwaves with clocks; symptomatic of impatience and our sloppy preference for immediacy over efficiency.

photon: the theoretical particle used to explain light.

photophobic: fear of light (or preference for darkness), usually used of insects and animals. The opposite, phototropic, means light-seeking.

photovoltaic cell: the proper name for a device manufactured to pump electricity when light falls on it.

photovoltaics: PVs or modules that utilize the photovoltaic effect to generate useable amounts of electricity.

photovoltaic system: the modules, controls, storage, and other components that constitute a stand-alone solar energy system.

plates: the thin pieces of metal or other material used to collect electrical energy in a battery.

plug-loads: the appliances and other devices plugged into a power system.

plutonium: a particularly nasty radioactive material used in nuclear generation of electricity. One atom is enough to kill you.

pn-junction: the plane within a photovoltaic cell where the positively and negatively doped silicon layers meet.

pocket plate: a plate for a battery in which active materials are held in a perforated metal pocket on a support strip.

pollution: any dumping of toxic or unpleasant materials into air or water.

polyurethane: a long-chain carbon molecule, a good basis for sealants, paints, and plastics.

power: kinetic, or moving energy, actually performing work; in an electrical system, power is measured in watts.

power-conditioning equipment: electrical devices that change electrical forms (an inverter is an example) or assure that the electricity is of the correct form and reliability for the equipment consuming it; a surge protector is another example.

power density: ratio of a battery's rated power available to its volume (in liters) or weight (in kilograms).

PUC: Public Utilities Commission; many states call it something else, but this is the agency responsible for regulating utility rates and practices.

PURPA: this 1978 legislation, the Public Utility Regulatory Policy Act, requires utilities to purchase power from anyone at the utility's avoided cost.

PVs: photovoltaic modules.

R

radioactive material: a substance which, left to itself, sheds tiny, highly energetic pieces that put anyone nearby at great risk. Plutonium is one of these. Radioactive materials remain active indefinitely, but the time over which they are active is measured in terms of half-life, the time it takes them to become half as active as they are now; plutonium's half-life is a little over 22,000 years.

ram pump: a water-pumping machine that uses a water-hammer effect (based on the inertia of flowing water) to lift water.

rated battery capacity: manufacturer's term indicating the maximum energy that can be withdrawn from a battery at a standard specified rate.

rated module current: manufacturer's indication of module current under standard laboratory test conditions.

renewable energy: an energy source that renews itself without effort; fossil fuels, once consumed, are gone forever, while solar energy is renewable in that the sun we harvest today has no effect on the sun we can harvest tomorrow.

renewables: shorthand term for renewable energy or materials sources.

resistance: the ability of a substance to resist electrical flow; in electricity, resistance is measured in ohms.

retrofit: to install new equipment in a structure that was not originally designed for it. For example, we may retrofit a lamp with a compact fluorescent bulb, but the new bulb's shape may not fit well with the lamp's design.

romex: an electrician's term for common two-conductor-with-ground wire, the kind houses are wired with.

root mean square: RMS, the effective voltage of alternating current, usually about 70% (the square root of two over two) of the peak voltage. House current typically has an RMS of 117 volts and a peak voltage of 167 volts.

RPM: rotations per minute.

R-value: Resistance value, used specifically of materials used for insulating structures. Fiberglass insulation three inches thick has an R-value of 13.

S

seasonal depth of discharge: an adjustment for long-term seasonal battery discharge resulting in a smaller array and a battery bank matched to low-insolation season needs.

secondary battery: a battery that can be repeatedly discharged and fully recharged.

self-discharge: the tendency of a battery to lose its charge through internal chemical activity.

semiconductor: the chief ingredient in a photovoltaic cell, a normal insulating substance that conducts electricity under certain circumstances.

series connection: wiring devices with alternating poles, plus to minus, plus to minus; this arrangement increases voltage (potential) without increasing current.

set-back thermostat: combines a clock and a thermostat so that a zone (like a bedroom) may be kept comfortable only when in use.

setpoint: electrical condition, usually voltage, at which controls are adjusted to change their action.

shallow-cycle battery: like an automotive battery, designed to be kept nearly fully charged; such batteries perform poorly when discharged by more than 25% of their capacity.

shelf life: the period of time a device can be expected to be stored and still perform to specifications.

short circuit: an electrical path that connects opposite sides of a source without any appreciable load, thereby allowing maximum (and possibly disastrous) current flow.

short circuit current: current produced by a photovoltaic cell, module, or array when its output terminals are connected to each other, or "short-circuited."

showerhead: in common usage, a device for wasting energy by using too much hot water; in the Gaiam Real Goods home, low-flow showerheads prevent this undesirable result.

silicon: one of the most abundant elements on the planet, commonly found as sand and used to make photovoltaic cells.

single-crystal silicon: silicon carefully melted and grown in large boules, then sliced and treated to become the most efficient photovoltaic cells.

slow-blow: a fuse that tolerates a degree of overcurrent momentarily; a good choice for motors and other devices that require initial power surges to get rolling.

slow paced: a description of the life of a Gaiam Real Goods employee . . . not!

solar aperture: the opening to the south of a site (in the Northern Hemisphere) across which the sun passes; trees, mountains, and buildings may narrow the aperture, which also changes with the season.

solar cell: see photovoltaic cell.

solar fraction: the fraction of electricity that may be reasonable harvested from sun falling on a site. The solar fraction will be less in a foggy or cloudy site, or one with a narrower solar aperture, than an open, sunny site.

solar hot water heating: direct or indirect use of heat taken from the sun to heat domestic hot water.

solar oven: simply a box with a glass front and, optionally, reflectors and reflector coated walls, which heats up in the sun sufficiently to cook food.

solar panels: any kind of flat devices placed in the sun to harvest solar energy.

solar resource: the amount of insolation a site receives, normally expressed in kilowatt-hours per square meter per day.

specific gravity: the relative density of a substance compared to water (for liquids and solids) or air (for gases) Water is defined as 1.0; a fully charged sulfuric acid electrolyte might be as dense as 1.30, or 30% denser than water. Specific gravity is measured with a hydrometer.

stand-alone: a system, sometimes a home, that requires no imported energy.

stand-by: a device kept for times when the primary device is unable to perform; a stand-by generator is the same as a back-up generator.

starved electrolyte cell: a battery cell containing little or no free fluid electrolyte.

state of charge: the real amount of energy stored in a battery as a percentage of its total rated capacity.

state-of-the-art: a term beloved by technoids to express that this is the hottest thing since sliced bread.

stratification: in a battery, when the electrolyte acid concentration varies in layers from top to bottom; seldom a problem in vehicle batteries, due to vehicular motion and vibration, this can be a problem with static batteries, and can be corrected by periodic equalization.

stepwise: a little at a time, incrementally.

subarray: part of an array, usually photovoltaic, wired to be controlled separately.

sulfating: formation of lead-sulfate crystals on the plates of a lead-acid battery, which can cause permanent damage to the battery.

superinsulated: using as much insulation as possible, usually R-50 and above.

surge capacity: the ability of a battery or inverter to sustain a short current surge in excess of rated capacity in order to start a device that requires an initial surge current.

sustainable: material or energy sources that, if managed carefully, will provide at current levels indefinitely. A theoretical example: redwood is sustainable if it is harvested sparingly (large takings and exportation to Japan not allowed) and if every tree taken is replaced with another redwood. Sustainability can be, and usually is, abused for profit by playing it like a shell game; by planting, for example, a fast-growing fir in place of the harvested redwood.

system availability: the probability or percentage of time that an energy storage system will be able to meet load demand fully.

T

temperature compensation: an allowance made by a charge controller to match battery charging to battery temperature.

temperature correction: applied to derive true storage capacity using battery nameplate capacity and temperature; batteries are rated at 20°C.

therm: a quantity of natural gas, 100 cubic feet; roughly 100,000 Btus of potential heat.

thermal mass: solid, usually masonry volumes inside a structure that absorb heat, then radiate it slowly when the surrounding air falls below their temperature.

thermoelectric: producing heat using electricity; a bad idea.

thermography: photography of heat loss, usually with a special video camera sensitive to the far end of the infrared spectrum.

thermosiphon: a circulation system that takes advantage of the fact that warmer substances rise. By placing the solar collector of a solar hot-water system below the tank, thermosiphoning takes care of circulating the hot water, and pumping is not required.

thin-film module: an inexpensive way of manufacturing photovoltaic modules; thin-film modules typically are less efficient than single-crystal or multi-crystal devices; also called amorphous silicon modules.

tilt angle: measures the angle of a panel from the horizontal.

tracker: a device that follows the Sun and keeps the panel perpendicular to it.

transformer: a simple electrical device that changes the voltage of alternating current; most transformers are inductive, which means they set up a field around themselves, which is often a costly thing to do.

transparent energy system: a system that looks and acts like a conventional grid-connected home system, but is independent.

trickle charge: a small current intended to maintain an inactive battery in a fully charged condition.

troubleshoot: a form of recreation not unlike riding to hounds, in which the technician attempts to find, catch, and eliminate the trouble in a system.

tungsten filament: the small coil in a light bulb that glows hotly and brightly when electricity passes through it.

turbine: a vaned wheel over which a rapidly moving liquid or gas is passed, causing the wheel to spin; a device for converting flow to rotational energy.

turnkey: the jail warden; more commonly in our context, a system that is ready for the owner-occupant from the first time he or she turns the key in the front-door lock.

TV: a device for wasting time and scrambling the brain.

two-by-fours: standard building members, originally 2 by 4 inches, now 1.5 inches by 3.5 inches; often referred to as "sticks".

U

ultrafilter: of water, to remove all particulates and impurities down to the submicron range, about the size of giardia, and larger viruses.

uninterruptible power supply: an energy system providing ultrareliable power; essential for computers, aircraft guidance, medical, and other systems; also known as a UPS.

V

Varistor: a voltage-dependent variable resistor, normally used to protect sensitive equipment from spikes (like lightning strike) by diverting the energy to ground.

VCR: videocassette recorder, a device for making TVs slightly more responsive and useful.

VDTs: video display terminals, like televisions and computer screens.

vented cell: a battery cell designed with a vent mechanism for expelling gasses during charging.

volt: measure of electrical potential: 110-volt house electricity has more potential to do work than an equal flow of 12-volt electricity.

voltage drop: lost potential due to wire resistance over distance.

W

watt: the standard unit of electrical power; 1 ampere of current flowing with 1 volt of potential; 746 watts make 1 horsepower.

watt-hours: 1 watt for 1 hour. A 15-watt compact fluorescent consumes 15 of these in 60 minutes.

waveform: the characteristic trace of voltage over time of an alternating current when viewed on an oscilloscope; typically a smooth sine wave, although primitive inverters supply square or modified square waveforms.

wet shelf life: the time that an electrolyte-filled battery can remain unused in the charged condition before dropping below its nominal performance level.

wheatgrass: a singularly delicious potation made by squeezing young wheat sprouts; said to promote purity in the digestive tract.

whole-life cost analysis: an economic procedure for evaluating all the costs of an activity from cradle to grave, that is, from extraction or culture through manufacture and use, then back to the natural state; a very difficult thing to accomplish with great accuracy, but a very instructive reckoning nonetheless.

wind-chill: a factor calculated based on temperature and wind speed that expresses the fact that a given ambient temperature feels colder when the wind is blowing.

wind spinners: fond name for wind machines, devices that turn wind into usable energy.

Why There's No National Eleectric Code in This *Sourcebook*

Previous editions of this *Sourcebook* have printed in full the Suggested Practices of the National Electric Code with respect to Photovoltaic Power Systems. This time around, to save paper, trees, and unnecessary costs and because the NEC is so easily accessible (and free!) on the Internet, we recommend you download it at: http://www.nmsu.edu/~tdi/Photovoltaics/Codes-Stds/Codes-Stds.html. If you don't have access to the Internet, you can purchase the most recent handbook (see p. 535).

Product Codes Index

GAIAM REAL GOODS

Product Index

PRODUCT INDEX

PRODUCT INDEX

Subject Index

SUBJECT INDEX

solar water systems. *see also* solar systems
  backup, 402–403
  circulation pumps, 351–352
  heating, 6, 388–402
  pumps/pumping, 102, 348–355
  supply questionnaire, 356
Solex multicrystalline module, 97
SolFest, 572–573
solid block carbon, 420
solid oxide fuel cells, 164
Southwest Windpower, 144
Spirit of the Sun, 78–80
springs, water, 342
standalone systems (battery backup), 167–171
Standard Test Conditions (STC), 182
state rebate incentives, 106
Steen, Athena and Bill, 73
stick-frame houses, 58–59
stone houses, 71–72
storage water heaters, 385–386
storm windows, 290
stoves, 294–295, 295–296, 301
straw bale building, 61–64
straw-clay building, 69–71
Stream Engine Turbines, 134
string ribbon photovoltaic (PV) cells, 97
submersible pumps, 345, 353
sun conditions, 101
sun-free zones, 56
Sunhawk, 78–80
sunlight, direct, 55–56
Sun-Mar composting toilets, 438
sunspaces, 48
surface pump, defined, 345
surface water storage, 344
sustainability, 32, 37–38
sustainable living workshops, 559–571
sustainable transportation product listings, 495–498
swimming pool collector solar water heaters, 394
swimming pools, 352
system sizing worksheets, 171–176, 579–580

T

table lamps, 331
tankless water heaters, 386–387, 395, 402–403
tanks, water, 344
tank-type water heaters, 385–386, 395
television interference, 253
temperature, photovoltaic (PV) performance and, 101
temperature conversion chart, 587
thermal mass, 46, 52, 53
thermal storage walls, 48
thermostatic control, 402–403
thermostats, 300
thin-film photovoltaic (PV) cells, 97
three-way sockets, 331
tilt angle/orientation, photovoltaic (PV) modules, 122
time and money plans, 76
time-of-use metering, 182
toilets. *see* composting toilets
tools, kitchen and home product listings, 475–476
tools, outdoor and yard, 469–470
total dissolved solids (TDS) in water, 416
towers, wind energy systems, 147

toxic chemicals and modern building, 34
toxic metals in water, 416
Toyota, 164–165
Trace/Xantrex, 189
tracking photovoltaic (PV) mounts, 123, 213
track lighting, 331
Trainer, Ted, 13
transportation
  alternative-fueled vehicles, 487
  battery-electric vehicle (EV), 491
  biodisel, 488–489
  biofuels, 488–489
  dependent economy, 16–18
  electric vehicles, 491–492
  ethanol, 488–489
  fuel cell vehicles, 164–165, 492–493
  global warming and, 3
  hybrid vehicles, 488–490
  hydrogen vehicles, 492–493
  The Hypercar, 494
  low-emission vehicles, 488–490
  and peak oil, 2–3
  plug-in hybrid (PHEV) vehicle, 491
  resources, 494
  sustainable, 485–494
Trombe walls, 48
troubleshooting guide, battery, 230–231
turbidity, water, 416
turbines, DC, 134–135
turbines, small wind, 144

U

ultrafiltration, water, 420–421
Uni-Solar amorphous module, 97
United States
  energy offices, 594–597
  magnetic declinations, 583
utility intertie systems
  batteries, 178–179
  with batteries, 105–106
  costs, 179–180
  inverters, 189–192
  inverters with back-up ability, 193–196
  legal issues, 177
  net metering, 177
  product listings, 189–196
  safety issues, 177–178
  without batteries, 102–103

V

vampires, 253–254, 292–293
vehicles. *see* transportation
ventilation, 54, 304
venting, 403
volatile organic compounds (VOCs), 417
volt, 98
voltage, battery systems, 170
voltmeters, elastic, 221
"voluntary simplicity" movement, 13
volunteer programs at The Solar Living Institute, 573

W

washers, clothes, 318–319
waste, landfill, and modern building, 33–34
water, bottled, 346–347
water conversions, 345
water crisis, 346–347
water heating and heaters. *see also* water pumps and pumping
  books and reading, 413

closed-loop antifreeze solar systems, 398–401
concentrating collector solar, 394
drainback solar systems, 395–398
electric supply, 403
emergency preparedness and, 270
evacuated-tube collector solar, 393–394
flat-plate collectors solar, 392–393
fuel choices, 404–405
pressure/temperature relief valve, 403
product listings, 406–413
solar hot-water circulation pumps, 351–352
solar systems, types of, 394–401
tankless, 386, 387, 402
types, 385–387
venting, 403
water pumps and pumping. *see also* water heating and heaters
  automation, 355
  books and reading, 384
  freeze protection, 353–354
  friction loss charts, 380, 586
  high lifter pumps, 352–353
  ponds, streams, and other surface water, pumping, 342, 350
  product listings, 359–384
  PV modules, 354–355
  ram pumps, 352, 377
  solar hot-water circulation pumps, 351–352
  solar-powered, 102, 348–355
  solar water supply questionnaire, 356
  submersible pumps, 353
  terminology, 345
  truths and tips, 354
  water-powered, 352
  well water, 349–350
  wind-powered, 353
water quality
  activated carbon filtration, 419–420
  bottled, filtered, or purified, 418
  emergency preparedness and, 270
  filtration and purification, 419–421
  importance of, 415
  mechanical water filtration, 419
  product listings, 422–427
  purification product listings, 422–427
  reverse osmosis, 420–421
  testing, 417–418
  treatment methods, 421
water-saving showerheads, 293
water-source heat pumps, 298
water supply systems
  books and reading, 384
  collectors, 391–394
  components, 341–342
  delivery systems, 345
  emergency preparedness, 270
  pressurization, 351
  storage, 343–345
  storage systems, 343–345
  tanks, 344
water use, daily, 344
watt-hour (Wh), 98, 172–175
watt (W), defined, 98
waveforms, 249
weatherization, 285–287, 303
wells, water, 342
wet-cell batteries, 228–229, 232–233
wholesale cost of producing electricity, 105

GAIAM REAL GOODS

GAIAM · REAL ☀ GOODS

ORDER 24 HOURS A DAY, 7 DAYS A WEEK

Send orders to: Gaiam, Inc., 360 Interlocken Blvd., Ste. 300, Broomfield, CO 80021
Send returns to: Gaiam, Inc., 5455 West Chester Rd., West Chester, OH 45069
Technical Offices: PO Box 593, Hopland, CA 95449
Customer Service and Technical Assistance: M-F 8:30am–6pm (MST)

ORDER BY PHONE
800.919.2400
TECH FAX: 303-222-3599

order online
www.realgoodscatalog.com
INTERNATIONAL ORDERS: 707-744-2100

ORDER BY FAX
800.482.7602
INTERNATIONAL FAX: 707-744-2104

SHIP TO	ITEM #	QTY	PRODUCT DESCRIPTION	SIZE	COLOR	GIFT BOX	ITEM PRICE	TOTAL
A B C								
A B C								
A B C								
A B C								
A B C								
A B C								
A B C								
A B C								
A B C								
A B C	37-0000		**Go Zero!** ADD $2 TO PLANT A TREE THROUGH THE CONSERVATION FUND AND ZERO-OUT THE ECO-IMPACT OF SHIPPING YOUR ORDER. DETAILS, OPPOSITE PAGE AND WWW.REALGOODS.COM/GOZERO					

PAYMENT INFORMATION
Cash and credit card checks are not accepted.
☐ Check
☐ Money Order
☐ Gift Certificate # _____
☐ American Express
☐ Visa
☐ MasterCard
☐ Discover Card

Credit card no. _____
Exp. date ___ ___ CID# ___
   M M Y Y

Signature of authorized buyer _____

†**SUPPORT ECO-EDUCATION!** Add a tax-deductible donation to your order to support the work of the Solar Living Institute (see p. 59).

SHIPPING & HANDLING Per address

SUBTOTAL	CHARGE
up to $25.00	$5.95
$25.01–$50.00	$7.95
$50.01–$70.00	$9.95
$70.01–$100.00	$11.95
$100.01–$125.00	$13.95
$125.01–$150.00	$14.95
$150.01–$180.00	10% of subtotal
$180.01–$200.00	9% of subtotal
$200.01+	8% of subtotal

Call for rates to AK, HI, U.S. Territories & Canada.

RUSH SHIPPING Must call by 12:30 p.m. EST. Rush charge is in addition to regular shipping charges.

SERVICE under 10 lbs.	CHARGE
2-day service	$9.00
Next day by 4:30 p.m.	$16.50

Applies to in-stock items only. Call for rates on packages over 10 lbs. availability and shipping rates to AK, HI and U.S. Territories. FedEx cannot deliver to P.O. boxes or P.O. zip codes.

Subtotal _____
ADD: sales tax (CO, CA, OH only) _____
ADD: shipping & handling charges per address _____
ADD: oversize shipping charges per item (amount listed in parentheses after price) _____
ADD: rush shipping charges per address _____
Donation to the Solar Living Institute † _____
ADD: $5 for each gift box _____
SUBTRACT: any gift certificates or coupons _____
TOTAL In U.S. dollars _____

A SOLD TO
Name _____
Address _____
City _____
State _____ Zip _____
Phone No. ( ) _____
E-mail address _____

B SHIP TO (if different from A)   ☐ THIS IS A GIFT
Name _____
Address _____
City _____
State _____ Zip _____
Gift message: _____

C SHIP TO (if different from A)   ☐ THIS IS A GIFT
Name _____
Address _____
City _____
State _____ Zip _____
Gift message: _____

THANK YOU FOR YOUR ORDER! PLEASE MAKE SURE YOU'VE INCLUDED YOUR COMPLETE NAME, ADDRESS, ZIP CODE, PHONE NUMBER AND PAYMENT INFORMATION.

REAL ☀ GOODS

Four Easy Ways to Order

Phone 800.919.2400

Order 24 hours a day, 7 days a week

Customer Service and Technical Assistance:
M-F 8:30am-6pm (MST)

Our sales and technical experts are ready to help with any questions you might have — including more information about any of our products, technical specifications, answers about your options for going solar, and much more.

Fax 800.482.7602

Please include daytime phone number and credit card information.

WEB

www.realgoods.com

Order via our secure website, where you'll also find much more information about renewable energy for your home or business.

Mail

Send your completed order form (left) to: **Gaiam Real Goods, 360 Interlocken Blvd., Ste. 300, Broomfield, CO 80021.** Include a phone number and check or credit card information. All checks must show a street address. **Checks with larger dollar amounts may be delayed up to two weeks for processing. No personal checks over $5,000.**

Debit Card Users

Your bank may remove or set aside the necessary funds to cover the order for a period of up to five business days. Gaiam, Inc. does not collect payment until the order has shipped.

International Orders

Phone: 707.744.2100

Fax: 707.744.2104

Please call or fax for a shipping quote. Funds must be in U.S. dollars, drawn from a U.S. bank. Credit cards not accepted on International orders.

We Use Recycled Packaging

In an effort to lessen our impact upon the planet, we reuse boxes and packing materials. New boxes and mailing envelopes are made of up to 100% recycled material. We use clean waste paper collected from local sources and 100% natural, corn-based, biodegradable "peanuts" to protect your order during shipping.

Shipping

Please allow 7 to 10 business days for standard delivery of in-stock items. Some larger or bulky items, or items shipped directly from a manufacturer, may require additional shipping charges and delivery time; any such charges are indicated in parentheses following product price. Freight items require a quote. We're always happy to provide estimated shipping cost and date of delivery; just call us at 800.919.2400. Please inspect all shipments at the time of delivery. If you discover any damage, please advise the carrier immediately; **once your order is signed for, Real Goods is not responsible for merchandise damage.** Be sure to save all shipping and packing materials.

Corrections

Merchandise and prices in this catalog are effective through April 30, 2008. While we make every effort to ensure the accuracy of our information, errors will occasionally occur. We reserve the right to make corrections and price changes.

Return Policy

You can return most unused products to us for a replacement or refund within 30 days of receipt. Items must be in new condition with all original packaging, paperwork and accessories supplied. For your protection, please send the return via a traceable method. Return freight and original shipping are non-refundable. Special orders, appliances, heaters, items returned without original packaging or packing materials, and items sent directly from the manufacturer may be subject to a restocking fee of 10–25%. Please call us for return authorization prior to shipping any return.

Our services and programs for you, the planet and our global community

Community & Culture

At Real Goods you'll find some artisan-crafted and fair-trade gifts that help sustain a rich diversity of social threads around the world, and the ecosystems to which they're so closely linked. We also give back through programs like our Earth Day for Kids program that introduces inner-city kids to solar by bringing them to the Solar Living Center for a day.

Carbon-Zero Programs

Through our partner The Conservation Fund, Gaiam has made a contribution that enables Go Zero to plant 1,430 trees, completely offsetting the CO_2 impact of 2006 operations at our headquarters and fulfillment center. We also make it easy to be a Hero of Zero by offsetting the eco-impact of shipping your Gaiam order – or giving Go Zero as a gift. Learn more at gaiamliving.com/gozero.

A Program of
THE CONSERVATION FUND

Meaningful Choices: Always a Great Gift

Gift Certificates and eGift Cards

Always appreciated, valid on any Real Goods purchase. Call for details or visit www.realgoods.com/egift.

Gift Registry

You know what you want to make your home and lifestyle more health- and eco-conscious. Let your friends and family know, too. Call for our registry packet or visit www.realgoods.com/registry.

Gift Packaging

Our beautiful gift packaging includes a card with your personal message. For mail orders, please call for number of gift boxes you'll need.

New! Give the Gift of Zero

... and help slow global warming. Add just $2 to your order and Go Zero will send a certificate to your gift recipient recognizing that you've planted a tree in his or her name — a tree that will absorb one ton of carbon dioxide from the atmosphere over its lifetime as it helps restore native ecosystems and animal habitat. Visit www.realgoods.com/gozero.

A Program of
THE CONSERVATION FUND

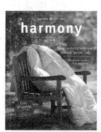 

Healthy People, Healthy Planet
Sign up for our other Gaiam catalogs

Visit www.gaiam.com/catalogrequest to receive:
Gaiam Living for the essentials of a beautiful, health-friendly, eco-friendly home.
Gaiam Mind•Body for our bestselling workout DVDs, eco-activewear and wellness solutions.
Gaiam Real Goods for Solar Power products and options for living lighter on the planet.

Join our Community.

You'll find people who share your passion for living greener and healthier lives, and ...

- John Schaffer's rants & revelations in the Community Blog
- Inspiring videos on alternative fuels, organic gardening & more
- $10 off your next Real Goods purchase

Realgoods.com/community